Handbook on Structural Testing

HANDBOOK ON STRUCTURAL TESTING

COPUBLISHED BY
SOCIETY FOR EXPERIMENTAL
MECHANICS, INC.
BETHEL, CONNECTICUT 06801

EDITED BY
ROBERT T. REESE, PH.D.
AND
WENDELL A. KAWAHARA, PH.D
SANDIA NATIONAL LABORATORIES

Published by
THE FAIRMONT PRESS, INC.
700 Indian Trail
Lilburn, GA 30247

Library of Congress Cataloging-in-Publication Data

Reese, Robert T.
   Handbook on structural testing / by Robert T. Reese, Wendell A. Kawahara.
   p. cm.
   Includes bibliographical references and index.
   ISBN 0-88173-155-2
   1. Structural analysis (Engineering) 2. Testing I. Kawahara, Wendell A. II. Title.
TA646.R36   1993   624.1'71--dc20   93-9439
                                                        CIP

Handbook On Structural Testing by Robert T. Reese, Wendell A. Kawahara.
©1993 by The Fairmont Press, Inc. All rights reserved. No part of this publication may be reproduced or transmitted in any form or by any means, electronic or mechanical, including photocopy, recording, or any information storage and retrieval system, without permission in writing from the publisher.

Published by The Fairmont Press, Inc.
700 Indian Trail
Lilburn, GA 30247

Printed in the United States of America

10 9 8 7 6 5 4 3 2 1

ISBN 0-88173-155-2   FP

ISBN 0-13-379306-0   PH

While every effort is made to provide dependable information, the publisher, authors, and editors cannot be held responsible for any errors or omissions.

Distributed by PTR Prentice-Hall, Inc.
A Simon & Schuster Company
Englewood Cliffs, NJ 07632

Prentice-Hall International (UK) Limited, London
Prentice-Hall of Australia Pty. Limited, Sydney
Prentice-Hall Canada Inc., Toronto
Prentice-Hall Hispanoamericana, S.A., Mexico
Prentice-Hall of India Private Limited, New Delhi
Prentice-Hall of Japan, Inc., Tokyo
Simon & Schuster Asia Pte. Ltd., Singapore
Editora Prentice-Hall do Brasil, Ltda., Rio de Janeiro

AUTHORS AND CONTRIBUTORS

NAME	AFFILIATION
Alan G. Beattie, Ph.D.	Sandia National Laboratories, New Mexico
Gilbert T. Blake	Wiss, Janney, Elstner, and Associates
Thomas G. Carne, Ph.D.	Sandia National Laboratories, New Mexico
Richard L. Crabb	Sandia National Laboratories, New Mexico
Robert Czarnek, Ph.D.	Concurrent Technology Corp
John W. Dreger	University of Wisconsin—Madison
Elizabeth A. Fuchs, Ph.D.	Sandia National Laboratories, California
Richard Grigat	Grumman Aircraft Company
Richard L. Hannah	Lockheed Aeronautical Systems (ret.), JP Technologies
Alan Humphrey	MTS Systems Corporation
David W. Larson, Ph.D.	Sandia National Laboratories, New Mexico
Lloyd J. Lazarus	Allied Signal Aerospace Company Kansas City Div.
William B. Leisher	Sandia National Laboratories, New Mexico
Donald R. Lesuer, Ph.D.	Lawrence Livermore National Laboratory
Albert N. Lin, Ph.D.*	National Institute for Standards and Technology
James W. Little	National Institute for Standards and Technology
Wei-Yang Lu, Ph.D.	University of Kentucky
Arup K. Maji, Ph.D.	University of New Mexico
Michael D. Messman	General Motors Proving Ground
Thomas Paez, Ph.D.	Sandia National Laboratories, New Mexico
Niel R. Petersen	MTS Systems Corporation
Richard Pettit, Ph.D.	Sandia National Laboratories, New Mexico
Alex Redner, Ph.D.	Strainoptic Technologies
Reginald Robinson	LTV Aircraft Products Division
Robert F. Rowlands, Ph.D.	University of Wisconsin—Madison
Robert C. Schwarz	Grumman Aircraft Company
Samuel H. Smith	Battelle Columbus Laboratory
William T. Springer, Ph.D.	University of Arkansas
Stuart Swartz, Ph.D.	Kansas State University
Frank J. Szafranski	General Dynamics Convair Division
Charles Taylor, Ph.D.	University of Florida (emeritus)
Benjamin A. Wallace, Ph.D.	University of Oklahoma
Ming L. Wang, Ph.D.	University of New Mexico

NOVEMBER 1991

*Dr. Lin died in 1992
(The protocol for naming Scandia National Laboratories changed in Oct. 1992.)

PREFACE

This handbook's purpose is twofold: first, it organizes and reviews fundamental types of structural tests, test methods, and procedures. These tests, test methods, and procedures can be applied to large civil engineering structures, aircraft and aerospace structures, mechanical assemblies, and miniature electronic components and parts. Second, since insight gained from a test is derived from the type and quality of the data obtained, the authors describe principles and restrictions underlying a variety of measurement and diagnostic techniques as applied to and used in structural testing.

This handbook is addressed to a diverse audience including novice and experienced practicing technical staff of engineers and technicians, management, students, and instructors. Description of broad concepts in structural testing is emphasized. Examples of applications are given throughout the text. Also included is a discussion of safety, health, and environmental considerations in structural testing.

Increased emphasis on improving national productivity goals and quality assurance incentives demand more and better information from structural tests. Yet, laboratory budget constraints continue to tighten. Hence judicious choices of testing methods, instrumentation, acquisition and allocation of other resources (e.g., test equipment, computer support) are required. These choices are suggested in this handbook as based on the technical material and the author's experiences. The authors present essential considerations in testing to inform the reader about advantages, limitations, and disadvantages of various methods, techniques, and types of tests that can be performed. In addition, this handbook also provides summaries of the operating characteristics of various sensors, transducers, and test equipment.

This book consists of six fundamental areas:

1. An overview and history of structural engineering and testing (chapters 1 and 2)
2. A description of why tests are performed and the types of structural tests that are routinely conducted (chapters 3 and 4)
3. The relationships between design loads and loading conditions used in laboratory and field tests on structures along with control systems used for test equipment involved (chapters 5 through 7)
4. A a presentation of the methods and techniques of experimental mechanics and nondestructive techniques applied to structural testing (chapters 8 through 12)
5. The safety, health, and environmental considerations involved in structural testing along with the measurement uncertainties found in tests (chapters 13 and 14)
6. Many examples of structural tests (chapter 15)

Unifying motives behind this handbook are: (1) the authors' concern for decreasing quantity and quality of laboratory and other courses where testing and experimental practices are introduced and the students experience working with test equipment and specimens and (2) the decreasing levels of understanding of testing among young engineers. By sharing their extensive experience and expertise, the authors help mitigate this imbalance caused by academic emphasis on analytical areas of engineering mechanics at the expense of experimental and testing techniques. The authors also desire to complement the information and materials available from the Society for Experimental Mechanics and other sources.

The authors thank many who have contributed to discussions guiding the work including members of the Society for Experimental Mechanics—Ian Allison, Clarence Calder, Susan Foss, William Fourney, Albert Kobayashi, and Robert Sullivan. Members of SEM's Structural Testing Division were particularly helpful in providing encouragement and direction. We also thank Professors Vagn Askegaard, J. Arcan, and D. Allan Firmage for their perspectives and ideas. Many members of the Western Regional Strain Gage Committee also gave many suggestions and contributed ideas and resources and we specifically thank Charles Wright, Udell Merritt, Larry Shull, Peter Stein, Michael Lemcoe, John Quinley, Don Roach, and Mike Tovey. The U.S. Department of Energy's Interagency Manufacturing Operations Group (IMOG) subgroup on Mechanical Testing also gave many ideas and insights into testing with specific thanks to Don Lesuer, Mike Stout, and Stanley Beitscher. The authors also thank members of Sandia National Laboratories for their help including:

Thomas Baca, Melvin Callabresi, Fred Cericola, Neil Davie, Daniel Dawson, Larry Dorrell, Terry Ernest, Vernon Gabbard, M. L. Heisler, John Keilman, Scott Klenke, John Korellis, Joseph Kubas, Rodney May, Michael Nusser, Ken Padilla, Jeanne Ramage, Bill Robinson, Marlene Shields, and John Totten.

The authors also give special thanks to Richard L. Hannah for his extensive review of the manuscript.

Readers and users of this handbook are encouraged to communicate with the authors through the Society for Experimental Mechanics, 7 School St., Bethel, CT 06801.

Robert T. Reese
Albuquerque, New Mexico

Wendell A. Kawahara
Pleasanton, California

TABLE OF CONTENTS
CHAPTERS AND MAIN SECTIONS

1. INTRODUCTION 1

    1.1 Overview
    1.2 Scope of Handbook
    1.3 Motivation for Handbook
    1.4 Technology Transfer
    1.5 Test Perspectives
    1.6 References

2. HISTORY 7

    2.1 Historical Perspectives on Structural Engineering
    2.2 History of Structural Testing
    2.3 Structural Engineering: Definition and Phases
    2.4 References

3. TEST PURPOSES AND TEST EVALUATION CRITERIA 21

    3.1 Introduction
    3.2 Test Purposes
    3.3 Test Evaluation Criteria
    3.4 Determining and Evaluating Structural Responses
    3.5 Inconsistent Calibrations
    3.6 Non-Structural Failure Modes and Evaluations
    3.7 References

4. TYPES OF STRUCTURAL TESTS 33

    4.1   Introduction—Test Conditions and Environments
    4.2   Acceptance and Quality Assurance
    4.3   Calibration
    4.4   Connections and Joints
    4.5   Development of Instrumented Test Units
    4.6   Dynamic Tests on Actual Structures
    4.7   Energy Absorption
    4.8   Environmental and Materials Compatibility
    4.9   Failure Loads and Modes (Design Margins)
    4.10 Full Scale and Model Tests
    4.11 Inertial
    4.12 Lightning
    4.13 Mass Properties
    4.14 Materials Testing
    4.15 Modal Analyses
    4.16 Pressure / Vacuum Tests
    4.17 Proof and Operational Tests
    4.18 Residual Stresses
    4.19 Service Life Extension
    4.20 Shock Tests

    4.21   Stiffness Testing
    4.22   Structure-Control Interaction
    4.23   Thermal Testing
    4.24   Time Dependent Tests
    4.25   Vibration Tests
    4.26   Wind Tunnel Testing
    4.27   Summary
    4.28   References

5.   LOADS 65

    5.1   Introduction
    5.2   Static and Dead Loads on Structures
    5.3   Approximated Live Loads / Equivalent Static Loads /
            Building Code Requirements
    5.4   Stored Energy Calculations
    5.5   Measured Service Loadings
    5.6   Thermal
    5.7   Load Spectra and Load Envelopes
    5.8   Probabilistic Loadings
    5.9   Examples of Probabilistic Loads
    5.10   Application of Loads and Conditions to Structures
    5.11   Summary
    5.12   References

6.   LOADING SYSTEMS 87

    6.1   Introduction
    6.2   Simulations—Actual Loads, Test Conditions, and Failure Situations
    6.3   Testing Concepts—Open and Closed Loop Systems
    6.4   Loading Equipment
    6.5   Examples of Laboratory and Field Static Loading Conditions
    6.6   Examples of Dynamic Loading Conditions
    6.7   Boundary Conditions and Test Fixtures
    6.8   Summary
    6.9   References

7.   TEST CONTROLS 115

    7.1   Editors' Introduction
    7.2   Servohydraulic Test Systems and Controls
    7.3   Applying Servohydraulic Controls to Test Applications
          7.3.1   Servohydraulic Characteristics
          7.3.2   Pressure Control Servovalves
          7.3.3   Actuator Piston Area Selection
          7.3.4   Rotary Servoactuators
          7.3.5   System Operating Pressure
          7.3.6   Maximum Actuator Dynamic Response
    7.4   Typical Servohydraulic Actuator Components
    7.5   Control Response
          7.5.1   Displacement Control Systems
          7.5.2   Load Control Systems

    7.5.3 Remote Strain Gage Control Considerations
  7.6 Hydraulic Supply and Return System
    7.6.1 Hydraulic Power Supply System
    7.6.2 Accumulators
    7.6.3 Hydraulic Contamination Control
  7.7 Control Electronics
  7.8 Safety and Interlock Systems
  7.9 Hydraulic Shutdown Methods
  7.10 Electric Power Loss
  7.11 Servohydraulic Test System Command Signal Sources
    7.11.1 Introduction
    7.11.2 Constant Amplitude
    7.11.3 Block Programming
    7.11.4 Variable Amplitude
  7.12 References

8. OVERVIEW OF EXPERIMENTAL METHODS AND TECHNIQUES 135

  8.1 Introduction
  8.2 Point Techniques—Gages, Transducers, and Sensors
  8.3 Full Field Techniques
  8.4 Nondestructive Evaluation Techniques
  8.5 Example—Pressure Vessel
  8.6 References

9. POINT TECHNIQUES, GAGES, TRANSDUCERS, AND SENSORS 141

  9.1 Introduction
  9.2 Strain Gages
  9.3 Extensometers
  9.4 Displacement Gages
  9.5 Accelerometers
  9.6 Temperature Sensing Devices
  9.7 Load Cells
  9.8 Pressure Transducers
  9.9 References

10. INSTRUMENTATION AND DATA SYSTEMS 159

  10.1 Introduction
    10.1.1 Rate or Time Dependency of Loading
    10.1.2 Accuracy and Resolution
  10.2 Transducers and Sensors
    10.2.1 Bridge Completion
    10.2.2 Wiring
  10.3 Data Transmission
    10.3.1 FM-FM Telemetry
    10.3.2 Slip-Rings
  10.4 Signal Conditioning
    10.4.1 Power Supplies
    10.4.2 Offsets
    10.4.3 Amplifiers

   10.4.4 Filters
   10.4.5 Calibration
 10.5 Recorders / Displays
   10.5.1 Analog Meters
   10.5.2 Digital Voltmeters
   10.5.3 Oscillograph Recorders
   10.5.4 Servo Recorders
   10.5.5 Magnetic Tape Recorders
   10.5.6 Oscilloscopes
   10.5.7 Hybrid Waveform Recorders
   10.5.8 Data Loggers
 10.6 Computer Based System
   10.6.1 Multiplexing
   10.6.2 Data Sampling
   10.6.3 A/D Converters
   10.6.4 Interfaces
 10.7 Summary
 10.8 References

11. **EXPERIMENTAL METHODS—FULL FIELD TECHNIQUES** 183

 11.1 Introduction
 11.2 Moiré Interferometry
   11.2.1 Introduction and Historical Perspective
   11.2.2 Basic Concepts
   11.2.3 Principles of Moiré Interferometry
   11.2.4 Moire Interferometry in Engineering Practice
   11.2.5 Interpretation of Moire Fringes
 11.3 Geometrical Moiré
   11.3.1 A Moiré System for Determining In-Plane Strains
   11.3.2 Application to Determining Residual Strains in Riveting
   11.3.3 Determination of Out-Of-Plane Displacements
   11.3.4 Application to a Typical Design Problem
 11.4 Photoelasticity
   11.4.1 Introduction
   11.4.2 Brief History
   11.4.3 Photoelastic Procedure
   11.4.4 Polariscopes
 11.5 Photoelastic Coatings
   11.5.1 Fundamental Relationships and Measuring Techniques
   11.5.2 Application of Photoelastic Coatings to Structures
   11.5.3 Separation of Principal Stresses
   11.5.4 Test Report Preparation Guidelines
 11.6 References

12. **EXPERIMENTAL METHODS—**
  NONDESTRUCTIVE EVALUATION TECHNIQUES 237

 12.1 Introduction
 12.2 Acoustic Emission, Principles and Instrumentation
   12.2.1 Principles of Acoustic Emission
   12.2.2 Acoustic Waves

    12.2.3 Sensors
    12.2.4 Acoustic Emission Signals
    12.2.5 Acoustic Emission Instrumentation
    12.2.6 Signal Processing Methods
    12.2.7 Advanced Analysis Methods
    12.2.8 Conclusions
    12.2.9 References
  12.3 Ultrasonic Methods
    12.3.1 Introduction
    12.3.2 Applications
    12.3.3 Basic Principles
    12.3.4 Apparatus
    12.3.5 Testing Techniques
    12.3.6 References
  12.4 Impact-Echo Technique
    12.4.1 Introduction
    12.4.2 Theoretical Background
    12.4.3 Performing the Test
    12.4.4 Instrumentation
    12.4.5 Other Concerns
    12.4.6 References
  12.5 Real-time Radiography
    12.5.1 Introduction
    12.5.2 Basic Principles
    12.5.3 Basic Equipment
    12.5.4 Effects of Scattered Radiation
    12.5.5 Image Processing Techniques
    12.5.6 Example I—Flaw Detection of Composite
    12.5.7 Example II—Internal Strain Measurement for Concrete
    12.5.8 References

13. TEST ORGANIZATION AND SAFETY, HEALTH, AND ENVIRONMENTAL REQUIREMENTS 295

  13.1 Introduction
  13.2 Basic Test Planning
  13.3 Test Preparation
  13.4 Safety, Health, and Environmental Considerations
  13.5 Methodology for Evaluating Hazards in Structural Testing
  13.6 Summary of Safety, Health and Environmental Considerations
  13.7 References

14. MEASUREMENT UNCERTAINTIES IN STRUCTURAL TESTING 319

  14.1 Introduction
  14.2 Uncertainties—Top Level and Working Standards
  14.3 Uncertainty Types
    Type A Uncertainties
    Type B Uncertainties
  14.4 Combining Uncertainties
  14.5 Sources of Uncertainty
  14.6 Uncertainty Statements

    14.7   Out of Range Calibrations
    14.8   Value of Uncertainty Analyses
    14.9   Uncertainty Analyses of Structural Test Laboratory
    14.10 Uncertainties Due to Strain Gage Misalignment
    14.11 References

15. **STRUCTURAL TESTING—EXAMPLE TESTS, CASE STUDIES, AND APPLICATIONS** 331

    15.1   Introduction
    15.2   Loads
          15.2.1 Thermal (Hydrocarbon Fire)
          15.2.2 Biomechanical Force-Platform Based on Strain Gages
          15.2.3 Field Measurement of Tension in a T-142 Tank Track
    15.3   Quality Assurance and Quality Control
    15.4   Calibration
          15.4.1 Instrumented Press System
          15.4.2 Load Cell Calibration for Cable Tension Meter
    15.5   Connections and Joints—Fixtures and Test Procedures for Miniature Threaded Fasteners
    15.6   Energy Absorption—Support Structure
    15.7   Environmental and Materials Compatibility—Nylon Screws
    15.8   Materials Testing
    15.9   Modal Analysis
    15.10 Nondestructive Testing
          15.10.1 Nondestructive Testing of Concrete
          15.10.2 Glass-to-Metal Seals
    15.11 Service Life Extension—Load Testing of the Huey P. Long Bridge
    15.12 Smart Structures
    15.13 Stiffness Tests
    15.14 Testing of Aircraft Structures
    15.15 Test Controls
    15.16 Time Dependent Tests
    15.17 References

INDEX 395

ABBREVIATIONS AND ACRONYMS USED

AASHTO	American Association of State Highway and Transportation Officials	lb.	Pounds
		LVDT	Linear Variable Differential Transformer
AC	Alternating Current	LVT	Linear Velocity Transducer
ACI	American Concrete Institute	L/d	Length to Diameter Ratio
A/D	Analog to Digital Converter (ADC also)	MAWP	Maximum Allowable Working Pressure
AE	Acoustic Emission	MIL	Military
AISC	American Institute of Steel Construction	MM	Micromeasurements Division of Measurements Group
amp	Amplifier		
ANSI	American National Standards Institute	MPa	Megapascal
AREA	American Railway Engineering Association	ms	Millisecond
ASCE	American Society of Civil Engineers	MS	Military Specification
ASME	American Society of Mechanical Engineers	MSDS	Material Safety Data Sheet
ASTM	American Society for Testing and Materials	MTS	MTS Systems Corp.
BIPM	International Bureau of Weights and Measures	NASP	National Aerospace Plane
		NDT	Nondestructive Test (Methods)
BLH	BLH Electronics	NFPA	National Fire Protection Association
BLEVE	Boiling Liquid Expanding Vapor Explosion	NIST	National Institute for Standards and Technology
Btu	British Thermal Unit		
CAD	Computer Aided Design	op amp	Operational Amplifier
CAE	Computer Aided Engineering	OPL	Optical Path Length
CAM	Computer Aided Manufacturing	OSHA	Occupational Safety and Health Administration
CCD	Charge Couple Device		
CCF	Cross Correlation Function	OTDR	Optical Time Domain Reflectometry
CFR	Code of Federal Regulations	Pa	Pascal
CMOS	Complimentary Metal Oxide Semiconductor	PC	Personal Computer
CMR	Common Mode Rejection	PC	Post Cure Requirements for Strain Gage Installations
CPU	Central Processing Unit		
D/A	Digital to Analog Converter (DAC also)	PCM	Pulse Code Modulation
dB	Decibel, $20 \log_{10}$ (ratio)	PDM	Pulse Duration Modulation
DC	Direct Current	PVC	Polyvinylchloride
DDC	Direct Digital Control	QA	Quality Assurance
DOE	United States Department of Energy	QC	Quality Control
DOF	Degree of Freedom	RV	Potentiometer/Variable Resistor
DVM	Digital Voltmeter	RF	Radio Frequency
EMAT	Electromagnetic Acoustic Transducer	RSS	Root Sum Square
EU	Engineering Units	RTD	Resistance Temperature Device
FEA	Finite Element Analysis	RVDT	Rotary Variable Differential Transformer
FEM	Finite Element Method	SCO	Subcarrier Oscillator
FFT	Fast Fourier Transform	SEAOC	Structural Engineers Association of California
FM	Frequency Modulation	SEM	Society for Experimental Mechanics
FRF	Frequency Response Functions	SESA	Society for Experimental Stress Analysis (now SEM)
ft.	Foot / Feet		
HBM	Hottinger Baldwin Measurements	S/H	Sample-Hole Circuit
HSM	Hydraulic Service Manifold	S/H/E	Safety, Health, and Environment
in.	Inch / Inches	STC	Self Temperature Compensated—Strain Gages
IA	Instrumentation Amplifier	STD	Standard
IES	Institute of Environmental Sciences	TM	Telemetry
JPT	JP Technologies	TM	Texas Measurements
K	Spring Stiffness F/L	TMF	Thermomechanical Fatigue
Kip	1000 lbs.	UNC	United National Coarse Thread

UNF	United National Fine Thread	V to F	Voltage to Frequency Converter
UNM	United National Metric Thread	WRSGC	Western Regional Strain Gage Committee
U.S.DOT	United States Department of Transportation		

GREEK AND OTHER SYMBOLS USED

α	Angle of Incidence
α	Angular Acceleration
α	Attenuation Factor—Ultrasonics
α	Coefficient of Thermal Expansion
β	Acoustoelastic Coefficient
β	Beat-Phase, Fiber Optics
β	Oil Compressibility, psi
γ	Shear Strain
δ	Optical Retardation
Δ	Deflection
Δ	Small Change
ε	Strain, $\Delta L/L$
ε_i	Strain in i Direction
θ	Angle of Diffraction
θ	Angular Displacement
λ	Eigenvalue
λ	Wave Length
μ	micro, $\times 10^{-6}$
μ	Shear Modulus, $E/2(1+u)$
$\mu\varepsilon$	Microstrain
μs	Microsecond
ν	Poisson's Ratio
ν	Frequency of Oscillation—Light or Sound
π_l	Longitudinal Piezoresistive Coefficient, (M2/N)
ρ	Radius of Curvature
ρ	Density
ρ	Specific Resistivity of an Electrical Conductor
σ	Normal Stress
τ	Shear Stress
ϕ	Angular Change
ϕ	Optical Retardation
ω	Angular Velocity
ω	Light Frequency
$°$	Degrees (Temperature)

ROMAN LETTER SYMBOLS USED

a	acceleration, length/sec/sec
a	amplitude of disturbance
a_r	radial acceleration
a_t	tangential acceleration
A	cross-sectional area 12
A	magnitude of electric vector
A	Area of Servohydraulic Actuator
AT	Type A Uncertainty
b	body force
bi	body force in i direction
B	Liquid Bulk Modulus
BT	Type B Uncertainty
c	Distance to Extreme Fiber
c	Constant of Integration
c	Speed of Sound
C_{ij}	Strain Optic Coefficients- Fiber Optics
C	Capacitance
C	Speed of Light
$°C$	Temperature—degrees Celsius
C'	Speed of Light in a Transparent Medium
C_i	Correction Factor for Photoelastic Coatings
C_i	Elastic Constant for Acoustic Wave
d	Depth or Thickness
db	Decibel, $20 \log_{10}$ (ratio)
e	Strain Vector
e	Thermal Emissivity of Surface
f	Frequency
E	Modulus of Elasticity
E	Energy Stored in a Structure
E	Excitation Voltage in Wheatstone Bridge
E_f	Energy Stored in a Fluid
Eg	Energy Stored in a Gas
E_s	Strain Energy Stored in a Structure
E(x)	Mean of a Random Variable—may be μ_x
f	Frequency of Grating
f	Fringe Value
ft	Triggering Rate for Oscilloscope
fy	Yield Stress of Steel
f'c	Yield Stress of Concrete
F	Force, Resultant Force
F	Frequency, Hertz
F	Gage Factor-Strain Gages
F	Desensitization Factor for Lead Wires
$°F$	Temperature-degrees Fahrenheit
[F]	Matrix of Applied Forces
g	Acceleration of Gravity
g	Distance Between Two Adjacent Planes of Maximum Intensity
g2/Hz	Energy Delivered to Structure in Random Vibration
G	Modulus of Elasticity in Shear
h	Convection Coefficient (heat transfer)
hc	Thickness of Photoelastic Plastic
hs	Thickness of Metal
H	Rate of Heat Flow
Hz	Frequency in Hertz

i	Index	P	Pitch of Moire Line Ruling
I	Moment of Inertia	P	Probabilities
I	Intensity of Light a^2	P	Signal Power (Acoustic)
I_{ii}	Moment of Inertia About i Axis	P_1	Pressure at Location 1
I_1	Stress Invariant	P_2	Pressure at Location 2
I_2	Stress Invariant	P_y	Critical Buckling Load
I_3	Stress Invariant	Q	Flow (through servovalve)—Gallons per Minute
I	Impact Factor for Bridge Loads = $50/(L+125)$	$[Q_{ij}]$	Material Stiffness Matrix
IE	Impact Echo	r	Radius of Gyration
k	Constant depending on units used	R	Electrical Resistance
k	Coefficient of Thermal Conductivity	R_i	Reaction Force at Location i
k	Strain Optic Material Constant	R_v	Potentiometer
ksi	1000/be/square inch	s	Seconds
Kips	1,000 Pounds	S	Section Modulus, I/c
K	Bulk Modulus	S	Optical Path Length
K	Overall Stiffness of Load Train	S	Sensitivity
K	x 1000	S_a	Strain Gage Sensitivity to Axial Strain
K	Range Factor-Converts Square Root of Variance to Probability Interval	S_s	Strain Gage Sensitivity to Shearing Strain
		S_t	Strain Gage Sensitivity to Transverse Strain
K	Compliance of Oil Column in Servohydraulic System, lbs./in.	SA	Strain Sensitivity for Materials
		SN$\rightarrow$	Stress Vector
K_t	Transverse Sensitivity for Strain Gages	SH	Horizontal Shear Wave—Ultrasonic
[K]	Stiffness Matrix	SV	Vertical Shear Wave—Ultrasonic
l	Direction Cosine for x	t	Time
L	Length	t, t_c	Thickness of Photoelastic Coating
L	Bridge Span in Feet	T	Period of Rotation (1/F)
L	Lines (number of)	T	Temperature
m	Diffraction Order	T	Slab Thickness (Concrete)
m	Direction Cosine for y	T	Transit Time
m	Mass	T_i	Temperature at Location i
m	Meter	T_0	Kelvin Temperature of Surface
mm	Millimeter	u	Displacement, Typically in x Direction
M	Applied Moment	U	Strain Energy Stored Per Unit Volume
M	$\times 10^6$	v	Velocity
MPa	Megapascals	v	Displacement, Typically in y Direction
M_o	Initial Applied Moment	V	Potential Energy Wave (ultrasonic)
M_p	Plastic Moment	V	Volume
n	Direction Cosine for z	V	Voltage
n	Number	V	Output Voltage in Wheatstone Bridge
n	Model Scale	V_i	Velocities of Acoustic Waves
n	Index of Refraction	w	Displacement, Typically in z Direction
n_i	Fiber Indices	w	Uniform Load
N	Force, Newtons	x,y,z	Orthogonal Coordinates
N	Number of Repetitions of Force Application	[x]	Displacement Vector
N	Fringe Number or Order (Moiré)	X	Desired Double Amplitude
N	Number of Characteristic Times Periods When Loads are Applied	X	Random Variable
		X_c	Probabilitistic Load
p	Point	Z	Plastic Section Modulus
P	Longitudinal Ultrasonic Compressive Waves	Z_i	Characteristic Impedance of Acoustic Waves
psi	Pounds per Square Inch	ZT	Overall Uncertainty
P	Pressure (internal/external)		

Chapter 1

Introduction

Robert T. Reese, Sandia National Laboratories,
Albuquerque, New Mexico
Wendell A. Kawahara, Sandia National Laboratories,
Livermore, California

1.1 OVERVIEW

This handbook describes the testing of structures. Structures are typically thought of as large in size such as buildings, bridges, ships, and aircraft. The principles of structural testing developed in this handbook apply to all types of structures ranging from these large sizes to small mechanical components and electronic assemblies. These principles also apply to structures made of virtually any material—metals, woods, plastics, ceramics, and composites (e.g., ranging from reinforced concrete to advanced aerospace materials). We define a structure as any mechanical system, electrical component, or assembly of structural members, beams, linkages, plates, shells, honeycombs, and cables. Testing of structures means that they will be subjected to loading conditions or other stimuli to determine how they respond, using the methods and techniques of experimental mechanics and other disciplines. Information on testing of structures can also be found in general descriptions of mechanical or environmental testing (Refs. 1-1 and 1-6). Structural testing covers the wide variety of structures and focuses on the purposes, methods, and procedures used in testing. The goal of structural testing is to determine how structures behave statically and dynamically.

Testing is one of the five basic phases of structural engineering. The other four phases are: analysis, design, fabrication and assembly (or production), and use. Each phase gives information to and places requirements on the other phases and receives guidance and constraints from them. These interactive relationships are shown in Fig. 1-1.

Those interactions in Fig. 1-1 involving personnel and engineering activities from each phase occur for one-of-a-kind projects as well as mass-produced structures and assemblies. These interactions are increasingly influenced by the developments in each phase. For example, developments in computer capabilities have expanded analytical techniques which, in turn, gives greater direction to test planning, diagnostics, and instrumentation.

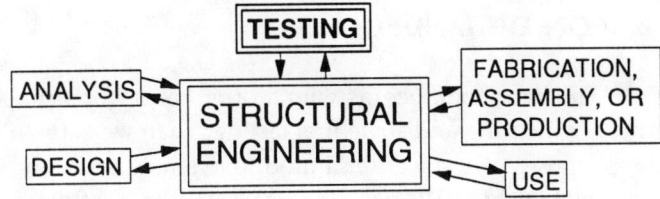

Figure 1-1. The Five Phases of Structural Engineering

A major intent of this handbook is to demonstrate and emphasize the value of testing in determining how structures respond. The purpose of the analysis and testing phases of structural engineering are the same: to determine how structures behave when loaded statically and dynamically. Analysis and testing should be complementary (Refs. 1-2 through 1-7). The concepts and descriptions in this handbook indicate the value of testing. We have developed this handbook into a one-volume detailed summary of structural testing that can be used by practicing engineers, technicians, new engineers, instructors, and others who have not been involved in testing such as students and project managers.

The authors of this handbook are involved in planning and conducting tests, training others to perform tests, developing test methods and techniques, and other related aspects of testing such as data acquisition, controls and software development. They have contributed to advances and developments of test equipment, computer systems, and diagnostic equipment which have permitted the introduction of new test methods and techniques and significant improvements in current testing capabilities. However, we share concerns regarding current trends in structural testing including the following: (1) reduced academic emphasis and instruction in laboratory courses, (2) reduction of funds available for laboratory personnel, equipment, and test operations, (3) a perception that testing offers less value to customers than in past years, and (4) increasing demands and requirements placed on test engineers for insight and understanding with particular emphasis on product integrity, liability issues, and safety and health concerns (Refs. 1-8 and 1-10).

The training of engineers and technicians in testing and laboratory activities has decreased significantly since the 1960's. Funding available for conducting tests, acquiring and maintaining equipment, and developing new techniques continues to decrease; yet more detailed information is expected from tests. With less academic training available and more definitive testing programs needed, there is more need than ever to document the techniques of structural testing (Refs. 1-9 and 1-10).

1.2 SCOPE OF HANDBOOK

The handbook begins with a brief history of structural engineering and structural testing. Then we explain the purposes of tests, criteria used to evaluate tests, and their relationships. We describe the various types of structural tests, and the application of forces, environments, and conditions to a wide variety of structural systems. We describe laboratory and field loading equipment for static, time dependent (creep, fatigue), shock, and vibration conditions. Next, control systems for structural tests are explained using servohydraulic systems for automotive testing as the example. We review some of the fundamentals of engineering mechanics with an emphasis on experimental measurements. We then present an overview of the various experimental methods and techniques in three chapters, each devoted to a major subject area (point techniques, transducers, and sensors; full or whole field methods; and nondestructive evaluation techniques). We explain the data acquisition and instrumentation systems used for structural testing including transducer and sensor operation, data transmission, signal conditioning, recorders, and displays. Next, the basics of planning and organizing tests are presented, including the increasing emphasis on safety, health, and environmental requirements. We explain measurement uncertainties that occur in structural testing conceptually and in applications. Finally, we present many examples that give insight into why and how structural tests were performed and how the tests results were interpreted.

A thorough discussion of structural testing involves many topics. Two key concerns underlie these discussions: (1) why tests are performed and what goals are expected to be met when the tests are completed and (2) the quality of the results. A basic goal in structural testing is to obtain information needed to certify that a structure or component is adequate to perform its intended functions. Certification may range from an agreement among a few individuals to formal procedures requiring evidence of strict adherence to federal regulations, governing national and local code requirements, and industry specifications. This handbook is a resource for ideas on how to perform "quality" tests. The quality of a structural test is defined by two basic factors: that information obtained is (1) sufficient and (2) valid. Some additional aspects of quality principles applied to structural testing are: (1) conformance to test requirements, (2) competent test performance, (3) adequate cost estimates prior to testing and financial controls during testing, and (4) being on or ahead of schedule.

1.3 MOTIVATION FOR HANDBOOK

This handbook is a ready reference for practicing test personnel and a basic text in structural testing for students. It provides insight and understanding to a wide variety of structural tests, and presents a detailed summary of testing not previously available. Many texts are available on specific fields of structural testing, including composites, fracture mechanics, materials testing, fatigue and cyclic testing, modal testing, and many others. However, there is a need for a basic book which brings together the fundamentals of structural testing common to virtually all types of structures and components.

Laboratory courses required in undergraduate mechanical and civil engineering curricula have been reduced by half or more since the 1960's as determined by discussions among the authors and others. These two engineering disciplines usually supply most of the people involved in structural testing. Current engineering graduates, however, usually do not have sufficient background and training in testing (Ref. 1-10). The situation is further compounded because emphasis on computer techniques attracts much of the time and interest of the students. It is difficult to interest students in a 1950's testing machine in favor of a state-of-the-art computer. One encouraging aspect of student interest in structural testing is shown by the increased involvement of electrical engineers because of the wide variety of electrically based sensors, transducers, data acquisition and instrumentation systems, test controlling systems, and software development.

Academic interest in testing has also diminished. More tests are performed in industrial settings rather than at universities. Test requirements have increased which need resources and facilities more likely available in industry rather than at universities. Tight time schedules and budgetary constraints also cause companies and private laboratories to rely on their own capabilities and resources. Industry often has proprietary interests so that tests conducted usually impose restrictions on access to test items and data. Government sponsored programs involving structural testing often have security restrictions as well.

Urgency to obtain answers often curtails research or projects previously conducted in university laboratories. Government support of structural testing activities has also reduced while compelling concerns of product liability and acceptance demand more comprehensive test capabilities and evaluation techniques. Support for test activities is likely to increase as a result of malfunctioning or defective products.

Most technologies associated with structural testing have been revised or improved in the past decade; several new techniques have also been introduced. Portable equipment has been developed permitting field investigation of a wide range of potential problems, with capabilities to perform complicated tests under adverse conditions. As a result, the methods of testing are considerably more powerful and, in many ways, less understood by those requesting test support, planning and designing test items, and specifying test instrumentation and data to be obtained. Computer systems available for laboratory and field use have greatly improved data acquisition and transmission, test system control, analysis of data, and overall test capabilities. This rapid development of test equipment, computer systems and software, and test methods and techniques makes it difficult to maintain current understanding about this quickly changing field. We give information which describes and organizes the various types of equipment and test techniques available so these will be more familiar to personnel needing to make decisions regarding test capabilities and equipment acquisition.

The motive for this handbook is to fulfill the current needs for understanding structural testing and for continuing development of expertise in testing. We intend that engineers new to or away from testing for a time, along with instructors and students, will receive sufficient preparation from this handbook in the various aspects of structural testing to be capable of specifying test requirements, understanding basic measurements, interpreting test results, and making competent decisions based on experimental evidence.

1.4 TECHNOLOGY TRANSFER

Technology transfer is a term describing the development of technology by one organization or group and its use by another organization. This handbook transfers structural testing technology from industry, test facilities and laboratories, government laboratories, and universities to readers.

Much of what an engineer learns is obtained through job-related experiences. The authors of this handbook have many experiences which serve as examples throughout. These include instruction to students, related research on fundamental principles and projects, and basic research and development of a wide range of structural systems, components, and assemblies to production of aircraft, farm equipment, highway bridges and other large civil engineering structures, automobiles, electronic parts such as computers, and many other types of structures.

Chapter 15 is devoted entirely to a series of examples of the authors' "favorite" problems and their solutions on a wide range of structural systems and assemblies. The transfer of structural testing technology often better understood in terms of example problems and how they were solved.

1.5 TEST PERSPECTIVES

Three basic perspectives for engineers and others involved in structural testing are:

1. Test Conductors: Those who perform tests on a continuing basis in laboratories and/or field locations,

2. Test Requesters: Those in project engineering or design organizations responsible for developing structures or components who need assistance in planning, conducting, and requesting tests to be performed by test laboratories or field test personnel,

3. Test Developers: Those in project engineering or design organizations working on their own structures and systems and do not rely on other organizations for help. They plan tests, design and build test facilities, and conduct structural tests themselves.

An individual involved in testing usually represents one of these perspectives—a test conductor, a test requester, or a test developer. These three different perspectives reflect the different viewpoints for the reader and user, and understanding these perspectives will aid in implementing structural testing technology. In each case there needs to be communication among test personnel, requesters, and users of test data so that each person involved (whichever perspective is represented) will know the needs of those requesting the test. Items such as schedules, costs, fixtures and interfaces, test purposes, test evaluation methods, diagnostic devices, and many other details need to be communicated and understood.

Each individual (conductor, requester, and developer) has different degrees of testing expertise, concerns, constraints, and priorities. Testing expertise indicates familiarity with equipment and safety practices, having personnel who are experienced and properly trained, having the instrumentation and data systems needed for testing, being able to provide the proper boundary and loading conditions so that the correct data are obtained, and completing test documentation. Testing concerns indicates considerations such as protection of test hardware, test costs, safety for personnel, equipment, and the environment, and proper operation of equipment and data systems. Testing constraints indicates the schedules that need to be met, the load paths that need to be used, limits on available applied forces and pressures and on resulting strains, accelerations, temperatures, and displacements. Testing priorities implies that certain minimum expectations need to be fulfilled; they may include items which would add information but may not necessarily be required as part of a successful test.

Test Conductors have greater testing expertise, experience with test equipment, familiarity with the instrumentation and data acquisition needs and capabilities, and will have an experience base of other tests to help them. As a result, they will likely be the most efficient in testing by obtaining the most information with the least amount of effort. However, since they are not as familiar with the test items and prototype units as the organization and designer requesting the tests are, then it is possible that some critical objectives can be missed in planning and conducting the tests.

Test Requesters understand their design and test hardware (e.g., prototype test units, handling equipment, or actual structures) along with its limitations but may not fully understand the capabilities and limitations of a test laboratory or field test facility. This limited knowledge can range from a lack of understanding of the experimental methods and techniques available to them to misconceptions about test capabilities particularly in recent developments and improvements. It is possible to have test requests which cannot be fulfilled completely or for which applicable experimental techniques are not available.

Test Developers may not be familiar with, but are responsible for implementing a test program for a project; they typically have many things to consider. These can include engineering designs, costs, schedules, and other developmental considerations. Other responsibilities can include planning a test program, designing and fabricating test hardware, and becoming familiar with experimental methods and procedures they will need for their project and often acquisition of test equipment and controls.

Test conductors are typically contacted by test requesters to perform tests in the conductor's facilities. The request usually includes some type of test plan, reasons for performing the tests, and desired information and data to be obtained. As a contrast, test developers are often left to their own initiatives and expertise. On occasion they may seek help from other organizations. Test developers are often faced with developing and designing test facilities, prototype test units, and planning, organizing and conducting tests.

Table 1-1 is a summary of these three perspectives and indicates how they can influence planning and conducting tests. This handbook addresses these perspectives and provides a basis for planning and conducting structural tests.

1.6 REFERENCES

1-1. _____, *Military Standardization Handbook.*, Metallic Materials and Elements for Aerospace Vehicle Structures, MIL-HNDBK-5D, Department of Defense.

1-2. M. Arcan and S. Balan, *Essai Des Constructions*, Editions Eyrolles, Paris, 1972.

1-3. V. Askegaard, "Testing of Structures and Structural Components," Proc. Society for Experimental Mechanics, International Spring Conference, Portland, OR, June 6-10, 1988, SEM, Bethel, CT.

1-4. V. Askegard, "Teaching and Research in Experimental Mechanics: Complementarity of Theory and Experiment," Internal Report, NR I 61, Technical University of Denmark, Lyngby, 1978.

1-5. E.O. Doeblelin, *Measurement Systems*, McGraw-Hill, New York, 1975.

1-6. C.M. Harris, *Shock and Vibration Handbook*, 3rd Ed., Chapter 25, McGraw-Hill, New York, 1988.

1-7. M. Hentenyi, *Handbook of Experimental Stress Analysis*, John Wiley & Sons, New York, 1950 (for additional background information).

1-8. J. Marin and J.A. Sauer, *Strength of Materials*, MacMillan Co., New York, 1954.

1-9. C.C. Perry and L.D. Lineback, "Why Management Won't Buy Experimental Mechanics," Proc. Society for Experimental Mechanics, International Spring Conference, Las Vegas, NV, June 9-14, 1985, SEM, Bethel, CT.

1-10. P.K. Stein and C.P. Wright, "Our Engineering Education—The Not-So Scientific Method," 13th Aerospace Testing Seminar, Aerospace Corporation and Institute of Environmental Sciences, Oct. 8-10, 1991, Manhattan Beach, CA.

Table 1-1. Three Basic Perspectives in Structural Testing

Personnel Degree of Familiarity with Structural Tests	TEST CONDUCTORS - Personnel in Full-Time Field or Test Laboratory Facilities	Project Engineering Groups	
		TEST REQUESTERS - Personnel Requesting Tests Performed by Other Organizations	TEST DEVELOPERS - Self-Contained Groups Performing Their Own Tests
Aspects Most Familiar With	• Experimental Methods Techniques, and Instrumentation • Test Controls and Capabilities • Analysis and Interpretation of Data and Test Results Related to Specific Test Equipment and Test Procedures • Minimizing Test Costs • Design of Test Fixtures • Test Organization • Performing Tests and Schedules	• Test Planning • Acquisition of Test Hardware • Definition of Test Purposes, Budgets, and Schedules • Allocation of Resources • Impact of Test Results on Projects • Analysis and Interpretation of Test Data Related to Specific Test Items and Prototypes • Machining and Fabrication Practices	• Test Planning • Design and Fabrication of Test Items and Hardware • Definition of Test Purposes, Budgets, and Schedules • Allocation and Use Personnel • Analysis and Interpretation of Test Data Related to Specific Test Items and Prototypes • Impact of Test Results on Projects
Aspects Somewhat Familiar With	• Behavior of Actual or Prototype Test Hardware, Fabrication of Test Items • Analysis and Interpretation of Test Data on a Specific Project	• Available Experimental Methods and Techniques • Analysis and Interpretation of Data for Test Controls and Equipment Operation • Test Organization and Scheduling • Calibration Requirements and Practices	
Aspects Less Familiar With	• Minimizing Overall Project Costs • Understanding Project Priorities and Constraints • Meeting Project Schedules	• Completeness and Thoroughness in Testing • Use of Optimum Test Methods or Techniques • Test Safety Requirements and Considerations	• How to Develop a Multiple Use Rather Than Single Use Test Facilities • Completeness and Thoroughness in Testing • Use of Optimum Test Techniques • Design of Test Fixtures

CHAPTER 2
HISTORY

*Robert T. Reese, Sandia National Laboratories,
Albuquerque, New Mexico*

2.1 HISTORICAL PERSPECTIVES ON STRUCTURAL ENGINEERING

The history of structural engineering is as old as primitive mankind and as varied as the imagination and abilities of modern man as evidenced in current technology. Early man learned to adapt to and cope with his surroundings and environments by initially providing shelter and substance. As civilizations developed, mankind began to build structures for religious, government, and commercial purposes as well as making improvements in their homes. Simple tools were invented to aid some aspects of life. Early man learned to use rock, timber, vines, soil (mud), straw, animal bones and hides, and other naturally available materials for their tools and structures. Man also became experienced in using the beam, the arch, and the column or post as basic structural elements. Principles of engineering mechanics were used without being analyzed and included the use of the inclined plane, buoyancy, friction, rollers and wheels, mechanical advantage (levers), and stored elastic energy (bows and arrows).

First reflections of early mankind are often of the impressive legacy of structures, buildings, and monuments that have lasted through the ages. The enormous pyramids of ancient Egypt were the tallest structures on earth for over three thousand years. The beauty of the Parthenon on the Acropolis in Athens has been copied in many public buildings. The greatness of these and many other civilizations was often measured by the size and grandeur of their buildings and monuments. Many examples of engineering masterpieces have survived the ravages of time and wars including the aqueducts, bridges, roads, and buildings of the Romans, the monuments and religious buildings of the Aztecs, Incas, and Mayas, the monuments of Abu Simbel along with the obelisks and the Sphinx in Egypt, the Great Wall of China, Stonehenge in England, the Indian cliff dwellings and burial grounds in North America, Anchor Wat in southeast Asia and the large stone structures like those on Easter Island in the Pacific Ocean. The early structural engineers responsible for these impressive projects probably learned their trade through some type of apprenticeship, indentured service, and on-the-job training. These projects often featured a high degree of skill on the part of stonemasons and other workers. Many of their skills remain unsurpassed today as exemplified by dressed stones fitted together without mortar. Structures built by these stonemasons have lasted through the ages on five continents. The rules developed and used by these master builders provided a stimulus for construction of many great projects (Refs. 2-10, 2-13, 2-24, 2-28, 2-29, and 2-32).

In this same period of early mankind there were equally impressive uses of metals, rocks, and soils. Man was interested in using metals long before recorded history. The evidence of these uses is spread over many continents and areas within them. Man probably learned first to use clays to make pottery and to use obsidian and other high strength rocks to make tools and weapons. The firing process used to make pottery led to the development of furnaces and kilns needed for extraction of metals from ores. Mineral ores were also used for pigments and their remains are found in petroglyphs and early drawings (Ref. 2-19).

Copper was first smelted in the Timna Valley in the Sinai desert about 3500 B.C. Evidence there shows that ores were placed in a pit and covered with wood. A fire stoked with air heated and melted the copper. The molten copper settled to the bottom of the pit and the ore was tapped off. This type of pit and heating system was repeated for most methods to extract metals from their ores (Ref. 2-19).

The next age of metals was the bronze age which began about 3000 B.C. Bronze is a combination of tin and copper. The early copper mines were not located near sources of tin. Traders moved the raw materials over land and by sea to these early smelters. This alloy of copper and tin offered some advantages to metalsmiths: lower melting temperatures, improved hardness, and easier casting including the development of the "lost wax" method used today. One may think that the metalsmiths were concentrated around the Mediterranean basin.

However, the largest casting of bronze (875 kilograms) found in antiquity was located in China and dates at about 1200 B.C. (Ref. 2-19).

Structural engineering and the development of iron and steel are inseparable. Iron foundries have been found in India and China in the second millennium B.C. Iron presented many problems to early metalsmiths because of higher melt temperatures and brittleness. The origin of "Wootz" steel was found in the Indus civilization around 1500 B.C. Wootz steel was made in small clay crucibles filled with wrought iron, some wood, and leaves from assorted plants. These crucibles were sealed and placed in a pit about one meter deep under a layer of charcoal. A fire was lit and combustion was aided by air blasts from bellows. The steel produced in this way was remarkably homogeneous and had the right amount and distribution of carbon. This steel was also traded to other parts of the world. As this civilization declined, the methods of steel making were lost for a time and then were rediscovered by Arab metalsmiths about 700 A.D. The strong, flexible steel swords of Damascus became famous weapons. With the decline of the Islamic civilization around 1000 A.D., the art of making high quality steel was lost again for a time (Ref. 2-19).

Acquisition of minerals usually was found in subsurface mines. Eventually mining processes encountered underground water which entered the mines through aquifers, springs, and rivers. The only available sources of continuous power needed to lift the water out of the mines were humans, animals, and waterwheel systems. Some animals used for this purpose actually lived their lives underground. The waterwheel system was more trouble free and developed most fully in the Rammelsburg mines in Germany and were used until 1910 (Ref. 2-19).

The modern development of the steel industry formed the foundation for the Industrial Revolution. The centers for development were near Coalbrookdale, England and Pittsburgh, Pennsylvania. Three important developments took place around 1850: first was the development of coke made from coal which made the increased temperatures needed to heat and melt the iron ores possible while preserving what was left of the depleted forests, the second was the development of a continuous source of power in the steam engine which made the removal of water and ores from the mines economic, and third was the construction of canals which enabled low cost and reliable transportation of the raw materials (Refs. 2-19 and 2-32).

A classic example of the development of metals, alloys, and castings is in the development of the printing process by Gutenberg. Many methods had been invented to make printed copies using characters carved from wood and formed from clay but they were not practical for large scale printing. Gutenberg's father worked in a mint and probably taught his son about metals and alloys. Gutenberg used an alloy of tin and lead (common solder) which was cast into the individual letters. The individual letters had a hole in the back of the type face which could be attached to a long rod. Gutenberg worked out a system using these individual movable letters which could be arranged to form words (Ref. 2-19).

Significant scientific investigations beginning about 1450 A.D. focused on answering many of the questions regarding nature and the world around mankind. Science began to describe many laws and occurrences in nature and these discoveries were applied. While these applications proceeded slowly when compared to developments in the modern era, many basic ideas were discovered and theories introduced in the period from 1450 to 1850 A.D. This four-hundred year period of the post renaissance era deserves special recognition because of the contributions, ideas, and theories developed by great thinkers (Refs. 2-13 and 2-24).

Leonardo da Vinci (1542-1619) is usually remembered for his famous art work. However, to the engineer he is likely to be as famous for his ideas of submarines, flying machines, bridges, and the practical statement of the moment of a force. Sir Isaac Newton provided the analytical basis for the work on structural analysis and engineering mechanics with the development of the three laws of motion and his formulation, along with Leibnitz and later Euler, of the basic relationships of calculus. Galileo (1564-1642) began the study of engineering mechanics by analyzing a beam in bending. While this analysis was incorrect, his work served to inspire others. There was great interest in the action of structural shapes and materials while being subjected to loading conditions. The work of the Bernoulli brothers (Jacob 1654-1705 and John 1667-1748) and Leonhard Euler (1707-1783) developed the theories of column buckling, elastic curves of beams, vibrations of beams and the theory of virtual work. Euler was the first to describe the motion of a body using differential equations with the results published in his famous book in 1736. Euler and Daniel Bernoulli corresponded during this period which promoted Euler's interest in the lateral vibration of bars and the differential equations governing these motions. Euler used variational calculus and developed Jacob Bernoulli's equation for elastic curves. Equation 2-1 describes the result which is not limited to small deflections (the modern version of the equation is in brackets). P is the applied force, x is the distance along the beam, and C depends on the elastic properties of the materials and the cross-section of the beam. (p in the modern equation is the radius of curvature which is proportional to the applied moment.) Euler integrated this equation using a series solution. The correct interpretation of C which for rectangular beams remained elusive for some time. C is proportional to the depth h of a beam as h_3 rather than h_2 as thought for many years. The major difficulty in determining the appropriate constants was the accurate description of the stresses in the beam cross-section (Ref. 2-28).

$$Px = C \frac{\frac{d^2y}{dx^2}}{\left(1+\left(\frac{dy}{dx}\right)^2\right)^{\frac{2}{3}}} \left\{\frac{1}{\rho} = \frac{\frac{d^2y}{dx^2}}{\left(1+\left(\frac{dy}{dx}\right)^2\right)^{\frac{2}{3}}}\right\} \qquad \text{Eq. 2-1}$$

Parent (1666-1716) was the first to determine how the stresses were actually distributed in the cross-section of a rectangular beam. Coulomb (1736-1806) was among the first to verify the behavior of a beam of rectangular cross-section subjected to bending. Coulomb used Hooke's Law (described later) to relate fiber stresses to location in the cross section, developed expressions for equilibrium, and correctly evaluated the stresses. He also considered the consequences of the beam materials yielding and contemplated the changes that could occur in the distribution of stresses. Coulomb also developed theories for the states of stress in the earth as affected by retaining structures. Timoshenko (Ref. 2-28) indicates that Coulomb contributed more than any other scientist (or engineer) to the understanding of elastic bodies during the eighteenth century. Among his important contributions were early experiments involving the tensile and shear strengths of materials and tests on beams with various supports. Navier (1785-1836) developed relationships for the bending of beams of various cross sections and formulated equations in a manner similar to the way they are used today in his publication of "Resume des Lecons de Mecanique," the first detailed book on engineering mechanics. He also analyzed arches and columns and evolved a general method of analyzing statically indeterminate structures. While Galileo is usually considered the father of engineering mechanics, Euler and Navier have the honor of being the initial great writers and organizers of the study of structures and could be considered the fathers of structural engineering. Coulomb was among the first to verify existing theories as an initial experimenter and could be the father of structural testing (Ref. 2-28).

It is important that these men be recognized, and it is also important for the reader to realize that structural engineering has a long and colorful history. (The reader is also referred to the references at the end of the chapter for additional information on history as well as each subject presented.) From these pioneering efforts three modern day engineering sciences have developed: engineering mechanics, structural engineering, and experimental mechanics.

The first truss structure in recorded history is credited to Palladio in Italy about 1550. While these structures were designed by trial and error procedures, they represented the first departure from the sole use of beams, arches, columns, and vine type suspension bridges used in most earlier buildings and structures. The modern period of structural engineering began about 1850 when Squire Whipple published the analysis of truss type structures. This work was likely based on the ideas developed by Stevin in Holland in 1586 which described the triangle for forces similar to a truss joint. The simple truss can be considered as the first engineered structure to be constructed. Beginning about this time, many advances occurred in each of these three sciences (mechanics, materials, and structural engineering) which began to develop and become distinct from each other. The most impressive characteristic of this entire period is the rapid rate at which ideas, theories, and practical applications were introduced and experiences were acquired. Among the important concepts developed was St. Venant's principle which has a particularly important place in structural testing. In modern terms it states that the effect of a force applied locally over a small area may be treated as simple statically equivalent force at a distance away from the point of application which is approximately equal to the thickness or width of the body (Ref. 2-18). Influence lines were introduced by Winkler in 1867 and have been used for analysis of moving and variable loads on beam and bridge type structures. Prof. Ritter introduced the analysis of trusses by method of sections in 1867 which is one of the initial analytical techniques taught to engineering students today. Maxwell introduced the analysis of redundant frameworks in 1864. Graphical analysis approaches were developed in this same time period along with methods of virtual work, elastic energy, and many developments in indeterminate structural analysis (Refs. 2-4, 2-10, 2-13, and 2-28).

Otto Mohr developed the conjugate beam method for analyzing beam deflections in the 1860's. Mohr also developed the circle of stress used to determine principal stresses and their orientations. This concept was also applied to strains and is used to interpret the results of structural tests. Many of the structural engineers in this era are credited with developing theories of failure including Rankine, Coulomb, St. Venant, Huber, von Mises, Hencky, Beltrami, and Haigh. These theories of failure are important for structural testing because they can define the type of individual test needed to characterize a structure, component or material (Ref. 2-28).

In the period since 1900, the development of new

theories, analytical techniques, and problem solving methodologies has shown an exponentially increasing growth. The analytical methods include moment distribution by Hardy Cross in 1930, slope deflection by Maney in 1915, and relaxation methods by Southwell in 1935-40. In the period following World War II and into the early 1960's, the advances in analytical methods included the development of plastic analysis for steel structures and ultimate strength design methods for reinforced and prestressed concrete. Handbooks were compiled summarizing basic analytical approaches, equations for many problems encountered in structural engineering, and many texts became available. Many methods were developed for analysis of complex structures (e.g., folded plate roofs, shell and shell-type structures, and rib stiffened plates) as described in texts by Timoshenko, and others (Refs. 2-4, 2-10, 2-17, and 2-24).

The 1960's also experienced the introduction of computers into the analysis and design process. The use of computers has greatly expanded the capabilities of the structural engineer to provide more thorough and detailed analyses as well as the ability to perform those analyses thought previously too difficult. The use of computers has changed the analysis procedure by emphasizing detailed numerical rather than analytically closed form solutions. The numerical techniques used now (finite element, finite difference, integration of differential equations, etc.) have opened many horizons to the structural engineer. The development of computer aided engineering (CAE), design (CAD), and manufacturing (CAM) techniques in the 1980's along with the great improvement in computer oriented graphics has contributed to enormous advancements in the abilities of the structural engineer.

Concurrent with the developments in analysis capabilities are the continuing improvements and discoveries in materials science. The most commonly used steels are now improved for resistance to corrosion, are produced by different processes, and are formed and worked using new procedures. The quality control procedures on materials processing has been improved. Many of the materials that are used in virtually every phase of life today were not used twenty and thirty years ago. A wide variety of wood and wood products (plywood, particle board, paneling, etc.) have been incorporated into construction practices. Plastics and other synthetic materials have been used in many types of structural members, parts, and assemblies. The most rapid developments have come in the use of composite materials. These materials have been utilized in aircraft, engines, tires, dishes, all manner of electronic parts, gears, and many other applications. It is difficult in this limited space to adequately describe the rapid advances and developments that have occurred in materials science. However, they have had an enormous effect on structural engineering.

Structural engineering has also been profoundly affected by the development of codes and specifications. Each major area of structural engineering has been defined in codes including structural steel design, reinforced concrete design, prestressed concrete design, pressure vessels and piping systems, bridges (both highway and railroad), pavements, practices and procedures for joining (soldering, brazing, welding, etc.), material properties through standardization of test procedures and results, and many other building, fabrication, and construction codes have standardized design and construction practices (Ref. 2-3, and 2-7).

Reference 2-7 contains a brief history of the problems of developing design codes for pressure vessels and boilers which culminated when the first code introduced was in 1905 for pressure vessels in response to a boiler explosion in a shoe factory in Brockton, Massachusetts. Since that time, design and construction codes were introduced by American Railway Engineering Association (AREA) in 1905, American Association of State Highway (Transportation) Officials (AASHO, now AASHTO) in 1914, the American Institute of Steel Construction (AISC) in 1921, and many others including mine safety and pipelines. The American Society for Testing and Materials (ASTM) has developed test procedures and specific equipment needed to test many types of materials used today. Concurrent with the development of codes was the use of specifications defining materials to be used, processes to be followed, inspection procedures, referenced test procedures, and many other standards and considerations which were organized, made available to, and required to be followed by the practicing engineers, designers, and contractors to initiate and complete various projects.

The term code implies the use of proven engineering practices and procedures for analysis, design, fabrication, use, and often for testing. Specifications developed and evolved so that details of specific engineering projects could be defined and that quality could be maintained. When computers were first being developed, programs were written for solving problems. The evolution and improvement of computer technology and software have lead to equivalent code status for computer based analysis and design techniques reflecting their importance in structural engineering.

Structural engineering has a long and involved history. The major contributions through the middle ages were largely those of construction and development of tools. Beginning about 1450 and lasting about four hundred years was a period in which science began to be incorporated into structural engineering. In the modern period since 1850, developments and discoveries were made in all areas at a continued accelerated pace. The development of computers has had an enormous impact on the analysis, design, testing, fabrication, and use of structures, mechanical systems, and electrical components. The development of codes has given proper balance to safety, conservation and wise use of resources, and wide acceptance of engineering practices and procedures.

2.2 HISTORY OF STRUCTURAL TESTING

Since this book is concerned with structural testing, it is appropriate to review some of its history. There are two parts to this history: the first is a discussion of the instruments and devices used to diagnose structural response and the second is a description of methods and techniques used to conduct structural tests.

Perhaps the first structural test was conducted when an important building or structure failed. Early builders were often required to lie down under a structure while the keystone of an arch or a beam spanning two columns was installed. The builders who lived were held in great esteem by the community and built structures that were considered safe by their users. There were many types of failures when structures were designed using trial and error procedures or even with some very limited accepted practices. The construction of large buildings, bridges, fortifications, and monuments often took a terrible toll in terms of human suffering. Much was learned from the problems encountered in construction processes enabling these early structural engineers to become master builders using trial and error procedures as a crude form of testing.

Da Vinci performed tension tests on iron wires using a test set-up shown in Fig. 2-1. Galileo also performed load tests on tensile and beam specimens. A sample of Galileo's tensile tests is also shown in Fig. 2-1. The wire supported a container for fine sand used to load the wire. The container was filled from the bucket shown on the right. The wire was attached to a spring which was also attached to the spout on the bucket. When the wire broke, the spring action would stop the flow of sand into the container. The weight of the sand was recorded and the failure load for the wire determined (Ref. 2-28).

Experimental measurements related to structural testing originated in England about 1670 with Robert Hooke (1635-1703). His experiments determined force-deformation relationships for axially loaded wires in tension. From the results of these tests he drew the following conclusion:

"It is very evident that the Rule or Law of Nature in every springing body is, that the force or power thereof to restore itself to its natural position is always proportionate to the distance or space it is removed therefrom..."

These experimental results led to the solution of many problems in structural engineering (e.g., the relationships of the normal fiber stresses in a beam in bending). These basic relationships are used today to described the elastic behavior of structures (Ref. 2-28).

The publications of experiments performed by Alphonse Duleau and Pierre C. Dupin about 1815 began the modern era of experimental mechanics. These experiments influenced engineering mechanics and structural engineers well into the 20th century. Duleau and Dupin were each faced with analysis and design of structures in which the behavior of the materials at small strain levels were important. Since no data existed describing the behavior of metals and wood at small strains, each undertook an experimental program to obtain data (Ref. 2-4).

Duleau was commissioned to design a forged iron bridge over the Dordogne river in France. He tested full-size structural shapes with various cross sections to determine their behavior in tension, compression, bending, torsion, and elastic stability. He found a satisfactory solution in linearity for small strains and deformations.

Dupin had responsibility to determine the changes in shapes of wooden ships after they were launched. He conducted tests on wooden beams and determined the actual deformed shape of the beams. From these experiments, a test procedure evolved which included dead-weight tests to determine stress and strain in solids. Duleau and Dupin are credited with introducing quasistatic linear elasticity because, for many years, those involved in analytical mechanics had expected nonlinear behavior of solids over the full range of deformations. The introduction and acceptance of linear elastic behavior for material behavior was a very important contribution in structural

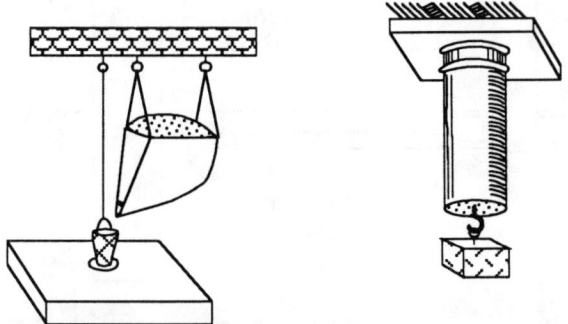

Figure 2-1. Tensile Tests Performed by Leonardo da Vinci and Galileo (From Ref. 2-28)

engineering

Prof. Bell (Ref. 2-4) summarizes the contributions of Duleau and Dupin in the following way:

"The sequel to their work in the 19th century is essentially devoid of controversy. In what might be termed the "Age of Design by Disaster," the relative simplicity of the linear theory of elasticity and its provision of a reasonable approximation in most instances provided a home base for applied engineering sanity."

The work of Duleau and Dupin laid the basis for structural testing. Their adoption of Hooke's law is fundamental to solid mechanics today.

Sometimes structural testing has been referred to as environmental testing. This does not mean that environments are tested but that structural systems and components are exposed to a potentially wide variety of environments. The term structural testing means that systems and components are physically tested. Analysis of structural systems and results of tests on structures should be complementary (Ref. 2-1).

Measurement Devices

The development of instruments and other diagnostic devices for structural testing followed three parts: first, their design and fabrication, second, the acceptance of the device, and third, its adoption and use by others. Many early tests appear to have used one-of-a-kind devices. Early pioneers would develop some type of device tailored for use in one test and then be faced with development of different types of devices for other tests. A single device or instrument was not adopted for use in structural testing and materials science until 1850 when Peter Barlow developed an extensometer for measuring surface strains in a structural member. This extensometer consisted of a single lever system which had a magnification of ten to one. Paine used a single lever extensometer with a magnification of 100:1 on New York City's East River Bridge project in 1883. Improvements were made in these mechanical gages by 1910 when magnifications of 2000:1 were available with gage lengths as short as 0.5 inches. These gages were self-contained and could be read directly.

Dynamic strains were measured using the deForest scratch gage. This type of gage made a scratch equal to the deformation over the gage length. When magnified, the scratch indicated the surface strain. Various types of these single lever type extensometers have been developed with an example of Capp's dividers shown in Fig. 2-2. The operation of the divider was through a change in length in the specimen which induced a rotation in the divider arm. The displacement of the tip of the arm corresponded to a change in length of the specimen per gage length. These single lever gages had mechanical limitations and were soon replaced by multiple lever systems shown in Fig. 2-3. The Huggenberger tensometer came in many sizes and configurations giving the suggestion that experimental stress analysis was possible and that many strain measurements could be performed on a test item. The next improvement in mechanical measurements of displacement came with the use of a rack and pinion system used in conjunction with a dial gage. Various types of dial gage systems were developed including Ames, Federal, Berry, Whittemore, and others with an example of a dial indicator shown in Fig. 2-4. These devices were also modified to measure angles of twist (Refs. 2-2, 2-4, 2-5, 2-11, 2-14, 2-15, 2-20, and 2-28).

Optical measurements of strain were developed when greater accuracies in mechanical extensometers were required and improvements in optical systems became available. The initial optical measurement systems consisted of

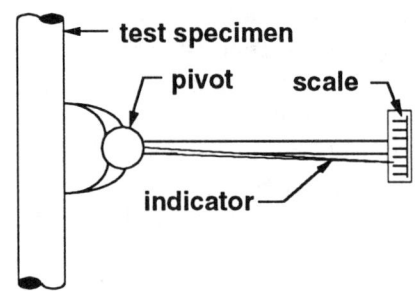

Figure 2-2. Single Lever Extensometer (Capp's Dividers)

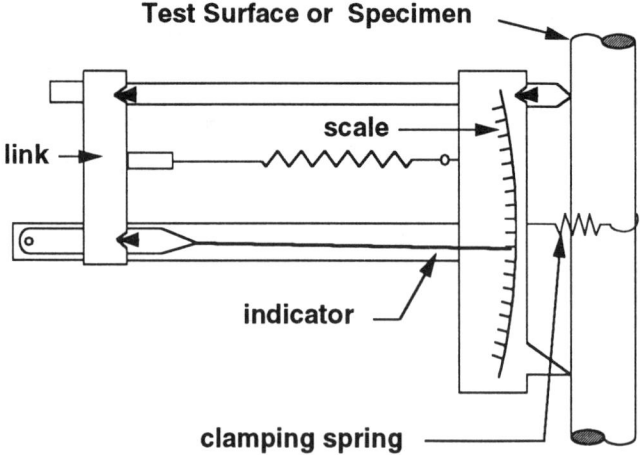

Figure 2-3. The Huggenberger Tensometer (multiple lever extensometer)

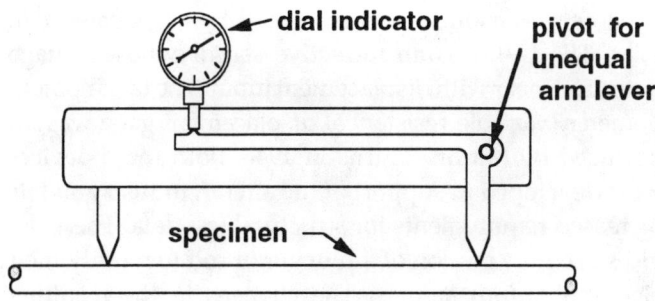

Figure 2-4. Dial Extensometer

both optical and mechanical features which were based on mirrors as the indicating device. A mirror would be installed in contact with a specimen. As the specimen changed length due to the applied forces, the mirror would rotate as shown in the Marten's extensometer in Fig. 2-5. The rotation of the mirror changed the reflection of the line of site as shown on the scale. With appropriate calibration and precision of manufacture, these optical devices could measure strains to a resolution of two microinches per inch. These mechanical and optical-mechanical devices for accurately measuring length significantly advanced the methods of determining the surface strain. However, they did have some disadvantages in that they were difficult to calibrate, to adjust, to attach and position, they were sensitive to vibrations, and they were typically bulky and had restrictions on where they could be located on a structure or specimen.

Another optical based method utilizes grids and fringe patterns. This method is commonly referred to as the moiré method. The term moire originally meant a weaving similar to fine cloth such as silk. The moire method uses as an indicator of strain the interference of superimposed geometric configurations including grids, speckles, and others. Photographs were taken of the test item subjected to incremental loading conditions. These photographs were superimposed making it possible to see interferences and gradients which were related to strains. The development of this basic technique began in the 1930's and was developed into a usable method as reported by Tollenaar in 1945. In the early 1960's, the actual interpretation of fringe patterns related to strains appeared in the literature. The use of moire fringe patterns to measure surface displacements was reported by Weller and Shepard in 1948. An example of a moire pattern is shown in Fig. 2-6 (Ref. 2-11).

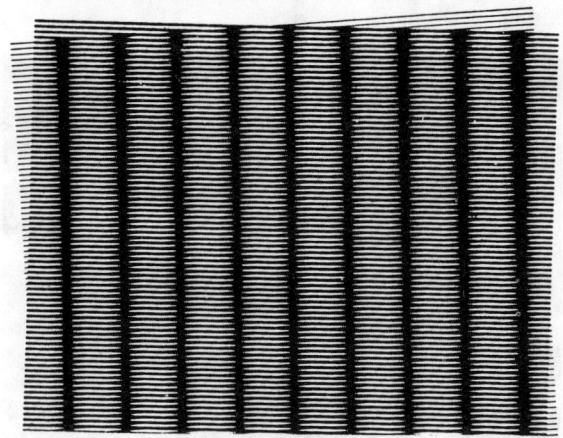

Figure 2-6. An Example of a Moire' Fringe Pattern

Another type of optical system is based on phenomena first investigated by David Brewster in 1814. He discovered that the optical properties of some transparent materials were modified when the material was subjected to stresses. This led to the development of photoelasticity. About 1912 photoelasticity became the principal tool of experimental stress analysis which lasted until about 1940. Photoelasticity is often used today with particular application to complicated assemblies and parts. While this method is usually considered to be used for solution of two-dimensional problems, it has been extended to three dimensions. Birefringent coatings have been developed which can be attached to an actual part or assembly. When a birefringent coated test item is loaded, stress patterns develop and can be viewed in the same manner as with photoelastic equipment. An example of a photoelastic pattern is shown in Fig. 2-7 (Refs. 2-9, 2-11, and 2-28).

Concurrent with the development of photoelasticity came the use of brittle coatings (Refs. 2-9 and 2-23). Researchers found that coatings could be applied to a structure would exhibit crack patterns when the test item was subjected to applied forces. It was also determined that the cracks that developed were typically normal to the principal strain direction. With suitable calibration, the use of brittle coatings developed into a valid experimental technique. An example crack pattern is shown in Fig. 2-8.

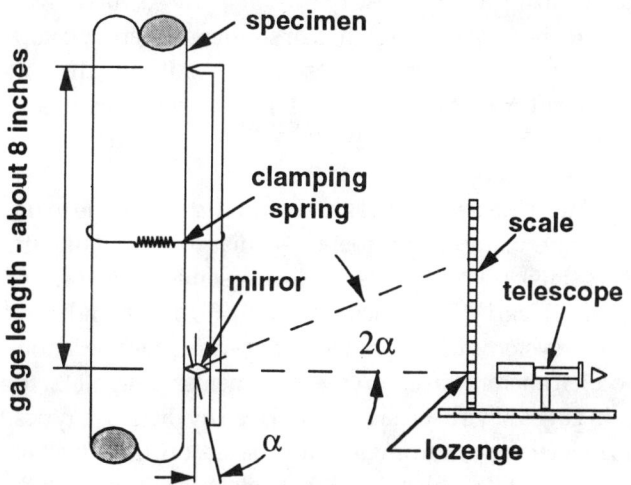

Figure 2-5. Marten's Optical Extensometer

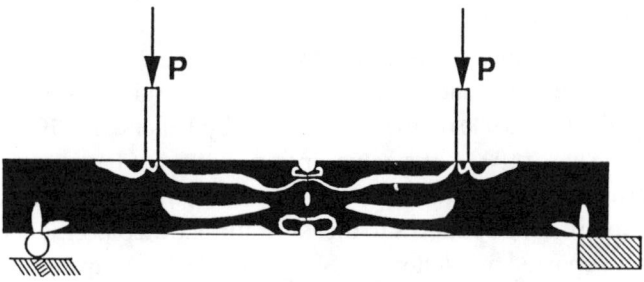

Figure 2-7. A Photoelastic Pattern on a Bending Specimen

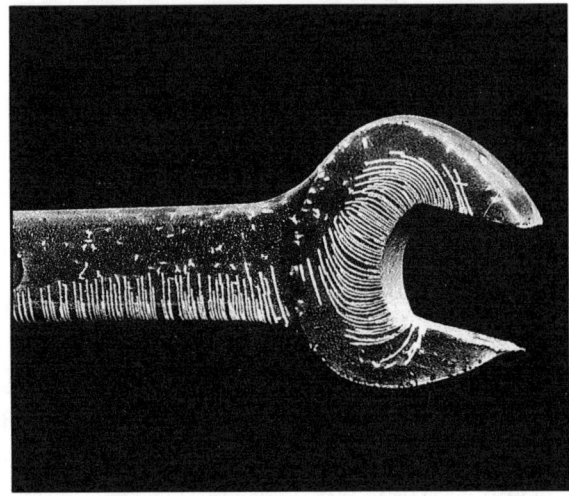

Figure 2-8. A Crack Pattern Generated Using Brittle Coatings

An extension and improvement of optical techniques was introduced in the 1970's by Gabor who won a Nobel prize for the development of holography. The application of holography to structural testing and measurements is in terms of holographic interferometry. Currently, holograms are made by using a laser light divided by a beam splitter. The collection of the original beam and the reflected light waves off a test object create patterns of interferences when compared using photographs or video camera techniques. These interferences can be calibrated and related to displacements. Another variation in optical strain measurements utilizes speckle patterns. These patterns are recorded through video or photographic techniques and are then compared to the patterns on a deformed structure. From the comparisons and appropriate calibrations, the surface strains and displacements can be determined.

Earlier descriptions of mechanical measurements of displacement showed the use of mechanical gages. The first electrically based displacement measurement system was the linear variable differential transformer (LVDT) patented by G.P. Moadley in 1936. The LVDT was developed as a transducer and marketed by M. Schaevitz in 1940. The LVDT is an inductive sensor whose voltage output is linear with displacement input. The linear potentiometer (variable resistance) displacement gage was introduced by Marlin Bourns in 1946. Both these devices were developed in support of the aircraft industry and its increased requirements for structural test data. These devices used the motion of a plunger or rod to modify their electrical output signals. The changes in the resulting electrical signal are calibrated and indicate accurate measurement of displacement. Improvements have been made in these devices so that measurements can be made to 10-5 inches. A variation in an optical technique utilizes a fiber optics technique introduced in 1980's. This device depends only on the amount of light reflected from a test surface. Calibration of this device can lead to measurement of displacements in microns and at very high frequencies. Other types of displacement measurement gages have also been developed (i.e., capacitance, eddy current, transmitted and reflected light systems, magnetostriction) (Refs. 2-6, 2-9, 2-11, 2-12, 2-14, and 2-28).

In 1850, Lord Kelvin attempted to make some exact measurements of electrical resistance. However, his experiments were affected by strain induced changes in the resistance of the wires. This idea was not pursued at this time but became the foundation of the electrical resistance strain gage. In the 1936 the electric resistance bonded strain gage was initially suggested by Edward Simmons. He suggested using a fine resistance wire be cemented to a test surface. Arthur Ruge, unaware of Simmons' suggestion, developed a wire strain gage in 1938. This wire type strain gage was originally mass produced by Charles Tatnall at Baldwin-Southwark. The production process included laying a fine wire around a small diameter tube. The tube was removed and the wire coil flattened. The deformed wire was then attached to a paper backing and this assembly was in turn glued to a structure or part. As loads were applied, the wire was stretched or shortened thereby changing its length and electrical resistance modifying the voltage output from the gage which is related to incremental strain (Refs. 2-2, 2-6, 2-8, 2-9, 2-11, 2-12, 2-16, 2-18, 2-20, 2-22, 2-23, 2-25, 2-26, 2-27, and 2-28).

The measurement of acceleration can be made in five basic ways using: (1) piezoelectric (manufactured or natural crystals), (2) piezoresistive (using semiconductors), (3) servo, (4) variable capacitance, and (5) mass and force balance systems used for measuring seismic accelerations. (Accelerometers based on measurements using metal foil strain gages have been largely replaced by these five types.) Piezoelectric accelerometers are based on the deformations of crystals which develop an electric charge with the charge being proportional to the acceleration. Piezoresistive accelerometers are based on the strain occurring in a semiconductor strain gage where the measured strain is proportional to the acceleration. Servo devices use a bar connected to a force restoring system in which the force required to maintain the bar in position is proportional to the acceleration. Capacitive devices consist of two plates of conducting materials separated by a gap. The acceleration is proportional to one of three changes: the gap material, the area between the plates, or the distance between the plates. The mass and force balance system consists of a single degree of freedom spring mass system enclosed in a frame which is rigidly attached to a base. When the base is excited from a seismic disturbance, the attached mass begins to accelerate. The measured force in the spring is proportional to the acceleration (Ref. 2-9).

Velocity gages have a similar history and makeup to accelerometers. These gages have also been made similarly to an LVDT with a moving coil in a stationary magnetic field. The voltage output is proportional to velocity. Various types of gages have also been made for measuring maximum velocities (e.g., the voltage output of an electrodynamic pickup is proportional to the relative velocity of a between the magnetic flux lines in the pickup and the coil.) Other techniques have been developed using optical methods which employ the Dgzsler effect, optical encoders, and high-speed photography. (Refs. 2-2, 2-9, and 2-12).

Recent developments in infrared cameras applied to thermoelastic behavior (Stefan's Law) make possible the determination of stresses by analyzing the thermal emissions. This technique was introduced in the 1970's and consists of detailed examinations of the relative temperatures on a cyclically loaded test item. The stress pattern intensities are related by temperature gradients. That is, the higher the temperature of the test item indicates higher stresses in the test item. This technique has particular application to in-plane forces and bending but has limited application in pure shear along with a limited temperature range (Ref. 2-15).

Experimental mechanics, historically referred to as experimental stress analysis, really began to have great impact on the design and fabrication of structures when the electric resistance strain gage, displacement gages, and other transducers based on these devices were developed. Strain gages and displacement transducers were used in many tests in development of World War II aircraft.

Testing Machines

Another aspect of the history of structural testing is the development of testing machines. Three main types of force testing machines have been developed and are in use today: the dead weight type, the screw driven, and the hydraulically actuated as shown in Figures 2-9 through 2-11. The dead weight type was developed first and it consisted of a series of mechanical levers used to multiply the forces generated by the dead weights. The screw type machine was developed next and it consists of two or more screws moving a cross head toward or away from a platen. The hydraulic machines generate forces using hydraulic pressures on a piston or ram pushing against or pulling on a crosshead. The development of the hydraulic machine by Amsler (Switzerland) in the 1920's gave capabilities for large forces. Improvements in each type of machine have resulted in their continued use today. These testing machines now have controllers today which can control the force, the stroke, the strain, and the rates at which they are applied, and provide opportunities for computer controls as well. Testing machines have changed dramatically since the development of the first dead weight system. These modern test systems can apply forces, pressures, torsion, shocks, vibration, fatigue, and creep loads and can be combined with temperature and other environments as well (Refs. 2-2, 2-8, 2-14, 2-15, 2-17, 2-27, and 2-28).

A testing machine consists of two parts: mechanical means for applying forces and methods and techniques for measuring the applied forces. The three mechanical systems used for applying forces were just described. The methods used to measure forces required some ingenious development and are described in the next section.

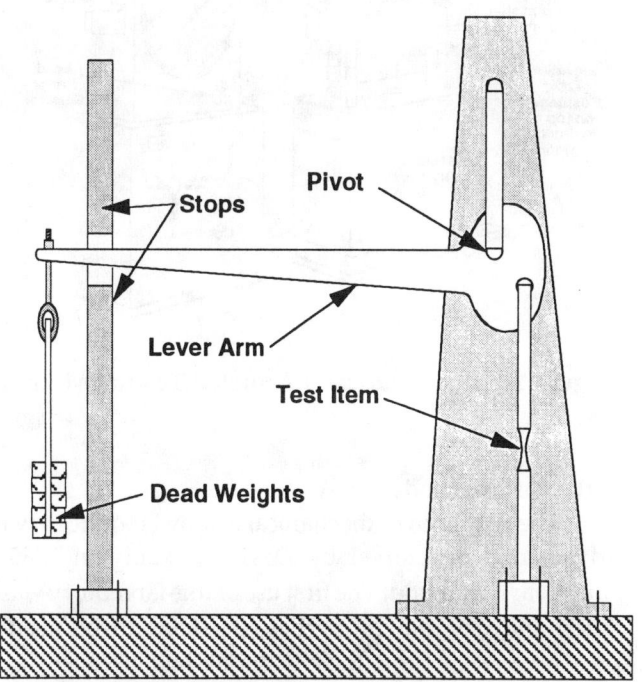

Figure 2-9. Dead Weight Testing Machine

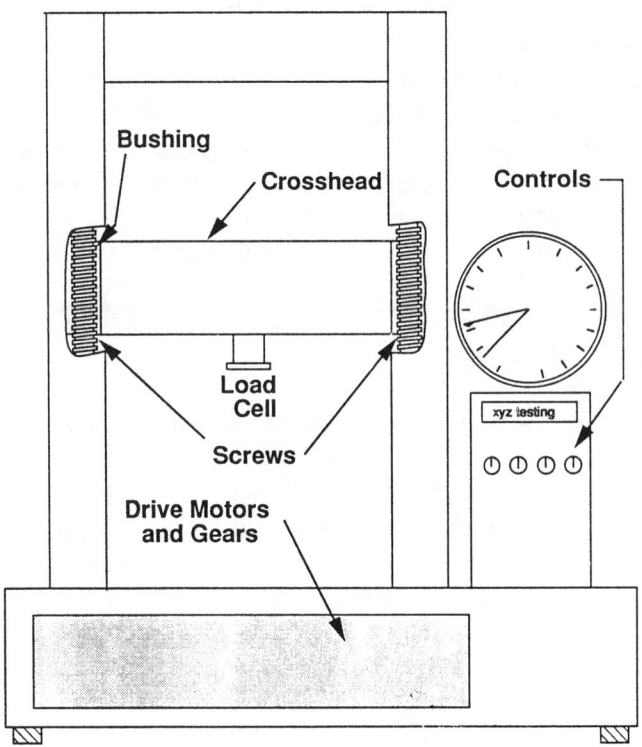

Figure 2-10. Screw Driven Testing Machine

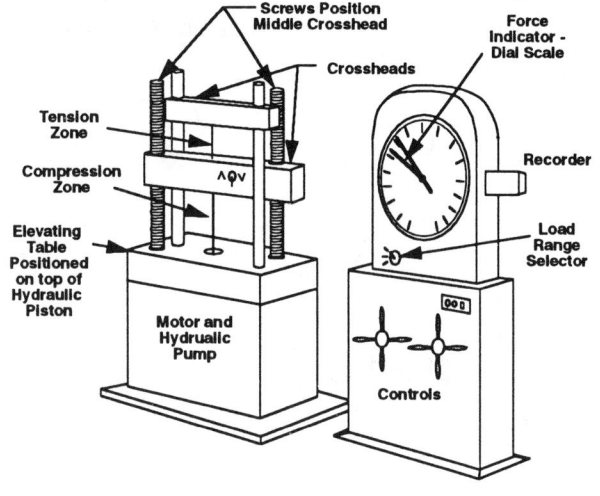

Figure 2-11. Hydraulically Actuated Testing Machine

Early Test Methods

The first known mechanical testing laboratory was a private one developed by David Kirkaldy in 1865 in Southwark, England. The first use of this laboratory was to determine material properties for steels to be used in pressure vessels. Since these materials had just been developed, it was decided that their properties needed to be ascertained. Many specimen were tested which led to the most complete description of material properties of steel and iron at that time. The laboratory used a dead weight and lever testing system along with some specially designed specimens. Many other aspects of material properties (e.g., effect of various shapes on strength, fatigue strength, effects of cold work on metals) were investigated. The first known publicly funded structural testing laboratory was developed in 1871 at the Polytechnical Institute of Munich. It had a testing machine with a capacity of 100 tons. Bauschinger, the director of this laboratory, developed a mirror extensometer which was also used in determining properties of various materials (Ref. 2-28).

Structural testing laboratories were developed so that tests could be performed on full size structural members and systems. The test items were often larger or of different geometry than could be accommodated in testing systems. These early test laboratories had stiffened floor systems which served as force platforms in which the test items could be secured to the floor and loaded, often in more than one direction. Whiffle tree systems were developed to distribute loads over an area or aircraft wing so that distributed loads such as snow or air pressures could be simulated. These systems used hydraulic or dead weights and a series of arms and levers which were in turn anchored to the structure being loaded. Other mechanical systems (e.g., reaction frames) were developed to apply forces, restrain test objects, and to support instrumentation systems.

Another contribution to structural testing comes in the development of techniques to measure forces. The beam balance system in the dead weight testers gave the forces directly. O. S. Peters developed the carbon pile strain gage/load cell in the 1920's. This device used an oscillograph to record the output voltage of the carbon gage as an electrical signal indicating force or strain. One unique feature of these older machines is the Bourdon Tube shown in Fig. 2-12. The Bourdon Tube is used to accurately indicate the applied force. This closed circular tube is designed so that it tends to straighten out when internally pressurized. A pressure line is connected from the hydraulic cylinder to the tube. As the pressure increases in the cylinder, that pressure is transmitted to the tube changing the position of its tip relative to its fixed base. The displacement is greatest at the end of the tube and the multiplying arm is attached there. The pointer on the dial gage is related to the displacement of the Bourdon tube. The testing machine is calibrated for force by relating the rotation of the pointer caused by the Bourdon tube to known forces measured by load cells or proving rings. Other methods have also been used to measure the forces in these machines: manometers are U shaped tubes which balance the pressures in either side of the tube. An unbalance in pres-

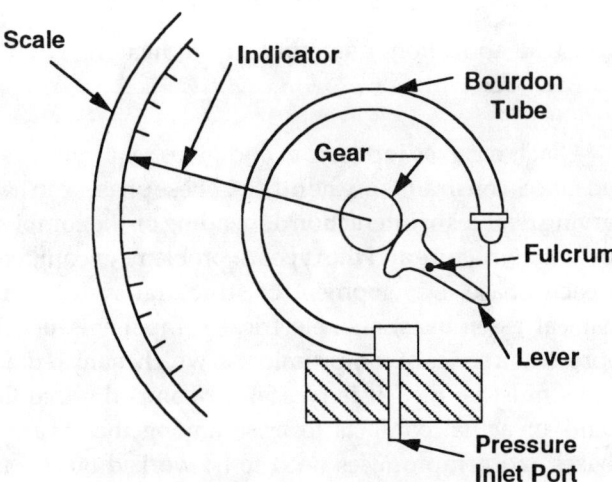

Figure 2-12. Bourdon Tube Used to Indicate Applied Forces

sures indicated by differences in elevations of the liquid on either side of the tube is a measure of the applied force. Dynamometers (spring balances) indicate the applied force by displacement of the spring. Pressure gages are also used to indicate force by using with the known area of the hydraulic piston multiplied by the measured pressure. The principal problem in the use of these pressure conversion systems is when air is trapped in the hydraulic lines making the systems respond out of phase with the applied force (Ref. 2-28).

Forces are now measured with load cells using strain gages or quartz type sensors, in addition to conventional dead weight systems. Forces can now be measured in other ways including ultrasonic detection of changes in lengths which can be related to forces by areas and mechanical properties as described in Eq. 2-2 (Ref. 2-18).

$$P = AE\Delta / L \qquad \text{Eq. 2-2}$$

where P = the force in a tensile member
A = cross-sectional area of member
E = modulus of elasticity
Δ = change in length
L = length of stressed portion of member

Many different types of structural tests have been introduced over the years. Among the first types was the model test. Scale models of various structures have been constructed and load tested in one type of test which is a direct model in which the necessary measurements to define the response of a structure are made directly or with the use of appropriate scale factors. The second type of model is called an indirect type used to establish force-deformation relationships so that influence lines can be developed. By utilizing appropriate scaling parameters and the Maxwell-Betti reciprocal theorems and the Muller-Breslau principle the appropriate similitude requirements can be established. G.E. Beggs introduced a deformator gage in 1922 so that influence lines could be determined from models. W.J. Eney introduced an improved version of a deformator in 1932. The reader is referred to reference 14 at the end of the chapter for additional information and ideas on structural models (Refs. 2-3, 2-4, 2-5, 2-6, 2-9, 2-11, 2-13, 2-14, 2-15, 2-18, 2-20, 2-21, and 2-28).

An often overlooked structural test is the determination of mass properties (weights, moments of inertia, center of gravity, etc.) Early ship builders were concerned about these properties; however, there has been no recorded historical evidence that these measurements were made. The first evidence of these measurements in the modern era came with the Wright brothers and their development of the airplane. Measurements of these physical properties (e.g., weight and center of gravity) were needed for design of the first airplane.

The development of the airplane led the Wright brothers to some interesting solutions of designing a structural system when the forces to be encountered were not known. They decided to invert an airplane and use gravity loading to destructively test the wings. Then, a ratio of actual wing strength to its dead weight capacity could be expressed in multiples of gravitational loading or g's. The performance of the airplane was then limited by imposed factors of safety based on these static tests. This type of test became a standard for aircraft design. An example of an early static test is shown in Fig. 2-13. As more information became available on aerodynamics, additional tests were performed to prove a particular design of an airplane. These static tests along with dynamic and fatigue tests are performed on aircraft today. Reference 2-30 gives results of a static test performed on the wings of modern passenger airplane.

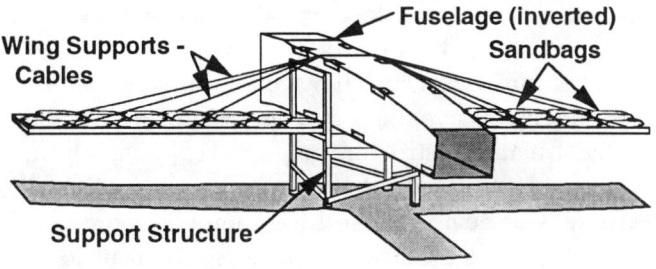

Figure 2-13. Albree-Fraser U.S. Army Scout Airplane Undergoing Static Testing at McCook Field (circa 1917). (This airplane was manufactured by Pigeon Hollow Spar Airplane Co.)

Among the relatively recent advancements in experimental devices is the development of the foil strain gage; displacement measurements using eddy currents, light transmission and reflection devices, capacitance gages, and holographic/interferometry systems; testing machines and pressure systems which are equipped with function generators for control of load, stroke, strain, cycling, etc.; data recording systems which can accommodate large strain rates and can record large quantities of data in microsecond time regimes (Refs. 2-15, 2-21, 2-23, 2-25, 2-26, 2-27, 2-28, and 2-31).

Structural testing has a more recent history than structural engineering. The development of methods and techniques of experimental mechanics for structural testing has its origins around 1800 and really accelerated in the World War II era in terms of equipment and measurement devices. The Society for Experimental Stress Analysis (SESA) was formed in 1943 and was renamed the Society for Experimental Mechanics (SEM) in 1984. This professional organization is dedicated to the advancement of experimental mechanics by organizing technical activities and serving as a forum to exchange ideas, and encouraging the development of experimental methods and techniques.

2.3 STRUCTURAL ENGINEERING: DEFINITION AND PHASES

A structure is an engineered or otherwise developed system or device composed of a single or interdependent parts required to fulfill some intended functions and to resist the applied loads, environments, and other conditions imposed upon it. For example, the frame and body of an automobile have the primary function of supporting the vehicle weight and its cargo while resisting the transportation shock and vibration conditions imposed. Another function of the frame and body is to provide protection for the occupants by absorbing energy in a collision. An additional requirement of the body is to have pleasing aesthetic qualities needed to make the product saleable. It is typical for a structure to have one or more major functions as well as other functions which it may be required to fulfill on a routine basis or as a single event over the expected lifetime of the structure. A structure may consist of a single beam or column, a complicated building or aircraft, a mechanical assembly or device, an electrical component, or virtually any other object which has some requirement to serve mankind (Ref. 2-10).

Structural engineering consists of five basic phases:

1. Design
2. Analysis
3. Testing
4. Fabrication, Assembly, or Production
5. Use

Each receives input from and gives requirements to and places constraints on the others. These phases can have varying degrees of interaction depending on the complexities of the structure and the types of problems encountered in each phase. Development of structural systems, mechanical assemblies, and electrical components usually represents a series of compromises in which many requirements must be met. It is typical for competing requirements or actual conflicts to arise among these various phases and compromises need to be worked out for the good of a particular project (Refs. 2-3 and 2-10).

The design process includes the following:

1. Identification of the intended functions or requirements that a structure, structural system or component must fulfill

2. Determination of the applied forces or other loading conditions to be imposed on the structure

3. Development of a system to satisfy the functions and resist the loads.

The design process is an iterative procedure during which various designs and ideas are advanced and prepared for review. Compromises are made by selecting features from the various designs and ideas and discarding those concepts which do not appear as favorable nor offer the necessary advantages. As a minimum, the design process should include choices of basic configurations, sizes, and shapes of the structure, assembly, or component, choices of materials, engineered drawings and often prototype hardware as well cost estimates.

The analysis phase consists of predicting how a structural system will respond to various loading conditions and then evaluating the adequacy of a particular design using the fundamental principles of engineering mechanics and procedures. These principles reveal the relationships between the applied forces and pressures and other environments and conditions and the response or behavior of the structure or component. The procedures reflect adoption of and adherence to recommended design values, knowledge and understanding of material properties and behavior, and adequate predictive capabilities describing structural behavior. Various analytical methods and techniques are used for predictions of behavior depending on the type of structure, the loading conditions and forces applied, and the kind of information needed to describe the response of the structure or part. Often, the level of effort or type of analytical approach is directly related to the complexities of the structural system or component and the loading conditions. Most structural analysis problems can be approached with more than one theory and method. The use of the various theories and methods will define or at least suggest limits to possible response of a structure as it is loaded. Differences in results are likely to occur using various analytical techniques because of the assumptions used in particular methods, the limitations of the solution techniques, and how these results are interpreted. It must be emphasized that these analytical techniques continue to improve and are more powerful, widely available, and accurate than ever before.

The fabrication, assembly, or production phase is that activity in which prototype or actual hardware, buildings, components, or other structures are constructed and assembled. Typically, most structures consist of a variety of elements or components joined together to form a system or component. There are many methods and techniques for joining structural members together (welding, soldering, brazing, adhesives, mortars, bolts, rivets, reinforcing steel, filaments, etc.). These attachments can be made in the field or on assembly lines. These joints are critical to a structure because they make it possible to fulfill the intended functions. Also, these joints are important in a structure because there is a modern day axiom which indicates that "the integrity of a structure is as good as its joints." This rule is not true in every situation. However, there is a high degree of correlation between degradation in structural performance and the reduction in integrity of joints and connections. The portions of structural systems and components receiving the most scrutiny in all phases of structural engineering are the connections or joints. Failures are often precipitated at or adjacent to a connection. The fabrication process can enhance or degrade the strength or behavior of connection often by subtle means. Therefore, fabrication is vital to the successful performance of a structure. Another important aspect of the fabrication process is the costs involved. Choices must be made on the materials (e.g., concrete, steel, wood, composites), the connections (bolted, riveted, welded, adhesives, etc.), and the other details of fabrication and construction, and minimizing costs influences the types of processes and procedures used in fabrication, assembly, and production.

The testing phase consists of obtaining experimental evidence describing how a structure responds to applied forces or loading conditions. The type of test conducted depends on the information needed, the intended use of the structure or component, and the capabilities available to perform a particular test. Types of tests include the following: full scale and models, calibrations, acceptance or quality assurance types, proof tests, materials characterization, behavior of joints and connections, stiffness or load-deformation tests, energy absorption, strength or failure load and mode determination. These tests can be performed in force, inertial, pressure, temperature, shock, vibration and other loading conditions. Each of these basic types of tests will be described in detail in Chapter 4. Tests on structures and components serve to verify analyses, provide information on design margins and factors of safety, and give evidence that a structure can perform its intended functions and meet the requirements imposed.

The actual use of a structural system is the fifth phase of structural engineering. Structures typically serve their users as they were designed and intended to be used. Occasionally problems requiring a structural solution are discovered after a structure is built. The knowledge of possible corrections can be used to modify hardware in production, to make modifications to existing structures, and to provide improvements in future designs. It is also common for a structure to be required to fulfill new, different or extended functions after it has been designed, analyzed, tested, and constructed. Therefore, the "use" phase of a structure is important because it gives the final verification of design and it can also provide new requirements. Some structures can be easily modified or adapted to new or different requirements; however, each case usually requires its own evaluation. Many examples exist of a structure designed for one use and converted to another: a warehouse converted to a laboratory where temperature and humidity controls are important, a roof beam required to support a solar heating system, a Boeing 747 airplane required to carry additional loads such as the Space Shuttle, and a computer system designed for home or laboratory use experiencing transportation shock and vibration conditions when used for data acquisition and test control in over-the-road testing of a new automobile.

These alternate uses for an existing structural system can present some significant problems to engineers including evaluation of the structure, determination of tests or other proof needed to certify that the new requirements can be met, performing the necessary analyses, and obtaining the available documentation and description of the existing structure. One way an engineer provides for the unexpected is the use of design margins. These margins or factors of safety have developed because there are often inadequate definitions of material properties, insufficient quality assurance controls on fabrication or construction processes, or less than adequate definition of the applied forces or loading conditions to permit a structure to be used at or near its limiting capabilities. Design margins provide for safe and adequate use of a structure over its lifetime of intended service.

2.4 REFERENCES

2-1. V. Askegard, "Teaching and Research in Experimental Mechanics: Complementarity of Theory and Experiment," Internal Report, NR I 61, Technical University of Denmark, Lyngby, 1978.

2-2. T.G. Beckwith and N. L. Buck, Mechanical Measurements. 2nd Ed., Addison-Wesley Co., Reading, Mass., 1973.

2-3. L.S. Beedle, et.al., Structural Steel Design. Ronald Press, New York, 1964.

2-4. J.F. Bell, "Experimental Solid Mechanics in the Nineteenth Century," the 1989 William M. Murray Lecture, Proc. Society for Experimental Mechanics Spring Conference, Cambridge, Massachusetts, May 28-June 1, 1989.

2-5. B.C. Boggs, The History of Static Test and Air Force Structures Testing, Air Force Flight Dynamics Laboratory, AFFDL-TR-3071, Dayton, Ohio, June 1979.

2-6. G.A. Brewer, Practical Solutions to Problems in Experimental Mechanics. 1940-1985, Vantage Press, New York, 1987.

2-7. R. Chuse and S.M. Eber, Pressure Vessels: the ASME Code Simplified, 6th Ed., McGraw-Hill, New York, 1984.

2-8. J.W. Dally and W.E. Riley, Experimental Stress Analysis. McGraw-Hill, New York, 1978.

2-9. R.C. Dove and P.H. Adams, Experimental Stress Analysis and Motion Measurement Merrill Books, Columbus, Ohio, 1964.

2-10. D.A. Firmage, Fundamental Theory of Structures, Krieger, New York, 1985.

2-11. M. Hetenyi, Handbook of Experimental Stress Analysis. John Wiley and Sons, New York, 1950.

2-12. E.E. Hurceg, Handbook of Measurement and Control. Schaevitz Engineering, Rev. Ed., Pennsauken, N.J., 1976.

2-13. J.S. Kinney, Indeterminate Structural Analysis. Addison-Wesley, Reading, Mass., 1957.

2-14. A.S. Kobayashi, Ed., Handbook on Experimental Mechanics, Prentice-Hall, New York, 1987.

2-15. A.S. Kobayashi, Manual on Experimental Stress Analysis. 4th Ed., Society for Experimental Mechanics, Bethel CT., 1983.

2-16. J. Marin, and J. A. Sauer, Strength of Materials. MacMillan Co., New York, 1954.

2-17. C.C. Perry, and H.R. Lisner, The Strain Gage Primer. 2nd. Ed., McGraw Hill, New York, 1962.

2-18. E.P. Popov, Introduction to Mechanics of Solids, Prentice-Hall, Englewood Cliffs, NJ, 1968.

2-19. R. Raymond, Out of the Firey Furnace. The Impact of Metals on the History of Mankind, Pennsylvania State University Press, University Park, PA, 1986.

2-20. R.J. Roark, Formulas for Stress and Strain. 4th Ed., McGraw-Hill, New York, 1965.

2-21. G.M. Sabnis, et. al., Structural Modeling and Experimental Techniques, Prentice-Hall, Englewood Cliffs, NJ, 1983.

2-22. C.R. Smith, "Bicycle Spokes—Their Use in Testing Aircraft Structures," Proc. of SESA, Vol. 3, No. 2, 1946.

2-23. P. Stein, T. Kemeny, and Havrilla, K., Golden Book of Strain Gages. Load Cells. and Brittle Coatings. to be published.

2-24. S. Strandh, A History of the Machine. A&W Publishers, New York, 1979.

2-25. J.E. Starr, "The Bonded Resistance Strain Gage—Its Golden Anniversary," Experimental Techniques, Vol. 9, No. 12, Dec. 1985.

2-26. J.E. Starr, J. Dorsey, and C.C. Perry, "Fifty Years of the Bonded Resistance Strain Gage—An American Perspective," pp. 1-21, Proc. Western Regional Strain Gage Committee, Jubilee Issue, June 1988, Portland, OR., SEM, Bethel, CT.

2-27. F.G. Tatnall, Tatnall on Testing. Reprinted 1979 by Department of Mechanical Engineering and Applied Mechanics, Univ. of Pennsylvania.

2-28. S.P. Timoshenko, The History of Strength of Materials. McGraw-Hill, New York, 1953.

2-29. P. Tompkins, Secrets of the Great Pyramids. Harper & Rowe, New York, 1971.

2-30. J.P. Wallace, "(Boeing) 757 Major Static Test Instrumentation Systems," Proc. of the Combined Meeting of Technical Committee on Strain Gages and Western Regional Strain Gage Committee, Las Vegas, NV, June 13-14, 1985, Society for Experimental Mechanics, Bethel, Ct.

2-31. A.L. Window and G.S. Holister, Strain Gauge Technology, Applied Science Publishers, London, 1983.

2-32. N. Hawkes, Structures, The Way Things Are Built, Macmillan, New York, 1990.

Chapter 3

Test Purposes and Test Evaluation Criteria

*Robert T. Reese, Sandia National Laboratories,
Albuquerque, New Mexico*

3.1 INTRODUCTION

Each structural test involves imposing one or more loading conditions on test hardware and obtaining some measurements and other indications describing structural response. Each test conducted should have one or more purposes or reasons why the test was performed. Included as possible reasons are (1) the thorough technical understanding of and the insights-comprehending how and why-a structure or assembly responds to applied forces and other conditions, (2) to satisfy certain requirements (e.g., codes, regulations), and (3) to obtain specific data needed (e.g., strength, accelerations, displacements). Evaluation criteria are also needed because their use provides the necessary evidence and assessments of test performance. These criteria enable test personnel to comprehend the underlying causes for structural responses, interpret test results, and select test techniques.

For example, when the purpose for testing a lifting fixture is to prove it is safe for use around people, evidence is needed that the necessary test was performed and how the structure responded. The test purpose in this example is a proof test. The evaluation criteria is the evidence indicating how the structure performed in the test (e.g., the lifting fixture had sufficient strength to support the required overload with no observable damage) and how the evidence should be interpreted (e.g., that the structure is safe to use around personnel).

Test purposes and evaluation criteria need to be established before a test is performed because they are essential in planning, organizing, and conducting a test. Test purposes and evaluation criteria provide the necessary guidelines for performing successful tests, defining and providing limits to test activities, and ensuring high quality tests (Refs. 3-3, 3-8, 3-11, 3-13, 3-16, and 3-21).

The real value and importance of structural testing comes when the responses can be understood and interpreted, insights are developed into how a structure responds, prototypes or actual structures are shown to be capable of meeting and exceeding its intended functions, or design modifications and improvements can be initiated based on insights gained from testing. The understanding and insight gained from structural tests may come in many ways including: (1) correlation with analytical models of structural behavior, (2) proving or offering evidence that structures can meet their intended strength, deflection, vibration, shock, or other requirements, (3) demonstrating that components can meet their functional requirements (e.g., electrical performance) while subjected to a variety of loading conditions, or (4) some other equally important requirements or performance expectations such as fatigue resistance or large plastic deformation capabilities in deep-draw metal forming. Test purposes and evaluation criteria are the necessary guides in planning and performing tests as well as the basis for interpreting the test results. Test evaluation criteria can also be considered to be the moderators between expected and actual response of a structure. Well-defined test purposes and evaluation criteria are very important aspects of performing valid tests.

Another important aspect of performing tests related to test purposes and evaluation criteria is the number of opportunities to gather information. Some test plans have many opportunities; that is, they have been developed with statistically significant number of tests, have many specimens or items available for testing, address the important variables, and will have extensive data analyses performed. Some types of tests (e.g., modal analysis) often require many cycles of testing to develop statistically sufficient insights into structural behavior. Other tests are considered single cycle; that is, there is only one opportunity to gather information. Destructive tests (e.g., static, dynamic) on prototype or actual structures are examples of a single opportunity tests because damage to the structure is usually considered non-repairable or the material properties have been altered significantly so that the structure cannot be tested again in the same manner. The test purposes suggested in the next section apply to the broad range of tests—the single opportunity through the statistically based test program consisting of many individual tests.

Two examples of investigations of structural failures are given to introduce why test purposes and evaluation criteria are so important. The first example is the problems that occurred in some of the 3,500 ships used for transport purposes in World War II. These Liberty and similar ships were built of preassembled sections which were then welded together. There was a great need for these ships because of losses occurring in combat and the need to move massive amounts of cargo and personnel great distances. These ships were constructed at a record breaking pace with one shipyard turning out one completed ship per day. Since the design of the ships was similar and sufficient number of failures occurred, it was possible to analyze the structural deficiencies in a statistical basis.

Many of the failures of these ships occurred in service. An investigative board was formed to evaluate design and construction methods used in all welded ships. This board reviewed many aspects of potential problems: materials, design, fabrication, welding, and inspection. This investigation resulted in the development of the relationships among the types of steel plates, weldments used, and the failures that occurred. From this investigation some specifications of materials were developed including the 15 ft-lb. Charpy Vnotch toughness criterion which has been used for many years (Ref. 3-18). The failures that occurred in these transport vessels were among the first instances of repeated brittle fracture of welded materials. Current engineering practice requires the characterization of critical materials and weld processes (Refs. 3-7 and 3-18). These practices are based in part on the problems encountered by these Liberty ships.

The second example is the dynamic behavior and failure of the Tacoma Narrows Bridge which collapsed after only four months of service. Prior to the collapse large amplitude vibrations (lateral, vertical, and torsional) were observed. The cause of these vibrations was interactions of the bridge superstructure with the wind loadings prevalent in the area. The bridge width (two lanes-less than 40 ft.) and cross-sectional shape made significant aeroelastic interactions along with it difficulties in shedding vortices. The Tacoma Narrows Bridge experienced repeated cycles of wind-induced vibrations resulting in a dramatic fatigue failure of the 2,800 ft. superstructure. The aerodynamic loads on this bridge were not considered to be the limiting design condition.

After the collapse of this suspension bridge, other suspensions bridges were investigated to determine if they were also susceptible to large amplitude vibrations caused by wind loadings. Models of these other bridges were typically subjected to gusts in wind tunnels. The other bridges evaluated did not have the detrimental characteristics of the Tacoma Narrows Bridge. A general rule of behavior was developed which indicated that wind-induced vibrations, if allowed to continue over long periods of time, can lead to sufficient numbers of load cycles to induce fatigue in critical portions of structures (Refs. 3-12 and 3-20). This bridge failure is an example of where only one possibility existed to obtain data. The behavior of this bridge was studied in detail and the results were successfully applied to many other bridges. This example indicates the level of effort that may be required to determine how a structure actually responds to the loads imposed.

3.2 TEST PURPOSES

In this section we focus on test purposes or the reasons why structural tests are performed. The use of test purposes makes it possible to determine what insights are needed and the outcomes expected. In the next two sections (3.3 and 3.4), we will introduce how test evaluation criteria can be used to understand and interpret structural test results and give specific criteria used in testing.

If the outcome of every structural test was known before tests were performed, then there would be little need for testing. Testing offers irrefutable evidence of the behavior of structures subjected to actual, simulated, overload, or in other needed conditions. Test results and data along with the associated correlations and interpretations indicate whether the performance of tested structures measures up to the desired standards or expectations. The reasons for performing structural tests should focus on and emphasize fundamental and objective understanding of structural behavior. Structural tests performed on complete systems, assemblies, components, and individual members or elements typically satisfy two fundamental needs: (1) obtaining enough information to determine how they behave statically, dynamically, in failure situations, or in any other conditions of interest or concern and (2) to obtain evidence to confirm or verify they satisfy requirements and specifications (e.g., codes, regulations, company rules, requester supplied conditions, or test expectations) (Refs. 3-1, 3-8, 3-11, 3-16, and 3-21).

Test quality was defined in Chapter 1 as consisting of two parts: (1) obtaining sufficient information to describe structural response and (2) that the information obtained is valid. This commitment to quality forms a consistent basis for test purposes because well-executed structural tests give objective evidence of structural behavior. The importance of clearly defendable test evidence describing performance and capabilities of structural systems (e.g., commercial aircraft, shipping systems for nuclear materials, pressure vessels and systems) is essential in developing compliance to codes, regulations, and design requirements as well as gaining public acceptance and trust.

Test purposes vary widely. The introduction to this chapter contained an example of a proof test on a lifting fixture. Test purposes may be considerably more complicated such as the certification of a commercial aircraft which includes a multi-step testing process. Satisfying federal requirements for air worthiness will require many tests during development of individual components and in the ground and flight tests on prototype and actual aircraft.

Test money is precious and the need to gain as much information as possible from each test is more important than ever before. The amount and requested detail of test results are greater than ever. Improved diagnostics, data analysis techniques, and controls for test equipment are developing more rapidly than ever. Those involved in structural testing need to consider in sufficient detail and thoroughness why a test is to be performed, what can be learned from it, how the response of the structure or component will be better understood, and how to take best advantage of test capabilities and opportunities. Test planners and requesters need to spend sufficient time and effort in defining and planning why structural tests are needed and how they should be performed.

In this age of interpretation of codes and regulations, management directives, and budgetary constraints, it is possible to perform a test without useful purposes. For example, suppose funding were allocated to perform proof tests on proposed permanently attached personnel safety railings on wings for commercial aircraft. These railing are to be a part of an improved safety program for maintenance workers. If the tests are not performed in a specified time period then funding will be lost. Such proof tests would have little real value as the implementation of these railings is a very unlikely reality because of the loss of airworthiness and structural integrity of the aircraft. Performing such tests is likely to be demeaning to a test organization, to the responsible engineers and technicians, and to the management which has been forced to complete this proof test. However, on occasion, this type of test may be required.

The shrewd structural test engineer will use the opportunity to test these personnel railings to learn something from each test performed such as—can the testing system be operated differently, can engineers and technicians expand their knowledge and expertise in the use of the test equipment, can different diagnostic capabilities and test techniques be incorporated into laboratory practices? Even in tests which may appear to have meaningless value, experience can be gained, training given, and test capabilities improved.

Test purposes can be considered from many perspectives; the usual reference is in terms of the structure being tested. Structural testing to be performed on prototype hardware or actual structures can be described in six basic parts as summarized in Table 3-1 (a more detailed flow chart for test organization is given in Appendix C and discussed in Chapter 14). Table 3-1 gives the relationships among test purposes, evaluation criteria, test activities, and the desired insights needed from structural testing.

Test purposes provide the initial, continuing, and lasting focus and direction on why structural tests are performed and guide the planning and organizing efforts. Their importance in test planning and organization efforts cannot be overemphasized.

Table 3-1. Six Basic Parts of Structural Testing Relating Test Purposes, Evaluation Criteria, and Insights Needed

Basis/Source of Test Activity	Six Basic Parts of Test Planning, Organizing, and Activities
Test Purposes	Determining Loading Conditions, Determining Boundary Conditions and Laboratory or Field Simulations
Test Purposes	Determining Test Methods and Procedures, Safety Procedures, and Basic Test Sequences
Test Purposes/Evaluation Criteria	Selecting and Installing Diagnostic Devices—Sensors, Transducers, and Gages
Test Purposes/Evaluation Criteria	Obtaining Data, Results, Observations, and Other Descriptions of Structural Responses
Evaluation Criteria	Analysis, Organization, and Presentation of Data, Correlation/Interpretation of Data and Test Results
Insights into Structural Behavior	Evaluated and Interpreted Data and Test Results, Analysis of Failures and Other Responses

3.3 TEST EVALUATION CRITERIA

Test evaluation criteria consist of methods and ways that test results can be interpreted and understood. The methods and ways may be commonly used (e.g., strength tests where maximum applied forces are measured) or may be unique to a particular test. Test purposes cannot be fulfilled unless the test data, results, and other indicators of response can be objectively assessed and evaluated.

Evaluation criteria are usually specified through methods and techniques of engineering mechanics and materials science. Criteria developed through the rigors of mechanics and materials have associated sensors, transducers, and other diagnostic methods plus accepted test techniques. These instrumentation systems typically interface into data acquisition systems and can be correlated with analytical methods and techniques. The data obtained from the associated measurement devices used in testing will make correlations and comparisons of theory and test data more direct and easier to understand. Criteria not described in terms of engineering mechanics typically make testing activities and data interpretation more difficult.

For example, a safety evaluation of an electronic component may require a crush test or exposure to an engulfing fire. The main emphasis in evaluating the performance of this component is often to determine if it functions electrically while it is being tested and/or after it has been crushed or burned. Planning and conducting tests for this component with the appropriate diagnostic devices from engineering mechanics can be very challenging tasks because the failure modes are really defined in terms of the lack or reduced electrical function of the component. Further, relating degrading electrical functions in terms of the diagnostics available through engineering mechanics is not straightforward.

It is not uncommon for tests performed in structural testing laboratories to have the responsible personnel adopt a wait and see (e.g., "play it by ear") approach to determine how a complicated assembly will respond. That is, the test evaluation criteria may be defined even as late as when the structure fractures prematurely in a test. Or, in rare instances, the evaluation criteria can be so removed from the realms of engineering mechanics that those performing the test are at the mercy of the requester. Recalling that the purpose of testing is to gather data and information so that understanding of and insights into structural behavior can be gained, it is important to realize that any test may produce surprises which have not been accounted for in the analyses or expected behavior of the structure under load. There is a need to compare and evaluate the response of a test structure with predictions of behavior.

People involved in structural engineering have usually been trained in theoretical analyses and computer generated predictions of structural behavior. Newly trained people or those not familiar with testing are often constrained by these analyses and sometimes do not understand or appreciate subtleties and complexities of how some aspects of real structural systems (e.g., threaded connections in dynamic conditions, residual stresses, and tolerance build-ups) can affect responses. They tend to be suspicious of or are unwilling to accept the test results, particularly when the results are different than expected because they do not match the predictions. Rarely do test results and theoretical predictions agree completely to the smallest measurement of strain, displacement, acceleration, or frequency. As the structures and components become more complicated there is even less likelihood that analytical results and test data will agree exactly. Some test engineers have said that tests on complicated structures involving many load paths and multiple failure modes may result in a significant lack (e.g., 30 percent or more) of agreement and test results.

Typically, a structural test will exercise prototype or actual hardware or components in one or more ways. However, it is difficult in one test to exercise all the main load carrying portions of the structure. Suppose a simply supported I-beam is subjected to two concentrated loads P_1 and P_2 applied on top of the beam perpendicular to the flanges and in the plane of the web as shown in Fig. 3-1. The applied loads contribute to only elastic behavior in the beam as the beam returns to its original position when the loads are removed. Both the undeformed and displaced configurations are shown in the upper right portion of Fig. 3-1. If an acceptance criterion was developed that limited the deflection of the beam to a certain value, then the performance of the loaded beam could be evaluated in a test or in actual use. If the evaluation criteria limited the displacement of the beam to the value shown by δ and the applied force caused a greater displacement, then the performance of the beam would be considered unsatisfac-

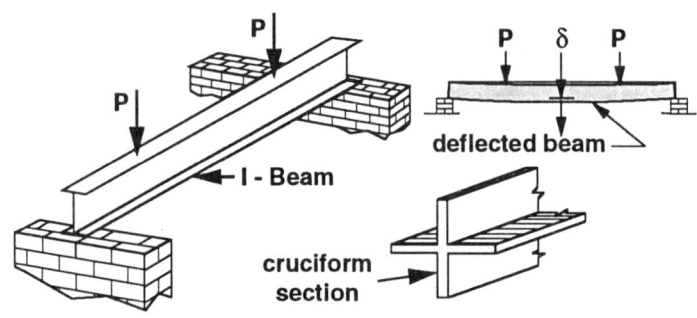

Figure 3-1. Simply Supported Beam

tory or even a failure although the behavior of the beam was elastic and no permanent deformation resulted. In addition, suppose there was a requirement for the beam to meet certain performance criteria for combined axial and bending loads, or for axial and lateral loads in two planes along with torsional loads applied about the longitudinal axis of the beam. Providing these loading conditions in tests could be difficult, particularly if the beam were very flexible in torsion (e.g., cruciform or X-shaped cross-section; flexible structures are difficult to test because of requirements for applied forces to maintain alignment to a structure which undergoing significant deformation).

The major point in this beam testing example is that the test evaluation criteria are needed to plan the test. If only in-plane loads are expected and tests are limited to these conditions, then one type of response can be determined. However, if the actual use of the beam includes both in-plane and torsional loads, then they both need to be considered and accounted for in test planning. Test evaluation criteria are particularly important when overload, failure, or dynamic behaviors of the structure are expected.

There are many other possible design considerations and corresponding test evaluation criteria for the simple beam shown in Fig. 3-1 as follows:

1. Stresses sufficient to cause yielding at any point between the applied loads.
2. Possible web crippling at each support.
3. Possible torsional buckling of a thin-walled beam.
4. Possible shear failure in the web where the loads are applied.

In the example of the beam bending test given above, it was indicated that the beam passed a strength requirement but did not pass a deflection condition. The same beam could also have difficulty with an axial compression test which could induce a buckling type of failure. The test involving the combination of beam buckling and bending can be performed. However, diagnostic information must be planned for in advance of the test.

When structural systems and components are tested rather than just individual elements as suggested in Fig. 3-1, adequate simulation of loads and how the loading conditions are applied are some of real challenges of testing. Typical setups for tests (both static and dynamic except for creep and fatigue) often consume about 90-95 percent of the time needed for testing. Much of the time needed for setting up a test is in attempting to comply with the test purposes and evaluation criteria to ensure that the loading conditions are properly applied and that the boundary conditions are realistic. There are times when the purposes and evaluation criteria can be simplified without compromising the importance of the test results. Simplified purposes (e.g., performing a proof test to demonstrate structural integrity rather than conducting a strength test to determine failure loads and modes) can result in test setups which are easier and less costly to accomplish.

After nearly every test there is a period of time in which a reconciliation between theory and test results occurs. This reconciliation can be straightforward if the test is well-planned and the evaluation criteria are well thought-out beforehand. However, this reconciliation process can be difficult if the test data and theoretical predictions differ widely. (One way to help this process is outlined in Chapter 4 where the various types of structural tests will be described. The titles given to each type of test either state directly or imply the basis of the evaluation approach that will be used to judge the structural behavior.)

On rare occasions when large discrepancies between theoretical predictions and test results occur, some tense situations can develop. The first question raised is the validity of the calibrations of the instrumentation, transducers, and gages used in a test. After the calibration issues have been adequately addressed and discounted, those contesting the test results question the signal conditioning, data acquisition, or other systems for what they perceive to be inappropriately acquiring, mishandling, or incorrect manipulation of the data. When these approaches do not bring about the satisfaction needed or realign the test results with theory, those questioning the test data then begin to doubt the capabilities of test personnel in terms of their competence, integrity, or any other areas of apparent vulnerability. The point is that in some situations people want the test results to agree with the answers they already have. It is not always possible for all people involved with a test to be objective with regard to the test results. In defense of those requesters, many are often under pressures (e.g., schedules, costs, self-induced, managerial) to deliver a tested prototype or actual structure which passes all test requirements. Sometimes satisfying all the expectations of test results is not possible. Those involved in structural testing who will not tolerate test results which differ from predictions of behavior invite failure or at least a less than successful test. It is almost always better for everyone involved in structural testing to discover problems as early as possible in laboratory or other controlled conditions to avoid more costly (and potentially embarrassing) situations and delays found later in production, assembly, and use.

There is a need for those involved in testing to temper and reconcile discrepancies between analytical predictions and test data. One essential way to accomplish this is to develop test evaluation criteria prior to conducting a test. The main points regarding evaluation criteria are the following:

1. evaluation criteria are needed for every test performed,
2. the criteria may be defined in terms of engineering mechanics,
3. the criteria may be defined in legitimate and appropriate non-structural terms,
4. the criteria are subject to change during a test,
5. the criteria may be defined during or after the test, and
6. how the data are recorded, displayed, and presented (e.g., tabulated, plotted, combined by channels) are often very important in understanding how a structure responds
7. that discussion prior to a test regarding the expected insights and data is usually essential in performing a quality test.

Failures of members and connections of systems and assemblies often occur in structural tests. Most structural tests are considered as potentially destructive. Therefore, failure is often thought of as the only criteria for structural testing. While this is not true and there are other criteria that are important for structural testing, the major source of interest in a structural test and the primary concern for conducting tests safely are in terms of some types of failure of the structure or component. Unfortunately, there is not a simple term to describe structural failures, reduction in structural capability, or limits on acceptable behavior. Structural tests do cover these possibilities.

Therefore, the evaluation criteria will be defined in three basic ways:

1. those described by relationships developed in engineering mechanics (e.g., yielding, fracture, frequencies) and through typical structural test methods and techniques
2. those defined in other ways (e.g., electrical malfunction, leakage) and by other disciplines, whether they are scientific, empirical, or subjective.
3. those occurring with inconsistent calibrations.

The following three sections (3.4, 3.5, and 3.6) describe how to determine structural response in tests. The first section (3.4) is devoted to understanding structural responses from an engineering mechanics and materials science viewpoints. The second section (3.5) describes how inconsistent calibrations can affect the understanding of structural response (e.g., where measurements made depend on the instruments used only to find that the instruments give understandable but incorrect indications). The third section (3.6) discusses evaluations of structural response from other viewpoints (e.g., electrical malfunctions of components subjected to applied loads).

3.4 DETERMINING AND EVALUATING STRUCTURAL RESPONSES

This section describes various ways that structural systems and assemblies respond when loaded. These descriptions are developed through the relationships developed in engineering mechanics, structural engineering, and materials science. The use of methods and techniques of experimental mechanics, nondestructive evaluation, and other disciplines provide the information and evidence of the structural responses. This section is focused on understanding how structural systems can respond and giving criteria needed to evaluate test performance.

There are some basic concerns in structural engineering: (1) adequate structural strength, (2) capabilities to meet the intended service functions over the projected lifetime of the structure including dynamic loads, (3) possible presence of undesirable responses (e.g., excess vibrations, deflections), and (4) adequate tolerance for possible damage from unanticipated loads (e.g., fire, vandalism, and disasters). This section supplies suggested means of addressing these concerns and gives criteria for evaluating structural behavior.

One distinct measure of degraded structural performance is when failure (e.g., separation of parts) occurs. There are some responses of structures which are considered as failures even though separation of parts does not occur (e.g., gross deformation). There are other responses which are not failures but can be considered as precursors to failure conditions or, at a minimum, produce undesired actions of the structure (e.g., excess vibration). Structural systems can respond in the following ways (listed below) indicating failure, reduced integrity, lessened capabilities, or compromises in performance. Each response listed will be described later in this section. It should also be pointed out that the responses listed are often not independent, they can occur simultaneously, and the actions from one response can initiate problems resulting in other detrimental actions to occur (e.g., vibrating at resonant conditions can cause degradation in threaded connections).

1. Buckling
2. Connections and joints—degradation, separation

3. Dynamic characteristics—mode shapes, frequencies, and damping that are not desired or out of limits

4. Elastic behavior—excessive deformation

5. Environmental incompatibilities—corrosion, embrittlement, moisture plasticizing, and annealing

6. Fracture, spall

7. Inertial characteristics-not meeting requirements or out of limits

8. Plastic deformation, yielding

9. Residual stress—formation, relieving, influence

10. Stiffness, compliance—too stiff, too flexible

11. Strength—insufficient

12. Time dependent behavior—anelasticity, creep, fatigue, fretting, wear

13. Undesired failure modes—failure of test fixtures, tire blow out, missile failure

Buckling: This type of failure is considered unique because of the action of the structure. There are four basic types of structural systems which can buckle: columns, plates, shells, and assemblies. Buckling implies that there is a displaced configuration of the structure which has a lower energy level associated with it than the previous unbuckled configuration. A simple rod which is pin supported on each end and axially loaded in compression will eventually buckle, particularly if it is long enough. Buckling may be initiated while the structure remains elastic or while the structure experiences plastic deformations as well. A simple example of plate buckling is the common effect of an oil can. Thin sheet material is commonly used for the sides and tops of instrumentation racks, computer systems, storage cabinets, etc. There are times when these systems are overloaded or displaced causing these panels to deform and buckle.

Another type of buckling failure is when external pressure is applied to a thin walled pressure vessel such as a water tank having its pressure relief valve frozen shut in winter when a sudden demand for water occurs. The release of the water creates a volume of reduced pressure within the tank. This pressure differential can buckle the water tank. Theoretical predictions of buckling do not involve estimates of material strength (Ref. 3-6). In actual practice, buckling usually occurs with a combination of loads applied to a structure. Columns with axial loads applied usually have some bending forces applied as well. Analytical predictions of buckling are usually modified to reflect the presence of other loading conditions (Refs. 3-5, 3-6). Buckling usually initiates in a small localized area. Buckling may be confined to that area or may propagate and affect the whole structure.

Connections and Joints: Earlier it was mentioned that a structure is "as good as its joints." A likely candidate for test and evaluation efforts to be focused is on problems involving connections and joints (Refs. 3-2, 3-7, 3-13, 3-16, 3-20, and 3-21). The purpose of a connection is to transfer forces among the joined parts. Any process which degrades the transfer of forces is detrimental to joint integrity. Test efforts in evaluating joint behavior are usually focused on trying to determine how a connection can degrade and what magnitudes, orientations, or cycles of loading are necessary to initiate the degradation of a joint.

There are many possible joining methods and techniques. There are also many possible failure mechanisms. For example, structural connections can be made with bolts, rivets, welds, adhesives, compression bonding, plating, soldering, brazing, etc. Often these joining processes require the addition of heat, sometimes in large amounts, in order to generate the bonds required. The added heat can degrade the material in or adjacent to the joining area. Connections are also often subjected to corrosive environments. Connections and joints usually receive added scrutiny in analyses and tests because of their susceptibilities to degradation because of how they were made or from their exposure to potentially harmful influences. It is very difficult to provide guidelines or to describe an exhaustive or comprehensive listing of the types of failures or degradations of joint behavior. Some examples of possible degradation are: possibility of joint opening (e.g., creation of gaps) which degrades dynamic response, flaws in welds or brazed joints which can cause cracks to form and propagate, imperfect alignment of joined parts (e.g., butt welds) which can create additional stresses. Whatever joining processes are used in test structures will affect the integrity of the connections particularly if structural failures or dynamic responses are possible modes of behavior.

Dynamic Characteristics: A vitally important aspect of the behavior of structural systems is their mode shapes, frequencies, and damping. Modal analysis is used to determine the dynamic characteristics of structures and assemblies. There can be many different combinations of responses which can lead to undesirable behavior of a structure or to actual failure or separation of parts. For example, acoustic coupling of highway road noise into a closed passenger section of an automobile can give an uncomfortable "beat" to the occupants. Another example of a potential problem is the coupling of frequencies in automobile suspension and steering systems. If the dynamic response of the tire/wheel/suspension system of a car combines with the steering system, then the driver could have some difficulties in controlling the car. Another example of potential dynamic response problems is with packaging and shipping systems. If the contents shipped in the protective packaging system combines with the dynamic response of the transport system (truck and trailer, railcar, aircraft, or ship), then damage to the contents or shipping system could occur. Examples of this potential damage include engine mounts on automobiles vibrations from trains or trucks used in transport and moving fluids in tanks and piping systems could damage the restraint structures supporting them. Modal analysis is used to determine the dynamic characteristics, their origins (e.g., automotive systems interacting with pavements, rotating machinery causing undesired excitations of buildings), and can often suggest what modifications can be made to reduce or eliminate undesirable effects.

Elastic Behavior: An example of elastic behavior which is detrimental to the integrity of a structure is when a beam deflected excessively (Refs. 3-5 and 3-18). It should be noted that structural damage was not the issue; that is, the structure would return to its original position when the loads were removed (e.g., elastic deformation). A functional damage threshold was exceeded because the beam did not appear to be capable of sustaining the loads. Most building codes have a requirement regarding the limits on deflections for floors (Ref. 3-9). While this type of restriction on deflection has appearance as its primary motivation, most civil engineering structures which have excessive deformations of its major load carrying elements or members can be considered inadequate for use.

An example of elastic degradation due to excessive deformation is in housing construction. Typical walls and ceilings are constructed with timber beams (e.g., nominal 2×4 in., 2×6 in., or 2×8 in. construction grade lumber). These beams and columns provide the nailing and attachment points for wall board, paneling, and other coverings. In opening and closing the doors to a room, small differential air pressures occur. These differential air pressures can cause small pressure loadings on the wall coverings. Repeated cycling of these connections can separate the wall coverings from the beams (rafters) or columns (studs). A functional failure threshold would occur when nail heads reappear through the wall board or other covering.

An example of a type of elastic deflection and the perception of the problems associated is in the aluminum skin of a B-52 bomber. Portions of this skin on the fuselage of this aircraft have a rippled appearance when the aircraft is at rest on the ground. This appearance does not detract from the structural integrity of the aircraft nor does it hamper aerodynamic performance as the rippled skin goes away when the aircraft is in flight when the wings are supporting the plane. These aircraft have the greatest longevity of any military aircraft built to date. However, if the skin of a commercial aircraft had the same rippled appearance, potential passengers could perceive the airplane is unsafe for flight. Regardless of the documented evidence of airworthiness and demonstrated ability to fly with ripples in the skin, a passenger aircraft exhibiting such appearance would likely be perceived as unsafe. This example points out the problem that a structure can be perfectly capable of meeting its design intent but the perception of its ability is suspect.

Environmental Incompatibilities: An example of an environmentally induced structural failure is corrosion (Refs. 3-16 and 3-18). There are many examples of how a structure can corrode: (1) direct exposure to acids, gases, saltwater air, and other potentially non-compatible materials, (2) indirect exposure by electrical conduction, acting as an antenna, etc., (3) other chemical reactions which in turn cause degradation. Some of the mechanisms which induce failures are understood and can be designed and tested for. Others mechanisms are only partially understood. For example, the actions of acids on metal structures is well understood and, with appropriate plating or with other means, the structural members and parts can be protected. However, the presence of hydrogen at a potential crack interface in high strength steels is likely to produce a brittle fracture for reasons not well understood. Some of these incompatibilities can degrade structures rapidly (e.g., hydrogen from an improperly controlled plating process which can embrittle high strength bolts overnight). However, it is typical to have these incompatibility problems occur after structural systems have been in use for some time.

These environmental incompatibilities usually are manifest in some type of changes in material properties (e.g., annealing). Determining the cause of environmental incompatibilities usually involves material sciences. Evaluating the impact of these incompatibilities on structural systems is a valid reason for performing structural tests.

Fracture. Spall: Fracture is possibly the most common type of structural failure. Fracture suggests a separation of parts (Refs. 3-7, 3-16, and 3-18). Fracture can occur in at least two modes: brittle and ductile. An example of brittle fracture is the failure of concrete test cylinders under a compressive loading condition. The friction between the crossheads of a testing machine and the test cylinder causes a triaxial state of stress in the concrete. When the applied force is large enough, then a shear failure is initiated. It is typical for this type of test to have the failed cylinder appear as two cones with their points touching and the remainder of the material fractured. While brittle fracture is known to occur in structures and has resulted in some spectacular and disastrous past failures (e.g., Liberty

Ships as described earlier in this chapter), the incidents now are quite few and current materials and analytical procedures have reduced this type of failure to a minimum.

An example of ductile fracture occurs when yielding of the structure takes place in a metal structure and the yielding continues until a crack is initiated. The crack continues to propagate and separation of parts occurs. There are several analytical approaches to fracture mechanics (e.g., stress intensity, crack tip deformation, and energy methods.) In addition, there are test methods developed to determine fracture toughness and crack growth in metals, plastics, composites, etc. Fractures which can be detected and are related to degradation in structural performance or capabilities are test evaluation criteria (Refs. 3-2, 3-6, 3-7, 3-10, 3-16, 3-18, and 3-21).

Spall is the creation of sufficiently large stresses within a portion of a material that plastic deformation and separation of the material occurs. These stresses may come from thermal shocks in which the differential temperatures cause stresses to form. If the thermal shock occurs rapidly before temperature equilibrium is reached, then thermal shock can cause spallation (removal) of material. Spall can also be caused by stress waves in the material. A compressive stress wave traveling in a structural member is reflected as a tensile wave when it reaches a free boundary or an interface between material layers. If the wave has stresses which are large enough and of the right shape and frequency to cause yielding and failure, then spalling of the material will result (Ref. 3-16).

Inertial Characteristics: Rigid body dynamic responses are controlled by inertial characteristics. When the inertial characteristics of test structures (e.g., mass, centers of gravity, and moments of inertia) do not conform to requirements, then operations of flight and vehicle structures will not meet expectations. Measurements of these dynamic characteristics is accomplished using methods and techniques of "mass properties." When these characteristics do not meet design requirements, some detrimental behavior can result. The center of gravity may not be in the desired location, the flight vehicle may be too heavy for the engines reducing performance, the moments of inertias may be too large or too small resulting in difficulties in maneuvering and flight. Designs of most flight structures or ground vehicles have desired limits on weights, locations of centers of gravity, and moments of inertia. A passenger vehicle may have too high a potential for rolling over when cornering because of inadequate inertial characteristics. Meeting these necessary physical performance conditions (e.g., weight, inertial, and locations of centers of gravity) are examples of effective structural test evaluation criteria (Refs. 3-8 and 3-12).

Plastic Deformation. Yielding: Another way in which an individual member or structural system can respond is defined in terms of plastic deformation or yielding (Ref. 3-18). In this context, yielding will be assumed to occur without rupture. Ductile materials will exhibit yielding before failure. This yielding may be confined to a very small location (e.g., a heat affected zone adjacent to an opening in a steel pressure vessel wall). This yielding may also affect a large amount of material (e.g., the whole wall of this same pressure vessel). A threshold of structural damage will usually occur when yielding of the material takes place thus describing one of many related structural test evaluation criteria. The quantitative criteria for relating yielding of materials under combined states of stress are not complete. Much of the data on yielding is obtained from tests on materials specimens. This data may have complete or only limited application to actual structural elements and systems. Performing tests on structural systems and assemblies where yielding and plastic deformation are expected is suggested to effectively evaluate structural behavior.

Residual Stresses: Another example of the way a structure can respond is the development and dissipation of undesired residual stresses (Refs. 3-5, 3-6, and 3-14). Virtually all machined and assembled parts of structural systems have some residual stresses. How these stresses influence structural response and the integrity of a structure can be critical in evaluating their performance in a test. Suppose that a hole was drilled in a metal using a dull drill. The heating that occurs in the drilling process can change the heat treatment of or plastically deform the material surrounding the hole. The dissipation of these stresses may or may not be performed as part of a heat treatment process. If left untreated, then these residual stresses can be the source of flaws, cracks, deformation (e.g., spring back), and other undesired features in a structure or component. The presence of residual stresses and how they are diminished needs to be considered in evaluating test results.

Stiffness: The force-displacement relationships are essential in understanding how a structure responds. The basic stiffness relationships are defined in Eq. 3-1. The stiffness matrix relates the displacements of the structure to the applied forces.

$$(F) = [K]\{x\} \qquad \text{Eq. 3-1}$$

where (F) = applied forces
$[K]$ = stiffness matrix
$\{x\}$ = displacements

The compliance of a structure describes the amount of deformation or displacement of a structure or component as a function of the magnitude of the applied loads

(Refs. 3-5 and 3-11). The stiffness/compliance relationships developed define how the structure will respond to vibration conditions and shock loadings. The evaluation criteria associated with stiffness tests may be as straightforward as that when is loaded exhibits excessive deformation or does not deform enough. Stiffness tests are usually investigations to determine how complex structures (e.g., with many connections and complicated load paths) actually deforms under load. Any values are usually acceptable unless some gross deformations or yielding occurs. Structural tests are typically configured to measure the deflections of structural systems at various points. These measurements are usually used to verify or modify analytical models of a structure. The evaluation criteria then focus on the values obtained to determine if they are reasonable (e.g., linear with applied load, follow predictable patterns as in the deflection of the beam example (Fig. 3-1) used earlier in this chapter in which the deflection measurements of the beam should be greatest at midspan and reach zero at the supports).

Strength: The first consideration in the design of most structural members and systems is to have sufficient strength. The actual strength of a test item can be determined only by a failure test. Strength tests are usually performed on prototypes or items and the results are applied to those items made and assembled like the failed test item. The strength of a structure, a member of a structure, or a component is related to the magnitudes and directions of the forces and conditions applied (Refs. 3-1, 3-2, 3-6, 3-10, 3-14, 3-15, 3-16, 3-18, 3-19, and 3-21). The rate at which the loading is applied, the manner in which it is applied (e.g., point, line) can also have an appreciable effect of the response of a structure.

In structural testing, load testing to prove that a structure is strong enough typically consists of applying forces to a prototype or actual structure which are in excess of the design conditions. These proof tests form the basis for one type of strength evaluation criteria for structures. If a structure is "strong enough" to pass a test involving the application of these larger than design loads, then it has passed a strength test. Given in Table 3-2 is a series of typical values for proof testing requirements.

Determining effective measures of evaluating the strength of members and systems is essential in an objective assessment of structural integrity. One problem associated with proof tests is the choice of factors (multipliers of applied forces) to be used in testing. Sometimes there is a notion that larger factors make a safer structure. Too large a factor can cause test conditions which can actually damage the tested structure.

Time Dependent Behavior: Examples of time-dependent tests include creep which is time dependent deformation, fatigue which is a cyclic repetition of loads, stress relaxation which is a reduction in stresses over time (Refs. 3-9, 3-15, and 3-21). All materials creep; it is a matter of degree. Some will creep so little that measurements of deformation can hardly be made. Some materials creep readily. Well-developed test procedures and techniques have been formulated so that materials and structural designs can be evaluated before they are implemented (Ref. 3-10).

Fatigue of structures and members usually results in one or more of the three types of failure: elastic, plastic (ductile), or brittle. Elastic failure would be defined in terms of excessive deformations or deflections which could result from a reduction in section modulus of the members. A plastic failure implies the potential for slow crack growth. Brittle behavior implies a fracture of the member or structure below its yield stress.

Stress relaxation occurs in many types of structural systems. Stress relaxation is the decrease in stress after a given time at the same strain or displacement. A common example is the force (stress resultant) in some threaded fasteners which decreases after initial installation.

Other examples of time dependent behavior include wear and fretting (Ref. 3-21). Wear occurs when two bodies are in contact and move relative to each other. Frictional forces resist the motion and must be overcome for relative

Table 3-2. Typical Proof Test Load Factors

Design /Structure	Proof Test Load Factors
Lifting Straps	1.0 to 2.0 Maximum Load Rating
Below the Hook Lifting Devices and Equipment	1.25 × Maximum Load Rating
Pressure Vessels (ASME)	1.5 × Maximum Working Pressure
Lifting Fixtures for High Explosives in Manned Areas	4.0 × Maximum Loads to be Lifted
High Altitude Balloon Gondolas—Payload Hook	10.0 × Payload Weight

displacement to occur. Changing the contacting surfaces caused by wear alters the frictional forces. Wear is a function of time, the number of cycles of loading, the materials, and environmental conditions. Fretting is the wear of two surfaces in oscillatory motion (Ref. 3-3).

Undesired Failure Modes: Undesired failure modes can be surprises which occur during tests. These failures may occur from problems that happened in fabrication or assembly or from inattention in the analyses in which theory and test results do not agree. For example, a proof test was performed on a test fixture to be installed on a centrifuge. The design of the fixture included the use of both bolted connections and welds. The designer assumed (erroneously) that the welds and bolts acting in both shear and bending would share the applied forces together and their combined strength would be sufficient to withstand the steady state acceleration forces found in the centrifuge. In performing a test it was found that the welds reacted most of the applied loads first as there was intimate contact in the welded parts and no slippage in the weldment was possible. After the welds fractured, then the test forces were transferred to and resisted by the bolts. Unfortunately, the bolts were not strong enough either. Therefore, an undesired failure mode occurred which differed from the analytical predictions. This unanticipated response of a structure is usually clearly understood after a test but is often disguised prior to a test.

This discussion was developed to show many of the possible structural test evaluation criteria that can be developed which are based on principles of engineering mechanics, structural engineering, and materials science.

3.5 INCONSISTENT CALIBRATIONS

Inconsistent calibrations describes those instruments and transducers which can no longer be calibrated or which fail to repeat their calibrations (Ref. 3-12). Many transducers and instruments used in structural testing experience rough handling, are exposed to harsh environments, and many times are damaged in service. One of the most subtle problems occurs when an instrument or gage has behavior which deviates from its former calibrated behavior while it is being used in a test. Problems of recognizing this deviant behavior and establishing limits and tolerances on acceptable performance is not as easy as its appears. Instruments and gages usually just do not suddenly fail. If they do, their erratic behavior is usually easy to detect. However, they sometimes just drift out of calibration. While the inability of a transducer or gage to pass a calibration may not be a part of a structural test, the evaluation of their performance along with the structures and components developed as part of a test setup or an instrumented test body suggest that inconsistent calibrations can play an important part in evaluating structural behavior. These inconsistent calibrations can be prevented by performing periodic and spot check calibrations on instruments and protecting the instruments while they are in service and in storage. Those involved in structural testing should be alert to the fact that the instruments used for measurements could be the source of indications of undesirable response of the test structures.

3.6 NON-STRUCTURAL FAILURE MODES AND EVALUATIONS

This section describes how the actions of a structural system can also be evaluated using other criteria which are not based on structural responses anticipated through engineering mechanics and materials science. These criteria usually relate to functional behavior rather than structural damage or failure. These criteria are best explained using examples. The emphasis in these criteria is that the test requester or product user may have reasons to evaluate performance in terms that are more direct or are more applicable to a structure or component than stresses, deflections, or temperatures.

The first example is that of a pressure vessel subjected to internal pressure. The purpose of a proof test on a pressure vessel is to show that the vessel is capable of containing a certain pressure without gross deformation. There are two basic kinds of failures which can result from a pressure vessel: rupture (separation of parts indicating a fracture) and leakage of the pressurizing gas or liquid. Leakage can be described by fluid flow theory to an extent (Ref. 3-4). Its description through irregular shaped openings, through long paths, and around seals and other obstructions will challenge analytical techniques. Therefore, predictions of leakage as a function of pressure as related to degradation of structural integrity makes this a very difficult analytical task.

In this example proof test, if rupture or leakage occurs before the desired pressure is reached, then the vessel fails the test; otherwise the structure passes provided that the materials and connections have not been degraded. Rupture of a metal walled vessel is amenable to analyses. Rupture of the walls of other types of materials used in construction (e.g., reinforced or prestressed concrete) is more challenging. However, leakage is more likely to be considered a non-structural failure, or it will be considered as a failure that can only be estimated with methods of engineering mechanics.

Another example of a non-structural failure are tests

performed on electrical devices (e.g., a flight recorder, combat aircraft electronic controls, data telemetry and recording systems) subjected to tests conditions simulating accidents or other worst case events. The major concern for these devices is the loss or confusion of output signals or improper functioning. There are two distinct subsets of this type of functional testing: (1) will the component or device be required to function just prior to and during the application of the loading conditions which cause the damage conditions or (2) will the test item be required to function prior to, during, and <u>after</u> the test. While this distinction may appear to be a small difference, in terms of test set-up and providing the input and output signals to a component undergoing tests in which posttest functionality is required can be very difficult and may create some challenging test opportunities. For example, a crush test on a critical electronic component may require that electrical wiring be protected from sharp edges of failed parts. Anticipating where these sharp edges may occur and how wiring should be protected can be difficult. Another example in this same crush test is when radiography is used as a diagnostic device to determine how the structure responds. The location and direction of the x-ray source depends on how the structure deforms so that the radiographs obtained are not obscured by the deformed structure.

Test conditions simulating accidents usually involve the application of large forces, thermal effects, or some other types of loading conditions. Predicting the relationship of structural response to altering the electrical output of one of these devices is difficult. Nonetheless, these tests are often performed to determine how these devices and systems would function in the worst conditions to which they may be exposed. This type of test is analogous to a test to determine failure loads and modes (see Chapter 4). The evaluation of mechanical failures in electrical component leading to electrical dysfunction requires monitoring indicators of the loading conditions (position, direction, and magnitudes of loads, pressures, and temperatures, etc.) and recording input and output electrical signals with particular emphasis on deviations from expected results. The evaluation of failure when changes in electrical output is affected is most likely to be accomplished in non-structural terms.

A somewhat poorly defined non-structural evaluation criteria is the appearance of a structure. Poor workmanship, process control, and inspection procedures can result in appearance deficiencies. There are times when a structure has been designed, tested and shown to be capable of performing an intended function except for the fact that its appearance suggests that it is not capable. This category is included in this discussion because untrained test requesters as well as practicing engineers will reject items that they feel will give an unsatisfactory impression to those using the structure. When tests are performed for those who do not have engineering or technical training, this criteria is more likely to be used that one might expect. Some examples of appearance deficiencies include poorly aligned bolt or rivet patterns, non-uniform protective platings or coatings, and poorly applied paint.

Those responsible for structural testing laboratories and have performed many tests over extended periods of time will attest to the basic fact that structural test evaluation criteria are often overlooked or not even considered in planning and conducting tests. It is very important that evaluation criteria be developed, examined carefully, and properly implemented to ensure that high quality structural tests are performed.

3.7 REFERENCES

3-1. _____, *Mechanical Testing. Vol. 8. Metals Handbook.* 9th Ed., American Society for Metals, Metals Park, OH, 1985.

3-2. _____, *Failure Analysis and Prevention. Vol. 10. Metals Handbook.* 8th Ed., American Society for Metals, Metals Park, OH, 1975.

3-3. _____, *Compilation of ASTM Standard Definitions*, American Society for Testing and Materials, Philadelphia, PA, 1982.

3-4. _____, ASTM Specification E 425-71 (1980), *Definition of Terms Related to Leak Testing*, American Society for Testing and Materials, Philadelphia, PA.

3-5. A. Blake, Ed., *Handbook of Mechanics. Materials. and Structures*, John Wiley & Sons, New York, 1985.

3-6. A. Blake, *Practical Stress Analysis in Engineering Design.* 2nd Ed., Marcel Dekker, Inc., New York, 1990.

3-7. D. Broek, *The Practical Use of Fracture Mechanics*, Kluwer Academic Publishers, Norwall, MA, 1989.

3-8. D.J. Ewins, *Modal Testing: Theory and Practice*, John Wiley & Sons, New York, 1985.

3-9. D.A. Firmage, *Fundamental Theory of Structures*, Krieger, New York, 1985.

3-10. F. Garafalo, *Creep and Creep Rupture of Metals*, McMillan Series in Materials Science, 1965

3-11. C.N. Gaylord and E.H. Gaylord, *Structural Engineering Handbook*, 2nd Ed., McGraw-Hill Co., New York, 1979.

3-12. C.M. Harris, *Shock and Vibration Handbook*, 3rd Ed., McGraw-Hill, New York, 1988.

3-13. H. Liu, *Wind Engineering*, Prentice-Hall, Englewood Cliffs, NJ, 1991.

3-14. R.E. Peterson, *Stress Concentration Factors*, John Wiley & Sons, New York, 1974.

3-15. E.P. Popov, *Introduction to the Mechanics of Solids*, Prentice-Hall, Englewood Cliffs, NJ, 1968.

3-16. W. Richards, *Engineering Materials Science*, Wadsworth Publishing Co., San Francisco, 1961.

3-17. D.A. Rigney, Ed., *Fundametals of Friction and Wear of Materials*, American Society for Metals, Metals Park, OH, 1981.

3-18. S.T. Rolfe and J.M. Barsom, *Fracture and Fatigue Control in Structures*, Prentice-Hall, Inc., Englewood Cliffs, NJ, 1977.

3-19. F.B. Seely and J.O. Smith, *Advanced Mechanics of Materials*, John Wiley & Sons, New York, 1965.

3-20. E. Simiu and R.H. Scanlan, *Wind Effects on Structures: An Introduction to Wind Engineering*, John Wiley, New York, 1978.

3-21. M.F. Spotts, *Design of Machine Elements*, 4th Ed., Prentice-Hall, Englewood Cliffs, NJ, 1971.

CHAPTER 4
TYPES OF STRUCTURAL TESTS

Robert T. Reese, Sandia National Laboratories,
Albuquerque, New Mexico
Wendell A. Kawahara, Sandia National Laboratories,
Livermore, California

4.1 INTRODUCTION—TEST CONDITIONS AND ENVIRONMENTS

The material in this chapter describes the various types of structural tests that can be performed. Some examples are: destructive to define failure modes and loads, pressure proof tests, random vibrations, impacts, and proof of compatibility with environments and surroundings. This compilation also aids in developing typical lists of instrumentation, test equipment, and auxiliary items needed in performing tests. Organizing structural tests by the types of tests that can be performed develops a consistent framework and makes presenting materials in subsequent chapters easier to understand.

Three basic descriptions of the structural tests will be developed in this chapter: (1) tests on actual, prototype, and unique (one-of-a-kind) structures can be performed, (2) tests can be performed using a wide variety of possible loading conditions (e.g., strength tests performed using forces, pressures, and thermal conditions), and (3) tests that have been developed using special techniques and equipment (e.g., impact tests using drop tables). Overlap exists among these basic descriptions of structural tests (e.g., stiffness relationships can be determined through static load-deflection tests, vibration, and modal analysis).

The first main point serving as a basis for this chapter is that virtually all types of structures can be tested. Many can be tested in laboratory conditions and some must be tested in the field. The laboratory offers advantages of test equipment, data systems, and having all the resources and capabilities available. Field testing often suggests having the structure available in their actual locations subjected to their service loads and conditions. Conducting tests in either laboratory and field locations have advantages and potential problem areas (see Chapter 14).

The second description is of types of tests which can be performed using a wide variety of test conditions. These test conditions need to be tailored to the requirements for obtaining the response or behavior of the test item. For example, proof or operational acceptance tests on mechanical assemblies or electrical components can be performed using pressures or vacuums, thermal gradients or temperature cycling, shocks, vibrations, or other loading conditions as needed to verify the test item will meet its design intent and will provide useful service. A single loading condition could be used for acceptance or proof testing or the test item could be required to successfully pass a series of tests using various loading conditions and even repeats of certain conditions.

The third description developed concurrently in this chapter is of types of tests using special techniques (e.g., both analytical and testing) and accepted methods and test procedures. Structural testing has also been described sometimes as environmental testing. These specialized types of tests use methods developed for subjecting items to particular environments. These environments are very difficult to generate except through the use of these special techniques and methods often used for simulating dynamic loading conditions.

An important idea in structural testing is that there is an overlap among these basic descriptions of tests. It can be difficult to define distinct boundaries among types of tests. It is also common for test items to be subjected to combined types of tests—strengths and stiffnesses as affected by temperatures and electrical function as influenced by shock and vibrations. The performance of test items in these conditions leads to the insight needed in terms of their anticipated integrity as they represent other similar items in the development of systems and assemblies. Further, it is more than likely that a more than one type of test can be performed on the same test items. For example, prototype hardware to be used for flight testing on a missile system can be subjected to a series of load-deflection tests to determine the stiffnesses of a structure or component, then subjected to proof tests simulating in-flight loadings due to maneuvers, and then used later in failure tests to determine design margins and failure loads and modes.

The following list of example test structures begins with the structural systems usually considered the most straightforward to test and concludes with the structure considered more difficult to test.

1. Structural system, prototype, or test item designed for and dedicated to testing in the laboratory.

2. Actual structure obtained and perhaps modified for test purposes available for testing in the laboratory.

3. Structural system, prototype, or test item made available for laboratory testing.

4. Actual structure not modified for laboratory testing.

5. Structural system, prototype, or test item designed for field testing

6. Actual structure awaiting field testing.

The reader will be introduced to categories of structural tests that are performed on structural systems, mechanical assemblies, and electronic components. These tests have diverse objectives and purposes and utilize virtually all aspects of experimental mechanics and nondestructive evaluation techniques. This chapter will cover the various types of structural tests and give the basic diagnostic techniques typically used along with the data usually obtained where this information is appropriate. The types of structural tests described in this chapter are given in alphabetical order for reference ease. This description of structural tests is intended to be an inclusive listing with no basic type of test being excluded.

Twenty-five types of structural tests are described in this chapter. Most structural tests performed can be categorized into one or more of the types described. Structural tests may be performed during design and development, during production, while fabrication and erection activities are in progress, or after a structure has been in use. Some types (e.g., acceptance and quality assurance, calibration, mass properties, materials testing) have been developed to evaluate systems, products, and materials for conformance to design requirements and product specifications. Other types of tests (e.g., energy absorption, failure loads and modes, environmental compatibility, instrumented test units, modal analysis, residual stresses, shock, vibration, wind tunnel) focus on supplying information for design and development activities. Other types of tests provide evaluations of structures so that further applications and uses can be made of them (e.g., pressure, thermal, proof and operational, and service life extension).

4.2 ACCEPTANCE AND QUALITY ASSURANCE

One type of structural test performed routinely in production of mechanical and electrical hardware is to determine whether an individual part or assembly part meets certain minimum requirements other than dimensional tolerance, finish specifications, appearance requirements, and other aspects related to machining and assembly. An example of an acceptance test is to show that a prototype pressure vessel can be pressurized to a certain minimum pressure. The procedure would be to extract a representative sample from a production lot, pressurize each vessel in the sample, and record results (e.g., pass or fail with some explanation of results). These activities are described as the quality assurance and quality control (QA/QC) tests.

Product development activities usually include detailed tests to ensure that products will function and meet their design intent. To ensure continued performance in production, the mechanical portion of the QA/QC tests are usually simplified versions of the more detailed structural tests formulated and performed in development activities. It is typical for a quality assurance type of test to have simpler diagnostic devices and transducers used, the test is usually set-up in an assembly or field test area rather than a structural test laboratory, the data obtained is usually substantially less, and the evaluation is often limited to acceptance or rejection of the item. Nonetheless, these tests are very important to ensuring continued quality and repeatability of products and assemblies. This type of test is essential in ensuring that a product delivered to a consumer will function as described in the following example.

Suppose that a manufacturer of electronic equipment desires to develop a high speed switch for use in controlling electrical current used in an electromechanical system. In development these switches would be subjected to cyclic operational checks while operating at ranges of temperatures and humidity levels. The QA/QC evaluation of these switches could include: (1) repeating the same tests on each switch (probably too costly), (2) performing tests on representative (or lot) samples from production, or (3) having no QA/QC testing program (probably not acceptable from a product liability viewpoint). The QA/QC evaluation could also include some failure tests on these switches to be compared with developmental test results. The results of the QA/QC testing program will provide a history of the product performance. These results are important because comparative information is supplied which can identify problems particularly if changes are made in production activities (e.g., materials, processing, assembly procedures).

QA/QC testing is usually directed at determining

the functionality of components and assemblies. QA/QC tests need to be designed to provide definitive information (e.g., pass or fail) while subjecting test items to representative and reasonable loading conditions while keeping test costs to a minimum. Virtually all organizations producing items for consumer use have a quality engineering and assurance programs and associated testing and evaluation requirements and procedures. The basis for the mechanical portion of these QA/QC activities is usually formulated on the results of structural tests performed in design and development activities.

4.3 CALIBRATION

Measurement processes depend on the calibrations of the instruments used. Calibration of the instruments relates the inputs they receive to the outputs they indicate. For example, a strain gage based load cell is subjected to an applied force. The force produces a strain in the sensing portion of the load cell. The output of the load cell's measurement system indicates a voltage change. The voltage change is proportional to the applied force.

Calibration is quantifying and defining the relationships between known inputs and observed outputs and the uncertainties associated with the two measurement processes. Calibration activities are essential in structural testing because they give confidence to the data taken and provide some type of certification or traceability to referenced standards. Among the basic types of calibrations performed in support of structural tests are: (1) force (and torque), (2) acceleration, (3) velocity, (4) displacement and length, (5) temperature, (6) pressure, leakage, and vacuum, (7) electrical charge, current, resistance, and voltage, (8) time, (9) frequency, (10) optical properties, and (11) radiation. In each case, it is necessary to have a known reference which is traceable to some accepted standard (e.g., National Institute for Standards and Technology—NIST), a calibration procedure which is repeatable and supplies the known condition, and a measurement device calibrated using that reference and accepted procedure. In some types of tests, relative information is all that is needed. However, structural tests require the use of measurement devices that are calibrated and are shown to be repeatable.

There are two basic approaches to calibration: (1) relating the operation of a transducer or device and its output to that of a known standard so the history of the instrument can be developed and documented and (2) modifying the equations or factors governing the constants used to convert the measured output (e.g., voltage) to the needed indication of behavior (e.g., strain, acceleration). The output of the transducer or device must agree with the known standard within acceptable uncertainties. (Measurement uncertainties will be described in detail in Chapter 15). The first approach is used in most standards laboratories so that histories can be kept on each individual transducer, instrument, or system used for calibration. The results of this type of calibration is usually a table summary comparing the measured values from the transducer and the values from the standard.

This historical approach is not as useful in most structural testing facilities. That is, a side-by-side comparison of output signals and other information is not nearly as useful as keeping a transducer or device in adjustment and calibrated so that its output can be read directly or its output signal can be fed into a data acquisition system. A structural tester is usually much more interested in the response of a structure or prototype test unit than they are over the long-term history of a particular transducer or device. Therefore, calibration without adjustment is useful, but does not usually meet the requirements of a testing operations.

Most of the transducers or devices used for structural testing involve the use of some electrical output (e.g., current or voltage) as related to the desired output (e.g., millivolts per volt per pound of force). There is an increasing motivation to use as many electrically based measurement systems as possible because of the ability to interface these signals with computer based or other types of electronic data acquisition, recording, and processing systems. While some measurements are made by observers recording data by hand, any tests requiring even small amounts of data are usually coupled with some type of data acquisition system. There is almost always an enormous increase in test capability when these electrically based transducers and devices are used. Most transducers require a known signal or excitation be applied (e.g., constant voltage or current) and that an output signal be accurately read and compared with an initial condition (e.g., to determine voltage or resistance change) (Ref. 4-9).

Force (or torque) can be measured with dead weights, electrical based load cells (resistance or piezoelectric), mechanical proving rings, spring scales, or beam balances. Pressures can measured using dial gages (force balances), Bourdon tubes (pressure balances), pressure transducers (strain gage or piezoelectric based), or with Pitot tubes. Typically, most forces, torques, and pressures are measured with load cells, torque transducers, and pressure transducers. These transducers require an input excitation and that the output signal be read by an electrical device (e.g., digital or analog voltmeter). Forces can also be determined from pressure measurements made on a known cross-sectional area of an actuator which is converted to a force.

Accelerometers, velocity gages, and displacement transducers are calibrated by placing known conditions on them and determining their output. For example, it is routine to calibrate an accelerometer in a centrifuge. By knowing the position of the accelerometer relative to the center of rotation of the centrifuge arm and the number of revolutions per unit time, the resulting acceleration is known, $a_r = r\omega^2$, where a_r = the radial acceleration, r = the radial position of the accelerometer, and ω = the angular velocity. A velocity gage can be calibrated by placing it in a constant velocity field or by placing it on a vibration machine with a reference accelerometer and integrating the acceleration time history. In a centrifuge, a velocity gage can be calibrated by using its position r and the angular velocity ω in the equation $v = r\omega$.

A displacement gage can be calibrated by using a precision micrometer, using an interferometer or some other optical device, or by transmission of frequency based signals such as light or sound.

Temperature calibrations are usually conducted by basing measurements made on a thermocouple performance compared to conditions indicated by precision devices typically constructed of platinum resistance elements. A thermocouple to be calibrated is placed in a temperature field along with the reference standards. The output of the thermocouple is related to temperature by the performance of the reference standard. Temperature measurements are also made with resistance temperature devices (RTD's). This device is a precision resistor whose change in resistance is proportional to temperature. Field measurements of temperatures can be made with thermocouples or RTD's using the manufacturer's calibrations.

Electrical measurements of charge, voltage, current, and resistance are essential because of the previous ideas introduced regarding current data acquisition, test operating, and processing systems. Electronic components such as voltmeters, power supplies, amplifiers, and other equipment used in data acquisition and recording systems will need to be calibrated individually or placed in a circuit so that they can be calibrated as a system using shunt calibration techniques. They can also be adjusted by the use of precision voltmeters. They can also be used without adjustment and have their output signals measured by these precision voltmeters.

Time periods for calibrations performed on transducers and gages can be some lengthy times (e.g., two years provided the device is not damaged or compromised) or the calibration may be required at time of use. Many recording devices (e.g., x-y recorders, oscilloscopes) need a known signal in order to establish the range of the instrument. Details of calibration will be discussed in Chapter 10 (Instrumentation and Data Systems) and in Chapter 14.

4.4 CONNECTIONS AND JOINTS

A suggested rule of evaluating structural integrity is that a structure "is as good as its joints." While this is not exactly true for all structures, it is true in a very large percentage of the assembled structures. A joint or connection can be made in many ways including: (1) bolting, (2) riveting, (3) adhesives and bonding, (4) welding, (5) compression bonding, (6) friction, (7) nailing, (8) plating, and (9) temperature effects such as shrink-fitting or interference fitting. These connections typically have large localized stresses when the joining process is completed. Others may affect the materials involved, either in terms of compatibility or by changing properties such as strength and ductility as in the heat affected zones in weldments.

The usual interests in this type of test are: (1) strength, (2) deformation, (3) potential failure modes and loads, and (4) other unique characteristics such as the loading at which joined parts initially separate, when color or appearance changes occur in plastics, etc. Since connections and joints implies that two or more elements or parts are attached or joined, the focus in this type of test will be on the joints. (Section 4.9 describes failure loads and modes.)

For example, suppose that a steel pressure vessel is constructed of a series of rolled plates which have been butt welded along their edges. If the vessel is pressurized, then the areas of interest and concern would likely be the welds and the heat affected zones adjacent to the welds. There are a variety of experimental techniques that could be used to determine the response of this vessel to internal pressure (Chapter 8). If this vessel has a large length of welds, then detailed examinations of these welds could become a critical part of the structural evaluation of this structure.

Suppose that two cylindrical pipes or conical sections are joined by a bolted connection. This connection is to be subjected to the combined action of internal pressure and pipe bending. In this prototype structural joint, a stiffness test is proposed. The example structure is shown in Fig. 4-1.

There are three basic concerns in this type of test: (1) what measurements can be obtained, (2) what measurands describe the failure of the connection, and (3) can the loading conditions be duplicated adequately. In this particular test the main question is whether the joint will allow leakage of pressure when subjected to internal pressure and moment as shown.

In this example, the diagnostics used are to determine the load-deflection relationships for the joint and correlate them with any leakage that occurs. Note the location of the port on the flange which allows internal pressure to be applied to the pipes or conical sections. The change in load-deflection characteristics is indicative of a

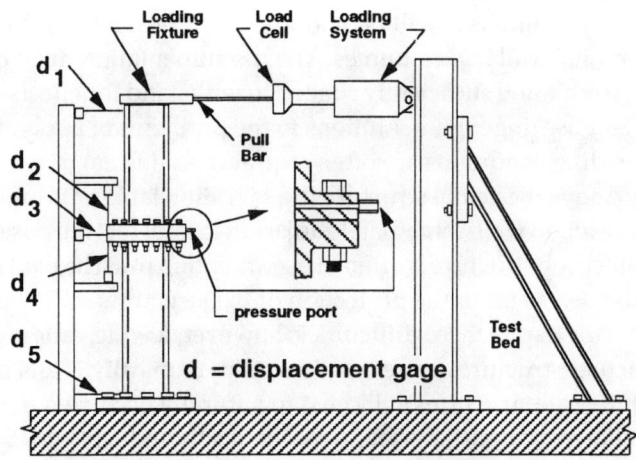

Figure 4-1. Example Structure for Connection Failure Test

reduction in clamping force across the seals allowing leaks to develop. As shown, gages capable of measuring very small displacements are positioned on the tension side of the joint (d_2 and d_4). These gages should detect if the sealing surface begins to separate which, in turn, should contribute to displacement of the O-ring seal and subsequent leakage. Other displacement gages are positioned to obtain the deformed shaped of the pipe joint under the loading descriptions described.

Another type of basic test typically involving connections and joints is slippage (relative displacement between joined surfaces using threaded connections). There are two basic concerns in this type of test: (1) determination of the applied forces when slippage occurs and ends and (2) accurate measurement of the displacements involved (typically very small). This evaluation of a bolted connection is particularly important to flight vehicles because joint slippages can provide a great deal of damping in vibration tests.

4.5 DEVELOPMENT OF INSTRUMENTED TEST UNITS

In the development phase of many projects, it is often necessary to design and fabricate a prototype structure which is instrumented to obtain its responses to various applied forces and loads. The development of these instrumented test bodies includes aircraft, vehicles, components, weapons systems, and a wide variety of possible structures. Basically the instrumented test body becomes a means to obtain detailed information about the component or structure. The data obtained from these test bodies is often telemetered back to a data station as in the flight tests for commercial and military aircraft or the data are recorded on board the test body and retrieved later.

The development of these test units involves the installation of gages and transducers and the verification that they are functional. It also includes operational checks of the instrumentation, signal conditioning, data acquisition systems, and telemetry or other recording systems.

An example of an instrumented test unit is the aircraft landing gear shown in Fig. 4-2. Figure 4-2 shows the landing gear in a laboratory testing program prior the measurement of forces in taxiing, takeoff, and landing simulations. The design of the landing gear involved both detailed analytical and experimental efforts. The placement of the strain gages and accelerometers was chosen so that correlation of the analytical predictions and test results could be accomplished. The landing gear was subjected to a series of simulated landings and takeoffs in a structural testing laboratory. For a landing gear passing these laboratory tests, confidence is given to the design. Successful completion of these tests are often required for certification of the landing gear prior to flight tests. After the prototype design for the landing gear was completed, then the aircraft was subjected to a series of landings and takeoffs with the strain and acceleration data recorded on board the aircraft. (The data could also be telemetered back to the data station.) Using this procedure, on-line data analysis techniques could be used and test conditions can be adjusted consistent with the capabilities of the aircraft and the loads placed on the landing gear. Comparisons were also made between the laboratory test results and the flight test data.

A significantly less sophisticated test than the landing gear test just described but one which is very meaningful in its own way is the development of an instrumented

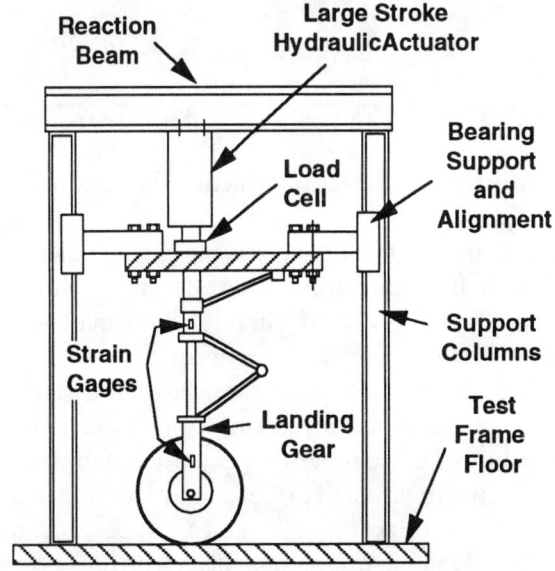

Figure 4-2. Landing Gear Tests

torque wrench shown in Fig. 4-3. Accurate measurements of applied torques are often required on installation of bolts and screws. The torque is influenced by many variables which are often difficult to quantify and assess the effects they cause (e.g., lubricants, thread fit, percent thread engagement, thread performance [galling], alignment and concentricity, etc.). The device shown is a torque wrench which has been instrumented with strain gages and then calibrated. The readout device also provides the power needed for the Wheatstone bridge circuitry. This wrench is just a sample of instrumented tools which are increasingly available for production as well as research applications.

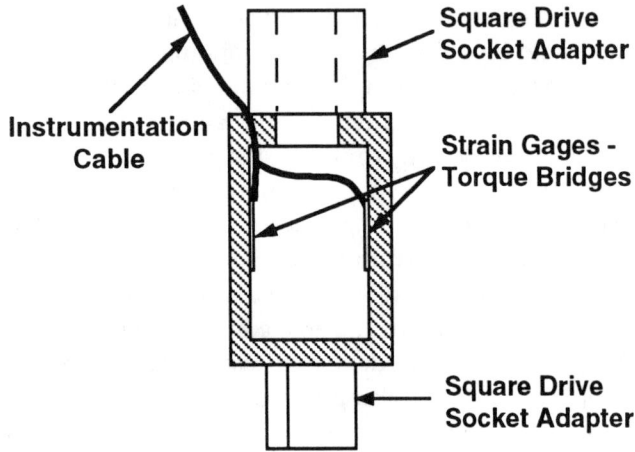

Figure 4-3. Instrumented Torque Wrench

Instrumented test units are essential for determining measured service loads. These loads are used in further laboratory simulations. It is important in this type of test to select the instrumentation needed that will give the desired insight and understanding of the behavior of the test item.

4.6 DYNAMIC TESTS ON ACTUAL STRUCTURES

Most civil engineering structures are unique. Testing these types of typically large structures presents challenges to the engineers and technicians because it is not possible to first build and test a prototype. Each structure is generally one-of-a-kind, although there may be a number of common or similar components.

Testing of these structures can be useful in advancing the state of the art of analysis and design for an entire class of structures. For example, the results of tests on one earth-filled dam can be applied to the design of future earth-filled dams even if the location, size, and soil types are different.

The dynamic testing of actual structures poses particular challenges to the structural test engineer because the structure is usually not located in or near a laboratory and its available resources. The instrumentation must be portable and sufficiently rugged to withstand the effects of field use. Ingenious solutions to the problems of applying loading conditions are often required, as it is not possible to move the actual structure to a vibration table or to build a reaction wall immediately adjacent to if for test purposes. Since the structure is often in use, attention must be paid to the users and to the protection of the structure.

Despite these difficulties, however, a wide variety of actual structures have been tested dynamically. Tests on full-scale structures subjected to naturally occurring transient vibrations as well as mechanically induced steady-state vibrations were made feasible by the development of the finite Fourier transform and frequency-controlled eccentric mass vibration generators.

Low-level transient vibrations consist of naturally occurring wind and microtremors as well as from actions of large crowds of people attending sporting events and concerts as they affect the response of stadiums and arenas. Tests simulating low-level vibrations have been conducted on buildings, bridges, earth-filled and concrete dams, fluid-filled storage tanks, tall chimneys, and hyperbolic cooling towers (Ref. 4-P23). The amplitude of vibration in these tests is very small, on the order of 0.001 to 0.0001 g's. Because the response of structures and supporting soils is nonlinear, the response frequencies found during these tests may be different from the response frequencies that occur during strong shaking. However, if the damage is not severe, the mode shapes will generally not change appreciably. Therefore, the response to ambient vibration can be used to detect possible changes in the dynamic characteristics of the structure.

Electrohydraulic actuators or eccentric mass vibration generators are used in forced vibration testing to produce significant motions. The usually small damping values of structures at resonance leads to large amplification factors resulting in response magnitudes on the order of 0.01 g's. Because the frequency and orientation of the exciting force can be controlled, these tests permit the excitation of specific motions, including torsion and higher mode responses. The steady-state excitation allows for detailed measurement of mode shape, including foundation response. Because the force level is adjustable at a given frequency, it is possible to determine the amplitude-dependent (nonlinear) characteristics of the structure. Other types of excitation have included man-excited vibrations, pull-back tests, and impulsive loads from rockets and small explosive cartridges. Force vibration testing has primarily been applied to building structures, although bridges, dams, intake towers, liquid storage tanks, and foundation structures have been tested as well.

An earthquake is the ultimate test of a civil engineering structure. The strong motion records obtained from bridges, dams, and buildings during actual events have yielded significant information about the high amplitude response of these structures. This has been particularly true for structures that have been subjected to low-level testing before and after the earthquake.

Rarely is there an opportunity to perform tests to destruction on an actual structure. In the 1970's, an 11-story reinforced concrete building was subjected to a series of small- and large-amplitude dynamic tests as the structure was prepared for demolition. A prestressed concrete bridge girder was tested to failure when it was removed after 20 years of service. Significant work on the repair of a steel bridge span was conducted after the bridge was demolished and the span moved to a test facility. (Ref. 4-P6, 4-P16)

4.7 ENERGY ABSORPTION

This type of test has the goal of defining how much energy a structure can absorb when subjected to some types of loads (typically impacts) which will cause large deformations of the structure or component. Many structural systems are designed to have deformable sections used to absorb and attenuate the forces due to impact. Examples of structures designed with energy absorption in mind include shock absorbing automobile components (e.g., bumpers, dashboards, and steering columns), packaging and shipping systems using foam, captive air bubbles, and other components used to transport a wide variety of products ranging from computers to medical organs to nuclear waste, and structures designed for air delivery (e.g., parachutes and energy absorbing pallets). The energy absorbed by these type of structures is related to three principal variables: the area under the load-deflection curve, the loading rate (strain rate sensitivities of the materials), and the axes of loading. The energy absorbed can also be described by moment-rotation (structures in bending) or by torsional characteristics. The examples here will focus on the load-deflection behavior.

The first example of energy absorption by a structure is of a 1/4-20 UNC stainless steel threaded fasteners in tension. There are many requirements for threaded fasteners, but the two most important are strength and ductility. Strength is important for design purposes and ductility is important for connection integrity. Lot sample tension tests on threaded fasteners give information on both strength and ductility. Three sample load-deflection curves for 1/4-20 UNC stainless steel fasteners engaged two, six, and eight threads into stainless steel are shown in Fig. 4-4 along with the value for the strength requirement from the

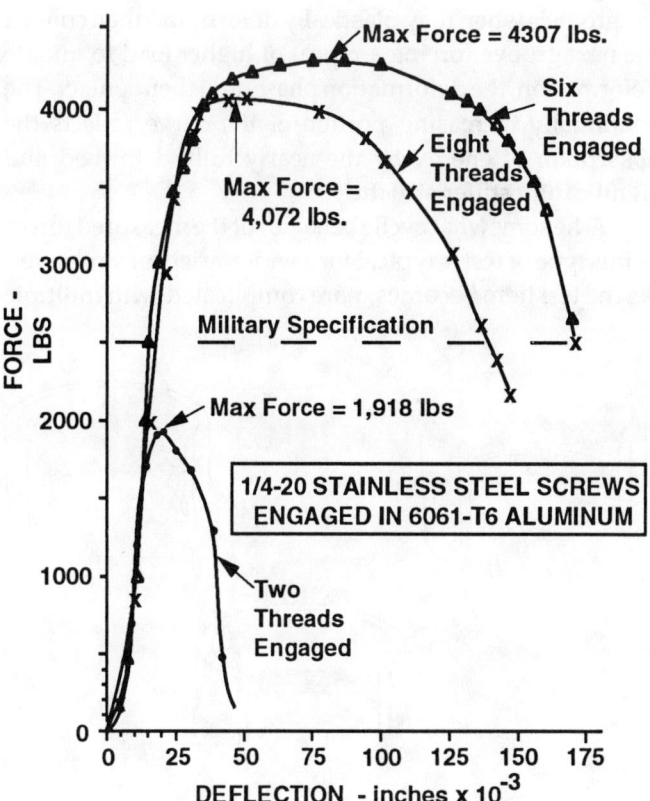

Figure 4-4. Load Deflection Test Results on Three 1/4-20 Screws

military specification.

The load-deflection curve shows a short linear (elastic) portion, a plastic portion adjacent to the maximum applied force, and then a decreasing portion of the curve when necking of the fastener commences through fracture. The displacement is measured by the break-away extensometer attached to the bolt/screw test fixtures. The attachment was made in this manner to minimize the compliance of the test equipment.

A second example shows the load-deflection curve for a shock mitigation system proposed for use on an axisymmetric shipping system for nuclear materials. The mitigation system is designed to attenuate the shock loads into the container. The mitigator design consists of materials (stainless steel with considerable available plastic deformation potential) and circumferential grooves designed to cause the structure to deform in a predetermined way regardless of the angle of impact. Three views of the shock mitigator and details of the grooves (undeformed, deformed from angled impact test, and deformed from an axial test) are given in Fig. 4-5. The resulting load-deflection curve for the axial impact test is shown in Fig. 4-6. This curve shows a series of approximate cyclic increases and decreases in applied forces as the structure deforms. These changes in measured forces correspond to the behavior of the grooves when they plastically deform and then contact the next groove forcing a repeat of higher load to initiate deformation, the deformation phase, and then contact. The continually increasing portion of the curve reflects the absorption of energy by the nearly fully deformed and significantly stiffer structure.

The somewhat cyclic behavior of the measured forces in this type of test is typical for a wide variety of structures. As the test item becomes more complicated with multiple load paths along with possibilities for buckling, then the measurement of the load-deflection curve is more difficult and must be done in real time.

For those dynamic tests in which automobiles or other structures to be impacted into stiff restraints, the measurement of displacements can be accomplished with high speed cameras. The measurements of forces applied to the test item can be performed by reacting the restraint with load cells.

Some considerations which are significant in the performing of this type of test are:

1. strain rate sensitivities and effects of materials in actual use versus laboratory simulation capabilities (e.g., do not fall for the trap that a material model exists for extrapolating static test results to dynamic behavior)

2. fixtures and boundary conditions

3. the structure or portion of the structure to be tested (e.g., a bumper versus the whole car)

4. how the energies will actually be measured and interpreted from the data available

5. the rates of loading

6. temperature effects and heat dissipation

The intent of an energy absorption test is usually straightforward. However, there are potentially many factors which can have varying influence on the results. Each of these factors need to be considered in planning and performing a test to determine how energies are absorbed by a structure or component.

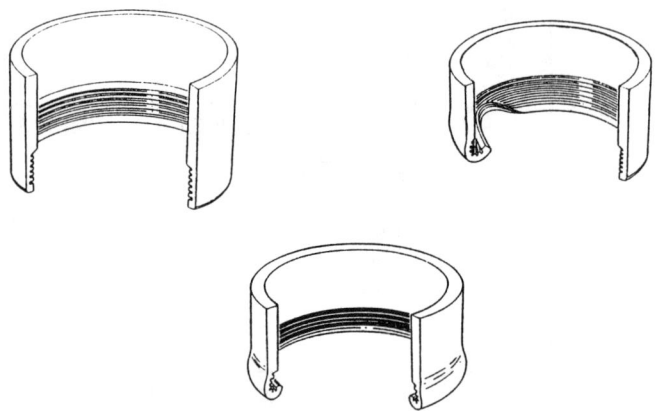

Figure 4-5. Crush Tests on Shock Mitigation System (Impact Limiter)

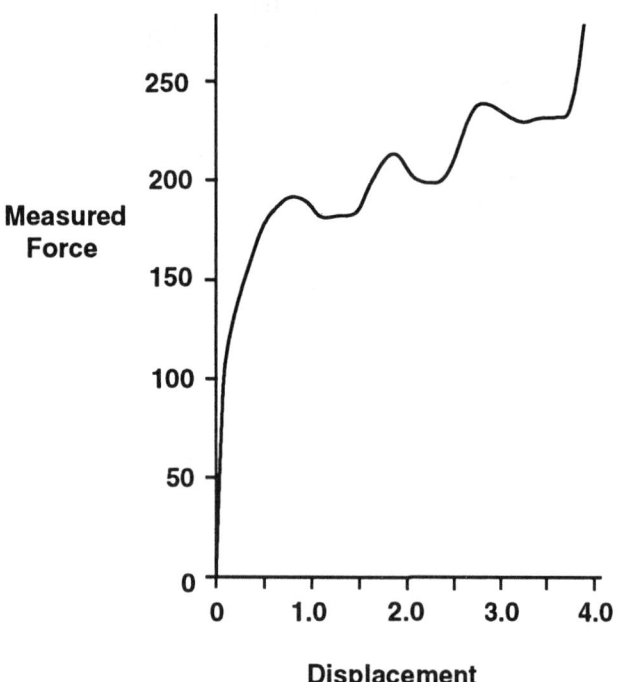

Figure 4-6. Load Deflection Curve for Axially Loaded Shock Mitigation System

4.8 ENVIRONMENTAL AND MATERIALS COMPATIBILITY

These types of tests can include tests in which a prototype or real structure is subjected to actual environments and conditions. These tests may include tests in simulated rain, sleet, hail, fog, temperatures (cycles and extremes), ice formation, steam, corrosives, sunlight, etc. The reason for including a compatibility type of test in this handbook is because the degradation of structural performance is often measured through methods of experimental mechanics. Materials science also has many techniques for evaluating the possible effects of environmental damage and evidences of materials incompatibilities (Ref. 4-16). The major focus of this type of test is to determine effects caused by reduced or a lack of materials compatibility with the surroundings. No specific guidelines can be given on how to plan structural tests because there are so many variables involved. However, a technically sound approach is suggested in this section.

Some types of environmental damage to structural systems is easily viewed in terms of the build-up of oxides (rust), the corrosion of surfaces, and the breakdown of lubricants causing wear in contacting surfaces. Some other types of environmental damage can be difficult to detect and are not yet even understood in terms of the complex mechanisms which govern their behavior (e.g., hydrogen embrittlement of steels). Corrosion can cause reduction in strength and ductility which can be determined by tests to determine strengths or other performance characteristics. Other types of surface disruptions can enhance crack propagation in metals. Reduction in the performance of lubricants can cause bearings to wear which can be detected as strains in housings. Each winter the freeze-thaw cycles of temperatures cause considerable damage to highway pavements. ASTM has many tests involving weathering effects (Ref. 4-2).

Structural tests performed to evaluate environmental and materials compatibility need to be formulated to focus on determining degradation in performance or inability to meet design intent. The structural tests performed typically need to be planned in conjunction with a materials evaluation as well. There are three principal sources or causes of reduced compatibility:

1. Presence of natural conditions (e.g., temperature cycles and extremes, humidity, salt spray) which will affect structural materials.

2. Presence of man-made chemicals and materials (e.g., solvents, cleaners, paints, acids, bases) which will cause degradation of the materials.

3. Presence of the combination of two or more conditions or materials where the synergistic effects will cause greater damage or where one acts as a catalyst for other reactions initiating even greater reduction in structural performance.

Materials incompatibilities usually occur in the surfaces of the exposed materials. For example, the presence of nylon around brass surfaces causes degradation of the brass. The nylon produces amines which interact with the moisture available in the air and cause surface cracks in the brass. Cadmium plated bolts protect high-strength steel bolts from corrosion. The plating process for the cadmium must be tightly controlled to ensure that hydrogen embrittlement of the steels does not occur. In one instance an aerospace structure was assembled using cadmium plated bolts. On returning to work the next day, the engineer found many of the bolt heads were severed from their shanks because cracks formed in the plating process had propagated under the stresses caused by assembly torques.

The evaluation of materials incompatibilities is usually thought of in terms of materials engineering. The proof of developing structures and components capable of withstanding potential incompatibilities is usually determined in structural testing. Planning structural tests involving possible material incompatibilities involves the following:

1. Controlled conditions (e.g., climatic, processes such as steam generation, concentrations of acids and bases)

2. Test procedures which can diagnose problems which will occur over time; most of the effects of incompatibilities do not develop rapidly

3. Techniques to evaluate degradation in the exposed areas of structures.

The basic approach involved in this type of structural test is to consider each situation as a separate case. The engineer should gather as much information as possible; there are often many subtle chemical or other reactions (e.g., phase changes) which initiate and significantly contribute to the response of the structural system or component as in the galvanic corrosion in carbon fiber composites.

4.9 FAILURE LOADS AND MODES (DESIGN MARGINS)

This category of structural tests is usually considered as requiring some types of destructive tests. Failure loads and modes are not identified until some type of failure criteria as defined in Chapter 3 are satisfied. Failure loads is a loosely defined term which can include pressures, shock pulses, and other conceivable loading conditions including multi-axis loading conditions. Failure modes usually imply when a structure loses enough of its integrity that it will no longer meet the design intent with typical examples of large deformations, broken bolts or rivets, and separation of parts (Ref. 4-10).

The major purpose of performing a failure test is usually to verify the design procedures and to determine if unexpected situations occur. There is typically some need to prove that a structure or component be stronger than the anticipated loads. The results of a failure test usually indicate the reserve strength and deformation capabilities in a structure.

The major request for data from this type of test is to define the conditions when failure occurs including the magnitude of the applied load or pressure, the displacements of important locations of the structure, and the behavior of joints and connections, and the modes of failure. Failures were discussed in general terms in Chapter 3. The structural and non-structural modes are important and the reader is reminded that both need to be considered in the evaluation of the test item or component. For example, suppose that an in-flight tape recorder for commercial aviation is in development and requires testing. Failure tests on prototype recorders could be conducted in terms of a crush displacement applied along the principle axes (x, y, and z). The loading conditions at the onset of nonlinear structural response would be of interest to correlate with analytical predictions. However, the designer would likely be more interested in what load and displacement history were required to cause the recorder to lose capability to record and playback vital flight information. Therefore, the failure criteria could include: displacements, loading conditions (forces and pressures from underwater exposure), temperatures, and electrical function.

Another example is a bolt in single shear. A typical test request would be to have a load-deflection test to failure; that is separation of the head and a portion of the bolt shank if it is so equipped. It is also possible to have failure through the threads if they have been loaded because of improper design procedures or if no shank is present. In conducting the test, if the threaded portion and head are separated, the threaded section will be firmly embedded in the test fixture making its removal difficult, often requiring the services of a machinist. If the threaded portion is failed in single shear, then it is virtually guaranteed it will be bound in the test fixture. (It may be single shear tests on threaded portions of bolts may require single use fixtures as it can be very difficult to preserve the fixture integrity.) If the test was terminated after the ultimate load was reached (the maximum point on the load curve) but prior to the point of separation, then the bolt could be separated easily from the test fixture as it is unlikely that the threads will be damaged. The question then becomes is the complete load deflection curve needed on each bolt or is it more important to identify the onset of failure and leave the test fixtures intact so other tests could be performed?

An example of a field failure test are the tests performed on the Bailey truss bridges used in World War II. These bridges were fabricated in pieces so they could be erected quickly under hostile conditions. They were used to support heavy loads (e.g., tanks and supply trucks) over rivers and gorges. These bridges were tested in Arizona prior to being used in combat conditions. The tests included adding individual bridge sections to a completed cantilevered portion until the bridge system could no longer support itself. The tested system failed when the lower compression chords buckled. The intent of these tests was to determine the number of bridge sections that could be safely erected in a cantilevered manner (Ref. 4-18.)

Design margins are often determined from the results of failure tests. With the modes and loads determined from failure tests, they can be compared to analytical predictions of structural performance and behavior in service and overload conditions or with other tests on similar hardware. As long as the loading and boundary conditions used in the structural tests adequately simulate the anticipated failure conditions, then design margins can be estimated. These margins are typically ratios of failure loads to service conditions. For example, pressure vessels are routinely designed with a predicted rupture of four times the maximum allowable working pressure (MAWP). Burst tests performed should achieve this pressure in order to be considered successful.

4.10 FULL-SCALE AND MODEL TESTS
Albert N. Lin, Ph.D., Research Structural Engineer, National Institute for Standards and Technology, Gaithersburg, MD

The testing of structures, components, and systems can take a variety of forms, depending on the objectives. In one form, information from the model or prototype test can be directly applied to find the solution. In the second form of structural testing, the test results indicate general trends and require further refinement prior to application.

Consider the following cases. In the construction of foundation structures for buildings and bridges, test piles are commonly constructed or driven into soil or rock and then tested to failure to determine the ultimate bearing capacity of the pile installed at the site. The results of these tests are used to verify the specific design at a particular construction site, which is normally a function of the anticipated loads and subsurface geotechnical conditions. Insufficient performance can be remedied immediately.

As a second case, the response of a structure to seismic loading is a function of the modal characteristics which are dependent upon the distribution of mass and stiffness. It is often difficult to assess the final mass and stiffness of a structure because of the non-structural elements (occupants and equipment). Proper design would require an iterative approach based on modification of the building structure, which is a logical impossibility. In this case, observation of prototype structural response (including possible failure) during earthquake conditions can provide valuable information for the analysis and design of similar structures in the future. This is the motivation behind the various local and national programs to instrument buildings, bridges, dams, and other facilities to record their response to earthquake forces and displacements. However, structurally significant earthquakes in a given region are infrequent. Therefore, experimental observation of the response of structures to ambient and forced vibration can be used to similar purposes.

In a third case, any large structure and many smaller ones also, no matter how unique they may appear, generally consist of a number of relatively common components—beams, columns, connections and joints, plates, shells, etc. Hence testing of such elements and these components assembled together provides useful data to describe the anticipated behavior of the components, which may then be combined to analyze the entire structure. Tests on these components and elements can be conducted on a full or reduced scale.

In general, full-scale testing of prototype structures is the preferred course of action. In such tests, material properties, construction and fabrication methods, magnitudes of loads, and other factors important in the service and ultimate conditions of the structure are representative of the actual case. However, practical limitations of cost, testing facilities, and time often preclude full-scale tests in favor of model tests.

Results of tests conducted on model structures and building components are often extrapolated to full-scale designs. If the scale reduction is not large (i.e., a large model) and the behavior being examined is quasi-linear (small strains and displacements and nearly constant material properties), such extrapolation is acceptable. However, when small models (large scale factors) using non-homogeneous materials (i.e., concrete, composites) are tested in the nonlinear range (large strains and displacements and changing material properties), serious questions will likely be raised regarding extrapolation to full-scale structures. Some of these questions can be addressed by conducting a series of identical tests at different scale factors to determine the linearity of the observed behavior with size.

Such a series of tests has been conducted as a part of the recently concluded United States—Japan Cooperative Earthquake Engineering Research Program. This program was designed to provide information on the response of structures and structural elements beyond what could be obtained from post-earthquake inspection or small scale tests. Particular questions involved the three-dimensional behavior of full size structures and components, the effects of structural damping, the behavior of joints and connections, and effect of equipment and furnishings. These objectives were met by a comprehensive test program consisting of the following elements.

First, a full-scale, seven-story reinforced concrete building was constructed and tested at Japan's Building Research Institute (BRI). Later, a full-scale, six-story steel framed structure was constructed and tested at BRI. Static and pseudodynamic tests were performed with the intent to damage the structure (greater than elastic strains were imposed) (Ref. 4-P16). Then the building was repaired and strengthened, non-structural elements were added, and the building retested. Both low level (linear elastic) and high level (damaging) force levels were used. Various full-scale components of beam-column connections and other structural elements were tested at the Ferguson Structural Laboratory at the University of Texas.

Second, a 1/5 scale model of the concrete frame-wall structure was constructed at the University of California-Berkeley Test Facility. The test results were compared to the full-scale test results and to the results of analytical models. Microconcrete and heat-treated steel were used to construct the model to give proper scaling for the aggregate in the concrete as well as the reinforcing steel used. Structural elements at a similar scale were also fabricated and tested.

Third, three models of the seven-story reinforced concrete building were constructed at 1/10 scale. Testing was conducted at the University of Illinois' single axis vibration (shake) table. Additional mass was added to the structure for proper scaling for gravity.

Fourth, models at a 1/12.5 scale factor were constructed of beam-column assemblies, an isolated shear wall, and a shear wall-frame unit. These models were tested at Stanford University's John A. Blume Earthquake Engineering Center.

A similar set of tests has been proposed to examine the behavior of bridge structures. In this program, full-scale test specimen and large models will be tested using structural vibrations, explosive simulation of earthquake ground motions, and a structural test (reaction) frame. The behavior of the material and supporting soil will be assessed.

4.11 INERTIAL
Dr. Albert N. Lin, Research Structural Engineer, National Institute for Standards and Technology, Gaithersburg, MD

Prototype and model tests conducted on the earth's surface are subject to an acceleration of 32.2 ft/sec or 9.81 m/sec due to the earth's gravitational field. Although such an environment is generally accepted and often not even considered in many structural tests, some areas of engineering, particularly in aeronautical and geotechnical disciplines, require an acceleration field greater than the 1 g gravitational field of the earth. The aeronautical considerations come from maneuvering loads created by aircraft as they turn, climb, and dive, the accelerations created by lift-off of rocket and missile systems and separation of their various spent portions or stages, and decelerations as bodies reenter the earth's atmosphere. Geotechnical considerations come from the forces involved in formation of rocks, sediments, and in ways to test engineered structures supported on soils or forces exerted on retaining walls.

The acceleration forces greater than 1 g can be generated in a centrifuge. A centrifuge generally consists of a test item attached to or supported by an arm which rotates in a horizontal plane. Usually a counterweight is attached to the opposite end of the arm to provide a balancing force. In such a configuration, the test specimen is subjected to a centrifugal acceleration described in the Calibration Section. If the centrifuge is rotating at a constant rate, then $a_t = 0.0$; otherwise, there is a component of acceleration in the tangential direction as well. Using this principle in a structural testing context, it is possible to create an environment in which the acceleration field can be controlled.

Centrifuges usually come equipped with some types of electrically conducting slip rings. These slip rings can be used to supply electrical power to and obtain electrical signals from test units attached to and undergoing tests in a centrifuge. For example, a typical strain gage circuit uses two electrical lead wires to supply a voltage to a gage and two leads to obtain the strain signal from a gage. Both low and high voltage and low and high current slip rings are available and their use permits instrumentation and other devices (temperature chambers, vibration shakers, etc.) to be installed in a centrifuge and controlled during a test.

Aeronautical Applications

Centrifuges in aerospace applications are used to simulate the high g force fields typically associated with the launch, flight, landing, and other sequences in the use of typical structural systems as described in Chapter 5. The test items are typically full-scale prototypes or components ranging from small electrical switches, gyroscopes, and mechanical assemblies to large structures such as actual aircraft. Specialized centrifuges have been developed and used for testing and training fighter pilots for high performance combat aircraft.

Geotechnical Applications

In contrast to the aerospace applications, the geotechnical centrifuge is used to simulate the 1 g earth gravitational field on small specimens. Geotechnical projects usually involve the construction of dams, foundations, retaining walls, and other relatively large structures. It is impractical to construct full or large scale test specimens. Yet, when constructed to a manageable scale (e.g., 1/50 to 1/250), similitude is no longer possible because the principal forces in such structures are gravity or time dependent. As a results, the use of a centrifuge to generate a gravitational field of "n" times the earth's gravitational acceleration, allowing for testing of models constructed at a scale of "1/n." Under these conditions, the following set of scaling laws may be derived as given in Table 4-1. It is noteworthy that the stress and strain are unchanged in the prototype and the model.

Centrifuge testing has been used to study many types of problems including retaining walls, vibration of pile structures, earthquake simulation, liquefaction of soils, vibration of footings, seepage and stability of embankments and earthen dams, and general constitutive properties of soils (Ref. 4-P2).

A typical geotechnical centrifuge consists of a pivoted basket containing the specimen with the basket rotating from a vertical position to the horizontal during operation. The basket is located at the end of the arm of the centrifuge. The rotating arm is powered by a variable speed electric motor. Electric power, pneumatic power, and data signals are transmitted to the bucket through sliprings and rotating unions.

An example of a geotechnical centrifuge is the California Institute of Technology machine with a capacity of 10,000 g-lbs. Other major centrifuge facilities can be found at the University of California-Davis, Princeton University, CESTA in France, and Cambridge University in England.

Table 4-1. Scaling Relations for Centrifuge Applications

Quantity	Full-Scale Prototype	Centrifugal Model at "n" g's
Linear Dimension	1	$1/n$
Area	1	$1/n^2$
Volume	1	$1/n^3$
Time (dynamic terms)	1	$1/n^3$
Velocity	1	1
Acceleration	1	n
Mass	1	$1/n^3$
Force	1	$1/n^2$
Energy	1	$1/n^3$
Stress	1	1
Strain	1	1
Density	1	1
Frequency	1	n

4.12 LIGHTNING

Laboratory simulations of the naturally occurring lightning and discharges of static electricity have been developed in some selected test facilities. These simulation facilities are used to test a variety of structures, components, and materials which are sensitive to or can be affected by lightning. Two basic types of facilities have been designed and built: a capacitive pulser and a triggered lightning technique. The structure to be tested has a significant affect on how the electrical pulse is introduced because of the inductance it creates and how it is connected to the test source. A critical feature in this type of test is the rise time of the test current that can be delivered to the test item. Another critical simulation feature is how the lightning is distributed over the test item. Lightning typically consists of a highly active flash (large currents and voltages) followed by a longer duration, relatively low amplitude continuing currents.

Among the structures, components, and materials that are sensitive to the effects of lightning are: electronic systems, electric wiring, cable, and power distribution systems, munitions and explosives, antennas, aircraft, and missiles (Ref. 4-P19). Testing any of these items or systems will require the selection of the appropriate test levels and interfaces needed for adequate simulation.

Sandia National Laboratories in Albuquerque has developed a portable system which is contained in a steel ocean cargo container which can be placed on flatbed trailer. This system measures 20 x 8 x 8 ft. It consists of an electric field monitor used to monitor ambient conditions, visual recording systems (video and photographic), data acquisition systems used to monitor the responses (typically currents) in the test item as well as the incident lightning, data processing systems, and related equipment (e.g., portable power generator, controllers, and a lightning detection network). The system used to measure the intensity and duration of natural lightning is described in Ref. 4-P19.

4.13 MASS PROPERTIES

A fundamental structural test is the determination of the weights, centers of gravity, moments of inertia. Definition of these parameters is essential in any component or system that will be subjected to dynamic loads, or will experience some type of structure-control interaction during its service life e.g., aircraft, some automobiles, and weapons systems).

The weights of individual components and structural assemblies are measured by using calibrated scales which may employ a dead weight system, mechanical or electronic techniques, or a load cell system.

The centers of gravities (c.g.'s) and moments of inertia (m.o.i's) of components and fully assembled structural systems can each be measured using the same type of test equipment. The example used here is for relatively small components (less than 1000 lbs.). The first step in this type of structural test is to isolate the measurement system from the world around it. This isolation is accomplished by placing the test item on a table which is supported by an air bearing. This air bearing supports a table and test items for gravitational loads but offers minimal resistance to unbalanced forces or rotations if off-centered loads are placed thereon.

The test procedures used to measure the c.g.'s and m.o.i's begins with the calibration of the system prior to its use. A support plate with known inertial characteristics is attached to the test system as shown in Fig. 4-7. The bottom of the air bearing has a slender rod extending below. The rod is attached to a small load cell. Measurements of the force needed to restore the rod to a vertical position are made as the table and rod are rotated about the longitudinal axis of the rod with the table positioned at 0, 90, 180, and 270 degrees. The product of the restoring force, F_1, at its known position, a_1 is equated to the product of the unbalanced force caused by the weight of the test item, F_2, and its location, a_2. The only unknown in this equation is a_2 which locates the offset from the center of the rod to the center of gravity of the test item in the x and y positions in this plane. By positioning the test item in the two other principal planes, the c.g's in each orthogonal plane can be determined.

The moments of inertia are determined by using the same test system. In this case the air bearing is allowed to

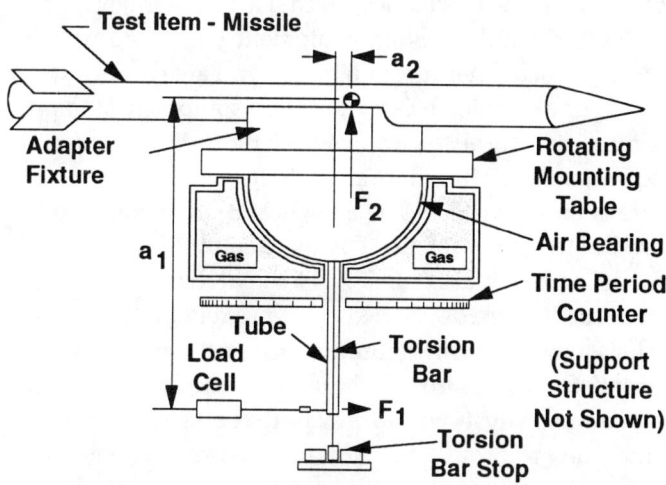

Figure 4-7. Air Bearing Support System for Mass Properties Measurements

rotated and is restrained by the thin bar with the end of the bar away from the bearing fixed. The period of rotation, T, is directly proportional to the moment of inertia of the combined test item and test system. By appropriate ratios of the periods of test system alone and the combined test item and test system, the moment of inertia of the test item can be found ($I = K T^2$).

The measured results are typically compared with the theoretical calculations and requirements. Most systems which have requirements for weights, centers of gravity, and moments of inertia will have tolerances associated with these measurements. Experience shows that identically made parts do vary in these properties, particularly when they are assembled into a system.

4.14 MATERIALS TESTING
Wendell A. Kawahara, Sandia National Laboratories, Livermore, CA

Introduction and Scope

Our modest intent for this topic is to offer structural testing personnel some motivation and rationale underlying a few selected materials test techniques, targeting an upper-division undergraduate level audience. Due to length restrictions, certain types of testing such as creep and fatigue are omitted; concepts from applied mechanics and materials science, essential for a fuller understanding of materials testing, are also kept to a minimum. A few preliminary remarks on materials testing are presented in this chapter.

Relating Materials Testing and Structural Testing

Materials testing and structural testing are two realms of experimental mechanics which share fundamental principles and similar testing equipment yet, paradoxically, relate to each other through theory and analysis.

For example, in cases where it is impractical to perform a structural test to evaluate the influence of each potential design iteration, an analytical model of the structure is indispensable. The analytical predictions, in turn, can be strongly influenced by the idealized mathematical model of material behavior. Such material or "constitutive" models are developed cooperatively by experimentalists who perform definitive materials tests, material scientists who ensure that the models have sound physical basis, and analysts who formulate the equations in a mathematically rigorous, kinematically consistent and numerically efficient format. Hence, structural testing is linked to materials testing through a desire to construct an accurate predictive analytical structural model.

Less frequently, given an accurate analytic model, structural tests may be used to extrapolate material properties. For example, the mode shapes and natural frequencies (eigenvectors and eigenvalues) provided by modal analysis have been used to infer the orthotropic elastic moduli of composite plates. The Taylor test is an example where some have attempted to extract high-rate stress-strain data from analyzing the deformed shape of a solid cylinder axially impacted onto a smooth rigid half-space.

Material test results quantify a material's intrinsic response to mechanical and thermal loadings regardless of the particular means that these loadings are generated in the structural application. By contrast, structural test results can be greatly influenced by details of the loading technique and geometry of the test item. Structural testing personnel can benefit from a familiarity with some basic materials testing concepts and techniques because:

1. When a discrepancy arises between structural test results and analytical predictions, the personnel may need to estimate the contribution of inappropriate models or materials data.

2. If existing material property data is inadequate for the structural analysis, the personnel may be required to define and perform those appropriate materials tests as well.

Some Resources on Materials Testing

Our presentation is limited to a general discussion of some techniques and concerns involved in conducting some common materials tests. Detailed descriptions of test method standards can be found in References 4-1, 4-2, 4-15, and 4-25.

Current developments in mechanical testing appear in periodicals such as SEM's Experimental Mechanics and Engineering Techniques, the ASME Journal of Engineering Materials and Technology, and the BSSM Strain Journal. Some compilations of metals test data are the Military Standardization Handbook, the Aerospace Structural Metals Handbook, the Damage Tolerant Design Handbook, the Metals Handbook, ASM's Atlas of Stress-Strain Curves and Atlas of Fatigue Curves and others given in References 4-3 through 4-6, 4-P7, 4-P10, 4-P12, and 4-P13.

Remarks on Specimen Selection

Specimens should be extracted from the same material to be analyzed, and preferably from identical stock because of potential influences such as heat treatment, cold-working, texture, and impurities. For example, material from a weld region may exhibit inferior strength compared to its adjacent base metal. Specimen geometries (e.g., tension, compression, torsion) should reflect the anticipated predominant deformation or failure mode because of possible material anisotropy (Refs. 4-1, 4-2, 4-15, 4-22, 4-25, and 4-P13).

Tensile specimens, typically straightforward to fabricate and test, are limited in strain for ductile materials because of necking. That is, a peak load is reached when the increase of applied stress (due to decreasing cross-sectional area) exceeds the increase of the material's sustainable stress (due to strain-hardening accompanying plastic deformation).

Since necking (localized deformation) develops when the specimen is stretched beyond the peak load, the usual load and displacement measurements no longer uniquely define the material's stress and strain. If a specimen is unloaded after necking but prior to failure, an approximation can be made for the local true stress and true strain state at the neck if the load data is used in conjunction with accurate neck radius-of-curvature measurements (Bridgman correction equations) (Ref. 4-P8). In this way, the stress-strain curve can be extended beyond peak load through the discrete pairs of stress-strain values (one pair for each unloading operation). Hence, the common term "Tensile Strength" (peak load per unit original area) is misleading for ductile materials because the area at peak load has decreased and furthermore the specimen has not yet failed.

In tensile testing brittle specimens, superior alignment of the test fixtures and hardware is essential so that a tensile bending stress caused by even small test system misalignments does not cause premature failure. Unintentional notches caused by machining or extensometry attachment can also cause premature failure in materials with low fracture toughness. Bend testing is often used instead of tensile testing because of easier specimen fabrication and testing requirements, although the influence of inherent flaw distribution may need to be accounted for in interpreting strength data.

Compression material test data is particularly useful in analyzing metal forming type processes which can generate large strains. Uniform true strains of -1.0 can be achieved in one continuous stroke, strains in excess of -3.0 can be attained by remachining between tests, although elevated temperature material recovery under no load during cooldown and heatup cycles between tests should be minimized (Refs. 4-P11 and 4-P12).

Compression specimens are typically right circular cylinders. Since uniform deformation is desired, alignment of the load train is critical. Friction between the specimen and loading platens can cause specimen "barreling," leading to erroneously high stress-strain curves based upon overall load and displacement. Often, a variety of concentric groove profiles are machined on the specimen loading faces to entrap lubricants and minimize barreling; this can be overdone, resulting in specimen "hourglassing" and erroneously low stress-strain curves.

Torsion testing can achieve equivalent plastic strain of 5 or more (Ref. 4-P9). Typically, short thin-wall specimens are used so that the shear stress can be proportionally related to applied torque. Long test sections may experience premature torsional buckling whereas very short sections can have undesirable stress and strain gradients from reinforcement at their ends. Alternatively, a stress-strain curve can also be deduced from a series of tests on solid rods having various diameters (ref. 4-P9).

Examples of Materials Test Data Misapplication

The following describes three common areas—extrapolation, isotropy, and equation of state.

Extrapolation: this means using material property data outside its domain of validity. Several examples are:

1. using static data for high-rate applications (i.e., ignoring the strain-rate sensitivity of materials such as polymers and ductile body centered cubic (BCC) steels.

2. using room-temperature data for elevated temperature applications (the temperature can be externally imposed or self-generated by adiabatic heating at high strain rates) (Ref. 4-P15).

3. Linear extrapolation of the tangent modulus to large strains from test data terminated at small strain; conversely, assuming that the elastic modulus of a tensile test is accurate if an extensometer calibrated to a full-scale strain of, say 50 percent or more was used.

4. Service environment is another concern. For example, nonconservative predictions can result from using virgin material handbook properties to analyze structures that may actually be embrittled by hydrogen (e.g., austenitic stainless steels, cadmium plated bolts) or oxygen (e.g., refractory alloys), or using "Dry As-Molded" stress-strain data for certain polymers in a moist service environment. Hydrostatic pressure can have a profound effect on the mechanical behavior of polymers and geologic materials. Pre-existing residual stress states are another set of conditions often ignored in analyses.

Isotropy: Many material models assume that the material behavior is the same in any direction. This would mean, for example, that tensile specimens extracted in random orientations from one region of a plate would all have the same stress-strain curve. Since properties in the rolling direction are often superior to those in the transverse and through-thickness directions, assuming isotropy can lead to nonconservative analyses. Isotropy can be especially unrealistic for continuous fiber composites, depending upon the constituents and layups (Refs. 4-22, 4-P3, 4-P4, 4-P5, 4-P14, 4-P20, 4-P21, and 4-P22).

Equation of State: $\sigma = \sigma(\varepsilon, \dot{\varepsilon}, T)$ This is an oversimplified idealization of material behavior. The instantaneous stress indeed depends upon strain, strain-rate, and temperature but is <u>not</u> in general a single-valued function of them. Just as for viscoelasticity, the history of deformation can influence subsequent material response. Since stress-strain curves are typically generated at a fixed strain-rate and temperature, an equation-of-state representation can be useful to summarize such a matrix of data. However, such a curve fit should not be construed as a material model, especially if nonproportional or reversed loadings and/or substantial temperature fluctuations are anticipated in the application. Paraphrasing Albert Einstein, "Everything should be made as simple as possible... but no simpler."

4.15 MODAL ANALYSIS
Dr. Thomas G. Carne, Experimental Mechanics Dept., Sandia National Laboratories, Albuquerque, NM

Modal analysis is a procedure that combines vibration test data and analytical methods to determine modal (natural) parameters including frequencies, mode shapes, and damping of a structural system. Modal analysis is a diagnostic tool used to understand how a structure responds dynamically. Modal analyses are performed for four major reasons:

1. To understand the structural dynamics of a system (deformation modes, modal frequencies, and damping).
2. Troubleshooting of structures experiencing problems in dynamic response.
3. Validation of analytical models (typically finite element models).
4. Contribute in the development of control systems for structures.

Because structures and components frequently store energy in the lower modes of vibration, modal analysis is particularly valuable because it identifies the frequencies and mode shapes beginning with the lower modes and increasing through as many as needed (typically less than 50, but as few as one). The major emphasis in modal analysis is to determine the frequencies and mode shapes. This information along with judgment and experience gives the insight into the dynamic response of a structure or component. Good design practice for structures subjected to dynamic loads requires that the modes not be excessively excited by the input conditions. Modal analyses provide the diagnostic approach needed to validate structures for dynamic inputs.

Modal analysis can be used for troubleshooting when problems have been encountered with a structure or component in production or use. For example, the installation of rotating machinery may cause excessive vibrations. The suspension system of an automobile is another example of how modal analysis can be used in troubleshooting. The road surface, tires, suspension, and steering systems of an automobile can interact in ways to cause considerable discomfort to a drive because a mode of vibration has been excessively excited. Modal analyses performed on these two example structural systems could reveal the cause of the problem. In addition, based on theses analyses, proposed remedies could be suggested and then evaluated.

Validation of an analytical model is an aim in many types of structural tests and modal analysis is frequently used to validate structural dynamic models. To develop a control system for structures, knowledge of the modal parameters is essential. Without that knowledge, the control system could cause an instability. Modal analyses give a major insight into structure-control interactions.

To perform a modal test, the test item must be excited in some manner. One can use artificial excitation such as a vibration shaker, an acoustic source, or an impact device (e.g., instrumented hammer shown in Chap. 6, Fig. 6-15), or a step-relaxation procedure in which a restraining force is suddenly released. One can also use natural excitations such as wind, road surface irregularities, wave action, or turbulent flight conditions. The response to the excitation must be measured using transducers (e.g., accelerometers, velocity gages, displacement gages, microphones, particles of sand, or strain gages). Most modal tests use accelerometers; ease of installation and their high frequency response are two compelling reasons. Accelerations and displacements have the same mode shapes while strains do not. Consider the free beam bending in the first mode as shown in Fig. 4-8. As the beam bends the acceleration and displacement mode shapes have two nodes along the beam and have their maximum values at the beam tips. However, the strains are maximum at midspan and zero at the beam tips.

A typical modal test includes the following: an array of accelerometers are placed on a structure connected to amplifiers and a data recording system, an excitation is

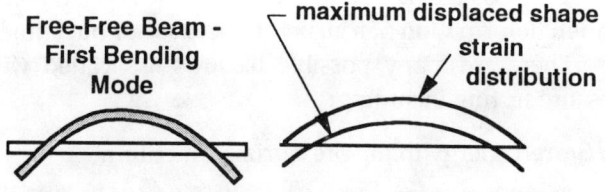

Figure 4-8. Free-Free Beam in First Mode Bending

applied, and the analog acceleration time histories and input forces are recorded for each accelerometer and each force input during the test. The data recording system would normally include anti-aliasing filters, analog-to-digital converters, and a digital storage device. The digitized data is typically converted by a digital Fourier transformation into the frequency domain which reveals the frequency content of the signals.

The frequency response function (FRF) is calculated or estimated using these data. The FRF is essentially the ratio of the response Fourier transform to the input Fourier transform.

In practice, several algorithms have been developed for calculating a "best" estimate of the FRF by reducing the effects of noise. These include H_1, H_2, H_v, and H_s (Ref. 4-11 and 4-17). All of these algorithms average the data from repeated tests. A FRF is developed for each accelerometer at each location where it is used to measure the response of the structure. Once the set of FRF's is obtained, then data analyzing techniques are used to extract modal parameters (frequencies, shapes, and damping).

Consider for example, a cantilever beam which has just three accelerometers attached. The FRF's and beam are shown in Fig. 4-9. The first three frequencies are indicated as ω_1, ω_2, and ω_3. The very light damping can be observed in the sharp narrow resonant peaks. The components of the mode shapes are determined from the FRF at each location on the beam axis. These shapes show the displacement of the beam if only one mode was excited. The shape for the first bending modes is given in the upper right of Fig. 4-9.

There are a large number of techniques to extract the modal parameters from a set of FRF's. These techniques vary in accuracy, speed of implementation, capability to handle less-than-perfect data, operator interaction, and requirements on the data. They work in both the frequency and time domains. The requirement of the data analysis and the quality of the data affect the choice of technique.

Some of the problems that are encountered in modal analysis are:

1. Obtaining "good" data (not obscured by noise, nonlinearities, or rattles).

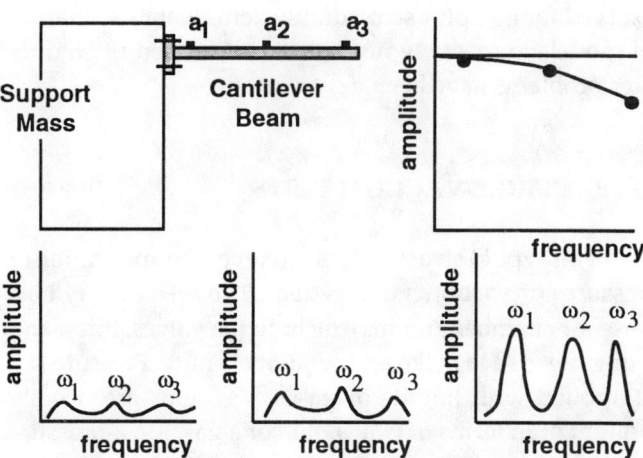

Figure 4-9. Cantilever Beam in Bending, Frequency Response Functions (FRF) for Accelerometers A_1, A_2, and A_3, and First Mode Shape (upper right)

2. Assuring proper analog signal processing (e.g., no overloads in any components, proper use of accelerometers, no clipping of signals, low DC drift).

3. Avoiding aliasing by using proper anti-aliasing filter before digitizing signals.

4. Avoiding test conditions in which the structure is not time-invariant. This is critically important if one uses the more automated mode extraction techniques.

5. Applying modal analysis techniques to structures which exhibit some or even considerable nonlinear behavior.

6. Identifying closely space modes.

7. Identifying lightly excited modes.

8. Defining how many and of what locations on the structure should be excited in order to excite all the modes of interest.

9. Deciding the locations to make measurements so that all the modes are observable and uniquely defined by this measurement set. The measurement set also needs to capture the principal kinematic energies and strain energies in the modes.

Modal analysis requires judgment and experience by those performing these tests. Among the advantages of modal analyses are that mode shapes, frequencies and damping are identified. This information is essential in understanding how a structural system responds dynamically. Among the disadvantages of modal analyses are the assumption that the structure can be described by linear or a nearly linear model, it is time consuming and is usually reserved for complex structural systems, detailed analytical models are usually needed, and it is used often only after problems have been experienced.

4.16 PRESSURE/VACUUM TESTS

This type of structural test involves the application of pressures or vacuums to a system. The system may be a vessel or chamber and may include the valves, lines, and fittings needed to make an operational entity. Pressure can be applied with liquids or gases. Vacuums are usually thought of in terms of evacuation of a gas. Most pressure and vacuum test hardware use some types of seals (e.g., metal-to-metal, O-rings, gaskets).

Pressure is defined as a force per unit area. The pressurizing medium provides a pressure to all parts of the structural system it can reach. Vacuum is defined as a space filled with gas at pressures less than atmospheric. Pressure units used are lbs/in^2 or N/m^2. Vacuum is defined in torrs (1/760 of a standard atmosphere). The following ranges for vacuum tests are:

1. Low   - 760 to 25 torr
2. Medium   - 25 to 1×10^{-3} torr
3. High   - 10^{-3} to 10^{-6} torr
4. Very High - 10^{-6} to 10^{-9} torr
5. Ultra High - 10^{-9} and less

There is one problem common to each test and that is leakage. No seal has been found which will not leak eventually. The question in conducting these tests is to determine whether the leaks are significant. That is, are the leaks large enough to inhibit the action of the pumps used, are the leaks a danger to personnel, equipment, and diagnostic devices used? The presence of leaks and their detection can be a significant challenge to those performing these tests. Leaks can be detected by changes in pressure or vacuum in comparison to expected or anticipated values when a test is being performed. Leaks can also be detected by noises, particularly if acoustic emission devices are used in conjunction with other diagnostic devices. Leaks can also be detected using a variety of techniques (e.g., soap bubbles or other similar liquids placed on the system, helium leak detection systems and other diagnostic systems developed for this purpose). The editors of this handbook chose to not include leak detection as a structural test. ANSI N14.5 gives a detailed summary of leak detection methods and standards. There are also many other available references under the general title of leaks and leak detection (Chap. 3, Ref 3-16).

The possible hazard in a vacuum test is the possible sudden deformation of a structure because of buckling.

There are many possible hazards associated with pressure testing including:

1. Stored energy in the pressurizing medium
2. Stored strain energy in the system
3. Sudden release of these stored energies
4. Containment of the sudden released energies
5. Possibility of high pressure leaks (do not search for leaks with your hands)
6. Possible high speed motion of pressure lines in the event of valve, fitting, coupling, vessel, or other system failure
7. Disruption of seals, sealing surfaces
8. Materials degradation—fracture, brittle failure, welds; even small flaws can cause significant damage potential to a structure
9. High localized stresses around holes, pass-throughs, attachment points.
10. Lifting, rigging, and handling pressure test hardware.

There are many examples of pressure testing in this handbook. Pressure tests are widely used in industry: aircraft fuselages for commercial and military aircraft, boiler and other vessel certification, air compressors, and other related systems.

Pressure systems are typically designed using ASME Section 3 and 8. These rules require systems designed with these codes have a working pressure which corresponds to a design margin of four on ultimate strength of the material. These vessels and systems are then to be proof tested to 1.5 times the working pressure. There also requirements for protection of the system so designed. These safety devices include relief devices including valves (typically spring loaded which open at pressures near the working pressure), burst disks which fail at a defined pressure, and other devices designed to relieve pressure. The fittings, valves, lines, and other hardware must be qualified for the pressures expected at operation.

There are usually rigorous requirements for assembly of hardware, selection and installation of fittings, flow diagrams to prevent unwanted pressurizations, operational schemes to vent pressure in the event of emergencies, and many other system operation and safety considerations.

Pressure tests should be undertaken only after training personnel, planning and taking adequate measures to ensure safety, following approved procedures, and having adequate test resources and equipment.

4.17 PROOF/OPERATIONAL TESTS

In the development of prototype hardware and components and its continuing certification for use, a typical test performed is a proof test. There are some formal procedures associated with proof tests of pressure vessels (i.e., ASME Sections 3 and 8 of the Pressure Vessel Safety Code). There are also some formal as well as informal procedures involved in proof tests on hardware. The term proof test usually implies: (1) a single test conducted once in the service life of a structure or (2) a test performed periodically to recertify that the structure is safe for continued operation (e.g., a pressure vessel, a repaired lifting sling, etc.). A corollary of this type of test is when a structure designed for one type of use and thought to be adaptable to use for another condition is certified for that condition by a proof test.

The term proof testing implies that the design loading conditions will be applied to the test item and it is typical that the loading is increased by some factor to allow for unknowns in the design process. It is assumed that a proof test will exercise a structure to conditions greater than expected in its service life. That is, a structure will be operated or used at levels lower than the proof test conditions. However, in a theoretical sense, it is difficult to actually perform a proof test that exercises each element, part, bolt, or connection to conditions greater than expected in service with typical margins of 1.15, 1.25, or 1.5 times the maximum service loads. Only by performing a rigorous and exhaustive set of tests could each part of a structure be tested to required proof test levels. In a practical sense, it is possible by conducting a limited number of tests, possibly even one, on a structure that will adequately demonstrate its ability to meet maximum service conditions plus an additional margin. The trade-off here between requiring tests on each part of the structure or a few or one key test is to require that the elements of the structure (e.g., bolts, welds, individual members and elements) have requirements to meet guaranteed minimum strengths and material certifications. These requirements are typical in the design process.

An example of a proof test is to apply loads to a structure to be used for flight tests. Suppose that an adapter is to mount a flight vehicle to a missile structure. This adapter should be subjected to loads simulating the maximum flight loads that are expected plus extra force for the design margin. Among the conditions that the adapter could be tested to include steady state (statically equivalent) values of accelerations, in-flight vibrations (sinusoidal, random, and transient), and shock loadings equivalent to various separation stages of missile operation. These test conditions could be applied individually or simultaneously in various axes.

Another type of proof test is for fixtures used in various tests. For example, fixtures are used in inertial devices (centrifuges) to support hardware and components to be tested. A static test can be used to place the expected loads on critical parts and joints of the fixture. This type of loading can exercise the joints and connections to the degree expected in a test. Additional factors (e.g., 1.5) can be used to multiply the effects of the applied forces. However, a static test cannot duplicate the body forces acting on each structural element found in the rotation of the centrifuge. There are many other types of fixtures used in structural testing. Their performance should be evaluated because they can become part of the problem (i.e., excess deformation, extraneous boundary conditions) rather than a means to complete the test.

Pressure vessels occupy a unique position in proof testing. A typical requirement for a vessel which will be operated with personnel in close proximity is to require the vessel to be pressurized to 1.5 times its working pressure before the vessel can be placed in service. This proof test should reveal major faults or flaws in the materials and construction of the vessel. This requirement is for hydrostatic tests using water or other fluid to convey the pressure. The ASME also requires a proof test to 1.15 or 1.25 times the working pressure if a pneumatic test using air or other gas to convey the pressure. (The reader is referred to the applicable sections of the ASME codes for specific requirements.) If a hydrostatic test is used, then the maximum allowable working pressure (MAWP) is then defined as two-thirds of its proof pressure. If proof levels higher than 1.5 times working pressure are requested, the potential for damage or degradation in structural integrity increases. For vessels designed with a factor of safety of 4.0, the proof test of 1.5 times working is considered optimum. Any greater proof test requirements could degrade the vessel. There will come a time when an in-service vessel may have its working pressure reduced or its service life shortened based on its performance in a proof test.

Proof tests are needed to satisfy requirements, national standards, and many company or organizational requirements. These proof tests give demonstrated evidence that the structure or prototype will meet its design intent and expected service conditions with some additional margins. Proof tests are commonly performed for commercial and military aircraft (static, dynamic, and pressure tests), lifting and rigging hardware, pressure vessels, test fixtures, and test equipment.

4.18 RESIDUAL STRESSES

These stresses exist in structures which are otherwise free from the effects of applied loads and from the constraints offered by boundary conditions. These stresses result from a variety of sources including machining operations (milling, drilling), joining processes (welding, brazing, gluing), formulation of composite materials and filament winding processes, and from interactions among the materials involved. Residual stresses can produce both advantageous situations for structures as well as introduce or cause detrimental effects. There are times when residual stresses are desired and there are times when they can initiate failures.

In terms of structural testing, residual stresses are difficult to measure and evaluate. Most of the experimental techniques used to measure residual stresses involve destructive evaluation of the test items. These techniques typically involve the measurement of strains, often in relative terms, so that estimates of the actual stress states can be calculated or otherwise inferred. The experimental difficulties are compounded by the unpredictable nature of these stresses and their abilities to combine with other effects such as material incompatibilities, stress corrosion, and time dependent behavior of materials. These combinations are often difficult to sort into individual conditions and to quantify each effect. As R.E. Rowlands explains in Ref. 4-21, "Although progress has been made in measuring residual stresses, extensive efforts remain to develop adequate nondestructive, cost-effective, and expedient techniques." Initially one might be discouraged with this introduction because of the apparent complexities and relative inaccuracies of the methods and techniques. However, there are many references and existing techniques developed for many problems. These techniques do require some effort to understand and use.

Residual stresses are widespread in structural systems and assemblies. Often they are ignored or considered only with minimal attention. Residual stresses are usually evaluated in terms of the context in which they occur and with particular emphasis on the materials involved. For example, the rolling operations for an I beam or wide flange shape results in differential cooling along the flanges and webs and through the thicknesses of the materials resulting in residual stresses (Refs. 4-10 and 4-26). The effects of these residual stresses on elastic behavior of this type of beam is minimal and even the failure conditions of the beams are not appreciably altered as long as they are loaded and act along their principal axes. The effects of residual stresses are attenuated because of the redistribution capabilities of steels, particularly structural grade materials. However, if the residual stresses occur near discontinuities or are oriented to abet formation and growth of flaws in materials, then the residual stresses can become vitally important in the testing of structures and components, particularly in fatigue.

It is suggested that prior to starting an experimental investigation into structures where residual stresses could play an important part in their behavior, that the references supplied at the end of the chapter be consulted. These references were chosen because they are widely available.

4.19 SERVICE LIFE EXTENSION

The purpose of this type of testing methodology and evaluation procedure is to extend the service life of a structural system. Many structural systems are designed with an expected period of service. This period may be calendar years, it may be the number of cycles of loading that must be endured, it may be the number of hours of expected operational use, or it may be when measurable limits of structural behavior are reached (e.g., deflected shape, elongation). It is often desirable to extend the life of structures beyond their initial design intent and this need has led to the development of structural test methods and techniques. The structures currently considered as candidates for extension of their service lives include aircraft (both military and commercial), petroleum and other piping distribution systems along with refineries and production plants, bridges (highway and railroad), and related structural systems which are often large and costly to replace (Ref. 4-P6).

In years past, these types of structures were replaced because newer improved structural systems were developed which made older systems obsolete. In recent years these older systems have proven very useful and are not being outdated. With economic resources not as plentiful as they once were, there is sufficient incentive to increase the working lives of older structural systems. The extension of service life of structural systems is an attractive alternative to developing new systems.

The challenge for extending the service life (often at considerable time intervals or cycles beyond the expected design life) has given significant opportunities for structural testing along with related detailed technical evaluations of structural systems. Much work has been initiated and many systems (particularly military aircraft) have had significant extensions of their service lives. Further development of the technology and evaluation procedures will be the focus of much technical work in the coming years.

The challenge focuses on six major aspects of testing with particular emphasis on nondestructive evaluation techniques:

1. Development of the full potential of nondestructive evaluation techniques.

2. Development of an industry commitment to fix the technical problems before political solutions are forced upon the industries.

3. Development other of test techniques and diagnostic tools which can be successfully applied to a wide variety of structural systems.

4. Development of information and data procedures to effectively evaluate existing structural systems.

5. Development of acceptable procedures for inspection, rework, retrofit, and recertification of structures.

6. Development of regulations and governmental controls which provide for and encourage the extension of service lives of structures.

Some of these challenges can and will be solved by the structural testing and nondestructive evaluation community.

One major technical challenge in service life extension is the development of technology which anticipates surprise behavior in structural systems. Many of the problems encountered thus far have been found only when parts of systems have been replaced. For example, the inspection of replaced parts in aircraft engines have shown indicated that about 80 percent of the defects found were discovered after the parts were replaced and not while the parts were in service. That is, prior inspection of parts to be replaced had not revealed the flaws and defects that were detected under closer scrutiny when the parts were removed from the engines and replaced.

As structures go through their service lives, it is common to identify some "hot" spots or critical areas which degrade faster than other areas. These critical areas are exemplified by wings of military aircraft which are subjected to high loadings from payloads and maneuvering forces. These wings often reach their limits (either cycles of use, strain levels in critical areas, etc.) before the rest of the aircraft reaches limits. Some military aircraft have had their wings replaced. Some bridges have had their main supporting members replaced or strengthened with added plates and stiffeners installed.

Specific methodologies have been developed for application to specific types of structures. A general approach to service life extension is described in the following sequence (Ref. 4-P6):

1. The identification of the critical locations in structures or assemblies where added stresses, wear, or degradation in materials have contributed potential or actual problems in structural performance.

2. The performing of representative and realistic failure tests on actual parts or pieces obtained from structural elements or components. A typical test is to use cyclic loading, temperature cycles, or vibration conditions.

3. Evaluation of the test data to determine actual failure levels and modes.

4. Overall evaluation of the remainder of the structure for comparative purposes so that estimates can be made of the increased service life expected if the critical locations are repaired, retrofitted, or otherwise improved.

5. Assignment of appropriate margins or limits on continued service to continue the integrity of the structure or system.

The principles of service life extension testing and evaluation will likely be applied to buildings, housing, automobiles, electrical transmission towers and poles, and other types of distribution systems (e.g., water and gas pipelines).

4.20 SHOCK TESTS

Shock tests can be grouped with many other types of dynamic or acceleration based tests. Shock tests differ from other time dependent tests because the major interest is the application of relatively large accelerations usually in the form of transient excitations. There are some types of shock conditions which involve transients impose on large steady state accelerations such as the lift-off of missile systems and then the separation of the staged sections (Ref. 4-20).

Most military hardware, many types of industrial transportation systems, and various electronic components are routinely required to survive and function after being shock tested. The shocks imposed replicate the loads encountered in service, handling, transportation, package delivery, and other environments where shock loads are present.

The testing of structures and components is accomplished through the use of many types of equipment. The general approach is to store energy in the test system and then release that energy rapidly so that a shock loading condition is imparted to the test item. Energy can be stored in the test hardware through the following means: (1) compressed gases, (2) compressed or tensioned springs, (3) hydraulic methods, and (4) through gravity and positioning of test equipment. The quick release of this stored energy provides the motions needed to create shock loadings on structures. The quick release can use locking pins, explosive separation devices (e.g., bolts, line or strip explosive charges), and bolts with known failure loads and modes (e.g., brittle).

The goals in shock testing are: (1) to provide an appropriate shock load (a peak acceleration level and velocity change through the needed acceleration-time history), (2) to provide repeatable tests for exercising actual and prototype structures and components, (3) to prove that structures will survive and function, and (4) to determine failure conditions for test items.

In compressed gas systems, a chamber stores compressed gas on one side of a piston. The piston is released from its restrained position and a pressure injection system (e.g., burst diaphragm) is activated causing the compressed gas to accelerate the piston. With the piston in motion, various techniques have been developed for shock loads to be imparted to the test item. Attaching the test item to the accelerating piston will cause a shock loading. By having the piston follow a specified path and impact into a mass is another way of developing a shock load. In this case the test item can still be attached to the piston or it could be secured to the reaction mass. The piston could also be secured to a mass initially and the collision of the piston-mass combination into another mass will also result in a shock load on the test item. By appropriate selections of pressures, piston sizes, masses, test items, and shock mitigation and tailoring materials among the interacting parts, the desired shock pulses can be achieved.

In mechanically loaded systems (e.g., springs), the energy is stored in the mechanical system. Various methods can be used to release the energy (e.g., explosive bolts which fail when a charge is fired, mechanical triggers). The test items are typically attached to a carriage. After traveling for a distance, the carriage is stopped. The acceleration-time histories can be imposed on test items in a similar manner as just described for the gas systems.

Shock testing equipment also includes a wide array of drop tables which are gravity actuated. A test table is hoisted into position and then allowed to free-fall onto a target. The acceleration-time history can be shaped by appropriate use of materials between the table and target. Higher accelerations can be achieved if the test table is pulled toward the target. Various pulling force systems have been developed ranging from the use of stretchable cords (bungees) to rocket sleds. These sleds have cables attached to test items directly with the cables passing through pulleys located adjacent to the target. The test item is separated from the cables just prior to impact so that free fall conditions can be achieved and additional forces from sled interactions are not imposed on the test item.

The art in conducting shock tests is selecting and having the appropriate equipment available for the test item, the acceleration-time history desired, and to obtain the test data. The data can be obtained through instrumentation cables attached directly to the test item, by telemetry, and through systems which allow post-test interrogation. Considerable effort has been directed at developing data systems which can record test data without having to be exposed to the high shock loads.

Among the potential problems in shock testing are:

1. Containment of the test item and any debris that may result from the test.

2. Rugged and reliable data systems which may be required to withstand shock loads as well.

3. Tailoring the shock pulse to meet the test requirements.

4. Maintaining alignment of the test table or test item and the target.

5. Test system safety for operators and equipment (e.g., interlocks, arming systems, firing systems)

6. Adequate transducers and data acquisition systems which accurately read the high frequency shock pulses and measure the responses of the structural system or component.

4.21 STIFFNESS TESTING

Stiffness testing is the experimental determination of the displacements of a structure when loads are applied (Ref. 4-27). The purpose of stiffness testing is to supply actual data needed to define or understand how a structure actually deforms. The analytical models (usually developed using finite element analyses—FEA) are used to convert the governing partial differential equations into more manageable and numerically efficient matrix based algebraic equations. This conversion process makes the modeling of complex structures (including plates, beams, shells) subjected to static and dynamic loads amenable to solutions using computer software. In any modeling process there are many assumptions that have to be made (e.g., structure geometries, material properties, and finite element representations). Models for complex structures (e.g., with joints and connections, multiple load paths, and complicated geometries) usually need some experimental verification. Stiffness testing supplies the information to verify these models.

Stiffness testing involves (typically static) loading of a structure elastically and measuring the resulting displacements and/or strains. Some common reasons for performing stiffness tests are to help validate deflection, fatigue, buckling, resonance, and flutter analyses. Joints can be especially challenging to model; stiffness testing can help develop a more realistic model. Conversely, if an accurate analytical model exists, stiffness testing can infer damage in a structure from the deviation of measured stiffness from that predicted for an undamaged structure. For example, finite element analysis compliance calculations are used to infer crack growth in fracture toughness specimens. Stiffness testing offers valuable insight into structural behavior.

Finite element analyses typically consist of many nodes and elements used to make a model. The displacement of each node does not need to be measured in order to have a valid stiffness test. These equations are closely coupled so that only a few displacement measurements are needed to ensure an adequate representation of the displaced shape of a structure. These measurements can often be used as inputs or known conditions which the model must satisfy. The stiffnesses can be adjusted so that the predicted deformations coincide with the measured displaced shape.

There are some fundamental considerations in stiffness testing regarding the following:

1. Applied forces
2. Measurements of forces, deflections, and other measurands
3. Practical aspects of testing

Some concerns in stiffness testing regarding how the applied loading reflects the actual application are:

1. Does the magnitude of the applied loading cause damage or nonlinear effects due to overloading? Are the loads large enough so that adequate measurements can be made but small enough so that the structure is not damaged?
2. Does the distribution of the applied load properly model pressure, gravity, maneuvering, acceleration, or other loads? Many loads are applied at convenient locations (sometimes the only location available) on the structure rather than at centers of gravity.
3. Does the direction of the applied load follow the desired path; are the loads required to change orientation because of deformations of the structure? (A very flexible structure often has large elastic deformations; flutter is an example of a follower force)
4. Does the rate of applied loading reflect any viscoelastic or dissipative conditions in the structure?
5. Is the duration of the applied loading long compared to actual loads? (At elevated temperatures is creep deformation large compared to the desired elastic deformation?)
6. Are there other effects which could influence how loads are applied or modify structural response (e.g., stiffness measurements prior to and following thermal cycling)

Some concerns in stiffness testing regarding the measurement of displacement are:

1. Measurements are accomplished in two ways: by instruments in contact with the surface of the structure or by non-contacting techniques. On compliant structures does the contact force from the displacement gage influence the structural response? (would non-contact gages or techniques be preferable?)
2. Is the resolution of the transducer appropriate for the range of displacement to be measured? Many of the displacements are very small values.
3. If the total displacements are measured, is the gross motion unduly large compared to any desired differential displacements among the points on the structure?
4. Are unexpected behaviors planned for (e.g., in asymmetric or real structures, it is possible to apply a force at a location and have small deformations occur in the opposite direction at nearby locations.)
5. Are residual stresses present and do they influence the deflections?
6. Are only elastic strains present?
7. Displacements are measured relative to some reference. Is the choice of the reference locations proper or will it also experience some deformation?

With the use of finite element analysis (FEA) so widespread, we have offered some fundamental situations which may lead to discrepancies between calculated stiffnesses and structural test results. A more complete discussion of finite element analysis is given in numerous other texts.

4.22 STRUCTURE-CONTROL INTERACTION

Structure-control interaction means the combination of sensing and control devices and techniques used to define structural response. For example, the controls on an automobile are exerted by the driver on the accelerator, transmission, brakes, and steering system. The interactions between the driver and the automobile define where the car is heading, speed, etc.

With the increased use of a variety of sensors, microprocessors, and other systems capable of reacting to electrical input, structure-control interaction is now a much larger and more complicated field.

Many manufacturing operations are conducted using robots and other instrumented systems used for handling, assembling, and testing a wide variety of products. Among the advantages of these systems is the relief of tedious repetitive behavior on the assembly line. A disadvantage of these interactive systems is in making them responsive and intelligent enough to complete their job, particularly when anomalies occur.

In Chapter 6 the concept of open and closed loop testing will be introduced. Part of this introduction contains a description of feedback. Feedback is the process of sending and receiving information, comprehending the meaning of the information, and then taking appropriate actions. In structure-control interaction, feedback is essential to successful completion of the interactions.

Feedback in structure-control interaction depends on sensing a change in the structure (e.g., acceleration, displacement, force, pressure, strain, temperature, velocity, etc.) and then developing appropriate characteristics in the control system to respond to these measurements. The art in development of these systems is to provide sufficient but not excessive instrumentation.

The possibilities for structure-control interactions are virtually limitless. Current developments include sensing materials that are part of the structure itself, improved microprocessors, miniaturization of components and making them more rugged and resistant to very harsh environments, and shielding of the small electrical signals.

4.23 THERMAL TESTING

Thermal testing consists of placing various conditions of heat energy on a structure. The heat energy may be in the form of a fire, air at elevated temperatures, radiation from a variety of sources, or the near absence of energy (cryogenic conditions) on a structure or test item.

Thermal testing is important when the test item is required to be used and survive temperature variations through thicknesses or walls, temperature cycles which cause differential expansions among the materials involved, and direct exposure to thermal effects. For example, thermal cycling tests can be important for electronic systems which experience temperature cycles while in use. These cycles can influence the integrity of solder and other interconnections because of detrimental effects caused by thermal expansion and contraction. Experience shows that thermally induced stresses and loads on electronic systems are a principal mechanical design requirement.

Power generation and distribution systems also have thermal design requirements. These design requirements are often coupled with the presence of hazardous or corrosive materials (e.g., steam under pressure). Thermal and environmental test evaluation of the materials and components of these systems can be essential prior to placing them in service.

Thermal testing can be accomplished in a variety of ways including:

1. Temperature chambers capable of holding relatively large items

2. Heaters and other specially designed equipment attached directly to the test item or structure

3. Burn facilities which can expose the test items directly to the effects of controlled fires

4. Radiant facilities which offer heat from electric lamps and other sources of radiant energy

5. Baths or containers of very cold liquids (e.g., nitrogen)

There are three important parameters associated with thermal testing: (1) maximum or minimum temperatures, (2) rates of temperature changes, and (3) spatial distributions of temperatures over the test item. These variables typically indicate the type of test facility needed, how the tests need to be controlled, and how instrumentation requirements will be defined.

There are three major problem areas associated with thermal testing: (1) data acquisition, particularly for very hot or very cold tests, (2) control of the temperatures and spatial distribution of thermal effects, and (3) test safety. For those tests conducted at or near ambient conditions, test control, data acquisition, and safety concerns are typically not difficult to accomplish. If thermal tests are to be conducted in conjunction with other types of tests (e.g., materials, proof, stiffness), then the constraints from the thermal tests can influence material properties (e.g., most properties—particularly strength and stiffness are altered with sufficient temperature change), make measurements more difficult to obtain as instruments must be able to operate at or be protected from the effects of increased temperatures, require added safety equipment to perform tests safely (e.g., thermal protection for personnel), and may require additional analysis to correct for modifications to the data (e.g., apparent strain in gages due to temperature changes). Performing tests at elevated temperatures often requires additional planning, equipment, and procedures. For those tests conducted at very low or elevated temperatures, these problem areas become very important and need to be considered in planning and conducting the tests.

4.24 TIME DEPENDENT TESTS

Creep

Nearly all materials and structures made of them exhibit time-dependent behavior to some extent, depending upon the test temperature and environment, load magnitude, and load duration. Even prestressed concrete beams and floors experience sagging under their own weight.

The total strain at any point consists of recoverable and permanent parts. Each part may have time-independent and time-dependent components, again depending upon the test conditions.

For recoverable strain, the elastic component is relatively instantaneously recovered while unloading; its magnitude is usually proportional to the load. In stiffness testing, the load magnitude and duration are limited so that this elastic component dominates the total strain. For polymeric materials (and metals to a lesser extent) the elastic strain is often accompanied by a time-dependent anelastic strain component, observed as an attempt by the structure to resume its original undeformed configuration for some time after unloading.

If the loading magnitude or duration (at a given test temperature) is sufficient, some of the total strain will not be recoverable even long after unloading; this is the permanent component of strain. The permanent strain can have a relatively time-independent component, the plastic strain, like that generated while yielding a steel tensile specimen at a constant velocity at room temperature. Or, it can be time-dependent such as creep strain accumulating in the same specimen under contact stress at elevated temperature.

So, the total strain consists of time-independent and time dependent recoverable and permanent parts. The proportion of these parts often change during a test. During a constant stress tensile creep test on a metal, the elastic strain remains about constant while the creep strain increases with time. In a stress-relaxation test, a specimen is quickly strained a fixed amount; as the strain is held constant the load is observed to decrease. This is analogous to the preload decay in a nylon or soft lead washer after the bolthead is torqued. The dropping load means the elastic strain is decreasing with time. Since the total strain is fixed, this elastic strain decrease is being balanced by an increase in anelastic and creep strain. For the nylon washer, it may be more anelastic than creep and vice-versa for the lead washer.

We briefly list some failure modes that structural testers should additionally consider when creep behavior is anticipated. They should also generally be alert when tests are requested at temperatures near or above one-half of the melting temperature (in degrees Kelvin) of any critical part under stress, especially for long loading durations (Refs. 4-12, 4-13, 4-14, 4-19, 4-24, and 4-P18).

Excessive creep deformation leading to interference of parts is a failure mode. In turbine engines, rotor blades creep under radial centrifugal loads at elevated temperatures; the blade tips may engrave into the engine housing. Insofar as the grain boundary sliding contribution to creep, one counter is to grow single-crystal blades; another is to form the blades superplastically (with very small grainsize) then follow with a post heat treatment to dissolve most of the grain boundaries and recover the creep resistance inherent in the alloy.

Temperature cycling can cause excessive deformation to accumulate due to incremental deformation. The source of this "ratcheting" can be time-independent plastic strains and/or time dependent creep strains. For the latter, the phenomenon of "enhanced creep" may arise, wherein the creep strain accumulated during cycling exceeds that which would be calculated from continuous operation at the higher temperature boundary.

Stress relaxation in threaded connections can result in separation of parts and fatigue failure or reduction or loss of initial tension (preload). In aluminum electrical wiring, the loss of preload can result in a poor connection, high local electrical resistance, heating, and sometimes even fires.

Creep rupture is a catastrophic failure mode. For example, boiler vessels and high pressure piping walls can reduce in thickness (a corrosive synergism may accelerate the process). Since the pressure is constant, the wall stress increases with thinning, thus accelerating further creep, thinning and finally rupture. Empirical relations (e.g., Larson-Miller, Monkman-Grant, Manson, Sherby) based on uniaxial tensile creep rupture data can be used to help avoid creep rupture although appropriate generalizations from uniaxial to multiaxial stress states is still a research topic.

Creep crack growth under load can occur as the material near the crack tip progressively ruptures due to void growth under high triaxial stress at elevated temperatures.

For slender members like columns and shell structures under compression, creep-induced buckling can be a catastrophic failure mode. That is, collapse can occur due to creep deformation which causes the structure to be increasingly imperfect. Consider an elastic column under compression, and imagine a plot of compressive load versus midspan deflection perpendicular to the longitudinal axis of the column. For a perfectly straight column, the load-deflection curve will trace the load axis up to the Euler buckling load. For columns with an initial curvature (imperfection), the load-deflection curve will start at zero load and a slight positive deflection. As the load increases, the load-deflection curve will rise, bend over to the right and follow a state of zero slope where the load-deflection values at this point define an unstable state corresponding to that initial imperfection. Similar points in the load-deflection plane exist for other values of initial imperfection. One can anticipate that the locus of all such points is a curve starting at the Euler buckling load and decreasing as a function of deflection. In creep-induced buckling, the structure is initially loaded to a point safely under this instability locus; however, as creep proceeds under constant load, the deflection can increase with time until the locus is reached.

Thermo-mechanical fatigue (TMF) is another creep-related failure mode. Consider, for example, a service cycle where strain and temperature are cycled between two levels with a hold time at the higher level, all the while being proportional to each other. If the ratio of strain range to temperature range is other than the thermal expansion coefficient of the material, a stress will be generated when the thermal expansion strains are exceeded. During the dwell time, some of the stress will relax. After sufficient repetitions, a hysteresis loop in the stress-strain diagram may stabilize; this behavior is analogous to low cycle fatigue involving plastic deformation where the number of cycles to failure can be significantly fewer than for high cycle fatigue involving only elastic behavior.

Another example of creep related testing is accelerated aging simulations. The purpose of these tests is to determine some limits to typically time-based evaluations of service conditions. Simple examples of aging tests to evaluate structural behavior repeated over many cycles of loading are closing and opening a mechanical locking systems (e.g., automobile doors) and exercising seals on pressure systems. In accelerated aging tests, if creep behavior is expected to occur, it can have a significant influence on the test results and needs to be included as part of the test. In the example of an elastomeric seal on a pressure system, the accelerated testing evaluating this seal would need to include those effects which could degrade the seal (e.g., permeation around and through the seal, sufficient time for the relaxation in the preload on the seal) in order to have a realistic test. As a contrast, if the preload was not relaxed, then there may not be a path around the seal and if there was not enough time for materials to permeate through the seal, then an nonconservative evaluation of the seal would occur on just one pressure cycle. If the seal is needed over many cycles of use, then it is essential that each cycle of the test be performed to exercise the seal as it will actually be used. If the seal was also required to maintain pressure over a long period of time, then it should be tested over that time period.

The evaluation of certain types of electrical components (e.g., glass or ceramic insulators) can involve accelerated aging tests. These parts are typically expected to withstand long periods of use with environmental effects (e.g., temperature variations, humidity, corrosives) having the major degrading influence on how they behave. These materials can deteriorate because of the condition of their surfaces. Small surface cracks can develop which weaken them and allow penetration of materials whose effects further weaken them.

As an example of the accelerated aging tests on these electrical components is to simulate the requirements for their insulators to experience temperature cycling. An example of how not to perform an accelerated aging test is in the evaluation of a lot sample of connector pins made with ceramic insulators. The connectors were required to withstand many cycles of temperatures ranging from -65 °F to 200 °F over a few minutes. An accelerated aging test was proposed which involved placing the connectors in a solution of dry ice and alcohol (about -65 °F) and then dropping them in boiling water with many repetitions of this cycle. When the ceramic insulators began cracking (even on the first cycle), it was determined that the connectors were not in thermal equilibrium. With a relatively short pause (about one minute) in the middle of each cycle at ambient temperatures so that thermal equilibrium could be reached or at least approached, then the thermal shocks were reduced and the connectors survived. The actual service conditions had a much slower temperature cycle (about 15 minutes from coldest to warmest conditions).

Accelerated aging tests are valuable tests and they will yield insights into structural behavior and capabilities if the tests adequately replicate expected service conditions.

Fatigue

Structures are often subjected to fluctuating or oscillatory loads. The behavior of these loaded structures can differ fundamentally from those loaded statically or even dynamically if fatigue considerations are present. Fatigue is a structural behavior which is characterized by four features: (1) reduced strength, (2) reduced ductility, (3) load ranges (minimums and maximums), and (4) number of load repetitions expected in the service life. The effects of fatigue can occur in widely different materials (e.g., metals, plastics, rubber, and concrete). Each material has some inhomogeneities; these imperfections (e.g., surface cracks, inclusions, voids, and interfaces in composite materials) are the source of fatigue. When a structure has some stress concentrations, the presence of these imperfections gives opportunities for cracks to grow. When the stress concentrations are combined with repeated loads, the resulting effect is to give greater emphasis to these imperfections. Characterizing these imperfections is important but a tedious process. Evaluating a structure (known to have some flaws) in a well-formulated fatigue test will integrate the effects and give insight into how a prototype or actual structure will respond.

Fatigue loads on structures may be cyclic as in rotating machinery, or random as in flutter on aircraft wings (Ref. 4-11). These loads may alternate between tension and compression, may increase or decrease around a mean value, and they may be uniformly to randomly distributed over a structure. Experience shows that loads alternating about a mean value but always in tension or compression produce a certain damage potential; loads alternating from zero to a peak value (in tension or compression) produce a greater damage potential; and, loads alternating from tension to compression produce even a greater damage potential. A common way of representing these stresses is through the stress ratio $R = \sigma_{min}/\sigma_{max}$; $R = -1$ for equally alternating tensile and compressive stresses, $-1 < R < 0$ for alternating tensile and compressive stresses, and $0 > R > 1$ for alternating tensile stresses (Refs. 4-7, 4-8, and 4-25).

The classic approach in fatigue studies is to generate a series of S-N curves for a specimen or portion of a structure exposed to cyclic loading. The ordinate S is for tensile stress and the abscissa N is for the number of cycles. Each point on the curve indicates the failure of a specimen at the number of cycles corresponding to the constant stress amplitude. These curves (for metals) typically exhibit a decreasing value for the failure condition until a horizontal asymptote is reached with the implication that the fatigue loads corresponding to this level of stress can be endured indefinitely.

Three areas of research have increased our understanding of fatigue loads and their impact on structural testing: (1) the development of relationships between plastic strain in materials and fatigue life, (2) the development of fracture mechanics to enable competent evaluations and assessments of structural designs, and (3) testing capabilities combined with statistically significant test methods giving better information and data for design purposes. It is now possible to assess the fatigue life of a structure or component based on crack-growth data, strain-life data, along with the stress-strain relationships for materials on very small scales. Areas or portions of structures which are "hot spots"—known to be subjected to large fluctuating stresses are often isolated and tested separately. Aircraft test laboratories often use many testing machines devoted to fatigue testing of component sections. These include uniaxial (tensile specimens), biaxial (specially designed items, tension-torsion specimens) and triaxial tests.

With the continued development and use of high strength materials, some composites, and minimum weight structures, the presence of fatigue conditions are more prevalent than ever. Fatigue testing can be essential in the development of any structure which will experience fluctuating loads including structures designed for land transportation systems, aircraft, and railroad use.

Fatigue and fracture mechanics are often linked together. Fatigue may not result in fracture (although degradation is usually measured in this manner). Fracture can occur without fatigue. They are found together in fatigue testing because specimens and components usually fail by fracturing. Occasionally other failure modes are found by fatigue tests but they are infrequent and do not result in a worst case or even typical failure condition.

4.25 VIBRATION TESTS

Many structural components, mechanical assemblies, and electrical systems are designed using static design principles. These principles use the estimates of maximum acceleration in a statically equivalent sense and apply the resulting force (F = ma) to the component or assembly. The component or assembly is then configured and sized to resist these applied forces along with some appropriate design margins. This design process usually focuses on the connections or joints in terms of the threaded fasteners used, welds, or other joining processes as well as the strengths required for individual parts in the component or assembly. Once the design has been chosen then prototype structures or assemblies are typically fabricated. The integrity of the design is verified, often as the most important part of testing, by the performance of the prototype structures when they are subjected to vibration conditions.

Vibration testing is the most widely accepted way of design verification for a wide variety of components and assemblies. This approach is used for many types of mass-produced components and assemblies ranging from computer parts to mechanical assemblies to larger structures such as aircraft. A vibration test is intended to serve one or more of the following purposes: (1) a proof test, (2) an overload or significantly greater exercise of the system, and (3) to determine the damping of the system, component, or assembly. From the results of vibration tests combined with experience in testing and analyses, estimates can be made of the possible degradation of the structural system being tested (Ref. 4-11, 4-20, 4-23, and 4-P1).

For vibration tests on a unique structure or one which cannot be degraded or destroyed by testing, then a different approach must be used. The intent in these tests is to learn as much as possible about how the structure behaves, when failure thresholds are likely to occur, and where weak points of the structure are located. It is typical in this approach to apply vibration conditions (see Chap. 6) and evaluate the responses. If the responses are small (e.g., do not exceed the elastic behavior of the materials), then the intensity of the vibration signal is increased and the responses are reevaluated. From this comparative type of approach the critical responses can be determined (Ref. 4-P23).

There are two basic types of vibration tests: (1) sinusoidal and (2) random. The sinusoidal test consists of certain acceleration level (g's) combined with a frequency sweep at a range from an initial frequency f_i to a final frequency f_o or a constant displacement over a portion of this curve. The objective of most sinusoidal tests is to find the resonant frequencies of the structure and then dwell on those frequencies in further tests. The sweeps typically consist of a 1 g acceleration, then repeat tests with incremental increases in magnitude until damage potentials are reached. The sinusoidal vibration test can contain significant amounts of energy, particularly when the structure is being excited at one of its resonant frequencies. The equipment limitations in this type of test are displacement (e.g., typically about 1 inch maximum travel), frequencies (e.g., up to about 3,000 Hz), or force depending on the size of the vibration equipment. The dwell portion of a sine vibration test is also used to evaluate and verify the performance of a component over its intended service life.

The fundamental difference between sine and random vibration testing are that the structure or component in a random vibration test will be exposed to energies at all frequencies in the bandwidth selected while a sine vibration test will output energy (amplitude) at only one frequency. This single amplitude is then swept (the frequency changes) over the frequency range of interest. The sine vibration test is typically specified as an amplitude (e.g., 5 g's) over a desired frequency range (e.g., 20 to 2,000 Hz). A random vibration test is specified in terms of the energy delivered to the structure (e.g., 0.1 g^2/Hz) in the frequency bandwidth (e.g., 20 to 2,000 Hz). A random test requires a statistical description in order to evaluate the energies involved.

Both types of tests can be required and can be necessary for a component. The necessity for both types of tests depends on the service requirements, acceptance criteria, and on the structure or component itself. Many components or assemblies are exposed to a predictable and repeatable sets of conditions (e.g., rotating machinery). These components or assemblies will likely have their performance evaluated in a sinusoidal vibration test which is essential to its acceptance. Another component could be exposed to widely varying vibrations conditions better replicated by a random test. Many environments (e.g., flight, over-the-road transportation, wind buffeting) exhibit a statistical nature in their description and are better described in random vibration testing.

4.26 WIND TUNNEL TESTING
P.J. Mole, F.J. Szafranski, and D.R. Booth, General Dynamics Corporation, Convair Division, San Diego, CA.

The design and development of an aircraft and an automobile usually depends on conducting tests on the aerodynamic shape in a wind tunnel. The purpose of these wind tunnel tests is to provide accurate measurements on the six force and moment components exerted on a model or full-scale aircraft or automobile by the wind. These forces are compared with analytical predictions. As the design of a prototype aerodynamic structure proceeds, other wind tunnel tests are conducted on revised shapes until a baseline design is established.

The six components of forces and moments measured in a wind tunnel test are the axial forces in the x, y, and z directions along with the moments applied in these same directions. This topic has been included as a separate section in this handbook because of the level of effort in developing and maintaining test capabilities along with performing these tests, the complex equipment involved, and the great detail involved in developing measurement systems. Aircraft testing will be used as the example (Ref. 4-P17).

The customary approach is to use a prototype model of an aircraft attached to a wind tunnel balance. A balance is a complex load and moment measuring system used to measure the six components of force acting on the model. The model is attached to the balance which in turn is attached to a sting or support secured in the wind tunnel. The sting is an aerodynamic structure which is installed in the wind tunnel and provides support for the test item and force measurement system. The sting also provides paths for the instrumentation and signal conditioning lines to be brought from the balance to the data acquisition systems. The model has to be configured in such a way that its attachment to the balance will result in the measurement of the forces and moments desired.

Types of Structural Tests

A wind tunnel is typically a large structure which accelerates air using propellers or turbine fans or compressed air, focuses the air with ducts which usually increase its velocity, controls the air velocities and flow conditions (e.g., laminar flow, free from vortices), and then directs the air around the model and balance. Considerable attention has been placed on the design of wind tunnels to ensure that the flow of air is at the velocities and conditions desired. Attention has also been paid to ensuring that boundary effects are minimized or eliminated in terms of their possible effects on the passage of air around the model.

There are two basic types of wind tunnels: one is a single pass type of system where air is introduced, accelerated, ducted, focused, and allowed to exit; the second is similar except that the air is cycled around through the propellers or turbine fans on a repeated basis. This cycling allows higher velocities to be reached as the incoming air is already moving at a higher speed. These systems which recycle moving air are usually more efficient (they take less energy to operate) than the single pass systems.

One type of balance is shown in Fig. 4-10. The model is attached to the forward taper section. The aft taper section is attached to the tunnel support or sting. The model is then cantilevered on the beam balance. From Ref. 4-P17:

"the normal force, pitch moment, side force, and yaw moments are determined by using four arm active bridges. These moment bridges are labeled forward pitch and yaw, aft pitch and yaw as shown. The axial force is measured using the strain gages placed on the axial webs. These gages are wired into four arm active bridges, and measure the web end moment which is directly related to the axial force. The rolling moment is measured with gages applied to the end of the roll webs, similar to the axial gages. These gages are also wired into four arm active bridges."

There are other types of balances. The design requirements for these balances are (1) to keep the costs down, (2) have high stiffness and dynamic (frequency response) characteristics, (3) minimize interactions and hystereses, and (4) have large strength capacities.

The testing concepts developed for and used in wind tunnels also apply to other similar types of tests where the loading conditions need to be focused on the test item. For example, development of ships, torpedoes, and other systems used in water are usually tested in some type of relatively large tank or basin. Unlike the wind tunnel, these model test items are attached to a sting and moved through the water. The energies associated with moving large volumes of water and establishing and maintaining the proper flow characteristics to properly simulate realistic interactions of the test item and water are difficult to achieve. The problems of developing instrumented force balance systems to measure the forces and moments on models under test are similar for these water systems.

4.27 SUMMARY

Various types of structural tests have been described. These types of tests range from those that are oriented toward specific conditions (e.g., shock, vibration, thermal, lightning) to those that can be performed in different conditions (e.g., quality assurance and quality control, failure, proof) to those that combine test and analytical capabilities (e.g., modal analyses). It was also shown that there can be considerable overlap among the various types of tests.

Table 4-2 summarizes the types of tests and the diagnostics, instrumentation, transducers, and sensors commonly used. While the instrumentation aspects of structural testing have not been formally introduced, it was felt that some initial relationships between instrumentation and the types of tests would be helpful and would help focus the discussion in the following chapters. Also included are various loading rates and a brief summary of nondestructive evaluation techniques used in structural testing in terms of the types of tests. The loading rates (static, low and high dynamic rates) basically indicate the following: static = independent of time, low = some dynamic response occurs with various modes of vibration excited, and high = exciting stress wave effects in solids with plastic deformations occurring.

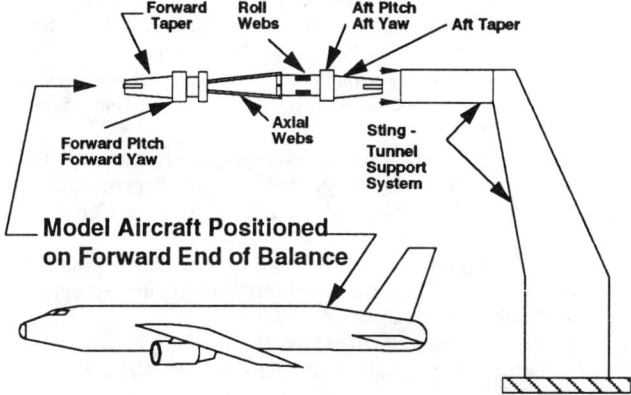

Figure 4-10. Wind Tunnel Balance

4.28 REFERENCES

4-1. _____, *Metals Handbook*, Ninth Edition, Volume 8, Mechanical Testing, American Society for Metals, Metals Park, OH, June 1985.

4-2. _____, *Annual Book of ASTM Standards*, American Society for Testing and Materials, Philadelphia, PA.

4-3. _____, *Military Standardization Handbook*, Metallic Materials and Elements for Aerospace Vehicle Structures, MIL-HNDBK-5D, Department of Defense.

4-4. _____, *Aerospace Structural Metals Handbook*, AFML-TR-68-115, Department of Defense Mechanical Properties Data Center.

4-5. _____, *Damage Tolerant Design Handbook*, MCIC-HB-01, Metals and Ceramics Information Center, Battelle Columbus Laboratories, Columbus, Ohio.

4-6. _____, *Metals Handbook*, American Society for Metals, volume 1, Properties and Selection of Metals.

4-7. P.R. Abelkis, and C.M. Hudson, *Design of Fatigue and Fracture Resistant Structures*, ASTM STP 761, Philadelphia, PA, 1982.

4-8. C. Amzallag, B.N. Leis, and P. Rabbe, *Low-Cycle Fatigue and Life Prediction*. ASTM STP 770, Philadelphia, PA, 1982.

4-9. T.G. Beckwith, and N.L. Buck, *Mechanical Measurements*, Addison Wesley, Reading, MA, 1973.

4-10. L.S. Beedle, et.al., *Structural Steel Design*, Ronald Press, New York, 1964.

4-11. J.S. Bendat and A.G. Piersol, *Random Data. Analysis and Measurement Procedures*, 2nd Ed., Wiley Interscience, New York, 1986.

4-12. G. Bernsaconi and G. Piatti, Eds., *Creep of Engineering Materials and Structures*, Applied Science Publishers, 1979.

4-13. H.E. Boyer, Ed., *Atlas of Creep and Stress-Rupture Curves*, ASM International, Chicago, IL, 1988

4-14. H.E. Boyer, Ed., *Atlas of Stress-Strain Curves*, ASM International, Metals Park, OH, 1987.

4-15. H.E. Davis, G.E. Troxell, and G.F.W. Hauck, *The Testing of Engineering Materials*, 4th Ed., McGraw-Hill, New York, 1982.

4-16. G.W.A. Dummer and N.B. Griffin, *Environmental Testing Techniques for Electronics and Materials*, Pergamon/MacMillan Co., New York, 1962.

4-17. D.J. Ewins, *Modal Testing: Theory and Practice*, John Wiley & Sons, New York, 1985.

4-18. D.A. Firmage, private conversation

4-19. F. Garafalo, *Creep and Creep Rupture of Metals*, McMillan Series in Materials Science, 1965

4-20. C.M. Harris, *Shock and Vibration Handbook*, 3rd Ed., McGraw-Hill, New York, 1988.

4-21. A.S. Kobayashi, Ed., *Handbook on Experimental Mechanics*, Prentice Hall, New York, 1987.

4-22. A.K. Mukherjee, J.E. Bird, and J.E. Dorn, *Transactions of American Society Metals*, Vol. 62, pp. 155-179, 1969.

4-23. D.E. Newland, *An Introduction to Random Vibrations and Spectral Analysis*, Longman, New York, 1975.

4-24. A.R.S. Ponter, and D.R. Hayhurst, Eds., *Creep in Structures*, Proc. 3rd Symposium, Leicester, UK, Sept 8-12, 1980, Springer-Verlag, 1981.

4-25. C.W. Richards, *Engineering Materials Science*, Wadsworth, San Francisco, 1961.

4-26. J.F. Throop and H.S. Reemsnyder, *Residual Stress Effects in Fatigue*, ASTM STP 776, Philadelphia, PA, May 1981.

4-27. J.S. Przemieniecki, *Theory of Matrix Structural Analysis*, McGraw-Hill, New York, 1968.

Papers and Proceedings

4-P1. _____, "Background of Vibration Testing," Spectral Dynamics Division, Scientific Atlanta, Atlanta, GA.

4-P2. _____, Compilation of Papers on Geotechnical Centrifuge Testing, International Journal of Soil Dynamics and Earthquake Engineering, Vol. 2, No. 4, October 1983.

4-P3. _____, Composite Materials: "Testing and Design (7th Conference)," ASTM April 2-4, 1984, Philadelphia, PA.

4-P4. _____, "Mechanics of Composite Materials Directory," Mechanics & Surface Interactions Branch, Nonmetallic Materials Division, Air Force Wright Aeronautical Laboratories (AFSC), Wright-Patterson AFB, Ohio.

4-P5. _____, "Strain Measurement on Composite Materials," Epsilonics, Vol. IV, Issue 2, October 1984, pp 14-15, Measurements Group Inc., Raleigh, NC.

4-P6. J.W. Baldwin and H.J. Salane, "Fatigue Tests of a Twin Girder Highway Bridge," *Proc. SEM Fall Conference*, Indianapolis, No. 7-8, 1988, SEM, Bethel, CT., 1988.

4-P7. S.J. Bless, T.C. Challita, and A.M. Rajendran, "Dynamic Tensile Test Results for Several Metals," Materials Laboratory, Air Force Aeronautical Laboratories, Wright-Patterson AFB, Ohio, AFWAL-TR82-4026, April 1982.

4-P8. P.W. Bridgman, Trans. Am. Soc. Met., Vol 32, 1944, p 553.

4-P9. G.R. Canova, S. Shrivastava, J.J. Jones, and C. G'Sell, "The Use of Torsion Testing to Assess Material Formability:, Formability of Metallic Materials-2000 A.D.," ASTM STP 753, 1982, pp 189-210.

4-P10. D.R. Christman, "A Selected Bibliography on Dynamic Properties of Materials," Defense Atomic Support Agency, Washington, D.C., DASA 2511, June 1970.

4-P11. S.S. Hecker, M.G. Stout, D.T. Eash, "Experiments on Plastic Deformation at Finite Strains," in Plasticity of Metals at Finite Strain: Theory, Computation and Experiment, E.H. Lee and R.L. Mallett, Editors, proceedings of Research Workshop at Stanford University, June 29-30, July 1, 1981.

4-P12. W.A. Kawahara, "Compression Materials Testing at Low to Medium Strain Rates," ASME Winter Annual Meeting, December 7-12 1986, 84-WA-MATS-15.

4-P13. U.S. Lindholm and R.L. Bessey, "A Survey of Rate Dependent Strength Properties of Metals," Air Force Materials Laboratory, Wright-Patterson AFB, Ohio, AFWL-TR-69-119, April 1969.

4-P14. A.M. Lindrose, COMPOS, Sandia National Laboratories Report, SAND78-0177, 1978.

4-P15. J. Lipkin, M.L. Chiesa, and D.J. Bammann, "Thermal Softening of 304L Stainless Steel: Experimental Results and Numerical Simulations," IMPACT '87: Inter. Conf. on Impact Loading and Dynamic Behavior of Materials, Bremen, FRG, May 1987.

4-P16. Mahin, S.A. and Shing, P.B., "Pseudodynamic Method for Seismic Testing," *Journal of Structural Engineering*, ASCE, Vol. 111, No. 7, July 1985.

4-P17. Mole, P.J., "Strain Gage Applications in Wind Tunnel Balances," General Dynamics Corporation, Convair Division, San Diego, CA., Feb. 1989.

4-P18. Mukherjee, A.K., "High Temperature Creep Review," Treatise in Materials Science and Technology, Vol. 6, pp. 164-221, 1975.

4-P19. Schnetzer, G.H. and Fisher, R.J., "The Sandia Transportable Triggered Lightning Instrumentation Facility," 1991 International Conference on Lightning and Static Electricity, April 1991.

4-P20. G.C. Sih and A.M. Skudra, Editors, "Failure Mechanics of Composites, in Handbook of Composites," Vol. 3, Elsevier Science Publishers, 1985.

4-P21. M.E. Tuttle and H.F. Brinson, "Resistance-foil Strain-gage Technology as Applied to Composite Materials," Experimental Mechanics 24 (1), pp. 54-56, March 1984 (errata June 1986 pp 153-154).

4-P22. J.M. Whitney and I. M. Daniel, Eds., Experimental Mechanics of Fiber Reinforced Composite Materials, Society for Experimental Mechanics Handbook No. 4, 1984.

4-P23. A.N. Lin, "Measurement of Prototype Cooling Tower Ambient Vibration," Proc. SEM Fall Conf., Greenlefe, FL, Nov. 17-20, 1985, Bethel, CT.

Table 4-2. Summary of Type of Tests and Diagnostics / Instrumentation Systems Typically/Frequently Used

S = Static Loading/Response
L = Dynamic Loading/Low Rate Response
H = Dynamic Loading/High Rate Response
A = Acoustic Emission
E = Eddy Current
M = Magnetic Particle
P = Penetrant
R = Radiographic
U = Ultrasonic

Type of Test and / or Technique	Accelerometers L H	Displacements S L H	Load Cells Pressure Transduc S L H	Strain Gages S L H	Thermocouples S L H	Velocities L H	Non-Destructive Techniques A E M P R U
• Acceptance QA / QC	x x	x x x	x x	x x x	x	x x	x x x x x x
• Calibrations	x x	x x x	x x	x x x	x x	x x	
• Connections and Joints	x x	x x x	x x x	x x x	x		x x x x x
• Development Test Units	x x	x x x	x x x	x x x	x x	x x	x
• Energy Absorption	x x	x x x	x x x	x x x	x x x	x	x
• Environmental Compatibility	x x	x	x	x	x	x	x x x x x
• Failure Loads and Modes	x x	x x x	x x x	x x x	x x x	x x	x x x x x x
• Full Scale & Model Tests	x x	x x x	x x x	x x x	x x		x x x x x x
• Inertial	x x	x x	x x	x x	x x x		
• Lightning	x x			x x		x x	x
• Mass Property	x x	x x					
• Materials Properties	x x	x x x	x x x	x x x	x x x	x x	
• Modal Analy.	x x	x	x x	x x	x	x x	
• Pressure/Vac			x x x	x x	x x x		x x x
• Proof and/or Function	x x	x x x	x x x	x x x	x x x	x x	x x x x x x
• Residual Stresses			x x	x x	x x		x
• Service Life Extension	x x	x	x x	x x x	x x x	x x	x x x x x x
• Shock	x x	x	x x	x x	x	x x	
• Stiffness / Compliance	x	x x	x	x x x	x		x x x x
• Structure-Control Interaction	x x	x x	x x x	x x x	x x	x x	
• Thermal			x	x x x	x x x		
• Time Dependent	x x	x x	x x x	x x x	x x	x x	
• Vibrations	x x	x		x x	x	x x	
• Wind Tunnel	x x	x x		x x	x x x	x x	

CHAPTER 5

LOADS

*Robert T. Reese, Sandia National Laboratories,
Albuquerque, New Mexico*

5.1 INTRODUCTION

Virtually all structures experience some type of forces during their service life. These forces range from gravitational loads on buildings and bridges, to thermal effects on heat exchangers, to in-flight accelerations from maneuvering aircraft and missiles, to loads on machine tools in cutting operations, to electronic switch chatter and many, many others. The forces and conditions applied to a structure or component are usually called "loads." This seemingly inexact term is used to describe and define conditions that are applied or imposed upon a structure and result in accelerations, deformations, internal stresses, and other responses of the structure, mechanical assembly, or electrical component. Loads may be applied statically (independent of time), dynamically (time dependent) and, when applied dynamically, are often classified as various forms of excitations or stimuli. Loads may be known (deterministic) or defined probabilistically in which statistical estimates of the loads (static, dynamic) are made. The term "loads" is used because it has simplicity, it represents a suitable compromise to describe the wide variety of applied conditions, and it is sufficiently inclusive to cover the wide ranges of these conditions when used with additional parameters to describe the applied forces and pressures, thermal gradients, vibrations, shocks, accelerations, and the wide variety of other conditions and environments which can be imposed on structures (Refs. 5-1, 5-4, 5-8, 5-9, 5-10, 5-11, 5-12, 5-13, 5-15, 5-16, 5-17, 5-19, 5-20, 5-21. 5-22, 5-23, and 5-24).

The Institute of Environmental Sciences (IES) uses the term environments to define the loads and conditions placed on structures and components. Some military standards (e.g., MIL-STD-810D) require the identification of the environments that military equipment will be exposed to in their service life. Many natural environments (e.g., climatic conditions over many years) have been gathered and extensive data banks are available. Other environments (e.g., lightning) have less information available describing them. It is not the purpose of this handbook to impose a redefinition of terms. The reader is alerted that phenomena producing structural responses need to be described and organized. We have simply chosen to use loads to describe these phenomena because it has wide acceptance and allows for description of virtually any stimuli.

The following relationships will be used in this handbook to differentiate between the loads used for design purposes and the forces and conditions to be applied in tests:

- Loads occur in service and are used for design purposes.

- Loading conditions are imposed on a structure or component in laboratory and field tests and are usually (not always) intended to approximate the service loads.

Loads can be considered in two basic types: static and temporary. Static loads typically do not change with time or change so slowly that they produce only static responses. Both static and dead loads are caused by gravitational attraction. Static loads can also be caused by assembly conditions, thermal heating and cooling, and rotational actions. Static equivalent loads are used to adequately represent and to simplify more complex loads. Exceptions for static type loads which can inflict structural damage are creep rupture or some type of environmental damage (e.g., corrosion). Temporary loads which vary with time do so in ways that affect structural responses. Loads will be covered in this chapter and loading systems and examples of laboratory and field testing methods will be described in Chapter 6.

This chapter emphasizes how and where loads occur in service. Detailed definitions are found in the references supplied. Since this handbook is directed at testing, the description of loads will reflect a viewpoint from testing considerations. The loading conditions used in laboratory and field simulations will need to adequately replicate the service or use conditions. (There are times when loading conditions may not reflect service or use conditions; loading conditions may be needed to exercise a structure under known conditions needed to understand and verify behavior of a structure. For example, unit loads for evaluating load-deflection relationships for finite element model correlations.)

There are many ways "loads" can be defined. Several will be described in this handbook. There are no specific rules involved in these descriptions except for the intent to obtain an accurate yet simple definition of the applied forces, conditions, and environments used for design purposes. This chapter contains a summary of loads along with key references (Refs. 5-1, 5-5, 5-6, 5-7, 5-8, 5-9, 5-10, 5-12, 5-13, 5-14, 5-15, 5-17, 5-18, 5-19, 5-20, 5-21, 5-22, 5-23, 5-24, 5-P1, 5-P2, 5-P3, 5-P6, and 5-P9).

Table 5-1 shows the relationships among the various types of loads. There are no distinct boundaries in these relationships. The descriptions of loads overlap among the various types. The descriptions of loads become more complicated as greater time dependence occurs. Static loads require magnitudes, directions, and points of application. Static loads can also require some indications of creep considerations. Dynamic loads require all the static descriptions plus the rates of application of loads, cyclic or random behavior, frequencies, and indications of fatigue actions (Ref. 5-8).

Gravitational induced loads is a broad category. Among these gravity caused loads are dead loads (those which always reside permanently with the structure—the weights of the various beams, floors, etc.). Static loads are induced by gravity or caused by assembly or other actions can temporarily reside in or on structures including the weights of office furniture, occupants, thermal effects on thin-walled storage tanks, and preloaded threaded connections. Equivalent static loads are used to adequately represent and simplify some of the complex dynamic loads that can be applied to structures earthquakes, wind, snow, waves. The simplifications were used initially because descriptions of loads and design guidance were needed. Experience has shown that these descriptions are adequate (Refs. 5-8, 5-9, 5-10, 5-22, 5-23, and 5-24).

Some loads are known directly by the weights of individual parts of a structure because their dimensions and densities (unit weights) are known (Refs. 5-8, 5-19, and 5-21). Some loads are defined by specific or other governing codes for design of highway and railroad bridges (Refs. 5-8 and 5-20). Other loads have to be estimated or approximated as in the maximum torques delivered by engines (Refs. 5-13 and 5-15). Some loads can be described only in terms of limiting conditions such as a completely full or empty water tank. Other loads need additional terms for describing their time durations (impacts), the area and intensity distributions of temperatures (early morning sunlight causing thermal expansion on a portion of an empty steel oil tank), flow rates (for pumping fresh concrete), and energy depositions (lightning impacts on radio tower antennas) (Refs. 5-1, 5-3, 5-5, 5-7, 5-8, 5-9, 5-10, 5-12, 5-13, 5-14, and 5-17). Some types of loads must be measured on instrumented scaled models or prototype structures before they can be adequately described such as aircraft (Ref. 5-6), automobiles in wind tunnels, and on model ships involved in ice breaking. Each structure and the anticipated loads need to be studied in some detail to determine how the structure will respond and how detailed a description of loads will be needed.

Loads are usually defined by the following terms: (1) forces, (2) pressures, (3) accelerations, (4) velocities (e.g., fluid flow), (5) displacements, (6) thermal gradients or temperature changes, (7) frequencies, (8) vibration excitations and environments, (9) shock conditions, or (10) by some process which results in loads being applied (e.g., encapsulation, packaging, freezing liquids). Forces are usually described in convenient units—newtons, pounds, kips (1000 lbs.). Pressures are defined in terms of forces per unit area. Accelerations are usually given in multiples of gravitational acceleration (g's) along with a duration and pulse shape for shock conditions. Vibrations are defined in terms of wave forms, energy levels, frequencies, durations, and whether the conditions are steady state, random, or transient. Thermal loading conditions are described by expansions, contractions, gradients, heat fluxes, changes, or absolute values of temperatures. Whatever methods and units are used, the description of the loads used for design conditions must be simple, concise, and capable of adequately describing the phenomena and environments applied to a structure or component (Refs., 5-1, 5-6, 5-8, 5-9, 5-13, 5-19, 5-22, 5-23, 5-24, 5-P1, 5-P3, and 5-P6).

Once the loads are known, approximated, or otherwise described, it is then necessary to know where they are applied to the structure and in what direction(s) they act. Since force is a vector quantity, it is necessary to know the magnitude and direction to complete its mathematical description. Loads can be applied at a point, along a line, distributed over various areas, or for inertial conditions per unit mass or volume. Some type of description is needed to define these parameters (Ref. 5-8, 5-9, 5-15).

It is also necessary to know if the loads are applied suddenly or gradually. Those loads which do not excite resonances, inertial effects, or other dynamic responses in the structure are considered static or independent of time. Those loads exciting inertial effects and initiating time

Table 5-1. Relationships Among Actual Loads, Design Loads, and
Laboratory/Field Loading Conditions and Simulations

Actual Loads Design		Loads/Conditions	Laboratory/Field Simulations
Dead Loads-Mass of Structure Static Loads Equivalent and/or Approximated Static Loads		Weights of Structures Supported; Weights of Occupants and Equipment Code or Other Static Approximations of Loads (Wind, Snow, Earthquake)	Forces, Moments Equal or Greater Than Weight of Structure, Occupants, and Equipment Forces and Moments Positioned to Cause Worst Case Moments, Forces, Pressures
Inertial Loads - Centrifuge/Flight		Weight of Structure Times Acceleration in g's	Steady State Acceleration or Applied Forces and Moments
Internal Pressure Systems/Vessels		Maximum Allowable Working Pressure—MAWP	Pressure—Gas or Liquid at 1.5 or other factor times MAWP
External Pressure Systems/Vessels		Pressure in Containment System (e.g., vessel)	Pressure
Live Loads-Change with Time—Snow, Wind, Earthquakes, Traffic on Highway Bridges, Diurnal Cycles (People, Tides), Thermal Forces		Uniform Load-Distribution of Snow, People, Traffic Distributed Pressures/ Loads-Wind, Waves Distributed Lateral Forces-Earthquakes, Explosives, Earth Forces Temperatures	Equivalent Static Forces Applied at Points or Distributed Equivalent Static (Peak Magnitude) Pressures Thermal Heating or Cooling
Time Dependent Behavior-Creep, Fatigue, and Other Material Degradation		Factors Address Loss of Strength, Ductility, and Relaxation as they affect service life, strength, and integrity	Numbers of Cycles of Loading Steady State Loads Applied for Creep
Dynamic Loads	Aerodynamic/ Hydrodynamic Pressures	Pressure Distributions, Time Dependencies	Pressure or Equivalent Force Distributions, Time Based Descriptions
	Vibration - Free/Forced Periodic Random	Steady State or Transient, Fixed or Free Boundary Conditions, Frequency Range, Amplitude, Duration Energy (Spectral Density)	Force-attached to exciter, Deterministic (known), Nondeterministic (random), or Transient Frequency Range, Amplitude, and Duration
	Shock - System/Component Response	Wave Form/Shape, Amplitude Duration, Shock Spectra	Acceleration, Frequency Content, Duration, Shock Spectra
	Shock - Material Response, Stress Waves	Wave Directions, Material Properties (elastic, plastic, hydrodynamic), Strain Rates, Boundary Conditions, Heating Effect	Longitudinal (dilatational) or Shear (distortional) waves, Elastic, Plastic, Hydrodynamic Waves, Heating Effects, Strain Rates, Various Test Methods

varying responses of a structure or component are time dependent and are typically called dynamic loads. The "gray area" between static and dynamic loads depends on the structure and how it responds in terms of resonances, accelerations, and inertial effects to the imposed conditions. The same loads applied to one structure may be considered static but when applied to another structure are treated as a dynamic loading. Dynamic loads also have other characteristics such as some type of oscillatory behavior in which the conditions are applied and removed (either totally or in part) and applied again as in steady state, transient, or random vibration. An example of dynamic loads which are applied, removed, and applied again is in cutting tool operations found in machining. It is common for some structures such as an aircraft to experience many different types of loads at nearly the same time (i.e., constant internal pressure on the fuselage, aerodynamic forces on the wings and tail assembly in flight and maneuvering, engine thrust and vibrations, etc.) (Refs. 5-2, 5-6, 5-8, 5-9, 5-10, 5-12, 5-13, 5-14, 5-17, 5-18, 5-19, 5-20, 5-21, 5-22, 5-23, and 5-24.)

Another broad subset of dynamic loads is described by shock loads. These loads excite time dependent responses in portions or in all of the structure or component. Shock loads are usually considered as impacts, impulses, energy depositions or combinations thereof. Impact typically indicates some type of collision involving two or more bodies. For example, an automobile colliding with another or with a stationary object are examples of impact loads. It is often required for structures involved in impacts to absorb large amounts of energy and experience significant amounts of deformation. An impulse is applied in a short time, often before a structure has time to respond. An example of a shock load often called an impulsive load is applied in less that 1/10 of the fundamental (lowest) period of the structure such as a bullet striking an object. Impulsive loads may be energy deposited in a structure from instantaneous heating, exposure to intense noise (i.e, pressure changes), detonation of explosives, etc. If these depositions of energy are rapid with respect to the natural frequencies of the structure and its components, then the inertial effects and resonances will be excited and the resulting loads will be superimposed on the statically applied forces (Ref. 5-10).

This chapter begins with a discussion of static or dead loads on structures from a civil engineering viewpoint. Statically equivalent live loads are then described. Structural testing involves storing energies in structures. A section describing the sudden release of this energy is given. Measured service loads are then described in terms of procedures and examples. Thermal loads are described in terms of the governing equations of heat transfer and practical temperature limits. Next, load and load spectra and envelopes are outlined. An example of probabilistic loadings is given. Next various other loads—temperatures effects, shrink fits, machine tools are outlined. This chapter concludes with a discussion of how loads are actually applied.

5.2 STATIC AND DEAD LOADS ON STRUCTURES

Dead loads are simply those loads which remain as part of the structure over its service life. These dead loads include the weight of the structure and any permanently attached equipment including the floors, ceilings, walls, plumbing, columns, and roofing of a building or the frame, the wings, tail assembly, engines, seats, and auxiliary equipment of an aircraft.

Dead loads can be estimated accurately by the volume of a structure and the densities (unit weights) of the materials used. Many structural members have their weights per unit length already determined. The unit weights for various materials are given in Table 5-2. References are also given for additional sources of densities (unit weights) at the end of the chapter (Ref. 5-8, 5-9, 5-19, 5-21. 5-22, and 5-23). Many handbooks contain sections of unit weights for various materials. The information presented here is a summary of the information available (Refs. 5-1, 5-7, 5-8, 5-9, 5-12, 5-14, 5-15, 5-20, 5-21, 5-22, 5-23, 5-24, and 5-25).

Static loads are closely related to dead loads. However, they can be applied to a structure and then removed. Or, they can be applied in sequence to a structure so that worst case conditions of forces and moments (stresses) will result. Static loads are applied to a structure gradually over a long time period and do not excite the dynamic behavior of the structure. Examples of static loads include bulk materials in storage bins and silos and any other materials, furniture, and items that can be placed in or on a structure. Other examples of static loads include assembly conditions (e.g., interference fits), preloads in threaded connections, preloaded springs, and stresses caused by mismatches in thermal expansion and contraction.

5.3 APPROXIMATED LIVE LOADS/ EQUIVALENT STATIC LOADS/ BUILDING CODE REQUIREMENTS

The loads described in this section (approximate, equivalent, design based) may be considered as temporary as contrasted with the permanent nature of the weights of the members and attached equipment. Temporary loads

Table 5-2. Densities (Unit Weights) of Materials
(Refs. 5-1, 5-8, 5-9, 5-16, 5-19, and 5-21)

Material	Kilograms per Cubic Meter	Pounds Per Cubic Foot	Material	Kilograms per Cubic Meter	Pounds Per Cubic Foot
Aluminum	2660-2830	165-175	Ash	600	28-38
Beryllium	1860	115	Cedar Western	420	18-29
Beryllium-Copper	8266	515	Cypress	460	22-25
			Douglas Fir	480	28-32
Brass	8490-8770	530-545	Elm	630	29-39
Bronze	8765-8820	545-550	Hickory	720	35-41
Cobalt	8320-9240	515-575	Mahogany	500	26-28
Copper	7820-8930	485-555	Maple	630	27-39
Gold	19360	1260	Oak	630	35-55
Iron, Cast	6930-7330	430-465	Pine		
Lead	11370	710	East White	350	21-23
Magnesium	1775-1830	110-115	Ponderosa	400	24-25
Manganese	7625	485	Poplar	420	19-21
Mercury	14150	900	Redwood	400	21-25
Molybdenum	10260-13870	640-865	Spruce	400	21-26
Nickel	8460-8900	525-555	Walnut	550	34
Hastelloy	8320-9240	520-575			
Inconel	8070-8460	500-530	Particle Board	415-1280	25-80
Monel	8460-8900	530-555			
Platinum	21450	1340	Hardboard	800-1440	50-90
Rhenium	21080	1310	Gypsum Wallboard		
Silver	10510	585			
Steel	7698	490	Paper	1140-1300	70-85
Tantalum	16640-16865	1040-1050			
Tin	7320-7770	455-485	Soils	1000-1975	60-120
Titanium	4380-4850	270-300			
Tungsten	19730	1200	Acrylics	1165-1420	70-90
Uranium	18950	1180	Epoxies	1130-2000	70-125
Vanadium	6380	400	Nylon	1140	71
Zinc	7160	445	Polyester	1390	86
Diecast	5025-6710	310-420	Rayon	1470-1555	92-97
Foundry	5025-6300	310-395	Concrete	1750-2410	105-150
Zirconium	6490-6575	405-410	Glass	2300-2520	140-165

include occupants, merchandise, snow, etc. This category also includes some dynamic loads which can be treated statically including wind, earthquake, moving vehicles and trains, ocean waves, etc. It is typical in this category to have additional multiplying factors which increase these statically equivalent loads to account for impacts, gusts, other rapid or transient conditions or load distributions. The term live loads in civil engineering is used to describe loads which will vary over the service life of the structure. In the design phases described earlier, these loads are typically applied to a structure in sets of worst combinations to produce maximum stresses and deflections in the structure. It is customary practice in structural testing to position applied forces to produce maximum responses of a test structure (Ref. 5-3, 5-6, 5-8, 5-9, 5-10, 5-11, 5-12, 5-13, 5-14, and 5-15).

Some live loads (e.g., dynamic loads caused by a moving truck passing over a bridge) are approximated as equivalent static loads. One typical factor used is twice the maximum static load. This factor of two is based on a one degree of freedom spring mass model. For most large civil engineering structures this factor does not exceed two. However, it is possible for the dynamic forces and structural behavior of a structure to interact in such a way that this load factor is actually greater than two (Refs. 5-8, 5-9, and 5-10).

One procedure used to determine statically equivalent loads is to estimate the maximum value of the dynamic load and multiply that value by a factor greater than one. Highway bridge designers use an impact factor to included the effects of dynamic loads multiplying the design load by $I = 50/(L + 125)$ where L is the length of the bridge span in feet which is loaded to produce the maximum stress in the supporting beams with I limited to 30 percent) (Ref. 5-20).

Reference 5-22 contains the requirements for minimum loads on buildings and other structures which describe these approximated live loads and equivalent static loads.

Building codes have been developed for the design of civil engineering structures (Ref. 5-23). These types of structures are governed by two basic requirements: safety and serviceability. A safe building will be designed and constructed to support all loads without exceeding the allowable stress or specified strength for the materials used in the construction. A serviceable structure will have sufficient stiffness to limit static deflections and dynamic responses in vibration so as to not adversely affect the function of the building or the comfort of its occupants.

It is extremely unlikely that an entire structure would be loaded with the maximum live and dead loads at any one time. The Uniform Building Code, for example, permits a reduction of the live load as a function of the tributary area of the element under consideration, by defining $R = 0.08 (A-150)$, where R is the allowable reduction in live load (percent) and A is the tributary area (sq. ft.) of the element under consideration. Of course, this reduction is not always used as in the case of a sports stadium or concert hall, where all the seats and public areas can conceivably be occupied (Ref. 5-23).

It would be impractical and prohibitively expensive to design and construct structures to withstand extremely rare events, such as large earthquakes, without damage. Therefore, the design philosophy of the Structural Engineers Association of California (SEAOC) Recommended Lateral Force Requirements is that a properly designed structure should be able to (a) resist a minor earthquake without damage, (b) resist a moderate earthquake without structural damage but with some nonstructural damage and loss of contents, and (c) resist a major (rare) earthquake without collapse or endangerment of the safety of the occupants. It is recognized that the structure may require replacement after such a major event. Also, the definition of minor, moderate, and major earthquakes will be dependent on the seismicity of the area (Ref. 5-24).

Some dynamic loads can be approximated by equivalent static loads. In the design of railway bridges impact factors are used to account for rolling actions of the engines and cars (increase of 10 percent) and for bridge lengths, number of tracks, and bridge type (beam, truss) (Ref. 5-7). This impact factor is applied to live loads and accounts for the dynamic effects of the moving traffic. Short span bridges are stiffer and are influenced more by moving loads than longer spans. Similar equivalent static load analyses can be used to account for dynamic lateral forces induced by wind or earthquake.

Civil engineering building codes provide for a minimum level of performance. Vitelmo Bertero, researcher in earthquake engineering, states that "building codes are necessary but not sufficient" to ensure adequate performance (Ref. 5-21). The experimenter responsible for testing a civil engineering structure or structural component should be aware of the functional requirements of the element under consideration, and should not allow mere conformance to building code requirements be the only criterion.

5.4 STORED ENERGY CALCULATIONS

Strain energies are stored in structures when they are loaded. The unplanned or unanticipated release of this stored energy can cause parts of the structure to become quite active. Bolts may have their heads or nuts severed and released with considerable velocity, parts of pressure vessels and related piping can be separated from the test item and loading equipment and move because of the sudden release of stored energy. The calculations of these energies are important in proof and failure testing or when a structure could fail. The important point for test engineers is to anticipate that these energies can be released suddenly and that the accelerated structural parts need to be restrained or directed to capture systems. An example of the energies associated with pressure tests is provided to illustrate these types of loads.

A pressure system can consist of vessels, pumps, pressure intensifiers, pressure lines, relief valves, other valves, and gages. When such a system is pressurized, energy is stored in the pressurized fluid and as strain energy in the structural parts. If the fluid is a gas or a two-phase fluid, much of the energy is stored in the gas itself. If the fluid is nearly incompressible (e.g., hydraulic oil, water), then much of the energy is stored in the structural system. A two-phase fluid is one where the same pure substance exists in liquid and vapor phases in equilibrium at the same temperature and pressure. Examples of these two-phase fluids are steam, carbon dioxide, liquified petroleum gas, and refrigerants (Ref. 5-P9).

The total energy involved is the sum of the energy stored in the fluid and the energy stored in the structural parts, or $Q = \Delta U + W$, where Q is the total energy, ΔU is the change in internal energy of the fluid, and W is the work done by the structural parts or elements.

The energy stored in a liquid system is given in Eq. 5-1.

$$E_f = \frac{P^2 V}{K B}$$ Eq 5-1

where
E_f = energy of compression of the liquid
P = system pressure
V = system volume
B = liquid bulk modulus (B = 1/β where β is the isothermal compressibility coefficient)
K = constant depending on the units employed

The bulk modulus of water ranges between 2.9×10^5 and 3.5×10^5 psi between 30 and 190 °F for pressures between atmospheric and 2,000 psi and increases to 9×10^5 at 100,000 psi. The strain energy stored in the structural parts depends on their configurations and wall thicknesses.

The energies stored in gas systems is significantly larger than for liquid systems. These energies can be calculated using the ideal gas law relationships and isentropic expansion as given in equation 5-2. The energies calculated with Eq. 5-2 are conservative because actual pressures measured for real gases are lower, particularly at very high pressures. Some corrections or factors are needed to determine actual pressures. One method using Amagat relationships is described in Ref. 5-25.

$$E_g = \frac{P_1 V_1}{(k-1)} \left\{ 1 - \left(\frac{P_2}{P_1}\right)^{\frac{(k-1)}{k}} \right\}$$ Eq. 5-2

where
E_g = energy of compression in the gas
k = ratio of specific heat at constant pressure, c_p, to that at constant volume, C_v (k = 1.404 for nitrogen)
V_1 = volume of vessel and system
P_1 = initial pressure, often atmospheric
P_2 = final pressure

The various specific heats, bulk moduli, and other parameters are available in materials handbooks. A comparison of the stored energies for liquid (water) and gas (nitrogen) stored in the same vessel (5000 cu. in or 2.89 cu ft) at different pressures is given in Table 5-3 (Ref. 5-P9).

Once the energies are calculated, assumptions need to be made about how this energy is transferred to the ejected parts of the structure as a result of a failure test. One method adopted for this type of calculation is the relationship of stored energy to that released from a mass equivalent explosive. More information is known about the destructive power associated with explosives in terms energies imparted to ejected parts.

Energy can also be stored in the strained mechanical

Table 5-3. Comparison of Energies for Pressurized Liquids and Gases Stored in the Same Pressure Vessel (Water or Nitrogen in a 5,000 cu. in. vessel)

Pressures psi	Energy Stored in Liquid, E_1 foot-pounds	Energy Stored in Gas, E_g foot-pounds (nitrogen, k = 1.404)	Weight (lb) of Equivalent High Explosive $E_g/1.55 \times 10^6$ ft-lbs/lb TNT	Ratio E_g/E_1
10	0.4	1,210	0.0008	3,025
50	3.24	15,310	0.0049	4,670
60	4.27	20,600	0.013	4,825
70	5.42	26,120	0.017	4,820
100	9.63	43,730	0.028	4,540
200	31.8	108,950	0.07	3,425
1,000	660.	725,120	0.470	1,100
2,000	2,543.	1,157,500	0.734	455
10,000	58,280.	8,800,800	5.67	150

parts (e.g., tensioned or compressed springs). The energy stored in these parts is governed by Eq. 5-3:

$$e_s = \frac{1}{2} k x^2 \qquad \text{Eq. 5-3}$$

where e_s = energy stored (e.g., in-lbs)
      k = stiffness (lbs/inch)
      x = displacement from rest or initial position (inches)

Many mechanical parts (e.g., bolts) can have large internal forces but do not store great amounts of energy because the displacements are small. However, the energy stored in a nylon parachute shroud line can be very large as the nylon can increase in length as much as 50 percent while under load. Mechanical springs can also store considerable energy.

5.5 MEASURED SERVICE LOADS

One common method of determining the loading conditions on a structure is to design and fabricate a prototype. This prototype structure or component can be instrumented with the appropriate transducers and gages and then tested to representative loading conditions in laboratory and/or field tests. For example, the design of a component for an aircraft often includes tests on a "fly around" instrumented prototype or initial production designed in the manner just described. The data obtained (e.g., accelerations, strains) are analyzed. Tests are then planned in a laboratory so that the response of the component is duplicated as nearly as possible. The output from the vibration testing machine or other type of test equipment is used as the excitation for the continued development of the component. Virtually all types of wheeled vehicles (e.g., automobiles, tanks, farm combines) have been instrumented to determine the actual service conditions they encounter.

There is an important distinction to be made here: the prototype structure or test unit has been designed and instrumented with the intent to measure its response and not to measure the applied loads. For certain classes of structures (e.g., aircraft) it is possible to obtain the force-time history transmitted through the landing gear. Wheel spindles are also instrumented so that the forces imparted to an automobile occurring as it travels on the highway can be recorded and used in later tests.

These measured responses can be duplicated in the laboratory or field tests by appropriate specifications for shock, vibration, and in facilities using servohydraulic or other controllable testing systems. The data obtained from these service load tests describe the response of the prototype or test unit to the conditions imposed.

For certain classes of structures (e.g., linear elastic) it is possible with mathematical techniques (deconvolution) to "work backwards" and determine the forcing functions applied to the structure which resulted in the measured responses. In order to determine the excitation or stimuli causing the response, the prototype or test unit will need to be mounted in a testing machine, vibration shaker, or shock tester which can approximate the applied loads, environments, and conditions. Then, the output from the testing machine can be adjusted until the laboratory response of the prototype or test item closely approximates the the original measured responses. Once the output from the testing machine is determined, the prototype design can be further evaluated, modifications and improvements to the design made, and further tests performed to qualify and prove that the component will meet its expected design conditions.

Measurement of service loads gives guidance for design and leads to detailed specifications and tests for many structural systems and components including automobiles, weapons systems, and many consumer products tested by Underwriter's Laboratory and many others. Safety engineering of automobiles is an example where over-the-road data and crash test data are used to improve the deformable car bodies and passenger restraint systems used to protect occupants.

Basically the design approach can be described as a cycle consisting of "storage-readiness-use-retirement" as described below:

1. In "storage," a component would typically need to be evaluated for cycles of temperatures, moisture, and material and environmental compatibility concerns.

2. In a "readiness" status, a component or system would need to have its capabilities to perform its intended functions verified and proven. The readiness cycle could include transportation and handling conditions imposed as the structure or component is moved or readied for use.

3. The "use" cycle consists of monitoring how structures or components respond to the actual conditions in which they are used. Also included in this portion of the design are considerations regarding safety of items for which product liability requirements must be met. These requirements often include evaluations of overload or other failure or near-failure conditions.

4. The "retirement" portion of the cycle consists of the necessary steps to remove an item from service by

eliminating and destroying hazardous materials present, disconnecting electrical power supplies, and rendering the component or structure to a harmless condition while preserving the environment.

For example, an automobile has a "storage-readiness-use-retirement" design approach. In storage an automobile is expected to contain the fluids and gases stored in it, to keep the power supply from leaking away, and remain as impervious to the storage environments as possible. In its readiness state, many of the functions of the engine operation, electrical systems, brakes, tires, and other important and necessary operational and comfort functions and abilities of the automobile are examined and tested. The readiness state may also require transportation and handling from the point of manufacture to the point of sale. In its use state, an automobile is usually examined carefully by the driver and occupants for smooth performance, comfort, and reliable operation, as well as safety considerations involving potential collisions, braking, acceleration, turning, etc. In a retirement state, car bodies eventually are scavenged for parts and returned for potential recycling purposes.

Continuing with the cycle of "storage-readiness-use-retirement," the following suggests many of the types of related structural tests that may need to be performed while a structural system or component is being developed, during its use, and when it is to be retired:

1. Storage: (temperature cycles, environmental conditions, etc.)

2. Readiness: (transportation vibrations and shocks, accident/crash loads, pressure, etc.)

3. Use: (flight and maneuvering loads, temperatures, countermeasures, pressure, overload conditions, etc.)

4. Retirement: (environmental concerns, disassembly/crushing/other dismantling procedures, separation of parts, etc.)

Performing tests using service conditions (storage, readiness, use, and retirement) require adequate replication or simulation of these service conditions. In complicated structural systems such as automobiles, aircraft, or computer controlled machining centers, the measured responses of the structural system will need to be approximated as closely as possible in tests conducted in the laboratory.

5.6 THERMAL

Thermal conditions are among the most challenging for engineering designs. Thermal conditions may be steady state, time-varying, and are dependent on the mechanical and thermal properties of the materials involved. The environments creating thermal loads may involve gradual heating, self-heating, frictional heating, rapid depositions of energy, or may result from internal heating on highly stressed parts. It is common practice in design for high temperature applications to limit the temperatures imposed on a materials to values at or below their service temperatures. It is also common practice in design for low temperature applications to keep the temperatures above minimum service temperatures (Refs. 5-P1, 5-P3, 5-P4, 5-PS, 5-P6, 5-P7, and 5-P8).

For applications where engineering designs include temperature considerations, it is customary to perform analyses to determine the magnitudes, durations, and spatial distributions of the temperatures. These quantities should be estimated by analytic methods, experience, etc. There are three basic methods of heat transfer: conduction, convection, and radiation. Conduction is the transfer of heat through materials, typically solids. Convection is the transfer of heat by motion of the hot material. Forced air furnaces and hot-water heating systems are examples of systems with motions of a gas or fluid. Radiation is the emission of energy from the surfaces of a body. The energy is radiant energy in the form of electromagnetic waves. These waves travel with the speed of light and are transmitted through vacuums. Most bodies are not transparent to these waves and absorb the energy as a result. This absorption process converts the energy to heat.

A comparison of the three governing equations for heat transfer are given in Table 5-4. The important relationships involved show that heat energy is linear in some methods of heat transfer while in radiation it is to the fourth power of the temperatures. The constants associated with the terms in the equations and values of terms used for specific materials are available in Ref. 5-P1.

Reference 5-P6 gives evidence that a large gasoline (hydrocarbon) fire produces maximum temperatures of about 2,000°F. Since the materials available for fires consist mostly of hydrocarbon materials, this temperature can be considered as an upper temperature limit for a very large percentage of fires. Structural tests can have thermal loads defined by each of the equations given in Table 5-4. The loads may be governed by one or more equations. The most difficult tests to perform usually involve radiant heat transfer.

Table 5-4. Comparison of Equations Governing Heat Transfer

Governing Equations for Heat Transfer		
Conduction	Convection	Radiation
$H = K A (t_2 - t_1)/L$ where H = rate of heat flow K = coefficient of thermal conductivity A = cross-sectional area $t_2 - t_1$ = temperature difference L = thickness or length in direction of heat transfer	$H = h A \Delta t$ where H = rate of heat flow h = convection coefficient A = area of surface Δt = temperature differences between gas or fluid surface of material	$R = e \sigma (T_o^4 - T_i^4)$ where R = rate of emission of radiant energy per unit area e = emissivity of surface σ = Stefan-Boltzman constant $= 5.7 \times 10^{-8}$ W/m^2 T$_4$ T_o, T_i = Kelvin Temperatures of surfaces

The National Aerospace Plane (NASP) involves the concept of designing and testing an aircraft which can land on long conventional runways while being able to reach space without added rocket booster systems. This proposed aircraft presents many problems to designers and materials scientists because of the extreme temperatures involved (e.g., as high as 4,000 °F) on leading edges and operating surfaces. Significant developments of materials, test methods, and design will be required for this aircraft. Hopefully, these developments can be achieved and the capabilities added to current technology for solution of very complex thermal problems (Ref. 5-P10).

5.7 LOAD SPECTRA AND LOAD ENVELOPES

Load spectra and load envelopes are methods used to summarize, organize, and provide limiting values or conditions for loads applied to structures, members, or components. These summaries often give limits to the combined actions of various loads (e.g., axial and lateral loads on columns). They usually include analytical predictions based on test results. The organization of loads gives information in forms useful for test purposes.

An example of a load envelope for static loads is given in a beam-column structure. These envelopes are also called interaction curves for axial force and moment. This beam-column structure can be loaded only in compression (column action) or laterally (beam bending). When this structure is loaded in both compression and bending, larger stresses occur than for just one loading condition. When the ranges of loads can span between zero to the yielding condition in compression or the full plastic moment capacity in bending, then an envelope of the combined interactions is needed (Ref. 5-1). The governing differential equation for such a structure is given in Eq. 5-4:

$$EI \frac{d^4v}{dx^4} + P \frac{d^2v}{dx^2} = 0 \qquad \text{Eq. 5-4}$$

where E = Modulus of Elasticity
I = Moment of inertia of the cross-section
P = Applied force

Reformatting this equation for the critical buckling load (P_y) and full plastic moment capability (M_p), Eq. 5-4 can be rewritten as (Ref 5-1):

$$\frac{P}{P_y} + \left(\frac{M_o}{M_p}\right)\left[\frac{Z}{S} \text{ secant}\left(\frac{L}{2r}\sqrt{\left(\frac{P}{P_y}\right)\left(\frac{\sigma_y}{E}\right)}\right)\right] = 1.0 \qquad \text{Eq. 5-4}$$

where P = applied axial force, lbs or Newtons
P_y = load at initial yielding of beam material, lbs or Newtons
M = applied moment, inch-pounds or Newton-m
M_p = plastic moment capability, inch-pounds or Newton–m

Z = plastic section modulus, inches^3 or meter^3
S = section modulus (I/c), inches^3 or meter^3
L = length of beam, inches or meters
r = radius of gyration, inches or meters
s = yield strength of material, pounds/in^2 or N/m^2
E = modulus of elasticity, pounds/in^2 or N/m^2

The load envelope for Eq. 5-5 is shown in Fig. 5-1. This load envelope defines the magnitudes of various combinations of axial and lateral loads that can be placed on this beam-column. In general, a load envelope consists of interaction diagrams defining combinations of loads that can be safely placed on specific structures or types of structures.

The curved portion of Fig. 5-1 gives the limiting conditions on applied loads (axially, end moments, or in combination) for the beam-column with equal and opposite end moments. For beam columns of reinforced concrete, many tests have been performed on various types of columns (e.g., square, rectangular, round, spiral reinforced, rectangularly reinforced, etc.). With appropriate design margins, these envelopes are used for design purposes to predict behavior of beam-columns.

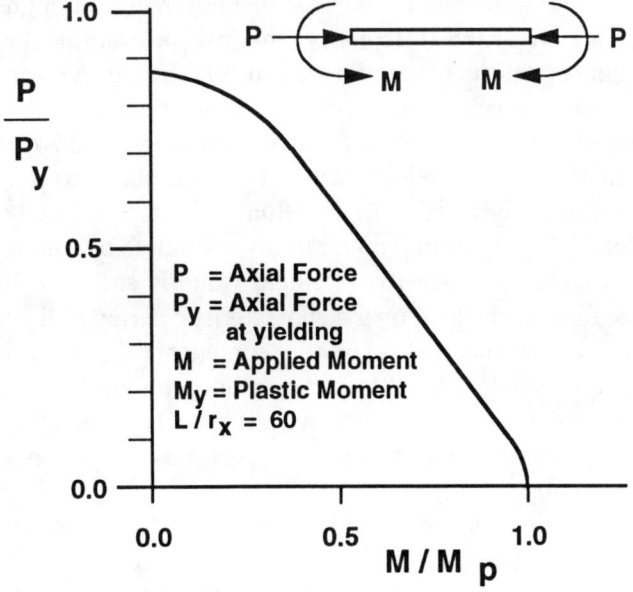

Figure 5-1. Comparison of Elastic Limit Envelope with Ultimate Strength for a Beam Column with Equal and Opposite End Moments (Ref. 5-1)

As dynamic loads are imposed on structures, the time based descriptions for the loads and the various responses of the structure enter into the problem. The definition of the load spectra and envelopes becomes more difficult for dynamic loads than for static loads.

The descriptions of dynamic loads are often based on results from tests on prototype structures and assemblies (Refs. 5-2, 5-6, 5-10, 5-12, and 5-14). These test results are combined with analytical definitions of loads. In the development of a electrical or mechanical component, it is common to specify an initial dynamic test (e.g., shock and/or vibration), instrument the test item (e.g., accelerometers), and perform tests. After confidence is gained in the prototype design, then instrumented prototypes are subjected to flight environments, over-the-road travel, and other anticipated service or other realistic conditions. These test units usually have important components and areas instrumented to gain information about structural response and to obtain definition of the load spectra needed for further laboratory tests.

Armed with additional insight and data, the design engineer and test personnel begin the second round in the design of components and systems. An initial vibration specification for a component is shown in Fig. 5-2 (a). This initial specification is typical because it provides an acceleration-time history which will govern a broad range of possible vibrations. After some initial testing, the required specification of an acceleration-time history which has sufficient excitation for product acceptance testing is given in Fig. 5-2 (b). This second specification is considerably reduced from the initial requirement. This reduced acceleration-time history was determined by testing and the analyses of data which showed that the higher initial specification was not needed for evaluation of this component in production.

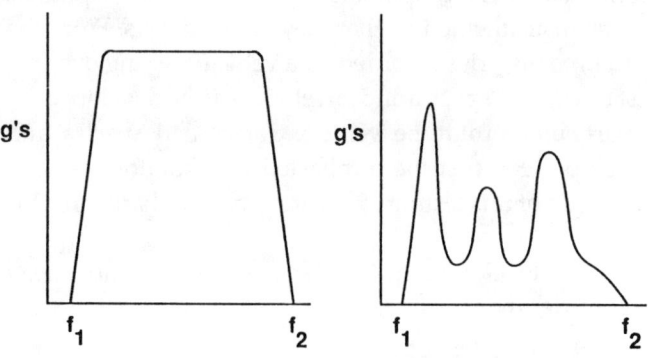

Figure 5-2. Comparison of Vibration Test Specifications

This iterative design and test procedure consists of establishing an initial test specification, testing and analyzing to redefine the test conditions, testing again to determine performance, etc. In actual practice this procedure is usually more complicated because important areas of the structure and key components are usually loaded in three mutually orthogonal axes, often simultaneously. It is rare that the design of a system or component restricts the excitations to only these three principal axes; that is, there is usually some excitation off-axis. Vibration tests conducted in the x axis can produce responses in the y and z axes. However, conducting dynamic tests that involve controlled simultaneous excitation in three principal axes is very difficult. The vibration envelopes can account for this off-axis responses from test measurements made off the principal axis of vibration. If many measurements can be made during product development and acceptance testing, then data bases can be developed defining vibration envelopes (for each principal axis) can be used to evaluate future products.

Dynamic multiple axis testing can be performed if various types of tests can be brought together. For example, if a structure is subjected to steady state acceleration forces and vibration conditions while in flight, one way to perform such a test is to mount a vibration shaker on a centrifuge. The test item could then be exposed to steady state accelerations from the centrifuge and to vibrations from the shaker. These test and evaluation techniques will provide improved definitions of the load spectra and envelopes needed for development of critical components and structural systems.

Automobile test tracks have been developed and used to exercise prototype automobiles with as many different over-the-road conditions as possible. The rates of accomplishing these purposes range as high as 99 percent in describing the conceivable loads that an automobile will experience. These measured forces can be determined from instrumented spindles (axle ends). These spindles can be instrumented so that they are load cells. From the load time histories recorded by a vehicle passing over the test track, load spectra and envelopes can be developed for a particular prototype vehicle. These load spectra and envelopes can then be duplicated in laboratory testing. Examples of this testing effort are given in Chapters 6 and 7.

The development of load spectra and envelopes need to consider the following:

1. The possible loads and where they will be applied to the structure or test item.
2. The possible distributions of cargo and equipment in the prototype structure (e.g., fuel in aircraft wing tanks, passengers in automobiles)
3. How the applied forces, accelerations, and other diagnostic information will be measured.
4. How the data will be stored and transmitted.
5. The resolution of data signals.
6. Schedules, costs, and other aspects of engineering development activities.

5.8 PROBABILISTIC LOADS
Thomas L. Paez, Ph.D., Experimental Mechanics Department, Sandia National Laboratories, Albuquerque, NM

It is acknowledged in the design of structures that material and structural characteristics can practically never be predicted precisely in advance of fabrication; neither can the loads (except perhaps dead loads) applied to a structure during its design life. Typically, structural characteristics display relatively little variation; therefore, they are considered deterministic, i.e., predictable and known in advance. On the other hand, structural loads, particularly dynamic loads, display a relatively high degree of variation, and these are often treated using a probabilistic approach. When loads are not accurately predictable in advance they are known as random (Refs. 5-2 and 5-10).

Both static and dynamic loads can behave randomly, and when they do, they require probabilistic characterization. A static load that is random is characterized in terms of a random variable. A dynamic load that varies randomly is characterized in terms of a random process. We consider first random static loads.

When we say that a static load varies randomly we mean one of two things. First, it may be that when a structure is fabricated a load that does not change throughout the life of the structure is random. For example, the weight of a girder in a reinforced concrete structure will not be known precisely in advance of fabrication. The other situation that can occur with respect to a static load is one with a time varying load that is treated as static for purposes of analysis varies in a random fashion. Specifically, a load can vary (though not rapidly enough to be considered dynamic) such that a single value is sufficient to characterize the load over a characteristic period of time. For example, the live load on one segment of the floor of an office building, related to personnel occupying the floor segment, may vary randomly, but may be adequately characterized by the average load over the time period of 30 minutes.

Random static loads are characterized as random variables. Random variables are entities that quantitatively characterized the results of a random experiment. For example, a random experiment might be described as follows. Measure the total quasi-static load caused by vehicles on the Golden Gate Bridge as a function of time between midnight on December 31 and midnight January 1, next year. A random variable related to this random experiment might be defined as the maximum value of the measured quasi-static load. This is a quantity between zero load and infinity. Clearly certain ranges of loads are more likely than others. Another random variable might be defined as the fraction of time during the day when the load exceeds 1000 tons in which case the time fraction between zero and one. The specific value that a random variable assumes as the result of a random experiment is known as the realization of the random variable. The range of possible values that realizations can assume when we perform a random experiment is known as the range of realization of the random variable.

The quantity that characterizes the chance that a random variable has of generating a realization in a specific range of values as the result of performing a single random experiment is the probability density function (pdf) of the random variable. Let X be a random variable and let $-\infty < X < \infty$ be the range of realization of the random variable. Random variables are usually denoted with capital letters while realizations of random variables are usually denoted using the corresponding lower case letters. The pdf of the random variable X is a function denoted $f_X(s)$, $-\infty < x < \infty$, and it is related to probabilities for the random variable X as follows:

$$P(a < X < b) = \int_a^b f_X(x) \, dx, \quad -\infty < a < b < \infty \qquad \text{Eq. 5-6}$$

$P(a<X<b)$ is the probability that when we perform a single random experiment the realization of the random variable X will fall in the interval (a,b). Probability is a number between zero and one and reflects the relative chance of occurrence of an event. The expression on the right of Eq. 5-6 is an integral and it defines the area under the pdf curve over the interval (a,b). Because the probability on the left must be positive, we require that

$$f_X(x) \geq 0, \quad -\infty < x < \infty \qquad \text{Eq. 5-7}$$

Because we must be certain (i.e., the probability equals 1) that each realization of the random variable X falls in the interval $(-\infty, \infty)$ we require that

$$\int_{-\infty}^{\infty} f_X(x) \, dx = 1 \qquad \text{Eq. 5-8}$$

A much simplified description of how we use the pdf for static loads is given in the following. In the situation where a static load on a structure is fixed at the time of fabrication and does not change over the life of the structure, we simply establish the probability that the load falls within a certain range by using Eq. 5-6. For example, the probability that the static load is lower than X_c is

$$P_c = \int_{-\infty}^{x_c} f_X(x) \, dx, \qquad \text{Eq. 5-9}$$

If we wish to design a structure to support the load in question with the probability of survival of P_c, then we design the structure to support the load X_c where the value of xc is implicitly evaluated from the above integral.

When a structural system is analyzed statically but the loads on it vary slowly in time, then the pdf of the load can be used to estimate the approximate number of characteristic time periods during which the load will surpass the given level. For example, let X be a random variable with pdf $f_X(x)$, $0 \leq x < \infty$. When we wish to estimate the number of characteristic time periods during which a variable load will surpass the level xc we use the quantity

$$N = \int_{x_c}^{\infty} f_X(s) \, dx \qquad \text{Eq. 5-10}$$

where N = total number of characteristic time periods during which the load is applied

Some important measures of the potential values random variables can assume are contained in their moments. First, the mean, E[X], of a random variable is defined as the centroid of its pdf (the first moment about the origin) as given in Eq. 5-11:

$$E[X] = \int_{-\infty}^{\infty} x \, f_X(x) \, dx \qquad \text{Eq. 5-11}$$

This is the single value most representative of the range of possible values of the random variable X. The mean, E[X], is also sometimes denoted by μ_X.

The variance of a random variable defines the variation tendency of its realizations. It is the second moment of the pdf about the mean and is defined as

$$V[X] = \sigma_X^2 = \int_{-\infty}^{\infty} (x - \mu_X)^2 f_X(x) \, dx \qquad \text{Eq. 5-12}$$

This is the average of the square of the deviation of random variable realizations from the mean, and has units of x^2. A measure of the average deviation of values of a random variable X from the mean that has the same units as X is the standard deviation. This is defined as

$$\sigma_X = \sqrt{V[X]} \qquad \text{Eq. 5-13}$$

The mean and standard deviation provides two fundamental measures of the predominant values and range of values of a random variable.

The characterization of a random load as a dynamic function of time is much more difficult than the characterization of a static load that is a single random variable or simple sequence of random variables. The reasons that a random process is conceptually defined as a sequence of random variables are (1) when time is continuous the random process consists of an uncountable number of random variables and (2) complete characterization of a random process depends not only on the behavior of the individual random variables, but also on the joint behavior of pairs of random variables, triplets of random variables, etc. It is beyond the scope of the present description to completely develop the definition of random processes, but some fundamental principles can be introduced.

Because the random process is a sequence of random variables, a first level description of random processes simply involves description of the individual random variables in the random process. We use the notation $\{X(t), t \varepsilon T\}$ to symbolize a random process and the meaning is this: a random process $\{X(t), t \varepsilon T\}$ is a function of a parameter (time t in this case) and it is only defined for certain values of the parameter. Those values are elements of the set T. For each value of time in T (denoted $t \varepsilon T$) the random process $\{X(t), t \varepsilon T\}$ has a random variable $X(t)$ defined. Each random variable has a mean, variance, pdf, etc. A fundamental description of a random process is the function that is the sequence of values that are the means of the random variables $X(t)$ for each $t \varepsilon T$. This is called the mean function of the random process $(X(t), t \varepsilon T)$ and is denoted by

$$\mu_{X(t)} = E[X(t)], t \varepsilon T \qquad \text{Eq. 5-14}$$

Likewise, a random process has a variance function and a standard deviation function.

$$\sigma_X^2(t) = V[X(t)], \quad t \varepsilon T \qquad \text{Eq. 5-15}$$

$$\sigma_X(t) = \sqrt{V[X(t)]}, \quad t \varepsilon T \qquad \text{Eq. 5-16}$$

and these functions define the sequences of variances and standard deviations of the random variables in the random process.

The most common form for the pdf of a random variable is the normal pdf because it arises frequently in nature. The normal pdf for a random variable with a mean of 10.0 and a standard deviation of 2.0 is shown in Fig. 5-3.

The description of a random process goes far beyond these measures to characterize the time and frequency features of random signal sources. The fundamental measure of some random processes is the spectral density. When the random process has a steady state character known as stationarity (i.e., its features do not change rapidly with time), then it possesses a spectral density. The

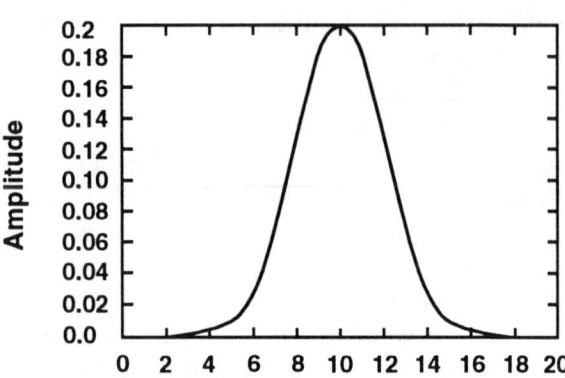

Figure 5-3. Probability Density Function (pdf) for Random Variable

spectral density of a stationary random process is a measure of the mean square signal content of a random source as a function of frequency. To describe this without the use of mathematical formulas we offer the following explanation of spectral density. (This description does not reflect the actual manner in which spectral density is estimated.) Refer to Fig. 5-4.

Every stationary random source can be broken into frequency components by the use of band pass filters. For example, the source that generates the signal shown in Fig. 5-4, on the left, can be filtered using a sequence of band pass filters that span the frequency ranges (0,100), (100,200), ..., (900,1000) Hz. (Assume that the source has no frequency content beyond 1000 Hz.) The filtered signals that result are those shown in the second column of Fig. 5-4. Each of the filtered signals has a mean square value (that can be estimated using statistical analysis) and the mean square value of the jth filtered signal is denoted by σ_j^2. (If the random process has a nonzero mean it should be subtracted algebraically to eliminate its influence on the signal component in the lowest frequency range.) The quantity σ_j^2 is the mean square value of the signal component with signal content in the frequency range $((j-1)100, (j)100)$Hz. The spectral density in the same frequency band is defined as the mean square of the signal component σ_j^2 divided by the bandwidth of the band pass filter used to generate the jth component (100Hz in the present case). When σ_j^2 is divided by the filter bandwidth the resulting quantity can be plotted as function of frequency, say, for example, the center value of the frequency band. The value of the spectral density so defined establishes the mean square character of the random source in the frequency band of the band pass filter.

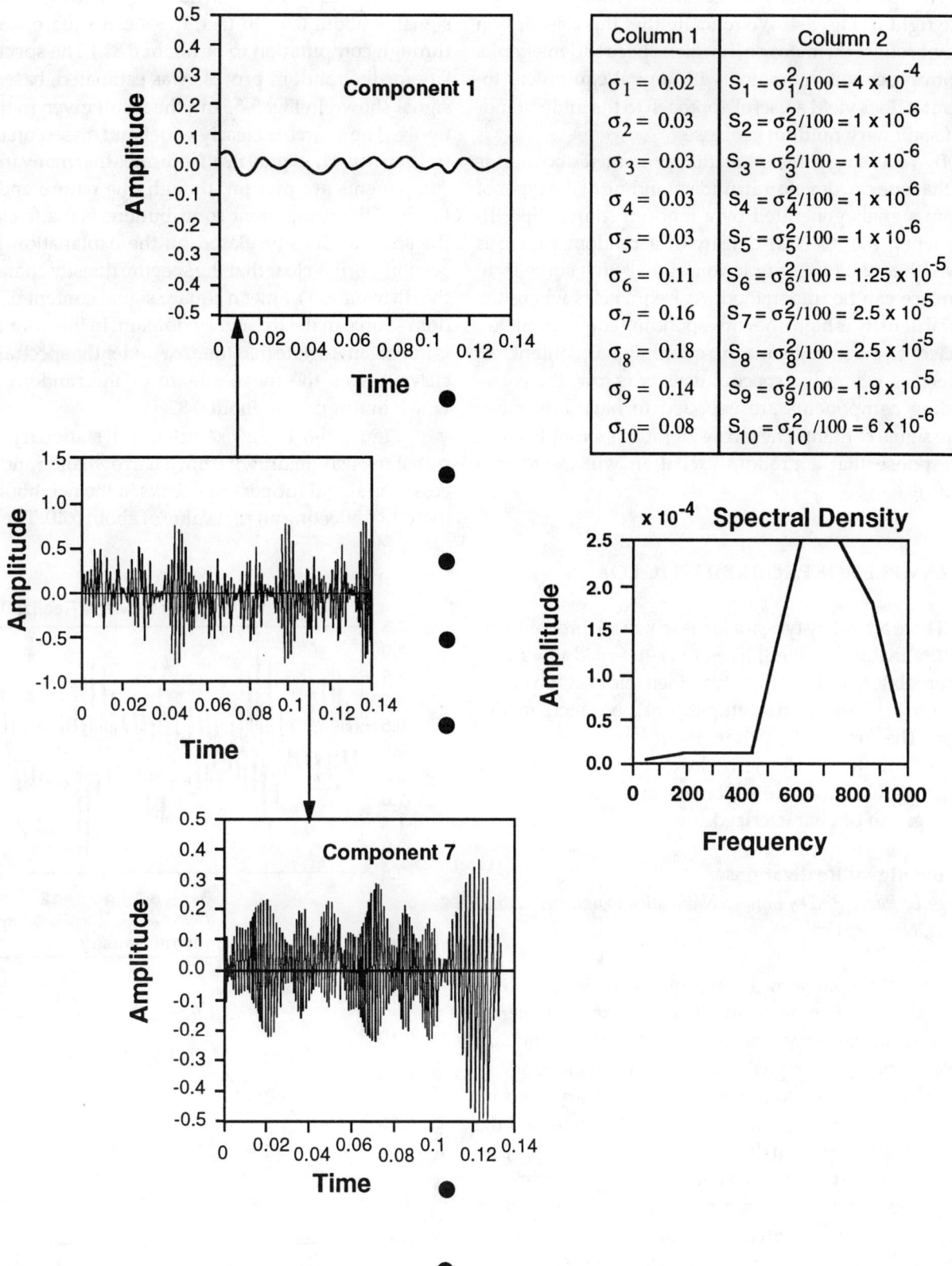

Figure 5-4. Random Signal Realization. Source on the Left, Component 1 at Top, Component 7 at Bottom, and Spectral Density at Right

The spectral density of the random source is shown on the right in Fig. 5-4. We reiterate that this description does not reflect the manner in which spectral density of a random process is estimated with digital equipment today, but it does yield a useful approach to the understanding of stationary random processes.

By reversing the explanation presented above, it is clear that one can develop an understanding of the types of random signals generated by a random source. Specifically, when the spectral density of a random source is known, then the types of random signals that come from the source can be understood. At frequencies where the spectral density is high, the corresponding components are expected to have high mean square signal content. At frequencies where the spectral density is low, the corresponding components are expected to have low mean square signal content. These considerations strongly affect the response that a random excitation will excite in a structure.

5.9 EXAMPLES OF PROBABILISTIC LOADS

There are many types of loads which are probabilistic in nature including wind, tides, over-the-road vibrations in automobiles, in-flight flutter, and seismic loads on buildings and bridges. Two examples are presented in this section. The first explains how the information given in Section 5.8 is used. The second example describes how seismic disturbances are described and how structural responses can be characterized.

Random Signal Realizations
Thomas L. Paez, Ph.D., Sandia National Laboratories, Albuquerque, New Mexico

This example shows some random signal realizations drawn from random process sources and their corresponding estimated spectral densities. A typical stationary random process has signal content over a wide range of frequencies. The signal content of a stationary random process is reflected in both the signal realizations and the random process spectral density. When a random process has signal content only over a narrow band of frequencies, then the source is called narrowband. When a random process has signal content over a wide band of frequencies (whether the signal content is uniform or variable), then the random process is called wideband.

Figure 5-5 is an example of a stationary random signal realization drawn from a wideband random process. The signal (upper) has peaks in the neighborhood of 2.0 to 2.5, and if we were to plot a histogram of the signal we would find that the root mean square (rms) value of the signal is about 0.9. (In fact, the mean square was found through computation to be about 0.82.) The spectral density for the random process was estimated, based on the signal shown in Fig. 5-5, and the result given in the lower figure. The source is clearly wideband, based on the spectral density. The signal itself indicates that many frequency components are present, though, the nature and importance of the component contributions is made clearer by the spectral density. Based on the explanation given in Section 5.8, it is clear that the spectral density characterizes the distribution of mean square signal content of the random source in the frequency domain. In this connection, it is important to note that the area under the spectral density curve equals the mean square of the random process, which in this case is about 0.82.

Figure 5-6 is an example of a stationary random signal realization drawn from a narrowband random process. The signal (upper) has peaks in the neighborhood of 0.9 to 1.0 reflecting an rms value of about 0.40. The spectral

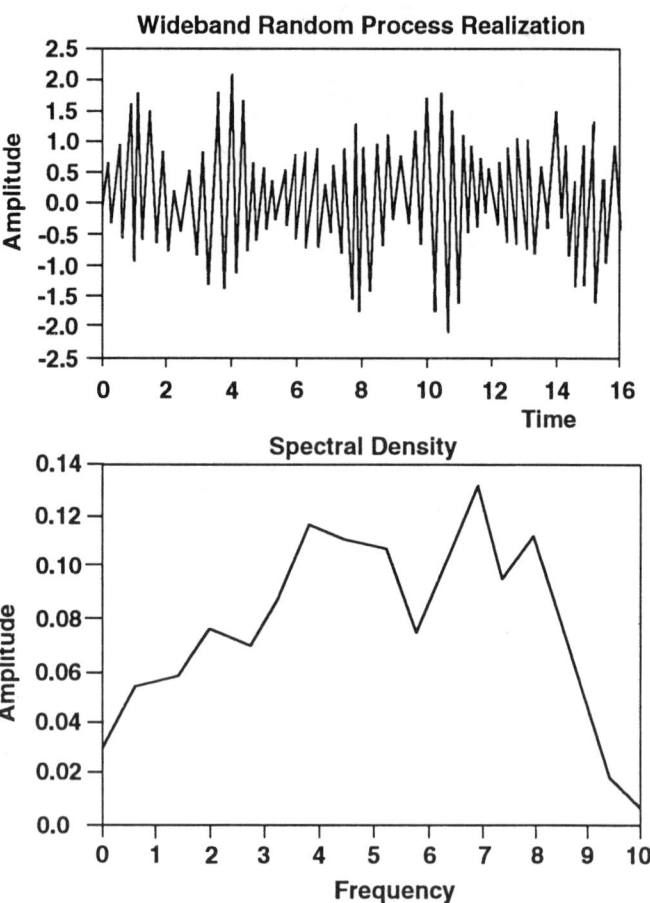

Figure 5-5. Stationary Random Signal Realization Drawn from Wideband Process. Source is the Upper Signal and the Spectral Density in the Lower Figure.

density of the random source was estimated as in the previous example, and the results are shown in the lower part of Fig. 5-6. It is clear that the random process is narrowband from its spectral density. The time domain realization also reflects this in that the time history looks like a sine wave with randomly modulated amplitude. This is the characteristic of a narrowband random process. It is important to note that when the random process becomes more complicated, it becomes difficult to discern its character from the time history.

Seismic Disturbances and Structural Responses

When an earthquake occurs, energy is released into the ground in both radial and vertical directions. Depending on the location of the structure relative to the epicenter or areas of wave reflections, the initial disturbances will excite nearby structures in both horizontal and vertical directions. The excited structure will respond according to its inertial and dynamic characteristics. That is, two structures each with different mass and stiffness distributions and other properties will respond in different modes, frequencies, and potential damage if the intensity of the disturbance is sufficient.

Civil engineers have developed approaches for sensible design of structures subjected to earthquakes. This sensible approach is to design a sufficiently strong structure to avoid collapse in the most severe earthquakes while accepting the possibility of some repairable damage. The sufficiently strong structure prevents loss of life from building collapse. Thus safety, costs, and structural integrity are all addressed in their proper perspectives.

This design approach consists of two essential parts: (1) descriptions of the seismic disturbances and (2) characterization of the responses of the proposed structure. Recording the evidence of an earthquake is in the form of a seismograph giving the acceleration time history of the motion of the recording station or the earth. This recording is a random acceleration time history. From this record, successive integrations yield the velocity and displacement histories. A classic example of this type of record is the El Centro, California earthquake of 1940 (Ref. 5-9).

By analyzing a proposed structure as a single or multiple degree of freedom dynamic system subjected to seismic disturbances, plots can be obtained of the response of the structure. One approach is to use the maximum acceleration and an area of the acceleration-time history giving the velocity change imparted to the structure. The other maximums (e.g., velocity and displacement) can be used as well. These maximums can also be converted to other representations such as the energy absorbed (e.g., $1/2\ mV^2$) and is called pseudovelocity.

Figure 5-7 shows the north-south component of ground accelerations in the El Centro Earthquake. This normalized plot shows accelerations in g's. Figure 5-8 shows the response spectrum for a typical earthquake. The figure is a tripartite logarithmic which indicates at one location the displacement, velocity, and acceleration values for an analyzed structure. The scales are positioned in the following manner: the horizontal scale is the fundamental vibration frequency for the structure, the vertical scale is the pseudovelocity as described above, the scale inclined toward the upper right corner is the displacement, and the scale inclined to the upper left corner is the acceleration. The undamped spectrum is for various types of analyzed structures indicated as points on this plot. Short, stiff structures having higher frequencies are shown as points a and b, multistory structures as points c and d. More compliant structures such as water tanks and suspension bridges are shown as points e and f (Ref. 5-9).

Response spectra as shown in Fig. 5-8 are dependent on the damping in the structure. Increased damping reduces the spectrum shown in this figure. The response

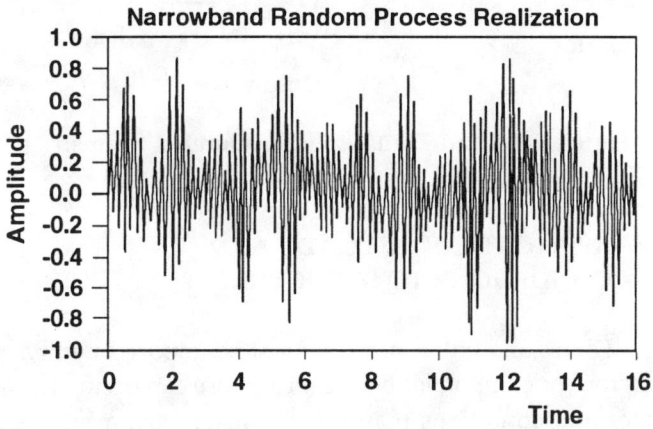

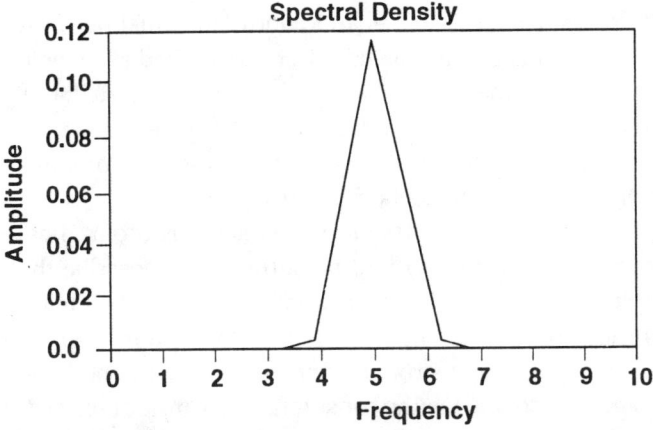

Figure 5-6. Stationary Random Signal Realization Drawn from a Narrowband Process. Source is the Upper Signal and Spectral Density is the Lower Figure.

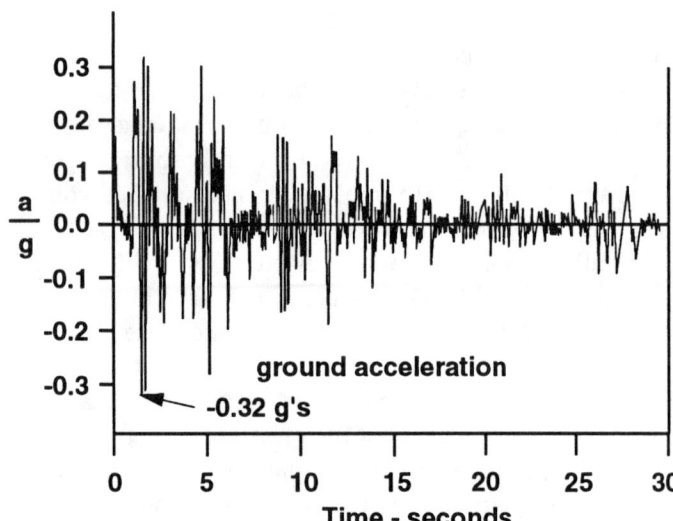

Figure 5-7. North-South Component of Ground Acceleration of El Centro Earthquake.

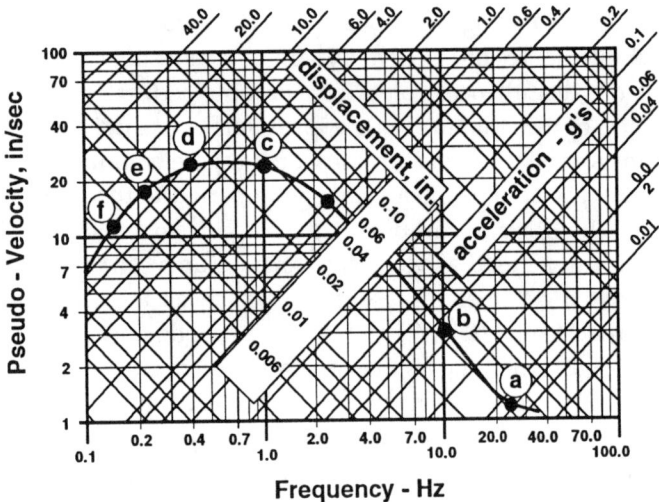

Figure 5-8. Response Spectrum for Typical Earthquake (Tripartite Logarithmic Plot) (Ref. 5-9).

spectra are also dependent on the types of soils involved and how they interact with the structure.

Since many structures exhibit some inelasticity when subjected to large forces, their behavior can be approximated as multistep model of an initial elastic region, an elasto-plastic portion of constant resistance, and an elastic unloading. Granted that this is an approximation but the experience shows that this procedures gives generally reasonable results.

This simplified approach outlined in this section can be improved by incorporating probabilistic descriptions of the motion histories involved (e.g., acceleration-time). Using more detailed models (multidegree of freedom will also develop more realistic assessments of the behavior of structures subjected to dynamic probabilistic loads (Ref. 5-9).

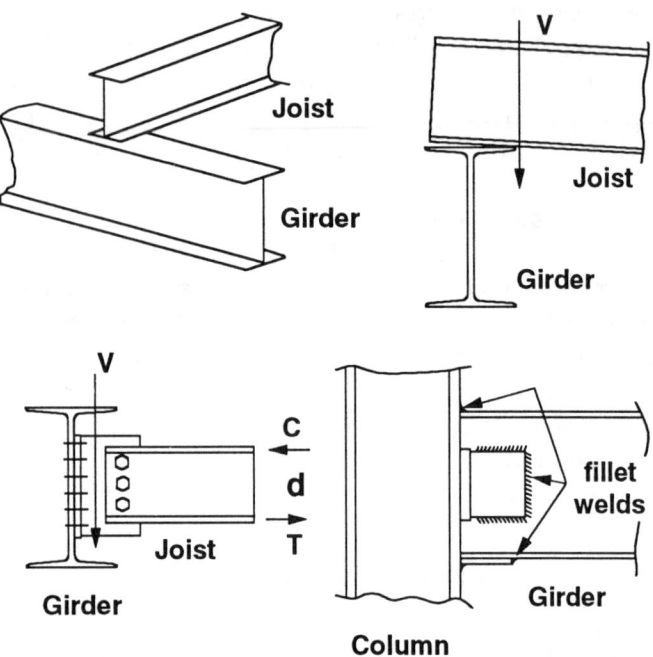

Figure 5-9. Load Transfer in Supporeted Beams.

5.10 APPLICATION OF LOADS AND CONDITIONS TO STRUCTURES

Forces and environments must be imposed upon a structure or component before a response can occur. The forces and conditions may be in contact (e.g., a bridge resting on its supporting columns), nearby (e.g., radiative heating from a fire), or in the path (e.g., fluid flow or blast overpressures). The forces and conditions must be sufficiently localized to be considered to be applied at a single point, or, they may be uniformly distributed along a line or randomly occur over a portion of a structure. They may also result from inertial effects or even stress wave motions within the structure (Refs. 5-8, 5-9, and 5-16).

Assuming that these four descriptions (point, line, distributed, or inertial) will be sufficient to describe the applied forces and other conditions (along with their magnitudes, directions, and time based descriptions) which can be imposed on a structure or component, then both the design aspects and testing considerations can be described.

Point or Concentrated

A point or concentrated load can be shown by beams resting on other beams as shown in Fig. 5-9. Technically, there is actually an area of pressure of unknown magnitude but whose resultant force (pressure x area) equals the weight of the beam and what it is supporting. (It would be very complicated to determine the actual pressure distribution under the beam flange because it depends on the areas in contact and the relative stiffnesses of the flanges.) It is usually sufficient to consider this type of load to be concentrated at the web of the beam. One could also argue that the load transfer is through a line of contact parallel to and directly under the web of the beam. This added complexity of describing the application of loads is rarely needed. The usual practice is to assume that the loads supported by the upper beam are transferred as a point load to the intersection of the planes of the two webs of the beams. This choice of a point load makes the analysis of the beams simpler as they would both act in bending. The beam would be subjected to both bending and torsion if the loads were transferred off this point of intersection. One can see that this choice is not necessarily conservative in terms of the superposition of stresses in the event (quite likely) that the loads are actually transferred somewhat outside the point of intersection of the webs of the beams.

There are many other examples of points loads including (Ref. 5-9):

1. Cranes are used to lift and maneuver a wide variety of objects. Most cranes use hooks. The lifting lines from the hook to the item are often reacted at specific points on the item.

2. Transport of items in trucks, rail systems, and in aircraft often use various types of tie-down systems. These tie-down systems also have the applied forces reacted at specific points. A related example is the cam-lock corners installed on shipping systems used for transport by ship and barge. These locking corners attach a container to the ship directly or to other containers.

3. Vertical tower structures are often restrained by cables which apply point loads to the tower.

4. Bicycle spokes offer support to a wheel rim.

5. Engines are mounted at two or more points to an automobile body or attached to a wing structure.

It is typical in structural testing to have a limited number of opportunities (often one) to attach a loading system to a test item. One force can be distributed through levers and load spreaders.

Line

Loads can also be applied along lines or paths. These lines or paths may be straight or curved. For example, an interior wall or partition in a building, positioned on an unsupported portion of a floor, would be considered a line load. A cylindrical tank transported on its side could be handled with straps to remove it from a truck trailer. These straps give a line loading to the tank over a portion of its circumference.

The forces acting on removable closures on pressure vessels (e.g., flat, elliptical) transmit the axial forces and moments distributed around the circumference of the vessel wall. These loads can also be considered as line loads acting around a circular or other path conforming to the geometry of the vessel. The line loads acting on a pipe flange is a related example of how forces are transmitted from one section of pipe to another.

Seals used to capture pressures in vessels and systems are usually preloaded into position. These preloading forces can also apply line loads to the structure where the seal is made.

Another example of line loads is the use of various techniques for rapid separation of parts. Such devices as focused explosive cutting charges are used to sever portions of structures. Sufficient explosive energy to cut even specially designed weakened portions of structures will cause line loads (and moments) to a structure.

High strength straps, chains, or other flexible systems are often used in structural testing to apply forces at or near desired locations (e.g., center of gravity) on complex structures. These line loads used in testing are used in lieu of more complicated fixtures.

Distributed/Pressure

Many types of loads are distributed over a portion or all of a structure. In a pressure vessel, the pressures are typically uniformly distributed over the structure. For a pressurized water tank where the weight of the water becomes an appreciable part of the design loads for the structure, there is a uniform pressure loading plus a distributed pressure loading due to the weight of the water (Ref. 5-3, 5-6, 5-7, 5-12, 5-13, 5-14, and 5-15).

The early morning heating of a cool empty oil storage tank described earlier is another example of a distributed load. The differential thermal expansions of these typically thin-walled vessels is an important part of their design.

Building codes require that certain uniform loads are placed on a structure for design purposes. These code requirements do give flexibility depending on the type of use the building will have. Many buildings have portions of their floors which are removable or portable to allow access to equipment, piping systems, electrical systems, and other reasons. These portable floors are usually designed with an allowable point load and a permitted uniform load. Mezzanine type structures have similar types of requirements (Refs. 5-22 and 5-23).

Loads on aircraft are another example of distributed loads. The pressure differential across the wing (bottom to top) must be sufficient to offer a force equal and opposite to the weight of the airplane. This pressure differential is distributed over the wing in a manner controlled by the aerodynamics of the aircraft and the shape of the wing cross-section (Ref. 5-6).

There are many other types of distributed loads. The intent here is to offer a few examples to help anticipate distributed loads in structural testing. Simulating distributed loads is not an easy situation in structural testing.

Inertial

Many types of mechanical assemblies and electrical systems use inertial devices. These devices typically use rotary motion to provide a known force field to control their operation. The resulting loads depend on the angular velocity and on the mass of the device. These loads (often called body forces) are internal to the item and are sometimes difficult to simulate in a test. For an item subjected to inertial loads, these loads are pervasive throughout a structure, and while they will vary with position and angular velocity, they will act at every infinitesimal point.

Rotating machinery is an example of mechanical assemblies which are subjected to inertial loads. A rotating shaft in which a bearing suddenly freezes is subjected to a torsional load over its length as it decelerates. A common household washer and dryer are two types of mechanical systems which experience inertial loads. These loads are most noticeable when an unbalance occurs in how the contents are distributed. The design of these items must allow for some inertial loads plus an unbalanced load (Ref. 5-3).

5.11 SUMMARY

Loads vary from simple, easy to understand conditions imposed on a structure to very complicated, time dependent, rapidly varying excitations. Most of the examples cited in how loads are applied were static in nature. When dynamic loads are imposed on a structure, they require additional descriptions (e.g., duration, wave form, etc.). The reader should be alerted to how important the descriptions of loads are in testing of structures, particularly when dynamic considerations are present at a requested or required test.

Those planning and organizing tests should carefully consider the factors used to multiply or otherwise affect the description of loads. Some factors have often been added to these loads so that the loading conditions used in a test can approach or even exceed a conservative or sound approach used in testing a structure. Pressure vessels proof tested beyond 1.5 times their working pressure could cause some permanent damage to the vessel. While factors increasing the test requirements to exceed service conditions are standard practice, there needs to be a proper balance between overtest requirements and actual service conditions so that continued structural performance is not compromised in testing. Those new to or less familiar with testing and even experienced test personnel sometimes are prone to add excess requirements and even specify significant overtests in the belief that the structure will be better having been through a more "rigorous" test. Excess test requirements can contribute to reduced structural integrity and should be avoided.

An effective test must comprehend realistic customer related load spectra. For many test items, a service history is the all-important concern. Determining the service history for the economical, cost effective design of automobiles, farm equipment, bridges, and any other structures usually rests on choosing the right loads for validation testing. Given competitive market pressures for engineered equipment, using realistic loads can lead to efficient products.

5.12 REFERENCES

Texts and Handbooks

5-1. L.S. Beedle, et.al., *Structural Steel Design*, Ronald Press, New York, 1964.
5-2. J.S. Bendat, and A.G. Piersol, *Random Data. Analysis and Measurement Procedures*, 2nd Ed., Wiley Interscience, New York, 1986.
5-3. A. Blake, *Practical Stress Analysis in Engineering Design*, 2nd Ed., Marcel Dekker Inc., New York, 1990.
5-4. R.E. Bolz, and G.L. Tuve, Editors, *Handbook of Tables for Applied Engineering Science*, Chemical Rubber Co., Cleveland, OH, 1970 or latest edition.
5-5. H.E. Boyer, and T.L. Gall, Editors, *Metals Handbook, Desk Edition*, American Society for Metals, Metals Park, OH, 1985.
5-6. E.F. Bruhn, *Analysis and Design of Flight Vehicle Structures*, TriState Offset Co., 1973.
5-7. R. Chuse, and S.M. Eber, *Pressure Vessels, The ASME Code Simplified*, McGraw-Hill, New York, 1984.
5-8. D.A. Firmage, *Fundamental Theory of Structures*, Kreiger, 1985
5-9. E.H. Gaylord, and C.N. Gaylord, *Structural Engineering Handbook*, 2nd Ed., McGraw-Hill, New York, 1979.
5-10. C.M. Harris, *Shock and Vibration Handbook*, 3rd Ed., McGraw-Hill, New York, 1988.
5-11. T.G. Hicks, Ed., *Standard Handbook of Engineering Calculations*, McGraw-Hill, New York, 1972.
5-12. H. Liu, *Wind Engineering*, Prentice-Hall, Englewood Cliffs, NJ, 1991

5-13. L.S. Marks, *Standard Handbook for Mechanical Engineers*, McGraw-Hill, New York, latest edition.

5-14. E. Simiu and R.H. Scanlan, *Wind Effects on Structures: An Introduction to Wind Engineering*, John Wiley, New York, 1978.

5-15. M.F. Spotts, *Design of Machine Elements*, 4th Ed., Prentice-Hall, Englewood Cliffs, NJ., 1971.

5-16. C.W. Richards, *Engineering Materials Science*, Wadsworth Publishing Co., San Francisco, CA, 1961.

5-17. J.A. Zukas, et. al., *Impact Dynamics*, John Wiley & Sons, New York, 1982.

5-18. _____, Guide to Safe Handling of Compressed Gases, Matheson Gas Products, Inc., Seacaucus, NJ., 1983.

5-19. _____, *Manual of Steel Construction*, American Institute of Steel Construction (AISC), Latest Edition, New York.

5-20. _____, Standard Specifications for Highway Bridges, American Association of State Highway and Transportation Officials (AASHTO), Washington, D.C., 1989.

5-21. _____, *Wood Handbook*, U.S. Dept of Agriculture—Forest Service, Washington, D.C., 1987

5-22. _____, Minimum Design Loads in Building and Other Structures, American Society of Civil Engineers, (Revision of ANSI A58.1-1982), New York, 1990.

5-23. _____, *Uniform Building Code*, International Conference of Building Officials, Whittier, CA, 1988.

5-24. _____, Recommended Lateral Force Requirements and Commentary, Seismology Committee, Structural Engineers Association of California, Sacramento, CA, 1990.

5-25. T. Baumeister, Mark's Mechanical Engineers Handbook, 9th Ed., McGraw-Hill, New York, 1958, p. 4-10.

Papers and Reports

5-P1. B.E. Bader, "Heat Transfer in Liquid Hydrocarbon Fires," Chemical Engineering Progress Symposium Series, Vol. 61, No. 56, p. 7840, 1965.

5-P2. R.K. Clarke, J.T. Foley, W.F. Hartman, and D.W. Larson, "Severities of Transportation Accidents," Sandia National Laboratories Report, SLA 74-0001, Albuquerque, NM, 1976.

5-P3. J.E. Kennedy, "Behavior and Utilization of Explosives in Engineering Design," 12th Annual ASME Symposium.

5-P4. J.J. Gregory, R.Mata, Jr., and N.R. Keltner, "Thermal Measurements in a Series of Large Pool Fires," Sandia National Laboratories Report, SAND 85- 0196, Albuquerque, NM, 1987.

5-P5. D.W. Larson, R.T. Reese, and E.L. Wilmot, "The Caldecott Tunnel Fire Environments, Regulatory Considerations, and Probabilities," 7th International Symposium on Packaging and Transportation of Radioactive Materials (PATRAM VII), New Orleans, LA, May 15-20, 1983.

5-P6. D.W. Larson and R.T. Reese, "Design Considerations for Severe Thermal Environments Based on Tunnel Fires," Proceedings Society for Experimental Mechanics, Savannah, GA, November 1987.

5-P7. L.H. Russell and J.A. Canfield, "Experimental Measurement of Heat Transfer to a Cylinder Immersed in a Large Aviation Fuel Fire," Journal of Heat Transfer, ASME, Series C, 95, pp. 397-407, August 1973.

5-P8. _____, "The Holland Tunnel Chemical Fire," The National Board of Fire Underwriters, New York, May 13, 1949.

5-P9. _____, "Pressure Safety Manual," SAND90-1818, Sandia National Laboratories, Albuquerque, NM, 1991.

5-P10. _____, "Structural Testing Technology at High Temperatures," Proc. SEM Fall Conference, Nov. 4-6, 1991, Dayton, OH.

Chapter 6
Loading Systems

Robert T. Reese, Sandia National Laboratories,
Albuquerque, New Mexico

6.1 INTRODUCTION

This chapter describes the loading systems for laboratory and field tests. The loads described in Chapter 5 are intended to be the best estimates of the expected conditions that a structure must survive and function in service, overload, and in one time occurrences. Loading conditions are imposed on structures and test items in the laboratory or field by loading systems. The loading systems and laboratory methods are expected to produce simulations of expected conditions which duplicate as nearly as possible the response of the structure to the actual or design conditions. Loading systems are also used to exercise a structure under known or defined static and dynamic conditions.

This chapter contains five basic sections:

- The overall concept of laboratory and field simulations needed to duplicate service, overload, or other known conditions producing realistic responses of the structures and components tested
- A discussion of open and closed loop testing systems and test controls
- Loading equipment used to impose the wide variety of conditions (forces, vibrations, shocks, temperatures, etc.) on structures
- Examples of laboratory and field loading conditions
- A description of boundary conditions and fixture design.

A chapter summary is given in Table 6-2.

6.2 SIMULATIONS—ACTUAL LOADS, TEST CONDITIONS, AND FAILURE SITUATIONS

This section emphasizes simulations of actual or expected loads, test conditions, and possible failure situations which act on, produce, or define the response of the structure or component. Structural testing requires some types of simulations in order to set-up, perform, and complete a valid test. Simulations will very likely be required for the following (Refs. 6-3, 6-4, 6-7):

1. Methods of applying the loads including: the locations, directions, and manner of application (point, line, distributed, or inertially); loading rates and other time-dependent descriptions of loadings, magnitudes (e.g., service or use, overload with factors of safety (proof), and/or failure conditions); substitutions of loading conditions in terms of the equipment used which will give useful results ensuring that other loading conditions (e.g., temperatures, displacements, gravity free) can be properly applied.

2. Boundary conditions including: basic structural support or reaction conditions of free, pinned, fixed, moment or shear restraint, and others with the intent of duplicating the actual or expected conditions found in service or the extreme conditions expected.

3. Substitution of laboratory techniques and methods in place of the actual or expected conditions so that diagnostic information can be obtained and the imposed conditions known within the limitations of the equipment and facilities used.

4. Evaluation of the procedures and data to ensure that a test performed is equivalent to the actual or expected conditions and gives the kinds of responses and insights of structural behavior needed.

The first important point from Chapter 5 is that almost all loads are estimates and that very few exact descriptions for loads are available. However, loads typically have many ways in which they can be closely approximated or estimated so that a structure or test item can be realistically exercised and the response determined. Some tests are performed in which known conditions are imposed on a structure and the response (e.g., accelerations, displacements, strains, temperatures, velocities) are measured. These test results can be used to verify analytical models or just give insight into how a structure responds. There are also a few exceptions to the basic premise that loads on structures are usually approximations. A pressure vessel is an example of a structure which is typically designed with a desired working pressure and a required proof pressure that is 1.5 times the working pressure. In this case, the loads are known accurately.

The second point which will be developed in this chapter is that any test performed is a simulation. Therefore, one way to describe structural testing is that it consists of obtaining and interpreting information from tests which can be described as "simulations of approximations." The approximations or estimates come from the descriptions of loads needed so that the structure can be designed and fabricated. The simulations come from the requirements which must be fulfilled so that a test can be specified, planned, and performed in the field or in a laboratory where known conditions can be imposed on a structure or component and responses measured and recorded.

The terms simulations and approximations could imply some degree of inability to perform meaningful tests or that the measurements might not be an accurate indicator of the performance of a structure or component. While there is seldom a completely exact description of the loads, structural tests can be planned and performed so that limiting values, boundaries, and other conditions can be applied which almost always provide more than adequate simulations satisfying the need to duplicate the expected events and requirements. These approximations and simulations are completely analogous to the assumptions made in performing structural analyses. It is standard practice in structural engineering to make some assumptions in order to understand and predict the response of a structure. The approximations and simulations required in structural testing are a necessary part of the work just as assumptions of material properties, geometry, and loads are necessary in structural analyses (Ref. 6-1, 6-2).

This section will address the simulations. Laboratory and field simulations that are needed for test purposes consist of the following which expand and clarify the list of four given above:

1. Test control in terms of open or closed loop; tests can be performed in either mode or in combinations which can adequately approximate the actual conditions.

2. Systems (equipment) used to apply loads and loading conditions.

3. Estimates of the loading conditions (from test plans and objectives) to be applied including their location, magnitudes, rates or functional descriptions, and direction(s),

4. Fixtures and devices used to supply boundary conditions (test setup), interfaces, and attachment to testing machines, frames, vibration shakers, etc.,

5. Installation and use of transducers and gages to determine the loading conditions and the structural response at the locations and directions which will give test results which can be understood, interpreted correctly, and correlated with analyses,

6. Data acquisition systems including signal conditioning, processing, and recording instrumentation which nearly duplicate the response of the structure or component to loading conditions.

7. Analysis and interpretation of the test data and evaluation of structural behavior to verify the adequacy of the simulation.

Test controls given in section 6.3 describe open or closed loop systems which govern how the loading conditions are controlled and how the feedback and diagnostic systems interact. The systems used to apply loads and loading conditions are described in the third section. Examples are given in the fourth section which outline how loading conditions have been applied to certain test structures. The art and science of developing boundary conditions through test fixtures is described in the fourth section. The utilization of transducers and strain gages is given in Chapter 9. Data acquisition systems are described in Chapter 10. The analysis and interpretation of test data are described in Chapter 15 through a series examples.

Simulations needed for structural testing are best described by examples. One example is a complex cantilevered beam/bracket supporting a component as shown in Fig. 6-1. The beam is enlarged to show critical features for structural testing purposes.

The component supported by this example beam could be a small critical subsystem for a missile, an air conditioning unit supported by two or more of these

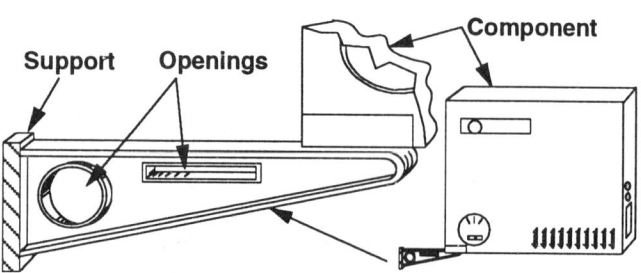

Figure 6-1. Complex Cantilevered Beam

beams, or other necessary item for a mechanical or electrical system. The beam has a series of openings needed for other functions such as electrical systems, hydraulic lines, or other controls. It is common in many types of structures (e.g., aircraft, missiles, weapons) to have major structural elements equipped with openings for access or passage of other essential functions. The openings are expected to reduce the beam's strength even though the areas around the openings have been stiffened. The design of this particular beam has been pursued with sufficient engineering analyses so that the prototype bracket is of minimum weight and optimum materials for machining, joining, and fabrication purposes.

Tests have been proposed to evaluate the strength and integrity of the beam and to proof test it for margins over its expected service conditions. This proof test could be part of a quality assurance effort if many of these beams were to be made. The beam is to be subjected to both static and shock loads. The shock loads are expected to be much larger in magnitude. The shock loads could come from stage separations in a missile system or from vertical accelerations in earthquakes. In either case, the beam and component are expected to experience accelerations or body forces proportional to their mass distribution.

One type of simulation is a static load test on the beam assembly. Static loads are not the same as the actual accelerations but they can be made equivalent. The laboratory static loading can cause a larger local shear force at the point of application which could cause larger localized stresses. Static loading would be applied slowly enough that rate effects in materials would not be involved. Another type of proof test would be to subject the beam to a shock test. This shock test would be important if the beam and component were expected to respond dynamically. If the expected acceleration pulse (velocity change) produces vibrations in the beam system, then vibration tests could also be used for proof test purposes.

The proof test could also be accomplished in a centrifuge so that the distribution of forces would more closely approximate the actual loads. Regardless of how the proof test is conducted, the following things must be accomplished to ensure that a valid proof test was conducted:

1. The first objective is the verification of joint integrity at the support.

2. The second objective is the verification of the structural integrity of the beam itself (e.g., the fabrication and joining processes).

3. A third possible objective is the dynamic behavior of the beam (e.g., resonances, amplifications/attenuations of shock loads and/or vibrations).

4. A possible fourth objective is to verify that the loading conditions used in the proof test did not compromise the integrity of the beam or create other unwanted responses.

Two examples of static proof tests are also shown in Fig. 6-2 and 6-3. In Fig. 6-2 shows how the beam could be attached to a test fixture and positioned in a testing machine so that the applied force would be reacted at one point (e.g., the load cell). The load simulating the accelerated component would subject the beam to two point loading conditions distributed according to its mass distribution as shown by the unequal forces supported by the load distributor. An advantage of this proposed test is that only one load path is needed to simulate the equivalent static force of the component and beam which makes setting up and performing the test simpler. A disadvantage of this proposed test is that larger shear forces could be applied to the beam around the rectangular opening.

In Fig. 6-3, the proposed test indicates how an equivalent moment and shear force could be applied to the beam. The application of this equivalent force system would produce different stresses around the openings but would exercise the support joint in the desired manner. An advantage of this test setup would be a more realistic distribution of stresses in the beam except where the loading fixture is attached. A disadvantage of this test setup is that three loading systems would be required which makes testing more difficult.

The proof test could address the support joint, the beam, the strains around the openings, the fabrication processes, or other suspected problem area. The designer/

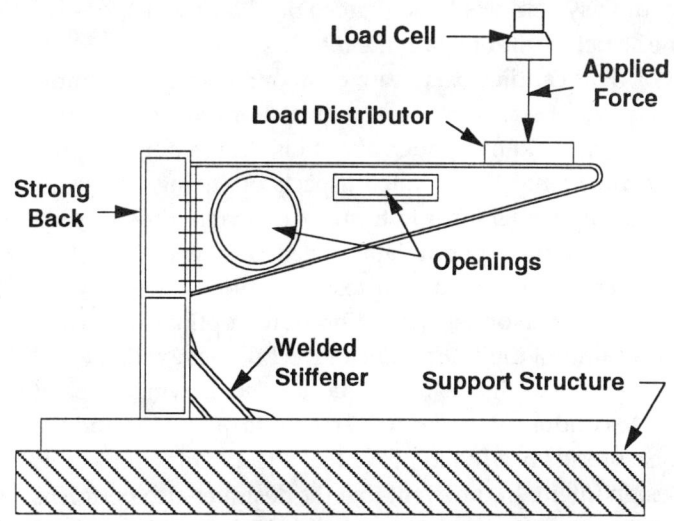

Figure 6-2. Example Setup for Tests on Cantilevered Beam

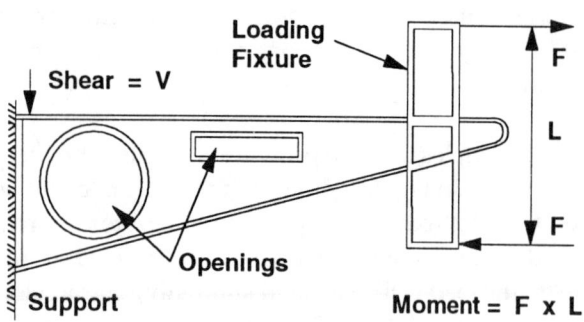

Figure 6-3. Cantilevered Beam Example Structure and Loading Systems

test requester will likely know the expected weakest parts in the prototype design test hardware. By having a good understanding of the expected response of a structure, tests can be better planned, the results needed can be obtained, and the necessary judgments and evaluations of structural performance can be made. A proof test performed on the structure may result in premature failure or permanent deformation.

It is likely that a test laboratory will have many types of equipment needed to perform tests but will probably not have all the different types of equipment needed to complete requested tests. Therefore, some compromises would probably be needed in order to set-up the testing system to be used for the test and some simulations are required in order to complete the test. Even in this relatively simple example of a cantilevered beam, there were many aspects of the proposed proof test which would require an evaluation of the possible simulations in the laboratory to determine if the test procedures, boundary conditions, and loading system were adequate to produce the response of the structure and to gain the insights needed.

This cantilevered beam component system example is typical of structural tests required or requested. Often the designer knows much about his design but often has not considered the detailed aspects of testing. These aspects may be deficient in terms of describing the loading conditions, the types of data needed, or how the structure will actually respond in a test. It is often necessary for laboratory personnel to describe the test options within the constraints of the equipment and techniques available and then let the designer/test requester decide which type of test to conduct along with the methods of loading and the measurands needed to prove that a test item will meet its design intent. In the event that the designer also performs the testing, then it is important that he look objectively at the test item and develop a test plan that will provide the necessary information. In these cases, compromises and simulations should be expected. As the structure or system to be tested becomes more complicated and consists of more parts and assemblies, then the structural tests will also become more complex, more details will need to be considered, and more alternatives and options available along with more compromises and simulations needed to obtain the test data.

6.3 TESTING CONCEPTS—OPEN AND CLOSED LOOP SYSTEMS

Dr. Stuart Swartz, Civil Engineering Dept., Kansas State University, Manhattan, KS.

There are many different ways to describe testing. Testing can be described in terms of the kinds of loading conditions imposed (e.g., dead loads, lift-off loads). Testing can also be described by the results of the tests performed (e.g., materials characterization, proof, stiffness). The manner in which tests are conducted is one basic way to describe testing. With the development of computer based systems, interactive controllers, and very responsive testing systems, current descriptions of testing usually use the term "open and closed loop testing." This basic way of describing testing is chosen because it reflects how the test personnel and equipment interact with the test item, the loading equipment, data acquisition systems, and the measurands obtained. In any test, there will probably be some parts that can or must be performed manually and some parts that can be more efficiently and precisely by a testing machine, its controllers, software and other features.

Open loop testing basically consists of performing a test in which the operator has control and provides the feedback to the test equipment. For example, in a static test situation, an open loop test suggests that the loading conditions, measurements, and data acquisition systems are all controlled manually or, if portions of these systems are automated, are controlled or that can be interrupted and redirected by manual commands. If the test data contributes information suggesting another increment of loading, added data to be taken, or to modify or enhance some other aspect of the test, then the decisions to make changes are made by people rather than automated controllers, computer systems, or software.

A closed loop system consists of performing a test in which a portion or all of the control of the test is given over to the testing machine, the computer systems, the built-in controllers, or other devices which are interactive with the testing machine, transducers, and gages. (These closed loop systems do have override controls for emergency situations. Closed loop control does not mean complete abdication of responsibility by the operator.)

In its basic configuration a closed loop testing system consists of a hydraulic actuator and servovalve assembly, pump, transducers, and a command and feedback controller. Such systems may be assembled for particular structural testing requirements or may be configured with a testing machine, testing frame, or test bed. As an example of the former case the Stevin Laboratory at the University of Delft in The Netherlands has a central pumping station from which hydraulic lines and outlet connections are distributed throughout the laboratory. (Central pumping systems are common to most servohydraulic testing systems.) For a particular structural test, one or more hydraulic actuator/servovalve assemblies are employed and coupled with one or more force and displacement transducers for feedback control and measurement along with appropriate controller(s) and output devices.

Theory of Operation

The controlled, closed-loop functions are outlined. The servo-controller accepts command and feedback signals associated with a controlled variable. For example, the physical positioning of the piston in the hydraulic actuator in the testing machine. The command signal represents a programmed or desired voltage typically as a percent of full scale and the feedback signal is the actual voltage associated with the measured quantity such as the actual position of the piston measured by an LVDT. If the command and feedback signals are not equal, then the servovalve is actuated permitting the flow of hydraulic oil from the pump and moving the piston to the desired location. The rate of movement is generally controlled by the machine but sometimes, depending on the type of feedback control or unintended effects, can be exceedingly rapid. The movement continues until control and feedback signal values are identical at which time the loop is balanced or "closed." In order to prevent serious damage to the system, test fixtures, or test structure or component, limits may be set on the control variable when reached or exceeded will shut off power to the machine.

The use of stroke control is analogous to using a deformation-controlled testing machine, e.g., a screw-driven machine described later. The use of load control is similar, but not identical, to using a hydraulically driven machine. In load control, a specific value of maximum applied load which is intended to be reached is programmed. The controller commands the servovalve to open and allow pressurized hydraulic oil to enter the piston in the appropriate direction until the load cell provides the output signal equal to the programmed value. This maximum load may be steady state or have upper and lower limits such as tension and compression or larger and smaller values of tension. Whatever the limits are, the servovalve will open and close to permit the passage of hydraulic oil into and out of the piston to achieve the applied loads programmed.

The use of strain control is basically unique to the closed loop system. In this the controlled variable is the strain which may be obtained from a strain gage mounted on a specimen or from an extensometer or displacement gage attached to the specimen. It is this mode that allows the evaluation of post-peak response, or strain softening, in a controlled manner to fracture of the specimen or test item.

Operation in Practice

The operation of a closed loop system for materials testing is discussed here but most of the ideas are applicable to a structural testing system.

The system illustrated in Fig. 6-4 may be imagined to be used to test a specimen to failure. The beam and its supports are called the test assembly. The system shown has three variables which may be measured, any one of which may be controlled. These are stroke (movement of the hydraulic actuator), load (applied force measured by the load cell), and strain or displacement (measured at some location on the beam). The following procedure is typical:

1. Turn on electrical equipment and establish that all input settings and voltages are correct. It is very easy to alter these settings and they may drift over long periods of time.

2. Warm up the system to the operating temperature of the hydraulic oil. Then, turn off the hydraulic power.

3. Place test specimen into the test assembly, place and install appropriate transducers, establish zero readings of transducer output including load cell, adjust output devices.

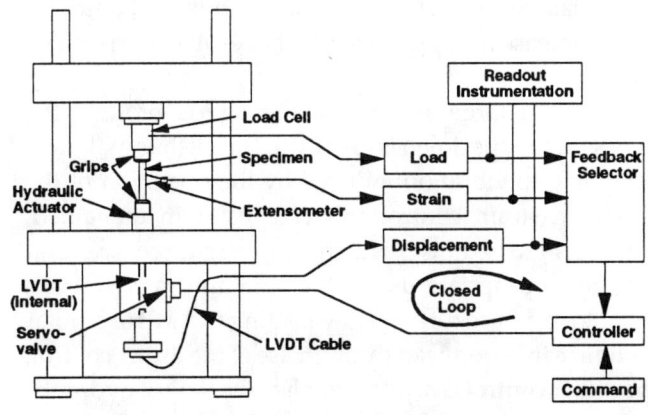

Figure 6-4. Closed Loop Testing System

4. Introduce prepared testing program, e.g., frequency, wave function, peak values, or other function described by computer software, etc.

5. Turn on machine to low hydraulic power and bring specimen to near contact with load cell (or its extension). Complete zeroing process with high power and zero value of controlled variable to "close the loop" at low, or zero, load.

6. Run program; operator intervention and re-zeroing may be necessary.

7. After completing program, disengage test assembly from contact with load cell and remove power.

Because of the very rapid response depending on the flow capacities of the pump and servovalves, a closed loop system is ideal for both fatigue studies and failure investigations in the post-peak load regime. At the same time it is potentially a very dangerous device and all precautions to ensure safety of the operator and attached equipment and fixtures should be made.

As a simple example of what can go wrong, consider a testing arrangement where a displacement is used as the control variable. If the polarity of the displacement transducer does not match that of the stroke transducer, failure will occur almost instantaneously as soon as the specimen starts to deform. Thus, in step 1 above, one of the important checks is to be sure polarities are in agreement between whatever control variable is being used and the stroke transducer.

Another situation that can lead to disaster (e.g., the unplanned test) is to program a peak magnitude of the control variable far in excess of that which can occur. Actually, if load control is used in a test to failure, a point will be reached where the command load is greater than the specimen can carry; the specimen starts to fail—thus further reducing its load carrying capacity which is sensed by the load cell as a decreasing load; thus the controller tries to increase the applied load. The result is a very rapid failure.

When a large amount of damage is present, the stiffness of the specimen is greatly reduced, thus making it susceptible to vibration induced by the operation of the system. In strain control this can result in unwanted, spurious signals (noise) which can be reduced or eliminated by appropriate use of the gain and rate controls. Unstable response during slow loading (RAMP or similar function) can be reduced through use of the dither control. (The dither control is used to regulate the flow of hydraulic fluid through the servovalve. The dither control is always maintaining the servohydraulic system at the levels requested. A characteristic of servohydraulic systems is that they are always anticipating any changes in levels and the "hunting" process to select the next correct level is regulated by the dither control.)

Finally, with respect to output and test results, careful logging of all variables, settings, etc., should be a part of the documentation and should be done before starting the test.

Open and closed loop tests are often separated on the basis of the time needed for the structure or component to respond. That is, dynamic tests in which the structural response is short (less than a minute) are usually conducted in a closed loop. Those tests requiring data to be taken in real time need to have the data acquisition and test control systems automated and are almost always conducted on a closed loop basis. Static tests are often conducted on an open loop basis. However, as soon as more than 3 to 5 channels of data are requested, then static tests are usually conducted on a closed loop basis as well. Those tests in which data are taken at the operator's or test conductor's discretion are typically open loop tests. One final observation is that control of tests is becoming more automated as electronics and feedback systems improve with the result that an increasing proportion of tests will be conducted on a closed loop basis.

6.4 LOADING EQUIPMENT

There are many systems used to apply loads to structures. The primary means of applying loads are through testing machines, structural test frames using hydraulic actuators and other devices, pressure systems and contained pressure in vessels, thermal systems, inertial devices such as centrifuges, vibration equipment, shock and impulse loading systems, blast and explosives, and energy deposition techniques. The equipment and techniques used are separated in this manner so that their descriptions could be simplified and more easily explained.

Testing Systems and Machines

The term testing machine can imply many different types of equipment to engineers. Testing machines are used to provide capabilities for a wide variety of structural tests and also to determine many different material properties. Testing systems and machines have been developed for the following five basic purposes (Refs. 6-1, 6-3, 6-4, 6-6, 6-7, and 6-8):

1. Measurements of hardness including resistance to indentation, abrasion and wear, scratching, and to determine machinability

2. Evaluation of the structure of materials both microscopic in terms of grain structure, inclusions, etc., and macroscopically in terms of appearance, absorptivity, etc. and other metallographic or equivalent features

3. Determination of properties often indicating uniformity such as thicknesses, integrity of welds and bonds, etc.

4. Determination of a wide variety of properties such as electrical resistivity, coefficients of thermal expansion, and other important physical properties and characteristics

5. Evaluation of strength, stiffness, stability, creep resistance, and fatigue characteristics.

In the context of structural testing, testing machines will have a more restricted definition and application in terms of determining strength, stability, stiffness, creep resistance, and fatigue characteristics by application of known forces, strains, or displacements. Testing machines are used to impose loading conditions on structures and test hardware with the machines equipped with sufficient transducers and other diagnostics so that the conditions imposed can be controlled and will be known within some small experimental uncertainty. Basically, the term testing machine describes the equipment that can be used to apply conditions (force, torque, displacement, rotation, strains, accelerations, temperatures, shocks, pressures, etc.) at points, along lines, or to a portion of or the entire structure in both static and dynamic loading conditions depending upon a machine's capabilities. Because of the interactive controls and multiple mode operations currently available, the term testing machines has been replaced by testing systems. These systems offer test control as well as data acquisition.

Testing machines are mechanical systems with various types of controls: manual, mechanical and electrical servovalves, computer software, and function generators which are now usually described as testing systems. There are two basic types of testing systems in terms of their abilities to impose loads and displacements—mechanical and hydraulic. The testing systems are described in these two basic categories so that capabilities for each type can be differentiated. (These two categories are not based on the control systems.)

The two basic categories are further divided each into two basic types. The mechanical testing machines consist of two ways in which loading conditions are applied: through dead weights and using mechanical advantage of screws driven by motors. The hydraulic machines consist of tension or compression (e.g., pilot valve controlled) and tension and compression systems (e.g., servovalve controlled). Examples of the three basic types of the early versions of testing machines were shown in Chapter 2, Figs. 2-9 through 2-11.

Torsional loading conditions can also be applied in the screw-driven and hydraulic testing machines. A linear motion hydraulic actuator or ram and a crosshead head motion provided by screw-drives can be replaced with a rotary actuator or a motor-driven gearing system to apply torsional loading conditions. A dead weight testing machine can also be modified or configured for use in a torsional test as well. Special testing systems can be combined with certain types of specimens to give many possible combinations for testing including:

1. Tension/Compression/Internal Pressure
2. Tension/Compression/External Pressure
3. Tension/Compression/Torsion
4. Tension/Compression/Torsion/Internal or External Pressure

The combinations of tension/compression/torsion indicates the use of multiple actuators or screw-driven systems. These combinations of loading conditions suggest that the test item is subjected to simultaneous loading conditions. To create a tension/compression/pressure system, a testing system would need to be coupled with a pressure system to provide the internal or external pressure. If external pressure is applied to a test item, then the pressure will need to be confined in a containment system (e.g., vessel).

An example of a structure subjected to tension, compression, torsion, and internal and external pressures is an underwater pipeline. The tension, torsion, and compression stresses would come from bending and maneuvering the pipe as it is lowered into the water. The internal pressures would come from the pressurized liquid traveling in the pipe. The external pressures would come from the hydrostatic pressures from the weight of the water above the pipe, particularly when the pipe is empty or when no pressurized flow is occurring.

Examples of a test involving tension/compression and pressure and another involving external pressure and tension/compression are shown in Fig. 6-5 (a and b) respectively. Figure 6-5 (a) shows a special test specimen which can be subjected to internal pressure and tension/compression forces. Creating these biaxial stresses can be important for some materials and to evaluate some types of structures. Figure 6-5 (b) shows a testing machine positioned inside a pressure vessel. Test items requiring loading conditions of external pressures and tension/compres-

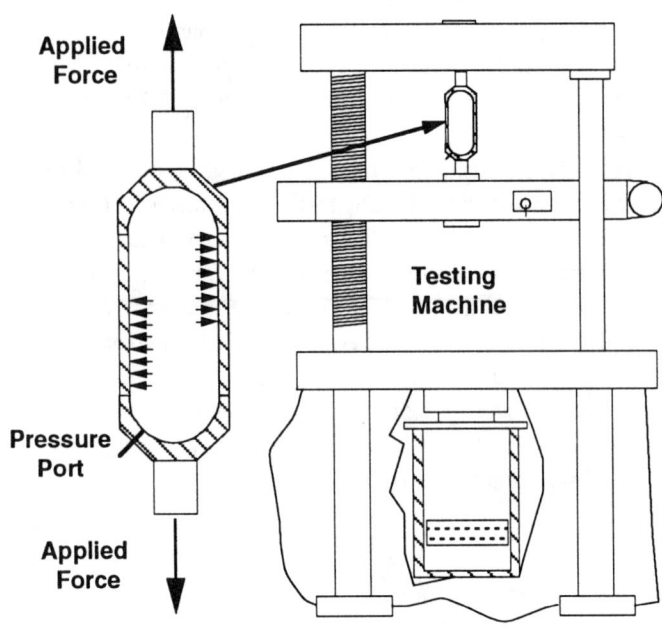

Figure 6-5a. Tension/Compression Pressure

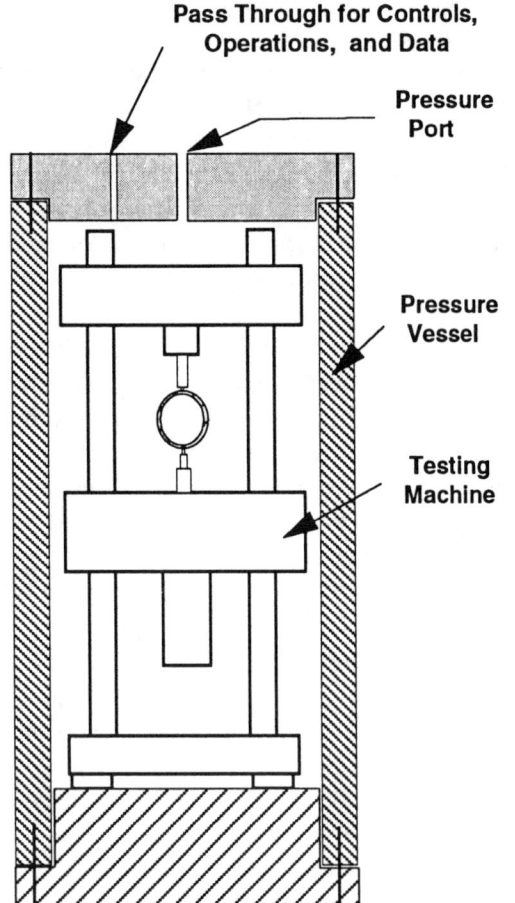

Figure 6-5b. Testing Machine in Pressure Vessel

Figure 6-5. Modified Testing Setups

sion forces could be tested in this system. There are some obvious additional complexities in performing tests with a testing machine placed in a pressure vessel—positioning test items, testing machine hardware subjected to external pressures, instrumentation and data acquisition systems capable of operating under the combined influences of pressure and pressure media (e.g., water), and seals to protect mechanical surfaces and parts. For example, the positioning of test items could require access ports in the wall of the pressure vessel near the working level of the testing machine or it may be necessary to remove the testing machine from the vessel, insert the test item, and return the testing machine with the test item to the vessel.

Multiple mode tests (1 through 4 above) will usually require some types of additional test fixtures with the added complexities of separating the effects so there is a minimum of interaction between the modes. In addition, the interfaces between portions of the actuators will need to be protected and sealed.

The testing systems with added capabilities have appearances nearly the same as the typical testing systems. The tension-compression (closed loop) system shown in Fig. 6-4 could have a torsional actuator added to the hydraulic actuator already present. Each of the testing machines could be modified so that the multiple loading systems would apply the conditions described.

Testing machines were developed during the first half of the twentieth century. While improvements were made during the time since then, it was the development of microelectronics that gave considerable increase to the machine capabilities primarily in terms of the controls, feedback, and computer interactions that are possible today. For example, the early mechanical fatigue tester basically consisted of a motor driven cam which imposed a known oscillating displacement on a test item. The motor has a revolution counter to record when failure occurred. The same type of machine could be used today except that it is likely that it would be modified with the following possibilities: (1) strain gages could be installed on the test item with the strain signal transmitted to a system producing constant strain on the test item even if cracks occurred, (2) an instrumented motor shaft which could sense the changes in compliance as the specimen cracked thus controlling the strain rate, and (3) computer controls on the closed loop testing machine or motor which could be used to cycle through service and overload conditions.

The possibilities appear virtually limitless for structure-control interaction and the loading conditions that can be imposed as well as responded to through use of computers, improved diagnostic devices attached to the test items, the testing machines, and, perhaps, materials which have built-in diagnostic capabilities.

Because the cost of new equipment sometimes exceeds operating budgets, many laboratories are faced with making improvements and modifications on existing equipment. Many laboratories which have the older tension or compression testing machines will find their capabilities have been greatly enhanced by computer controls and feedback devices. Installation of improvements in older equipment can be a cost effective performance enhancement alternative to acquisition of new equipment.

Among the hazards associated with this type of equipment (whether old, refurbished, replacement, or new) are:

1. The sudden release of large forces and energies

2. The many pinch points, and related hazards to hands and arms.

3. The high pressure hydraulic systems—fittings can become damaged, hoses become worn; sudden release of these high pressures will create personnel hazards and environmental concerns (the release of hydraulic oils into laboratory plumbing drains).

4. The mechanical gears and electric motors need to be isolated from people.

5. Lifting and rigging heavy items into and out of the test spaces within these testing systems because spaces are not readily accessible with overhead cranes. Portable boom cranes, fork lifts, and elevating tables are commonly used to lift items into position.

6. Often these testing systems are placed so that the areas around them are congested, have trip hazards present in the form of hydraulic and control lines, power cords, data acquisition and instrumentation cables.

Dead Weight Testing Machine

The dead weight machine consists of a series of weights which are applied directly to the test item or having the forces induced through one or more lever arms which multiply the forces. Typically, these machines use lever arms to achieve the required forces on the test items. A typical use for dead weight testing machines is in creep testing in which the forces must be maintained accurately over long periods of time. Another example of the use of dead weight testing machine is in force calibration (Ref. 6-1).

In Fig. 6-6, a lever type dead weight testing machine is shown. The weights are stacked on the elevator on the right side. The weights apply a tension force to the train and lever arm above. The arm is pivoted about the fulcrum. The resulting moment is resisted by the tensile force in the

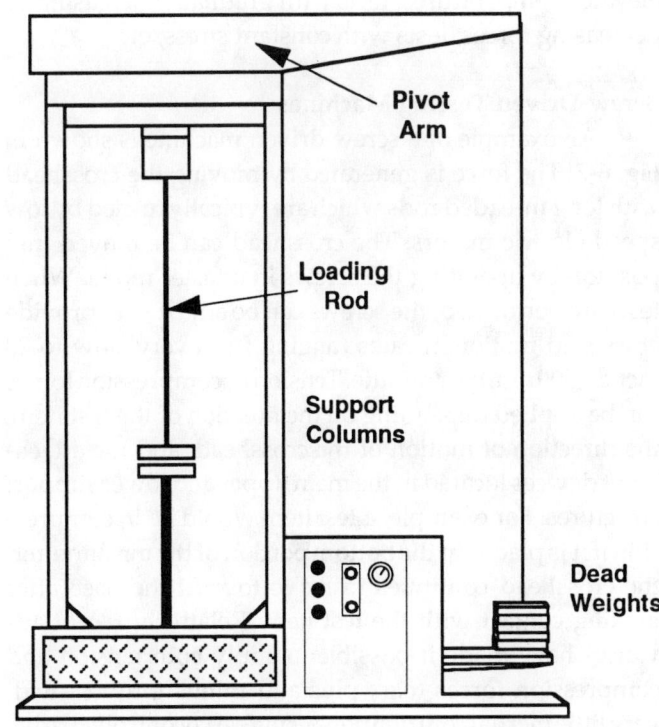

Figure 6-6. Dead Weight Testing System.

left load train. The test item or specimen can be subjected to tension or compression forces depending on how they are attached to the load train. Tension forces are applied by direct attachment to the load train. The applied tensile forces are resisted by the base plate and columns. Compression forces are applied through fixtures attached to the load train. A typical example of a fixture is a "compression cage" in which the test item is placed between two plates. The lower plate is attached to the upper tension rod. The upper plate is attached to the lower tension rod. These two and other similar plates have hole patterns which allow the load paths to pass freely.

To operate the machine, the test item is positioned in the machine when there is no load on the right load train. The elevator is positioned above the foot on the right load train. The weights typically have a hole in their center and they are positioned around the load train. The desired amount of weight is placed on the elevator which in turn is lowered past the foot transferring the weights to the right load train. The force applied to the right load train is multiplied by the ratio of the distances from the pivot point to the test specimen giving the force applied to the test item and left load train.

The weights can also be applied directly to the test item in a different type of machine without multiplication of the forces. Variations of these machines also include capabilities to perform stress relaxation tests, low and elevated temperatures, tests with gradually increasing or decreasing forces, tests with constant stress, etc.

Screw-Driven Testing Machines

An example of a screw-driven machine is shown in Fig. 6-7. The force is generated by moving the crosshead with long threaded rods which are typically turned by low speed electric motors. The crosshead can be moved into position by operating the screws in a faster mode. When tests are performed, the screws can be adjusted to provide crosshead motion at rates ranging from very slow to 20 inches (500 mm) per minute. Tension or compression forces can be applied depending on the position of the test item, the direction of motion of the crosshead, and the attachment devices located in the main upper and lower support structures. For example, a test item would be in compression if it is placed in the bottom portion of the machine and the crosshead continues to move toward the base after making contact with the test item. Relatively recent advances have made it possible to apply both tension and compression forces (e.g., plus and minus torques) with very little inertial contributions using servocontrolled drive motors and linear direct current amplifiers.

The motors can be manually controlled with potentiometers, with function generators, and with interactive computer software. The motors and testing frame are protected with limit switches so that the movable crosshead does not become bound against either end and to prevent jamming the screw threads. Other principles of good mechanical design have also been included such that the use of backlash nuts which prevent added frictional loads in the screws and preloads used when the frames are assembled to reduce the compliance of the system. The screws are often protected with flexible covers to keep the work area clean and to prevent entanglement of parts.

Tension or Compression Hydraulic Testing Machines

These machines were originally called universal testing machines because they supply tension and compression forces one at a time but cannot cycle between each force mode. The term universal testing machine is obsolete now because the newer machines are capable of tension, compression, and have other improvements (e.g., feedback/control systems). However, the machines are not obsolete because they were so well built that they will provide many more years of additional service. Further, with upgrades and improved controls, these machines can be retrofitted to provide very useful and competent test capability with an example shown in Fig. 6-8. Figure 6-8 shows a typical two post system. The hydraulic cylinder provides a force to the table. The table is attached to the upper crosshead by long columns and the applied forces can be transmitted to the upper crosshead. The movable

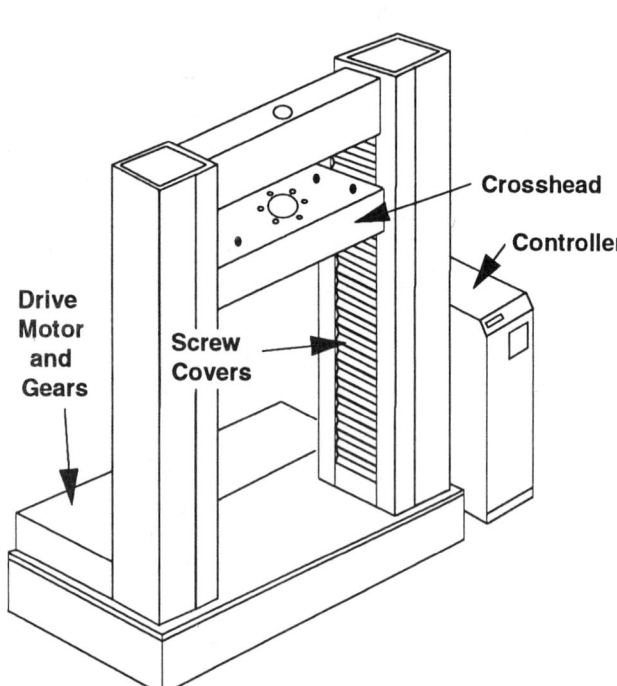

Figure 6-7. Screw-Driven Testing System.

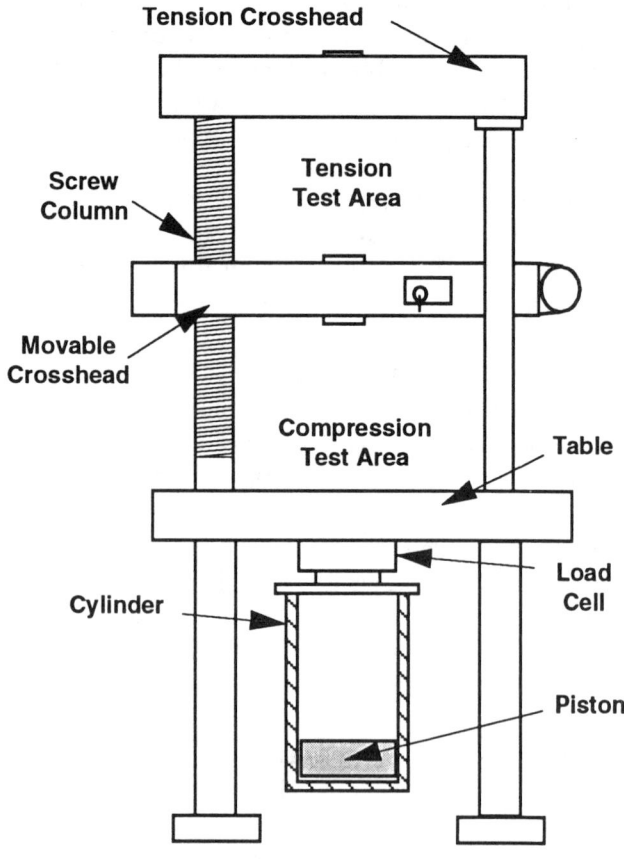

Figure 6-8. Hydraulic Testing System

crosshead is positioned for the test and, when stationary, provides the reaction point for the forces generated by the hydraulic cylinder. The tension space above indicates that the upper crosshead is moving away from the positioned movable crosshead during a test. The compression test area indicates that the table is moving toward the positioned movable crosshead. These machines typically have jaws and chucks imbedded in their crossheads. These jaws and chucks have tension force capabilities equal to the maximum forces that can be generated by the machine.

These machines usually have 3,000 psi hydraulic cylinders located in the base of the machine. The hydraulic cylinder is machined to a tight tolerance within its case or capsule. A certain amount of fluid is allowed to flow through the tolerance zone to lubricate the cylinder. The fluid thus used is returned to the reservoir to be filtered and recycled into the pump. The cylinder travels in one direction only and since it is not sealed, it can apply only compression forces on the top or table which precludes the ability to produce cyclic tension and compression forces on a test item.

The control of these machines is typically through manual operation of pilot wheels which control the flow of pressurized hydraulic oil into and out of the cylinder. One wheel is used to control the loading and one wheel is used to control unloading. Newer versions of these machines have been equipped with controls complementing manual operation so that load, stroke, and strain capabilities can be obtained.

Servohydraulic Testing Systems

These testing systems are similar to the tension or compression testing machines in that the applied load is transmitted hydraulically. The use of servocontrols provides added test features when using this equipment. The hydraulic cylinder is sealed so it is capable of displacement in two directions so that both tension and compression forces can be applied. These systems are also equipped with built-in displacement gages (usually LVDT's) so that the displacement of the actuator can be defined as well. Feedback controls are also available so that the load, stroke, or strain on a test item or specimen can be measured and used as part of the machine control. A servohydraulic example of this machine is shown in Fig. 6-9. The crosshead can be moved manually, by using the hydraulic actuator and positioning blocks, or by hydraulic means. The hydraulic pump (not shown) may be built into the testing system or located at some other place for easier servicing, noise reduction, or space limitations.

These testing systems are equipped with load cells, displacement transducers, and feedback systems so that interactive control of load, stroke, and strain can obtained.

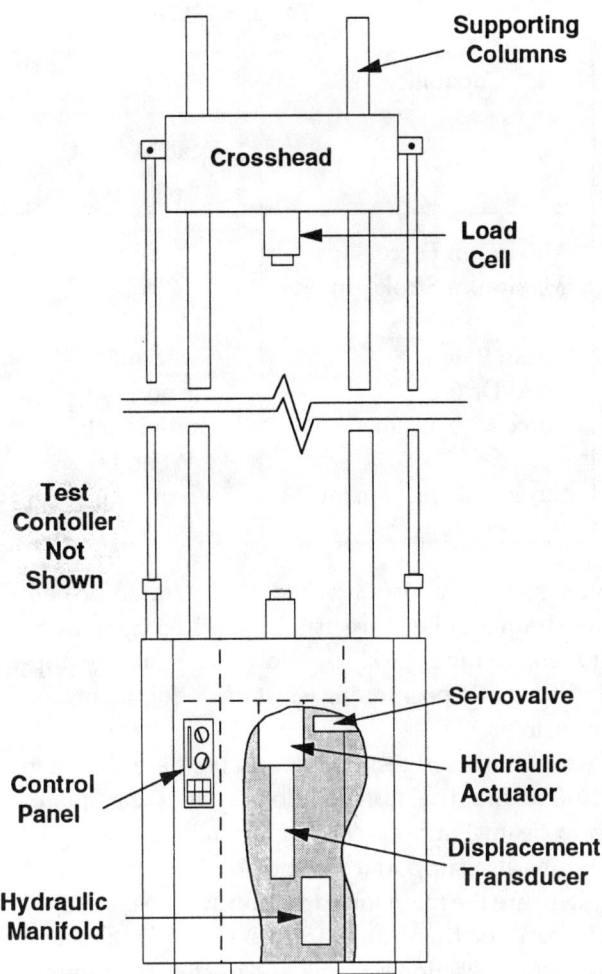

Figure 6-9. Servohydraulic Testing System

Strain control is obtained through extensometers or strain gages attached to test items. Some of the machines have the hydraulic actuator positioned in the crosshead while others have the actuator in the base of the machine as shown in Fig. 6-9. These systems are capable of closed loop operations because of the interactive controls, servovalves, and transducers. They can also be controlled using function generators or computer software interfaced with the hydraulics. A brief comparative summary of testing system capabilities is given in Table 6-1.

Structural Frames and Test Beds

Structural test frames and beds are used to accommodate large test items, test hardware having multi-axis loading requirements, large off-axis loading requirements, and other unique test requirements. A test frame or bed may be relatively small (e.g., desk top size) or large (e.g., building size). Test items and hardware are attached to the bed or frame. These beds or frames usually have a stiffened back or force reaction system. This back may be permanently attached or portable. Other beams, columns, and

Table 6-1. Comparison of Testing System Capabilities

Capability	Type of Testing Machine			
	Dead Weight	Screw-Driven	Tension or Compression	Tension and Compression
Maximum Force, kips	50	200	10,000	10,000
Maximum Stroke, in.	5	Machine size minus crosshead	16 - 20	16 - 20
Strain Rate	variable	low	low	high
Load Drift	none	+/-0.1%	+/-0.5%	+/-0.2%
Force Measurement	Individual Weights	Load Cell	Bourdon Tube or Load Cell	Load Cell
Stroke/Displacement	Test Item	Cross-head	Cross-head	Actuator

reaction systems are also attached to the frame or bed. These test frames or beds are usually serviced by an overhead traveling crane, forklifts, and other related equipment for lifting and positioning test items, equipment, and loading systems.

Two examples of test frames and beds are shown in Fig. 6-10. The first is a stiffened floor system (e.g., plates welded to beams) supported by four columns. The two rear columns also support a stiffened back. This back and floor system are the main force reaction systems. The only movable parts of this frame are the side rails and any portable beams, reaction systems, and loading equipment. The floor, beams, stiffened back, and columns have through or threaded holes on a uniform spacing to permit bolted attachment of test items and hardware. This type of frame has the appearance of a giant erector set and is used many different ways to accommodate test items and equipment. It is not uncommon to have this type of frame modified and added to in many ways. Many of these frames have had extra plates, stiffeners, and beams welded to them, holes have been drilled for attachment of loading equipment, test items, and other modifications for test purposes.

The second example consists of a stiffened floor system. This floor can have through holes, threaded inserts, slots, rails, or tracks embedded in its floor for attachment of test items, equipment, and loading systems. Testing machines and systems have also been integrated into these floor systems. The example floor system with an integrated testing machine is shown in Fig. 6-11. These larger stiffened floor systems have been used to test full scale bridge beams and structural sections as shown in Fig. 6-11, aircraft wings, fuselages, and tail sections, and a variety of other large structural systems. An example of an aircraft wing subjected to vertical loading conditions is shown in Fig. 6-12.

Pressure Systems and Vessels

Pressures for test purposes are typically applied through two primary media: fluids and gases. In either case, pressures must be applied to the fluid or gas which in turn transmits the pressure directly or indirectly to a test item. Pressure systems consist primarily of pumps, intensifiers, accumulators, controls, and safety devices. A pump is a relatively simple mechanical device in which low pressure (e.g., ambient air pressure) is applied to one side of a piston while the other side of a piston acts on a

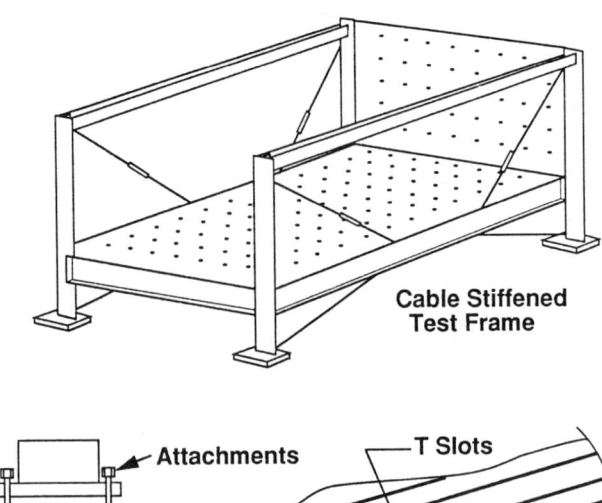

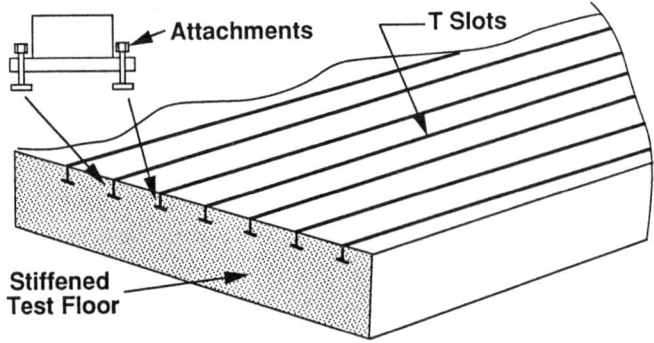

Figure 6-10. Test Frames and Test Beds

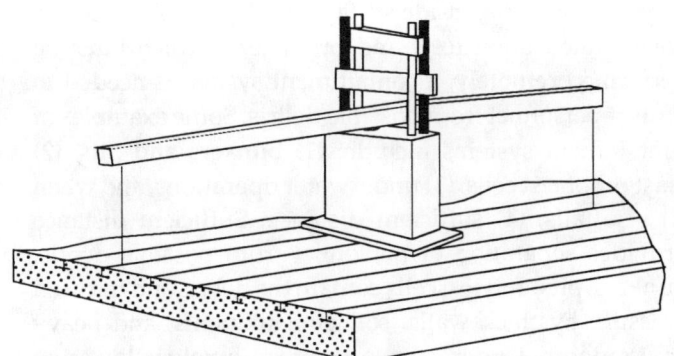

Figure 6-11. Testing Machine Positioned in Stiffened Floor System

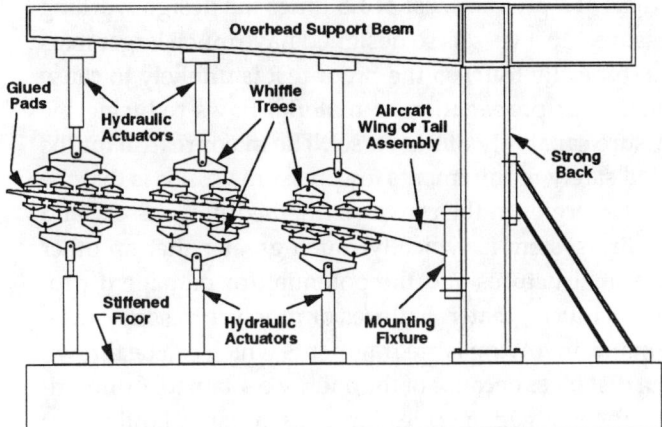

Figure 6-12. Aircraft Wing Subjected to Load Tests on Stiffened Floor; Vertical Forces Applied Through Whiffle Trees.

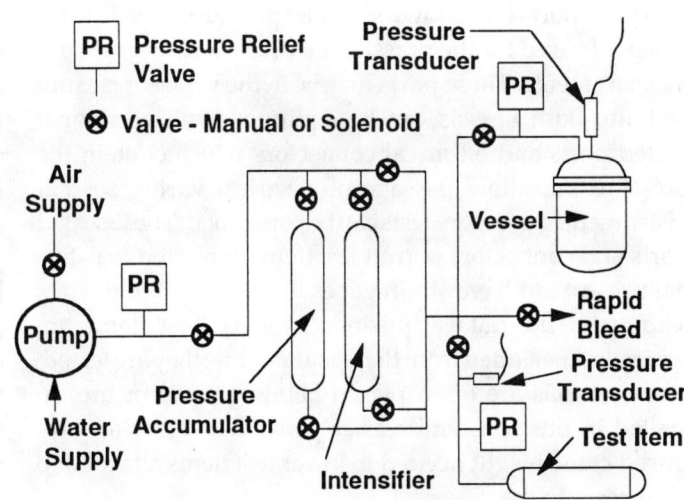

Figure 6-13. Pressure System Layout

relatively small volume. The piston is moved by pressure acting on the low pressure side creating high pressure downstream. A check valve allows passage of the high pressure. A bleed valve or opening in the low pressure chamber of the pump relieves the driving pressure allowing the pump to recycle. Each cycle of the pump increases the pressure on the outlet side. The pressure is restrained by a check valve. The area ratio of the inlet side over the outlet side provides an increase in pressure on the outlet side of the pump. This cycle can be repeated until very high pressures (e.g., 150,000 psi) can be reached. An example pressure system is shown in Fig. 6-13.

Each cycle of a pump produces a small step function increase in pressure depending on the volume of the pump. Intensifiers and accumulators can be installed on the outlet side of the pump. The purposes of these devices are to obtain higher pressures than the pump can deliver and to smooth out the steps as the pressure increases. An intensifier is similar to a pump but there is not an automatic cycling in each step. Input pressure can be brought to one side of the intensifier and the area ratio principle used to increase the output side of the intensifier. An accumulator is a type of pressure reservoir which can be store and then release a supply or pressure to a test item.

Control of pressure systems can be through manual valves, through air or electrically operated servovalves, and can be adapted to computer controls and function generators. It is possible for the output of a hydraulic testing machine (the pressure in the hydraulic system) to be the input (low pressure side) of a pump so that higher pressures could be obtained. (Typically, the area ratio for input to output sides of a pump or intensifier is 10:1.)

The pressure vessels are used in two basic ways: (1) to provide a pressure source for processes, operating and testing systems, and other applications requiring an uninterrupted pressure supply and (2) to contain the pressures applied to test items. Examples of commercially available pressures sources which have pressure vessels include air compressors and systems using U.S. D.O.T. gas bottles. There are many different types of pressure vessels; two are shown in Fig. 6-14. On the left is a vessel with a bolted closure with two ports (not shown) in the head of the vessel. The removable closure permits access to the vessel so test items can be placed inside. The ports are through holes in the vessel wall or closure. The ports are needed for the inlet pressure supply, an outlet for removing air from the vessel, and for pressure measurement.

Figure 6-14 also shows a vessel with an elliptical removable closure (head). This closure is lowered onto a set of O-rings. The closure is tightened by pulling the tapered clamping rings toward the vessel circumference using the ratcheting turnbuckles. This clamping action seals the vessel along the O-ring surfaces. This vessel will also need ports (not shown) for inlet pressure, bleeding air if water is used for the pressure medium, and for pressure measurements. These ports may be in the vessel or closure or both. Both vessels can be equipped with additional sealed ports and electrical connectors which contain the pressure but allow passage of electrical wiring so that instrumented pressure tests can be performed. These sealed ports and connectors permit test items subjected to external pressure to have strain gages installed with the wires leading to the data acquisition system. Test items are typically suspended from the closures while they are tested. These vessels are often placed below ground or are installed in pits to permit easier personnel access and reduced crane height needed to lower test items attached to closures into the vessels.

The design and certification of pressure vessels for manned operations comes under the guidance of the applicable sections (III and VIII) of the ASME pressure vessel codes or equivalent requirements. Many commercial suppliers offer fittings, tubing, valves, manifolds, and other equipment needed to build pressure systems. The equipment supplied comes in various operating pressure ranges including the very high pressure ranges in excess of 100,000 psi.

In terms of containment, there are two concerns: pressure and debris from tests in which failure occurred. Normally, a vessel used to contain pressure would not be used as a containment system for failure tests particularly if debris from a failed test item is expected. Typically, failure tests would be conducted in some other type of containment system. However, debris from tests could impact the wall of the vessel and initiate cracks and other degradation of the inside surfaces of the vessel can occur. For planned burst tests and proof tests which must be performed remotely, a containment system is needed to protect personnel from possible debris. Some examples of containment systems include: (1) bunkers and pits, (2) blast proof test cells, (3) underwater operations, and when all else fails, (4) sufficient distance. Sufficient distance provides separation of personnel from possible debris. Bunkers, pits, and test cells contain the debris and released pressure by thick walls, soil embankments, and heavy access doors. Underwater operations involves lowering test items into the water and using the water to absorb the energies associated with ejected parts and released pressure.

Certification of pressure systems and vessels usually consists of a pressure test at 1.5 times the design working pressure. Most vessels so designed have much larger margins (typically four) so the proof test is unlikely to cause initiation or propagation of material flaws reducing the pressure capability of the vessel. The major reason for the added safety requirements for pressure vessels is that the stored energies in the pressurizing medium, vessels, and pressure system is typically much greater than in other types of structures and the potential for damage if ruptured is much greater. Failures of pressure vessels can be dramatic, involving large fragments which can be thrown great distances because of the energy contained. Comparisons of contained energies for a gas versus a liquid were described briefly in Chapter 5, Table 5-3. Further, there have been enough failures involving vessels that they require greater consideration in design and certification than other types of structures. For those systems (typically above 3,000 psi) which do not have design margins of four based on ultimate strength, then additional procedures (e.g., detailed analyses, material tests results and certifications, certified fabrication procedures with special emphasis on machining, forming, and welding, and testing) are needed to certify a pressure vessel for manned operation.

Pressure systems are protected using pressure relieving devices. A variety of devices are used: (1) burst disks designed to fail at specific pressures, (2) spring loaded valves which allow passage of fluid or gas when the force induced by the pressure overcomes the preload in the spring, and (3) other types of mechanically operated systems designed to open at or above specific temperatures. These devices should be set, installed, used, and periodically inspected to ensure safe operations involving pressure systems and vessels (Refs. 6-10 and 6-11).

The nemesis of pressure systems and vessels is the presence of leaks. Leaks can occur in a myriad of places—welds, seals, connections, fittings, etc. Leakage involves the location(s) and rates. Finding the location ranges from

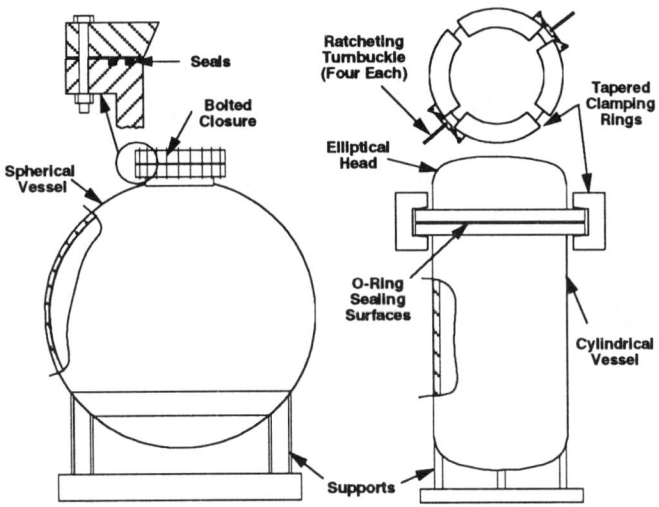

Figure 6-14. Pressure Vessels: Spherical and Cylindrical with Elliptical Ends

the obvious hole to a leaking fitting to sophisticated techniques (e.g., helium leak detection). The importance of the location ranges from a fitting needing tightening (easily remedied) to a hole in a vessel wall (with remedies ranging from rewelding to redesign). A leak can be observed by a drop in pressure while attempting to maintain a constant pressure. Leak rates can be determined by the drop in pressure over time to more sophisticated methods involving accurate flow measurements.

The occurrence of leaks during a pressure test cause the following problems: (1) an evaluation of the locations—ranging from a repairable loose fitting to a critical weld which may not be amenable to repair, (2) once located and evaluated, should attempts be made to pump past the leak and assume that the leaks (in noncritical areas) can be repaired later, and (3) modify the test requirements based on reduced performance of the vessel and/or system during a test. Repair of leaks can range from tightening a fitting to repeated grinding and rewelding and retesting.

Performing pressure tests should involve the following:

1. Determine the vessel and/or system to be tested and the test objectives.

2. Perform energy calculations using Eq. 5-1 and 5-2. These energies indicate the levels of protection and the type of facility needed in order to perform the tests safely (e.g., barriers, remote operations, capability of containment systems).

3. Evaluate options—test using liquids whenever possible to reduce the energy in the pressurizing medium (see Chapter 5, Table 5-3) and facilities needed versus required.

4. For units externally pressurized, provide some type of posttest pressure relief device (valve, removable plug) to vent any pressure accumulated during testing.

5. Check for leaks when the test item cannot sustain pressure—do not use your hands except for very low pressures (e.g., less than 30 psi). One possible procedure is to pressurize the system to some value, then reduce the applied pressure by fifty percent and check for leaks.

6. Provide other safety measures—personnel access denial, restraints on pressure lines in case of failure of test item, fittings, etc.

Thermal Systems

Thermal systems for structural tests basically involve three types of systems: temperature controlled ovens for steady state or programmable temperature conditions, radiant heat systems for rapid temperature rises and temperature gradients, and burn facilities to create maximum temperatures and durations of fires. The ovens rely primarily on convection, conduction, and radiation as the means of heat transfer to the test item. The radiant heat and burn facilities rely primarily on radiation as the means of heat transfer.

Temperature controlled ovens typically operate between −100 °F and 1000 °F. They can control temperatures within a few degrees for nearly steady state conditions. The ovens can also be programmed so that temperature cycles can be applied to test items. These ovens are typically heated with electrical resistance heaters, heat lamps, etc. Ovens can be used by themselves, installed in testing machines surrounding a test item, or on test frames and beds. In addition, most ovens can have other conditions introduced such as humidity, salt fog, or other environmental or climatic conditions (Ref. 6-5).

Radiant heat facilities typically use some type of quartz lamps which heat up rapidly when activated electrically. These facilities are capable of heat fluxes as large as 400 Btu/ft-sec and in rise times of two or three seconds. The amount of heat needed for a test depends on the test item, its size and materials of construction including emissivities, and on the design or service requirement that is to be duplicated.

A burn facility implies the combustion of flammable materials, typically hydrocarbons. Various types of burn facilities exist including: a chimney like facility, an open pool, an enclosed pool, and an air curtain burner. These facilities can have the supplies of air and fuel mixed so there is an ideal burn (stoichiometric), fuel rich, or air rich. Rich meaning there is an excess supply of fuel or air in comparison with ideal burn conditions.

A chimney facility is constructed much like an ordinary fireplace with the test item positioned above the point where the fuel and air are combined and ignited. A pool facility consists of a lake or pond of fuel with the test item positioned near or above the surface of the fuel. An enclosed pool is similar except that a barrier surrounds the pool limiting or controlling the amount of air that can combine with the fuel vapor near the surface of the pool depending on the position of the barrier. The temperatures in these burn facilities range from 1,400° to 2,400°F with a variation of about +/− 100°F. The rate of heat production is limited to about 20-30 Btu/ft -sec. The temperatures are determined with strategically placed thermocouples on the test item and in the fire. A typical fuel used is JP-4 for jet aircraft.

Modal Analyses

The purpose of this type of equipment is to excite a structure to determine its mode shapes and frequencies.

The loading conditions must be dynamic so that relative motions are created within the structure or component. Among the types of equipment used are: instrumented hammers, vibration exciters, forced or constrained displacements, etc. An instrumented hammer consists of an ordinary hammer with a piezoelectric force transducer installed as shown in Fig. 6-15. The force transducer is calibrated so that the impact force-time history can be measured. The hammer tip is interchangeable so that different frequencies can be excited. With accelerometers and strain gages strategically placed on the test item, the excitation caused by the hammer blow will cause the structure or component to respond dynamically which is measured by the gages. Analyses of the data will yield the mode shapes and frequencies.

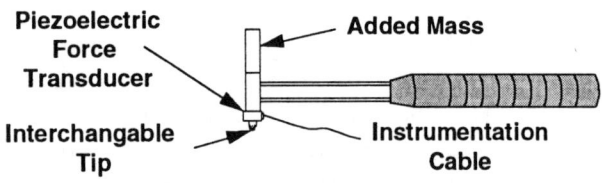

Figure 6-15. Instrumented Hammer Used for Modal Analyses

The structure can also be excited by an appropriately mounted vibration shaker. Similar instrumentation would be used and the analysis of the data gives the mode shapes and frequencies. A forced or imposed displacement can also cause a structure to respond. For example, a cantilever beam could be preloaded into a deflected shape. If the preload is transmitted through an explosive bolt or some other type of restraint which can be released rapidly, then the structure will respond dynamically until the vibrations damp out.

Modal analyses also consist of extensive data analyses using computer based software systems. These analytical methods used to interpret the data give the mode shapes and frequencies. The details of modal analyses are given in Chapter 4 (Ref. 6-4).

Inertial Systems (Centrifuges and Spinners)

Centrifuges are used to create uniform acceleration conditions or gradients along a test item. Centrifuges are devices which have a rotating arm attached to a drive mechanism. The inertial forces are created by the combinations of the radial and tangential accelerations given by:

$$a_r = \omega^2 r \text{ and } a_t = r \alpha. \quad \text{Eq. 6-1 and 6-2}$$

where a_r is the radial acceleration and a_t is the tangential acceleration, ω is the angular velocity, r is the radius to the point measured, and α is the rate of change of angular acceleration. For nearly steady state conditions, a_t is zero and the inertial force in directly proportional to the mass of the test item and the radial acceleration (angular velocity).

Centrifuges come in a wide variety of sizes for testing structures and components ranging from a radius of nearly zero to a maximum of about 35 feet. The accelerations vary from slightly over one g to more than 2,000 g's. An example centrifuge used for structural testing is shown in Fig. 6-16. This centrifuge has the drive motors located under subflooring in the centrifuge pit. A containment structure is built around the perimeter of the centrifuge. A dynamic balance compensator is built into the two arms to allow for differences in the calculated balances required for operation. This compensator permits some vertical movement of the support arms while the centrifuge is rotating. The centrifuge also has slip rings so that data acquired during a test can be transmitted to data acquisition and recording systems. These slip rings also provide a means to operate test items (e.g., controls, actuators, electrical systems). Larger centrifuges are usually serviced by an overhead crane.

Spinners are usually air-driven turbines placed in a fixture. The fixture is usually positioned over a sealed pit. The pit can be evacuated so that little or no air resistance is encountered while the spinner is in operation. Test items are attached to the turbine drive motor through fixtures and adapters. Spinners can operate at no-load speeds up to 40,000 rpm and greater. Spin testers are often used to evaluate the performance and operation of high speed rotating assemblies and systems (e.g., artillery shells, jet and turbine engine components—blades, compressors, and drive motors).

Among the assumptions made in using centrifuges and spin testers is that the test item will not fail. In order to

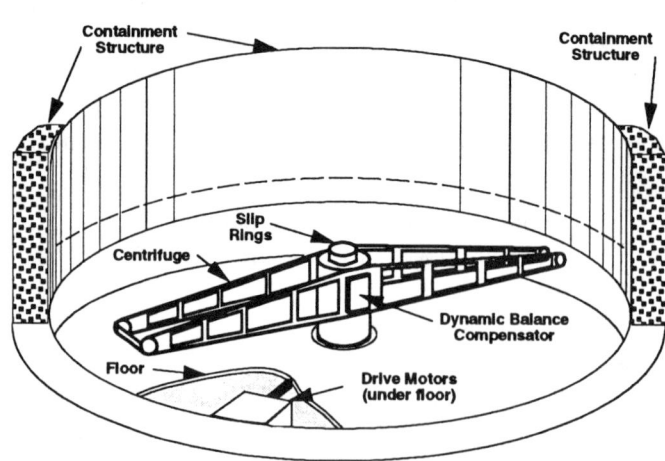

Figure 6-16. Centrifuge

rotate, centrifuges and spin testers must be dynamically balanced. If failure occurs, then the dynamic balancing of the arm allowing rotation is upset when the structure deforms or changes position as it fails. When the dynamic balance of a centrifuge or spinner is upset, there is a danger of damaging their bearings, containment systems, slip rings, counter systems, drive system, etc. A few centrifuges have been designed to tolerate ejection of or at least gross displacement of test items. However, this is not the usual case so that centrifuges usually need to be protected from displacement of test items. Among other features which can be incorporated in centrifuges and spinners include temperature chambers, vibration shakers, control and diagnostic information transmitted through slip rings. (Details on slip rings are available in Chap. 10.)

Vibration Systems

Vibration equipment consists of three basic types of exciters: electromagnetic, hydraulic, and mechanical. Examples of each of these vibration test systems are given in Fig. 6-17. These exciters can produce steady state, transient, and random vibrations depending on the controls used. Each type of equipment has advantages (e.g., frequency ranges) and some disadvantages (e.g., test item size) as will be explained in this section. There are two ways vibration test equipment are used: (1) with items attached directly to a shaker and (2) an exciter used to vibrate large structures such as buildings or bridges. Test items are typically rigidly attached to a plate or mounting fixture which is connected directly to the shaker. The motion of the shaker is transmitted to the fixture and the attached test item. For test items much larger than the plates or mounting fixtures, portable exciters and driver systems (power sources) have been developed which can be installed at various locations on the structure or test item. Clever placement of these portable shakers can cause vibrations and resonances in the structure.

The mechanical exciters use cams or a variety of other devices (e.g., roller cranks, variable springs, eccentrics, unbalanced masses) to induce small varying displacements to the table supporting the test item or directly to the test hardware. The hydraulic systems are similar to a testing machine with the motions of the support or base plate caused by moving the hydraulic fluid rapidly through servovalves (Refs. 6-3, 6-4).

The electromagnetic exciters move a magnetic field coil suspended in a driver coil. The driver coil is attached to the shaker support structure through flexures to minimize interactions of the field coil and the support structure. As the magnetic field changes, small displacements are induced in the field coil which cause vibrations in the attached test items. The motion of the field coil is very similar to voice coil arrangement in a loud speaker system. Vibrations can also be imposed on structures by acoustic systems which can operate at 150 dB or more. Many flight structures have requirements to pass acoustic induced vibrations simulating missile lift off and engine noise.

Steady state vibration consists of sinusoidally varying displacements of the plate or mounting fixture. Transient vibration is defined by a sinusoidal excitation multiplied by an exponentially decaying signal. Random vibration is defined by the superposition of many sinusoidal signals and exponential increasing and decreasing signals. Examples of these three basic types of acceleration-time histories are given in Fig. 6-18.

Vibration testing consists of four basic parts:

- Specification of the vibration environments (sine, random, frequencies, magnitudes, duration)

- Attachment of the test item or structure to the properly selected vibration machine through the use of fixtures and preloading techniques

- Selection of the diagnostic devices and their installation at critical and/or informative locations

- performing the test including the data reduction and analyses.

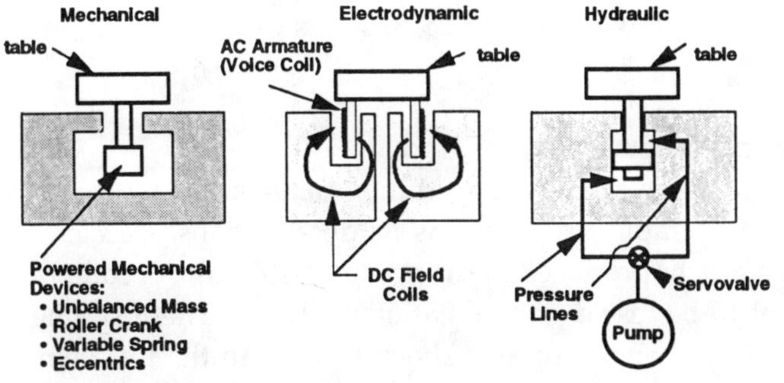

Figure 6-17. Vibration Testing Systems

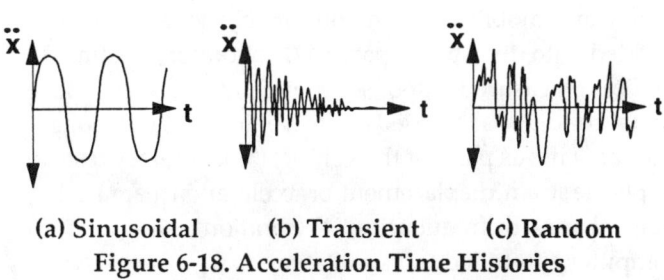

(a) Sinusoidal   (b) Transient   (c) Random

Figure 6-18. Acceleration Time Histories for Vibration Environments

These four parts depend in varying degrees on the equipment: the input depends on a particular machine and/or laboratory capability, the attachment depends on the basic machine, modifications made to it, and proper fixtures, the diagnostic devices chosen depend on the data acquisition and control systems, and the analysis of the data depends on the type of data and the computer systems and software available.

Vibration test equipment can be summarized in the following manner:

1. The frequency range delivered by the shaker to the test item is inversely proportional its the force capability.

2. Mechanical shakers can be used for low frequency tests (up to 100 Hz).

3. Hydraulic shakers can be used up to 2000 Hz.

4. The upper frequency ranges (well above 2000 Hz) can be reached only with electromagnetic shakers which cover all ranges of frequencies that can be achieved by test equipment.

5. The largest forces can be delivered by hydraulic shaker systems.

6. Electromagnetic shakers are typically limited in displacement to about one inch.

7. Hydraulic have more displacement capability than electrodynamic shakers.

8. Mechanical shakers have the most displacement capabilities—about 6 inches.

9. Hydraulic and mechanical shakers perform sinusoidal vibration tests but the precise controls on complex random or transient wave forms is difficult. Advances in electronic controls and servovalves have improved these controls will facilitate random vibration tests on future hydraulic vibration systems.

10. Electromagnetic shakers and acoustic systems can produce complex random motions in test items.

An example of vibration testing (including fatigue) is contained in the description of loading equipment used for testing automobiles. Automobile or vehicle testing can be divided into five basic parts: (1) laboratory testing to replicate measured responses from field tests (reproduce actual conditions imposed on the vehicle), (2) component tests on various parts of the vehicle, (3) laboratory tests to duplicate strain, displacement, or acceleration magnitudes, pulse shapes, or frequencies, (4) vibration testing, and (5) computer controlled test simulations.

An example of a testing system used in the development of automobiles is shown in Fig. 6-19. The automobile is supported on four spindles. These spindles are coupled to hydraulic actuators which supply forces to the spindles. The force time histories can replicate the force-time histories measured in over-the-road conditions. The limitations of this type of test are found in three basic areas: (1) the validity of the measured test data, (2) the capabilities of the servohydraulic systems (e.g., force capability, frequency response, displacement, acceleration, and velocity requirements), and (3) the ability to compromise between the test expectations and capabilities.

The test capabilities shown in Fig. 6-19 can be expanded to include forces applied in other directions to simulate braking and turning. Tires could be added to the vehicle to include their effects on the response of the vehicle. The evaluation of the developed automobile is in the hands of a potential customer. The customer's feelings toward the response and ride characteristics are subjective at best. These test capabilities ensure that the vehicle can be tested to determine realistic responses and that potential problem areas can be identified and corrected prior to production.

Figure 6-19. Vehicle Testing Example (Courtesy MTS Systems Corp.)

Shock Systems

Shocks can be induced in structures and test hardware by rapid accelerations or decelerations. Decelerations can be produced by dropping the structure or component to be tested onto a stationary target. The resulting velocity change causes a shock loading to the test item. Accelerations can be produced by placing the structure or component in a type of gun. With a rapid creation of pressure from the detonation of explosives or the sudden release of pressurized gases, the structure or component is subjected to a rapid increase in forces causing an acceleration. Shocks can also be produced by preloading a structure into a deflected position and then suddenly release the preload (Ref. 6-4).

Shock testing machines are manufactured by many companies and many laboratories have designed and built special and general use systems. They basically consist of an elevating table for mounting the test item. The table can move to a position above a contact surface. A shock loading is produced when the table is dropped along guide rails onto the contact surface. The shock pulse can be tailored according to the shock absorbing materials used on the contact surfaces, the velocity change produced, and the masses of the table and the test item. Greater velocities can be achieved if long elastic cords attached to the base of the machine are tensioned to provide a pull-down force to the table and test item.

An example of shock machine is shown in Fig. 6-20. The test item is located on the mounting plate. The lifting winch is used to raise the mounting plate to an elevated position. The mounting plate impacts onto a programmer which is specially selected materials (e.g., felt pads, honeycomb) designed to supply a certain shock pulse (acceleration time history) to the component.

Gas guns, horizontal actuators, and other similar devices rely the sudden release of energy (typically pressure) which is captured behind a movable attachment system (e.g., piston) and test item. The release of the stored pressure causes rapid movement of the piston. The resulting acceleration can be the desired shock depending on its rise time. The moving piston and test item can also be impacted into a stationary target, a movable mass, or other type of target depending on the type of shock pulse needed. An example of a gas gun is shown in Fig. 6-21. In this example the test item is attached directly to the piston. The shock pulse experienced by the test item is the same as that given to the piston.

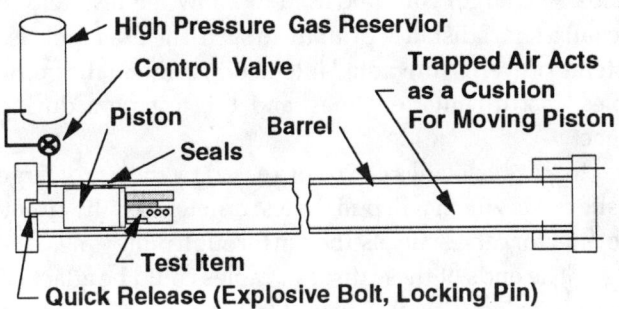

Figure 6-21. Gas Gun

Shock tests are also performed using drop towers and cable suspension systems. A shock testing system used for large structures is shown in Fig. 6-22. This facility consists of a cable stretched between two hills or towers. An instrumented structure to be tested (shown in greatly exaggerated size) is raised up toward the cables. The instrumentation cable travels with the test item. The test item can also have data systems which can be recovered after the test or telemeter data to a receiving station during the test. The left tower also shows a cantilevered section which also has a target at its base.

The test item is typically allowed to free fall. The disadvantage in free fall is that the test item may be acted on by lateral aerodynamic loads depending on the height

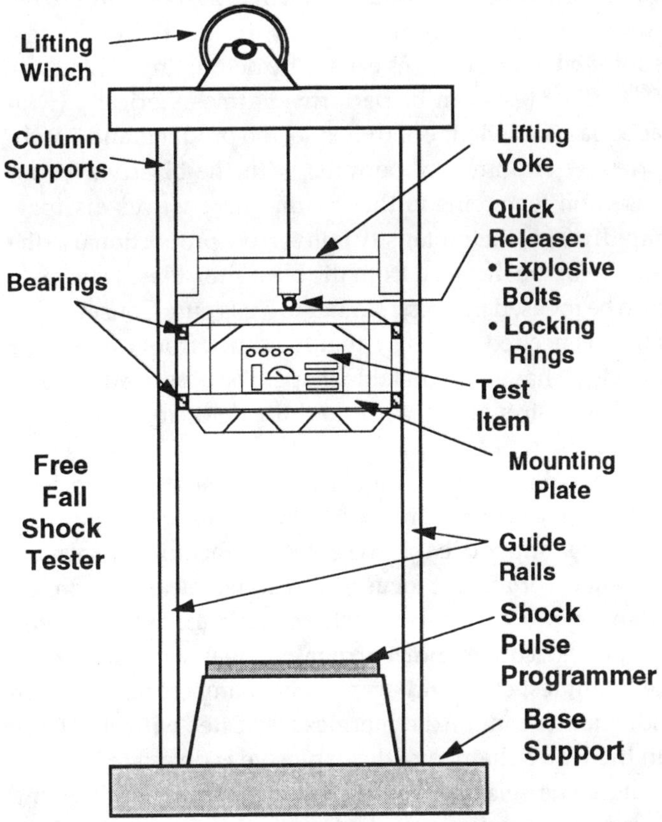

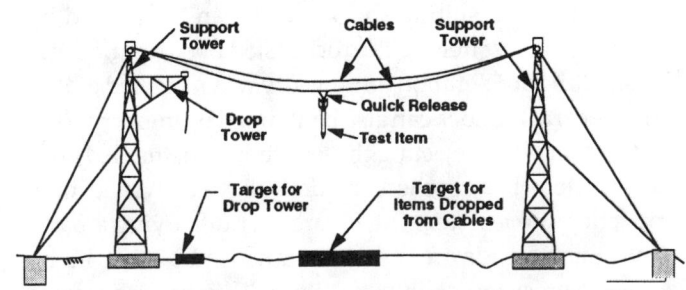

Figure 6-20. Shock Test Machine.

Figure 6-22. Drop Tower/Cable Shock Test Facility

of the tower and prevailing weather conditions. These lateral loads will cause the test item to deviate from its course to the target. These shock tests often use high speed photograph or video coverage which are directed to a specific impact location. Test items which impact at locations other than where the cameras are aimed reduce the effectiveness of the test. The test item can also be attached to two cables (not shown) which are located near the impact point. These director cables guide the test item to a predetermined location on the target in a similar manner to the guide columns in a shock machine. The impact velocity will be reduced slightly in this type of directed free fall because of frictional forces between the vertical cables and the attachment guides. Release systems (e.g., explosive bolts which fracture on detonation of a small directed explosive charge) could be used to allow the test item to free fall a short distance prior to impact. The use of release systems prevents unwanted interactions among the guide cables, instrumentation lines, and the test item during impact.

Impact velocities can be increased by the use of large elastic cords which will pull the test unit toward the target. The director cables can also be run through pulleys near the target. The ends of these director cables could be attached to a rocket sled. This sled would travel along a horizontal path away from the target area. The acceleration time history of the rocket sled could be designed so that it provides the desired tension force in the cable needed to pull the test item to the target at the additional desired velocity without whipping or overloading the cable. The test item would also be released (e.g., explosive cutters) from these guide cables to permit free fall into the target at or very near the desired location for camera coverage. The velocities achieved in these free fall assisted tests are limited only by practical limits of height, forces in elastic cords, or the accelerations delivered by a rocket sled.

Shocks can also be supplied by rocket sled track systems—a sort of open air gas gun. This type of loading system produces accelerations to test items attached to rocket sleds. The sleds are accelerated down a single or dual rail track by the ignition of rocket engines. Velocities of 5000 ft. per second can be achieved on tracks as short as 5,000 feet. The resulting shock pulses can be tailored by attaching the test item to the rocket sled and experiencing the resulting acceleration time history when the sled moves along the track. Shock can also be attained by impacting the sled into an object, ejecting the test item from the sled after the test item has reached the desired velocity and then impacting the test item into a target or deploying a parachute. Shock loads on a test item can also be achieved by accelerating a target into the item. This approach of a moving target allows the test item to be hard wired which usually aids instrumentation and data acquisition. This "reverse ballistic" technique was developed by Sandia National Laboratories.

The purpose of these shock generation systems is to give enough energy or velocity to the test item to expose it to required or representative service, proof, or destructive conditions.

Blast and Explosives

Blast waves can be produced by the detonation of explosives, from overpressures caused by boiling liquid expanding vapor explosions (BLEVE) which have occurred in transportation accidents involving flammable materials, and explosions involving grain dusts, mists of hydrocarbon liquids, etc. Blast waves are created by the detonations and result in overpressures from the expansion of the explosive products. The expansion increases the air pressure surrounding the detonation site. The increased air pressure creates dynamic loads on structures in its path. Tests involving explosions where pressure transients are the important part of the loading condition can best be replicated by other controlled explosions. The rapid rise times and pressure decay are difficult to simulate with other types of dynamic loading conditions. The concern for explosions has prompted other types of investigations of storage and containment systems for explosive liquids and gases. As a result, there have been many types of test structures (e.g., bunkers, missile silos, residential buildings) subjected to blast waves caused by detonating explosives.

Blast tests can be performed atmospherically (with adequate knowledge of the explosion products and appropriate environmental permits) with the blast waves focused on test items in their path. These waves dissipate rapidly (e.g., their intensity is inversely proportional to the square of the distance from the blast site). Blast waves can also be focused in a pipe similar to a gas gun. Considerable effort is needed to ensure that the shock fronts are planar or at least have a predictable shape. The pipe needs to be of sufficient strength to withstand the detonation of the explosives.

Loading systems for blast waves and explosive detonations usually require special facilities. Even for very small quantities of explosives, rugged facilities are needed to contain or at least focus or direct the released energies. Many types of bunker systems such as earth covered reinforced concrete or corrugated steel structures, blast resistant test cells, and even water chambers have proven adequate to contain the energies associated with blast tests. In the water chambers, the explosive is detonated underwater. The released energy is transformed into moving large amount of water which mitigates the release of energy to the surroundings.

Energy Deposition

Energy can be deposited in structures from natural sources such as lightning, from weapons effects such as shrapnel, bullets, implosions from collapsing air bubbles on turbine blades, and the release of a wide range of energies from nuclear effects. The energies released can be simulated in a variety of ways. Lightning can be releases of large amounts of various forms of electrical charges. The frequency ranges can be very large, approaching 50 GHz, the durations can be as long as 200 microseconds, and the electric field can be as large as 5,000 volts per meter.

Specialized facilities have been built for testing structures and components subjected to the deposition of a wide variety of energy forms. There are many unique or nearly unique facilities for these tests. They are often found as part of weapons development programs as well as some industries.

The effects of energy deposition is a two part problem: the first is the effects of how the energy content interacts with the structural materials involved (e.g., absorption, reflection, spallation), and the second is how the structure responds to the reactions the materials have to the exposure to energy sources (e.g., dynamic responses, differential displacements). Testing of structures subjected to rapid deposition of energies would need to address both the materials effects and structural response. Some forms of deposited energy impose high frequency loads on structures and can excite many higher modes of vibration.

6.5 EXAMPLES OF LABORATORY AND FIELD STATIC LOADING CONDITIONS

Parachute Lines

This example shows the main support bridal for a parachute system for the crew capsule of a modern military aircraft. The parachute system ready for testing in a structural test frame is shown in Fig. 6-23. The main line supporting of the parachute is attached to a point very near the center of gravity of the crew capsule. The six upper lines are divided into pairs with each pair connected to an individual parachute. A total of three parachutes control the descent of the crew capsule. Suppose a test request was received to perform a proof test and then a failure test on the main line and the six smaller lines. The proof test demonstrates the adequacy of the current design and the failure test indicates the margins and can suggest possible improvements in the design depending on the failure loads and modes.

The actual loads on parachute lines are dynamic and are different each time the parachute is deployed. Some estimates by computer models indicated that maximum force of about 40,000 lbs. on the main line was expected. A

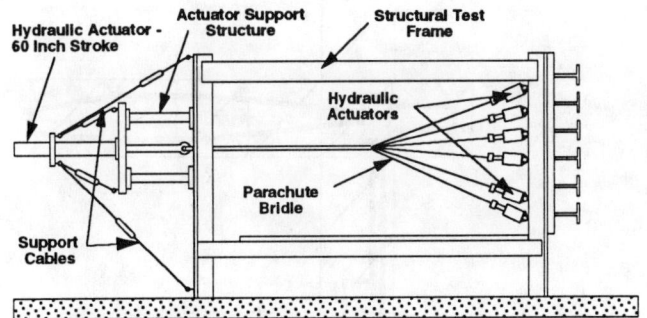

Figure 6-23. Parachute Bridal System Strength Test

static equivalent test was chosen because it is easier to control the loading conditions, to measure the forces in each part of the parachute bridle, obtain the data, and the dynamic loads do not significant change the strength or stiffness of this material in the strain ranges expected. It is assumed in the design that each of the smaller parachutes are required to resist the same maximum force. At the end of each parachute line is a hydraulic ram and a load cell. In the proof test on the lines, the major portion of each increment of applied force was made through the hydraulic ram on the main line. Then, the forces in each smaller line were brought to the same magnitude. This cycle was repeated until the proof load was reached.

In the failure test, the same loading procedure was used. Failure was noted by the slope of the load deflection curve and by the noise from the breaking of individual fibers in the lines.

Space System Payload

A second example of the application of static and dynamic loads is shown in Fig. 6-24 consisting of a space system payload. The payload consists of a strong and relatively stiff truss structure containing particle sampling system. The flight into space consists of a series of lift-off, maneuvering, and separation sequences which create loads on the payload system. In the design of this particular payload, it is customary to perform a series of tests on the prototype designs prior to flight tests, production, and actual flights. A critical part of these tests is the adapter hardware needed to attach the truss structure to the flight structure.

The lift-off and maneuvering sequences produce approximately steady state accelerations, vibrations, and some shocks as stages of the missile separate. In this example, only the simulations for the steady state accelerations will be described. Figure 6-24 shows the payload system ready for testing in a structural frame. These static tests are intended to determine stiffnesses, proof of design, and perhaps failure tests needed to determine design

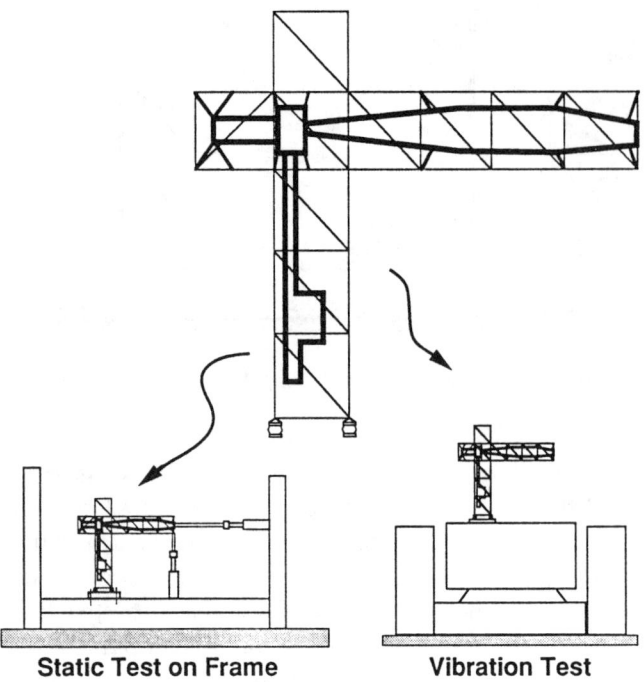

Figure 6-24. Space Payload Ready for Static and Vibration Tests

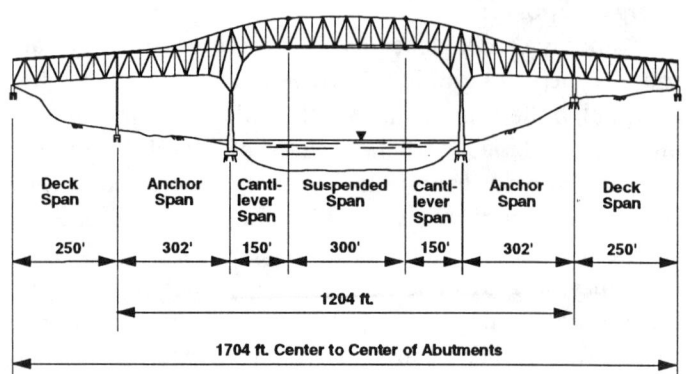

Figure 6-25. Summit Bridge over Chesapeake and Delaware Canal—Seven Spans—Decks, Anchors, Cantilevers, and Suspended.

margins. The prototype sampling device would be included in these static tests only if the sampling structure added significant stiffness and strength to the truss structure or if the sampling structure interaction with the truss structure caused some effects that needed to be evaluated with tests. Two loading conditions are shown in the static test in Fig. 6-24 which could simulate the forces found in flight.

The prototype design unit is also shown ready for tests in a vibration exciter. The purposes of these tests are similar to the static tests: (1) stiffness determination (the mass of the prototype is needed for dynamic response of the test item), (2) proof of design that it can meet the anticipated dynamic loads plus overload conditions, and (3) possible dynamically induced failure conditions ranging from structural joint degradation to anomalies in the operation of the sampling system.

6.6 EXAMPLES OF DYNAMIC LOADING CONDITIONS

Highway Bridge

The first example is the loading for tests on a highway bridge. Bridges are typically designed with the American Association of State Highway and Transportation Officials' H20-S16 loading. The Summit Bridge over the Chesapeake and Delaware Canal is shown in Fig. 6-25. The bridge consists of a five span truss with the structural test involving the three center spans which have an overall length of about 1200 ft. In a study of traffic flow induced vibrations of the supports for the Summit Bridge, the bridge was excited by the side-by-side passage of two trucks traveling at 5 and 30 miles per hour. An example test truck is shown in Fig. 6-26. Table 6-2 below gives the summary of the axle weights actually used and compares them with the bridge design loads. A comparison of the mean or any set of applied loads versus the design conditions shows differences in magnitudes with the greatest percentage difference in the front axle. Note that the AASHTO loading is for a three axle truck and the trucks used had five axles. Further information on the load testing of bridges is given in Chapter 15.

Parachute System Development Testing

The second example of loading conditions used in laboratory tests involves the development of a parachute system for an aircraft. Fig. 6-23 showed the parachute lines being proof and failure tested. To determine how the parachute behaves in service, a series of tests are proposed which simulate the worst case deployment situations. These tests are proposed for a rocket sled track. Prototype parachutes are packed and loaded on board a rocket sled. The sled is equipped with an ejection system (air powered

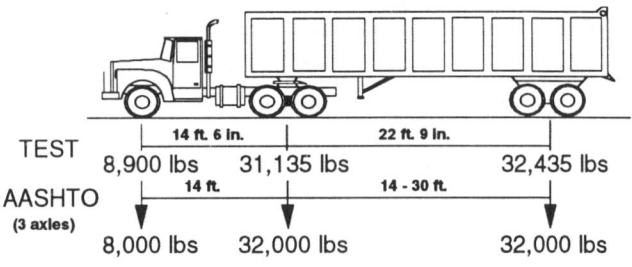

Figure 6-26. Truck Used for Load Tests on Summit Bridge

Table 6-2. Summary of Axle Loadings and Spacings

Truck Number	Front Axle Spacing ft. in.	Rear Axle Spacing ft. in.	Front Axle Weight lbs.	Drive Axle Weight lbs.	Rear Axle Weight lbs.
1	16'6"	23'6"	9,000	30,900	32,700
2	16'6"	23'6"	9,100	31,100	32,600
3	14'6"	22'9"	8,500	31,500	32,600
4	16'6"	23'6"	9,000	31,000	31,500
5	16'6"	22'9"	9,000	31,000	32,400
6	14'6"	23'6"	8,700	31,300	32,000
AASHTO	14'0"	variable 14 to 30'	8,000	32,000	32,00
Mean of Applied Forces			8,900	31,135	32,435

piston) which removes the scaled aircraft or other flight system containing the prototype parachute system and attachments from the sled as shown in Fig. 6-27.

In a typical test, the sled is accelerated along a track to the desired velocity (e.g., greater than the deployment velocity of the prototype parachute), a signal is given to the air piston which releases the scaled aircraft and contents. The parachute is deployed and is tested at conditions similar to the service requirements involving high speed ejection from an aircraft. The speed of the rocket sled can be adjusted by the drive motor and the mass of the sled. High speed video or motion picture coverage diagnostics are usually used during deployment to determine how the parachute actually behaves. If positional measurements of the test item are made as a function of time, then the force-time history acting on the item can be estimated.

Since the force-time history is supplied by the parachute and aerodynamic drag on the test item, estimates of the deployment forces from the parachute can be estimated. These force estimations are important because a parachute behaves somewhat differently each time it is deployed. Having estimates of actual forces significantly aids the parachute system development.

The sled shown is supported on two rails. The sled is stopped by the water brake which picks up water stored between the rails and redirects the water forward.

Vibration Test

An example of vibration testing is that different types of excitations that can be given a single threaded fastener. The purpose of this type of test was to determine what vibration excitation (e.g., sine or random) and the levels that were needed to degrade a threaded connection. A single #4-40 UNC screw attached an annular mass (1.8 kg) to a vibration table. (It would be very unlikely that this large a mass would be attached with just one screw.) Various excitations were applied to the screw in order to determine what levels would be needed to reduce preload in the threaded fastener that had been torqued to 10 inch-lbs (about 80 percent of the guaranteed minimum strength of the screw). It was determined that random vibration

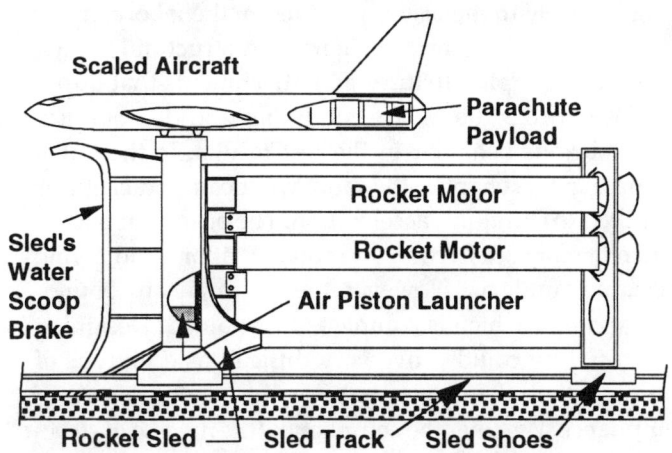

Figure 6-27. Sled Ejector System for Parachute Flight Tests

excitations from 2 to 3000 Hz did not supply sufficient energy to reduce the preload. A sine vibration was then applied. The frequency was varied so that resonances were found. The preload could be reduced by dwelling at the resonant frequency for the threaded connection/mass system. Indication of preload reductions were changes in the frequency response curves.

Pseudodynamic Testing (Ref. 6-9)

Because of the large size and mass of civil engineering structures, it is often impossible to subject large-scale and full-sized to significant dynamic loadings from vibration systems or external forces. Yet, the response of these structures in the dynamic inelastic regime is quite often the objective of the test program.

The pseudodynamic test method has been developed to overcome these disadvantages. The method has been used to test a variety of structures and structural components, ranging from seven-story reinforced concrete frame building to tubular steel frames. Requiring nearly the same equipment as conventional quasi-static testing, it is now possible to subject large, strong, or heavy structures to dynamic loadings. Essentially a computer-controlled experimental technique, pseudodynamic testing uses direct time-step integration to solve idealized equations of motion for the structural specimen. The resulting displacement response is then imposed quasi-statically onto the test specimen, and the resulting forces developed in the specimen are measured and used as the basis for the next calculation. Therefore, the stiffness properties are revised and updated in each time step.

A drawback of the method appears to be a sensitivity to experimental uncertainties and numerical calculational errors that can lead to significant distortion in the final results. However, if sufficient care is exercised in both areas, the results of pseudodynamic testing compares favorably with the results of tests performed on vibration systems, with far less equipment and expense.

Quick-Release Tests

The dynamic characteristics of some structures can be determined easily by pull-back or quick-release tests. In this method, the structure is displaced from its equilibrium position and suddenly released. The displacement time history can be used to estimate the fundamental natural period and damping of the structure. If the initial displacement is large enough, the response of the structure can exceed the linear range. It is also possible to have a large enough force applied that other modes of the structure are excited; that is, the mode shapes are a function of the initial preload.

Ambient Vibration

The response of an actual structure to its normal dynamic loads, such as low speed wind, can provide valuable information that can be used to predict its response to more unlikely events, such as earthquakes. The normal loads on many structures contain many sources of dynamic input, including microtremors, winds, wave action, and man-made loads such as automobile traffic. If the frequency content of these ambient vibrations is sufficiently broadbanded, the response of the structure can be analyzed to determine some of the natural frequencies and mode shapes of the response. Ambient vibration has been used to obtain estimates of the dynamic properties of buildings, pipe lines, bridges, earth-filled and concrete dams, tall chimneys, hyperbolic cooling towers, and offshore structures.

The use of ambient vibration data to identify dynamic characteristics requires a certain number of cautions: (1) the input motions must be sufficiently random and broad-banded; otherwise the structure will respond to the input as steady-state response, (2) because the response amplitudes will generally be small so that the structure will be excited within the linear range of behavior, (3) sufficient response data must be collected to increase the signal-to-noise ratio in the resulting averaged frequency spectra of response, and (4) if the data are collected in digital form, the sampling rate, low-pass filtering frequency, and the maximum Fourier frequency must be determined carefully so as to not introduce aliasing uncertainties. (Chapter 10 introduces the topics of aliasing, filters, etc.)

6.7 BOUNDARY CONDITIONS AND TEST FIXTURES

Boundary conditions which will constrain or permit the intended structural responses need to be duplicated as far as possible in the tests conducted in the laboratory or field. Some of the great challenges in structural testing come in developing fixtures and attachments that duplicate the expected boundary conditions and do not introduce spurious behavior in the test results. In the initial example in this chapter, the cantilever beam/bracket was subjected to a uniform acceleration. The beam was shown as fixed against deflection and rotation at one end giving the name cantilever. However, there is rarely any boundary condition which is completely restrained. Designing such a fixture could prove very difficult as each type of equipment used for test purposes has some degree of compliance. Whether the compliance (the inverse of structural stiffness) is important depends on many variables including the test equipment and fixtures plus the measurement capabilities. In simple terms, the design of test fixtures is often critical to performing a successful test. The design process usually requires basic engineering as well as considerable skill in developing something that is easy to use and produces the required results. There is probably about equal shares of creativity or "art form" in design of fixtures as there is detailed engineering. Clever fixture design comes from experience including some mistakes (Ref. 6-2).

There are basically two types of test hardware or structures to be tested: those which have been designed with testing in mind and those which have not been designed so. Hardware which has been designed anticipating that tests will be performed is likely to interface with laboratory hardware and equipment. That hardware which has not been designed with structural testing in mind or, perhaps anticipating a minimum amount of testing, will likely require fixtures and methods of attachment to testing equipment. Many types of structures and components will not have been designed with testing in mind. It is often customary for testing to be considered after other parts of the design have been agreed to, often after prototype hardware has been fabricated, or even after production problems have been encountered. It is important for the structural test personnel to be involved early in projects or for the responsible engineer to consider the possibility of testing in their design efforts.

A classic example of an incorrect boundary condition is a compressive test on a solid right circular cylinder specimen as shown in Fig. 6-28. The undeformed test specimen is shown on the left. The unlubricated deformed test specimen is shown in the center. The lubricated tested specimen is shown on the right.

As the force is applied to the unlubricated specimen, the specimen tends to assume a barrel shape. The triaxial stresses resulting from resistance to radial deformation near the top and bottom of the specimen cause the barrel shape. These stresses are caused by friction forces which prevent expansion of the base and top of the specimen. Simple lubrication (e.g., petroleum jelly) of the base and top can reduce or eliminate the restraint caused by friction. Simple lubrication and concentric grooves in the ends of the specimen capture the lubricants and effectively eliminate the frictional forces. The results of tests performed on specimens with and without restraint are considerably different because the triaxial stresses on the bases of the specimen will add considerable stiffness and bias the test results in a nonconservative way.

A second example of the need to carefully consider the boundary conditions is on the test of a tube or pipe in bending as shown in Fig. 6-29. The test item is placed in a fixture so that a "pure" moment can be applied through failure of the pipe or tube. That is, a "pure" moment applied to the joint indicates that there is no shear force acting across the connection. The test item consists of two sections joined by bolted flanges. The larger section contains an opening on one side (not shown) so that it is not an axisymmetric structure and is more representative of typical test hardware. As the forces are applied through the hydraulic jacks, the tube or pipe deforms elastically and then plastically as the displacement of the jacks increase.

Note the base of one of the pivoting arms is fixed but that the base of test fixture is allowed to displace along the floor of the test frame while the pipe section deforms. Restrained bases on both sides of this test fixture could result in added moments and shears even if the test structure was completely symmetric.

In summary, simulating boundary conditions depend on the following:

1. The magnitude and direction of the loading conditions

2. The rate and other parameters describing the loading conditions

3. How the loading conditions can be applied versus how it will be applied in service or overload conditions

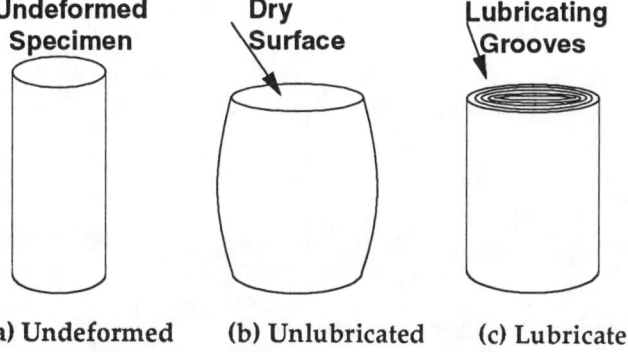

(a) Undeformed (b) Unlubricated (c) Lubricated

Figure 6-28. Compressions Specimens

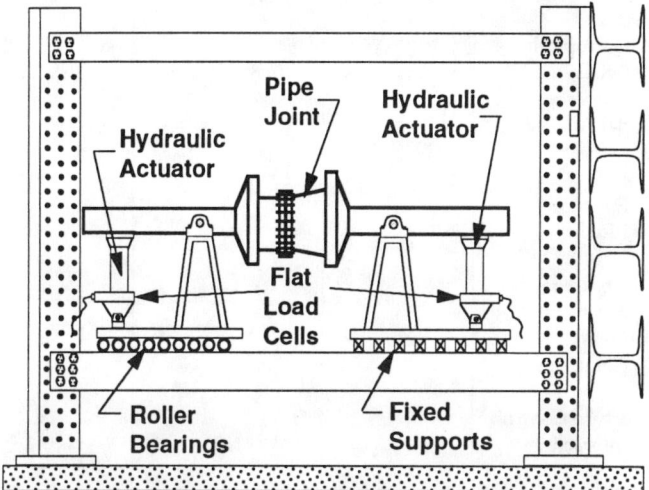

Figure 6-29. Pipe Joint Subjected to Pure Moment

4. The condition to be duplicated (e.g., free, pinned, fixed, etc.)
5. Laboratory equipment available or that can be designed and fabricated

6.8 SUMMARY

This chapter described the various types of loading systems, equipment, and test methods typically used in structural testing. The information in this chapter relates the description on types of tests (Chapter 4) and loads (Chapter 5) into simulations used in performing laboratory and field tests. The major points in this chapter are: (1) structural testing consists of simulating actual or expected overload conditions, (2) many different types of loading equipment are available or can be developed for conducting tests, and (3) the boundary conditions used in testing have significance in test results.

Table 6-3. Summary of Types of Tests and Applicable Loading Conditions

Type of Test	Loading Conditions															
	Static		Modal Analyses			Vibrations			Inertial	Shock			Thermal			Other
	Force	Pressure	Hammer	Shaker	Other	Sine	Transient	Random		Impact	Blast	End Deposit	Burnt	Oven	Radiant	Icing, Sound Waves, Electric Current, Voltage
Acceptance QA/QC	x	x	x	x	x	x	x	x	x	x		x		x	x	x
Calibrations	x	x				x			x	x				x	x	x
Connections & Joints	x	x				x	x	x	x	x	x	x	x	x	x	x
Development Test Units	x	x	x	x	x	x	x	x	x	x	x	x	x	x	x	x
Energy Absorbed	x	x				x	x	x	x	x	x	x	x			x
Environment Compatible	x	x				x	x	x	x	x	x	x		x	x	x
Failure Load and Modes	x	x				x	x	x	x	x	x	x	x			x
Full Scale & Model	x	x				x	x	x		x	x	x	x	x	x	x
Inertial	x					x			x							
Mass Properties	x	x				x			x							
Materials Properties	x	x							x	x		x	x	x	x	x
Modal Analysis	x		x	x	x	x				x						x
Proof & Function	x	x	x			x	x	x	x	x	x	x	x	x	x	x
Residual Stresses	x	x														x
Service Life Extension	x	x				x	x	x	x							x
Stiffness/Compliance	x	x	x	x	x	x	x	x	x							
Structure-Control Interaction	x	x				x	x	x	x							x
Time Dependent	x	x				x	x	x	x							x
Wind Tunnel	x	x	x													x

6.9 REFERENCES

6-1. _____, *Metals Handbook. 9th Ed. Vol. 8. Mechanical Testing*, American Society for Metals, June 1985.

6-2. A. Blake, *Handbook of Mechanics. Materials. and Structures*, John Wiley and Sons, New York, 1985.

6-3. T.G. Beckwith and N.L. Buck, *Mechanical Measurements*. Addison-Wesley, Reading, MA, 1973.

6-4. C.M. Harris, *Shock and Vibration Handbook*, 3rd Ed., McGraw-Hill, New York, 1988.

6-5. C.M. Herzfeld, Ed., *Temperature. Its Measurement and Control in Science and Industry*. Vols. 1, 2, and 3, Reinhold Publishing Corp., New York, 1962

6-6. C.W. Richards, *Engineering Materials Science*. Wadsworth, San Francisco, 1961.

6-7. G.M. Sabnis, *Structural Modeling and Experimental Techniques*, Prentice-Hall, Englewood Cliffs, NJ, 1983.

6-8. L.H. Van Vlack, *Elements of Materials Science*, Addison-Wesley, Reading, MA, 1974.

6-9. S.A. Mahin and P. B. Shing, "Pseudodynamic Method for Seismic Testing," *Journal of Structural Engineering*, ASCE, Vol. 111, No. 7, July 1985.

6-10. _____, "Safety and Relief Valves, Test Performance Codes," ASME/ANSI PTC 25.3-1988, ASME, New York, 1988.

6-11. _____, "Repair of ASME and National Board Stamped Pressure Relief Valves," The National Board of Boiler and Pressure Vessel Inspectors, Columbus, OH, 1988.

Supplier Catalogs:

The reader is referred to buyers' guides available in many periodicals. These buyer's guides have the most recent information available from companies manufacturing and supplying testing equipment. Much of the information in this chapter was found in these catalogs. These catalogs are available to industrial users at little or no cost.

CHAPTER 7
TEST CONTROLS

7.1 EDITORS' INTRODUCTION

There are many ways that the subject of test controls could be introduced and described. There are many types of control and feedback systems used in structural testing. The editors have chosen to use two examples in this chapter. These examples describe servohydraulic control systems. These systems have the most active development work and are the most widely used in structural testing.

This chapter is composed of two sections. The first section gives a description of test controls viewed from the complicated design aspects of servohydraulic test equipment, selecting the proper equipment for the tests, typical problems encountered in trying to match the test equipment to test requirements, and supplying the equipment to customers. The second section consists of a description of the possible sources for test system command signals with application to servohydraulic test equipment.

7.2 SERVOHYDRAULIC TEST SYSTEMS AND CONTROLS
Niel R. Petersen, MTS Systems Corp.,
Minneapolis, MN 55424

Early dynamic loading tests were accomplished with various machines including motor-driven screw jacks, crank-and-flywheel machines, and rotating-beam assemblies. They were used to experimentally obtain dynamic material properties and/or structural responses. Tests requiring multiple coordinated force inputs required multiple fatigue machines, resulting in complex and inflexible mechanically-coupled major test fixtures. Consequently they were restricted to only a few input loading points. Realistic service loading could only be crudely approximated by constant-amplitude, or at best, block cycle dynamic loading. Pneumatic cylinders could be used in low force cyclic tests, but the cycle rates were inadequate to obtain high-cycle fatigue information. Magnetically driven resonant testing machines could accomplish high frequency loading, but their physical size prevented them from being used in multiple input situations and the test controls severely limited the ability to vary the load input spectrum to resemble service conditions. In addition to these load control problems, data acquisition had to be accomplished with a lab notebook, and test monitoring was done with cycle counters and skilled test technicians.

Dynamic loading techniques were revolutionized with the development of servohydraulics for aerospace flight surface control industry in the 1950's. Further advancements were made by incorporating computers in the 1960's. These achievements, taken with improved servovalve flows and frequency responses, have enabled modular closed loop servohydraulic loading systems to expand dynamic test possibilities. As a result, the most complex, powerful, high performance, closed loop servohydraulic systems are now found in the materials characterization and structural testing disciplines. These achievements have required test lab engineers to become reasonably proficient in closed loop theory and practices so that they may obtain maximum utility from this technology. Today, state-of-the-art testing laboratories routinely perform sophisticated multiple input service history simulations, with automatic data acquisition and test monitoring.

The purpose of this chapter is to address several important aspects of servohydraulic test system hardware selection and application, with the goal of giving users insight into test planning and execution and to obtain better test results.

7.3 APPLYING SERVOHYDRAULIC CONTROLS TO TEST APPLICATIONS

Closed loop servohydraulic control systems are very powerful, robust devices with excellent response and small signal resolution. For most test applications, they are used as an element in a closed loop servo system. The actuators are motion generating devices that are quite insensitive to the reacted forces. The operating principles and characteristics of closed loop servohydraulic control systems which use these actuators, are ideal for operation in position control, where the dynamics of the control loop are dominated by the actuator parameters.

However, most test needs require, or at least benefit considerably, from controlling input parameters other than displacement. The use of feedback transducers, such as load cells, strain gages, pressure transducers, and even calculated parameters, permit many new test possibilities. However, the use of servohydraulics with control variables other than position feedback considerably complicates the control scheme since the control system dynamics are now dominated by a much more complex specimen transfer function. In addition, when compared to other many non-testing servohydraulic applications, obtaining the maximum dynamic response consistent with accuracy from the servo system is usually much more important. Simple stability and static accuracy are no longer the dominant system criteria, as users look to optimize closed loop dynamic response in order to increase testing fidelity and speed.

Theoretical analyses of closed loop servo systems are very complicated and often beyond the immediate needs of the test engineer. An appreciation of the subtleties of dynamics and servo control however is beneficial for successful test planning and execution. The test equipment industry has created a number of ways to apply high speed servos to a wide variety of applications, and standard electronic controllers appropriate for most materials characterization or structural testing needs have evolved. These controllers possess sufficient flexibility and control power to accomplish most test objectives with minimum setup and operating attention.

7.3.1 Servohydraulic Characteristics

Servohydraulic structural loading systems most often use flow-control servovalves connected to double-acting hydraulic cylinders. These servovalves are an electronically-commanded, variable-opening, four-way closed-center spool valves which operate from a remote fixed pressure hydraulic source which is commonly set at 3,000 psi. An example valve is shown in Fig. 7-1. Servovalves contain one or more stages of internal hydraulic amplification and sometimes even internal electronic feedback, enabling them to smoothly modulate considerable power with only a small input current.

A simple mechanical model of a servohydraulic actuator assembly assumes it as a relative-velocity generator which moves at a velocity proportional to the current applied across the servovalve. More complex models will include the oil column spring rate and piston rod mass.

In each direction of actuator motion, 3,000 psi hydraulic fluid is passed through two metering orifices inside the servovalve and through the actuator. Pressure drop occurs inside the actuator, depending on the reacted force,

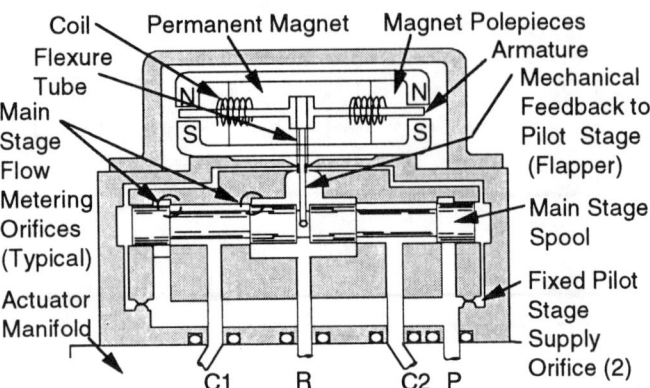

Figure 7-1. Typical Two-Stage Servovalve Schematic

but no matter which way the actuator is commanded to move, the metering orifices cause a volume of 3,000 psi supply oil to be displaced by the actuator piston. A similar volume is simultaneously returned to the hydraulic pump reservoir at atmospheric pressure.

The servovalve orifice pressure-drop process generates considerable heat in the oil. For example, the 3,000 psi supply pressure will cause the oil temperature to increase about 20°F in each pass through the system. This heat must be removed with a large heat exchanger (usually an oil-to-water type) with a continuous heat flux capability equal to that of the main pump motor's power output. Ordinary industrial hydraulic power supplies intended for open center valves are not rated for this amount of heat rejection capability. Very severe and dangerous oil overheating can occur from the use of an unsuitable pump assembly.

The current applied to the servovalve is closely related to the orifice opening for flow control servovalves, and the result is an actuator output velocity without much regard for the applied force. In other words, the servovalve current does not significantly relate to the actuator output force. With flow control servovalves, the actuator output force becomes only a consequence of impeding the motion of the actuator. It is widespread intuition error that "servovalves admit pressure into an actuator." Such thinking may occasionally lead to erroneous conclusions concerning the nature of load control servohydraulics. For example, quick motion disturbances to a load control system may cause considerable load reflection and load error, unless the control loop reacts very rapidly. The advantage of flow control servovalves is the robustness of the resulting output motion, the very fine control resolution, and dynamic range that can be obtained.

Operating efficiency is not a virtue of a hydraulic servo. Large hydraulic servo systems consume considerable amount of power. They will, however, give unequalled controllability with remarkable open-loop output flow resolution as low as 0.1 percent. This resolution can be further improved for static use with a small high frequency oscillating current applied to the servovalve to keep it from sticking from super fine contamination. The remote pump also allows the concentration of considerable dynamic loading power in a small package, without a local heating problem. Thus, a number of powerful loading actuators can be readily concentrated around a test structure.

7.3.2 Pressure Control Servovalves

There are a few hydraulic servovalves which, for low frequencies, are made to be pressure control devices instead of flow control servovalves. They generate a differential pressure across the control ports approximately proportional to the applied current, with comparatively little flow sensitivity (at least for steady state situations). This alternate type of control action might seem better for load control situations, but the resulting lack of actuator motion output robustness makes them unable to reject the stick-slip effects of actuator seal friction, creating a number of additional performance difficulties. All fixed-pressure-source servovalves, whether flow or pressure control, must rely on zero-lap four-way spool valve as the main stage. It is only the internal mechanisms of the pilot stage that make the servovalve sensitive (pressure control) or insensitive (flow control) to the control port differential pressure. Such internal mechanical feedback mechanisms will tend to have a fixed, mechanical, spool-movement response to pressure-error gain. Thus, pressure control valves limit the use of the electronic derivative and integral terms of control loop optimization present in electronic servo-controllers. Actuator friction effects, which can usually be ignored for flow control servovalves, will frequently cause annoying limit cycling for a pressure control system.

Control loops with pressure control servovalves may fail in a more benign manner than with flow control valves, providing the internal pressure feedback control is working. However, modern control systems use redundant overload and loss-of-control interlock detectors to sense loss of control or feedback, and provide the ability to unload safely.

7.3.3 Actuator Piston Area Selection

The actuator piston area determines its force/velocity response to a particular servovalve flow. The maximum force is obtained from a servoactuator when the actuator is stalled, at which point there is little flow or pressure drop in the servovalve. Equation 7-1 relates the supply pressure, the required force, and available piston area.

$$F = P \cdot A \qquad \text{Eq. 7-1}$$

where F = Output force, pounds

A = Piston area, square inches
P = Differential design or supply pressure, lbs/inch^2

The available force versus velocity characteristics of the actuator are determined by the orifice-flow relationship inside the servovalve. Figure 7-2 illustrates the rather complex flow versus actuator differential pressure for various valve openings.

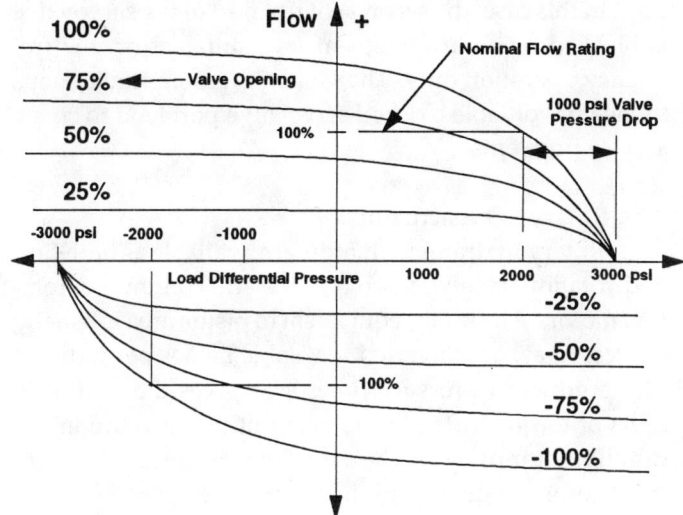

Figure 7-2. Flow Versus Pressure Drop for Different Valve Openings

The actuator sizing process can be simplified with assumptions and rules-of-thumb. For example, actuators will have a static nominal force rating which uses about 95 percent (about 2,900 psi differential pressure) of the supply pressure. For fatigue testing applications, it is common to use a pressure rating about 85 percent (2,500 psi differential pressure) of the supply pressure. As can be seen from Fig. 7-2, these choices are slightly arbitrary, but for most applications, they are reasonable. Occasionally, other area-choosing criteria may be used. One selection criteria is the actuator differential pressure rating for maximum actuator power output (a maximum instantaneous force * velocity product) for a given fixed servovalve or servovalve opening. Manipulation of the orifice flow equations will show that this condition occurs when two-thirds of the available supply pressure is dropped across the actuator, and one-third across the servovalve. For this reason actuators used to drive energy absorbing loads such as shock absorbers are frequently sized for 2,000 psi pressure drop at maximum specified force.

It is not necessary for an actuator to have equal area on both sides of the piston. Single ended actuators are frequently used for moderate-frequency, long-stroke tests when the force demands many be unequal, or where the additional length of the double ended actuators becomes a setup problem. Single-ended actuators must be carefully used in long-stroke applications to avoid buckling problems near maximum extension. With an unequal area actuator the control system response tends to be different in two directions, although the consequences of this are usually minimal.

Sometimes servovalves are used as three-way devices during an actuator arrangement that has only one port. In this case, the second control port of the servovalve is blocked. Such arrangements are suitable for low frequency operation only. The violent pressure fluctuations in the unavoidable blocked servovalve port tend to create service problems.

7.3.4 Rotary Servoactuators

Rotary hydraulic actuators are available as both limited-rotation vanetype (see Fig. 7-3) and continuous-rotation motors. The rotary equivalent to piston area is usually expressed as, or converted to, "cubic inches per radian." The product of pressure times area gives the output in units of torque instead of force. Continuous rotation hydraulic motors may also be operated in torque or displacement control with appropriate transducers, or in velocity control with tachometer feedback. Compared to electric drives, very high angular accelerations and dynamic ranges are possible from hydraulic rotary devices since very high torques are generated with minimum associated actuator inertia. The control possibilities of rotary systems should be similar to linear systems although control problems for rotary systems tend to be slightly more common. Typical test lab rotary setups have higher inertias, greater deflections, lower natural frequencies, and low inherent damping, when compared to typical linear setups. The same configuration guide rules used for linear systems apply even more strongly for rotary systems.

7.3.5 System Operating Pressure

Although 3,000 psi has emerged as a de facto standard servo-system pressure, there is no firm requirement to operate a servo test system at this pressure. Other supply pressures are used, and many be deliberately chosen for reasons such as component life, noise, safety reasons, compatibility, and temporary test needs. Generally, very large systems may benefit from higher pressures and very small systems may benefit from reduced pressures. Virtually all systems can be operated at reduced pressures to improve stability and give smoother, quieter response if the output force needs are reduced accordingly.

7.3.6 Maximum Actuator Dynamic Response

The maximum frequency response of a servovalve determines its ability to deliver changing flow in response to a changing current. It is established by the design of the servovalve pilot stage and its ability to dynamically position the main stage spool. The flow versus frequency of some common servo-valve models are show in Fig. 7-4.

Maximum dynamic actuator performance is established with a servovalve/actuator combination. Items of concern to the test planner while selecting an actuator are: the piston area, the internal volume of oil, the end attachment configuration, compliance and stiffness of the actuator mechanical assembly. It is also very important to provide a sufficient number of degrees-of-freedom to the actuator end attachments in an effort to protect the actuator from excessive side loads and to maintain the test validity.

7.4 TYPICAL SERVOHYDRAULIC ACTUATOR COMPONENTS

Commercially available linear actuator piston areas vary from about 0.5 square inches (for about 1,500 lbs maximum output force) to well over 1,000 square inches for multimillion pound output force. Servovalve sizes range from about one gallon per minute (gpm) maximum flow to over 400 gpm (this valve is a 200 pound steel block

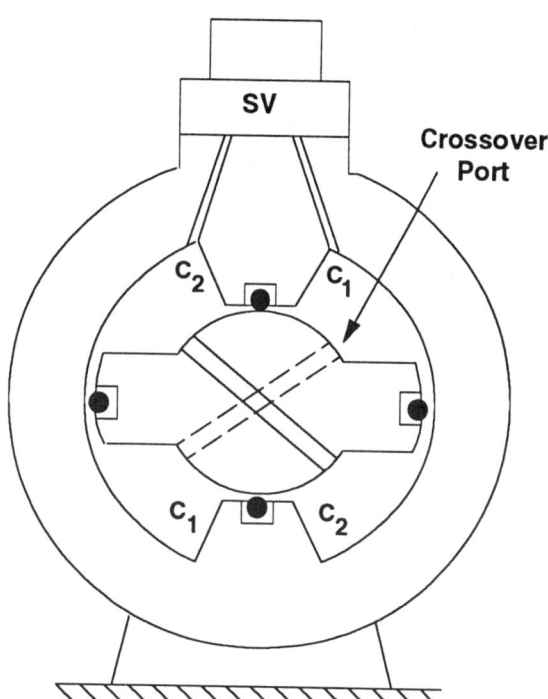

Figure 7-3. Limited Rotation (±50°) Vane-Type Rotary Actuator

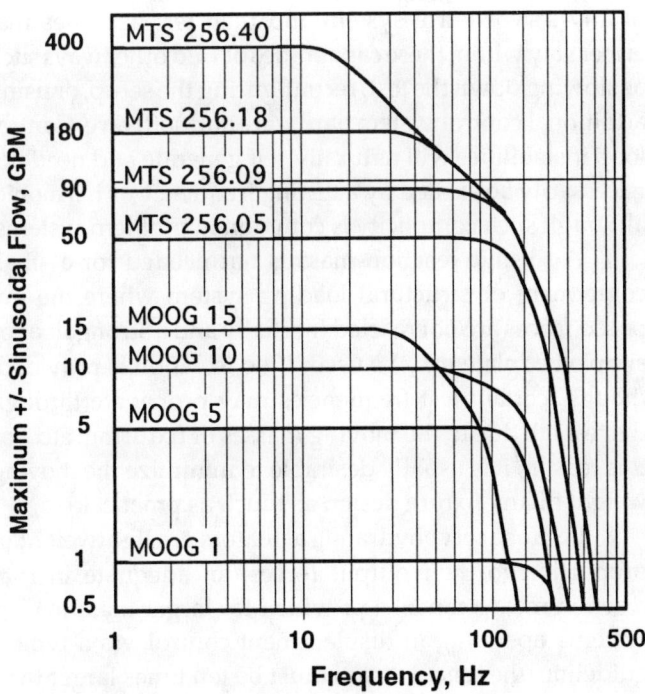

Figure 7-4. Typical Maximum Servovalve Flows Versus Frequency

requiring very stiff three inch outer diameter supply and return hoses). Frequency response extends into the low hundred Hz range, although there is substantial reduction of maximum available dynamic response accuracy, and especially the ultimate flow resolution, as the size of a servovalve increases.

For static testing needs, small servovalves can be used with large actuators, with consideration for the internal leakage of the actuator. On the other end of the scale, servovalve/actuator combinations have been built which have output velocities over 300 inches per second. Full scale aircraft structure tests typically operate form 0.1 Hz to 2 Hz and component tests from one to 25 Hz. Vehicle simulation tests routinely include frequency content up to 50 Hz. Depending on the setup for the test structure and component selection parameters, maximum closed loop operating frequencies may extend to as much as 100 Hz. Open loop operation has been accomplished to well over 500 Hz on certain material characterization and vibration test applications.

The servovalve flow required to achieve a desired actuator velocity can be calculated using the actuator piston area (remembering that there are 231 cubic inches per gallon). A simple equation to determine the maximum servovalve flow for sinusoidal operation, which is related to servovalve flow, actuator size, output sinusoidal amplitude and frequency is given in Eq. 7-2.

$$Q = F * A * X / 1.23 \qquad \text{Eq. 7.2}$$

where Q = Servovalve peak flow, gallons per minute (gpm)
 F = Operating frequency, Hz
 A = Piston area, square inches
 X = Desired double amplitude displacement, inches peak to peak

For applications involving loading a stiff structure, this simple flow versus output amplitude equation does not take into account the additional force-train deflection resulting from the compliance of the oil trapped inside the actuator. For example, at maximum dynamic operating force, about 0.5 percent of the actuator stroke (X from Eq. 7-2) will be lost due to oil compressibility. Test systems operated at high forces and frequencies, and short strokes (e.g., stiff structures) will have substantially reduced amplitude performance due to oil column compliance. Detailed performance calculations for stiff structure operation involve a number of additional factors and are best accomplished with computer programs available from system suppliers.

However, the oil column compliance is easily calculated. It can be added to the structure and force-train compliance to approximate the total double amplitude deflection that must be delivered by the servovalve to the actuator piston area. Since there are two chambers in an actuator, and actuator compliance calculation should use two parallel oil spring rates to account for both sides of the piston. But, for a centered actuator, where the two volumes are equal, the oil column compliance is given in Eq. 7-3:

$$K = 4 A^2 * \beta / V \qquad \text{Eq. 7-3}$$

where K = Oil column compliance, lbs/inch
 A = Piston area, square inches
 β = Oil compressibility (also called bulk modulus), lbs/inch^2 with a typical value of 180,000 psi
 V = Total amount of oil contained inside the actuator including servovalve porting, cubic inches

The oil column compressibility "constant" (β) will vary with the hydraulic fluid in use, and even with the same operating condition. With clean, clear petroleum based hydraulic fluid, the bulk modulus (β) may be as high as 200,000. Operation of a poorly designed hydraulic system, however may cause the oil to become entrained with air, reducing the bulk modulus to less than 100,000 psi and shortening oil life drastically.

7.5 CONTROL RESPONSE

Whereas the previous discussion has been used to define the maximum performance capabilities of a simple actuator-servovalve combination, the small signal dynamic response to a small signal dynamic command requires that other elements in the control loop respond as well. Each element can be described as a block with a certain limited response in a closed loop system as shown in Fig. 7-5. When all of the elements are connected together as a loop, the possibilities of instability are substantial, especially with more dynamically complex load control applications. Control loop stability commonly limits the servo loop performance to well below the maximum actuator response. Commanding the system with varying frequency low-amplitude sine waves and measuring the feedback response is a way to evaluate the stability and usable frequency of a system. Test system resonant frequencies can be established along with the nature of the resonance. A properly performing control loop will not have serious amplitude response peaks or valleys over the range of expected operating frequencies.

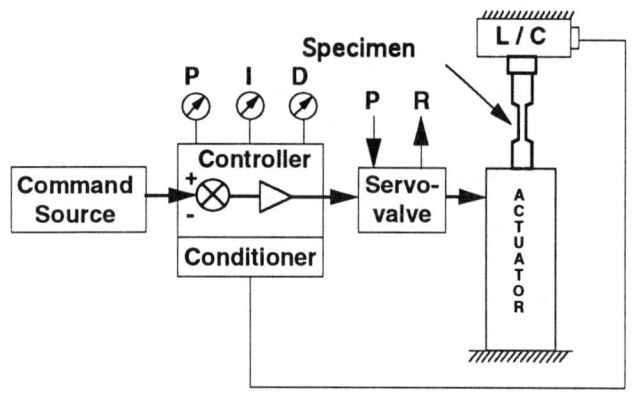

Figure 7-5. Closed Loop Block Diagram

Like all servo-systems, feedback response to a command signal is not instantaneous. As a control loop element, the damped response of typical servovalves makes an ideal loop component, but it must be remembered that other less well behaved elements are present. These less well behaved elements include the hydraulic actuator, feedback transducers, fixturing, and of course, the all-important test structure. Taken altogether, these blocks determine the overall system response to a desired command. In most systems, and especially for load control, it is the less well-behaved factors which ultimately determine the maximum achievable system response fidelity.

The test engineer must recognize the finite response of any servocontrol system and plan test activities that either stay within these capabilities or find other ways such as slowing down the test, reconfiguring the setup, or using additional control system hardware and software. Control loop instabilities will naturally self generate and need not necessarily be excited by a testing frequency or harmonic, although such complicity is common in problem systems.

A seismic reaction mass is not needed for a small component or structural loading system where the imposed forces are not reacted inertially and a strong floor or simple bedplate may be used. If no isolation is provided, though, certain test frequencies may propagate through the test lab due to the moving masses of fixturing, etc. For this reason, it is usually desirable to minimize the moving weight of any fixture design as much as practical.

Because servohydraulic actuators are lightweight in proportion to their output forces, an adequate inertial reaction means must be provided for larger mass-loaded systems operating in displacement control. As a typical guideline, the seismic mass must be ten times larger than the maximum unbalanced inertial output force anticipated. A more detailed analysis may allow much smaller mass or may even require a larger reaction mass. Typically, servoactuators themselves are capable of withstanding 5 g's of body acceleration, but most times the test structure or the surroundings establish a more stringent need for seismic isolation. Foundation experts can establish reaction needs once the nature of the setup inertial forces versus frequencies are determined.

7.5.1 Displacement Control Systems

A servohydraulic actuator makes a very robust control device for displacement control applications. Displacement feedback is commonly obtained from an LVDT mounted coaxially within the actuator for displacements up to +24 inches. Displacement measuring devices such as ultrasonic sensors, linear encoders, or even large displacement potentiometers at other locations can be used for displacement feedback, although at some point the structural dynamics of the setup or transducer, will create a more complex control situation requiring special treatment. Otherwise, little special stabilization effort is usually needed for stroke control applications. An exception, however is if a heavy mass is able to resonate on the oil column compliance within the frequency response bandpass of the servovalve. If such as resonance should be encountered, stabilization and compensation is readily accomplished with an additional transducer measuring actuator differential pressure or mass acceleration, and electronic networks in the servo-controller.

The oil column resonance of a mass-loaded equal-area double-acting actuator at the center of its stroke is given approximately by Eq. 7-4:

$$f_n = 2500\, A / \sqrt{W*V} \qquad \text{Eq. 7-4}$$

where f_n = Oil column natural frequency, Hz
  A = Actuator piston area, square inches
  W = Total directly couple moving mass including the piston rod, pounds
  V = Total internal oil volume of the actuator including ports to the servovalve, cubic inches

The natural frequency obtained from Eq. 7-4 will vary as the bulk modulus of the oil changes. The mass term should include direct coupled mass, but not mass which is isolated from the actuator by structure compliances such as tires, soft structures, etc. If this frequency is within the response capability of the servovalve, the actuator response will have some tendency to ring at this frequency, depending on the internal leakage of the actuator. This ringing can be most effectively damped out electronically with the differential pressure stabilization referred to.

7.5.2 LOAD CONTROL SYSTEMS

System configuration becomes very important when planning load feedback tests. Some guidelines include:

1. *Reaction Path*: There must be a statically coupled force reaction path for the structure, feedback load cell, and loading actuator force train. It is impractical to directly operate a simple load control system into a freely floating specimen. A static reaction path must be provided or otherwise the actuator will quickly run into the end of its stroke attempting to create a desired force or even a slight unavoidable load offset. Figures 7-6, 7-7, and 7-8 show examples.

   There are iterative ways, described later, which will enable replication of the dynamic components of a desired load versus time history into a floating structure. These test methods involve a supervisory digital control system to command an actuator operated usually in displacement control.

2. *Load Cell Position*: The feedback load cell should be placed in a location where the load cell measures the structure forces with a minimum of intervening mass-acceleration error. The fixed load cell control arrangement offers the greatest readout and control accuracy. Whenever possible the load cell should be placed as shown in Fig. 7-6, on the fixed or non-moving reacted side of the test structure to minimize the mass acceleration load sensing errors.

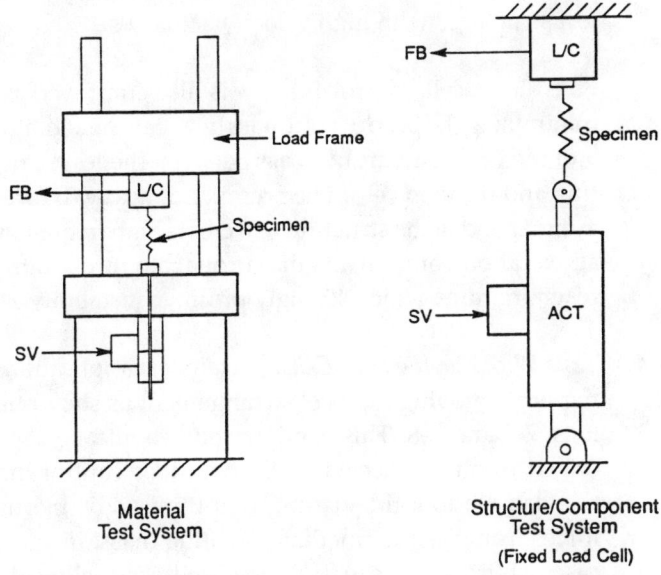

Figure 7-6. Typical Fixed Load Cell Control Setups

3. *Testing Stiff Structures*: With typical stiff structure load control applications, a small amount of actuator motion will result in full load feedback. Such load control systems have high mechanical gain (e.g., there will be a major load feedback response for only a tiny current input to the servovalve). Unfortunately, the mathematics of stability analysis will only allow a certain amount of total gain in closed loop system. If the mechanical feedback gain is high, the electronic controller forward gain must be correspondingly low, resulting in excessive control response errors.

The following equation is a simple approximation that can be used to predict the degree of control difficulty for fixed load cell control with stiff test structures:

$$T = 1.5 * 10^6\, A^2 / K * Q \qquad \text{Eq. 7-5}$$

where T = Transit time, milliseconds
  A = Actuator piston area, square inches
  K = Total overall load train spring rate, lbs per inch
  Q = Servovalve maximum flow in gpm

Equation 7-5 roughly approximates the time for the load feedback to travel its full range at maximum servovalve flow. If this time is greater than about 10-20 milliseconds, the control should stabilize readily. If the time is substantially less, it may be appropriate to either reduce the servovalve size or increase the area of the control actuator. Equation 7-5 scales the approximate controllability of a fixed load cell base. It will identify practical combinations of high compliance test structures with high velocity actuators.

Fixed-load cell control fidelity is also improved by minimizing the product of: (mass between the actuator and the test structure) * (mass between the test structure and the load cell). The presence of heavy fixtures at either end of the structure will create high frequency acceleration components that propagate through the reaction frame which strongly promote instability.

4. *Tests With Moving Load Cells*: Certain test applications require a moving load cell arrangement as shown in Figs. 7-7 and 7-8. This configuration should be used only if absolutely necessary. Both the load control and the load readout fidelity are compromised by inertia forces from the intermediate isolating mass. In some cases it is impossible to fix the load cell, especially with multiple loading inputs into a single test structure as shown in Fig. 7-8.

An easily calculated indicator of potential moving load cell control problems is to determine the resonant frequency of the intermediate isolating mass on the compliance of the test structure, or even the first resonant frequency of the test structure itself. Above this resonant frequency the load cell/mass combination is really behaving as an accelerometer, where feedback is sensitive to motion rather than force. Substantial er-

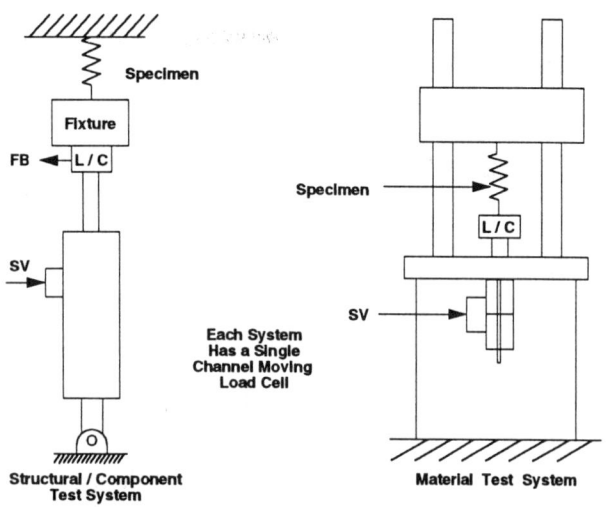

Figure 7-7. Single Channel Moving Load Cell Control Examples

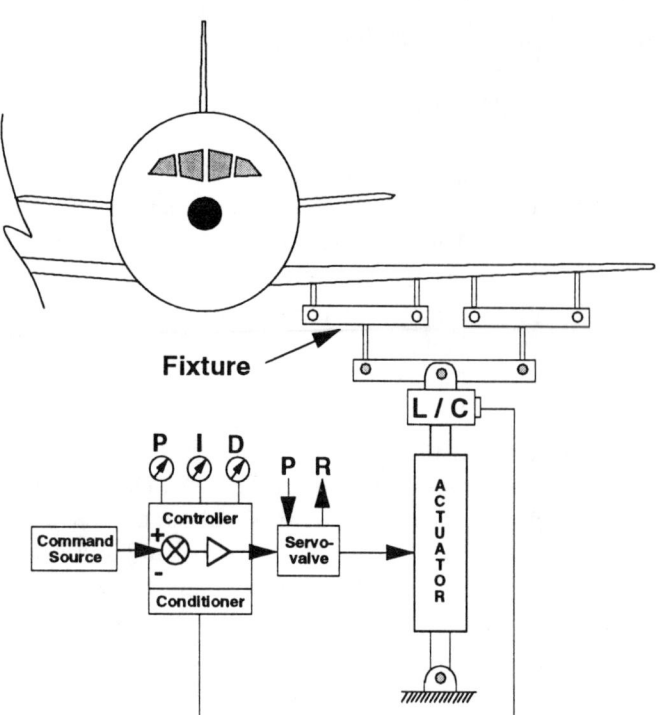

Figure 7-8. Multiple Moving Load Cell Control Example

rors creep into both the force control and the force readout system at frequencies above about one-fourth of this frequency. For this reason, moving load cell control systems are not practical at high frequencies. The control and readout accuracy problem is made worse by increasing amounts of fixture mass and becomes especially severe with compliant structures.

A most common control and test validity problem is that inertia forces are not considered when comparing a moving load cell output with the real forces present in a test structure. An example of a moving load cell output is shown in Fig. 7-9 which is a multiple input control system for vehicle testing.

For complex reasons, a moving load cell control loop will tend to naturally become unstable at high frequency, even though the calculated problem resonant frequency is really much lower. The problem frequency identified by the fixture mass resonating on the structure spring is a frequency at which the feedback goes to zero, and above which the load feedback transducer becomes an accelerometer. It can no longer measure the forces in the structure.

By deliberately limiting the bandwidth of the control, stability of the moving load cell control loop can be

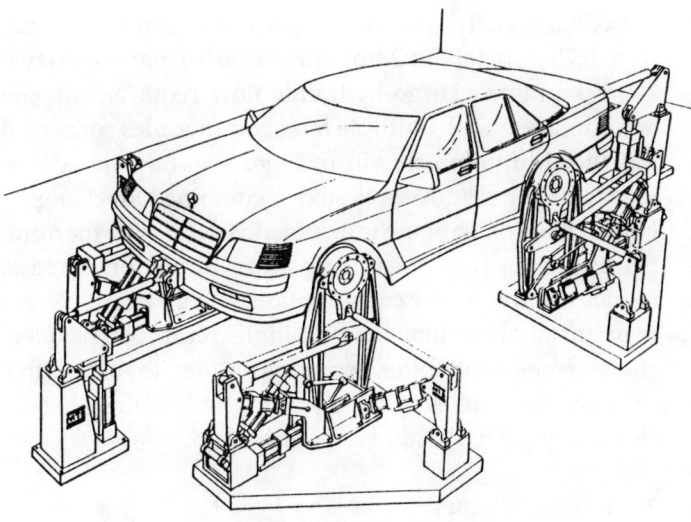

Figure 7-9. Typical Multiple Input Moving Load Cell Control System

restored, but not the response. A modern electronic controller will include the capability of adjustable control bandwidth.

Surprisingly, test setup combinations that provide better moving-load-cell control include a stiff structure, a more compliant oil column in an actuator (for example, a long static stroke), and a moderate response servovalve. However, the most important configuration item is an absolutely minimum weight between the load cell sensing element and the structure compliance.

Use of Differential Pressure of Control Cylinder: Under some low performance circumstances the differential pressure of the control cylinder can be used to approximate the output force. A differential pressure feedback transducer will also be sensitive to actuator bearing and seal friction. Using it as a control feedback imposes an additional piston rod mass into the control system making it similar to a moving load cell control situation. If a double acting hydraulic cylinder is of unequal area, the calibration factors of each half of the differential pressure cell must be proportionately scaled since pressure on both sides of the piston vary simultaneously during all types of operations.

Actuator differential pressure control should be considered only with careful skepticism, especially in low-force or compliant-structure situations where the seal friction or flow loss errors may be comparatively very large.

Tests with Load Cell Mounted Underneath the Actuator Base: A misapplication sometimes encountered has the load cell mounted underneath the actuator base. This is rationalized by thinking "it gets the feedback close to the actuator." In reality, this is not a useful strategy, as it really acts to "get the feedback even farther from the test structure." This configuration has many negative consequences including inertial effects which create an even more difficult control situation, and added possibilities for major loading errors. Furthermore, the base-mounted load cell configuration adds the moving weight of the piston rod, the high frequency vibration of the control actuator body, and servovalve hose forces to the load sensing and control errors. It may expose the load cell to substantial bending or shear loading-induced errors. The base mounted load cell configuration definitely should be avoided.

7.5.3 Remote Strain Gage Control Considerations

At times it may be desirable to control from a special transducer such as a strain gage bridge mounted on a test structure. This type of control can be expected to work quite well, as the inertial load components of a particular test setup are minimized. Care must be used, however, to be absolutely certain of the control bridge integrity because a failed feedback transducer will cause the servoactuator to go hard-over, traveling to its end stops at maximum velocity, smashing whatever is in its way.

The calibration of structure-mounted strain feedback bridges should be checked very carefully to ascertain that the control loop is actually controlling the desired variable without adding variations of its own. It is best to check a control bridge (or any critical bridge for that matter) not only for its sensitive axis calibration, but also for the effects of the five extraneous loading axes. For example, a bridge designed to measure tension-compression should be checked for sensitivity to moments about transverse axes, the moment about the longitudinal axis, and shear sensitivity about both lateral axes. Performing a simple analysis will help assure the test validity in the presence of reasonably anticipated cross-talk amounts. Remember also, that the errors of the transducer will not show during operation as the servo loop actually controls the strain gage bridge output signal without regard to the thermal drift and cross-talk effects. Using full strain gage bridges for all feedback transducers helps minimize these effects.

7.6 HYDRAULIC SUPPLY AND RETURN SYSTEM

A hydraulic servo-system requires a constant pressure source and a free return reservoir for hydraulic fluid. Servo-systems are sensitive to fluidborne contamination. Low quality, poorly constructed, or infrequently maintained systems will quickly have severe and very expensive reliability problems. Figure 7-10 shows a schematic of a commercial hydraulic service manifold (HSM) that is commonly used to provide hydraulic services to an individual control channel (or even a set of channels with a similar function). The HSM includes hydraulic accumulators to provide the necessary pressure and return fluid compliance, and a filter to protect the servovalve from catastrophic particles that may be in the hydraulic system. This in-line filter is exposed to considerable vibration and cannot really be expected to control the system contamination level. Hydraulic service manifolds also provide an On-Off function, and many include a hi-low pressure mode of operation to soften the initial start-up transient of the system. The HSM will typically have a shutdown time on the order of 100 milliseconds, allowing the system to use an interlock chain to provide automatic hydraulic shutdown in the event of a system fault or test structure failure.

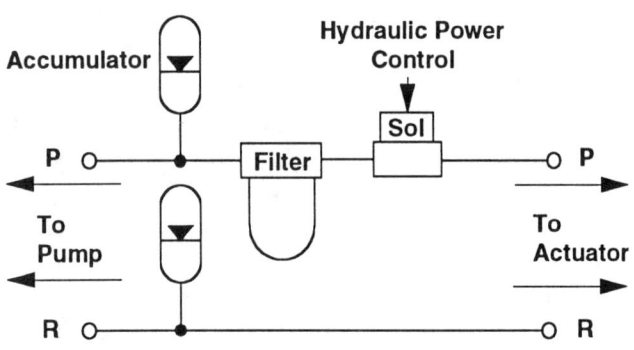

Figure 7-10. Typical Hydraulic Service Manifold (HSM)

For larger installation, a steel (or stainless steel) hydraulic hardline system is commonly used to connect a remotely located pump to the test lab or system. Short hydraulic hoses connect the hydraulic power supply (HPS) and the HSM to the hardline. Three lines are usually provided: (1) a 3,000 psi rate pressure line, (2) a large diameter 3,000 psi pressure rated return line, and (3) a moderate sized low pressure drain line for most installations.

Hydraulic hoses may be viewed as contamination generators especially when subjected to fatigue applications involving much hose jostling or internal fluid cavitation. Long hose runs should be avoided, but if required, actuator mounted last-chance filters should be provided to protect the servovalve.

7.6.1 Hydraulic Power Supply System

A wide range of hydraulic power supply assemblies are commercially available, with flow ratings typically from 0.75 gpm (about 2 hp) to 190 gpm (usually with two 200 hp motors). Large hydraulic flow requirements are accomplished with multiple reservoir modules connected together. Multiple reservoir arrangements have the advantages of being readily separated or combined for changing lab test needs. Changing test demands determine the number of pumps that must be operated at any one time to reduce electric power consumption. A very large, permanent installation may use a single reservoir/heat exchanger/reservoir filter assembly, rather than multiple reservoir assemblies so that operating and cooling efficiencies under partial load may be enhanced.

7.6.2 Accumulators

Hydraulic accumulators as shown in Fig. 7-11 are used on the pressure line to furnish short-term, high-flow demands to the servovalve. The nitrogen precharge inside one or more accumulators becomes a giant oil flow spring for high system flow demands. Pressure-side accumulators should be located nearest the servovalve, which is the source of the flow transient to be suppressed. For convenience, the pressure accumulator system can be broken up into small units immediately adjacent to the servovalve, and larger units are located remotely or even at the hydraulic power supply. Accumulators are also used near the servovalve on the return line since identical flow transients are also present in the return line. A hydraulic service manifold will contain both pressure and return accumulators for general purpose testing. However, the built-in pressure accumulators will not be sufficient for large actuator low-frequency test applications. Additional large accumulators will be needed remotely.

The use of accumulators to average system flow demands reduces the maximum hydraulic pump flow necessary to support typical actuator operation. For fast sinusoidal test programs, only about 63 percent of the peak servovalve flow from Eq. 7-2 must be furnished as steady

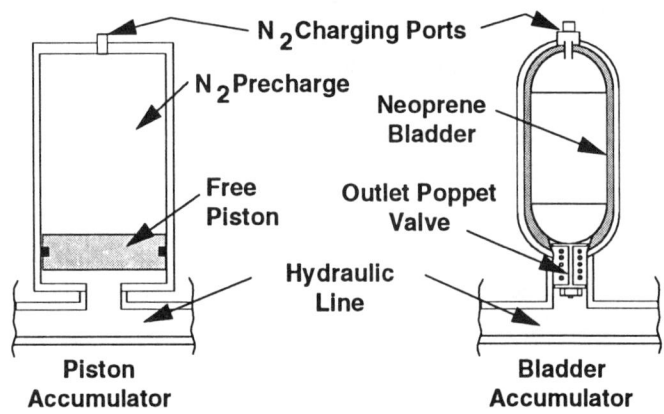

Figure 7-11. Hydraulic Accumulators

state flow by the hydraulic power supply, and the remainder can be obtained from one or more accumulators. For higher frequencies the necessary accumulators can be quite small but they need to be in close proximity to the servovalve, even to the point of being mounted directly on the servovalve manifold itself. Low testing frequencies allow the accumulator to be quite remote, but the necessary accumulator size increases greatly as the test frequency decreases.

For typical random program input tests, the steady state hydraulic power requirements may be only 20 to 35 percent of the peak hydraulic flow capability, with the balance being made up with accumulator capacity. It is not practical to precisely size or guarantee the performance of a particular accumulator/actuator/pump combination without knowing the exact time history demands of the test system program. Sometimes a simple rearrangement of the test program sequence can be used to allow the system hydraulic pumps to keep up with the system flow demands. Large accumulators are necessary to take full advantage of the difference between peak and average flow demands. Many times a short peak flow demand can be met with additional accumulators rather than additional pump capability.

Consideration must be given to adding the quiescent pilot flow and null leakage of each servovalve when sizing hydraulic pump flow needs. This is typically on the order of 0.5 gpm for each small servovalve increasing to about four percent of the main stage flow for larger servovalves.

Larger hydraulic power supply assemblies frequently use a variable volume pump with a large accumulator bank to conserve energy and reduce the system cooling needs, although the pressure regulation of such systems is not as precise as fixed volume pumps with an outlet relief valve. Other very large installations may use an "unloading valve" scheme on each fixed volume pump to cycle the flow into the pressure line, or bypass it a low pressure back to the reservoir. Costs of electric power to operate large systems are considerable and additional attention must be paid to allocating space, water and air cooling, maintenance, etc. These should be carefully figured into the initial configuration options for most cost effective testing. It is also important to use large system accumulators with a variable volume pump installation so that the pump variable volume feature is not exercised excessively. Otherwise the pump assembly itself may be subject to excessive fatigue as its internal variable volume control is cycled twice per system loading cycle.

Excessive flow transients in the return line of a system should be avoided as there is some evidence that ongoing water-hammer effects cause premature fluid breakdown. Back pressure relief valves on the return line near the pump reservoir may be necessary in severe cases to provide an adequate return ambient pressure to prevent fluid separation.

7.6.3 Hydraulic Contamination Control

Oil contamination control is essential on any servosystem and a wide variety of expensive service problems will usually be encountered before contamination problems are suspected. Although servovalves have been traditionally regarded as contamination sensitive, in reality it is the high pressure piston pump that will fail in the most disastrous manner in a contaminated system.

It is strongly recommended that hydraulic fluid samples be taken regularly for analysis, with careful monitoring for fluid degradation with time. Fluid contamination control is an evolving art, but modern hydraulic test systems rely heavily on low pressure hydraulic pump reservoir fine filters to minimize system fine particles. It is not economically practical to adequately filter the main pressure line of a fatigue operating servoactuator sufficiently to prevent contamination-related wear. In-line pressure filters are subjected to so much vibration that the trapped particles migrate through the filter media after a short time. Even the filter media itself is subject to fatigue due to flow variations. If it should fail due to fatigue, a massive slug of contamination will be released potentially jamming or failing the servovalve. For that reason, modern systems include only coarse but robust in-line filters in the HSM.

Large, high-capacity, depth-type, low pressure filters should be located in the hydraulic pump reservoir. These should be changed regularly in conjunction with a fluid monitoring program to maintain system reliability. It is a major benefit to grossly oversize the filter system as the total contamination capacity of a filter is inversely proportional to the flow through it. Most filter systems are catalog-rated only for clean flow capability. Gross oversizing allows the elements to be operated disproportionately longer, reducing service costs.

The operating environment for the hydraulic fluid in a fatigue system is much more severe than in a typical industrial hydraulic application. The test system hydraulic fluid is continuously sheared at high pressures over the orifices of a servovalve or relief valve, and may be subjected to cavitation abuse in the return line under cyclic conditions. However, certain industrial hydraulic fluids have proven themselves superior through years of experience and for reasons not well known. The variable volume piston pumps used in test systems are very sensitive to the proprietary anti-wear additives to the extent that the system suppliers recommendations should be followed carefully. Even apparently equivalent industrial oils of other well known brands have suddenly been found to create problems with certain applications. A fatigue test servosystem is definitely not a typical industrial application. Fluid selection and changes should be made very carefully.

For example, Mil-H-5606 hydraulic fluid has excellent low temperature properties, but these are rarely needed by a test system. However, the poor film strength and low viscosity of this fluid decreases test system component life and increase leakage problems.

Fire resistant fluids should not be considered unless absolutely mandatory. They tend to result in an unpredictably short life for many system components, and are very expensive and difficult to install and maintain.

7.7 CONTROL ELECTRONICS

A number of types of commercial servo-controllers are available which combine the functions of program generation, feedback conditioning, control, limit detection, interlock, and other miscellaneous system needs. Both analog and digitally based systems are used for programming and for control. Computers commonly assume the data acquisition and command generation functions. The newest generation servo-controllers are digital, although analog systems are still very common and have a number of redeeming features for certain applications. The primary advantages of direct digital control (DDC) are its almost instant reconfigurability and its setup repeatability.

Servo-controller feedback conditioning is accomplished with either a DC excitation and instrumentation-quality low-drift gage amplifier, or an AC based carrier conditioner for an LVDT stroke transducer. Other feedback devices commonly used include strain gage pressure transducers, strain extensometers, angular position and torque transducers, velocity transducers, and tachometers with controller modifications. It is important the feedback conditioners have good stability and noise rejection characteristics as such artifacts on the conditioner output become a substantial system noise problem with high response servovalves. Otherwise, power line transients, electronic drift, poor electrical grounding of the test structure, and signal ground loops can cause serious overloads to the test structures or create undetectable errors.

Some multi-actuator applications contain so many inter-actuator cross-talk possibilities that the feedback, command, and controller/actuator combinations are best electronically redefined in a static "control matrix" or "degree of freedom" form for ease of operator interface and control loop setup. Despite the apparent complexity, multiple actuator setups respond well the these interconnections. The control loop interactions and chances of operator error are greatly reduced, since the new electronically defined loops handle severe cases of redundant control where, in an earthquake simulator for example, more than three actuators may be connected to a single axis of a rigid platform.

7.8 SAFETY AND INTERLOCK SYSTEMS

In many test setups considerable damage may occur to the test structure or even to the test system if the program command becomes excessive or if control is lost. Although it is rare that a good quality actuator is damaged from excessive force or velocity, the test structure is frequently the most expensive part of a test and is most vulnerable to damage. Further, the value of the test structure increases as the test progresses. To protect test structures and testing systems, servohydraulic test systems rely on daisy-chained interlocks which automatically indicate, and shut down the hydraulic system for protection. Such interlock systems must include logic which allows the operator to identify the particular interlock which initiated the shutdown, even though additional interlocks may have been tripped during the shutdown transient. The feature is especially important for diagnosing large system or unattended shutdowns.

Because most servohydraulic actuators are motion generating devices, care must be used to detect the failure of any servo-loop to prevent the actuator from stroking to its maximum static force or the end of its travel. For example, loss of feedback signal during a dynamic command can be particularly violent, as the actuator may thrash through its entire stroke range. Various interlocks such as feedback or monitor transducer limits, servo-loop error detection, mechanical limit switches, etc., are used to automatically remove hydraulic power from the system to protect the test structure in the event of a control fault. Some servohydraulic systems though, are so fast that considerable damage can still be done in the very short time required for a solenoid HSM to shut down. With high performance actuators, some failure modes, such as loss of feedback, are difficult to detect with the system error detectors. Of particular difficulty is whenever there is a loss of feedback with near zero command, or high velocity system where the control loop gain must be set to low that considerable error must be generated to adequately drive the servovalve fully open.

Great care should be exercised by the test engineer to verify the installation safety of transducer feedback and servovalve drive cables. Experience indicates that these simple system elements are responsible for most unexpected loop failures. Operation of the system must not expose control-critical cables to excessive or concentrated flexing.

7.9 HYDRAULIC SHUTDOWN METHODS

Simple, dedicated-pump systems are usually controlled from a system control console in which the interlock chain can also shut down the pump. Multiple test systems run from a central pump station and require hydraulic service manifolds (HSM) for hydraulic control. Although operation with manual shutoff valves is theoretically possible, such a setup is not able to use an interlock chain to any safety benefit. The HSM allows shutdown of a test station or single actuator, while leaving a central lab hydraulic pump in normal operation.

Interacting overload/shutdown effects should be considered very carefully in multiple actuator situations. Specimen coupling may cause greater forces to be reacted at certain points as the load is shed by other loading inputs. Larger multi-channel structural loadings systems often use a pair of solenoid "lock" valves which shut off the servovalve ports to an actuator, in combination with "dump" valves which connect both actuator control ports to the return line, releasing all forces.

Other overload prevention schemes include mechanical/hydraulic "load limiter" relief valves mounted on the actuator which positively prevent more differential pressure to be generated than is set on the valve. These valves operate independent from electrical power and instrumentation, and are common for setups with very valuable test structures. Positive load limiting requirements involving unequal area actuators, and non-zero load limit requirements can be achieved with a more complex hydraulic load limiter valve. Adjustable units are available which will positively limit the actuator output force without any control electronics. These units are internally compensated for actuator area-ratio effects and may have multiple stages to optimize the set-pressure accuracy and full-flow operation. If one loading direction must be inhibited, and if the actuator has equal area in both directions, a check valve across the actuator provides a reliable anti-compression (or anti-tension) function. A series manual valve also allows it to be operated as a regular tension-compression actuator.

7.10 ELECTRIC POWER LOSS

Digital programming and control systems are sensitive to power faults and power loss and may cause hardover actuator response. For this reason, an uninterruptible power source (UPS) is frequently used, especially when a direct digital control system is involved. The UPS power source must be a quality unit that includes power line surge protection and a comprehensive battery health monitoring feature. Reliable, proven UPS systems are emphasized since simple battery backup systems have a way of being symptomatically functional but not performing when needed.

Some newer analog servo-controllers include a built-in rigorous line power integrity monitor and a very short term capacitively-based emergency backup power scheme. This feature maintains control loop integrity for about one second during a power loss, allowing a typical HSM to vent the system hydraulic pressure accumulators.

7.11 SERVOHYDRAULIC TEST SYSTEM COMMAND SIGNAL SOURCES
Alan R. Humphrey, MTS Systems Corp., Minneapolis, MN

7.11.1 Introduction

The purpose of this section is to give the user an overview of possible sources for test system command signals providing an understanding of the resources required to design and execute test profiles.

Once the servohydraulic test system has been setup and closed loop control has been established throughout the desired frequency bandwidth, the test system is ready for operation. The system may now be driven by summing a command signal into the control loop, thus commanding the servovalve and causing subsequent actuator motion or output force. The command signals cause forces or motions which the operator determines necessary to investigate the test specimen characteristics. Typical tests may include cyclic material property measurements, strengths, ductilities, static and dynamic ratings, modal analysis, durability fatigue testing, and noise and vibration measurements.

The magnitude of the desired force commands can be determined in many ways. A good detailed test specification will define exactly what is required for either input forces or achieved loads at key points on the specimen. Theoretical or calculated forces may be used to setup the test if a specification does not exist, or if the test is relatively simple. For more detailed testing, a specimen may be instrumented and calibrated, allowing actual data to be collected under typical or worst case operation. The test can then be accelerated using editing to remove minimum damage inputs, to develop the test profile.

The command or drive signals can range in complexity from a simple constant amplitude sine-wave created by a function generator, to a complex computer developed drive signal, optimized to reproduce exact specimen responses at remote measurement positions. However, to simplify this discussion, typical commands will be divided into three major categories: constant amplitude, block programming, and variable amplitude as shown in Fig. 7-12.

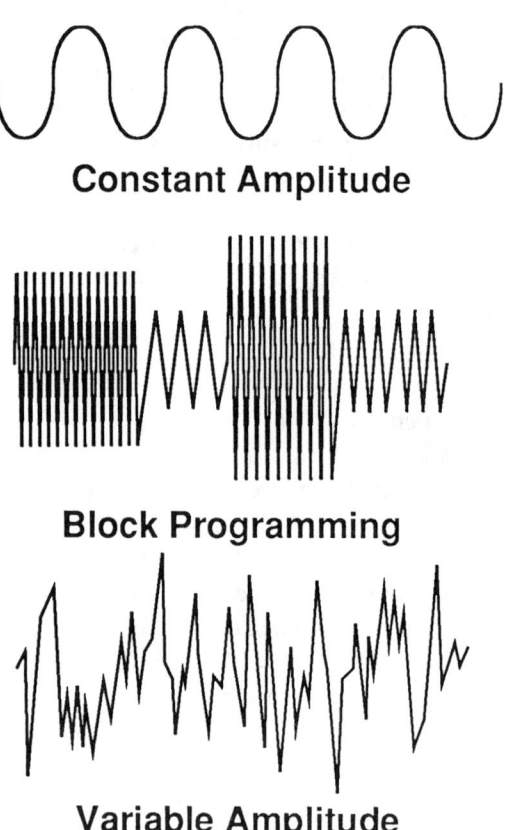

Figure 7-12. Three Major Categories of Command or Drive Signals

7.11.2 Constant Amplitude

This type of input is frequently used on simple specimen tests where measurements are being taken to categorize the specimen or where one amplitude and frequency combination are determined to be adequate to test the specimen. To generate the drive signal, a basic function generator equipped with an elapsed cycle counter would be adequate; however, more systems are now beginning to use personal computer-based systems equipped with digital to analog signal converter cards to provide the function generation capability, as well as provide a simple monitor and data manipulation system. The amplitude of the command or drive signal would be controlled via the analog span control and a mean offset could be applied to the test via a set point control.

A typical example of where constant amplitude cycles are used to drive a servohydraulic system is a materials testing machine. Here, a standard specimen is tested to failure by cycling the applied force sinusoidally at a fixed frequency. The test is then repeated on new specimens for a selection of strain range amplitudes. The number-of-cycles to failure information is then used to construct a strain versus life curve for specimen material, for fatigue life prediction on components subjected to a range of strain amplitudes during normal service (see Fig. 7-13).

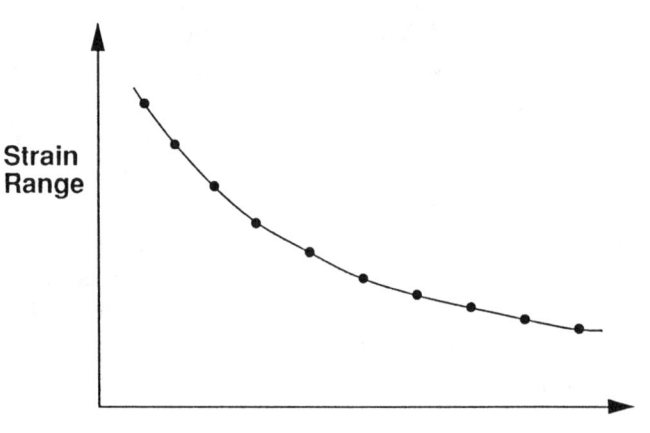

Figure 7-13. Strain Range Versus Cycle Life for Test Specimens

A second example is an automotive component which responds excessively, or resonates due to vertical inputs to a vehicle running over rough road surfaces, causing premature failure. The component could be instrumented with either an accelerometer or strain gage bridge at a critical location and response data collected while driving over a standard test surface. If the resonant frequency or possibly harmonics are dominant in the response transducers auto spectral density, the test operator may decide to drive the test system at this frequency to develop an accelerated durability test for this particular component. The magnitude of the drive signal would typically be determined by exciting the system while monitoring the response transducer feedback, adjusting the excitation level until the response approximated the peak levels measured on the original surface.

7.11.3 Block Programming

This input is typically used on specimens subjected to uniaxial or multiaxial phased amplitude loading. The block cycle excitation signal consists of a number of different amplitude sections of drive signals with either sinusoidal or ramp waveforms, where each block may also have an independent man offset. The blocks are generally created on a personal computer based system which allows profiles to be generated using a spreadsheet data entry format to specify the drive profiles. The block content is normally based on the measured specimen response time history during operation. The time history cycle counting information obtained from rainflow or level crossing analysis is then used to determine the drive profile characteristics for each block in the sequence. Sometimes blocks are developed from an industry standard load spectrum for the specimen being tested.

A test sequence is formed when all blocks have been created. This is fed to the analog controllers in the test system via digital to analog converters. The test will run until the sequence of blocks has been completed, or a monitoring or analog limit has been exceeded, typically indicating specimen degradation or failure.

Block cycle testing is a good way to simulate a large range of test amplitudes versus cycles. It is relatively simple to generate the drive profile for each block and the blocks can also be modified for repeat or similar test applications. The blocks require only minimal specification and storage requirements. Multiple tests can be stored on a personal computer based system with only limited disk capacity. Since cycles of equal magnitude are grouped into blocks, this technique causes some distortion of the sequencer effects which the specimen would actually see in normal operation, hence the fatigue life of the component may be affected. Also, tests developed based upon level crossing counting will tend to give conservative test results due to the technique used by the reconstruction algorithm.

The test can be optimized during operation to ensure that the applied forces are correct. The simplest way to do this is by scaling the applied analog drive signal with the span control until the achieved peaks are correct. More sophisticated personal computer systems are capable of performing such peak assurance optimization automatically. By monitoring the feedback from the servosystem, null pacing optimization will cause the system to achieve every commanded end level in the drive profile within a user specified tolerance band, before proceeding with the next end level. The operator specifies the error integrating gain factor, which is a measure of how much to increase the command signal per unit of time to achieve the specified end level. This technique has the effect of extending the test duration as the drive profile is held until each end level is achieved as shown in Fig. 7-14.

Amplitude control optimization boosts the amplitude of the command signal at each end level in the block

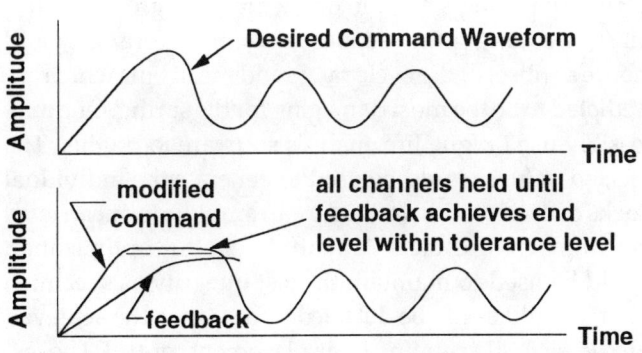

Figure 7-14. Desired and Modified Command Signals

cycles to compensate for limited servosystem response and bring the feedback up to the desired end level. The command is increased by a gain factor multiplied by the DC analog error at each end level over a series of cycles in a block until the desired end level is achieved. This level of command signal is then maintained for the duration of the block (see Fig. 7-15). When a new block is started, the optimization algorithm is restarted. The amplitude control technique boosts amplitude and maintains frequency content, hence it is the preferred method when test time is crucial.

A vehicle coil spring primarily subjected to vertical loading is a good application example for block cycle testing. The test system would incorporate a restraint fixture for the spring, a servohydraulic actuator mounted to cycle the spring along its axis, instrumentation and a block cycle programmer. The spring could be instrumented for either vertical displacement or strain. Response data would be collected while running over the accelerated

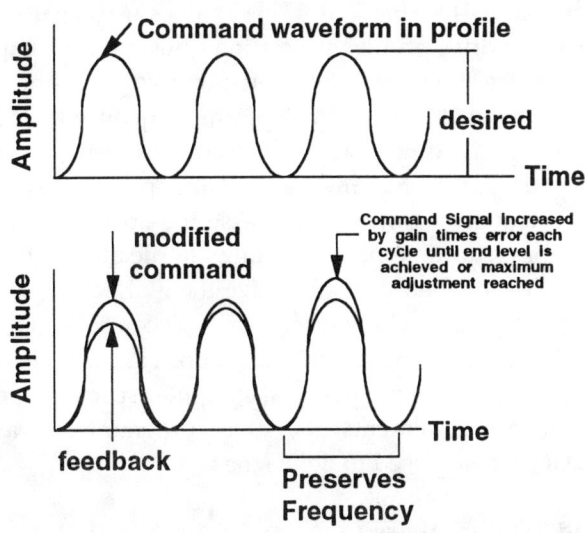

Figure 7-15. Modification of Command Signal

durability testing schedule on the proving ground. After rainflow counting the response data, the operator would choose a subset of the cycle range and mean bins which are predicted to be the most damaging for the spring. One may possibly use fatigue life analysis software to do this. The selected values would be used to generate the individual blocks of cycles, each defined by an amplitude, mean level and number of cycles. Amplitude control optimization would be used to maintain the test integrity as specimen failure would easily be detected by a drop in the achieved force level while running in displacement control. The test would consist of repeating the sequence of blocks until either the total number of desired cycles was exceeded or the specimen fails.

7.11.4 Variable Amplitude

Many techniques exist for developing a variable amplitude drive signal. They are typically applied to systems where the complexity of the test, or the required accuracy of the desired results cannot be met by either constant amplitude or block programming methods. The determination of which variable amplitude technique to use for each application is also dependent upon the above factors as well as possible equipment, budget, and time limitations.

Reconstructed Peak-Valley Excitation

Typically used as an alternative to block cycle programming, the reconstructed peak-valley technique has the advantage of maintaining the sequence effects that the specimen would see under normal operation, hence the test simulation becomes more realistic. The same test equipment is used, but rather than breaking the specimen response signal into blocks, the response time history peaks and valleys (and optionally times), are stored sequentially for each channel (see Fig. 7-16). These values are then used to generate a drive profile by inserting sinusoidal or ramp waveforms between adjacent points, thus constructing a drive signal which maintains the same sequence as the original response time history. If the time between points was also used, the resulting signal will approximate the original frequency content, otherwise the operator may specify the delta time between points to dictate the frequency. This method can still utilize the null pacing optimization technique; however, amplitude control can no longer be used as the amplitude is not constant. This technique requires more personal computer specification and storage requirements than block programming, as more data is being used to define the test.

Shaped Random Noise

For specimens subjected to variable amplitude load-

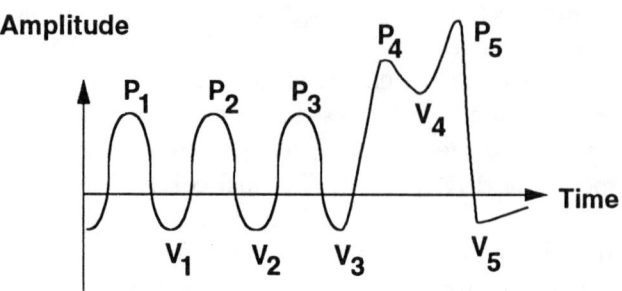

Figure 7-16. Sample Response Time History for Each Channel

ing in either uniaxial or uncorrelated multiaxial conditions, where the test data is stationary and Gaussian, it may be possible to use a shaped random noise drive signal to perform the test. The test specification parameters that may define this type of testing are: mean, standard deviation, and spectral shape (typically autospectral density of the desired response data). Otherwise, response data would be collected from the specimen to establish these parameters. The advantages of this method are that it is easy to specify, a wide variety of natural vibrations can be simulated, and in simple cases, the drive signal can be created without a computer. The main disadvantages are that it is not applicable to non-stationary signals and the operator has little control over the achieved end levels. If the control computer has the capability, the test may be compensated manually by adjusting the waveform definition parameters and retrying the test, or an iterative technique could refine the drive signal until the response auto spectral density matched the drive spectral density. This technique is typically used on simple vibration systems or when specifically called-for in the test procedure.

Real Time Direct Excitation

This technique involves measuring responses from the specimen and then playing these time histories into the test system servohydraulic controllers as command/drive signals, using either a magnetic tape playback deck or a computer system equipped with digital-to-analog converters. The magnitude of the drive signal can be modified by adjusting the span on the analog controller or by scaling the drive signal externally. Possible applications would be for specimens where the response data is non-stationary or non-Gaussian in distribution. This simple method has the advantage of minimal processing and equipment requirements; however, in most applications, it may result in a poor simulation test. This is because it does not compensate for specimen nonlinearities, the limited servohydraulic performance characteristics, and other factors which may influence the data. When setting up a test of this nature, the initial specimen response is normally monitored and the drive signal is sealed until the peak levels are being achieved with reasonable accuracy. This frequently leads to an inaccurate test of the specimen as the peak events are typically nonlinear, hence the general data levels would not be achieved correctly. Except for very simple or linear specimen test setups, this technique is normally superseded by an iterative type of approach.

Double Integrated Acceleration

If a response acceleration signal is available for the specimen, it may be possible to double integrate this signal to provide a displacement signal which could then be used as a drive signal for an analog displacement control loop as shown in Fig. 7-17. This could be used to run an inertially reacted displacement control test where the response data was non-Gaussian or nonstationary. This would require equipment to perform the double integration, scaling, and high pass filtering operations to eliminate any offset caused ramp functions induced by this operation. This technique is very similar to real time direct excitation and is subject to the same setup and accuracy limitations as shown above. Accelerometer dynamic range and offset considerations typically limit the upper/lower frequency range of the double integration process to about 10:1, although with care it can be extended to perhaps 30:1. The limited response capability of the the system displacement control loop must be considered also.

Artificial Excitation

In situations where the specimen response is unknown or cannot be measured, an artificial drive signal can be generated to develop an approximate test for the specimen or to verify an analytical model. The operator takes the available information pertinent to the event to be simulated to create a drive profile using a computer based segment generation program. The developed drive signal is then output to the test system analog controller via a digital-to-analog interface card. For example, for a vehicle driving over a pothole, the required information would be the x-y coordinates for all intersection points on the cross section of the pothole, the vehicle speed and the wheelbase of the vehicle. From this, a drive signal could be approximated for a tire coupled simulate, assuming that the wheel follows the contours of the pothole perfectly and without deformation. The speed and profile of the drive signal could be adjusted until the obtained motion was acceptable to the test operator. This technique obviously has many potential problems; however, it does provide a quick method of getting a simple test running.

Time History Reproduction

The time history reproduction technique is used to reproduce field measured specimen responses in the laboratory with a high degree of accuracy. Since the phase relationship between channels is maintained, this method can be used for multiaxial testing where the inter-channel phase relationship can be critical to accurate test development. It can be applied to a diverse range of test systems with varying levels of complexity, with current applications ranging from a simple single-channel component test used for noise and vibration testing to a sixteen channel spindle coupled road simulator used for full vehicle durability testing. Simple systems normally used a personal computer based software package where digital-to-analog and analog-to-digital converter interface cards are mounted in the backplane of the personal computer. This makes the system compact and portable, allowing use on multiple test systems if required. Complex systems with more than four control channels use a more powerful computer system to speed the numerous calculations and give the operator more flexibility to run the system and trouble shoot problems. This system is typically dedicated to one particular test as the complexity of the test normally requires sophisticated specimen monitoring in all stages of the test. Both systems use the same operating principles and the overall process can be divided into six major phases summarized below:

Data Acquisition: The specimen to be tested must be instrumented with sufficient channels to allow a drive signal to be developed. As a minimum, this would be one channel of response for every channel of drive signal in the proposed test system, where each response transducer is primarily, but necessarily entirely, sensitive to one axis of loading. For complex test applications, response transducer redundancy is common to ensure that the full frequency range of interest may be simulated and to check the correlation at other positions on the specimen which were

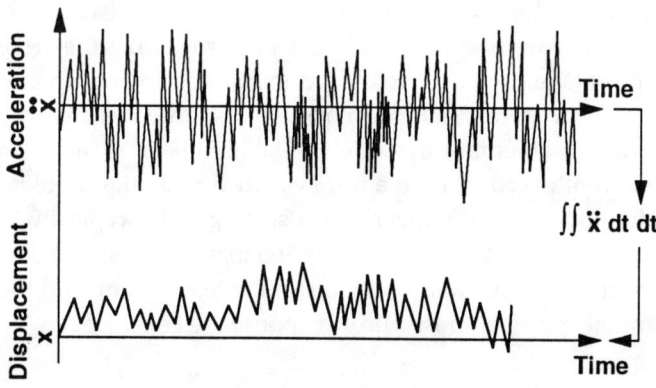

Figure 7-17. Sample Double Integration of Acceleration History

not directly controlled during the simulation process. Data is typically recorded in either analog or digital format for all test surfaces or events which are to be simulated on the set system as shown in Fig. 7-18.

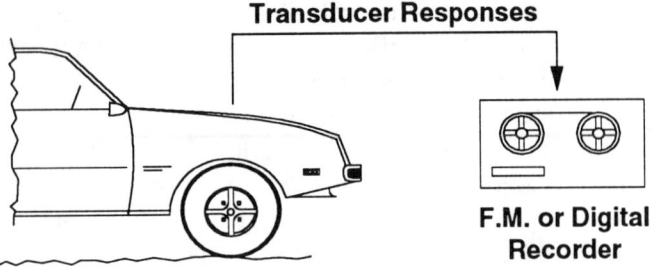

Figure 7-18. Data Recording Sequence for Individual Transducers

Data Editing: The test data is transferred into computer files where it is first checked for validity and then edited to remove unwanted portions such as periods of undamaging data or instrumentation noise spikes (see Fig. 7-19). Editing unwanted data is a very important process as it allows the user to concentrate only on significant events, greatly decrease the overall test time, thus allowing accelerated testing to be carried out. This process is done using visual aids or statistical analysis; however, modifications are made across all channels simultaneously in time to ensure that the correct phase relationship is maintained between all response channels. Data is typically edited in short blocks, with automatic routines being used to smoothly splice new data blocks together.

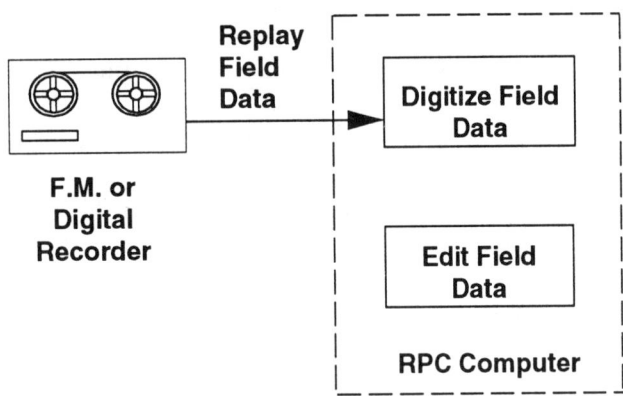

Figure 7-19. Recorded Data Validation and Removal or Unwanted Signals

Frequency Response Function Measurement: The frequency response function of the test system is a linear mathematical model of the system, which incorporates the characteristics of all components within the system such as the analog console, actuator performance, fixture cross coupling, specimen characteristics, and anti-aliasing filter characteristics. Hence, when this is used to develop drive signals for the system, all of these linear effects are compensated for, thus improving the accuracy of the test. To measure the frequency response function, a frequency shaped random noise drive signal is used to excite all drive channels in the test system. While the system is running, the responses of all the specimen transducers are sampled so that the linear response-to-drive relationship can be determined for all frequencies excited by the white noise drive (see Fig. 7-20). The frequency response function of the test system is calculated from these two signals. This may then be analyzed to evaluate the controllability of the specimen response transducers. The potential control bandwidth for the system will normally be determined at this time.

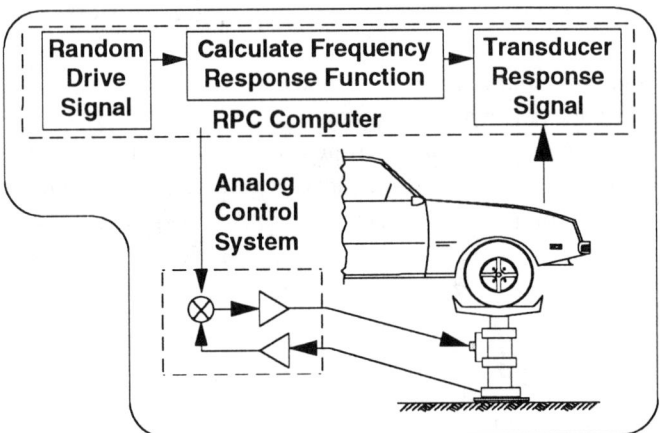

Figure 7-20. Frequency Response Function Calculation Schematic

Initial Drive Estimate: By using the desired specimen response signal which was created by editing the response data and the system frequency response function, an initial drive estimate for the system can be calculated (see Fig. 7-21). This drive signal usually only approximates the drive time history required to reproduce the specimen response in the laboratory, as the frequency response function is a linear mathematical model for a nonlinear system. To prevent overdriving the test system, a potential problem due to system nonlinearities, a scaling factor typically in the range of 0.4—0.8 can be applied to the drive signal. The initial drive signal is used to run the test system, and the initial specimen transducer response are collected.

Iteration: To compensate for differences between the desired responses and the initial responses obtained above, an iterative process is used. The initial responses are sub-

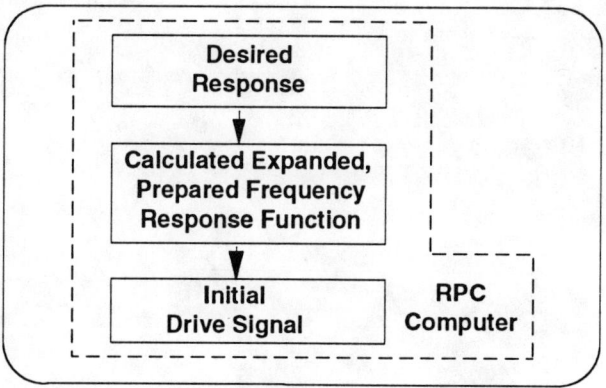

Figure 7-21. Calculation of Initial Drive Estimate Schematic

tracted from the desired responses on a point-by-point basis to generate an error signal. This is then used to calculate a correction drive signal, again scaled by a safety factor, which is then added to the initial drive estimate. The modified drive signal is then used to run the test system and new response signals are collected as shown in Fig. 7-22. This iterative process is repeated until the error between the desired and achieved specimen responses is judged to be insignificant.

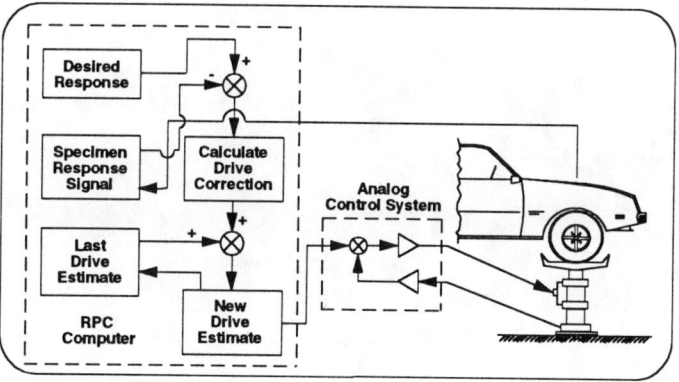

Figure 7-22. Modified Drive Signal and System Response Schematic

Durability Testing: The iterated drive files are used to build a sequence consisting of passes and repeats of drive files to meet the test requirements. This sequence is then used to drive the test system until the sequence is completed or a failure occurs. While the test is running, the computer can be used to monitor the test rig, the specimen, or operator specified external events to ensure that the test integrity is maintained as shown in Fig. 7-23.

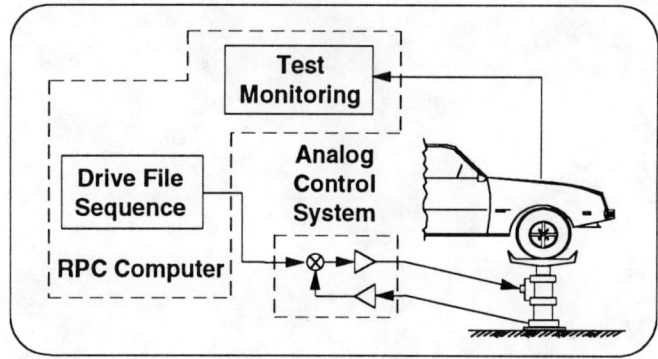

Figure 7-23. Drive, Test Control, and Monitoring Systems

7.9. REFERENCES

7-1. Bendat, J.S. and Piersol, A.G., *Random Data Analysis and Measurement Procedures*, 2nd Ed., John Wile and Sons, New York, 1986.

7-2. Fisher, D.K., "Theoretical and Practical Aspects of Multiple Actuator Shaker Control," Proc. of the IES, pp. 153-174, June 1973.

7-3. Ramirez, R.W., *The FFT Fundamentals and Concepts*, Prentice-Hall, Englewood Cliffs, NJ, 1985

7-4. Nolan, S.A. and Linden, N.A., "Integrating Simulation Technology into Automotive Design Evaluation and Validation Processes," SAE 87194, presented at Passenger Car Meeting and Exposition, Dearborn, MI, Oct. 1987.

7-5. Lund, R.A. and Donaldson, K.H., "Approaches to Vehicle Dynamics and Durability Testing," SAE 820092, Presented at SAE International Congress and Exposition, Michigan, Feb. 1982.

7-6. Donaldson, K.H. and Dabell, B.J., "Decision Path Guidelines for Applying Digital Testing Techniques in Fatigue Service Simulation," presented at SEE, London, May 1985.

7-7. Herraty, A.G., "Selection Criteria for Simulation Tests Systems," Presented at Inst. Mech. Engrs. Prediction and Simulation of In-service Conditions Conference, London, May 1985.

7-8. Isaacson, R.D., "Measurement of Service Environment," Paper 820684, Presented at SAE Fatigue Conference, Michigan, April 1982.

7-9. Bendat, J.S. and Piersol, A.G., *Engineering Applications of Correlation and Spectral Analysis*, John Wiley and Sons, New York, 1980.

7-10. Crochiere, R.E. and Rabiner, L.R., *Multirate Digital Signal Processing*, Prentice-Hall, Englewood Cliffs, NJ, 1983.

7-11. Otnes, R.K. and Enochson, L., *Applied Time Series Analysis: Basic Techniques—Vol. I*, Wiley-Interscience, New York, 1978.

7-12. Barber, A.J., "Durability Test Technology Training Course Notes," MTS Systems Corp., Minneapolis, MN.

7-13. Ericson, C., "Remote Parameter Control Training Course Notes," MTS Systems Corp., Minneapolis, MN.

CHAPTER 8
OVERVIEW OF EXPERIMENTAL METHODS AND TECHNIQUES

Robert T. Reese, Ph.D., Sandia National Laboratories, Albuquerque, NM
Wendell A. Kawahara, Ph.D., Sandia National Laboratories, Livermore, CA

8.1 INTRODUCTION

In any structural test, it is likely that a variety of experimental approaches, methods, techniques and equipment such as transducers, gages, sensors, and other measurement devices could be used. Some of the real pleasures of working in the experimental mechanics field are to take advantages of new technologies and the continuing improvements in existing techniques and equipment which have greatly expanded experimental capabilities. Unfortunately, there is a divergence that continues to develop between the increasing rate of development of new techniques and improvements in existing methods, and the diminishing amounts of formal instruction introducing these methods (Refs. 8-2 through 8-6).

This chapter provides an overview to the next three chapters which describe experimental methods and techniques used in structural testing. These following chapters condense and summarize information giving the needed overviews and details previously unavailable or arrived at with considerable effort. This information will help bridge this divergence and offers practicing engineers and others the summaries and details they need in performing structural tests.

There are four categories for experimental methods and techniques: (1) point techniques, (2) whole or full field techniques, (3) nondestructive evaluation methods, and (4) transducers, sensors, and other measurement devices. These categories have been used for many years and still offer the basic distinctions needed. Point techniques, sensors, and transducers will be combined into one chapter. Another chapter describes full or whole field techniques. A separate chapter is devoted to nondestructive evaluation methods. The emphasis in these following three chapters is on aspects relating to structural testing and to complement SEM's Handbook on Experimental Mechanics (Ref. 8-5).

There are some unique aspects of each of these four categories and there are areas where they are quite similar. Some structural tests may require the use of only a single gage (e.g, an extensometer to measure strain.) Other tests may involve the use of many transducers, gages, and diagnostic devices. Some techniques involve considerable equipment, require special fixtures, need computer controls and data acquisition systems, require carefully controlled conditions, and often result in great amounts of data. Other techniques will need little instrumentation and have very simple evaluation requirements (e.g., pass or fail). Regardless of the relative complexity of a particular test, the measurements made in a test should give insight to the response of the structure. Choosing the right experimental methods and measurement devices to do the job is essential in performing high quality tests.

There is usually more than one method or technique which can be successfully applied to a particular structural test. One potential concern in structural testing is to be "boxed in" in terms of only having one available method for a specific test. When only one approach is available, there almost always seems to be difficulties when tests are performed. Those involved in structural testing do not need to be limited in testing capabilities if they will understand the methods and techniques available, how they can be utilized to give the insights needed, and how various methods can yield equivalent results and understanding of structural behavior (Refs. 8-3, 8-5, 8-6, 8-8, and 8-9).

The remainder of this chapter gives a brief overview of point techniques and transducers, whole field techniques, and nondestructive methods. An example of a structural test utilizing these various methods and techniques follows.

8.2 POINT TECHNIQUES—GAGES, TRANSDUCERS, AND SENSORS

Some point measurements can be made with mechanical means such as dial or digital displacement gages, torque wrenches, micrometers, etc. Measurements made with mechanical systems are typically self-taught or can be learned with the instructions usually provided with these devices.

Measurements made in structural tests using point techniques, sensors, and transducers interface with electronic and computer systems. Electronic systems convert the physical quantities sensed by the gage or transducer into signal voltages which can be measured directly, amplified and measured, or can be made into suitable form for indication of behavior.

A point technique is an experimental approach in which measurements are made at one or more locations on a structure. These point techniques often use a measurement device such as a strain gage, load cell, accelerometer, or thermocouple. Measurements at points on a structure define responses at specific locations on a structure. The structural responses are typically accelerations, displacements, forces, velocities, temperatures, or strains. A gage or transducer is attached to or is positioned in close proximity to a desired location on a structure. The response of the gage or transducer indicates the behavior of the structure at that point (Refs. 8-1, 8-2, 8-4, 8-5).

A clear distinction among gages, sensors, and transducer would be appropriate at this point. Unfortunately, clear differences do not exist; however, a transducer is usually considered to be made of one or more gages protected in some type of housing or enclosure. These transducers are calibrated to convert the physical quantities sensed into electrical signals. For example, a load cell measuring forces can use strain gages in certain configurations to detect strains in the active part (flexure) of the load cell (e.g., force transducer). These strains are converted to electrical signals through the gages and associated Wheatstone bridge electrical circuitry. With a force calibration, a relationship is established between force and signal voltage output. Transducers can be much more complicated than a relatively straightforward conversion of a strain induced voltage signal to a known force. Transducers can use almost any electromechanical principle to convert measured quantities into electrical analogs. For purposes of organization of these chapters, gages, sensors, transducers, and other devices have been combined in this chapter. A major reason for this organization is because it is difficult to describe distinctive differences for the various gages, sensors, and transducers used in structural testing.

The point technique is based on the correlation of the information gathered at specific locations on the structure and the descriptions of structural behavior provided by the analytical models. This correlation and comparison process will give insights into the behavior of the whole or at least a portion of the structure being tested.

8.3 FULL FIELD TECHNIQUES

Whole or full field techniques describe the behavior of a portion or all of a structure typically in terms of strains, displacements, stresses, or temperatures. This category of techniques has similarities because they rely on some type of optical, lens, or camera based system which gives capability to view a portion or all of a structure. The use of one of these techniques (e.g., photoelastic coatings, brittle coatings, moiré), usually does not permit the concurrent use of another of these techniques in the same test.

The whole field techniques are based on the behavior of light in terms of its reflection, refraction, diffraction, interference, and image formation. Full field techniques are usually thought of in terms of the behavior of the surfaces of a structure that can be exposed or otherwise made available to cameras and optical methods. However, techniques have also been developed in which slices of a structure (e.g., three-dimensional photoelastic models of automobile engines) can be made and their behavior evaluated using full field techniques. Among the whole field techniques are: photoelasticity, moiré, holography, hic interferometry, brittle coatings, thermal emissions, and modal analyses.

8.4 NONDESTRUCTIVE EVALUATION TECHNIQUES

Nondestructive methods can give both point and whole field response of the structure. However, they differ from the first two methods because nondestructive methods are not supposed to impair or damage the structure. (This does not imply that the use of point or whole field techniques do inflict damage on a test structure or keep it from useful service.) Another way of describing the differences between nondestructive methods and point or whole field techniques is that nondestructive methods involve application of some type of energy to the structure (e.g., X-rays, ultrasonic signals, electrical currents, etc.) or rely on some action of the structure (e.g., sound generation) as the diagnostic indicator. Loading conditions are used to exercise a structure and the responses are indicated by point or whole field techniques. Nondestructive techniques rely on the use of energy sources (e.g., X-rays) to determine structural response.

Nondestructive methods can be grouped into about ten basic categories:

1. Acoustic (both sonic and ultrasonic)
2. Electrical and electrostatic
3. Electromagnetic induction

4. Magnetic
5. Penetrant
6. Photometrics
7. Pressure, flow, and leakage
8. Radiography
9. Thermal
10. Optical

Nondestructive methods are usually thought of in terms of detection and characterization of flaws in a material or structure. The flaws cause interruptions in the energy applied and detected by the method used. Recent developments have expanded some of these methods to give structural responses directly (Refs. 8-6 and 8-9).

8.5 EXAMPLE—PRESSURE VESSEL

The use of these four categories of experimental techniques and methods is best described by a proposed test. Suppose that a thin-walled steel pressure vessel shown in Fig. 8-1 is subjected to a pressure test. This vessel has been designed according to the applicable sections of the ASME Pressure Vessel Code. Similar vessels could be used for a variety of purposes; in this example it is a 1/8 scale model of a nuclear reactor containment structure (Ref. 8-7). This model was tested to failure (internal gas pressure) to evaluate the predictive capabilities of analytical methods beyond the range of design loads. The vessel has the following features:

- Openings simulating those found in actual containment structures with thickened sections around each opening. (The thickened sections had extra materials added at a distance around each opening.)
- Thin wall construction with circumferential stiffeners.
- Elliptical base section.
- Hemispherical top section.
- Drain port to release condensation.
- Pressurization system (not shown except for the inlet port).

All four categories of experimental methods could be used on this test as follows:

- Strain gages for strain measurements.
- Displacements of the specific locations on the wall of the vessel and around the stiffened openings.
- Pressure transducer to determine the gas pressure and possible leakage.

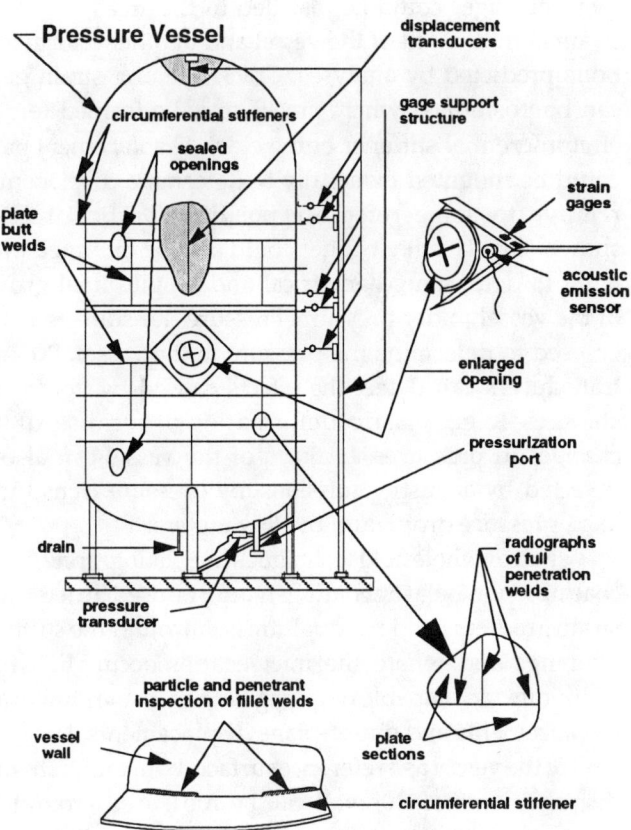

Figure 8-1. An Example Pressure Vessel Instrumented with Transducers and Sensors from the Four Basic Experimental Mechanics Methods

- Full field techniques (e.g., moiré, photoelastic, or triangulated surveying coordinate system) for in-plane strains and displacements and out-of-plane displacements.
- Radiography to evaluate the welds used in fabrication.
- Penetrants to evaluate the fillet welds
- Ultrasonics to evaluate potential flaws in the base materials and stiffened sections.
- Still photography and high speed video and cameras to record structural behavior at failure.
- Thermography to map the temperature distributions around the openings.
- Acoustic emission used to indicate the growth of flaws and mapping of the areas where failures initiate and propagate.

Various parts of the surface of the vessel would be appropriate places for a strain measurement. If a detailed closed form or finite element analysis were available, then the gages could be installed in the areas of greatest strain at midheight of the vessel and at other critical locations predicted by analyses. Crack-opening strain gages can be positioned where cracks may be formed (e.g., at circumferential stiffener butt welds). Displacement gages could be mounted externally to determine displacement relative to some reference position (vertical rod) as shown. Displacement gages could also be mounted internally to determine diametrical and longitudinal growth of the vessel under pressure. Pressure transducers would be used to determine the pressure in the vessel. Pressure transducers can detect the effects caused by volumetric changes (e.g., plastic deformations) because of the changes in pressures. Leakage of the vessel can also be detected by acoustic emissions, by pressure transducers (e.g., pressure drop), and by flow meters.

Full or whole field techniques (e.g., holography, brittle coatings, photoelastic coatings) could be used to determine strain gradients and residual stresses around the stiffened openings and where the intersections occur. Full field techniques are capable of giving both in-plane strains and displacements and out-of-plane displacements. Using the wall of the vessel as a reference surface, both the strains and displacements of the wall can be measured around the circumference of the deformed vessel along with diametrical growth. The residual stresses present from differential cooling in the welding processes can also be evaluated with whole field as well as point techniques.

In the area around one of the larger openings in the vessel, a moiré setup could be used to determine the in-plane strains and the out of plane displacements of the vessel wall. Acoustic emission transducers can be installed on the vessel in an array so that the location of any cracks that may initiate and propagate will be heard by the sensors (microphones) used and their location determined. In the event that a crack does form, then the data from the acoustic emission test will show the approximate location of the crack by overlaying the responses of each sensor in the array, a pressure vs. acoustic count summation will show when the crack was first detected, and give positive evidence that crack was initiated.

Ultrasonic inspection of the vessel can reveal flaws in the vessel walls, in the thickened portion around the ports, etc. X-rays can be used to perform detailed radiographic inspection of the welds used to join the vessel sections along with the welds used to join the thickened areas around the ports. Section 8 of the ASME Pressure Vessel Code requires radiographic inspection of welds in certain classes of pressure vessels. Performing a pressure test on this failure where rupture could occur will be constrained by safety considerations and data requirements. One option is to perform the pressure tests remotely. Transducers can be hardwired into the data acquisition system. Transducers that can convert measurements into electrical signals (e.g., most transducers) can have these signals transmitted by telemetry or recorded on-board the test item. Remote testing of pressure vessels (e.g., possible burst tests can be performed safely because the data can be acquired without exposing personnel to possible hazards of shrapnel and direct exposure to released pressure in gases or liquids used).

If shock or vibration loads could be imposed, then accelerometers could be located at strategic points so that dynamic response of the structure could be determined. If a similar vessel was designed for a water distribution system and was mounted near the top of a multi-story structure, then the dynamic response of the pressure vessel and distribution system could be important (e.g., behavior in earthquakes, waterhammer effects), particularly if some resonant frequencies could be excited in the structure.

By using this example, the complementary nature of these four basic methods of experimental measure have been shown. Each method can give similar indications of structural response but each approach also has unique features. It is a great advantage to the experimentalist to have access to these four basic methods and approaches to testing as each will give additional insight into the response of a structure. It is also important to recognize that virtually no laboratory will have all of the techniques and methods available. Therefore, equivalent methods and techniques will need to be considered and evaluated for each proposed test.

There are some considerations that should be identified and evaluated before proceeding with a structural test. Among the items and questions to consider are the following:

- What information and insights are needed from a particular test versus the choices of techniques available to perform the test? It is typical to have an imperfect match of test capabilities (equipment, trained personnel, etc.) with the test requested.

- What type(s) of test will be performed?

- What ranges of measurands are expected (how large will the strains, accelerations or displacements be?)

- Are the principal directions or axes known? Measurements are often made in these directions.

- What requirements exist for the accuracy and repeatability of the data as requirements will vary widely?

- What materials will the gages and transducers be attached or adjacent to? Material compatibility may need to be verified and potential problems identified.

- What is the orientation of and the accessibility to the surface or component to be instrumented? The usual rule is that most structures are not designed with ease of testing in mind. Sometimes structural tests are not even considered in development of a particular component or system.

- Will the gages and transducers be required to function in subsequent tests? Will some type of protection, electrical isolation, etc. be needed?

- Are there hazards to test personnel and observers and equipment involved (i.e., materials, handling, fixturing, etc.)? Are failures expected or could they occur? Proper planning for safety prior to a test has saved the career of many engineers and technicians.

- What time is available to complete instrumentation, fixture fabrication, and test setup? Often the schedules do not allow for sufficient time to properly prepare for and conduct tests.

- What is the experience of others who have performed similar tests? What problems have been encountered, can data and insight be obtained, and can the test be performed safely?

- Can laboratory tests be duplicated in the field, or can laboratory and field tests be conducted remotely (i.e., at ocean depths, in space, or deep in the earth's crust?

- What operating conditions will be present when the test is conducted? Extremes of heat, cold, moisture, dust and other pollution affect the operation of gages, calibration of transducers, and operation of instrumentation systems.

- What means will be used to transmit, record, and display the data? For hard wired systems, what is the distance from the gage or transducer to the recording device? For other systems, will on-board recorders be used or will the data be telemetered? In each case, considerable care must be taken to ensure that data obtained is free of spurious signals and accurately describes the response of the structure.

8.6 REFERENCES

8-1. I.M. Allison, I.B. MacDuff, and P. Stanley, *Monograph: Methods and Practice for Stress and Strain Measurement*, British Society for Strain Measurement, Newcastle upon Tyne, England, 1977.

8-2. J.W. Dally and W.F. Riley, *Experimental Stress Analysis*, McGraw-Hill, New York, 1978.

8-3. C.M. Harris, *Shock and Vibration Handbook*, 3rd Ed., McGraw-Hill, New York, 1988.

8-4. M. Hetenyi, Ed., *Handbook on Experimental Stress Analysis*, John Wiley, 1950.

8-5. A.S. Kobayashi, Ed., *Handbook on Experimental Mechanics*, Prentice-Hall, New York, 1987.

8-6. W.J. McGonnagle, Nondestructive Testing, Gordon and Breach, New York, 1975.

8-7. R.T. Reese and D.S. Horschel, "Design and Fabrication of a 1/8 Scale Steel Containment Model," NUREG/CR-3647, SAND84-0048, Sandia National Laboratories, Albuquerque, NM, Feb. 1985.

8-8. G.M. Sabnis, H.G. Harris, R.N. White, and M.S. Mirza, *Structural Modeling and Experimental Techniques*, Prentice-Hall, Englewood Cliffs, NJ., 1983.

8-8. _____, *Nondestructive Testing Handbook*, 2nd Ed., American Society for Nondestructive Testing, Columbus, OH, 1982.

Chapter 9

Point Techniques, Gages, Transducers, and Sensors

Robert T. Reese, Ph.D., Sandia National Laboratories, Albuquerque, NM
Wendell A. Kawahara, Ph.D., Sandia National Laboratories, Livermore, CA

9.1 INTRODUCTION

This chapter describes the point techniques and transducers used in experimental mechanics. This chapter has been assembled to complement the available texts and product information. Summaries and comparisons have been made in as many areas as practicable. Commercial suppliers of gages and transducers have excellent product manuals, offer many technical aids and helps in using their products; many offer technical consulting assistance, and most of this information is available at little or no cost.

9.2 STRAIN GAGES

There are two basic types: (1) those that are in contact with the surface and use a change in dimension per unit length of gage to monitor the strain such as a bonded electrical resistance strain gage and (2) those that are not in permanent intimate contact over the whole gage surface such as extensometers, optical devices, clip gages, and other devices which measure a change in length over a gage length. Unfortunately there is not a clear agreement on the terms used by the experimental mechanics community with regard to differentiating between these two types of strain gages. Therefore, this book adopts the following convention: a strain gage will be defined as in the first definition given above. That is, a strain gage measures a change in length over the length of the gage and translates that dimensional change into strain. The other devices, sometimes called strain gages, will be described in a separate section on extensometers which bridge the measurement capability between strain and displacement (Refs. 9-1, 9-3 through 9-11, 9-12 through 9-19).

Derivation of Strain Gage Equations

The strain gage is one of the most widely used devices in structural testing. When the term strain gage is used, it is typical for one to think of a bonded electrical resistance gage. The electrical resistance strain gage is based on two principles first discovered by Lord Kelvin in 1856. In his experiments on axially loaded copper and iron wires in tension, he determined that both the resistance and geometries of a conductor vary with increasing strain. Beginning with the basic relationship for resistance R of a uniform conductor is given by:

$$R = \rho L/A \qquad \text{Eq. 9-1}$$

where R = resistance in ohms
 ρ = specific resistivity
 L = length of conductor
 A = cross-sectional area

The total differential of Eq. 9-1 gives:

$$dR = d\rho\, L/A + dL\, \rho/A - \rho L\, dA/A^2$$
$$dR = (\rho A\, dL + A L\, d\rho - \rho L\, dA)/A^2 \qquad \text{Eq. 9-2}$$

The volume of the conductor is the product of the cross-sectional area and the length (V = AL). By differentiating this volume relationship,

$$dV = L\, dA + A\, dL \qquad \text{Eq. 9-3}$$

For small elastic volume changes, dV can approximate ΔV which is equal to $V_{final} - V_{original}$, then

$$dV = L_{final}\, A_{final} - L A$$

where L_{final} = $L(1+\varepsilon)$ and $A_{final} = A(1-\nu\varepsilon)^2$, then
 dV = $L A (1+\varepsilon)(1-\nu\varepsilon)^2 - L A$
or dV = $L A [\varepsilon(1-2\nu) + \varepsilon^2(\nu^2 - 2\nu) + \varepsilon^3(\nu^2)]$

$$\text{Eq. 9-4}$$

Ignoring the higher powers of the small strain terms (ε^2 and ε^3) in Eq. 9-4, recalling that engineering strain $\varepsilon = dL/L$, and equating Eqs. 9-3 and 9-4, then

$$L \, dA = -2\nu A \, dL \qquad \text{Eq. 9-5}$$

By substituting Eq. 9-5 into Eq. 9-2 and dividing by R, then

$$\frac{dR}{R} = \frac{dL}{L} + \frac{d\rho}{\rho} + 2\nu \frac{dL}{L} \qquad \text{Eq. 9-6}$$

or,
$$\frac{\frac{dR}{R}}{\frac{dL}{L}} = 1 + 2\nu + \frac{\frac{d\rho}{\rho}}{\frac{dL}{L}} = 1 + 2\nu + \frac{\frac{d\rho}{\rho}}{\varepsilon} \qquad \text{Eq. 9-7}$$

Equation 9-7 can be rewritten for strain sensitivity, S_A (resistance change per unit of initial resistance divided by the applied strain = $(\Delta R/R)/\varepsilon$ as follows:

$$S_A = 1 + 2\nu + (d\rho/\rho)/\varepsilon \qquad \text{Eq. 9-8}$$

Equation 9-8 is the mathematical statement of the discovery made by Lord Kelvin. The first two terms of Eq. 9-8, $(1 + 2\nu)$, are the resistance change due to geometrical effects and the last term, $(d\rho/\rho)/\varepsilon$, is the change in specific resistivity.

The gage factor F for a strain gage is defined as:

$$F = (\Delta R/R)/\varepsilon \qquad \text{Eq. 9-9}$$

F is equal to S_A and is the common measure and comparison for strain gages. Strain is typically determined by measuring accurately the change in resistance and then dividing this measured change in resistance by both the original resistance of the gage and the gage factor as shown in Eq. 9-10:

$$\varepsilon = (\Delta R/R)/F \qquad \text{Eq. 9-10}$$

Early Developments of Strain Gages

Strain gages were first introduced commercially by Charles Tatnall in 1938. These gages were made of wire attached to a thin paper backing. These gages were bonded to a structure with nitrous cellulose cement (Duco). Virtually every aspect of the strain gage including its construction and installation have been improved and modified extensively since their initial introduction. The major improvements are: (1) use of foil sensing elements rather than wire, (2) encapsulation of the gage for moisture protection and wear resistance, (3) use of plastics and other pliable and tough materials for the gage backing, (4) use of improved adhesives for bonding, (5) development of a wide variety of geometries, grid patterns, and tab layouts (electrical connections), and (6) improvements in lead wires and attachments. It is estimated that over 200,000 various types of gages are currently available.

The wire gage was the first strain gage developed concurrently by Simmons and Ruge in the 1930's. The wire used was constantan or nichrome. The initial backing was paper. This gage has been improved by the use of more resilient backing materials and is available in many sizes and configurations.

Foil strain gages were introduced by Saunders and Roe in England in 1952. They were developed as a new technology concurrent with improvements in etching and plating. The strain sensing element for a foil gage is manufactured by photoetching a metal film attached to a backing. Since this type of gage is the most commonly used strain sensing device, it will be described in detail later in this chapter.

Besides foil gages, other strain gages have been developed for special purposes such as high temperature applications and large signal outputs. The six basic types of strain gages (capacitance, foil, free filament, semiconductor, weldable, and wire) have been developed and are shown in Figure 9-1 and are summarized in the following paragraphs. A summary is given later (Table 9-4) which describes comparative advantages and disadvantages of each type of gage.

Other techniques and methods have been applied to measure strain including the capacitance gage, the free filament gage, the semiconductor gage, and weldable gages. The capacitance gage consisting of a parallel plate capacitor, was first introduced in 1966 by Hughes Aircraft Co. in

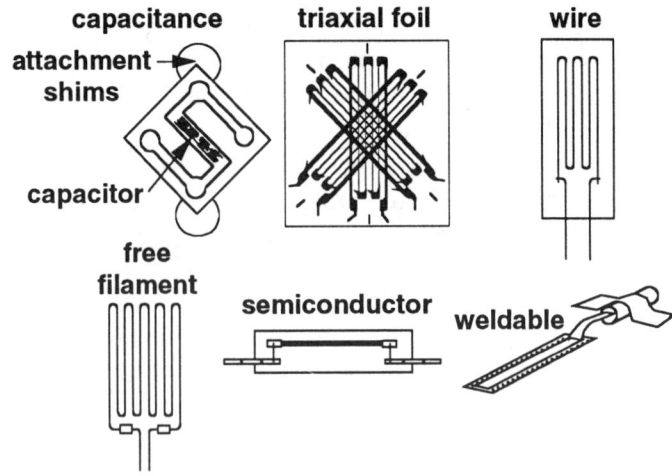

(1) Capacitance type (4) Semiconductor
(2) Foil (5) Weldable
(3) Free Filament (6) Wire

Figure 9-1. The Six Basic Types of Strain Gages

response to the need for reliable high temperature gages. This gage is attached to a specimen and an input voltage is applied. As strain occurs and the plates move relative to each other, the capacitance changes. Calibration of the gage relates the change in capacitance to actual strain measurements.

The free filament or element strain gage consists of a single filament of wire. The wire is attached to the surface with a sprayed or painted on with a brush on ceramic adhesive. This gage is an integral piece of the test item. These gages have special application to high temperature work in turbine engines and similar installations (Refs. 9-3 and 9-15). This type of gage was first developed by Boeing Aircraft Corp.

The semiconductor strain gage is described by Eq. (9-11) and consists of adapting the piezoresistive properties of silicon and germanium to strain measurement (Refs. 9-3 and 9-5). The piezoresistive effects were discovered by Lord Kelvin and used in wire and foil element gages. The piezoelectric materials such as crystalline quartz can also be used to sense strain because a change in electronic charge across the face of the crystal occurs when it is stressed. This change in charge can be proportional to strain.

A semiconductor is typically a rectangular filament of single crystal (usually silicon) which is mounted on a backing for handling and installation purposes. For strain gages made from semiconductor materials, the gage factor is given in Eq. 9-11 (Refs. 9-3 through 9-5, 9-9, 9-10, 9-12, and 9-13):

$$F = 1 + 2\nu + \pi_1 E \qquad \text{Eq. 9-11}$$

where F = gage factor
$1 + 2\nu$ = geometrical changes
π_1 = longitudinal piezoresistive coefficient (m^2/N)
E = modulus of elasticity of the gage material
$\pi_1 E$ = change in resistivity with strain $- d\rho/\rho/\varepsilon$
(this is the dominant term in Eq. 9-11)

Adapting semiconductor materials to strain measurement involves choosing the correct crystallographic orientation, the impurity levels, and calibration. The choice of P or N type silicons results in gage factors of 100 to 175 for P-type and -100 to -140 for N-type.

Semiconductor strain gages can be nonlinear. Nonlinear behavior of these gages is compensated for in their manufacture (by controlling the impurity levels in the crystals), by electrical circuitry using Wheatstone Bridge techniques (Chapter 11), and by limiting the strain ranges measured.

The vibrating wire gage consists of a tensioned wire positioned between two restraints with a coil of an electromagnet positioned midway between the restraints. A short current pulse through the coil gives a transverse vibration to the tensioned wire. As the tension in the wire changes due to relative displacement of the two restraints, the vibration frequency of the oscillating wire changes in proportion to the displacement. With the end restraints located as the ends of a strain gage, then the frequency changes measured by the coil are directly proportional to the differential displacement detected by the gage. With the differential displacement divided by the distance between the restraints, then proportional strain is measured. This type of gage is widely used in the United Kingdom with applications to concrete where the gage is embedded and protected. The vibrating wire gage more closely approximates an extensometer as is not included in the description of more conventional strain gages. It is described in detail in Ref. 9-15.

The weldable strain gage can be one of two types: (1) a strain gage attached to shim stock or (2) a conducting wire embedded in oxide powder encapsulated in a housing which is attached to shim stock. In either case, the gage is positioned so that the strain sensing element is exercised by the strain in the shim stock or in the housing. The periphery of the shim stock is spot welded to the surface of the structure. Since the gage attached to the shim stock is typically a wire or foil gage, only the embedded conductor gage is shown in Fig. 9-1. Some examples of the various types of foil strain gages are shown in Fig. 9-2.

The thin film strain gage consists of gaging materials deposited (often using vapor techniques) in very thin layers (0.0003 in.) on a structure. These gages were devel-

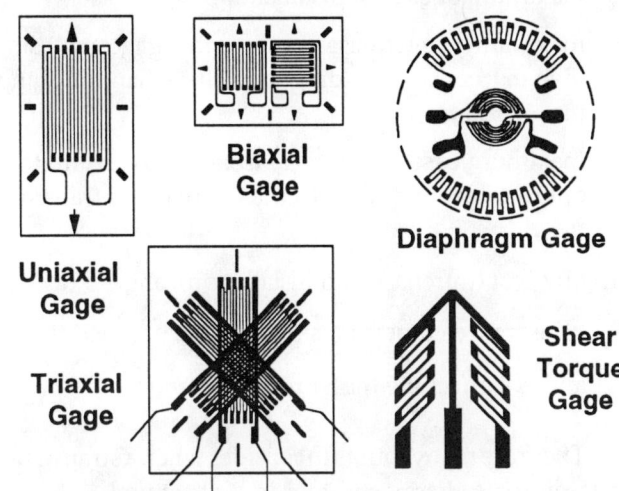

Figure 9-2. Some Examples of Various Types of Foil Strain Gages

oped in the 1960's for special applications such as strain measurements in the root of turbine blades for jet engines. Many aircraft engine companies have developed manufacturing and deposition techniques for this type of gage. Pressure transducers made with these gages are commercially available.

Applications of Strain Gages

The emphasis in this section will be on foil gages and will feature the following: (1) gage characteristics, (2) critical factors in gage selection, (3) gage construction, (4) gage descriptions and specifications, and (5) gage adhesives. (Strain gage circuitry will be described in Chapter II.) Summary charts and tables are presented to simplify and compare many aspects of selecting and using strain gages and adhesives.

The ideal strain sensing element and gage should have the following characteristics:

1. high strain sensitivity (maximum electrical output signal for a given strain),
2. maximum resistivity per unit size,
3. minimum or no hysteresis for accuracy and repeatability,
4. linear strain sensitivity in the elastic range for a structure as well as much of the plastic range as possible,
5. matching coefficients of thermal expansion for self-temperature compensation,
6. the largest possible range of operating temperatures,
7. excellent fatigue life for applications to cyclic loadings,
8. maximum of ease of installation,
9. maximum ruggedness to ensure that gages will survive extremes of loading as well as handling and installation,
10. maximum ease of electrical isolation to ensure that only strained induced electrical signals are transmitted,
11. low cost (purchase and installation), and
12. little or no electrical noise (generated or sensitive to).
13. and not affect the measured strain

There are many things to consider when a strain gage is chosen for a diagnostic tool in a structural test. The following critical factors need to be considered so that a gage installation will deliver the data required:

1. While the installation of a strain gage involves considerable skill and application of science, the process is considered more an art form than an applied science. Their use depends on installation by skilled technicians.

2. The complexity of and the time needed to perform a structural test often influences the choice of options available in strain gage installations. Structural tests can range in complexity from "quick and dirty" to agonizingly thorough. The quick tests typically need gages which can be installed in minimum time. The longer, more thorough tests may require gages to be installed and used some time later (e.g., many years in some cases). The numbers and locations of strain gages used in a test can range from a few to many.

3. Once the strain gages have been installed, there is a need to verify that the gages are functional. There is more involved than performing electrical continuity checks and verifying that the resistance to ground is very large. The most subtle form of a gage installation problem is with a gage that has an imperfect bond. This gage installation will detect strains; however, the measured strains will be different from the actual strains because there is not a perfect transfer of strain across the bond line between the gage and the structure surface. This problem can be eliminated by placing a known loading condition (e.g., static force) on an instrumented structure and measuring the output of the strain gages. With the applied force producing strains in the instrumented test item, the measured strains should compare with analytical predictions of strain within a few percent. If the results of this verification test show agreement between predicted and measured strains, then the installation is valid. However, if there is disagreement between measured and predicted strains, then it is likely there is some problem with gage installation that was not detected by the resistance to ground and electrical continuity checks.

4. There will likely be compromises, constraints, and some limitations involved in the installation of gages.

5. The magnitude of a strain signal is very small when compared to ordinary electrical signals found in a laboratory. It is essential that the strain gage circuitry be electrically isolated and insulated.

6. The cost of a strain gage is usually small in comparison with the labor costs of installation and costs associated with testing.

7. Sometimes tests are required to be performed in a minimum of time. The installation time for gages can

be the critical item in completing a test. The time of installation of a strain gage varies from a minimum of about 30 minutes for easily accessible location which has been properly prepared to to 8 hours for complicated situations where difficult surface access, preparation, cleaning, bonding, wiring, and other installation problems are encountered. These installation times do not include the time needed for curing the adhesives which ranges from one minute to 72 hours depending on the adhesive used. These times also do not include the time for installing the wiring and protective coatings.

Foil Strain Gage Construction

Strain gage construction can play an important role in how a gage is used. The most common strain gage currently used is the metal-foil bonded to a backing. A photoetching process provides a thin layer of metal on a relatively stiff but pliable backing. The backing provides electrical insulation and resistance to handling and installation forces. These gages can be manufactured to tight specifications while minimizing costs. Gages are available in 120, 175, 200, 350, 500, and 1,000 ohms (and greater) with lengths ranging from 0.2 mm (0.008 in.) to 100 mm (4.0 in). The most commonly used gages have resistances of 120 and 350 ohms and lengths between 1.5 mm (0.059 in) to 12.7 mm (0.5 in) (Refs. 9-1, 9-3 through 9-6, 9-9 through 9-11, 9-12 through 9-19).

A currently manufactured metal foil strain gage consists of four parts: (1) the strain sensitive electrical resistance element, (2) the backing material, (3) gage encapsulation, and (4) the electrical connections or tabs.

The strain sensitive resistance material is made of a variety of alloys. Table 9-1 describes these alloys and S_A from Eq. 9-8. The actual gage factors are determined from information supplied by each manufacturer for each lot of gages made.

Differential expansions due to temperature changes will occur in a strain gage installation unless the various parts of the gage sensing material, gage backing, adhesive, and protective coating have similar thermal expansion characteristics. The combination of these resistance changes occurring with temperature can be interpreted as mechanically induced strain. However, since they originated with temperature they are called "apparent strain" and are the source of one of the largest uncertainties in using strain gages. Some of the alloys in Table 9-1 exhibit small changes in resistance with temperature (e.g., Armour D, constantan, karma) while others have large changes (e.g., isoelastic, platinum-tungsten, nichrome). These alloys can be processed using heat treatment which adjusts the temperature coefficient of resistance to compensate for the resistance change in the combination of materials including the test item. This compensation procedure results in gages which are referred to as self-temperature compensated (STC) gages. This temperature compensation process makes it possible for these few alloys to be adjusted to cover the wide range of thermal expansion coefficients (α) in materials ranging from quartz ($\alpha = 0.5 \times 10^{-6}/°C$) to plastics ($\alpha = 65 \times 10^{-6}/°C$) over the operating temperature ranges for the gage and adhesives explained later in this section.

A strain gage actually has two gage factors for each uniaxial sensing elements: one gage factor is described in Eq. 9-10 which is for the major axis of the gage major

Table 9-1. Strain Sensitive Alloys and Strain Sensitivity, S_A

Strain Sensitive Alloy	Composition by Percentage	S_A
Armour D	70 Iron, 20 Chromium, 10 Aluminum	2.0
Constantan or Advance	45 Nickel, 55 Copper	2.1
Isoelastic	36 Nickel, 8 Chromium 0.5 Molybdenum, 55.5 Iron	3.6
Karma	74 Nickel, 20 Chromium 3 Aluminum, 3 Iron	2.0
Nichrome V	80 Nickel, 20 Chromium	2.1
Platinum Tungsten	92 Platinum, 8 Tungsten	4.0

direction of the sensing element and the other gage factor is for strain perpendicular to the major axis (K_t – transverse sensitivity). When a bonded gage is subjected to biaxial strain, the response is given by Eq. 9-12:

$$\frac{\Delta R}{R} = S_a \varepsilon_a + S_t \varepsilon_t + S_s \gamma_{at} \qquad \text{Eq. 9-12}$$

where ε_a = axial strain
ε_t = transverse strain
γ_{at} = shearing strain
S_a = gage sensitivity to axial strain
S_t = gage sensitivity to transverse strain
S_s = gage sensitivity to shearing strain

The gage sensitivity to shearing strain is small and is neglected. With $K_t = S_t/S_a$, Eq. 9-12 can be rewritten as:

$$\frac{\Delta R}{R} = S_a (\varepsilon_a + K_t \varepsilon_t) \qquad \text{Eq. 9-12}$$

where K_t = transverse sensitivity factor for the gage.

The values of K_t are small, they can be positive or negative, and are usually given in percent.

Backing materials for strain gages usually consists of one of the following: cellulose paper, epoxies, laminated epoxies, phenolics, polyimides, polyester resins, and teflon which is strippable. Backing materials serve the following functions: (1) provide the means to support the gage for handling and installation purposes including attachment of electrical connections, (2) provide a bondable surface so the gage can be attached to structures and components, (3) provide electrical isolation between the gage sensing material and the structure, and (4) provide for a linear and repeatable transmission of the strain from the surface of the test item to the sensing element.

Protective coatings range from moisture barriers to complete electrical isolation so that solder connections do not function as small antennas to a physical barrier against the effects caused by direct exposure to high pressure steam, acids, and other corrosives. Information on protective coatings is best located in manufacturer's catalogs. Reference 9-15 does offer a detailed discussion of gage protection (pp. 115-131). The use of protective coatings is essential in a good strain gage installation.

There are many geometries of electrical tabs available. There are also many options available for gage configurations. The best information is contained in the manufacturer's catalogs.

Strain Gage Descriptions and Operating Characteristics

There is no standard format for describing a strain gage. Among the features generally described are: (1) grid dimensions, (2) overall dimensions (length and width), (3) grid geometry, (4) resistance, (5) gage factor, (6) transverse sensitivity, Kt, (7) options (tab geometry, lead wires attached, (8) lot number, and (9) other features (standard gage, special order).

There is no standard format for ordering a strain gage. A method is presented here to compare a typical gage available from manufacturers in the United States. The comparison given in Table 9-2 shows how the gage is listed in a supplier's catalog and gives a comparison of the relevant features of this typical gage. The combinations of the geometries, configurations, sizes, and number of manufacturers result in the large number of gages available (about 200,000 currently). Readers selecting strain gages are encouraged to obtain the supplier catalogs. These catalogs are the best source of the gages actually available. Among the items to consider are: (1) is the gage a standard or special order type as some manufacturers can handle special orders and others must make added provisions, (2) would an alternate gage be acceptable, and (3) the options available for a particular type of gage.

In Table 9-3, a matrix summary of the strain gage types and various operational characteristics is presented. This chart was developed by SEM's Western Regional Strain Gage Committee. In the chart, X means that the gage has capability to perform the intended function. For example, a free filament gage is intended for use at high temperatures while most other gages cannot meet these operating requirements. A semiconductor gage has limited temperature operating ranges but can give much larger output than other gages (Refs. 9-P1 through 9-P4).

Figure 9-3 shows a typical gage with some of the many options available (solder tabs and locations, encapsulation—gage protection, lead wires attached).

Strain Gage Adhesives

Adhesives attach the gage backing to the surface of the structure. The characteristics of an ideal bond of this type are: (1) high shear strength, (2) high elongation, (3) matching thermal expansions of gage and adhesive, (4) provide maximum operating temperatures (cryogenic to very high), (5) minimize creep and hysteresis, (6) have material compatibility with gage backings and surfaces of structures, (7) have long shelf, pot, and working times, (8) require a minimum of clamping pressure, and (9) be easy to mix, use, and apply in thin layers with no voids or contaminants (Ref. 9-P4)

Various types of adhesives used for strain gages are listed below. Ceramics and flame spray (Rokide) are used

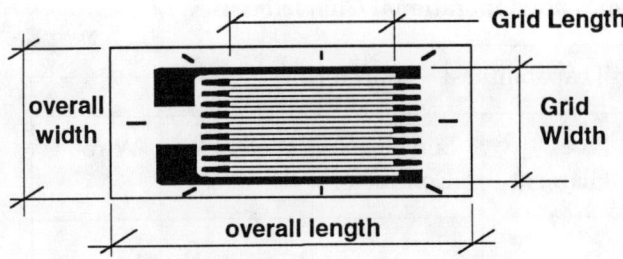

Sample Gage Characteristics:
  Resistance 350 Ω
  Grid Length = 6 mm (.25 in)
  Constantan Alloy
  Encapsulated
  Polyimide Backing
  Compensated for Steel

Some Options:
  lead wires attached
  solder tabs installed
  gage configurations
  encapsulation

Figure 9-3. Representative Strain Gage

In Tables 9-4 through 9-7 adhesives are characterized for strain ranges, temperature ranges, and curing times. Table 9-4 gives the operational temperature ranges (–400 °F to 1600 °F) for the various adhesives. The "recommended" range is within the specifications given by all manufacturers listed. The extended range is given by some of the manufacturers. Readers are encouraged to consult manufacturer's instructions and recommendations for specific adhesives. Note the discontinuity in the temperature scale at 1200 °F. Most of the operating temperatures range is between –100 °F and 400 °F (Ref. 9-P4).

Table 9-5 gives the strain ranges for the various adhesives with the same two scales of suggested performance (recommended and extended) used in Table 9-4. One percent strain is the same as 10,000 μin/in. Log strain is used in the table in order to condense the strain ranges into a one page summary. Most adhesives have upper limits of about 20,000 μin/in. High elongation adhesives are available and the reader is referred to product literature.

Table 9-6 shows the curing times for the various elevated temperature cured adhesives. The (Φ) indicates the time required for curing. The first entry in each category has the minimum recommended curing time for the particular adhesive in category. Other possible curing times are given on the same line in the table. One would need to know the temperature capabilities of the test item

for free filament gages. Cyanoacrylates ("super glue"), epoxies, epoxy phenolics, phenolics, polyesters, and polyimides are used for the other gages. Spot welding is not listed but is used to attach gages bonded to shims and weldable gages.

1. Ceramics
2. Cyanoacrylates
3. Epoxies
4. Epoxy Phenolics
5. Flame Spray
6. Nitro-Cellulose
7. Phenolics
8. Polyesters
9. Polyimides

Table 9-2. Manufacturer's Designations for Representative Gage

Gage Manufacturer	Typical Gage Designation	Grid Length (mm)	Grid Length (in)	Grid Width (mm)	Grid Width (in)	Overall Length (mm)	Overall Length (in)	Overall Width (mm)	Overall Width (in)
BLH	FAE-25-35S6E-L	6.35	.25	3.18	.13	8.89	.35	6.35	.25
HBM	3/350 LY 11	6.35	.25	2.9	.11	12.8	.50	6.3	.25
JP Technologies	PA-06-250AG-350-L	6.35	.25	6.35	.25	10.8	.43	6.35	.25
Measurements Group	EA-06-250AE-350-LE	6.35	.25	6.35	.25	10.8	.43	6.35	.25
PEC Indust.	BA-06-125-AB-250-CL	3.18	.13	3.18	.13	6.35	.25	3.18	.13
Texas Measurements	QLFA-6.350 11	6.0	.24	3.2	.13	—	—	—	—

Table 9-3. Summary of Strain Gage Types and Operational Characteristics

Operational Characteristics	Strain Gage Type (x = full capability, st = short term)					
	Capacitance	Foil	Free Filament	Semi-Conductor	Weldable	Wire
TEMPERATURES (1)						
Cryogenic	x	x			x	st
Low	x	x			x	x
Below Ambient	x	x		st	x	x
Ambient	x	x		x	x	x
Above Ambient	x	x		x	x	x
Elevated	x	x		st	x	x
High	x	st	x		x	
Very High I	x		x		x	
Very High II	x		x		st	
Very High III			x		st	
Extremely High			x			
EXCITATION (2)						
Current		x	x	x	x	x
Voltage	x	x	x	x	x	x
STRAIN RANGES (3)						
Small	x	x	x	x	x	x
Elastic	x	x	x	x	x	x
Plastic	x	x			x	x
ENVIRONMENTS (4)						
Hostile	x	x	x	x	x	x
Fatigue	x	x	x		x	

(1) Temperatures:
Cryogenic   = −273 to −173 C (−460 to −279 F)
Low   = −173 to −40 C (−279 to −40 F)
Below Ambient = −40 to   0 C (−40 to 32 F)
Ambient   =   0 to   50 C (32 to 122 F)
Above Ambient =   50 to 150 C (122 to 312 F)
Elevated   = 150 to 300 C (312 to 572 F)
High   = 300 to 480 C (572 to 896 F)
Very High I   = 480 to 600 C (896 to 1112 F)
Very High II   = 600 to 815 C (1112 to 1494 F)
Very High III = 815 to 1000 C (1494 to 1832 F)
Extremely High = > 1000 C (> 1832 F)

(2) Excitation
Currents   - Steady State
           - Pulsed
Voltages   - Steady State
           - Pulsed

(3) Strain Ranges
Small < 20 microstrain
Elastic ≤ 2000 microstrain
Plastic ≤ 200,000 microstrain

(4) Hostile Environments:
- in presence of live animal tissue and other natural corrosives
- subject to acids and other formulated corrosives
- required to resist steam and other water vapors
- installed in fresh concrete/ subjected to exothermic reactions
- subject to compressed gases
- exposed to sunlight or other high energy sources (radiation)

relative to the temperatures and times needed for curing a potential adhesive. The ∇> symbol indicates a post temperature cure usually at a different temperature from the initial cure. In Table 9-6 the abbreviations for the manufacturer's are: BLH = BLH Electronics, HBM = Hottinger Baldwin, HT = HITEC Products, JPT = JP Technologies, MM = Measurements Group, and TM = Texas Measurements. Rokide is sprayed in place and is not included in curing temperatures and times.

Table 9-7 gives the curing times for room temperature cured adhesives using the same manufacturer's designations given in Table 9-6.

The accuracy of Tables 9-4 through 9-7 is limited by the scaling used in personal computer word processing procedures rather than graphics. These tables are intended to summarize the information contained in many catalogs and manufacturer's specifications. The intent of Tables 9-4 through 9-7 is to present the capabilities and limitations of current strain gage use.

Factors to Consider in Using Strain Gages

This section summarizes the factors that need to be considered when a strain gage is chosen for a test diagnostic. The choice of the most appropriate gage depends on many variables as summarized in Table 9-8. This summary was prepared from manufacturer's information and data sheets along with many different textbooks. Each of the basic categories can be considered a variable in the gage selection process. Implied but not listed are the other variables such as gage size, orientation, and lead wire attachment.

The reader is reminded that there are also many variables associated with the surfaces and materials where the gage will be installed. There are important considerations including: (1) material compatibility or lack thereof, (2) presence of contaminants (e.g., out-gassing), (3) roughened or irregular surfaces, (4) difficulties in access, and (5) curvatures or other limiting conditions restricting the placement of the gage. There are over 200,000 different types of foil strain gages alone. Coupled with the other variables listed above gives over 600,000,000 combinations for strain gage installations that could occur. The reader is also reminded that there is an art to installing strain gages. Proper installation of strain gages should be performed by a skilled technician.

Summarized in Table 9-8 are the variables associated with a strain gage installation including gage types, gage backing materials, gage sensing alloys, adhesives and attachment methods, lead wire insulation (which requires material compatibility), protective coatings, and gage geometries. These variables are important in terms of the training and expertise of the installer, how well the gage

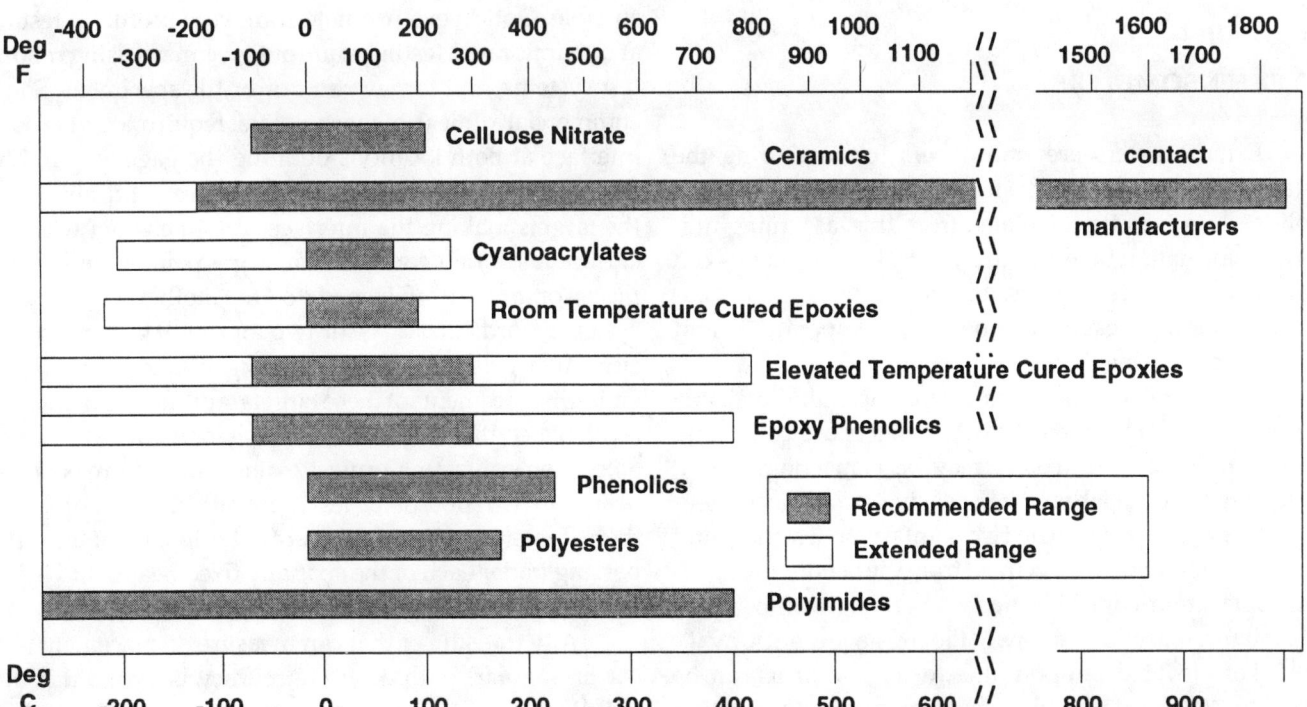

Table 9-4. Strain Gage Adhesives — Operational Temperature Ranges

Table 9-5. Strain Gage Adhesives — Operational Strain Ranges

Adhesive	Recommended Range (Log Strain - Percent at Room Temperature)	Extended Range
Celluose Nitrate	0.1 – 10.0	
Ceramics	0.1 – 0.5	
Cyanoacrylates	0.1 – 6.0	6.0 – 10.0
Room Temperature Cured Epoxies	0.1 – 3.0	3.0 – 10.0
Elevated Temperature Cured Epoxies	0.1 – 2.0	2.0 – 10.0
Epoxy - Phenolics	0.1 – 3.0	
Phenolics	0.1 – 3.0	
Polyesters	0.1 – 3.0	
Polyimides	0.1 – 2.0	

installation can meet the test requirements, and a host of other considerations specific to a particular test, test items, materials involved, signal conditioning, and data acquisition systems used.

9.3 EXTENSOMETERS

Extensometers are transducers for measuring the relative displacement ("extension") between material points: strain can be calculated from this, assuming uniform deformation between the points. Extensometers can be categorized several ways, for example by range, resolution, frequency response, operating temperature, and whether contacting or non-contacting.

There are several types of commercially available extensometers: mechanical extensometers are contacting while optical extensometers may be contacting or not. "Clip gages" are relatively inexpensive mechanical extensometers which fasten on to two points of the test piece and infer their relative displacement from the bending strain in a flexible strain-gaged elastic member. High temperature capacitance-based variations of the clip gage are also available. For elevated temperatures, extension arms can be used to stand off the clip gage (or a variety of other displacement transducers) away from a hot test area (Refs. 9-1, 9-3, 9-4, 9-7, 9-9, 9-17, 9-19, and 9-P1).

Optical extensometers are particularly attractive for elevated temperature testing through a suitable viewport and for testing where mechanical extensometers are unsuitable. Optical extensometers are very useful for testing at dynamic rates, testing multiple specimens using robotic test systems, and for very soft or brittle specimens. Some commercial optical extensometers require a light/dark interface at both locations defining the gage length. The displacements of the interfaces are tracked optically. Only the targets making the interfaces are in contact with the surface and often consists of a stripe painted on the test piece, or a "flag" fastened to it. Another video-based system records the image or two (or more) from reflective circular targets pasted to the moving test piece. The differential displacements of these targets indicates strains which is calculated during post-processing for strain calculation. Some non-contacting optical extensometers project laser beams on two spots of the test piece but, instead of tracking these material points, measures the length of material passing under each of the spatially fixed beams and infers strain by subtracting these measurements.

Any transducer that can measure displacement (linear or angular) with desired accuracy is a candidate for building an extensometer. Linear displacement transducers include LVDT's (linear variable differential trans-

Table 9-6. Curing Times for Adhesives—Elevated Temperature Cured

LOG TIME - HOURS
(Scale: 0.1, 0.2, 0.3, 0.5, 0.7, 1.0, 2.0, 4.0, 6.0, 10.0, 20.0, 40.0, 100.0)

Category	Source	Description
Celluose Nitrate	(BLH)	Φ 30 min @ RT plus 2 hr @ 120°F plus 2 hr @ 160°F or 30 min @ RT plus 2 hr @ 120 - 140°F
Ceramics	(HT)	Φ @ 400°F (precoat 30 min @ 200°F/repeat; Yellow Cerro)
	(MM)	Φ @ 600°F (precoat 15 min @ 350°F-M-Bond GA-100)
	(BLH) (CER 1000)	Φ air dry 30 min, 1 hr @ 220°F and 1 hr @ 600°F
	(JPT)	Φ @ 600°F (final coat) (PBX)
	(JPT)	Φ @ 600°F (final coat) (H Cement)
	(HBM)	Φ @ 350°F (CR 760)
Epoxies	(JPT)	Φ @ 75°F for 24 hrs plus PC @ 250°F - 2hrs (Epoxylite 810)
	(MM)	Φ @ 200°F, Φ @ 125°F, or Φ @ 75°F (M-Bond GA-2)
	(MM)	Φ @ 200°F to Φ @ 125°F (M-Bond AE-15)
	(BLH)	Φ @ 150°F (EPY-150)
	(JPT)	Φ @ 175°F (BR-104, RTC)
	(JPT)	Φ @ 350°F (Denex No. 3)
	(BLH)*	Φ @ 75°F plus PC @ 200-250°F (see note below)
	(JPT)	Φ @ 140°F (Hysol Epoxy)
	(MM)	Φ @ 165°F (M-Bond A-12)
	(MM)	-->350°F Φ to Φ @ 400°F (M-Bond 43-B)
	(TM)	Φ @ 284°F (A-2)
	(HBM)	Φ @ 350°F (H-3)
	(JPT)+	Φ @ 350°F plus PC 1 hr @ $\Delta 25$°F
	(JPT) (BR-22)	Φ @ 250°F plus PC 1 hr @ $\Delta 25$°F
	(HBM) (EP 310)	Φ @ 205°F plus 0.5 hr @ 400°F
	(MM)	Φ @ 250°F (M-bond GA-61)
	(HBM) (EP 250)	Φ @ 205°F plus 0.5 @ 390°F (EP250)
Epoxy-Phenolics	(MM)	Φ @ 300°F to Φ 175°F (M-bond 600)
	(JPT)	Φ @ 300°F plus PC @ $\Delta 25$°F (BAP-1)
	(JPT)	Φ @ 275°F 1 hr PC @ $\Delta 50$°F (BR-610)
	(MM)	Φ @ 375°F to Φ @ 212°F (M-Bond 610)
	(MM)	Φ @ 350°F to Φ @ 175°F (M-Bond 600)
Phenolics	(TM)	Φ @ 260°F then 1 hr @ 392°F (C1)
Polyimides	(BLH)	Φ @ 500°F (PLD 700)
	(JPT)	Φ @ 250°F ---> Φ @ 500°F (P adhesive)

LEGEND
Φ = Time in Hours Given By Manufacturers
PC = Post Cure
Δ = Added Temperature

Notes: * = BLH EPY 500, QA 500, QA 550, and QA 600   + = JPT Epoxylite 813

Table 9-7. Curing Times for Adhesives — Room Temperature Cured

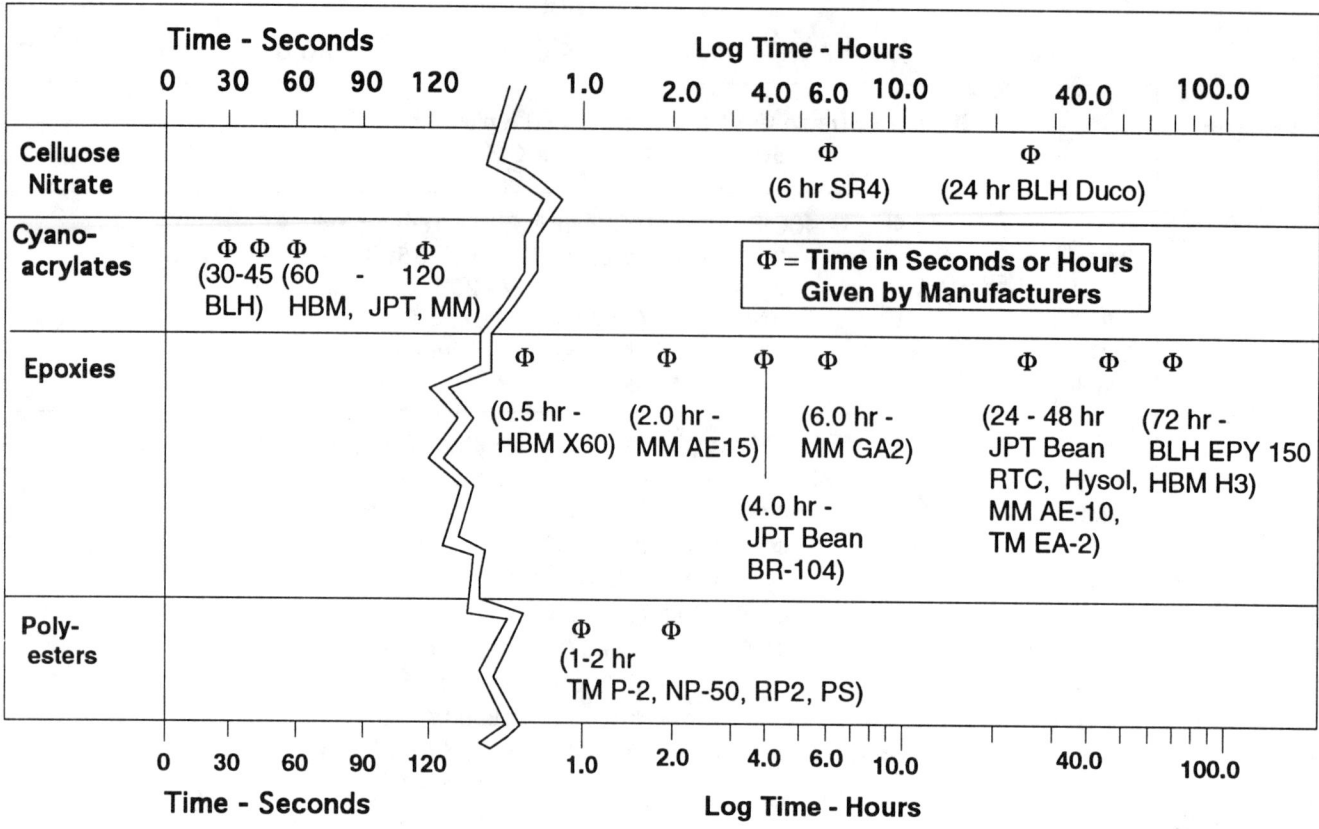

Table 9-8. Summary of Variables in Strain Gage Installations

Basic Categories of Variables in Strain Gage Installations						
Gage Type	Gage Backing	Gage Sensing Elements	Adhesive Attachment Materials	Lead Wire Insulation	Protective Coatings	Geometries
↓ Capacitance Foil Free Filament Semiconductor Weldable Wire	Epoxy Paper Phenolics Polyester Polymides Reinforced Fiberglass	Armour D Constantan Iso-elastic Karma Nichrome Platinum/ Tungsten	Ceramics Cyanoacrylates Epoxies Epoxy-Phenolics Flame Spray Polyesters Polyimides Nitro- Cellulose	Fiberglass Heatshrink Nylon Polyimide Polyurethane Polyvinyl Chloride Polytetra- flouro- ethylene Teflon Varnish	Bees Wax Epoxies Microcry- stalline Wax Nitrile Rubber Polyimide Lacquer Polysulfide RTV Silicon Rubber Toluene Acrylic	Uniaxial Biaxial Rosette Equiangular Rectangular Stacked (layered) Helical Diaphragm Strip or Chain Chain

formers), magnetostrictive transducers, capacitive or optical proximity sensors, and optical diode/photodetector position sensors. For higher frequency applications, for example, velocity data obtained from mechanical linear velocity transducers (LVT's) or laser vibrometry, can be integrated to obtain displacements. Rotary displacement transducers include those based on resistive potentiometers, precision differential capacitors, and encoders (optical, magnetic, or laser-based (Ref. 9-P2). The digital output nature of encoders makes them attractive candidates for digitally-based test system controllers, since they avoid drawbacks such as drift and noise associated with conventional analog transducers and their signal conditioners.

9.4 DISPLACEMENT GAGES

Displacement gages were developed in response to the need to accurately measure deformations of a structure. Mechanical gages (e.g., dial indicators) were the first instruments used to measure displacement. These early instruments required a person to be available within eyesight distance to record the data. For most applications today, displacement is converted to an electrical signal which can be used in data acquisition systems (see Chapter 11).

The transducers developed to measure displacement are based on the relative motion of one part of the transducer with respect to another part which changes the electrical output of the transducer. Each transducer is calibrated to convert the electrical output to a measured displacement. The transducers currently in use appear to be limited in their ability to make measurements only by the precision of the calibration and the fixturing needed to support and stabilize the transducer while it is being used. Displacement gages can be calibrated using gage blocks, precision micrometers, and other types of scales (Refs. 9-1, 9-7, 9-19, and 9-P1).

The first electrically based displacement measuring system was a linear variable differential transformer (LVDT) in 1936. LVDT's were used extensively in World War II aircraft and weapons work. The device consists of a rod surrounded by a transformer coil. The motion of the rod relative to the coil changes the electrical output of the transformer. The displacement of the rod produces a nearly linear change in output of the transformer. Improvements were made in LVDT's which produced better linearity, allowed for applications to vibrations, decreased their size which makes mounting and installation easier, and made possible measurements of force, torque, velocity, and acceleration (Ref. 9-7).

Linear wire wound potentiometers were the among the next transducers developed. Their characteristics are similar to an LVDT but they differ in the way the measurement is made. A linear potentiometer changes in resistance as the motion of the rod detecting displacement occurs. Initially, these transducers had a sort of stepped calibration relationship as the rod and contact moved from one winding of wire to the next. The resolution was limited to the diameter of the wire used. Conductive plastics have been developed to replace the wire windings so that there is a true linear relationship between displacement and voltage output. These transducers currently have two basic configurations: linear and rotary. The linear devices are used primarily for smaller measurements—150 mm (6 in) or less while the rotary transducers (sometimes called deflection boxes) have been developed to have ranges up to 15m (600 in) and more. The accuracy of the linear and rotary displacement transducers depends on how they are calibrated. They are capable of measuring 0.0002 mm (tens of microns) when calibrated using interferometric systems.

Examples of displacement transducers are shown in Fig. 9-4.

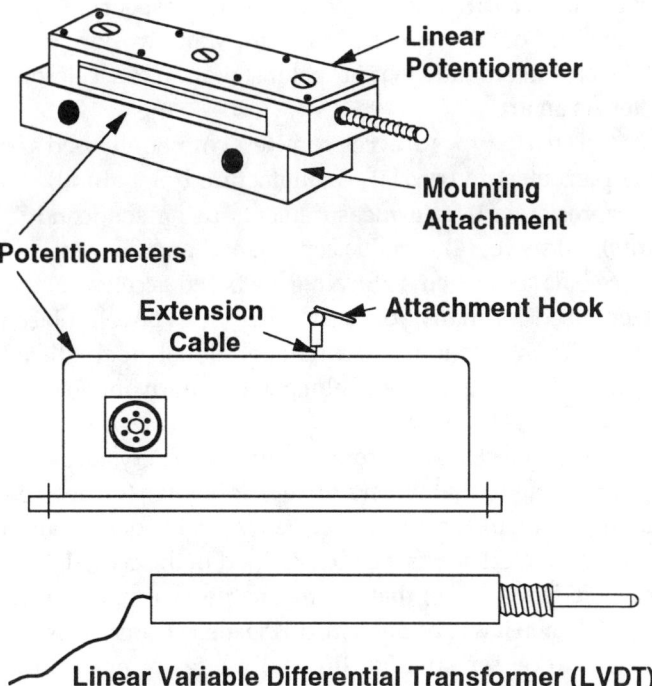

Figure 9-4. Linear Variable Differential Transformers (LVDT's) and Potentiometer Based Displacement Transducers.

9.5 ACCELEROMETERS
W. B. Leisher, Sandia National Laboratories, Albuquerque, NM

Accelerometers measure the time rate of change of velocity. Measurements of acceleration are needed to characterize the dynamic response of a wide variety of structures, components, and assemblies. A typical design procedure for components is to use static design principles to layout and size members and connections. These principles use statically equivalent forces to be resisted by attachments. After the initial design for strength considerations is completed and prototype assemblies are built, then shock and vibration tests are usually performed to verify the integrity of the component and to determine failure conditions. These tests usually rely on acceleration measurements. Some of the acceleration measurements are converted into forces acting on the component. These measurements may also be used to determine the frequency responses, mode shapes, and damping (Refs. 9-1, 9-4, and 9-6).

In order to measure acceleration, accelerometers must be mounted on a component or part of the structure that will respond dynamically. Vibration nodes of the structure do exist and an accelerometer placed on a node can indicate little or no acceleration. Placement of an accelerometer at a location where the dynamic response is excessively amplified can give invalid data also. Mounting an accelerometer in a proper location is the most important aspect of installation and obtaining valid test data. Many working with accelerometers state that proper installation of accelerometers is an art.

The five types of accelerometers commonly used are: (1) piezoelectric (crystal—manufactured or natural), (2) piezoresistive (strain measurements using semiconductors), (3) servo, (4) variable capacitance, and (5) mass and force balance systems. (Strain gage based accelerometers were used for many years but other types have replaced them.) Table 9-9 summarizes the operational capabilities of each type of accelerometer along with the advantages and disadvantages of each.

Piezoelectric accelerometers consist of a crystal placed in a housing so that when a force is applied to the device a deformation of the crystal occurs. When the deformation occurs, an electric charge is developed in the crystal. (The reverse is also true; that is, if a charge is applied to the crystal, then it will be deformed.) Piezoelectric crystals can be cut to be sensitive to different modes of mechanical deformation (e.g., shear, bending, compression). There are no moving parts in a piezoelectric accelerometer.

Piezoresistive accelerometers use the deformation of a mechanical part as indicated by the strain occurring in a semiconductor strain gage. The strain is proportional to the acceleration and can be calibrated.

A servo accelerometer consists of bar connected to a force restoring system. When an acceleration occurs, the bar is deformed. The force restoring system maintains the bar in its at rest position. The force required to maintain the bar at its rest position is proportional to acceleration.

Capacitive devices consist of two plates of conducting material separated by a distance. The plates are of common area and the gap between them is filled with a dielectric (insulator). These devices function based on one of three changes: in the dielectric material, in the area between the plates, or by the distance between the plates. A capacitive accelerometer is typically based on the changing distances between the plates as affected by the applied forces (accelerations).

Another type of accelerometer is a mass and force balance system. A large mass is supported by a spring or the mass rests on a captive fluid. Measurement of the force in the spring or the pressure change in the supporting fluid is proportional to acceleration. An accelerometer used to sense seismic disturbances is typically a mass and force balance system.

9.6 TEMPERATURE SENSING DEVICES

Thermocouples are devices (contacting or non-contacting) used to measure temperatures. T.J. Seebeck discovered in 1821 that if two dissimilar metal conductors were joined at their ends but separated in between so that the junction at one end could be maintained a temperature different than the junction at the other end, then a current (emf) would flow between them as long as the temperature difference is continued. The voltage output is a function of the temperature difference between the two metals (one is the reference and the other is the measuring junction). Using this discovery and the need to measure temperatures in a wide variety of conditions and environments, the science and art of temperature measurement has developed many methods and techniques for measuring temperatures (Refs. 9-1, 9-8, and 9-19).

A second type of temperature sensing systems is related to electrical resistance changes in materials which occur in a reproducible manner with changes in temperatures. Materials used for resistance temperature measurements are two basic types—conducting metals and semiconductors. A typical metal used is high purity platinum wire which is nearly linear in resistance change over a temperature range of –300°F to 1500°F. The conducting metals used for temperature measurements are usually called resistance temperature devices (RTD's). The semiconductor based sensors are usually called thermistors.

Table 9-9. Operational Capabilities, Advantages and Disadvantages of the Five Basic Types of Commonly Used Accelerometers

Accelerometer Type and Ranges	Frequency Ranges		Advantages	Disadvantages
	Low Hz	High Hz		
Piezoelectric 10 g's to 200,000 g's	2,000	200,000	less cost, less temperature sensitivity, can be hardened for hostile environments, stable over long times no moving parts, no external power source needed	cannot read very low frequencies, usually need charge amplifiers to reduce/eliminate electrical losses little means of damping
Piezoresistive 20 g's to 200,000 g's	D.C.	30,000	high output, DC response, may not need an amplifier easily damped	moving parts, not as rugged, need power source
Servo 2 g's to 100 g's	D.C.	1,000	very accurate, flat & linear, some have built-in amplifiers,	limited frequency and acceleration levels, need power source, some temperature effects
Variable Capacitance 2 g's to 100 g's	D.C.	1,000	low cost, low frequency, low acceleration, easily damped	limited frequency and acceleration levels, need power source
Mass and Force Balance	D.C.	1,000	low cost, low frequency, low acceleration, easily damped, rugged construction	limited frequency and acceleration levels, need power source

accelerations are measured in g's (9.8 m/sec^2, 980 cm/sec^2, 32.2 ft/sec^2)
typical temperature range: –30 °F to 150 °F
maximum temperature range: –65 °F to 250 °F

Both need to be in contact with the surface where the measurements are made.

The sensing elements for RTD's can be made in a variety of different forms. These forms depend on how they will be used, temperature range to be measured, exposure to harsh environments, and space and surface constraints. Flat grids are available for measuring temperatures on surfaces of test structures. Thin films of platinum can also be attached to the surfaces of test items.

Bridge circuitry (for details see Chapter 11) is used to determine the temperatures from the measurements of resistance change. These bridge circuits can be excited by a-c (alternating currents) and d-c (direct current) voltages. The RTD's range in resistance from a few ohms (~10 ohms) to 25,000 ohms. The RTD's with larger resistances are less affected by lead-wire and contact resistance variations.

The semiconductor sensors (thermistors) also are available in many forms, typically in probes, rods, or disks. They have wide use in electronic systems such as time delay elements, voltage and power controls, and temperature compensating devices.

A third type of temperature sensing devices are based on radiation from sources. These non-contacting devices sense the infrared radiation emitted from the source. Some terms given to these devices include radiation pyrometer, radiation thermometers, and optical pyrometers. Some other methods and techniques used for temperature measurement include thermal imaging techniques, electron beam fluorescence, and fusible indicators which have known solid-liquid transition temperatures.

Some factors to consider when making temperature measurements are:

- the measurement device should be capable of the temperature ranges expected
- the thermal emf must be large enough to obtain sufficient accuracy
- the relationship of emf to temperature should be linear or be capable of being defined by equations
- the measurement device should be resistant to heat induced corrosion or other degradation (e.g., formation of oxides)
- the measurement devices should be able to meet the other constraints of the test (e.g., geometry, low cost, etc.)
- whether the measurements need to be made in contact with the test surface or if they will by non-contacting means.

Thermocouples are typically described by seven basic types (S, R, J, T, K, E, and B) as defined by the National Institute for Standards and Technology (NIST). These types of thermocouples reflect the kinds of dissimilar materials joined together are are summarized by:

Type S (90 percent platinum vs. 10 percent platinum/rhodium)
Type R (87 percent platinum vs. 13 percent platinum/rhodium)
Type J (iron vs. constantan)
Type T (copper vs. constantan)
Type K (chromel vs. alumel or any others representing similar characteristics)
Type E (chromel vs. constantan)
Type B (tungsten vs. rhenium)

9.7 LOAD CELLS

Transducers such as spring scales, Bourdon tubes, and strain-gaged load cells infer a load from the displacement or strain of an elastic member. Among the considerations in the use of load cells are accuracy, linearity, hysteresis, capacity, mechanical configuration, environmental capability, overload protection, and limited frequency response. The elastic member is often called a flexure. In spring scales, the load is inferred from the deflection in, say, a linear or torsional spring. A commercial variation of this involves displacing an infrared beam impinging upon a semi-conductor position-sensing device whose output is amplified then converted to a voltage. In a Bourdon tube, hydraulic pressure applied to a structural actuator simultaneously pressurizes a curved thin-wall tube; the resulting deflection as this tube straightens infers the force applied by the actuator on the test piece, neglecting actuator sealing ring friction.

Strain gages provide signals when attached to items experiencing small deflection; hence the elastic members to which they are bonded can be very stiff. This improves frequency response, linearity, hysteresis, fatigue life, overall test system stiffness and control stability. High gage-factor semiconductor strain gages can be used in even stiffer load cells, for example, cyclic elastomer testing in the 1 KHz regime, although the temperature sensitivity of such gages must be considered. A series of design considerations for strain-gaged load cell is found in Ref. 9-12. Full four-arm Wheatstone bridge load cells come as small as 0.5 in. in diameter by 0.11 in. length with useful ranges from 1-20 KHz. Strain gage load cells exist for use to 700 °F; higher temperatures are achievable via capacitive gage technology with the requirement that the dielectric properties of the elements must be stable (Refs. 9-1, 9-17, 9-18, 9-19, and 9-P5).

High frequency applications up in the 10-100 KHz regime often employ a piezoelectric force transducer. The compressive force applied to a quartz disk induces an electric charge signal which is transformed into a voltage via a charge amplifier. Charge amplifiers tend to bleed off over long time periods, yet measurements spanning several minutes are possible. For tension, the disk is precompressed; tensile force is then sensed as a decrease in preload. For dynamic tests, the mass of any adapter between the test piece and disk should be minimized to reduce the amplitude of ringing superposed upon the desired load signal; in addition, an accelerometer can be mounted to the adapters to subtract this force component from the piezoelectric transducer's signal.

The force in a bolted joint can be measured by a miniature strain-gaged washer installed under the bolt head, but this changes the actual joint stackup and potentially the torque-force relationship being sought. Another technique is to measure the axial strain in the bolt, by a strain gage installed on the wall of a hole machined down the center of the bolt, or by an ultrasonic sensor which converts change in transit time of high frequency sound to mechanical strain, and when corrected for temperature effects on sound and material behavior and sound speed changes in the stressed portion of the bolt, provides an accurate measure elongation in the bolt. The ultrasonic sensor can be held in place with a magnet, by fixtures, or by hand. The elongation is proportional to the force in the bolt (Ref. 9-2).

Sensing films (about 0.005 in. thick) are relatively new. One is a thick film resistor sensitive to force; its electrical resistance decreases logarithmically as the force is increased. With a sensing film attached to the test hardware, a force is applied to a test item with the accompanying strain transferred to a sheet coated with a resistive polymer which shunts conductive "interdigitating fingers" on an adjacent sheet. Another uses a conductive grid whose intersection points are separated by a pressure-sensitive semiconducting layer; sensing cells as small as 0.010 in. square are available to measure load distribution profiles. Piezoresistive stress gages and piezoelectric films employed as pressure sensors for planar shock wave impact experiments are quite specialized and are not discussed here.

Torque cells are not as widely used as load (force) cells in most structural test laboratories partly because torsional actuators are more expensive that axial ones, and because a moment can be imposed on a structure by an equivalent force couple. However, torque and combination force-torque cells are commonly used in the straingage (static) and piezoelectric (dynamic) measurement of torques. A consideration in force-torque transducers is the amount of cross-talk between two channels, e.g., an apparent torque signal when a pure force is applied, and vice-versa. This factor should be accounted for in the control command signals as well as in data reduction. Other types of torque transducers include a differential transformer mounted circumferentially on an elastic shaft (i.e., torque and arc length travel instead of force and displacement) or non-contacting devices employing the magnetoelectric effect. For rotating shafts, slip rings or FM transmission are often used to extract the torque signals.

Load cells are found in many industrial applications (e.g., weighing, production control). Overall accuracies of 0.10 percent are routinely specified for commercial load cells. Precision load cells can have accuracies of 0.05 percent or better. These accuracies include the uncertainties associated with repeatability, nonlinearity, hysteresis, temperature changes, creep, etc. Long term stability of load cells is consistently within one percent and mostly less than one-half percent for measurements when used in normal environments. More data are needed to evaluate load cell performance in the following conditions: fatigue, temperature cycling, and continuous loading (Ref. 9-P3 and 9-P5).

Figure 9-5 shows examples of commercially available load cells.

1 = Low Profile, Tension/Compression, Test System Application

2 = Sealed Superbeam-Weighing Applications (all weather protection)

3 = Sealed Super-Ministructural Testing and Weighing Applications

4 = Sealed Superbeam-Allweather Applications

5 = Super Mini Load Cell - Precision Force Measurement

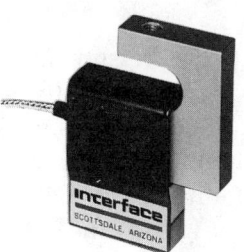

Figure 9-5. Load Cells (Courtesy of Interface Corp. Scottsdale, AZ)

9.8 PRESSURE TRANSDUCERS
William B. Leisher, Sandia National Laboratories, Albuquerque, NM

Pressure, force per unit area, can be measured in three basic ways: absolute, differential, or gage. Absolute pressure, such as a barometric reading, is measured relative to a very good vacuum. Differential pressure is the difference across a boundary measured relative to the pressure on one side of the boundary. If the pressure on one side is atmospheric, then differential pressure is the same as gage pressure. Gage pressure, such as a tire inflation pressure, is measured relative to the local atmosphere.

Pressure measurements required for structural testing may be either dynamic or steady state (e.g., static equilibrium). Usually, steady state measurements are sufficient, but it the structure is subjected to a flowing fluid, such as an explosive blast, then dynamic conditions prevail.

Steady state pressures may be measured with a variety of instruments ranging from simple Bourdon tube gages to complex pressure transducers with diaphragm deflections measured by a capacitance change or semiconductor or metal foil strain gages. Readouts may be visual dials or numbers or obtained using an electrical voltage/current which has been related to pressure during the calibration or even a combination of both. Stock instruments are available from many manufacturers in ranges from a fraction of a psi to 125,000 psi.

Dynamic pressures are normally measured with a flush diaphragm type of transducer. The diaphragm may be mounted parallel or perpendicular to the flowing stream. If the stream velocities are large, the mounting direction is important since the parallel diaphragm will sense true steady state pressure and the perpendicular diaphragm will measure total or stagnation pressure (steady state plus velocity head). A diaphragm moving with the fluid will sense the steady state (static equilibrium) pressures. Diaphragms positioned at angles between parallel and perpendicular give an unknown combination of steady state and stagnation pressures.

The device used to measure the deflection of the pressure-sensing diaphragm for dynamic conditions may be strain gages (either semiconductor or foil), an LVDT, a capacitance probe, or piezoelectric crystal.

Most pressure transducers will require some signal conditioning and a power supply. Some can drive a recording device directly without further amplification. The piezoelectric transducers can be used without external power supplies.

The recommendation for selecting a pressure transducer consists of two parts: (1) knowing or conservatively estimating the pressure conditions (magnitudes, rates, temperatures, and compatibilities with the fluids or gases) and (2) using the information and performance capabilities given by the manufacturers (Refs. 9-1, 9-14, 9-19, and 9-P1).

9.9 REFERENCES

9-1. T.G. Beckwith and N.L. Buck, *Mechanical Measurements*, Addison-Wesley Co., Reading, MA, 1973.
9-2. A. Blake, Ed., *Handbook of Mechanics. Materials. and Structures.* Wiley Interscience, New York, 1985.
9-3. J.W. Dally and W.F. Riley, *Experimental Stress Analysis*, 2nd Ed., McGraw-Hill Co., New York, 1978.
9-4. R.C. Dove and P.H. Adams, *Experimental Stress Analysis and Motion Measurement*, Merrill Books, Columbus, OH, 1964. (out of print).
9-5. R.L. Hannah and S. Reed, Eds., *Strain Gage User's Handbook*, Elsevier, London, 1992.
9-6. Harris, C.M., *Shock and Vibration Handbook*, 3rd Ed., McGraw-Hill, New York, 1988.
9-7. E.C. Herceg, *Handbook of Measurement and Control*, Schaevitz Engineering, Pennsauken, NJ, 1976.
9-8. Herzfeld, C.M., Ed., *Temperature, Its Measurement and Control in Science and Industry* Vols. 1, 2, and 3, Reinhold Publishing Corp., New York, 1962
9-9. M. Hetenyi, Ed., *Handbook of Experimental Stress Analysis*, John Wiley & Sons, New York, 1950.
9-9. A.S. Kobayashi, Ed., *Handbook on Experimental Mechanics*, Prentice-Hall, Englewood Cliffs, NJ, 1987.
9-11. A. S. Kobayashi, Ed., *Manual on Experimental Stress Analysis*, Society for Experimental Mechanics, Bethel, CT., 1983.
9-12. C.C. Perry and H.R. Lissner, *The Strain Gage Primer*, 2nd Ed., McGraw-Hill, New York, 1962.
9-13. J. Pople, *BSSM (British Society for Strain Measurement) Strain Measurement Reference Book*. BSSM, Newcastle upon Tyne, England, 1979.
9-14. Stein, P.K., *Measurement Engineering*, Vol.1, Stein Engineering Services, Phoenix, AZ, 1964.
9-15. A.L. Window and G.S. Holister, *Strain Gauge Technology*. Applied Science Publishers, London, 1982.
9-16. _____ , *Monograph — Methods and Practice for Stress and Strain Measurement*, British Society for Strain Measurement (BSSM), Newcastle upon Tyne, England, 1979.
9-18. _____ , Modern Strain Gage Transducers. Their Design and Construction (Part IV: Load Cells..Chapter 3)." *Epsilonics*, Vol. 2, Issue 3, Dec. 1982, pp. 5-7, (Measurements Group, Inc., P.O. Box 27777, Raleigh, NC).
9-19. _____ , *Strain Gage Based Transducers*, Measurements Group, Raleigh, NC, 1988.
9-20. F.S. Tse and I.E. Morse, *Measurement and Instrumentation Engineering*, Marcel Dekker, New York, 1989.

Papers:

9-P1. Brendel, A.E., "Transducer Tutorial: From the User's Viewpoint," Society for Experimental Mechanics, Fall Conference, Greenlefe, FL, Nov. 1985.
9-P2. Gardner, D., "Encoders for Quick Precision," Design News, Feb. 11, 1991, pp. 88-92.
9-P3. Ormond, A.N., "Choosing and Using Flexures," Ormond Inc., Santa Fe Springs, CA.
9-P4. R.T. Reese, "A Working Paper on Strain Gages and Adhesives-Operational Characteristics," Proc. Western Regional Strain Gage Committee, Society for Experimental Mechanics, Bethel, CT, Aug. 1990.
9-P5. Yorgiadis, A., "Long Term Stability of Strain Gage Load Cells," Proc. Western Regional Strain Gage Committee, Phoenix, AZ, Feb. 12-13, 1985, SEM, Bethel, CT.

In addition to these references many manufacturers of transducers and gages make technical reports, notes, and other publications available to potential users. These technical notes offer excellent basic instruction, detailed analyses of potential problem areas, and give excellent ideas and recommendations to enhance the use of strain gages, accelerometers, displacement gages, etc.

Chapter 10
Instrumentation and Data Systems

*John Dreger, Engineering Mechanics Department,
University of Wisconsin, Madison, WI*

10.1 INTRODUCTION

It has often been stated: "A test is only as good as its data." In the introduction to this handbook, it was stated that obtaining the insight and information (data) describing the behavior of a structure is the goal in structural testing. Therefore we must pay close attention to the means of obtaining and recording that data. The term "instrumentation" encompasses all the devices used to: sense, connect, condition, display, and record the data. Figure 10-1 shows a basic schematic for an instrumentation system including an active transducer.

In the instrumentation system to be described here the device used to sense the measurand will be a transducer or sensor. The measurand is defined as the quantity of the physical changes occurring in the structure due to applied excitation (e.g., force, acceleration, temperature) to the structure. The transducer converts that physical change into an into electrical change, usually by altering voltage or current. The subsequent components in the system will be devoted to communicating these changing voltages or currents to the engineer in meaningful terms.

Many things have to be considered in putting together an instrumentation system for a given structural test. Is it going to be a static test, where the loads are to be constant or very slow to change, such as a time dependent or creep test? Perhaps there may be rapidly changing loads or transients, such as vibration, cyclic loading, or impacts. What do we wish to sense or measure, what accuracy, and what kind of resolution do we need? Are parts of the structure moving relative to each other or the test facility? Are there environmental concerns such as elevated or reduced temperatures, electrostatic or electromagnetic noise? How will we observe and document the data? Is one channel of information sufficient or will we need multiple channels? Simple numbers read out on a meter may be adequate, but, more than likely, we will want to permanently record the data in some fashion. The next step would be to input the results into a computer, either real-time or from the recorder medium at a later date.

Ideally, all the components of the instrumentation system should be selected based on the requirements of the test. However, other than specialized transducers, most of the components will probably be commercially available instruments that are versatile and adaptable to most test application and procedures. As described in previous chapters, the structure can be excited in various ways: through force, displacement, acceleration, temperature changes. For simplicity, force (load) will be the assumed excitation of the structure for discussions in this chapter (Refs. 10-1 through 9).

10.1.1 Rate or Time Dependency of Loading

One of the first considerations in designing the instrumentation system is whether the test is to be static or dynamic in nature. This factor will have the greatest influence in determining what type of display and/or recording device is to be used. (The reader is referred to Table 10-3, page 177, for equipment used to acquire and record data including frequency response ranges.) In general, the transducer and signal conditioner are more flexible with respect to rate of change of the data signal and this will be discussed later. To define the domains of static versus dynamic in terms of frequency is difficult due to the type of structures, environments, and whether the loads are repetitive or transient in nature. Strain rate or displacement rate usually place limitations on transducers. Generally, rates less than 1 Hz are approaching static as far as recorders are concerned. Dynamic would therefore be anything faster. Transient loading, however, usually is expressed in

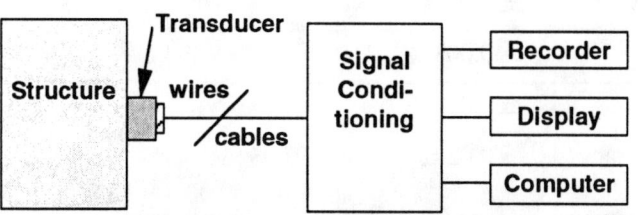

Figure 10-1. Basic Schematic for an Instrumentation System

terms of time to level rather than frequency and causes problems in capturing a true representation, particularly in digitizing systems such as computer data acquisition systems and digital oscilloscopes. Most large civil engineering structures have responses in the 1 Hz to 10 Hz range, which is not too fast for most transducers and recording systems. Other mechanical systems can respond at much higher frequencies limiting available sensors and recording systems.

10.1.2 Accuracy and Resolution

Many have fallen into the trap of confusing accuracy with resolution. One may have a load cell with 20,000 lb. capacity and have the output read on a 5 digit digital display or a digital oscilloscope with a 12 bit resolution, which both give a resolution of 1 lb. That amounts to a resolution of 0.005 percent of capacity. However, that load cell may have a nonlinearity of ±0.15 percent, which at a given load the reading may be off by as much as 30 lbs. In addition, the meter accuracy could be + 0.02 percent of reading, which could add another 4 lbs. of uncertainty in the measurement. There are many sources for uncertainties in the system and they can be additive. Table 10-1 gives examples of the accuracy and resolution of various instruments (Refs. 10-2 and 10-8) and a more detailed discussion of measurement uncertainties found in structural testing is available in Chapter 14.

As can be seen in Table 10-3, the recording method can possibly have the greatest influence on overall system accuracy and resolution. In high rate dynamic tests we may be restricted to an analog oscilloscope and a camera to record the data from the screen. This usually gives at best 1-3 percent accuracy as well as resolution. For static work a strip chart, x-y or servo recorder can give better than 1 percent results. Digitizing systems can give excellent accuracy and resolution; however, one must be careful in selecting the right type of A/D converter and in interpreting the data (Ref. 10-T1). Excessive noise in the data signal can appear as data and will be discussed later.

10.2 TRANSDUCERS AND SENSORS

Sensors/transducers have been covered in Chapter 9. However, since they are part of the instrumentation package and they will affect the type of signal conditioning to be used, we must discuss them here also. Transducers can be divided into two basic groups: self-exciting or needing external excitation. Excitation here refers to a voltage or current either produced by the device from the phenomena being measured or being supplied to the transducer for its operation as summarized in Table 10-2 (Refs. 10-2, 10-7, and 10-T2).

Table 10-2. Types of Transducers and Sensors

Self-Exciting	Needing Excitation
Thermocouples	Thermistors
Piezoelectric	Piezoresistive
Velocity Coils	Resistance Strain Gages
Photoelectric Devices	Linear Variable Differential Transformers (LVDT's)
	Capacitive
	Potentiometers

Thermocouples, resistance temperature devices (RTD's), and thermistors measure temperature changes;

Table 10-1. Example of Instrumentation Accuracies and Resolution

System Component	Parameter	Accuracy % of Full Scale	Resolution
Transducer (Load Cell)	Linearity Hysteresis Repeatability	± 0.15 % ± 0.15 % ± 0.05 %	
Signal Conditioner	Linearity	± 0.003 %	
Calibration Resistor	Tolerance	± 0.02 %	
Digital Panel Meter (4.5 digit)	Overall Accuracy	± 0.02 %	0.005 %
Total Uncertainty (worst case) =		± 0.397 %	0.005 %
For 20,000 lbs. Load Cell =		± 78.6 lbs.	1 lb.

however, while thermocouples produce a voltage proportional to temperature change, thermistors and RTD's are resistors that change their resistance value due to temperature change. Therefore, an external voltage source must be supplied to the thermistor or RTD, such that when properly wired in a half-bridge configuration, a voltage change can be measured proportional to a change in temperature as shown in Fig. 10-2.

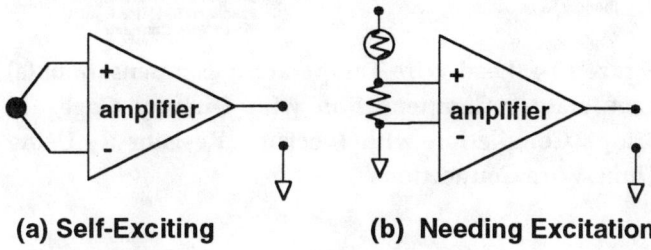

(a) Self-Exciting   (b) Needing Excitation

Figure 10-2. Two Types of Transducers

Another example of the self-exciting transducer is the piezoelectric based device, represented mostly by accelerometers and some load cells or pressure transducers. They use a crystal which produces a voltage or charge when stressed. One important characteristic of the piezoelectric device is that it is limited to dynamic forces greater than 0.5 Hz. Another factor that affects the associated instrumentation is that it has a very high impedance, which necessitates the use of connecting cables matched to the transducer and the amplifier in the signal conditioner. The amplifier is either a charge amplifier or a capacitively coupled very high input impedance amplifier, such as a voltage follower (Ref. 10-T9).

In contrast to the piezoelectric sensor, the piezoresistive type consists of a semi-conductor crystal that changes resistance under mechanical stress. It does not produce a charge or voltage output and therefore needs external excitation voltage for its operation. It has the advantage of sensing static forces as well as dynamic and has a lower impedance, which means conventional cabling and amplifiers can be used. Due to its high sensitivity, 10-100 times that of wire or foil strain gages, in many applications amplification is not needed and the transducer's output can be coupled directly to the recording/readout device. The disadvantages of piezoresistive elements are poor linearity over their ranges and thermal drift. When used as a strain gage for strain measurement, it is difficult to match the gage with the coefficient of thermal expansion for the material being tested, as can be done with metal or foil gages. They are also difficult to shunt calibrate as described in Section 10.4.

Probably the most widely used sensor in transducer applications is the metal foil strain gage. It has the greatest linearity and can be easily compensated for thermal effects. They are powerful tools in the sensing of mechanical changes which occur in structural testing. However, how they are connected and interfaced to the instrumentation system is of great importance (Ref. 10-T10).

10.2.1 Bridge Completion

Since the metal foil strain gage is also an excellent temperature sensor, we must be able to compensate or adjust for the thermal effect when measuring strain or other phenomena when using strain gage type transducers. This temperature-induced resistance change is referred to as apparent strain. As described in Chapter 9, the strain gage changes resistance proportional to the change in strain of the material it is mounted on. The resistance change is very small and an ordinary resistance meter is not sensitive enough to resolve the desired equivalent strain. The Wheatstone bridge is the most popular device used today to convert the small resistance changes into voltage changes which then can be amplified sufficiently to drive the recorders, displays, and computer data acquisition systems. A Wheatstone bridge circuit is shown in Fig. 10-3 (Refs. 10-T11, 10-T13, and 10-T14).

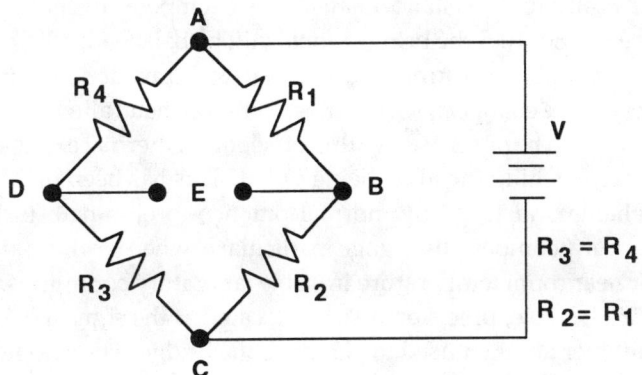

Figure 10-3. Wheatstone Bridge Circuit

With an excitation voltage, V, applied to the bridge at AC, the voltage drop across R_1 is

$$V_{AB} = \frac{R_1}{R_1 + R_2} V \qquad \text{Eq. 10-1}$$

and the voltage drop across R4 is

$$V_{AD} = \frac{R_4}{R_3 + R_4} V \qquad \text{Eq. 10-2}$$

The output voltage of the bridge, E, is

$$E = V_{BD} = V_{AB} - V_{AD}$$

Substituting Eqs. 10-1 and 10-2 into Eq. 10-3 and simplifying gives

$$E = \frac{R_1 R_3 - R_2 R_4}{(R_1 + R_2)(R_3 + R_4)} V \qquad \text{Eq. 10-4}$$

The bridge will be balanced and the output will be zero when

$$R_1 R_3 = R_2 R_4 \qquad \text{Eq. 10-5}$$

In measuring strain with one strain gage (referred to as the active gage), resistance can be represented in Eq. 10-5 by R_4, with R_1, R_2, and R_3 being equal and fixed. With a change in strain an equivalent change in resistance of R_4, will occur, causing an output voltage E. However, if a temperature change occurs as well, there will also be a resistance change and an output due to temperature (apparent strain). If a second strain gage (dummy gage) is mounted on a part of the structure under test such that it will be in the same temperature environment as the active gage but will see no strain during the test and is wired into the bridge as R_3 as shown in Fig. 10-4(a), then temperature compensation will occur. Ideally, both R_3 and R_4 would have an equal resistance change due to temperature and as seen in Eq. 10-5 the bridge would still be balanced, allowing the imbalance to occur due only to the strain dependent resistance change in R_4. For most common metal alloys foil gages can be purchased with coefficients of thermal expansion matching the alloy being tested. This has been quite reliable and it is now normal practice to eliminate the dummy compensating gage, particularly when working at or near room temperature in stable laboratory conditions. Fixed, stable, precision resistors located at the signal conditioner are then used to complete the bridge. The bridge circuit shown in Fig. 10-4(b) is a quarter bridge configuration which is the most common method used for discrete strain measurement.

In situations where bending strains are present one can mount two active gages on opposite sides of the bending member and achieve both excellent temperature compensation as well as doubling the voltage output of the bridge. An example of a half bridge configuration for measuring bending strains is illustrated in Fig. 10-5.

Since R_3 and R_4 would have the same resistance change, but of opposite polarity, by connecting them to the bridge as shown in Fig. 10-5 the result would be additive and a doubling of the output occurs.

For forces other than bending, such as axial or torsional, the half bridge can be used as shown in Fig. 10-6. As shown in Fig. 10-6(a), gage R_3 will measure Poisson strain

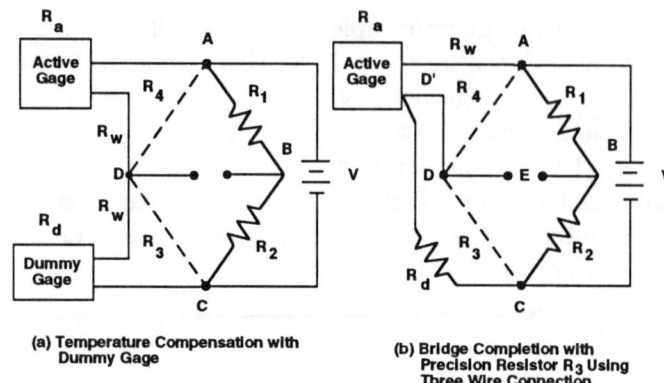

Figure 10-4. Lead Wire Temperature Compensation: (a) Temperature Compensation with Dummy Gage, (b) Bridge Completion with Precision Resistor R_3 Using Three-Wire Connections

and R_4 axial strain. Assuming Poisson's ratio of 0.3 and that the strain at R_3 will be negative relative to that at R_4, one can connect R_3 in the adjacent arm of the bridge and get an effective output (E) equal to 1.3 times the axial strain. The torsional load in Fig. 10-6(b) will result in R_3 and R_4 measuring equal magnitudes of strain but of opposite polarity and by connecting R_3 and R_4 in adjacent arms the result will be an effective output (E) equal to two times the torsional strain, similar to the result in the bending beam example.

By mounting two additional gages on the beam as shown in Fig. 10-5, R_2 adjacent to and oriented in the same direction as R_4, and R_1 adjacent to and oriented in the same direction as R_3, the output would be double that of the half bridge configuration as shown in Fig. 10-7.

The full bridge obviously offers several advantages over quarter and half bridge configurations; however, they are primarily the domain of transducers such as load cells, pressure transducers, and extensometers. Since their effective outputs usually represent something besides strain, they must be calibrated by applying the desired mechanical change to the device to determine their sensitivity (Refs. 10-1, 10-T5, 10-T6, 10-T13, and 10-T14).

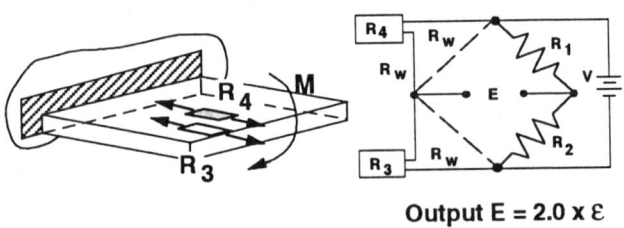

Output $E = 2.0 \times \varepsilon$

Figure 10-5. Bending Beam Example Using Half Bridge

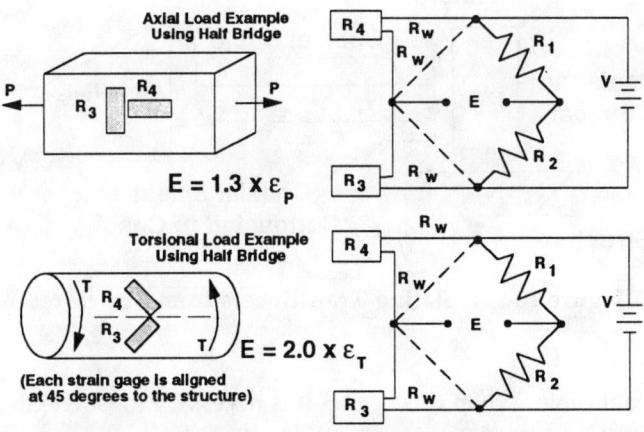

Figure 10-6. Half Bridge Examples

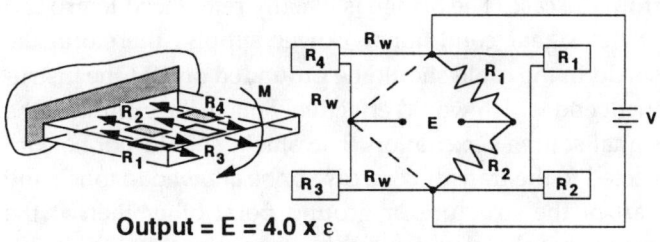

Figure 10-7. Bending Beam Example Using Full Bridge

10.2.2 Wiring

Always a major factor are the actual connections to the bridge and the wiring itself. Due to temperature changes during a test apparent strain can be encountered in the form of resistance changes in the lead wires in addition to those in the strain gage. Particularly vulnerable are the quarter and half bridge configurations. Since the lead resistance is in series with the strain gage, its resistance change can easily have as great an effect as the gage's resistance change due to strain. In the half bridge all leads going to the active and dummy gages must be of the same gauge (diameter), material, and length. In other words, all four leads must have the same resistance and be of the same material so that the resistance change due to the environment will be the same in all leads and their effects cancelled out in the bridge. Referring to Fig. 10-4(a), R_w and R_a are part of R_4 and R_w and R_d are part of R_3. From Eq. 10-5 it can be seen if R_4 and R_3 would change equally the bridge would remain balanced.

In the quarter bridge, Fig. 10-4(b), a third lead wire to the active gage is used to compensate for temperature. All the lead wires of equal length, gauge, and resistance, and are routed together in the same environment between the gage and signal conditioning. Since R_w and R_a are part of R_4 and R_w and R_d are part of R_3, they again will cause no imbalance of the bridge due to temperature change.

The added resistance of the intra-bridge lead wires in the active arms of the bridge will also cause desensitization of the gages in those arms. Since the lead wires' resistance does not change when the gage is subjected to strain, their resistance does not contribute to the strain related change in bridge output. In a quarter-bridge where two lead wires are used to connect the active gage to the bridge completion in the signal conditioner, as shown in Fig. 10-4(a), the desensitization factor F is:

$$F = - \frac{2R_w}{2R_w + R_a} \qquad \text{Eq. 10-6}$$

compared to:

$$F = - \frac{R_w}{R_w + R_a} \qquad \text{Eq. 10-7}$$

for a three lead wire connection as in Fig. 10-4(b). The third wire in Fig. 10-4(b) effectively moves the bridge point D to point D′ and is not considered an intra-bridge lead wire and thus has little effect on sensitivity. As can be seen from Eqs. 10-6 and 10-7, by using the three lead wire method the attenuation factor can be reduced by approximately one-half. Additional compensation for lead wire desensitization can be accomplished through proper calibration procedures to be discussed later.

Other environmental factors are also of concern in the wiring and connections of the gages or transducers to the instrumentation. Probably the most troublesome is electrical noise. A structural test lab or field location has many sources for electrical noise. Generally this includes noise from power lines, motors, transformers, fluorescent lighting, arc welders, mechanical relays, and solenoids. Transients on the electrical power lines can find their way into the instruments through the line cord. Power line filters and isolation transformers can be used to minimize problems transmitted through power lines, but do not help when the noise is inductively picked up by the transducer and/or cables/wires. This "inductive pickup" is called electromagnetic coupling and can be reduced by twisting the wires from the transducer closely together to electrically cancel the external magnetic field. Electrostatic coupling is another source of noise in the cabling/wiring connections. It is due to capacitive coupling to adjacent circuits. It is most critical in high impedance signal sources such as accelerometers and other piezoelectric devices. The best method to minimize this type of electrical noise is to use shielded cable. Co-axial cable is commonly used with accelerometers and high level signal sources; however, caution must be used as it affords no protection from electromagnetic interference and their use can create ground loops if the shield is connected to ground at both ends. With low impedance and low level signal sources, twisting the wires together with a shield wrapped around the wires and connected to ground offers the optimum protection. When using a shield it is of great importance as to how and where it is connected to ground. The shield essentially intercepts the capacitively coupled interference voltage and "drains" it to ground. An improvement over the braided type shielded cable is a foil wrapped cable with a drain wire which would be connected to ground. Examples of wiring to reduce electrical noise are shown in Fig. 10-8.

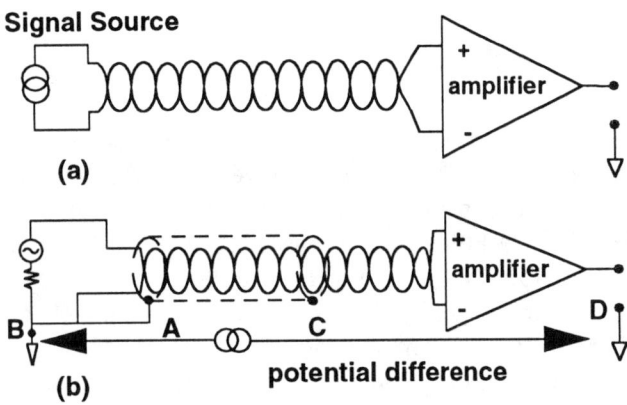

Figure 10-8. Wiring to Reduce Electrical Noise

When using a shield in the signal cable one must be very careful where the shield is connected to ground. Referring to Fig. 10-8(b), the best place to connect the shield ground is at the sign ground "B." Never connect the shield at both "B" and "D" as that would provide a current path from A, C, D, to B which would be coupled as noise into the signal leads inside the shield. This noise introduction to the system is referred to as a ground loop. The major rule in avoiding ground loops is to connect all grounds in a signal circuit at one common point, usually at the power supply ground in the signal conditioner. But, in some cases the signal source may be a self-exciting transducer such as an accelerometer, then the ground point of the shield would be at the amplifier/signal conditioner as shown in Fig. 10-9. With accelerometers the case is usually connected to the shield of the co-axial cable; therefore, one of the signal paths to the amplifiers would be grounded through the structure and to earth ground if the structure were of conducting material. This could form a ground loop through the building or laboratory ground to the signal conditioning and back to the accelerometer via the shield in the co-

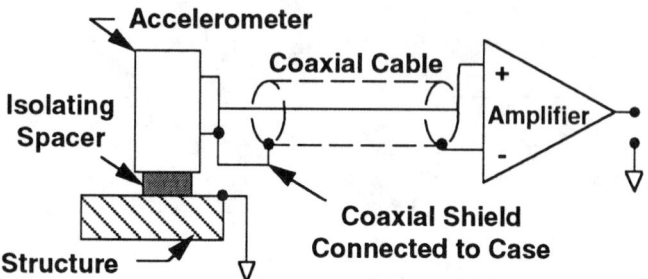

Figure 10-9. Isolating Transducers from Structures

axial cable. When this occurs it is necessary to isolate the accelerometer from the structure with an insulating spacer (Refs. 10-T4 and 10-T8).

With strain gage bridges and bridge type transducers the gages and sensing elements are electrically isolated from the case. The bridge is usually referenced to ground at the signal conditioner power supply; therefore, the shield in the cable should be grounded only at the instrument end which would effectively make the ground at the signal source's excitation. The shield should not be connected to the transducer's case or be allowed to touch any part of the structure or ground point other than at the power supply/signal conditioner as shown in Fig. 10-10.

With discrete strain gage measurement, quarter bridge, the three-wire leads should also be twisted or braided together as shown in Fig. 10-11. In electrically noisy environments flat ribbon cable should be avoided. Wire lengths should be as short as possible, and never have excess leads coiled, as that invites inductive noise coupling much like a transformer. Always avoid routing wires and cables parallel to, and in close proximity to power lines and cables carrying high currents and voltages (Ref. 10-T7).

For strain gage based transducers there has been a standard established in the United States defining the color coding of the cables and wires as shown in Fig. 10-12. It also defines the pin assignments, relative to the bridge "points," for most connector applications between the transducer and the signal conditioning (Ref. 10-3).

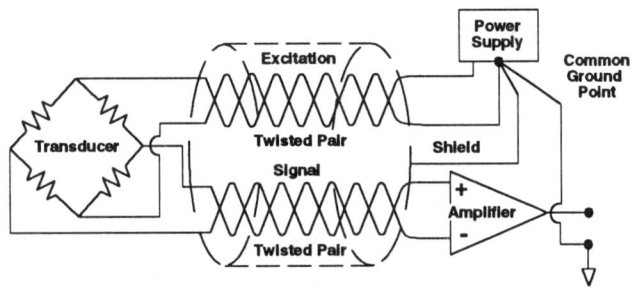

Figure 10-10. Proper Grounding Procedure

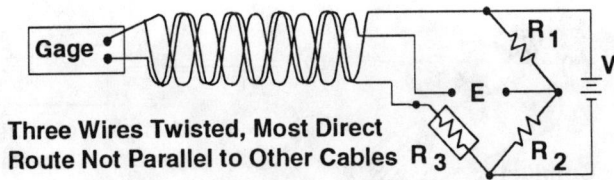

Figure 10-11. Quarter Bridge Wiring

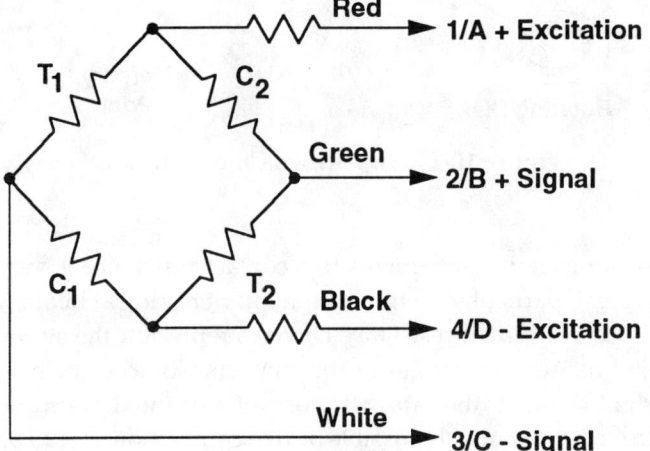

Figure 10-12. Standardized Color Coding and Connector Pin Designations for Strain Gage Based Transducers Specified by the Western Regional Strain Gage Committee and ANSI-SEM-1-1984.

10.3 DATA TRANSMISSION

In many test situations the components of the instrumentation system may be located remotely from one another, such as when there would be hazards involving the response of the structure, both to the personnel involved and the delicate and sensitive recorders and/or computers used. The transducers must be mounted on or physically close to the structure, and therefore, would require long cabling to connect to the associated signal conditioning and recorders. Long lines can attenuate data signals, particularly high frequency signals, due to the resistance and capacitance of the cable. Even short cables (under 10 feet) can attenuate signals with high impedance devices such as piezoelectric transducers.

One way of diminishing these effects is to digitize the data signal at the transducer before it is sent over the long lines. Digitizing basically chops the data signal up into a series of pulses. The data value will be related to the absence or presence of pulses rather that the amplitude or voltage level. Therefore, if due to cable resistance there is attenuation of the amplitude of the pulses or electrical noise causing amplitude fluctuations in the pulse amplitudes there will be no degradation of the data. Two of the most popular methods are called pulse-code modulation (PCM) where the number of pulses in a sequence is proportional to the voltage level and pulse-duration (PDM) where the pulse widths are proportional to the voltage level (Ref. 10-7).

10.3.1 FM-FM Telemetry

Related to the pulse modulation systems is the FM-FM telemetry which can be used when it becomes impractical or impossible to connect the transducers to the instrumentation via "hard wire." This technique can transmit data across small distances with the structure itself, across the laboratory, or great distances as with satellites in space. Another feature of this concept allows multiple channels of data to be transmitted over one RF (radio frequency) link. Each varying voltage signal from its respective transducer is converted to a proportionally varying frequency. This is done with voltage to frequency converters called subcarrier oscillators (SCO's). Each data channel would have a designated SCO and its center frequency would be different from the others in the system. In most standardized systems as many as 18 channels representing 18 different variables may be transmitted simultaneously. The 18 SCO center frequency bands range from 200 Hz to 70,000 Hz. This does limit the frequency response range of the data somewhat; ranging from 6 Hz for the 400 Hz SCO to 1 kHz for the 70,000 Hz SCO. The SCO frequencies are too low for practical transmission to the receiving site so the SCO outputs are "mixed" (added) then the composite signal modulates another oscillator of higher frequency and power such that its output can be coupled to an antenna and transmitted at an allocated radio-frequency (RF) channel. For low power, within the laboratory or structure, the standard FM broadcast band (88 to 108 megahertz) can be used. However, for outside the laboratory and longer distances, allocated frequency bands must be used. At the receiving end of the system an antenna and FM receiver are used to separate the "mixed" SCO frequencies from the RF. The "mixed" SCO frequencies then pass through band-pass filters which separate the various SCO frequencies. These is a band-pass filter for each respective SCO frequency. From there each SCO frequency goes to an FM demodulator which separates the data from the SCO frequency. A lowpass filter completes the job outputting a varying voltage duplicating the transducer's output as given in Fig. 10-13 (Ref. 10-2).

If the transducer is mounted on the moving part some means of couplings the data signal to the associated signal conditioning and recorder must be provided. One solution may be the FM-FM telemetry link just described. The SCO,

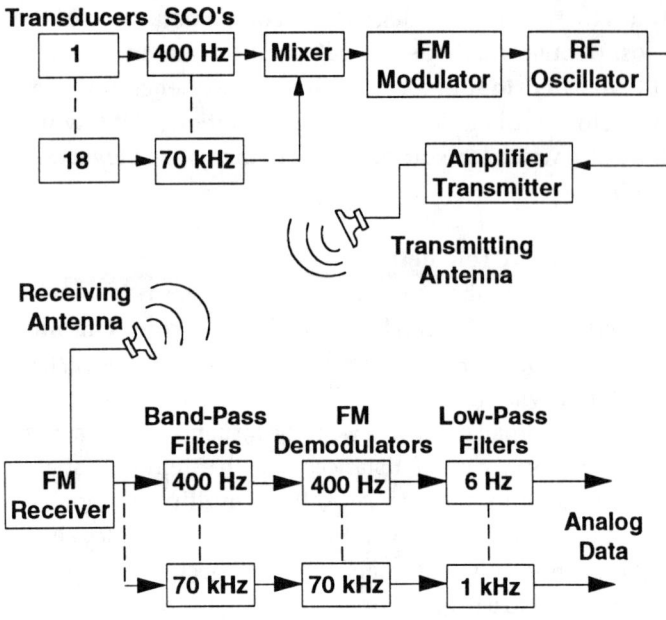

Figure 10-13. FM-FM Telemetry System

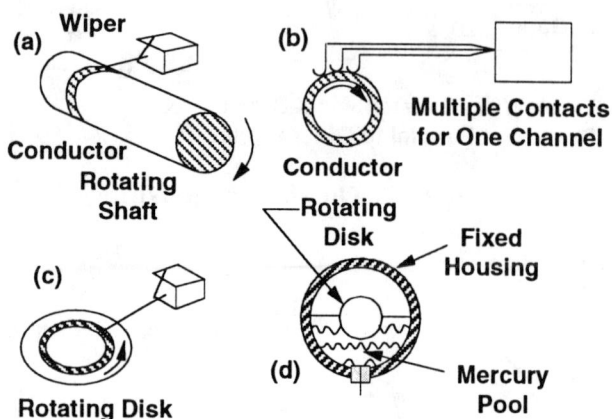

Figure 10-14. Slip-Rings Configurations

modulator, transmitter, antenna, and battery power source would be mounted with the transducer on the moving part and the receiving antenna and FM receiver with demodulators and filters would be on the fixed part of the structure. Another way of transmitting data from a moving structure to the recording system is through slip-rings.

10.33.2 Slip-Rings

With rotating members of structures slip-rings have been used quite successfully at speeds up to 100,000 rpm. Slip-rings usually consist of a rotating shaft or disk with a continuous electrical conductor attached to its surface. A mating wiper or slider on the fixed body makes continuous contact with the moving conductor. Examples of slip-ring configurations are given in Fig. 10-14. In many cases part of the structure being instrumented may be moving with respect to the rest of the structure (e.g., rotating shafts, centrifuges) (Refs. 10-2 and 10-8).

The major problem with slip-rings is the contact resistance between the conductor and the wiper. Because of this resistance, contact surfaces are usually made of gold, silver or other noble metals or alloys, making them expensive. Many schemes have been developed to alleviate this problem. Often multiple wipers will be used for each channel as shown in Fig. 10-14 (b). Another type uses a liquid conductor, such as mercury, to form the contact link between the rotor and stator (fixed housing) as given in Fig. 10-14 (d). It may have a contact resistance as low as 0.005 ohm with variations of ±0.00025 ohm compared to the 0.05 ohm variation with sliding contact types. With strain gage measurements this contact resistance is very crucial, particularly when using quarter bridge circuitry. Since very small resistance changes represent the strain, any resistance changes in the contacts would contribute significantly to the data in the form of unwanted strain due to electrical noise. If possible bridge completion should be accomplished at the measuring point ahead of the slip-rings. Bridge completions here put the slipring contact resistance in series with the excitation voltage leads and the output signal leads of the bridge which will have little desensitizing effect on the data signal.

10.4 SIGNAL CONDITIONING

In an instrumentation system the signal conditioning consists of components which can manipulate and alter (condition) the data signal from the transducer. It encompasses, but is not limited to: excitation source for the transducer, zero suppression, impedance matching, amplification/attenuation, linearization, differentiation, integration, filtering, and calibration. It is not the scope of this handbook to show how to design or build a signal conditioner because there are many commercially available with many of the capabilities discussed above. However, it is important to point out the differences between various types of conditioning (Ref. 10-P2).

10.4.1 Power Supplies

Sometimes the power supply or excitation source, for a transducer that needs external excitation, may be physically separated from the signal conditioner. However, the excitation is a variable in the system and its level and type affects the output of the transducer; therefore, for this discussion it is included in signal conditioning. For devices needing a stable excitation, such as strain gage based transducers, there are two basic types of regulated power supplies: constant-voltage or constant-current. Most signal conditioning systems currently available for use with resistive and bridge type transducers use a constant-voltage excitation. In the past constant-voltage power supplies were less costly and more readily available and, in many cases, batteries were used which are basically a constant-voltage source. Traditionally, most resistive type transducers are calibrated and have their Sensitivities stated in terms of millivolts per volt of excitation. Therefore, with a constant-voltage excitation it is more straightforward to establish desired gain levels in the system using manufacturer's supplied calibration data for the transducer. The Wheatstone bridge does have inherent nonlinearities, particularly when only one strain gage is active and at higher strain levels (Ref. 10-T5). There also can be desensitization of the bridge due to lead wire resistance when using constant-voltage excitation. This is due to a voltage drop between the power supply and the transducer. This can be overcome by using an extra pair of wires going to the transducer to sense the voltage at the gages in the transducer. This measurement of the voltage at the bridge is used to "adjust" the voltage output at the power supply to correct for voltage drop. Most nonlinearities are relatively small and are usually corrected numerically.

Despite the constant-voltage method being more popular, constant-current excitation offers several advantages. Probably the strongest case for constant-current is the elimination of the effect of the resistance in the power lead wires going to the bridge. As explained earlier, in full bridge transducers that may be of some distance from the signal conditioning, attenuation of the excitation voltage can occur which proportionally drops the output of the transducer. Also, in varying temperature environments the lead-wire resistance will change, causing some additional errors. With constant-current there will be no change due to the resistance of the lead wires, plus the two sensing wires are not needed. The nonlinearities in the bridge are not totally eliminated, but are reduced to as much as one-fourth (depending on ratio of active gage resistance to bridge completion resistance in a single active gage configuration) of the error as with a constant-voltage supply. A word of caution: many strain gage based transducers have modulus compensation and span set resistors which become useless when using a constant-current supply, plus the transducer would probably have to be recalibrated with the constant-current supply (Ref. 10-P1).

A four-wire constant-current configuration can be very beneficial when making strain measurements with a single gage, particularly with dynamic strains. The circuit is quite simple in that bridge completion is not necessary as shown in Fig. 10-15.

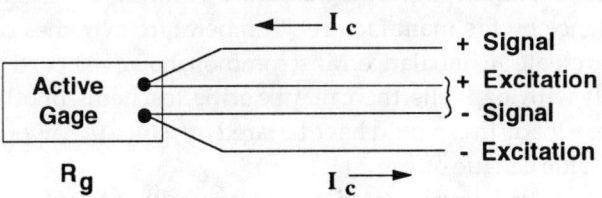

Figure 10-15. Single Strain Gage Without Bridge Using Four-Wires with Constant Current Excitation

The voltage drop across the gage resistance R_g with zero strain would be:

$$V = I_c R_g \qquad \text{Eq. 10-8}$$

A change in strain would produce:

$$V + DV = I_c (R_g + \Delta R_g) \qquad \text{Eq. 10-9}$$

As can be seen by Eq. 10-9, there will always be a voltage output regardless of strain level. Since there are no bridge or compensating resistors to provide a zero output at zero strain, we must deal with the "offset" (V) created because this could easily saturate most amplifiers. This can be eliminated by "injecting" a voltage of appropriate level and polarity into the amplifier to null out the offset. Another method would be to AC couple the signal leads (S+, S–) to the amplifier by putting a capacitor in each line. Since the offset would be a DC voltage (static) it would be blocked by the capacitors and any changing voltage would pass through them to be amplified. Obviously this technique limits its use to transient/dynamic signals. Also, a temperature compensating dummy gage cannot be used, so self-temperature-compensating gages are a must. However, with dynamic signals these gages are not as important. Another benefit from this method is that the signal produced is nearly four times that of a constant-current Wheatstone bridge where all four gages are of equal resistance. Where long lead wires, lead wires with high resistance, or slip-rings must be used, constant-current excitation can give superior results. But, in most measuring situations, particularly with full bridge transducers, existing voltage instruments give very adequate results. There are few instruments available which offer the choice of current or voltage (Ref. 10-T6).

10.4.2 Offsets

Offsets from transducers can be caused by various means. With strain gage based devices there can be slight imbalances in the bridge due to the gages not being perfectly matched in resistance (nominal gage resistance tolerance might be ±0.15 percent), although most commercially available strain gage transducers are trimmed for resistive balance by the manufacturer. Temperature extremes can also create an imbalance. Most common, however, particularly with load cells, there may be grips, test items, or other mean loads that would have be tared off, since they would provide a static offset.

As previously stated, transducer voltage offsets are a factor that usually are compensated for in the signal conditioning. If they are not too great they sometimes can be "adjusted" numerically after conditioning when analyzing the data or in the computer, if one is used, with software. Also, with low offsets, many recording devices have sufficient "zero suppression" adjustments available to balance out the "zero offset."

Most signal conditioners usually provide one of two types of offset adjustment. Until recent years the most popular for strain gage bridges and transducers has been a ten turn potentiometer (R_v) wired across the excitation point of the bridge with a fixed resistor (R_c) in series with the wiper and one of the signal points. Figure 10-16 shows methods to electrically balance Wheatstone bridges.

The external method shown in Fig. 10-16 (a) can have a "loading effect" on the bridge and can cause a slight error in signal especially with precision strain gage transducers. Most modern signal conditioners use the voltage injection method where the initial imbalance voltage is sensed and a voltage of equal magnitude, but of opposite polarity is added at the input stage of the amplifier to null the imbalance without any resistance circuitry added to the bridge. Besides avoiding the loading of the bridge it offers the ability to have automatic balancing in multi-channel systems. Variations of this allow manual adjustment of the injection voltage. In some cases where high precision is necessary and there are large strain levels (>10,000 microinches), it becomes necessary to know the initial imbalance at zero strain so that the appropriate mathematical corrections can be made with the data, due to the inherent nonlinearities of an imbalanced bridge. To do this one must be able to defeat or remove the balance circuit, not matter which type, so that the zero strain output can be read.

A special case dealing with offsets concerns thermocouples. All connections (junctions) in the wiring of thermocouples to the amplifiers or readout devices, generate voltages proportional to the temperatures at those points. Therefore, some provision must be made to cancel these voltages as they will be added to the thermocouple voltage which is the one of interest. Since the voltage developed at the connection points usually represents the ambient temperature and if the temperature is stable at all connection points and is known, the equivalent value can be subtracted from the data or stated as referenced to ambient temperature. One way to remedy this "offset" is to reference the thermocouple output to 0°C by placing a second (reference) thermocouple in an ice bath or a device that simulates 0°C and wiring it in series with the measuring thermocouple. This results in a differential measurement relating to the measured temperature to a stable, known reference. Another version of this cold-junction compensation uses a temperature-to-current integrated circuit that is placed in intimate thermal contact with the reference junction and it adds voltage to the measuring circuit equal to the junction temperature but of opposite polarity, giving the desired reading of the measuring junction.

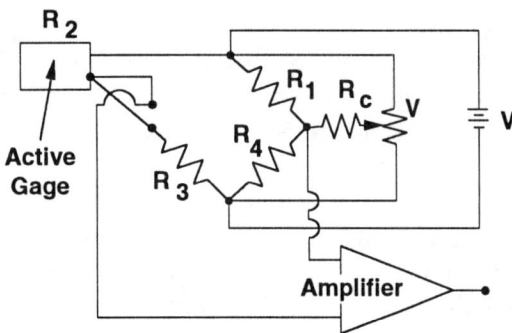

(a) External Bridge Balance Circuit

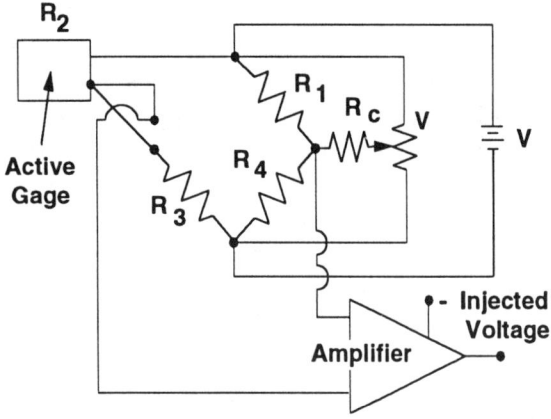

(b) Voltage Injection to Compensate for Bridge Imbalance

Figure 10-16. Bridge Balance Methods

10.4.3 Amplifiers

Most of the transducers and sensors that are used in structural testing are devices that have output signals that are of very low voltage levels. For these signals to be recorded or displayed they need to be amplified to levels sometimes as much as 10,000 times their original value. With gains of this magnitude, there are critical requirements in selecting amplifier used in the signal conditioning, including but not limited to: high input impedance, high common-mode-rejection, low drift, linear over large frequency range, adjustable gain, and low output impedance (Ref. 10-2 and 10-8).

Since the "ideal" transducer should have a low output impedance the amplifier should have a high input impedance so that it does not load and affect the transducer's output. As stated previously, this creates special considerations when using piezoelectric devices since they have very high impedance. Even the highest input impedance amplifiers can sill cause loading and errors in the data. A charge amplifier or a capacitively coupled amplifier is generally used in these situations.

Common-mode-rejection (CMR) is the ability for the amplifier to reject unwanted voltages (noise) "riding" on the data signal common to both signal leads coupled to the amplifier inputs. Most instrumentation amplifiers used in signal conditioners have a differential input configuration, meaning they amplify the "difference" in voltage between the two input terminals and "cancel" any voltage that appears simultaneously at both terminals as shown schematically in Fig. 10-17.

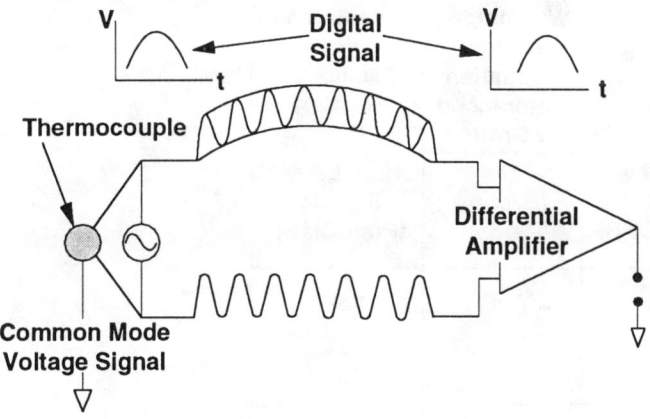

Figure 10-17. Common-Mode-Rejection

Drift (a change in amplifier output voltage without a corresponding change in input signal) in an amplifier is usually due to thermal effects, either from ambient environment or the self-heating of the internal components in the amplifier. For many years high gain DC amplifiers were thermally unstable and quite expensive. Therefore, most high gain systems used AC coupled amplifiers which were inherently stable. The problem was they were useless for DC signals and static testing. What was commonly used when DC signals were to be amplified, and is still used in some applications, was a AC carrier system much like the FM-FM subcarrier telemetry system discussed earlier. The transducer, using AC excitation, would modulate a carrier frequency such that the amplitude of the carrier would correspond to the measurand value while the frequency remained constant. After amplification the combined signal would be demodulated by filtering out the carrier signal leaving the lower frequency data signal. The high frequency response of such a system is limited by the frequency of the carrier. Usually the data signal is limited to one-tenth the frequency of the carrier. Today carrier amplifiers are primarily used with transducers that require AC excitation such as LVDT's, RVDT's, and capacitive based transducers. A carrier amplifier system is shown in Fig. 10-18.

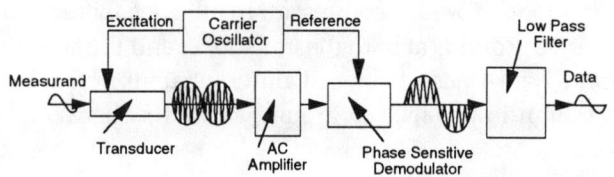

Figure 10-18. Carrier Amplifier System

Many of the LVDT's used in recent years incorporate the carrier amplifier in the transducer package, requiring only a DC power supply and producing a high level output (5-10 volts for full scale) that can drive recorders or displays directly. Again, dynamic response with these devices is limited to one-tenth the carrier frequency (Ref. 10-5).

Another way to amplify DC signals with a high stability AC amplifier is to use a "chopper" which chops up the incoming signal into pulses of constant width and frequency with the amplitudes representing the signal level. A high gain (as much as 100,000) AC amplifier is used and its output goes to a phasesensitive demodulator and low pass filter much like the carrier amplifier in Fig. 10-18. A hybrid version called a chopper-stabilized amplifier combines the drift-free characteristics of the chopper amplifier with the high frequency capabilities of a DC amplifier. By a using chopper amplifier at the "front-end" to amplify only the low frequency signal (the most prone to drift problems) while amplifying all the frequencies with a DC amplifier, a sizable reduction in overall drift is accomplished.

Since the developed of the integrated circuits, "pure" DC amplifiers have dominated the instrumentation field. They offer excellent thermal stability along with high frequency response, high input impedance, low output impedance, and easily adjusted gain. The basic DC amplifier building block is the operational amplifier (op-amp) which for noncritical situations can be used for amplifying transducer signals in low-cost signal conditioning. By far the most preferred amplifier is the instrumentation amplifier (IA). Most IA's use three or more op-amps in their design. In addition to basic op-amp qualities, they have superior common-mode-rejection, a true differential input, and the gain is set by a single, user selected resistor. For most low level signals, such as foil strain gages or strain gage based transducers and thermocouples, gains of 10x to 5,000x are common. The input configuration of the amplifier is determined by the transducer or signal source configuration. Fig. 10-19 shows several amplifier configurations (Ref. 10-T3).

Since the single-ended grounded amplifier configuration has no commonmode-rejection it should not be used with low-level signals or signals referenced to ground as ground-loops would be present. In all cases there should never be grounds at both the transducer and the amplifier input. The IA has a balanced, differential input that has a very high impedance to ground as well as between the input terminals. If used with floating signal sources such as a thermocouple, a bias current path must be provided at the input as given in Fig. 10-20.

The isolation amplifier is a special instrumentation amplifier that has its signal input circuit isolated from signal output and power supply circuits. In medical applications this is very important for safety to the patient as it

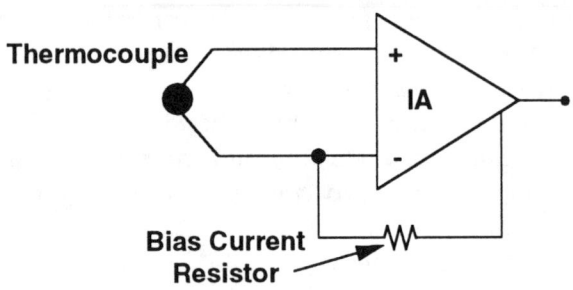

Figure 10-20. Floating Source with Bias Current Provision for Instrumentation Amplifier

Transducer Configuration (example) / Amplifier Input Configuration (Problem Areas)	Single-ended Grounded — Accelerometer	Single-ended Floating — Thermocouple	Balanced Grounded — Wheatstone Bridge, Excitation Referenced to Ground	Balanced Floating — Wheatstone Bridge Excitation Floating
Single-ended, Grounded	No (Ground Loop)	High Level Signals (Poor CMR)	No (Ground Loop)	High Level Signals (Poor CMR)
Balanced to Ground	No (Ground Loop)	Yes	Yes	Yes
Single-ended, Floating, Shielded	Yes	Yes	No (Poor CMR)	Yes
Balanced, Floating Ground	Yes	Yes (May need bias path to ground)	Yes	Yes (May need bias path to ground)
Isolation Amplifier	Yes	Yes	Yes	Yes

Figure 10-19. Amplifier/Transducer Compatibility

does not allow leakage currents from the power supplies to back through the transducer to the patient. Also, for small signal levels riding on a high common-mode voltage there can be a great improvement in CMR as well as safety as ground loops are eliminated.

Other than providing gain and impedance matching between the transducer and recorder/display, there are various specialized amplifiers that could be included in the signal conditioner that would: correct for nonlinearities in thermocouple and strain gage circuits, integrate or differentiate the data, and filter unwanted noise from the signal.

10.4.4 Filters

After following proper lead wire and connecting procedures and using shielding and guarding, there still may be excessive noise in the data signal. If the data and noise frequencies are significantly different, electronic filtering can be employed to attenuate the noise while passing the desired data signal. Normally this is done after amplification of the transducer's signal. Since most practical transducers do not have the desired low output impedance, an introduction of a filter circuit between the transducer and the amplifier would attenuate the input signal and distortion of the data signal. The basic types of filters used in signal conditioning are: low-pass, high-pass, band-pass, and band-reject. The simplest versions of these would be passive RC circuits shown in Fig. 10-21 which consist of resistors and capacitors connected in appropriate configurations.

The passive circuits shown in Fig. 10-21 are considered first-order systems that attenuate the input signal with respect to frequency at 6db/octave. To improve on that, second and third order filters can be used, consisting essentially of cascaded states of resistance and capacitance. A major consideration with any filter is attenuation of the desired signal as well as the unwanted (e.g., noise). The is most problematical with the low-pass RC circuit because the series resistance will provide a voltage drop for the DC signals due to the input current of the recorder. With the advent of the op-amp, the active filter has become commonplace. By configuring the op-amp as a voltage-follower, it buffers the passive RC elements from both the signal source and the recording devices. An example is shown in Fig. 10-22.

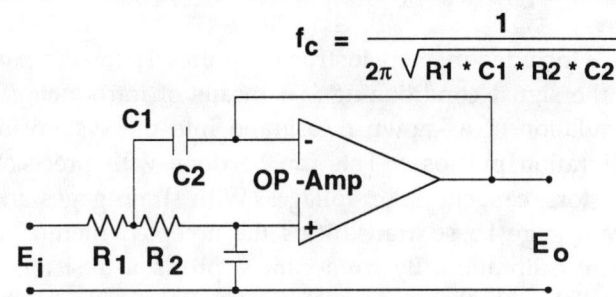

Figure 10-22. Low-Pass, Active, Second-Order Filter

The second-order active filter (anti-aliasing filter) shown in Fig. 10-22 produces an attenuation of 12 db/octave, which amounts to an order of magnitude improvement over a first-order filter.

It must not go unmentioned that all filters introduce a phase shift which will be seen as a delay between the variations in the input signal and the corresponding variations in the output signal. In complex signals (having more that one frequency component) there will be distortion as they pass through the filter due to the delay not being constant over the whole passband. An additional problem encountered in multichannel systems where the signals are sampled simultaneously, with each channel having a filter, there will be skew in the data unless all filters have identical delay characteristics.

10.4.5 Calibration

The discussion to this point has been in terms of resistances, voltages, charges, and currents. To relate the values of these to the appropriate measurand values one must have a means of calibrating the system. Ideally, the best method is to apply a known measurand (e.g., force, strain) to the transducer and observe the output at the recorder or display and compare it to the known input. This is not always practical, however, as one may not have a reference standard available on a day-to-day basis or the transducer may be at a remote location from the instrumentation. The transducers usually are calibrated independently by the manufacturer and their supplied calibration values are usually sufficient to set the sensitivities,

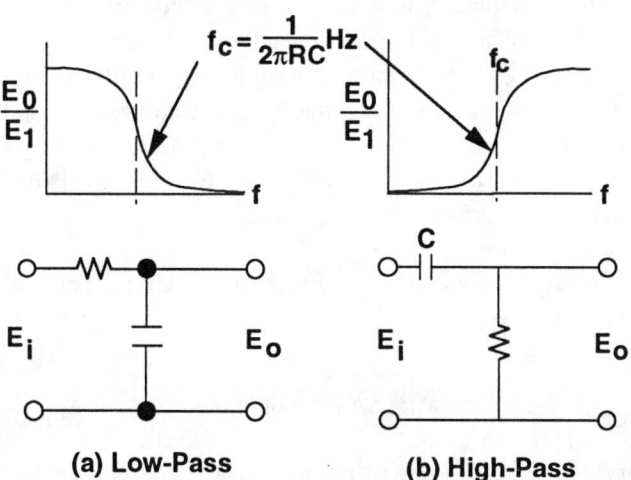

Figure 10-21. RC Passive Filters

excitation levels, and ranges of the signal conditioning and recorders. The transducer can be periodically returned to the manufacturer for recalibration or can be calibrated in-house using a "standard" transducer that is traceable to the National Institute for Standards and Technology. When calibrating the transducer alone, it is usually sufficient to accurately measure the input voltage and the output signal from the transducer. These voltages can be accurately measured using precision voltmeters (typically digital). It can also be a requirement in some test laboratories to calibrate power supplies, amplifiers, and recorders/displays.

Included in many instrumentation systems as a part of the signal conditioning is a means of introducing a simulation of a known measurand into the system for calibration purposes. This can be done with precision resistors, capacitors, or voltages. With strain gages and strain gage based transducers the accepted method is shunt calibration. By connecting a precision resistor in parallel with the resistance of one of the arms of the Wheatstone bridge, a resistive imbalance will result, thereby simulating a stain gage behavior. By shunting the appropriate arm and knowing the values of the strain gage and shunt resistor, the tensile or compressive value of simulated strain can be calculated and the excitation and/or amplifier gain can be adjusted to give the desired output as shown in Fig. 10-23 (Ref. 10-T12).

The formula for calculating the equivalent strain for a given shunt resistor is:

(simulated strain) $\mu\epsilon = Rg/GF(Rsh + Rg)$ Eq. 10-10

where Rg = resistance of shunted arm = resistance of active gage
       GF = gage factor of active gage
       Rsh = resistance of the shunt calibration resistor

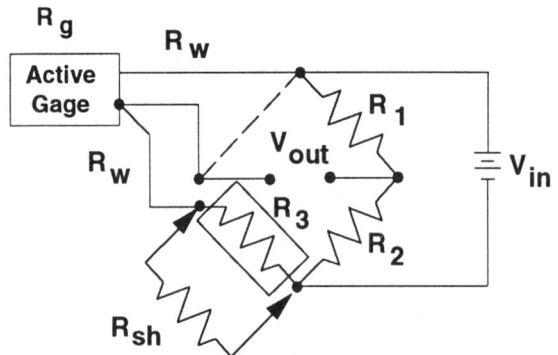

Figure 10-23. Shunt Calibration for Quarter Bridge, Three-Wire Connection

Ideally, to compensate for the lead wire desensitization discussed earlier, it is best to shunt the active gage directly at its remote location. However, it is usually more convenient to shunt the adjacent dummy arm at the instrument. If the three lead wire technique is used, lead wire desensitization will be compensated for with this method as well. When using half-bridge configurations as in Figs. 10-5 and 10-6, the number of active gages will necessitate an adjustment in the instrument sensitivity so that the output represents the actual surface strain sensed by the primary active gage. Where only one gage senses strain the number of active gages (N) = 1. If two gages are mounted with their respective axes perpendicular to each other (see Fig. 10-6 [a]), N = 1 + v where v is the Poisson's ratio of the test material. When two gages are mounted on opposite sides of a bending beam (Fig. 10-5), N = 2. Accounting for the number of active arms, Eq. 10-10 becomes:

$\mu\epsilon = Rg/GF (N) (Rsh + Rg)$ Eq. 10-11

Since shunting an arm of the Wheatstone bridge causes a resistive imbalance in the bridge, the nonlinear behavior of the bridge output at higher strain levels may require the data to be corrected numerically. For strain levels under 2000 $\mu\epsilon$, the nonlinearities are usually neglected.

For strain gage based transducers the shunt calibration method is still viable even though strain may not be the measurand. Usually the manufacturer of the transducer will calibrate the device with a known measurand, such as an applied force to a load cell, and provide a calibration resistor that when shunted across and arm of the bridge along with the designated excitation voltage, will give an output equal to a specific measurand that is based on the original calibration. The end-user can then use that resistor to adjust the sensitivity of the signal conditioner to get an output matching that of the manufacturers. As stated earlier, most strain gage based transducers have their sensitivities stated in terms of millivolts of output per volt of excitation at a rated capacity. If the signal conditioner has adjustable excitation and/or amplifier gain, they can be present by the user to calibrate the system by using the manufacturer's sensitivity value.

Output = Sensitivity × Excitation × Amplifier Gain
Eq. 10-12

or $V_{out} = \frac{millivolts}{V} \times V_{ex} \times V_{out}/V_{in}$ Eq. 10-13

For example: a 1,000 lb. load cell has a stated sensitivity of 2 millivolts/volt at 1,000 lbs. with an excitation voltage of

10 volts and an amplifier gain of 500 volts/volt at 1,000 lbs., the signal conditioner output would be:

$$V_{out} = \frac{0.002\,V}{V} \times 10V \times \frac{500\,V}{V} = 10V \qquad \text{Eq. 10-14}$$

10.5 RECORDERS/DISPLAYS

The transducer and relevant signal conditioning present the data from the test in form of a voltage. Unfortunately, the human observer cannot see voltage, so a means of displaying and recording such voltages must be provided so that the data can be interpreted. Probably the most important factor in selecting the appropriate device is its frequency response. As mentioned earlier, the data could consist of static, dynamic, or transient voltages. Sometimes the capability to get high frequency response will come at the expense of accuracy or poor noise rejection, so this discussion will focus on comparing recorders/displays in terms of frequency response, accuracy as well as other strengths and weaknesses.

10.5.1 Analog Meters

Probably the oldest and simplest device used to measure and display voltage is the D'Arsonval movement meter (Ref. 10-8). It is actually a current sensitive device but can measure voltage by keeping the circuit resistance constant. It primarily measures DC but through rectifier circuits can measure AC as well. However, it generally will read the RMS or average value of the AC voltage as long as it is sinusoidal in nature. For rapidly changing or transient signals it is of no value, with its response limited to < 1 Hz. Full scale accuracy is usually at best 1-3 percent. Even though they can measure very low voltages, they usually are not connected directly to transducer outputs because of their relatively low impedance.

10.5.2 Digital Voltmeters

For much more precision (0.01 to 0.1 percent full scale) digital voltmeters offer an excellent means of displaying data numerically. Again they are primarily static devices in that they must sample the data periodically and do the analog to digital conversion before the data can be displayed. Usually the rate of sampling varies between 1 to 4 samples per second. Also the data must be displayed long enough to be interpreted by the observer. Many options are available, including ability to read RMS, average or peak values, and interfaces to printers so that hard copy can be made. They have high input impedance and amplification which in many cases allows direct connection to the transducer. Many have the capability to allow the user to set scale factors such that the readout displays engineering units rather than volts.

10.5.3 Oscillograph Recorders

Related to the D'Arsonval movement meter is the galvanometer oscillograph. Where the meter has limited dynamic response the galvanometer version of the D'Arsonval movement has been designed to get rotational natural frequencies as high as 10,000 Hz. By attaching an ink pen to the arm of the movement and have it move proportional to the signal over a moving chart paper, a permanent record will be produced. Some oscillographs use a heated stylus in contact with heat sensitive paper instead of the pen and ink. In both types, due to the added mass of the pen and the friction of the contact with the paper, their frequency response is dropped to 200 Hz or less. By using a light beam instead of the pen the natural high frequency response of the galvanometer can be better realized. A very small mirror is fastened to the moving coil of the galvanometer and a light beam is reflected from it and focused as a spot on light sensitive paper that moves past at a fixed speed. As the galvanometer responds to the input voltage a trace is recorded on the paper representing the varying data signal. The major frequency limitation of the light-beam oscillograph has been the response speed of the paper, but that has been improved to the point that frequencies as high as 5,000 Hz can be reproduced. Early versions had the galvanometers driven directly by the transducers; however, since they were a current driven device with low impedance they required close matching via added series or parallel resistance. New instruments employ current amplifiers which can also introduce proper matching and damping. Accuracies have been improved from two percent for earlier recorders to 0.5 percent for many currently available.

10.5.4 Servo Recorders

Servo or self-balancing recorders are probably the best choice for static and time dependent tests. They have the best precision of any analog recorder including: high accuracy and resolution (0.2 percent), low hysteresis, and good repeatability and linearity. Their major limitation is frequency response, which is less than 2 Hz, but is usually stated in terms of slewing speed, with pen speeds of 30 inches per second being quite common. The pen moves along a linear slider which may 10 inches in length, and is pulled by a cable that is routed through pulleys and driven by a servo motor. Simultaneously, a potentiometer that is also coupled to the pen, measures the position and feeds back a voltage to the servo amplifier. The input signal voltage is also amplified and fed to the servo amplifier to be compared with the feedback voltage and thus provides the precision of a closed-loop servo system. In strip-chart versions voltage is recorded versus time which is represented by the time motion of the chart paper. In x-y servo recorders two voltages can be plotted with respect to each other rather than time. A second servo motor will move the slider or carriage with the pen instead of the chart paper moving. Most x-y recorders have a time base option which allows the carriage to move a a uniform rate of speed rather than proportional to a second voltage signal thereby recording voltage versus time. Accuracies and response times are similar to the strip chart versions. In recent years the servo motor has given way to a stepper motor in many recorders and plotters. The stepper motor is essentially a digitally controlled motor. The analog signal voltage is converted to a series of pulses that will drive the motor in very small, precise increments or "steps" with the number of steps being proportional to the signal voltage amplitude. The need for the feedback potentiometer is eliminated thus simplifying the system and providing better reliability and endurance of the recorder. Digital printers and plotters use this scheme.

10.5.5 Magnetic Tape Recorders

Tape recorders are primarily used in dynamic testing, but most versions have static or DC voltage recording capability also. The inherent nature of the magnetization of the tape by a rate of change of flux developed at the recording heads necessitates an AC or varying voltage signal fed to the heads. This is called the direct recording process and is the same method audio tape recorders use. The frequency range is approximately within the band of 50 to 1,000,000 Hz. The high frequency is primarily dependent on the gap distance in the heads, with the finer the gap the higher the frequency, and the tape speed, with 120 inches per second being the highest practical speed and giving the highest response. Direct recording is not the most accurate, due mainly to the poor signal-to-noise ratio caused by imperfections in the tape which yield varying voltage levels, which in audio systems in the main cause for "hiss."

When desiring to record DC levels as in static or time dependent testing, and to improve accuracy, the FM system is used. Much like the FM-FM telemetry system discussed earlier, the data signal frequency modulates a carrier frequency which is recorded on the tape in the normal way. Upon playback the modulated carrier is demodulated and the signal passed through a low-pass filter to reproduce the original data signal. The signal to noise ratio is normally an order of magnitude greater than the direct recording, producing accuracies of better than one percent. The frequency response is quite flat (±1 db) over a potential range of DC to 80,000 Hz. For a response of 80,000 Hz the carrier frequency would have to be 432 kHz and the tape speed of 120 inches per second.

Usually tape recorders are used as an intermediate recording medium in that the signal voltages still must be displayed and interpreted by some other means. The playback signals can be recorded on an oscillograph, servo recorder, displayed on an oscilloscope, or interfaced to a computer. Another function for the tape recorder in the test laboratory is for simulation testing. By using a portable tape recorder to capture data from action operation of vehicles, aircraft, or other structural systems in their actual environments, the test structures could be brought into the laboratory with the recorded load histories being played back to drive testing systems and/or vibration equipment duplicating the loads on the structure.

10.5.6 Oscilloscopes

These are used primarily as a device to observe dynamic signals. The analog oscilloscope has excellent high frequency capability (DC to 109 Hz) and is especially good for transient signals. It is rather limited in accuracy and in providing a hard copy record of the event. Accuracy and resolution are generally in the three percent range, similar to analog meters or older oscillographs. But, since an electron beam essentially replaces the galvanometer and pen combination, there is zero mass and friction, allowing the beam to move proportional to the input signal and "write" a trace on the phosphor of the cathode-ray tube (CRT) screen rather than on paper, giving a much higher frequency response capability. For one-time events or transients, the "scope" can be triggered by the event itself, with the signal being deliberately delayed such that the whole event can be captured and displayed without loss of data. External triggers can also be used that could precede the event by a known time interval, thereby providing a "window" to ensure data capture. To provide a permanent record of the event a special camera can be used to photograph the trace on the screen. In the case of transients, the camera can be "triggered" synchronously with the scope so that it will also capture and store the event. Storage versions of the analog oscilloscope provide for temporary (<1 hour) storage of signals on the CRT screen by use of variable persistence phosphors or statically charged grids behind the screen's surface. It is a useful method for holding one-time or transient events for observation and analysis. However, frequency response generally is lower that the conventional scope and the inherent inaccuracies of the analog scope are still there. Most oscilloscopes offer tow channels of input that are displayed as voltage with respect to time; however, one input channel can be substituted for the time base such that, one input signal can be displayed versus the other, much like an x-y recorder.

Many special plug-in input modules are available that can perform dedicated conditioning of the input signals or interfacing to various types of transducers. Some examples are: multiple channel inputs, strain gage/transducer conditioning, carrier amplifiers, spectrum analyzers, and high gain amplifiers.

The digitizing oscilloscope came on the scene in the early 1970's and offered a significant improvement in accuracy, resolution, and storage capability. An analog to digital (A/D) converter, usually one dedicated to each input channel digitizes the input signal voltage and stores the bits in a solid-state memory much like a computer. With a 12 bit system the data is measured 4096 times over a designated time period. The "block" of 4096 data points is then displayed on the CRT screen much like in an analog scope. The data is continually digitized and the memory is updated and succeeding "blocks" will be displayed replacing each preceding "block." The horizontal scale of the CRT screen is not in terms of frequency, but is calibrated in terms of time per point of data. If the A/D converter sampling rate is set at 1,000,000 samples per second, the time per point would be one microsecond and the full width of the CRT screen would display 4096 microseconds of data. Each succeeding sweep or "block" of data would be of the same time frame. At any time the observer can "freeze" the display by storing the currently displayed "block" in a second memory buffer which keeps that "block" of data displayed on the screen and succeeding "blocks" would not replace it. By moving cursors across the displayed data the observer can select a particular point of data and have its voltage and time from trigger annotated on the screen in digital form. The resolution of one part in 4096 equates to approximately 0.02 percent which is two orders of magnitude better that the analog scope. Also, one can "zoom" about a section of the data and expand the vertical and horizontal visual resolution of the display such that each of the 4096 points can be observed. As example display is given in Fig. 10-24.

At any time, from this "stored" mode, the data can be routed from memory through a digital to analog (D/A) converter and recorded permanently on a servo or other type recorder. The sampling speed of the A/D converter is the limiting factor for determining frequency response. It is a trade-off with the number of bits of resolution desired. Digital oscilloscopes are presently being offered with resolution of 8, 10, 12, 14, or even 16 bits. Examples of sample rates and practical frequency response are: for an 8 bit A/D—200 mega-samples per second or 10,000,000 Hz, for a 12 bit A/D—10 mega-samples per second or 500,000 Hz. A practical frequency response determination usually requires a minimum of 20 points to define a sine wave. A/D conversion and sampling concerns will be discussed in more detail later in the chapter.

10.5.7 Hybrid Waveform Recorders

By combining the features of the digital storage oscilloscope and the servo recorder or digital plotter, a versatile data acquisition system results. As discussed earlier, the servo recorder or digital plotter version have limited dynamic recording capability. However, their high precision is compatible with the 8 or 12 bit A/D converters and provides a good hard copy method for the digitally stored data. With the A/D converter providing moderate dynamic capability and having the data stored in a buffer memory, it can be down-loaded from the buffer to be recorded at a rate that falls within the frequency range of the recorder/plotter. Data stored in the buffer can be recorded as voltage versus time (y/t) or one input versus another (x/y). Other features include the ability to annotate the hard copy at any data point, much like the digital storage oscilloscope. Many have built in software that can provide: signal averaging, scaling in engineering units,

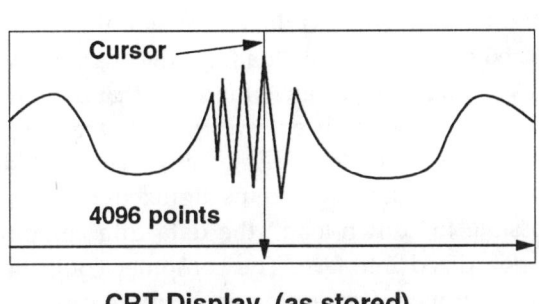
CRT Display (as stored)

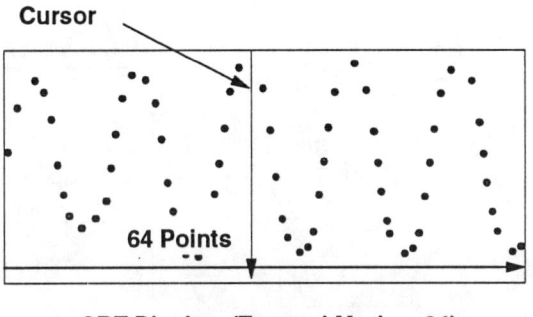
CRT Display (Zoomed Mode, x64)

Figure 10-24. Digital Oscilloscope Display

digital filtering, and numerous mathematical functions. Some give the option to record "direct" where the digitized signal goes directly to the recorder/plotter without being stored in the buffer. This essentially gives "real time" data recording, but restricts the frequency domain of the data to static or very low frequencies (<1 Hz). In the buffered mode the frequency response is dependent on the sampling rate of the A/D converter which is determined by the memory size and the total time that data is to be acquired. A typical recorder may have a frequency range of DC to 25 kHz with 32 kilobytes of memory for each of 8 channels at 8 bits of resolution. An example hybrid waveform recorder is given in Fig. 10-25.

10.5.8 Data Loggers

For acquiring and recording multiple channels (>8) of data, the data logger type of instrument effectively bridges the gap between multi-channel recorders and sophisticated computer based data acquisition systems. A sample data logger is shown in Fig. 10-26. There are many variations of this concept, with the discussion here emphasizing the basic features, strengths, and weaknesses. Generally, they are restricted to static or slow rate data, since the input signals are multiplexed or scanned before they are conditioned, amplified, and digitized. By using only one signal conditioner and analog to digital converter as shown in Fig. 10-26, higher quality and higher precision (up to 16 bits) versions can be used, while keeping the physical size small and costs low. (Multiplexing and sampling will be discussed in greater depth in the next section.) However, the multiplexing or switching of the multi-inputs sequentially into the single signal conditioner and A/D does contribute to the major limitation of the data logger which is their ability to capture only slowly changing data. Usually the number of channels the instrument is expandable, with most starting at 16 channels, and by adding boards or modules can be increased in increments of 16 up to hundreds of channels. Obviously, the greater the number of channels that share the single A/D the

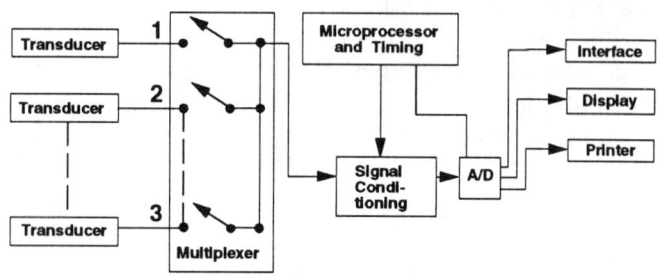

Figure 10-26. Data Logger System

greater the time between the samples of each individual channel. Scanning rates usually are in the range of 10 to 1000 samples per second. With 100 channels and 1000 samples per second rate, each channel would be sample 10 times per second.

By using a micro-processor and built-in software programs, scaling, offsetting, linearizing, and other special processing of the data are possible. For example, many can do rain flow counting of the data or provide time at interval information. These are powerful tools for obtaining load histories of structures and vehicles in particular, very compact in size and are truly portable acquisition systems. Many operate off the test vehicle's electrical power or battery and can accumulate data for long periods of time, even months perhaps, by storing the data in a solid-state memory for later analysis in the laboratory. Some have printers built in that can give real-time results and/or have serial or parallel interface ports for communicating with and transmitting data to a computer.

Table 10-3 contains a summary of the various methods and equipment used for recording and displaying data.

10.6 COMPUTER BASED SYSTEM

The instrumentation described up to this point has been primarily stand along systems that result in capturing, conditioning, storing, recording, or displaying the data. For analysis and data reduction the engineer is required to interpret the data from the "hard copy" recordings or find some means of feeding that data into a computer where he can use powerful software programs that can put the data into more meaningful terms. Many of the previously described systems digitized the data making it possible to "down-load" the data to a computer via a standardized interface. The computer could be a large, powerful main-frame type or the more readily available personal computer (PC). Due to their relatively low cost and popularity in word processing, computer aided draft-

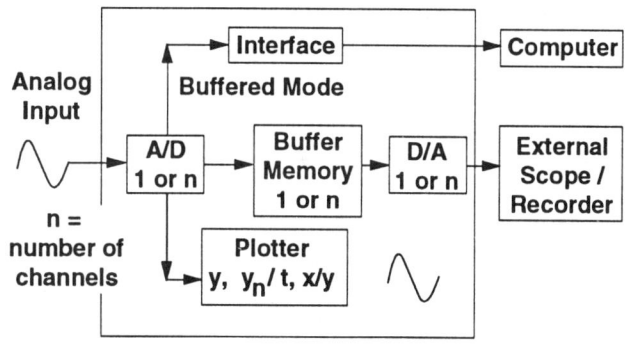

Figure 10-25. Example of Hybrid Waveform Recorder

Table 10-3. Comparison of Recording/Displaying Methods

Acquire	Record	Strengths / Advantages	Weaknesses/ Disadvantages	Frequency Response	Accuracy Percent of Full Scale (Resolution)
Computer	Printer Plotter Disk/Tape	• High Accuracy • Can Compute and Massage Data • Data Recording in Report Form • Multi-Channel	• Cost • Needs Some Computer Expertise • Low Frequency Data Rate	DC to 2,000 Hz (Depends on Number of Channels Sampled and Sample Rate)	(8 - 16 Bit) 0.4 - 0.0015 %
Conditioned Signal	Data Logger/ Scanner	• Multi-Channel • Can Massage Data • Data Printed in Digital Form • Portable	• Primarily Used for Static Tests • No Graphics • Time Skew Between Channels	Low (Depends on Number of Channels Sampled and Sample Rate)	(8 - 16 Bit) 0.4 - 0.0015 %
	Hybrid Recorder with A/D, Storage and Plotter	• High Accuracy • Dynamic with Buffer • Alpha/Numeric • Interface to Computer • Annotated Data	• Moderate cost • Limited Number of Channels (1-4)	DC to 25kHz	(8 - 14 Bit) 0.02 %
	Digital Storage Oscilloscope → Camera	• High Accuracy • Digital and Graphic Display • Can be Interfaced to Recorder or Computer • Fast • x-y, y-t	• Cost • Limited Number of Channels (1-4)	DC to 100 kHz	(12 Bit)
	Servo Recorder, x-y, Strip	• Accurate • Lower Cost • Can be Used to Record Digital Oscilloscope Data	• Static Data Acquisition • Limited Number of Channels (1-3)	2 kHz (DC to 30/sec)	0.2 %
	Tape Recorder	• Moderate Accuracy • Good Dynamic Recording • Lower Cost • Multi-Channel (1-14)	• No Graphics • Limited for Static Data • Calibration More Elaborate	50 to 10 Hz (Direct) DC to 80 kHz (FM-FM)	1.0 %
	Digital Meter	• Moderate Accuracy • Easy to Read • Inexpensive	• No Hard Copy • No Graphics • Static Only	Static	0.01 - 0.1 %
	Oscillograph Recorder - Ink Thermal, Light	• Good Dynamic Recording • Multi-Channel (1-8)	• Lower Accuracy • Moderate Cost • Expensive Paper	DC to 10,000 Hz	0.5 - 2 %
Analog Oscilloscope → Camera		• High Frequency Response • Good Graphics • x-y, y-t	• Lower Accuracy • Moderate Cost • Photo Only Record • Limited Number of Channels (1-4)	DC to 10 Hz	3.0 %
Analog Meter		• Low Cost • Easy to Use and Read	• Lower Accuracy • No Permanent Record • Static Only • Low Impedance	Static < 1 Hz	3.0 %

ing and finite element analysis they widely used in test laboratories. With the advent of the data acquisition plug-in boards shown in Fig. 10-27, they now can offer much more to the test engineer (Ref. 10-P3).

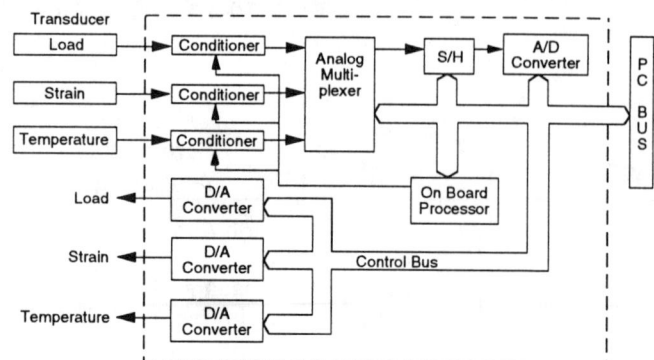

Figure 10-28. Separate Signal Conditioning A/D Board with Feedback Control

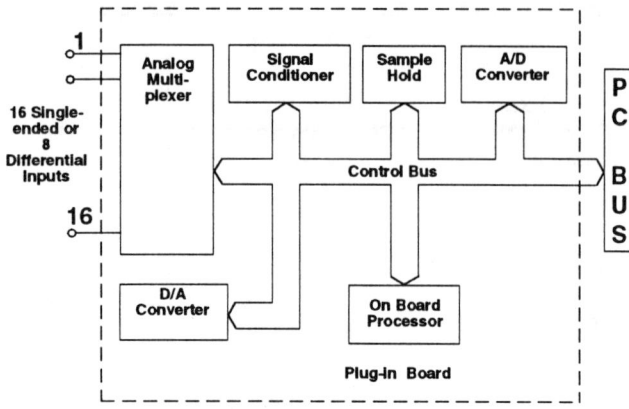

Figure 10-27. Data Acquisition Board for Personal Computer

Most of the data acquisition boards available today for the PC can acquire the data signals directly from the transducer, condition and digitize the data, and enter them into the PC's bus. With current 32 bit processors one can then view in real-time rapidly changing data on the CRT screen using the powerful graphics capability of the PC. Line plots, bar graphs, or pie charts can be displayed to give the user an immediate understanding of the data. Some examples of the many programs available for data analysis are: curve fitting, spectrum and frequency domain analysis, digital filtering and averaging, and statistical routines. There may be the need to use the PC to control some part of the test procedure such as maintaining a temperature environment controlling an oven or furnace by virtue of something learned from the data acquisition portion of the PC. By using a digital to analog converter (D/A) which many data acquisition boards employ, one can have the PC interpret the input data (measured temperature at points on the structure) and make the necessary adjustments to the heating device, as the D/A will output an analog voltage which will be compatible with the heater controller. An example of this configuration is shown in Fig. 10-28.

In selecting the appropriate data acquisition board there are many features and specifications that must be considered. However at this point it must be emphasized that it is very difficult to be very specific about describing state-of-the-art features, as this area of technology is changing so fast. Nonetheless, important considerations are, but not limited to: type of PC compatibility (IBM PC/XT/AT, Macintosh, Microvax, or Unibus), data signal rate and computer speed (sampling rates and type of A/D), amplitudes of input signals (A/D ranges needed), accuracy (A/D resolution), number of channels (is multiplexing needed?), ability to control some parameter of the test or test system (temperature—is a D/A needed?), amount of data to be accumulated over time (memory requirements), and will specific signal conditioning be necessary?

10.6.1 Multiplexing

Whenever more than one channel of data is to be acquired using only one signal conditioner and/or A/D converter, there must be a means of switching (multiplexing) between the input channels. Depending on the application, the multiplexer can be placed at different locations in the system. For most multichannel systems it is located at the "front end" and is connected directly to the signal source as shown in Fig. 10-27. However, this arrangement always produces time skew between the channels as the data is sampled and digitized sequentially. In dynamic tests, or where transient events may occur, the relative timing of the data of the various channels can be critical. Some systems put a sample-hold (S/H) circuit at each channel input ahead of the multiplexer as shown in Fig. 10-29. The sample-hold circuit is a device that can acquire the signal voltage and hold it for a period of time so that the multiplexer can sample and switch that stored signal to the A/D converter after all the channels have been sampled simultaneously. Then the signal can be digitized sequentially, with the end result being not time skew, as all the data points were digitized in one "scan" of the channels and were sampled at the same point in time.

An alternative for obtaining data points simultaneously is to have a discrete A/D converter for each channel and eliminate multiplexing. Each channel's A/D conversion would be "triggered" in synchronism by a common clock as the simultaneous sample-hold method. For a small number of channel (<8 for this example), this may be a practical approach; however, these systems could

Instrumentation and Data Systems

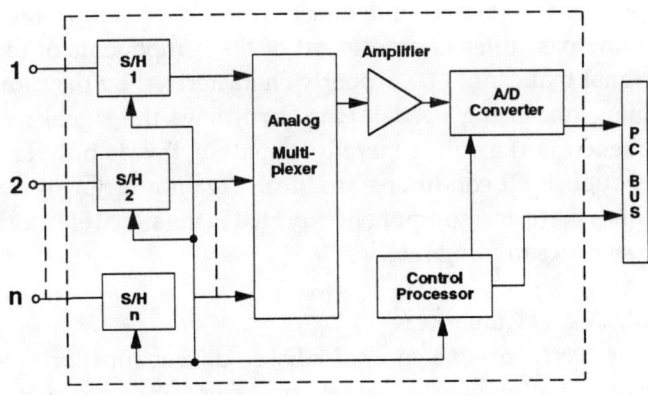

Figure 10-29. Simultaneous Sample-Hold System

become complex and costly for many multi-channel needs.

Another variation in the multiplexed system involves the situation where each input channel is from a different type of transducer, thus requiring separate signal conditioning for each channel as shown in Fig. 10-28. Here the individual signal conditioners would be at each channel input ahead of the multiplexer. This configuration can use the PC's benefits in that the gains, sensitivities, and offsets can be adjusted and controlled from software, thus providing an automated data acquisition system. Again, time skew and system costs could still be limiting factors.

There are two basic types of multiplexers: electrical/mechanical relays or solid-state CMOS switches. The principal factors in deciding which is better suited for a particular application are the switching or scanning speeds necessary and the signal level from the transducer or signal conditioner as given in Table 10-4. For low level signals (thermocouples) and low sample rates (<1 kHz), the relay has been the preferred type due to its low contact resistance. For high rate sampling the CMOS switch is superior. However, the CMOS switch does have some drawbacks in that it has some effective contact resistance, but with high input impedance amplifiers of today that problem has been minimized. Also, the maximum voltage that can be switched is limited to the supply voltage of the multiplexer drive circuit.

10.6.2 Data Sampling

Whenever digital handling of data is to take place it is important to talk about the issue of sampling. Sampling of the data may be done as part of the sensing, transmitting, recording, or presenting of the data signal. The signal is characterized by finding its value at a series of points. Generally these sampling points are evenly spaced in time. Its important that the frequency of the sampling is sufficient to adequately define the response of the structure being tested.

Often at this point the discussion turns to the Sampling Theorem and the conclusion is drawn that to completely characterize a signal one must sample at a rate of at least two times the highest frequency present in the signal. These discussions are more practical when sampling in preparation for conversion to the frequency domain by the Fourier transform. Sampling data for conversion by Fourier transform present special problems beyond the scope of this discussion (Ref. 10-4).

Commonly it is important to get a good time domain representation of the signal for a given purpose. The purpose might be to find the maximum value of a signal or to compare the signal to one acquired under different conditions. Finding peaks and valleys of the signal so that a complete signal and fatigue life analysis can be performed is also a common task. In these cases the decision on how fast to sample should be made with the application of the data to be obtained in mind.

Consider the simple example of looking for the maximum signal level during a transient event. We might actually be looking for a force or g-load during an impact, or pressure spike during combustion or some hydraulic event. How fast will we need to sample to get a good measure of the maximum value of the transient? Suppose the transient signal has the shape of the top half of a symmetrical triangular shaped wave and ranges from 0 to 100 in amplitude as shown in Fig. 10-30. Furthermore, assume that the ramp up and back to zero takes 20 milliseconds. Say sample takes place once every 10 milliseconds. If the sampling is badly time we might have two reading both turn out to be 50. So our maximum reading is 50 but the

Table 10-4. Comparison of Multiplexing Switching Methods

	Relays—Low Resistance	CMOS Switches—Higher Resistance
Speed	1-100 msec	10-100 nanoseconds
Contact Resistance	< 1 ohm	5-100 ohms
Driving Circuit	Needs Coil Power	Needs Driver Circuit
Signal Voltage	No Limit	Limited to Driver Voltage

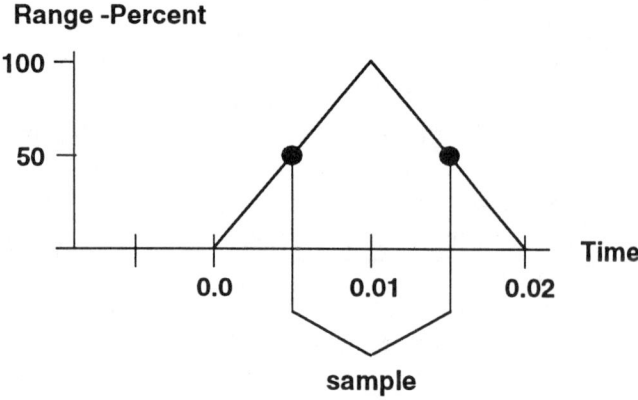

Figure 10-30. Sampling Rate Versus Signal Rate

signal went to 100. So, how fast should the samples be obtained? A suggested rule is: "sample at least ten times faster than the fastest signal you expect to see." This is just one example. The sample rate should be based on what the data are going to be used for. Each use, such as Fourier analysis, finding minima and maxima, or checking the shape of the signal will require different sampling rates.

The Nyquist criterion describes effective sampling rates. The frequency for sampling data is the maximum frequency that can be contained in a signal that can be sampled without introducing aliasing errors. The Nyquist frequency is equal to one-half the sampling frequency, However, even with electronic filtering it is impossible to assure that there will not be a signal above the Nyquist frequency in a given record.

Another important topic to cover when talking about sampling is the problem of aliasing. Consider a 100 Hz signal sampled at some frequency between 100 and 200 Hz as shown in Fig. 10-31. In reconstructing the data from the sampled signal one will not reconstruct the signal but an alias of it. To avoid being fooled by too low a sample rate, a low pass filter is used to rid of the components of the signal that cannot be properly characterized by the sampling rate being used. Again, two times the signal frequency is the rule generally state, but this is only fast enough if all conditions are ideal. Again a factor of ten times the fastest component expected in the signal is a more practical sampling rate.

10.6.3 A/D Converters

The heart of a computer based data acquisition system is the analog to digital converter. Since the signal coming from the transducers and signal conditioners are in analog form, they must be converted to a digital or coded format for the computer to accept, store, and manipulate them. The intent of this discussion is not to describe the theory and design of the various types of A/D converters, but to examine the differences in performance of two most common versions. The characteristics of primary importance are: speed, resolution, and linearity. For moderate to high speed applications, by far the most popular type uses successive-approximation technique. Essentially it is a servo-loop and in successive steps it compares its output to the input and keeps completed n-steps, where n is the resolution of the converter in bits. This process is quite fast and many of today's successive approximation types can do a 12-bit conversion in less than 2×10^{-6} sec. A sample-hold is needed ahead of this type of converter because it needs the analog voltage to be constant at its input while the conversion is being made. Also, high frequency noise or transient "spikes" can cause errors in the conversion, necessitating low-pass filtering of the analog signal in the signal conditioning.

The integrating A/D converter, of which the dual-slope type is the most popular, as its name implies, integrates or averages the input analog voltage over a fixed period of time. Since it averages the varying analog voltage over time, a sample-hold circuit is not used with this type, and inherently it is not as affected by noise or "spikes" as is the successive-approximation type. It also results in very small nonlinearity errors. However, its principal limitation is that it is considerably slower in its conversion rate. These characteristics make it the ideal A/D converter for digital panel meters, voltmeters, and many data loggers. Table 10-5 shows resolution versus accuracy and compares conversion speeds for the two A/D converters discussed here.

10.6.4 Interfaces

Communication among any of the components of the computer based systems requires some type of interfaces. Even within the PC itself there are many interfaces incorporated, such as between the data acquisition board and

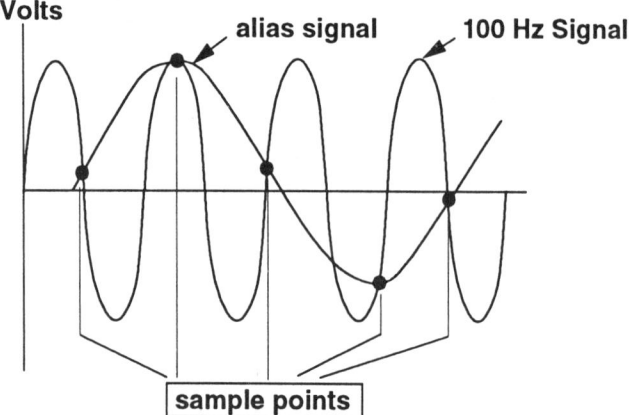

Figure 10-31. Aliasing of Data Signal

Table 10-5. Comparison of Performance of A/D Converters

Resolution Bits	Accuracy Percent	Conversion Time	
		Integrating A/D Converter Full Scale A/D Converter milliseconds	Successive-Approximations microseconds
8	0.4	0.3	0.8
10	0.1	1.0	1.0
12	0.02	5.0	2.0
16	0.0015	250.	15.0

the PC main bus. However, the main interest in this discussion involves interfacing to peripherals outside the PC, such as auxiliary disk and tape drives, plotters, printers, and modems.

There are many interfaces available, as plug-in boards, for various dedicated purposes. They are all summarized in two basic types: (1) serial or (2) parallel. The serial interface involves the sending of data one bit at a time over a single communication line. The parallel interface requires at least as many lines as there are bits in a word being transmitted (a minimum of 12 lines are needed for a 12-bit word).

For long distances (> 20 ft.), the serial transmission is usually the preferred method as the data can be sent over a single twisted pair of wires. By using a modem (a modulator/demodulator device that converts the serial data into audio tones), standard telephone lines become an excellent means of transmitting data over great distance. The most common serial interface used is the RS-232-C standard. The number of wires in the cables used can be as few as 4 and as many as 25. The IBM PC AT, for example, uses a 9 pin connector while most others use the standard 25 pin version.

Where higher speed for data transfer is needed, the parallel interfaces are usually incorporated, particularly where the distances are short, such as within the PC itself, and in connecting to peripherals that are in close proximity (<20 ft.). Until recently most printer interfaces were of the parallel type, but with the increased baud (a unit of data transmission speed equal to the number of bits per second) rates of the serial interfaces (as high as 100,000 bits/sec), many are now using the serial type.

One of the most used parallel interfaces with PC's is with the digital printers and plotters that are a necessary part of the data acquisition system. Compared to their analog counterparts discussed earlier, these devices provide digital annotation and recording as well as graphical presentation of the data. Therefore, the inputs to them must be in digitized form requiring the use of an interface.

10.7 SUMMARY

By first determining the rate of loading to be present on a structure under test, knowing the desired accuracies, and taking into account the test environment, one can set out to "design" the data acquisition system needed for the test. A list of the basic topics needing consideration when establishing an instrumentation and data system for a structural testing laboratory or field test situation are:

- Rates of load, stimuli, or excitation to the test structure

- Test environments (particularly those that can adversely affect the data)

- Choice of proper transducers and sensors, their characteristics and limitations

- Interconnections, interfaces, and signal transmissions

- Signal conditioning to match transducer and signal rates

- Recorders and displays to match signal rates and desired accuracies and resolutions

- Will a computer based system be needed to acquire and analyze the data?

10.8 REFERENCES

Texts and Handbooks:

10-1. J.W. Dally and W.F. Riley, *Experimental Stress Analysis*, 2nd Ed., McGraw-Hill Co., New York, 1978.

10-2. E.0. Doebelin, *Measurement Systems: Application and Design*, Revised Ed., McGraw-Hill, New York, 1975.

10-3. R.L. Hannah and S. Reed, *Strain Gage User's Handbook*, Elsevier, London, 1992.

10-4. A.S. Kobayashi, Ed., *Handbook on Experimental Mechanics*, Prentice Hall, Englewood Cliffs, NJ, 1987.

10-5. K. McConnell, "Notes on Vibration Frequency Analysis," SEM, Bethel, CT.

10-6. D.H. Sheingold, Ed., *Transducer Interfacing Handbook*, Analog Devices, Norwood, MA, 1981.

10-7. B. Tanner and E. Campbell, "Fundamentals of Digital Data Acquisition Systems," Campbell Scientific, Logan, UT, 1985.

10-8. F.S. Tse and I.E. Morse, *Measurement and Instrumentation in Engineering*, Marcel Dekker, New York, 1989.

10-9. E.L. Zuch, Ed., *Data Acquisition and Conversion Handbook*, Datel-Intersil, Inc., Mansfield, MA, 1980.

Technical Notes and Product Information:

10-T1. ———, "Analog Circuit Design Seminar—Notes," Analog Devices, Norwood, MA, 1982.

10-T2. ———, "The Application of Filters to Analog and Digital Signal Processing," Rockland Systems Corp., Rockleigh, NJ, 1980.

10-T3. ———, *Data Acquisition and Computer Interface Handbook and Encyclopedia*, Omega Engineering, Inc., Stamford, CT, 1988.

10-T4. ———, "Elimination of Noise in Low-Level Circuits," Gould Inc., Cleveland, OH, 1969.

10-T5. ———, "Errors Due to Wheatstone Bride Nonlinearity," Tech. Note TN-507, Measurements Group, Inc., Raleigh, NC.

10-T6. ———, "A Four-Wire Constant-Current Measurement Circuit," Experimental Stress Analysis Notebook, Issue 13, Measurements Group, Inc., Raleigh, NC, April 1990.

10-T7. ———, "Isolation and Instrumentation Amplifiers Designers Guide," Analog Devices, Norwood, MA, 1978.

10-T8. ———, "Noise Control in Strain Gage Measurements," Tech. Note TN-501, Measurements Group, Inc., Raleigh, NC.

10-T9. ———, "Piezoelectric Accelerometers—Instruction Manual," No. 101, Endevco Corp., San Juan Capistrano, CA, 1977.

10-T10. ———, "Piezoresistive Accelerometers—Instruction Manual," No. 121, Endevco Corp., San Juan Capistrano, 1978.

10-T11. ———, "The Practical Basis of Strain Gage Instrumentation," Experimental Stress Analysis Notebook, Issue 7, Measurements Group, Inc., Raleigh, NC, Nov. 1987.

10-T12. ———, "Shunt Calibration of Strain Gage Instrumentation," Tech. Note TN-514, Measurements Group, Raleigh, NC

10-T13. ———, "Wheatstone Bridge Primer—Part I," Experimental Stress Analysis Notebook, Issue 8, Measurements Group, Raleigh, NC, April 1988.

10-T14. ———, "Wheatstone Bridge Primer—Part II," Experimental Stress Analysis Notebook, Issue 9, Measurements Group, Raleigh, NC, September 1988.

Papers:

10-P1. C.A. Bowes, "Variable Resistance Sensors Work Better with Constant Current Excitation," Paper 16.9-4-66, ISA Conference Proceedings, October 1966.

10-P2. G. Klier, "Signal Conditioners: A Brief Outline," Sensors, January 1990, pp. 44-48.

10-P3. R. House, "PC-Based Data Acquisition Systems," Sensors, February 1989.

Chapter 11

Experimental Methods—Full Field Techniques

11.1 INTRODUCTION

This chapter describes full or whole field techniques used to examine a portion or all of a structural system being tested. These methods and techniques can be used to determine in-plane strains and displacements and out-of-plane displacements. These methods are typically based on the various properties and interactions of light or other types of waves with materials. Understanding these properties and interactions gives experimental capabilities which can be used in a wide variety of conditions and on many types of structures.

Among the obvious advantages of these methods over point techniques is the capability to evaluate the response of an area or all of a structure. Critical areas of structures such as welds, holes, stiffeners, turbine blades, and other complicated portions of structural systems can be examined in detail.

The methods and techniques described in this chapter are or are expected to be widely used in structural testing. The material in this chapter is intended to complement similar information given in Refs. 11-1 through 11.

11.2 MOIRÉ INTERFEROMETRY
Robert Czarnek, Ph.D., Concurrent Technology Corporation, Johnstown, PA.

11.2.1 Introduction and Historical Perspective

The idea of using diffraction gratings for measurements of displacements was proposed by Lord Rayleigh more than a century ago when he published his work on the fringes produced by crossed diffraction gratings in 1874, Ref. 1. These fringes are known as moiré fringes; the word moiré comes from the French word for "watered" silk. For a long time, however, the phenomenon of moiré fringes did not receive much attention. Slowly engineers started implementing the interference of relatively crude types of gratings in the form of coarse moiré, which is also known as geometric or traditional moiré. The development of new methods of manufacturing gratings made moiré interferometry more practical to use as a measuring tool.

High sensitivity moiré interferometry was introduced by Guild in 1956 as a powerful experimental technique for measuring deformations of solid bodies (Ref. 2). In its original form it was a method of measuring displacements in one in-plane direction on a flat surface of a solid body. A high frequency diffraction grating was replicated on the surface of the body and deformed with it. An optical system called an interferometer was used to measure the deformation of the grating and at the same time the deformation of the surface of the specimen to which it was applied. Guild's innovation extended the sensitivity of traditional moiré interferometry by two orders of magnitude, from about 25 µm per fringe order down to 0.25 µm, which is near the theoretical limit of sensitivity for interferometric techniques in the visible range. In the 1960's, when lasers became widely available, the method found practical applications in high-sensitivity measurements of displacements. One such application was described by Gerasimov and his colleagues in a series of papers dated 1963-1967 (Refs. 3 and 9). They used moiré interferometry with a reflective "specimen" grating and fringe multiplication by a factor of two to precisely control a ruling engine. (A ruling engine is a machine which cuts precisely spaced grooves in a diffraction grating.) In 1974, Wadsworth, Marchant, and Bishop described a moiré interferometer for measurements in two orthogonal directions (Ref. 5). In 1980, McDonach, McKelvie, and Walker published a more advanced three-beam moiré interferometer for two-directional measurements with high sensitivity. Although this interferometer was relatively large, it was portable and allowed measurements of deformation of specimens loaded in a testing machine (Ref. 6).

Since then moiré interferometry has found numerous applications in the testing of a wide variety of materials, from simple isotropic types to exotic high-performance composites. Some modifications have been made to the interferometers and different methods of specimen grating replication have been used, but the basic idea and physical background have remained the same. For a number of years most measurements were made with a simple one-mirror system for unidirectional measurements. This limited the number of strain components that could be extracted from an experiment to two normal strains. In 1983 the author introduced a new type of interferometer that allows moiré interferometry measurements of displacements to be made in two or three directions in a simple and routine way (Refs. 7 and 8).

Although moiré interferometry is only one of many methods of experimental mechanics, it has several unique features. A good understanding of its principles, advantages, and limitations is necessary to decide the most appropriate method to obtain the information and insights desired.

Moiré interferometry is based on the interference of light. A very close analogy can also be drawn between high sensitivity moiré interferometry its predecessor—geometrical moiré (Refs. 1, 9 and 10) even though the mechanism of producing the interferograms in these two methods is different, the equations relating the fringe patterns to the deformation that causes them are the same. This explains why both methods are categorized as moiré methods.

The advantages of moiré interferometry over other experimental techniques are numerous. Although it is a relatively simple method, it offers very high sensitivity, and it is a nondestructive technique that gives full-field information about deformation in the form of a contour maps of displacement components. Perhaps is most significant characteristic, however is that it can be applied to practically any solid engineering material. It should be emphasized that measurements are made directly on the material of interest, not on a material modeling the surface as in the case with photoelasticity.

Moiré interferometry does have its limitations, however. To date it is limited to flat surfaces of solid bodies. Until recently it has been confined to a laboratory environment and room temperatures, limiting its range of applications. Work is in progress to overcome these limitations.

An introduction of some basic concepts of optics is necessary to gain a thorough understanding of moiré interferometry. Sections 11.2.1 through 11.2.3 have been written accordingly; by closely following the definitions and derivations set forth in the theoretical portion the reader can expect to be able to use moiré interferometry effectively in a variety of experimental situations. The reader with a more advanced background may wish to go directly to the section 11.2.3 on Moiré Interferometry.

11.2.2 BASIC CONCEPTS

Nature of Light

Moiré interferometry is based on two physical phenomena: diffraction and interference of light. Before these two phenomena can be discussed, some concepts and definitions related to the nature of light must be introduced.

Since both diffraction and interference are associated with wave propagation, the so-called electromagnetic wave theory of light must be considered. According to this theory, a beam of light is defined to be a train of electromagnetic waves propagating in space. Consider the most fundamental component of such a beam consisting of a train of regularly spaced disturbances that vary with time and the distance measured along the direction of propagation. The strength of the disturbance can be viewed as the strength of either an electric or magnetic field at a point in space. That is, the magnitude of either an electric or a magnetic vector which is described below in Eq. 11-1 below. To avoid confusion it has traditionally been assumed that "A" symbolizes the magnitude of an electric vector.

$$A = a \cos \left[2\pi \frac{(Ct-Z)}{\lambda} \right] \qquad \text{Eq. 11-1}$$

where a = amplitude of the disturbance (constant for a parallel beam)
λ = wavelength (the distance between neighboring maxima of the disturbance)
C = velocity of light
t = time
Z = distance measured along the direction of propagation

The term $2\pi (Ct - Z)/\lambda$ is called the phase of the disturbance. In the case of a parallel beam, the phase is constant at any cross-section. Any surface along which the field strength is maximum (i.e., where the value of $(CT-Z)/\lambda$ has an integral value), is defined as a wavefront. The fact that the phase of the disturbance is constant along any cross-section of a parallel beam implies that the wavefronts are always plane cross-sections of this parallel beam, as illustrated in Figure 11-1.

The field strength in any given point of space varies through one full cycle in the time interval

$$T = \lambda / C \qquad \text{Eq. 11-2}$$

where T is called a period. Its reciprocal,

$$v = C / \lambda \qquad \text{Eq. 11-3}$$

is the frequency of oscillation.

Beams of light may also acquire irregular shapes as well (see Figure 11-2), for example, by reflection from an

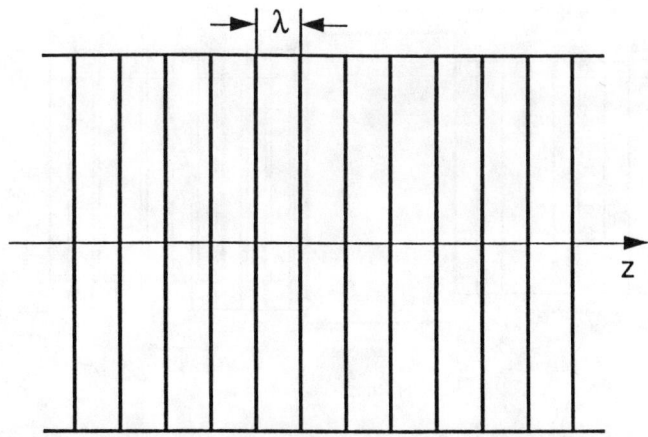

Figure 11-1. Collimated Light Beam Propagating Along the z-Axis. Wavefronts are Represented by Straight Vertical Lines and λ Indicates the Wavelength of Light.

irregular surface. The wavefronts of irregular beams constantly change shape during propagation, and in extreme cases, they may even fold as shown. This is a consequence of the rule that light propagates in homogeneous media in the direction of the wave normal; that is, in the direction perpendicular to the wavefront. Wavefronts of irregular shapes are called warped wavefronts.

The properties of light defined thus far correspond to a single wave train. In nature this kind of "pure" light does not exist. All natural light sources emit very complicated mixtures of wave trains varying in shape and wavelength. Light that can be collimated to a parallel beam at a single wavelength is called coherent light. The only source that is close to this "ideal" source is a laser. Since basic moiré interferometry works with coherent light only, any reference to light throughout the rest of this chapter should be assumed to be coherent light, unless stated otherwise.

It was mentioned before that Eq. 11-1 describes the strength of an electric field which is a vector quantity and the values of its components must be defined separately. Since the electric field vector is always perpendicular to the direction of propagation, the component Az is equal to zero. The other two components, Ax and Ay, can differ in amplitude and phase. For linearly polarized light the electric field vector oscillates only in one plane, and only the component parallel to this plane is defined by Eq. 11-1. The component perpendicular to the plane of polarization is equal to zero. A beam of light of any other polarization can be obtained as a combination of two linearly polarized beams having proper phases and amplitudes and mutually perpendicular planes of polarization.

When a wave train crosses a fixed point in space, its frequency of oscillation in visible light is of the order of 10 to 15 power cycles per second. No existing detection meters can measure individual cycles in this frequency range. Instead, receivers such as the human eye, photographic emulsion, and photoelectric cells respond to time-integrated energy and power.

The power per unit cross-sectional area of a beam of light is called its intensity. Electromagnetic theory shows that the intensity of light is

$$I = a^2 \text{ (as defined in Eq. 11-1)} \qquad \text{Eq. 11-4}$$

It can be seen that intensity is a time-averaged quantity and is independent of the wavelength and phase of the light beam.

Light propagates in space with astounding but finite velocity C, which in a vacuum has a magnitude of 300,000,000 m/sec. The velocity of light in a transparent material C', is less than that in a vacuum. The relationship between C and C' is

$$C' = C / n \qquad \text{Eq. 11-5}$$

where n = a material constant called the index of refraction.

Thus, when a wave train propagates in space, the time required for a wavefront to travel between two points,

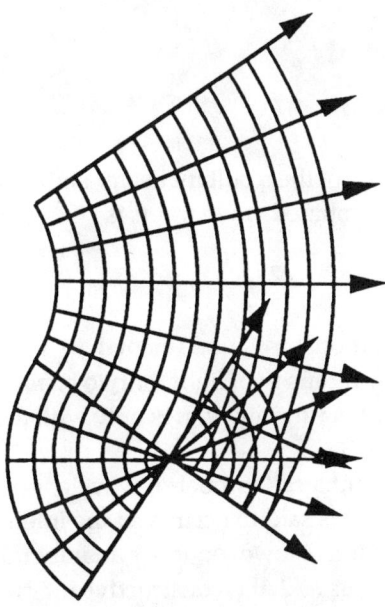

Figure 11-2. Irregular Beam. Due to the Warpage of the Wavefronts Part of the Beam Folded Over.

A and B, along its path depends not only upon the distance between them but also upon the indices of the refraction of the materials in the path.

Specifically, if on the way from point A to point B a wave has to go through k media with indices of refraction n_1, n_2, l_k, and the actual path length through each material is l_1, l_2, l_k, the time required to traverse the distance is

$$t = (n_1 l_1 + n_2 l_2 + \ldots + n_k l_k) / C \qquad \text{Eq. 11-6}$$

The optical path length (OPL) is defined as the distance a beam would travel in a vacuum compared with the time what would be required to travel through actual materials. Thus,

$$OPL = Ct = n_1 l_1 + n_2 l_2 + \ldots + n_k l_k \qquad \text{Eq 11-7}$$

Since the frequency and time are the same for both cases the number of cycles experienced by the light in the real path is identical to the number of cycles in the corresponding OPL in a vacuum.

Another property of light which must be well understood is interference. Since this property is the basis for all interferometric techniques including moiré interferometry, the following section is devoted to this subject.

Interference of Light

Interference is defined as the superposition of at least two coherent beams of light. An optical apparatus that produces two-beam interference is called an interferometer. A moiré interferometer is only one of many types of interferometers.

The principle of an interferometer is shown in Figure 11-3. A collimated beam of coherent light with wavefront W_0 enters the interferometer. As the beam B_0 travels through the apparatus it is divided into two separate beams, B_1 and B_2 with wavefronts W_1 and W_2 respectively. W_1 and W_2 have evolved from W_0 In the apparatus one of the beams, say B_1, is caused to travel a longer optical distance than the other. Due to this difference, S, in the the paths at the exit from the apparatus, the two beams recombine with a phase difference $2\pi S/\lambda$ radians.

In the simplest case the intensities, I_1 and I_2, of emerging beams B_1 and B_2 are equal. This case is called pure two-beam interference. Both beams travel through point P and assume that the wave equation for the longer path for wave train B_1 is

$$A_1 = a \cos 2\pi v t \qquad \text{Eq. 11-8}$$

Since the second beam (see Figure 11-3) had to travel a shorter distance, its phase at this point must be greater by

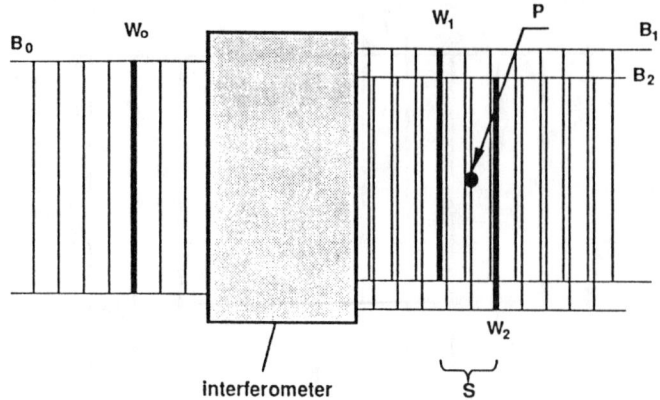

Figure 11-3. Principle of an Interferometer. Beam B_0 is Split in the Interferometer into Beams B_1 and B_2 After Recombination These Beams Interfere, Creating a Fringe Patterns That is a Contour Map of the Separation S Between Wavefronts W_1 and W_2

$2\pi S/\lambda$ and its wave can be described by

$$A_2 = a \cos 2\pi (vt + S/\lambda) \qquad \text{Eq. 11-9}$$

The resultant beam created by the recombination of beams B_1 and B_2 can be described by

$$A = A_1 + A_2 = a [\cos 2\pi vt + \cos 2\pi (vt + S/\lambda] \qquad \text{Eq. 11-10}$$

which, after simplification, becomes

$$A = 2a \cos\left(2\pi vt + \frac{\pi S}{\lambda}\right) \cos\frac{\pi S}{\lambda} = Q \cos\left(2\pi vt + \frac{\pi S}{\lambda}\right)$$

$$\text{Eq. 11-11}$$

where $Q = 2a \cos(\pi S/\lambda) \qquad \text{Eq. 11-12}$

is the amplitude of the resultant beam. The resultant intensity of the recombined wave trains is

$$I = Q^2 = 4 a^2 \cos^2(\pi S/\lambda) \qquad \text{Eq. 11-13}$$

This relationship, fundamental in interferometry, shows that the intensity of light for two beam interference is independent of time and varies only with the difference in path length, S.

If S is an integral number of wavelengths, constructive interference is said to occur and the intensity reaches a maximum value. However, if S is an integral number of wavelengths plus $\lambda/2$, then destructive interference takes place and the intensity reaches a minimum, which is zero.

The preceding calculations were made on the assumption that the interfering beams had equal intensities.

In practice, impure interference occurs, in which the intensities of the two interfering beams are different, and the destructive interference is incomplete. This means that the contrast of the interferograms, (i.e., the patterns created by the interference), is not optimal. Fortunately, the contrast, defined as

$$\text{contrast} = (I_{max} - I_{min})/I_{max} \quad \text{Eq. 11-14}$$

which turns out to be a relatively insensitive to the impurity of interference. For example, if the intensities of two interfering beams have the ratio of 4:1, then the resultant contrast is good and can be shown to be 89 percent.

It must be noted that since interference is based on adding two oscillating electromagnetic fields, the directions of their electric field vectors must be parallel. That is, the two interfering beams must be polarized in the same plane, otherwise full destructive interference cannot occur.

In this discussion on interference it has been assumed that the two interfering beams were collimated; that is, had flat wavefronts and were propagating in the same direction. In this case, if the beams were protected on a screen, an observer would see uniformly illuminated surface where the intensity of illumination would be a function of the optical path difference S.

In general, the optical path difference for interfering beams varies across the cross-section of the space in which they intersect. Specifically, this occurs when one or both of the beams have wavefronts that deviate from flatness; that is, when the wavefronts are warped. In the space where the two beams interfere the intensity of light varies continuously as a function of S according to Eq. 11-13. If a photographic plate is positioned in the plane intersecting this space it records the variation of the intensity across the plane in the form of an interference fringe pattern. This fringe pattern is a contour map of the distance between the wavefronts or, in stricter terms, of the phase difference between the two interfering beams. The wavefronts can be assigned integral numbers as they the leave the source. Since the wavefronts propagate away from the source, the wavefront numbers decrease as the distance from the source increases. The difference between the wavefront numbers of the wavefronts of the interfering beams is defined as the fringe order. If the wavefronts are smooth, fringe orders must vary in a smooth, continuous manner, and thus the changes of fringe order between adjacent fringes must be either one or zero.

A special case of interference is two beam interference, where two mutually coherent collimated beams of light intersect in space. Consider an interferometer that splits a beam of light into two beams, B_1 and B_2, that are brought together again as shown in Figure 11-4. B_1 and B_2 propagate on the x-z plane at angles α and $-\alpha$ with respect

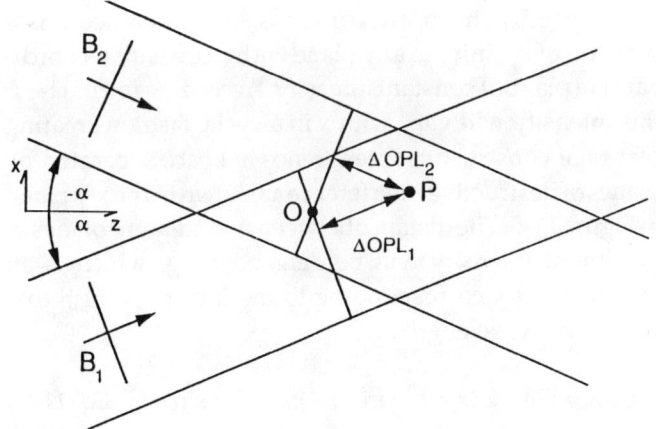

Figure 11-4. B_1 and B_2 are Two Interfering Beams. The Intensity of Light at P Depends on the Location of this Point in the Space Where The Beams Intersect, and Consequently on the Difference in Optical Path Lengths (OPL's), Measured from the Source Along the Paths of these Beams.

to the z-axis, respectively. In the space where they intersect they interfere. The intensity of light at point P depends on the location of this point in the space where they intersect, and consequently on the difference in optical path lengths (OPL's), measured from the source along the paths of these beams. The analysis of the intensity distribution in space follows.

The optical path lengths measured along the interfering beams to point O are OPL_1 and OPL_2 respectively. Point P, with coordinates x_p and y_p, is a point in their interference space. The difference between the optical path lengths at point O and at point P are measured in the direction of propagation, which means that for beam B_1, ΔOPL_1 is

$$\Delta OPL_1 = \overline{OP}(\cos\alpha; \sin\alpha) = z_p\cos\alpha + x_p\sin\alpha$$
$$\text{Eq. 11-15}$$

and for beam B_2,

$$\Delta OPL_2 = \overline{OP}(\cos\alpha; \sin(-\alpha)) = z_p\cos\alpha - x_p\sin\alpha$$
$$\text{Eq. 11-16}$$

As we already know from Eq. 11-11, the intensity of light for two interfering beams is a function of the difference between their respective optical path lengths, S. In this case,

$$S = OPL_1 + \Delta OPL_1 - OPL_2 - \Delta OPL_2 \quad \text{Eq. 11-17}$$

and after substitution from Eqns. 11-15 and 11-16,

$$S = OPL_1 - OPL_2 + 2x_p\sin\alpha = S_0 + 2x_p\sin\alpha$$
$$\text{Eq. 11-18}$$

where S_0 is the OPL difference at point 0.

Since S_0 and α are constants, the difference S is a function of x_p only, so any plane with a constant x-coordinate is a plane of constant intensity. According to Eq. 11-11, this intensity will vary with x in a cyclic fashion creating planes of constructive interference in space separated by planes of destructive interference as shown in the x-y plane in Figure 11-5. The distance between two adjacent planes of maximum intensity will be denoted by g, which is an increment of x corresponding to the increment of S that equals the wavelength λ.

$$\lambda = [S_0 - 2(x+g)\sin\alpha] - [S_0 - 2x\sin\alpha] \quad \text{Eq. 11-19}$$

and, after simplification,

$$\lambda = 2g\sin\alpha \quad \text{Eq. 11-20}$$

from where

$$g = \lambda/2\sin\alpha \quad \text{Eq. 11-21}$$

and the frequency, f is

$$f = 1/g = 2\sin\alpha/\lambda \quad \text{Eq. 11-22}$$

Two-beam interference can be viewed on a screen or through a microscope. It can also be recorded on a photographic plate. The plate is positioned in the virtual grating, usually in a plane perpendicular to the planes of interference and is then exposed. On the lines of intersection between the plate and the planes of constructive interference photographic emulsion gets exposed and the rest of the emulsion remains unexposed. After developing such a plate, the recorded pattern has the form of uniformly spaced parallel lines. That is, the plate becomes a diffraction grating. This method is widely used in the production of so-called holographic gratings. Depending on the emulsion and development process used in production, transmission, reflection, phase, and amplitude gratings can all be created. These gratings will be discussed in the section on diffraction gratings.

It is important to remember that one of the necessary conditions for interference is equal polarization of the interfering beams. Since the photographic emulsions are sensitive to electric rather than the magnetic vector component of the electromagnetic wave, the electric vectors of the two interfering beams should be parallel. This can happen only if the beams are polarized in the direction perpendicular to the plane parallel to their directions of propagation, in our case the x-z plane. This is especially important when the angle of intersection 2α is close to 90°. If this condition is not satisfied the contrast can diminish to zero.

Diffraction Gratings

A diffraction grating is a surface with regularly spaced lines, bars, or furrows. Diffraction gratings can be classified as either transmission gratings or reflection gratings (Figure 11-6a), based on their ability to transmit light, and as either phase gratings or amplitude gratings, based on their effect on electromagnetic waves (Figure 11-6b). Gratings can be either uni-directional or multi-directional. The most common multi-directional) grating used in moiré interferometry is a cross grating (Figure 11-6c).

The most significant parameter of diffraction grating is its frequency, which is the number of lines per unit length and will be denoted by the letter f. The distance between two neighboring grating lines in called the pitch of the grating and will be denoted by the letter g. Notice that g is the reciprocal of f.

Every set of uniformly spaced lines, furrows, or bars is a diffraction grating. However, at frequencies below about 40 lines/mm (L/mm) it is still possible to neglect the diffraction effect and use a simple geometrical approach. In the case of moiré interferometry out interest will lie in relatively high frequency gratings, from about 100 L/mm to as high as 4000 L/mm.

A transmission grating is able to transmit light so that diffracted light appears on the opposite side of the diffracting surface to the illuminating, or incident, light. It is usually made of a transparent material with two optically flat surfaces with the furrows or lines replicated on one of the surfaces, but it can be as simple as a set of straight, parallel wires uniformly spaced in one plane.

A reflection grating has one flat surface and can be made of any solid, homogeneous material. Diffracted light appears on the same side of a reflection grating as the

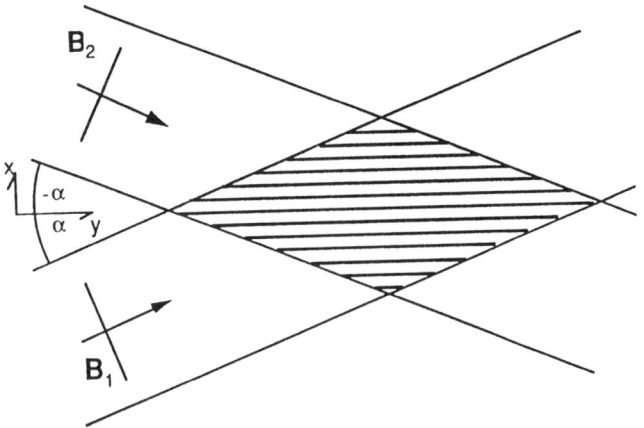

Figure 11-5. Two Interfering Beams Create Planes of Constructive Interference Separated by Planes of Destructive Interference, Indicated by White Spaces and Thick Horizontal Lines, Respectively.

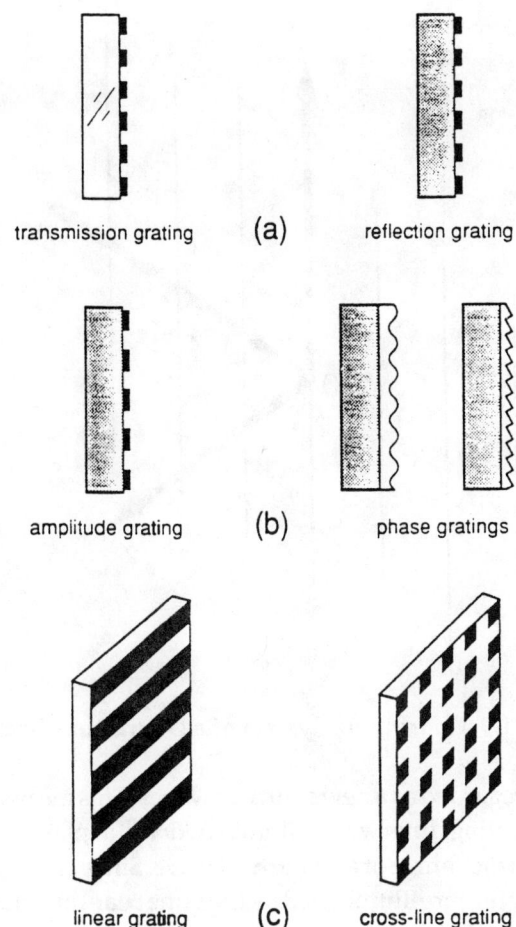

Figure 11-6. Diffraction Gratings Classified According to: (a) Their Ability to Transmit Light, (b) Their Effect on the Waves, and (c) the Distribution of the Lines on their surfaces.

incident light. It is important to note that since all materials reflect some light any transmission grating can be used as a reflection grating; its efficiency as such may be quite low, however.

An amplitude grating consists of either opaque bars and transparent spaces of reflective bars and non-reflective spaces. It is so named because the bars periodically alter the amplitude of incident light. Phase gratings have corrugated surfaces with either a symmetrical and non-symmetrical furrow profile; they alter phase of the incident light in a regular, repetitive way.

In general, phase gratings have higher efficiency that amplitude gratings, which is why they are the most commonly used interferometry. The shape of the furrows of a phase grating defines the light distribution between the diffraction orders, but does not have any influence on the shape of the wavefronts of these diffracted beams in the macro scale. The only case when the shape of a phase grating becomes an issue is when one tries to maximize the efficiency of an optical system utilizing a diffraction grating.

A diffraction grating divides an incident wave train into a number of wave trains of smaller intensities. These new wavefronts emerge in certain preferred directions, as illustrated in Figure 11-7. When a parallel beam is incident at angle α from the left, the grating divides it into a series of beams which emerge at preferred angles: $\theta_{-1}, \theta_0, \theta_1, \theta_2$. These beams are called diffraction orders and are numbered in sequence beginning with the zero order, which is an extension of the incident beam itself. Counter-clockwise diffraction orders are considered positive.

Reflection gratings and transmission gratings diffract light in the same way, except the incident beam of a reflection grating is the mirror image of that for a transmission grating, as indicated by the dotted line in Figure 11-7. For transmission gratings all the angles are measured counterclockwise from the direction normal to the grating surface and the diffraction orders are counted counter-clockwise from the 0-order. The 0-order is simple an extension of the illuminating beam. In the case of the reflection gratings, the angle of incidence is measured in the clockwise direction. This convention makes the notation consistent, assuring that all derivations done for transmission gratings apply to reflection gratings as well.

In order to simplify the analysis of the diffraction of light on a diffraction grating, consider a two-dimensional

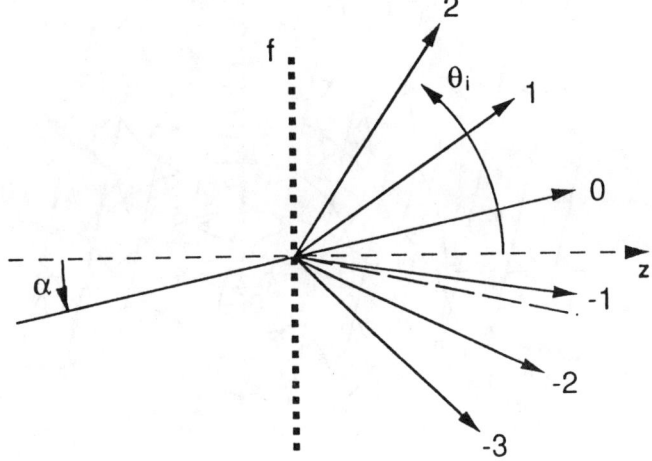

Figure 11-7. Diffraction Orders Created by a Transmission Grating. Angle α is the Angle of Incidence, θ_i are Angles of Diffraction and f is the Frequency of the Grating. The Diffraction Orders are Counted Counter-Clockwise from the 0-order, which is Simply the Extension of the Incident Beam. In a Reflection Grating the Incident Beam, Indicated above by the a Dashed Line, Would Correspond to the Mirror Image of that for Transmission, and Consequently a Positive Angle of Incidence Would be Measured Clockwise.

representation of a grating as shown in Figure 11-8. The grating lines are represented by points and wavefronts by lines. The interference of these waves creates the diffracted beams. Further, introduce an x-y-z coordinate system fixed with respect to the illuminating system and position the grating so that the line normal to it coincides with the z-axis, the lines of intersection of the grating plane with the wavefronts are parallel to the y-axis and the x, y, and z axes form an orthogonal right-handed system (Figure 11-9). The grating lines are numbered starting from the lines coinciding with y-axis and increasing along the x-axis and the wavefronts of the illuminating beam will be numbered in the order in which they left the source so that the wavefront numbers increase toward the source. This analysis is for a collimated beam, but it can easily be extended to the general case.

A collimated beam of light illuminates the grating from the left at some angle α. To simplify the analysis, assume that the grating lines are parallel to the y-axis. According to Huygens' principle each point on a wavefront

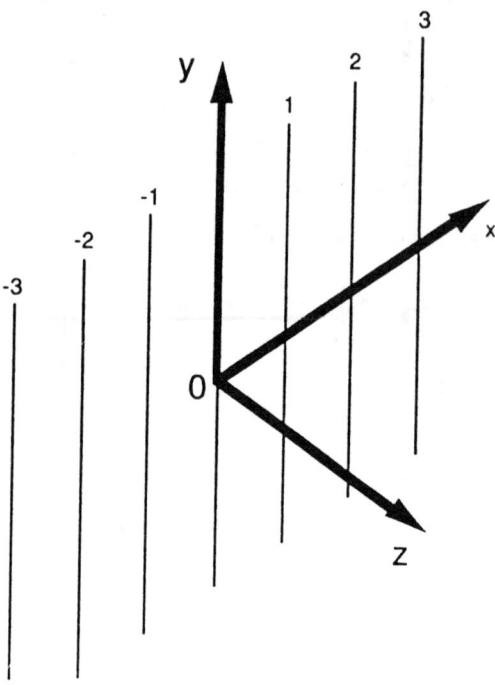

Figure 11-9. Coordinate System of a Diffraction Grating.

may be regarded as a new source of waves. This means that every grating line when illuminated by a plane wave becomes the source of a cylindrical wave. Since all the lines of the grating are illuminated by the same beam of light, the waves they generate are perfectly synchronized and at any instant of time their relative phase difference is the same. As a consequence all the cylindrical waves generated by the grating create a steady state of interference in space.

Since we are dealing with a steady state of interference, the intensity distribution in space does not change with time. This means that we can analyze the interference of light at some instant of time and the conclusions will be valid at all times. Consider the moment when the w-th wave front of the illuminating beam crosses the 0-th line of the grating. From the geometry of the system it can be shown that the wave number of the wavefront crossing the K-th line of the grating is equal to

$$w_K = w - \frac{k\,g\sin\alpha}{\lambda} = w - \frac{K\sin\alpha}{\lambda f} \qquad \text{Eq. 11-23}$$

where w_K = wavefront number of the cylindrical wave generated by the K-th line
w = the wave number of the wavefront of the illuminating beam.

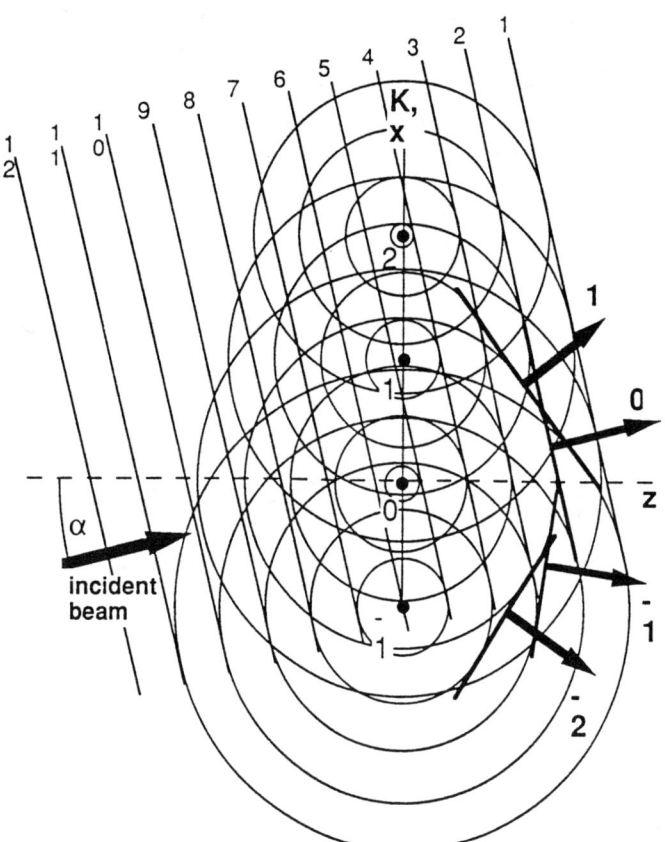

Figure 11-8. Diffraction of Light. As a consequence of Huygens' Principle the Lines of the Grating Represented Here by Points Become Sources of Cylindrical Waves Synchronized in Phase. The Interference of These Waves Creates he Diffracted Beams.

Notice that w_K need not be an integer; when it is not it will be called a fractional wavefront number.

All the waves generated by the grating lines interfere, creating diffracted beams. It can be proved that the wavefronts of these beams correspond to the envelopes of the interfering wavefronts, indicated in Figure 11-8 by the thick lines tangent to some of the circles with the arrows pointing in the direction of propagation. The diffraction order assigned to each of the beams is the number of full cycles by which two interfering wavefronts coming from line K and line K+1 differ.

Figure 11-10 shows the paths of two rays of light passing through lines L_K and L_{K+1}. After diffraction they interfere. Only the rays that create the m-th diffraction order are shown. The wavefront of the illuminating beam that at this instant is crossing line L_K is indicated in the figure by a thicker line, P_1L_K. Similarly the wavefront of the diffracted beam crossing the neighboring line L_{K+1} is indicated as line $L_{K+1}P_2$. Line $L_{K+1}P_2$ can represent a wavefront of the m-th diffraction beam only if the condition that the two interfering beams differ by an integral number of wavelengths is satisfied. Mathematically it can be written in the form of Eq. 11-24:

$$L_K P_K - P_1 L_{K+1} = m\lambda \qquad \text{Eq. 11-24}$$

now
$$P_1 L_{K+1} = g \sin \alpha \qquad \text{Eq. 11-25}$$

and
$$L_K P_2 = g \sin \theta \qquad \text{Eq. 11-26}$$

Substitution of Eqs. 11-25 and 11-26 into 11-24 leads to Eq. 11-19, which is known as the fundamental diffraction equation:

$$\sin \theta_m = m \lambda f + \sin \alpha \qquad \text{Eq. 11-27}$$

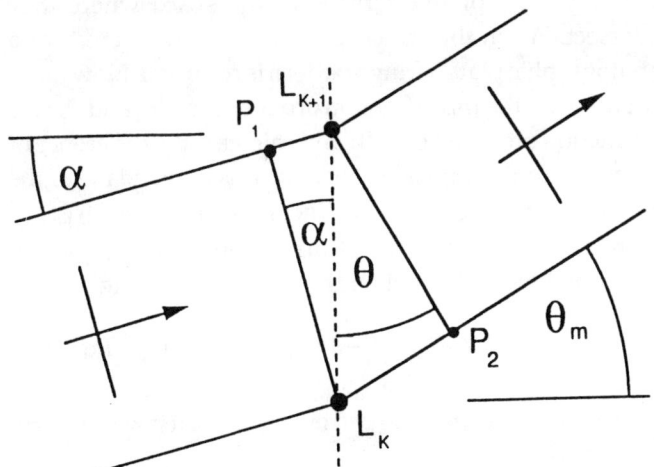

Figure 11-10. Angle of Incidence, α, and Angle of Diffraction, θ. L_K are line numbers, P_1 is a Point on the Wavefront of the Incident Beam, and P_2 is a Point on the Wavefront of the Diffracted Beam. The difference Between the Distances $L_K P_2$ and $P_1 L_{K+1}$ must be an Integral Number of Wavelengths.

where α = the angle of incidence
θ = the angle of diffraction
m = is a diffraction order
λ = the wavelength of light
f = frequency of the grating

Consider now one of the diffraction orders (i.e., m-th order) and derive an expression for the wavefront number at the surface of a grating as a function of the position of a point on the grating. Assuming that the grating is flat and uniform and the illuminating beam is collimated, the number of the wavefront of the illuminating beam at the K-th line is given by Eq. 11-23.

Assume that at line number 0 the wavefront number of the diffracted beam is equal to the wavefront number of the illuminating beam, in this case w. According to our derivation from above, in the formation of the w-th wavefront of the m-th diffraction order, light passing through line 0 has to travel a distance $m\lambda$ longer than light passing through line 1. Similarly, for light passing through line K the difference will be $Km\lambda$. Combining this information with Eq. 11-23, we can say that the wavefront number of the m-th diffraction order at the K-th grating line is

$$w_{Km} = w_K - mK = w - \frac{K \sin \alpha}{\lambda f} - mK \qquad \text{Eq. 11-28}$$

This information will be used later in the section devoted to moiré interferometry.

11.2.3 Principles of Moiré Interferometry

Moiré interferometry uses changes in the shape of the wavefronts of two beams diffracted from a grating formed on the surface of the specimen to measure displacements of the points of this surface. As was explained in the section on Interference of Light, an interference pattern is a contour map of the phase difference of the interfering beams. To be able to relate the moiré fringe pattern to the deformation of a specimen grating, consider the change of the wavefront number of a single diffracted beam at the K-th line as a function of the displacement of this line with respect to the illuminating beam. Then the phase difference of the two diffracted beams as a function of displacement of a point of a grating will be calculated. Each component of displacement will be considered separately.

Translation along the y-axis is the simplest type of displacement to analyze. In this case the grating points move in the direction parallel to the wavefronts of both illuminating beams. Since, as defined earlier, wavefronts are surfaces of constant phase, the phase of the diffracted beam does not change with translation in the y-direction, and thus no change in the interference pattern can be observed. This can be mathematically stated in the form of Equation 11-29:

$$\Delta w_{Kym} = 0 \qquad \text{Eq. 11-29}$$

where Δw_{Kym} is the change of the wavefront number of the m-th diffraction order due to displacement in y-direction.

When a point of a grating moves in the x-z plane, the corresponding change in the wavefront number will be equal to the change in the optical path length measured between the source and the point under consideration. Figure 11-11 shows the case where line K is translated by a distance U in the x-direction to K'. From the geometry we can calculate the change in OPL.

$$\Delta OPL_x = U \sin \alpha \qquad \text{Eq. 11-30}$$

The corresponding change in the wavefront number is simply ΔOPL_x divided by the distance between the adjacent wavefronts and taken with the negative sign, i.e.,

$$\Delta w_{Kxm} = -\Delta OPL_x / \lambda = -(U \sin \alpha)/\lambda \qquad \text{Eq. 11-31}$$

Figure 11-12 represents the translation in the z-direction or, in other terms, the W-displacement. The change in OPL is in this case

$$\Delta OPL_z = W \cos \alpha \qquad \text{Eq. 11-32}$$

and the corresponding change in the wavefront number is

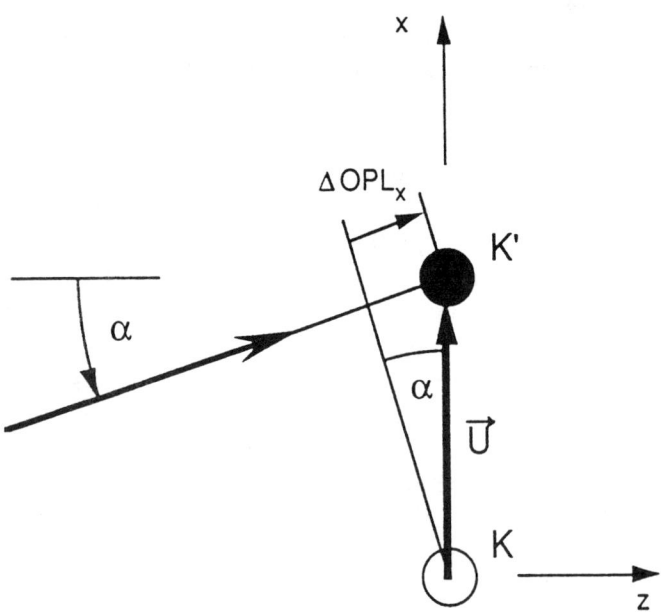

Figure 11-11. Change in Optical Path Length (OPL) Due to a Displacement in x-Direction.

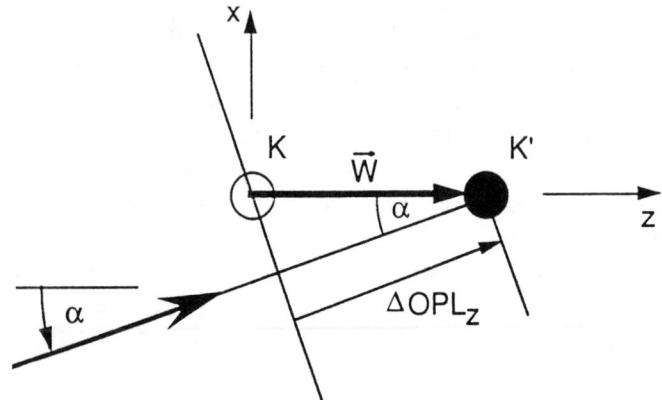

Figure 11-12. Change in OPL due to a displacement in z-direction.

$$\Delta w_{Kzm} = (W \cos \alpha)/\lambda \qquad \text{Eq. 11-33}$$

The total change in w_K is

$$\Delta w_{Km} = \Delta w_{Kxm} + \Delta w_{Kym} + \Delta w_{Kzm} = -\frac{U \sin \alpha - W \cos \alpha}{\lambda}$$

Eq. 11-34

Now illuminate the specimen grating with two mutually coherent beams B_1 and B_2 at angles α_1 and α_2, respectively. Each of the illuminating beams diffracts on the grating and creates diffraction beams with their respective wavefronts. To keep the notation consistent let the wavefront number of the m-th diffraction order of the i-th beam at the K-th grating line be $\Delta w_{KB_i m_i}$.

Each pair of the diffracted beams interferes, creating a steady state of interference in the space where they intersect. When this space is intersected by a screen or a photographic plate a fringe pattern is recorded. Now focus attention on the m_1 diffraction order of beam B_1 and the m_2 diffraction order of beam B_2 and calculate the difference of the wavefront numbers of these beams in the plane of the diffraction grating at the K-th line. This difference $\Delta w_{Km_1 m_2}$, represents the fringe order of an interference fringe pattern as defined in the section on the interference of light.

$$\Delta w_{Km_1 m_2} = \Delta w_{KB\ m_2\ 2} - \Delta w_{KB\ m_1\ 1} \qquad \text{Eq. 11-35}$$

After substitution for $w_{KB_i m_i}$ from Eq. 11-34 it takes the form of Eq. 11-36

$$\Delta w_{Km1m2} = \frac{-U(\sin \alpha_2 - \sin \alpha_1) + W(\cos \alpha_2 - \cos \alpha_1)}{\lambda}$$

Eq. 11-36)

To simplify the notation let the fringe order be denoted by N and its change by ΔN from now on. Equation

11-36 represents the general case where the angles of illumination and the interfering diffraction orders are arbitrary. This equation is relatively complicated and this general case does not have much practical value. The fringe order is a function of both the U- and W-displacements and an analysis of such a fringe pattern would be lengthy and require knowledge of the relation between these two components. The case can be greatly simplified if beams B_1 and B_2 illuminate the grating from symmetrical directions. In this standard moiré interferometry configuration

$$\alpha_1 = -\alpha_2 = \alpha \qquad \text{Eq. 11-37}$$

and Eq. 11-36 simplifies to

$$\Delta N = (2U \sin \alpha)/\lambda \qquad \text{Eq. 11-38}$$

How should this equation be interpreted? The term ΔN is an increment of the fringe order at a point on the specimen grating subjected to a displacement U. Notice that in the symmetrical configuration the W-displacement does not influence the magnitude of ΔN. This means that an observer looking at the interference pattern would see that during displacement the intensity of light at this point, for a given U, would go through ΔN cycles (from bright to dark and back to bright). Since the equation was derived for an arbitrary point on the specimen surface, it is true for a grating of any frequency, even a grating of variable frequency or simply a diffusing surface, as is used in holography. This phenomenon can and is used for measurements of displacements. With minor modifications it is widely used in measurements of fluid velocity. An electronic detector measures the frequency of the changes in the intensity of light diffracted from a particle suspended in a fluid and moving with it. In solid mechanics, the direct implementation of this phenomenon is not very attractive since it requires expensive automatic image analyzers. A much more practical method, moiré interferometry, has evolved to provide the information about these changes in the form of a contour map.

Basic Moiré Interferometer

As mentioned above, moiré interferometry is based on two-beam interference. A high frequency diffraction grating replicated on the surface of a specimen is illuminated by two mutually coherent beams from directions symmetrical with respect to the normal to the surface of the grating. The angle of incidence of the illuminating beams is defined by the diffraction equation (Eq. 11-27) with the condition that the first diffraction order for one beam and the minus first diffraction order for the other emerge along the normal to the specimen surface. The two diffracted beams interfere, creating an interference fringe pattern. A schematic diagram of basic moiré interferometer is illustrated in Figure 11-13. Two beams, A' and B' diffracted from the specimen grating are captured by the camera. Their interference creates on the camera back a fringe pattern that is a contour map of the separation of wavefronts of the interfering beams and at the same time a contour map of the displacement field of the surface of the specimen.

If the interferometer is tuned properly, a photographic camera focused on the plane of the undeformed grating will record the pattern containing zero fringes. This pattern is known as the "null field." In most practical applications this the desired initial pattern.

Any optical system that can produce illuminating beams that satisfy the conditions stated above can be used as a moiré interferometer. Not every system is convenient in real applications, however. When designing a moiré system one must keep in mind that the method is usually very sensitive and requires high stability. The system should be easy to tune and the illuminating system should not interfere with the recording camera.

A Simple Interferometer for Unidirectional Measurements

A simple moiré interferometer used in practice for measurements of displacements in one direction is shown in Figure 11-14. A laser beam is expanded by a microscope

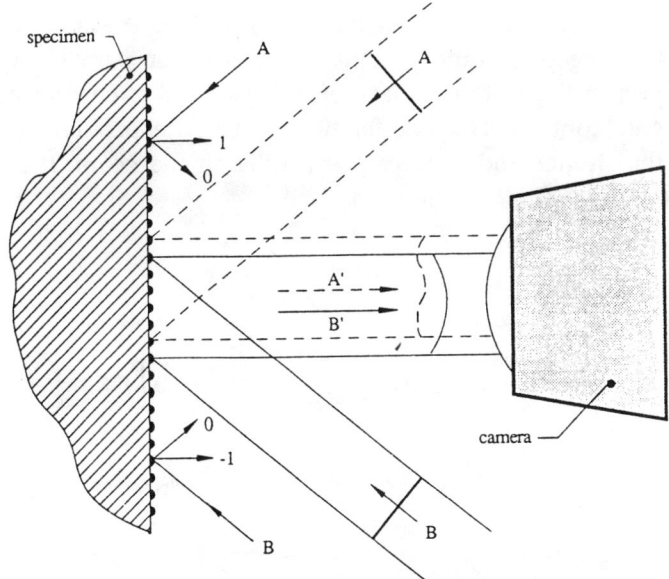

Figure 11-13. Basic Two-Beam Moiré Interferometer. Two Beams A' and B' Diffracted from the Specimen Grating are Captured by the Camera. their Interference Creates on the Camera Back a Fringe Pattern that is a Contour Map of the Separation of Wavefronts of the Interfering Beams and at the Same Time a contour Map of the Displacement Field of the Surface of the Specimen.

lens and collimated by either a lens or a parabolic mirror. If a high quality pattern is desired a spatial filter located in the focal plane of the microscope beam should be used. The collimated beam is directed by a plane mirror toward the specimen grating. Half of the beam illuminates the specimen grating directly and the other half after reflection from a plane mirror located close to the specimen and perpendicular to its surface. A large diameter lens collects the light diffracted from the specimen grating and projects it on the camera back. Usually the shutter controlling the exposure time is located between the laser and the beam expander.

To build a simple moiré interferometer, it will be necessary to have the following equipment:

- a flat, rigid surface (insulated from vibrations if possible)
- a laser with a polarized beam and at least a few centimeters of coherence length
- a beam expander, which can be just a 10× to 40× microscope objective
- a collimating lens or mirror
- a flat mirror the size of the collimator, attached to an adjustable mirror mount
- a flat mirror, a little larger than the specimen grating and attached to an adjustable mirror mount

The first step is assembly is to position the laser so that the polarization of the beam is either parallel or perpendicular to the plane of the system. This ensures a good contrast of the interference pattern. After positioning the shutter and the beam expander in the beam, it is necessary to locate the collimating element so that the resulting beam is precisely collimated. The easiest way to position the collimator is by the use of so-called auto-collimation. A flat mirror is located in the collimated beam, reflecting light back through the collimator to the beam expander. If the collimator is properly positioned, the reflected light will be focused in the focal plane of the beam expander lens where the spatial filter is located. The focused light can either be observed on the surface of the spatial filter, or if the filter is not used it is convenient to position a small screen in the focal plane of the microscope lens for this step and for the rest of the tuning procedure. This screen can be simply a piece of paper with a hole in it. The position of the collimator with respect to the beam expander should be adjusted so that the size of the bright point reflected back through the system reaches a minimum size. If possible, the beam expander should be located on the axis of the collimator, or at least close to it to avoid aberrations.

Once the beam of light is collimated, the specimen should be located in position and illuminated by the collimated beam reflected from the large flat mirror. The angle of illumination should be adjusted by rotating the specimen so that the first diffraction order emerges approximately along the normal to the specimen grating. Now the smaller flat mirror can be located close to the specimen grating and perpendicular to its surface. So far the elements are positioned only approximately. The system is now ready for fine tuning.

The tuning procedure is based purely on optical effects and will not require any extra measuring instruments. Two angles will have to be precisely adjusted: the right angle between the specimen grating and the flat mirror, and the angle of incidence, α.

If the angle between the specimen grating and the mirror is exactly 90°, then the light illuminating the two will be reflected by the grating and the mirror back in the direction it came from, and will be focused by the collimator on the beam expander. This reflected light will form two dots on either the spatial filter or the screen used in the auto-collimation. The 90° mirror (the one located near the specimen) should be adjusted so that the dots overlap with the focus of the beam expander.

The angle of incidence α should be adjusted so that the two beams diffracted from the specimen grating emerge along the normal to the specimen grating. Clearly, if this angle is properly adjusted, the two beams are parallel. The parallelism of these beams can easily be checked with the help of a positive lens. Usually the camera lens is used for this purpose. The two beams are focused by this lens to a point at the focal plane of the lens. If they are not parallel, then two separate dots will be seen in this plane.

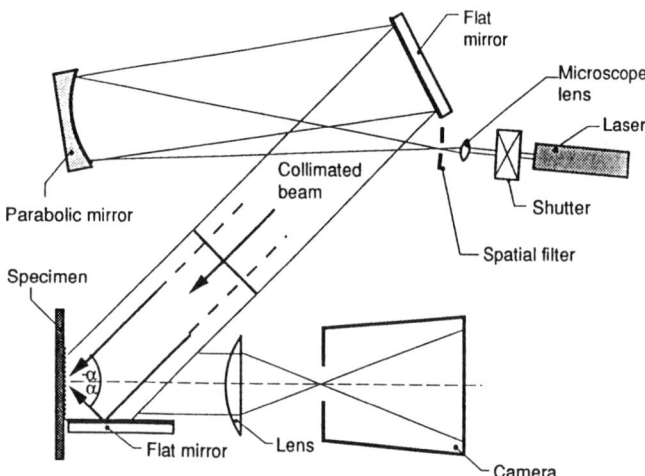

Figure 11-14. Simple Two-Beam Moiré Interferometer used in Unidirectional Measurements of Displacements. (This System is Intended for use on a Holographic Table.)

If the whole system is built on a table in a horizontal plane, then a horizontal separation of the dots is due to a deviation from the angle α. This should be corrected by adjusting the position of the (large) flat mirror directing the collimated beam. A vertical separation of the dots is due to deviation of the direction of the lines on the specimen grating with respect to the vertical direction. This can be corrected either by rotating the specimen, or if this is not possible, by adjusting the 90° mirror. Once the dots are brought together it should be checked that the light reflected back goes back to the beam expander. If not, the directing mirror should be adjusted. Usually a couple of iterations is sufficient to precisely position all the equipment.

The last part of the interferometer is the camera system. It is important that the camera is precisely focused on the specimen surface. The position of the camera lens and the camera back can be optically tuned using the principle of camera action. If an imaging lens is properly positioned, every ray of light coming from a given point will come to the same point on the camera back. By misaligning the 90° mirror or rotating the specimen we can provide two beams of light emerging from the surface of the specimen. These two beams will create two separate images of the specimen on the camera back unless the system is in focus. The camera lens and the camera back can be positioned very precisely by bringing the two images to a common location. Usually the specimen grating contains some small imperfections such as dots or scratches. Bringing them together ensures that the camera is adjusted properly. Now the interferometer must be brought back to the proper position by bringing the dots in the focal plane of the camera lens together.

The final tuning is guided by the number and direction of the fringes on the screen in the camera back. As mentioned earlier, the desired starting point is a field without fringes, called the null field. This can be accomplished by precisely tuning the position of the 90° mirror until a minimum number of fringes can be seen on the screen. It is unrealistic to expect zero fringes, but usually a good initial pattern contains half to one fringe in the field of view. Once the null field is achieved, the optics should be fixed in position throughout the whole structural test.

When the system is ready, measurement can begin. A force is applied to the specimen and the specimen deforms. Since the diffraction grating is securely cemented to the specimen surface, the grating follows the specimen's deformation exactly. Different points on the grating move in different directions by varying amounts, but as long as the grating is continuous the distribution of the displacements must be continuous as well. If so, the intensity of light produced by the interference must vary in a continuous manner. The result is a fringe pattern that is a contour map of displacements in the direction parallel to the plane of illumination and perpendicular to the bisector of the angle between the illuminating beams.

Three-Mirror, Four-Beam Interferometer

In order to define fully the state of in-plane strain, it is necessary to measure two orthogonal components of displacement. It is possible to measure displacements in more than one direction with the two-beam interferometer discussed in the previous section. To do so, the specimen and the loading system must be rotated with respect to the interferometer. This procedure, however, is not only time-consuming, it is inaccurate since the amount of rotation between measurements cannot be determined precisely. For several years two directional measurements were made in this way, but shear strains had to be ignored, or the arbitrary assumption that the displacement field was symmetric had to be made. Clearly, results of such tests contained errors.

A more accurate way of making two-directional measurements is possible: use of the two-beam interferometers described above and make the measurements without changing the position of the specimen or any of the optics. This is much more complicated, however, and quite costly to implement. In 1983 the author developed a three-mirror, four-beam interferometer that can be used to measure simultaneously, or nearly so, two orthogonal in-plane displacement fields (Refs. 11-7 and 11-8). It is an extension of the two-beam system.

Configuration of the Interferometer and Principles of Operation

The three-mirror, four-beam interferometer is illustrated in Figure 11-15. A large collimated beam of laser light illuminates the specimen and the three mirrors—B, C, and D. This beam is divided into four parts. Parts A' and B' illuminate the specimen grating in the horizontal plane in exactly the same way the two-beam interferometer discussed earlier does. Their diffraction orders interfere, producing a fringe pattern that represents the horizontal component of the displacement field, U. Parts C' and D' illuminate the specimen grating in a vertical plane after reflection from mirrors C and D, respectively. These two mirrors are positioned at 45° to the plane of mirror B, and perpendicular to the surface of the specimen. Due to the geometry of the system, the angles of illumination for beams C' and D' are exactly the same as those for beams A' and B'. If the specimen grating is a cross-line grating, beams C' and D' diffract and their diffraction orders interfere, producing a fringe pattern representing the vertical displacement field, V. The two patterns can be recorded by a single camera.

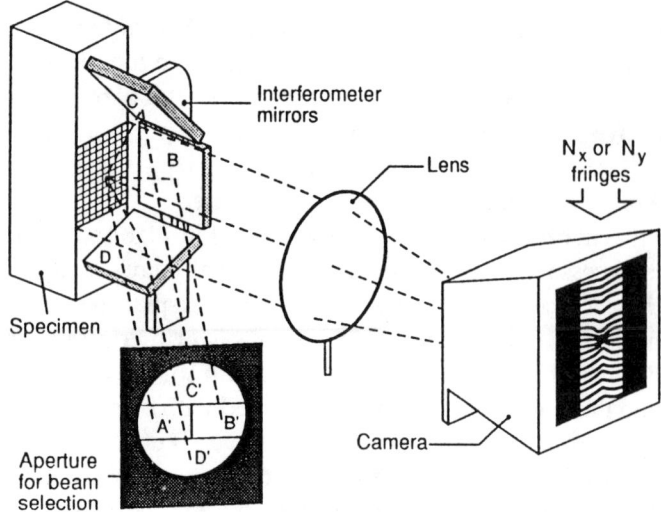

Figure 11-15. Three-mirror, four-beam moiré interferometer used to measure displacements in two orthogonal directions. This interferometer was developed for measurements on a holographic table.

They can be separated by blocking parts C' and D' of the collimated beam for the U pattern, and A' and B' for the V pattern.

If the system is properly tuned to the null fields in the undeformed stage, after deformation the tow recorded patterns are contour maps of the two orthogonal components of displacement. It should be noted that if any rigid body rotation occurs during deformation its contribution to the two patterns is exactly the same, meaning that shear strains can be extracted from the patterns without uncertainties in the measurements. That is, all three components of the strain tensor can be precisely measured on the surface of the

We also indicate that if the collimated beam coming from the laser is properly polarized for the U-field, its direction of polarization is either vertical or horizontal. In this case, beams C' and D' illuminate mirrors C and D with polarizations that are directed at 45° with respect to the plane defined by the direction of propagation and the normal to the surface of the mirror for each one of these beams. This configuration of beams has advantages, but it also has some disadvantages.

One advantage is that the polarization of the beams producing the U-field interference pattern is perpendicular to those producing the V-pattern, allowing easy separation of information with a polarizing beam splitter, and thus simultaneous recording of both fields. This can be important in applications to dynamic loads, where there is no time to switch the beams with an aperture. An application of this technique of simultaneous measurements is demonstrated in Ref. 11.

The coefficient of reflection for metals is strongly dependent on the angle of illumination, the wavelength and polarization of the incoming light (Ref. 11-12). In the three-mirror system beams C' and D' change their polarization from plane to elliptical. For some combinations of specimen grating frequency and wavelength of used light the elliptical component is negligible. For others, however, it is so strong that the contrast of the V-field pattern diminishes almost to zero. In these cases separation by polarization is difficult, and an aperture is necessary for field selection. The contrast of the V-field can be brought back to normal by introducing a plane polarizer in the camera system.

In dynamic applications there is no choice; simultaneous exposures with a polarizing beam splitter is essential. In static or quasi-static applications, where the exposures are taken in sequence, the aperture is the preferred method of field selection. It eliminates noise related to the other field and is much cheaper, since only one camera is needed and a polarizing cube is not used. In the author's laboratory, the two apertures are usually mounted on a rotating fixture and can easily be changed in a darkened room in a matter of seconds.

Tuning Procedure

As already mentioned, the three-mirror system is based on the two-beam interferometer. In the section on the simple interferometer a description was given of the procedure of constructing and tuning a practical two-beam configuration used for unidirectional measurements. All the steps of putting the three-mirror system together are exactly the same for the horizontal displacement field. In order to eliminate beams C' and D' during the tuning procedure the U-aperture should be used.

Once the U-field is tuned the remaining two mirrors responsible for the V-field can be aligned very quickly. When the aperture is removed from the collimated beam the specimen is illuminated by all four beams. Beams A' and B' are already in the right place; their diffraction orders are focused on one dot in the focal plane of the camera lens. The other tow can be tuned by adjusting the two 45° mirrors until the dots formed by the two remaining beams meet the first dot.

As before, the fine tuning of the system is guided by the fringe patterns on the camera back, one at a time. Small adjustments of the vertical mirror for the U-pattern and one of the 45° mirrors for the V-pattern allow the two null fields to be obtained.

Achromatic Interferometer

The two moiré interferometers presented (simple and three-beam) are systems that require some very strict conditions to be satisfied. The light used in a structural test must be both spatially and temporarily coherent, and the position of all the components of the system must be fixed in space during the test. In practice, this means that a high quality laser should be used and the experiment should be performed on a holographic table. These restrictions either limit the range of applications of the method to relatively small test items and low load levels, or result in very expensive tests. The achromatic moiré interferometer introduced by the author in 1984 eliminates most of the restrictions given above and allows moiré interferometry to be used outside the laboratory environment (Refs. 12 and 13). This means that not only small samples but also large specimens and even structures can be tested. The configuration of optics in this system is illustrated in Figure 11-16.

The design of the achromatic interferometer makes the optical path lengths of the illuminating beams equal for every point of the specimen grating, permitting the use of a chromatic source of light in this compact moiré system. A grating called a compensator provides order in the angles of incident light for each separate wavelength, so that the diffracted beams interfering in space create exactly the same fringe pattern for every wavelength.

The polychromatic light illuminating the compensator must be spatially coherent and collimated. As a consequence, the size of the emitting area of the light source remains a critical parameter in this design. For perfect collimation it should be a point source. Another important characteristic of light is its bandwidth; the broader the bandwidth, the larger the elements of the system must be. A laser diode seems to be the best answer. It produces light that is spatially coherent, and although its coherence length is very small, the bandwidth of light it produces is very small compared to that of other commonly used sources, providing a practical way to keep the system compact.

Probably the most important feature of the achromatic interferometer is its low sensitivity to vibrations. It has been proven theoretically and demonstrated experimentally that small deviations from normal incidence in the direction of illumination at the compensator do not cause a noticeable motion of the fringes. The two mirrors and the compensator can be mounted to a rigid frame and attached to the specimen. During the experiment the interferometer moves together with the specimen, eliminating the problem of relative motion and instability of the fringes. Experiments show that a fringe pattern can be comfortably observed during the load application on a testing machine.

The achromatic system can also be used with coherent light coming from a laser. In this case, advantage can be take of the insensitivity of the system to vibrations and of the high power levels at shorter wavelengths offered by gas lasers.

Specimen Grating Preparation

Before any measurements can be made with any interferometer, a diffraction grating must be produced on the surface of the specimen. In routine applications the most common method of specimen preparation is by replication from a special mold (Ref. 14). The mold is a phase holographic reflection grating with a transferrable reflective coating. It can be made of a high-resolution holographic plate exposed to a virtual grating created in space by the interference of two collimated laser beams (Ref. 15). During the development process the unexposed emulsion shrinks, creating a phase diffraction grating. After the grating is developed it is coated with a release agent. Once the plate is dry a thin metallic layer is evaporated on its surface. A mold prepared in this way is ready for replication.

During the replication process the prepared mold is cemented to the surface of a specimen. Depending on the specimen material, different cements can be used. The most common in these applications are two-component epoxy resins. After the adhesive fully hardens, the mold is pulled off the surface of the specimen. The reflective coating separates from the plate, giving high diffraction efficiency to the specimen grating. The whole process of making a specimen grating is illustrated in Figure 11-17.

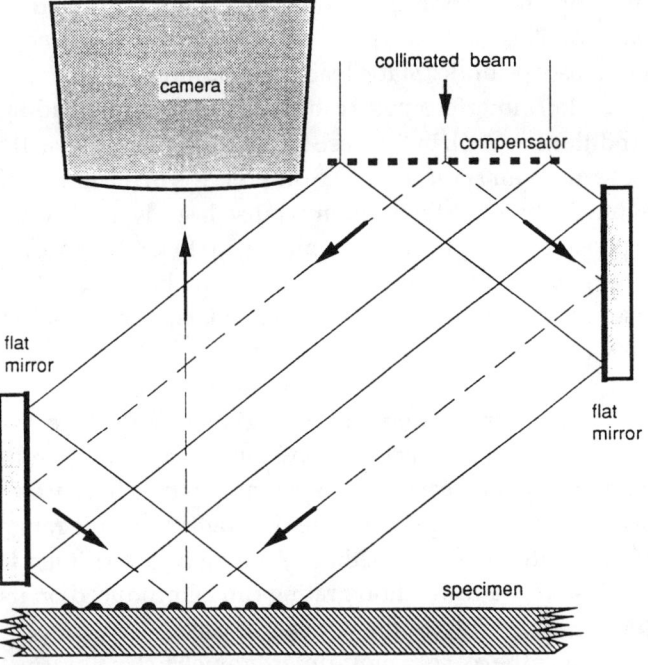

Figure 11-16. Achromatic Moiré Interferometer. Its Small Sensitivity to Vibrations Allows it to be used on a Testing Machine in a Typical Materials Testing Laboratory.

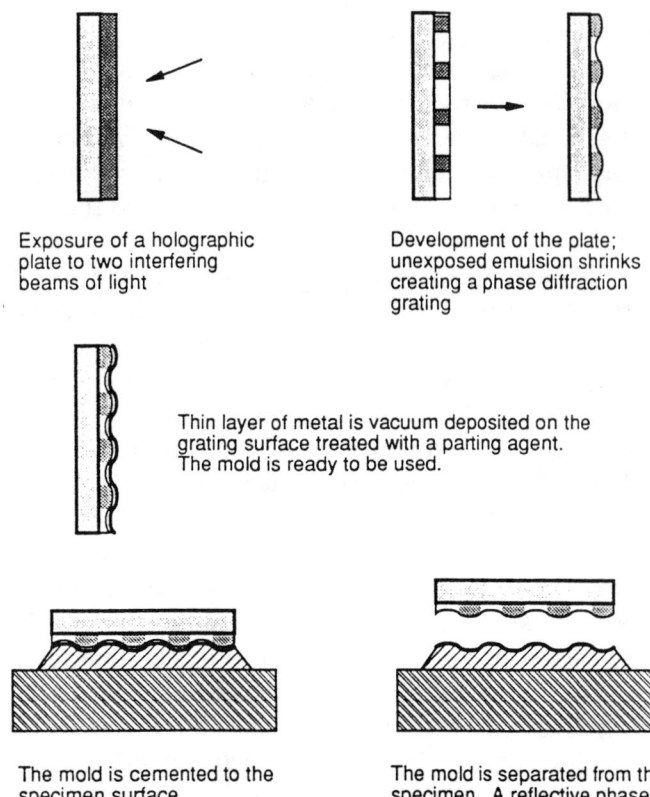

Figure 11-17. Process of Making a Specimen Grating

In special applications the mold can be made with photoresist or cut with a ruling engine. The quality and efficiency of such molds can be much higher, but these methods are expensive and difficult, and not recommended in simple, routine cases.

Special Techniques

Moiré interferometry as presented so far allows measurements of displacements with a sensitivity of about 400 nm per fringe order and with a resolution of approximately 100 nm. The size of the specimen grating, while theoretically unlimited, usually ranges between 1 and 30 mm. In perhaps 99 percent of all applications this is sufficient. Sometimes, however, when moiré interferometry is used in micromechanics studies, or when due to the small size and small deformation level, the number of fringes in the field is small, higher resolution of measurements is desirable. In recent years some new techniques allowing high-precision measurements have been introduced.

The simplest to apply is the technique of a carrier pattern. The use of the carrier pattern is analogous to FM radio transmission, where a high frequency electromagnetic wave is modulated by a low-frequency signal. In this case straight, high-frequency, uniformly space extra fringes are introduced either by a small rotation of the specimen or by a small change of the angle of incidence of the beams illuminating the specimen grating. Changing the angle of incidence corresponds to the introduction of an apparent uniform strain to the specimen, which must be taken into consideration during analysis. Rotating the specimen is preferable since rigid body rotation does not affect strains, and more importantly, does not require misalignment of the system. Thus it is easy to return to original patterns. A method of manual data extraction with an application of carrier patterns is present in (Ref. 16).

Since the carrier pattern is a contour map of linear function, the apparent displacement it represents can be easily subtracted from the displacement field represented by the fringe pattern that contains it. The high-resolution method of extracting data from interference patterns by using a computer interpretation of fringe patterns with carrier patterns combined with nonlinear processing is described in (Ref. 17). This technique allows full-field data collection with a resolution of about 20 nm, almost an order of magnitude better than before.

In cases of small fields of observation, it is not recommended to introduce carrier patterns of reasonable frequency using the methods described above. Separated beams coming to the camera lens might be subjected to different aberrations, which would introduce significant errors. In these situations, the sensitivity can be increased by an order of magnitude by the use of the holographic wavefront warpage multiplication method. The theory and implementation of this technique are presented in (Ref. 18). This method is practical for measurements under a microscope under static loading conditions.

Moiré interferometry can be used in dynamic loading conditions as well. In this case the optical system is exactly the same as for static tests. The difference is in the source of light. To "freeze" the motion a pulse laser is used. With expensive, sophisticated equipment, up to four exposures in sequence can be recorded during one load cycle. The laser most commonly used in dynamic applications is a ruby laser.

11.2.4 Moiré Interferometry in Engineering Practice

As was stated earlier, moiré interferometry is a null-field method of measuring displacements on a flat surface of a test item. This general statement defines a wide range of applications of the method. As long as a specimen is solid, so that a diffraction grating can be produced on its surface, moiré interferometry can be used.

Since the cost of a moiré interferometry structural test is a little higher than that for more conventional ones, moiré interferometry should be used only where it is needed. For a standard measurement of the elastic modulus for a homogeneous material it is cheaper and faster to use strain gages or extensometers, but if the behavior of the tested material is hard to predict, or the strain gradients are high, then this full-field technique should be considered. As a simple rule of thumb: whenever ten or more strain gages applied to a relatively small flat are considered necessary, it is wise to use moiré interferometry as long as environmental conditions permit it.

Currently routine measurements can be performed at room temperatures in reasonably stable conditions. Research is being done on the extension of the range of application into elevated temperature regimes. Recently in the author's laboratory, fringe patterns were recorded at 980°C at full sensitivity (Ref. 19). At the moment, however, such measurements are difficult to make and are far from routine. With time, high-temperature, high sensitivity moiré interferometry will be introduced to practical use, but it will require more expensive equipment, it will probably be limited to a few specialized laboratories.

Below are presented a few typical applications of moiré interferometry. All the patterns were produced with a three-mirror, four-beam interferometer. The frequency of the specimen grating in al the cases was 1200 lines per millimeter with a corresponding sensitivity of 417 nm per fringe order.

The first set of patterns illustrated in Figure 11-18 represents deformation of a material in an aluminum coupon with a crack grown in fatigue loading from an artificial notch. The measurements were made after a few thousand load cycles, with the specimen removed from the testing machine. Some variation in the direction and frequency of the nonuniformities in the plastic deformation of the material in this zone. If the data had been obtained with a number of strain gages it could be strongly affected by the resulting variations in strain gradients, and in the close vicinity of the crack uncertainties of the order of 50 percent would be likely.

The next set of patterns, illustrated in Figure 11-19, represents the displacement field on the edge of a one mm thick graphite/epoxy coupon with an embedded optical fiber (Ref. 20). Fibers like this are used in "smart" materials and structures as sensors of strain, damage, and other parameters. They have been developed for in-service monitoring of these materials. The cross-section of the sensor can be seen in the center of the picture in the form of an ellipse. The patterns presented demonstrate the effect of these sensors on the strain distribution in the material. In this particular experiment, where the optical fiber is perpendicular to the surrounding carbon fibers, the undesirable effects were the strongest and the strain concentration factor measured in the U direction was as high as 14. The zone where this high strain level occurred was a small

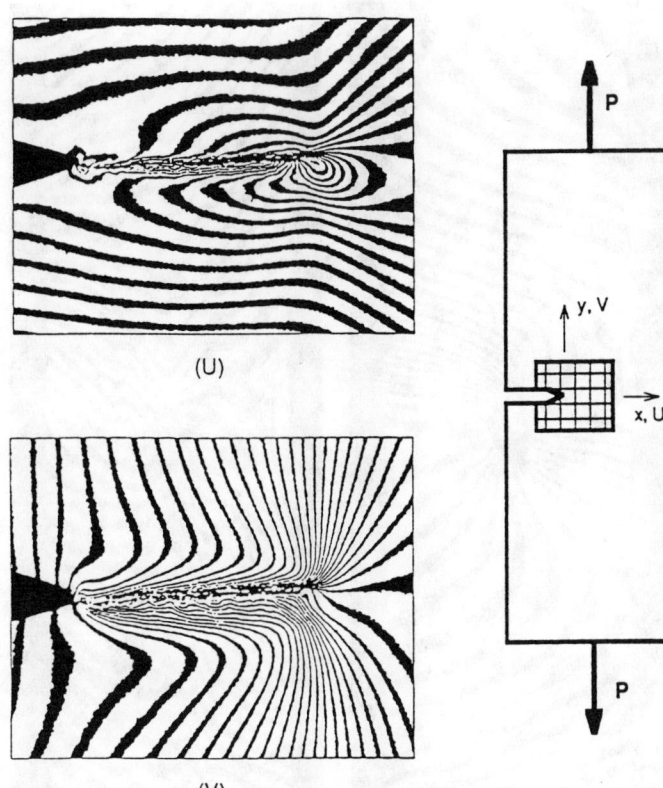

Figure 11-18. Residual U- and V-Displacement Fields Around a Fatigue Crack in a 7075-T6 Aluminum Coupon.

distance from the cross section of the optical fiber. The size of this zone and the very high gradients of strain would make it impossible to to detect by any conventional method.

The last example is an application of moiré interferometry to measurements of residual strains in thick composites (Ref. 21). A diffraction grating was replicated on the cross-sectional surface of a graphite/epoxy composite cylinder and the material was cut together with the replicated specimen grating to a depth of a few millimeters. The released stresses caused concentrated variations of strain at the edges of the cut that in Figure 11-20 can be seen in the form of fringes of varying frequencies and directions. The specimen was a section of a thick-walled graphite/epoxy composite cylinder. Part of the V-pattern with a carrier pattern of extension is demonstrated for the area in the dashed box. The small loops and zigzags along the edges correspond to variations from ply to ply. The data were collected with the help of carrier pattern techniques.

It can be noticed that although the specimens in the presented examples were very different, the experiments had one important thing in common. The deformation fields were highly nonhomogeneous and unpredictable. In order to find the highest value of strain a large number of data points had to be analyzed. The locations of these points were chosen according to the shape of the fringes

(U)   (V)

Figure 11-19. U- and V-displacement fields in a graphite/epoxy composite coupon with embedded fiber loaded in tension. The sequence of plies was [$90_2/0_2$/fiber/$0_2/90_2$].

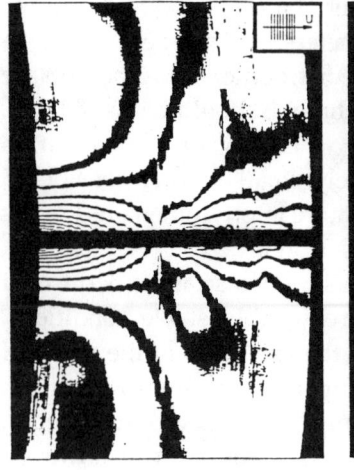 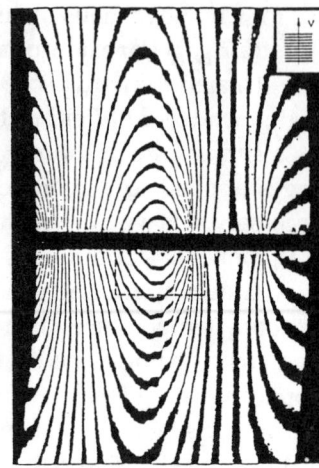

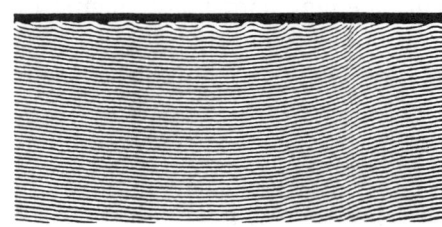

Figure 11-20. U and V Displacement Patterns Representing the Deformation of the Specimen Surface Due to the Release of Stresses in the Vertical Direction. The Specimen was a Section of a Thick-Walled Graphite/Epoxy Composite Cylinder. Part of the V-Pattern with a Carrier Pattern of Extension is Demonstrated for the Area in the Dashed Box.

recorded during the experiment and not in advance, as is the case with strain gages and similar point techniques. High-quality information can be obtained at practically any point of the field of measurements.

The range of possible applications of moiré interferometry is enormous. The presented examples show its advantages over the alternatives, if such alternatives even exist. They should also help the readers decide if their particular application it is the the method of choice.

11.2.5 Interpretation of Moiré Fringes
Extraction of Displacement Data

A moiré pattern is a contour map of a displacement component on the surface of a test item. It naturally comes without any description of the fringes. The only parameter that is known is the contour interval, which is defined by the frequency of the specimen grating and the diffraction orders that created the pattern. In a typical configuration with 1,200 line-per-millimeter specimen grating and interference between the first and the minus-first diffraction orders, this contour interval is 416.7 nm.

The first problem that one experiences looking a a fringe pattern is determining the fringe orders, or in other words, numbering the fringes in proper sequence. Usually moiré interferometry is used to measure deformation; that is, the gradients of the displacement field. In such cases it is necessary to determine the displacements relative to an arbitrarily chosen point. For a selected fringe a convenient number can be assigned. It can be zero, or for example, 100 if it is desirable to have all fringe orders positive. At this point it is convenient to define a coordinate system associated with the specimen grating.

The next step is to number the rest of the fringes. Usually something is known about the deformation of a test item. For example, it may be loaded in tension with a corresponding positive axial strain. In such a case, the fringe numbers will increase in the positive direction of the corresponding axis of the coordinate system. Neighboring fringes can differ by one or zero only. Knowing this information one can properly number all the fringes. Figures 11-

18 and 11-19 are examples of such a simple case.

In instances where the deformation of the specimen is difficult or impossible to estimate, as with residual stresses, a different technique of numbering is needed. Two approaches are most common. The first one, described in detail in (Ref. 16.) is based on the direction of the shift of the fringes due to a small rigid body displacement of the test item. In this approach the test item is shifted between exposures in a known direction by less than half of the fringe interval. The two patterns are then carefully compared and the sign of the fringe gradient can be extracted. this method is useful only in very stable systems in truly static cases. It was originally developed for coarse moiré, where stability is usually not a problem.

A much more practical approach is based on the introduction of frigid body rotation in a known direction or on a small mistuning of the optical system which results in the addition of a constant fringe gradient to the original interferogram. This fringe gradient is called a carrier pattern. The frequency of such a carrier pattern can be easily adjusted to allow comfortable data collection. This approach was used in an analysis of the residual deformation in a composite specimen as illustrated in Figure 11-20. The first two patterns are representing horizontal and vertical displacement fields directly. Below them is a magnified area of the V displacement pattern modified with a frequency type carrier pattern. This area is indicated on the original pattern by a dashed box.

Once a carrier pattern of proper frequency is introduced the sign of the fringe gradient is constant, making the fringe orders easy to determine. An additional benefit of this method is the effective increase of spatial resolution and accuracy of measurements, even though the sensitivity is the same. (Ref. 17) describes an analysis using this approach. It also shows improvement in the resolution of measurements by almost an order of magnitude.

It must be noted that a carrier patter introduced by rotation is much easier to used and analyze. It does not require misalignment of the optics as is the case with the carrier pattern of extension. Both variations of this method are useful, however. The choice often depends on the preferred direction of the fringes in the pattern to be analyzed.

If very high resolution is not an issue, two exposures can be taken, one without and one with a low frequency carrier pattern. The latter is used to determine the fringe orders in the exposure without the carrier pattern.

In more precise measurements it is more beneficial to use for deformation analysis only the interferogram with a carrier pattern. In this case once the data is taken from the picture displacements can be easily calculated by adding a linear function of coordinates of the data points to the measure values. This linear function corresponds either to the introduced rotation or to the apparent uniform strain or to a combination of both. The calculations can be easily performed using a personal computer or even a programmable calculator. A home-made program can be written or a commercial package can be purchased for plotting the reconstructed contour maps of displacements.

In recent years since personal computers have become widely available, considerable work has been done on automating the above-described procedure. In most cases a television camera is used as an input device while some researchers use scanners. Television camera offer a wide gray scale, allowing interpolation between fringes. However, they suffer from relatively spatial resolution, significantly limiting the frequency of fringes that can be analyzed. Scanners offer about four times higher resolution, but they require a hard copy of the fringe pattern. This makes the procedure more labor intensive, and due to the huge amount of data scanners generate, they require fast computers with large memories. Some of this work has been very successful, but at the present manual digitization of data in many cases is as efficient as fully automated one and in the cases where high gradients of strain exist the results are much more reliable.

Extraction of Strains from Displacement Patterns

So far only extraction of displacement data has been discussed. At first glance it might seem obvious that strains can be calculated through simple differentiation of displacement fields. This can be done but the process is not trivial. All experimental data contains some uncertainties. If the strains are numerically calculated as the increment of the relative displacement of two points divided by the difference between their coordinates, the calculated strain will be affected by an uncertainty proportional to the uncertainty in the displacement data and to the reciprocal of the difference between the coordinates called the gage length. The displacement uncertainty for high-sensitivity moiré is usually on the order of 100 nm with an manual analysis or 20 nm if a high-resolution procedure is used. For a 1 mm gauge length this translates to 100 and 20 microstrain of uncertainty, respectively, in the strain measurements. Further increase in the density of the collected data can lead to scatter of the same magnitude as the measured deformation, and therefore unacceptable. In such cases an effective smoothing technique must be employed.

Reference 22 describes a method that allows high-quality data processing with a very high data density without sacrificing the sensitivity of measurements. It is based on a hybrid approach combining the finite element method (FEM) with experimental data used as boundary conditions. A finite element mesh is automatically generated by a computer. It uses the digitized points as nodes of the mesh built out of triangular elements. The noes are loaded by measured displacements through springs that allow a reduction of scatter. Two approaches have been tried, one with springs in the plane of the mesh and the other with the springs perpendicular to the mesh. The discrepancy between the deformation of the FEM model and the experimental data can be controlled by a proper ratio of the stiffness of the springs to that of the mesh. The criteria chosen for selecting this ratio is that the discrepancy is kept below the resolution of the method, in the case of high resolution data reduction below 20 nm. It must be emphasized that the FEM model represents a mathematical function and not a real material and that is what justifies the freedom of tuning the stiffness of its components. Once the displacements are calculated at all the noes of the mesh, the same FEM analysis package is used to determine strains, and if the material properties are known, stresses.

11.2.6 References

1. Lord Rayleigh, "On the Manufacture and Theory of Diffraction Gratings," Scientific Papers, 1, 209; Phil. Mag. 47, 81-93 and 193-205, (1874).
2. J. Guild, "The Interference systems of Crossed diffraction Gratings; Theory of Moiré Fringes," Oxford at the Clarendon Press, (1956).
3. F.M. Gerasimov, V., P. Sergeev, I.A. Teltevskii, V.V. Sergeev, and B.V. Marichev, "The Use of Moiré Fringes to control the Ruling of Diffraction Gratings," Opt. i Spektroskopiya 19, 270 (1965); Opt. Spectry. 19, 152 (1965).
4. F.M. Gerasimov, "Use of Diffraction Gratings for Controlling a Ruling Engine," *Applied Optics*, Vol. 6, No. 11, (Nov. 1967).
5. N. Wadsworth, M. Marchant, and B. Billing, "Real-Time Observations of In-Plane Displacements of Opaque Surfaces," Opt. Las. Techno. 5, 119 (1973).
6. A. McDonach, J. McKelvie, C.A. Walker, "Stress Analysis of Fibrous Composites Using Moiré Interferometry," *Optics and Lasers in Engineering*, (1980).
7. R. Czarnek, "New Methods in Moiré Interferometry," Ph.D. Dissertation, Virginia Polytechnic and State University, Blacksburg, VA, (1984).
8. R. Czarnek, "Three Mirror, Four Beam Interferometer and Its Capabilities," *Optics and Lasers in Engineering*, Vol 15, pp. 93101, (1991).
9. A.J. Durelli and V.J. Parks, *Moiré Analysis of Strain*, Prentice-Hall, Englewood Cliffs, NJ, (1970).
10. F.P. Chiang, "Moiré Methods of Strain Analysis," Chapter VI, Manual on Experimental Stress Analysis, SESA, Rev. 3rd. Ed., A.S. Kobayashi, Ed. (1970).
11. M.S. Dadkah, F.X. Wang, and A.S. Kobayashi, "Simultaneous On-Line Measurements of Orthogonal Displacement Fields by Moiré Interferometry," *Experimental Techniques*, Vol. 12, No. 7, pp. 2830, (July 1988).
12. E. Hecht, *Optics*, Addison-Wesley Publishing Co., pp. 112-113, (1987).
13. R. Czarnek, "High Sensitivity Moiré Interferometry with a Compact Achromatic Interferometer," *Optics and Lasers in Engineering*, A Special Issue on Moiré Interferometry, Vol. 13, No. 2, pp. 99116, 1990.
14. J.W. White and W.A. Fraser, "Making Optical Elements," U.S. Patent 2,464,738, March 15, 1949.
15. P.M. Boone, "Laser Produced Moiré Gratings," *Strain*, Vol. 4, No. 89, p. 43, April (1969)
16. Y. Guo, D. Post, and R. Czarnek, "The Magic of Carrier Pattern in Moiré Interferometry," *Experimental Mechanics*, Vol. 29, No. 2, pp. 169-173, June 1987.
17. R. Czarnek, J. Lee, and T. Rantis, "Moiré Interferometry with Enhanced Resolution," *Experimental Techniques*, Vol. 14, No. 4, pp. 24-28, July/August 1990.
18. R. Czarnek, "Super High Sensitivity Moiré Interferometry with Optical Multiplication,": *Optics and Lasers in Engineering*, A Special Issue on Moiré Interferometry, Vol. 13, No. 2, pp. 87-98, 1990.
19. R. Czarnek, J.J. Wu, S.Y. Lin, and J. Lee, "High-Temperature, High Sensitivity Moiré Interferometry," *Experimental Techniques*, (in press).
20. R. Czarnek, Y.F. Guo, K.D. Bennett, and R.O. Claus, "Interferometric Measurements of Strain Concentrations Induced by an Optical Fiber Embedded in a Fiber-Reinforced Composite," Proc. SPIE - the International Society for Optical Engineering, Vol 986, Paper 09, pp. 43-54, (Sept. 1988).
21. J. Lee, R. Czarnek, and Y.F. Guo, "Interferometric Study of Residual Strains in Thick Composites," Proc. of the 1989 Society for Experimental Mechanics Spring Conference, pp. 356-364.
22. S.Y. Lin, J. Lee, and R. Czarnek, "Integration and Processing of High-Resolution Moiré-Interferometry Data," Proc. of the 1991 Society for Experimental Mechanics Spring conference, (June 1991).

11.3 GEOMETRICAL MOIRE
Robert C. Schwarz, Grumman Aircraft Corp., Bethpage, NY

The moiré effect results from the interference of two very similar arrays of equally spaced lines, concentric circular rings and rays, or patterns of dots. If two transparent sheets of identical patterns are overlaid and slightly offset, then the interference results are shown in Figure 11-21. Through suitable calibration procedures, the patterns of interference can be related to in-plane surface strains and out-of-plane displacements. In actual practice, the patterns given by an unloaded and loaded structure are compared, usually through photographs.

Moiré interferometry is an example of a laboratory technique which requires the making, recording, and processing of many test measurements. The use of this technique requires the ability to make many measurements and to make them rapidly. Moiré interferometry can be used to measure several geometric parameters. This section will describe how moiré methods have been applied to

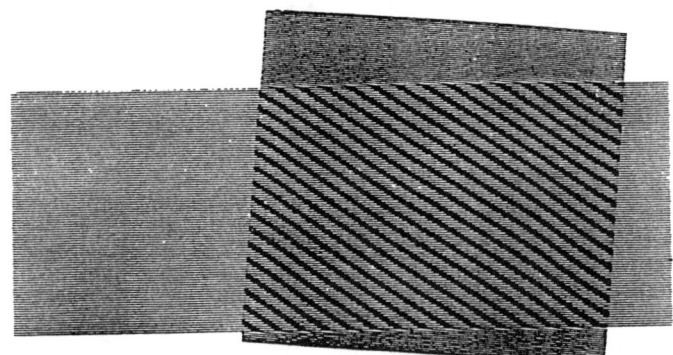

Figure 11-21. Example of Moiré Fringe Pattern

the problems of finding in-plane strains and out-of-plane displacements. Many of the principles and techniques employed are not new, and have been previously conceived of by other experimentalists. However, ideas have been implemented, and are currently being used to form an improved laboratory capability, that will be helpful to the industrial laboratory community and to educators (Refs. 1 through 7.)

11.3.1 A Moiré System For Determining In-Plane Strains

Specimen Preparation

Specimen instrumentation has two distinct phases. The first phase consists of maintaining a stock of moiré grids which can later be bonded to test specimens as required. The second phase consists of the actual bonding of the grid to the specimen. The significant details of these two phases follows.

The bondable grid method of specimen instrumentation was chosen because it accommodates specimens ranging in size from a simple tensile coupon to an integral structural element of a complete airframe. Also grid quality is easier to assure since the photographic reproduction process can be conducted under controlled conditions in a suitable equipped darkroom. Grids can be generally made efficiently at the convenience of laboratory personnel in large enough quantities to maintain an inventory. Our system employs a 10 × 10 cm (4 × 4 inch) array of black squares on a white background with a density of about 200 squares per cm (500 squares per inch). This density was selected to be well within the resolving limits of the small aperture 75 cm (30 in.) focal length objective mounted in the recording camera. In addition, the low density pattern enables a simple contact printing process to be employed in the making of secondary master grids to be used in the printing process. Thus, preserving the purchased master grid from handling damage. Commercially available stripping film (3M Corp. film number TS-5) is used as the bondable grid material. It is purchased in the standard 20 × 25 cm (8 × 10 in.) sheets and cut into quarters to yield 10 by 12.5 cm (4 × 5 in) rectangles that accommodate the 10 × 10 cm (4 × 4 in) ruled portion of the secondary master grids, still leaving a small amount of finger space for handling (Ref. 6). The secondary master grid is an amplitude type grating made on holographic glass plate (Agfa Scientia 10E56). It contains a pattern of sharply defined transparent squares in a black field. The secondary master grids are positive copies of the purchased primary grid (Measurements Group 10 × 10 cm - 4 × 4 in., 200 lpcm - 500 lpin, crossed grating, unmounted). After exposing, developing, and drying, each stripping film grid is checked for quality using a 400× microscope. Acceptance requires sharply defined boundaries throughout, good clarity where the pattern calls for transparency and a high density in the opaque field. It is also helpful at this time to cut one edge of the stripping film grid exactly parallel to the grid pattern. This cut edge is used later when positioning the grid on the test specimen for bonding. The finished grids are stored in plastic envelopes bearing their identification.

Experiments conducted in this laboratory have shown the superiority of a white surface behind the transparent areas of the stripping film grid when depending on diffuse reflection for photographically recording the grid pattern. For this reason the emulsion side of the stripping film grid is initially prepared with two very thin coats of white paint (Spruce Brand Gloss White Enamel 98-2). The choice of paint is important since many paints are discolored by the epoxy adhesive use to bond the grid in place. Next, the grid is bonded, emulsion side down, to the specimen with the grid oriented properly, using an epoxy adhesive (Tra-Bond BA-2115). A smooth parting material, such as drafting mylar, is placed under a thick glass plate, and a weight sufficient to produce a bonding pressure of 6.9 kPa (1 psi) is applied.

After curing, the support sheet of the stripping film is peeled off, leaving the very thin acetate film containing the moiré grid. For those laboratories which do not possess a darkroom capability, it should be noted that comparable grids and grills, ready for bonding, can be purchased from several manufacturers of experimental mechanics equipment.

Data Recording

The literature describes many optical arrangements for creating useful moiré fringe patterns when an undeformed master grating and deformed specimen grid are available. Of the various configurations commonly employed, the technique of photographically recording the specimen grid and then comparing this replica with a master grating was selected for implementation. The choice was based on the following characteristics of this method which are considered important in an industrial laboratory:

- The grid on the specimen surface need not be physically contacted. Only visual access is required as in an environmental chamber.

- Surface protrusions such as plugs, pins, and fasteners can be permitted in the grid zone since contact with a master grating is not required.

- The photographic replica of the specimen grid enables fringe multiplication and fringe sharpening techniques to be employed.

- Various mismatch techniques to control gage length can be easily applied at the time of recording the specimen grid.

Contrasting with these assets is the major disadvantage of the method, namely any uncontrolled camera or specimen movements will substantially affect the accuracy of these measurements. A change in the camera-to-specimen distance along the optic axis causes a magnification change in the recorded grid image and hence a change in grid pitch that can misinterpreted an in-plane specimen deformation. To preclude, or minimize the effects of such disturbances, the camera and supporting platform shown in Fig. 11-22 was designed for photographing specimen grids. Design features include the use of a process quality 75 cm (30 in.) focal length, objective. Also, the camera is mounted on a rigid platform that derives reference to the floor by means of screw adjustable friction pads. Finally, when experiments are performed in the laboratory, the specimen is prevented from moving out of plane by a vertical guide fixture secured to the floor. This support can also be seen in Fig. 11-22. It has been very effective in preventing motion normal to the plane of the specimen yet it does not interfere with the application of in-plane load.

Integral with the camera is a second optical device that enables the optic axis of the camera to be set normal to the plane of the specimen grid. This system is of particular value when the camera is being used outside the laboratory environment. In addition, the system provides a simple method of verifying the normality of the camera during application of the specimen loads. The alignment can be checked easily at any time. Figure 11-23 provides a schematic diagram of the alignment system. As shown, a 5 MW Helium Neon laser mounted on the top of the camera provides a small diameter beam having very little divergence over the distances covered. The beam is directed exactly along the camera centerline, through a small aperture plate installed in place of the lens. A second surface mirror on the specimen surface is used to return the beam to the aperture plate. Correct alignment is indicated when the reflected beam coincides with the out going beam, passing once again through the aperture, and illuminating the ground glass plate at the image plane of the camera. Misalignments cause the reflected beam to impact the aperture plate away from its center, and do not pass through. Corrective adjustments to the camera pitch and yaw are made by means of the camera's base mounted screw jacks and hydraulic table.

Once camera alignment has been achieved, further camera adjustments are made with the assistance of a real time moiré fringe pattern observed on a camera mounted

Figure 11-22. Recording Camera and Specimen Support Fixture

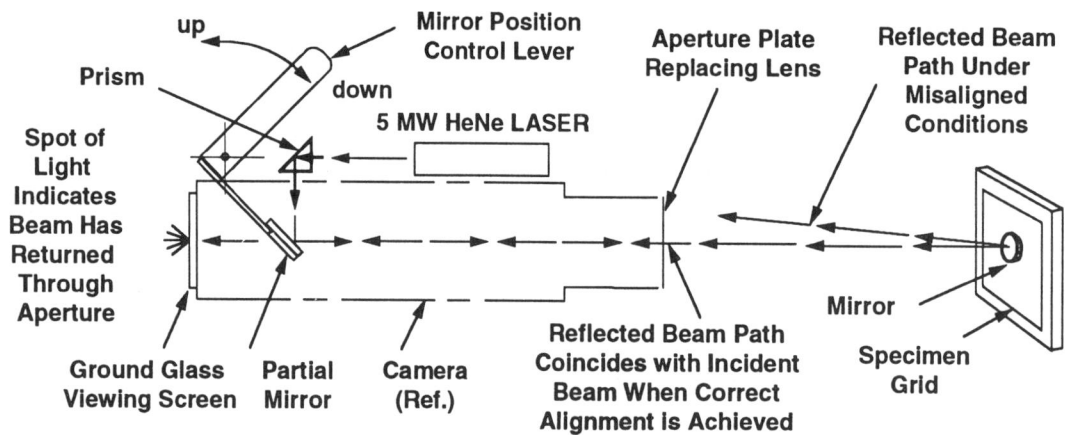

Figure 11-23. Recording Camera Alignment Schematic

viewing screen. The fringe viewing screen is one of several interchangeable accessories that can easily be installed at the image plane of the camera. Consistent installation accuracy, with respect to the image plane, is accomplished by interfacing each accessory at a common machined surface in the holding fixture which is permanently attached to the camera. The viewing screen assembly used for the purpose of adjusting camera focus, and mismatch fringe quantity consists of a 200 lines per centimeter (Lpcm) (500 lines per inch-Lpin) glass plate master grill held in contact with an image forming ground glass screen. (A grill is a uniformly spaced line ruling.) Slight mismatch, equivalent tensile or compressive field fringes (later referred to as Tare Fringes), can be made by changing the lens to specimen distance. Adjustable stops on the lens control rod are provided in both directions of movement to enable lens position either side of a clear field condition to be set initially, with the viewing screen installed, and then repeated later when photographically recording the specimen grid.

Illumination of the specimen grid is provided by two inexpensive commercial slide projectors (Edmund Scientific, 500 watt slide projectors), that have been modified by adding a supplemental lens to reduce the size of the illuminated field, and by the addition of a dimmer control in the lamp circuit to adjust the intensity level for best fringe viewing.

The grid image is photographically recorded on glass plate (Kodak Ortho Plate PFO) to assure a flat surface for the recording and to minimize dimensional changes of the image during developing. Handling the unexposed plates under normal room lighting conditions has been simplified by using laboratory designed plate holders. As with all the other camera accessories, the plate holders derive their image plane reference from the machined interfacing surfaces of the camera, but otherwise are similar to commercial sheet film holders.

After developing, the image on the glass plate is checked under a microscope for uniformity of exposure and line sharpness. Adjustments to the illumination and focus are made until an acceptable plate image is obtained.

Fringe Multiplication

The sensitivity afforded by the standard grid pitch of 200 lpcm (500 lpin) would be inadequate for most applications if the fringe patterns were derived by the classic moiré method. Therefore, a fringe multiplier based on a method developed by D. Post (Refs. 8 and 9) was assembled and is currently being used to provide multiplications of 2, 5 and 10. For the purpose of discussion, the arrangement of the optical elements is shown schematically in Figure 11-24. A photograph of the actual components is presented in Figure 11-25. The slotted mask in front of the camera lens allows only a single diffraction order of the master line grill to pass, and is positioned to select the most appropriate order to reach the camera as determined

Figure 11-25. Moiré Spatial Filter and Fringe Multiplier

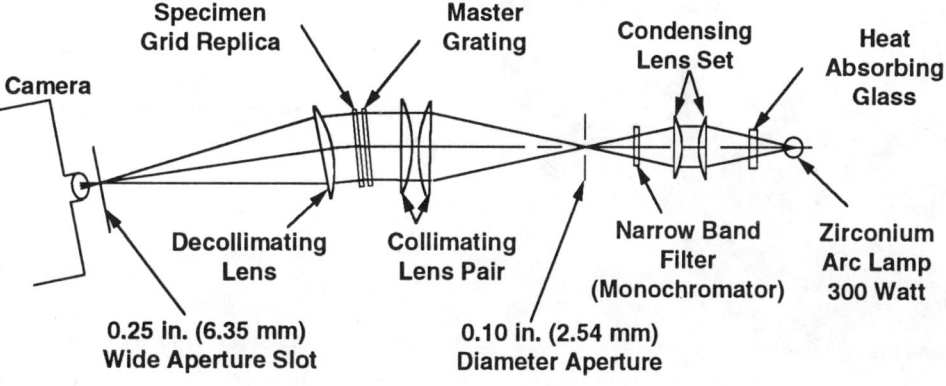

Figure 11-24. Fringe Multiplier Schematic

by the quality of the multiplied fringe pattern observed at the focal plane of the camera. The fringes are usually recorded on Polaroid film since this provides a quality print quickly.

The fixture used to hold the master grill and the specimen grid replica within the collimated light field is also shown in Figure 11-26. Important features of the fixture design provide for the interchanging of master grills, the simple installation of specimen grid replicas, and the adjustment of the master/replica parallelism, gap and rotational alignment. Interchangeability of several master grills is achieved by fitting each master with an aluminum plate frame having a standard hole pattern. Each master grill is adhesively bonded to the machined frame with its emulsion or profiled face away from the frame. Care is take to ensure that this outside face is flat and unencumbered, since reference is made to this face when adjusting the replica for parallelism and gap. An aluminum frame is also used to mount, one at a time, each specimen grid replica. Mounting is easily accomplished since a recess in the frame centers the glass plate and the depth of the recess is such that the emulsion face of the replica will lie only slightly above the frame surface, thereby allowing a strip of masking tape along two opposite edges to suffice for holding the replica in place. When installing the frame assembly in the spatial filter, two finger tight clamps are used to maintain contact with the fixture while rotational orientation is established by a single adjustable rail supporting the lower edge of the frame. Small rotations of the replica to achieve alignment with the ruling of the master grill are produced by adjusting the micrometer head controlling the support rail inclination. Two adjacent edges of the frame are accurately machined at right angles so that once the inclination of the support rail is set, the u and v displacement field fringes can be derived by a simple 90 degree rotation of the frame. Except for small rotational adjustments, the replica is fixed in space, and is the criterion for positioning the master grill. The portion of the fixture devoted to holding the master grill frame uses an assembly of commercially available modular positioning devices (Modern Optics Corp., Precision Vertical Translator VT02B, Translator Stages TS01B, Angular Orientation Device A002b) which enable adjustments while checking with a paper feeler gage at the four corners of the gratings for indications of a uniform gap. The masking tape holding the replica grid to its frame is approximately 0.010 cm (0.004 in.) thick, and when the face of the master grill is just contacting the tape a satisfactory gap results for these multiplications.

It should be noted that the recording camera of the spatial filter must be carefully set and adjusted to provide a sharply focused image with one-to-one magnification ratio.

Data Processing

The large quantities of fringe numbers and coordinates needed for computing strains led to a system which would support the use of a computer. The task of extracting the fringe coordinate data from the fringe photographs obtained from the spatial filter, and recording it in a form compatible with the data processing computer is accomplished using an electro/mechanical/optical fringe reader device constructed in the Grumman laboratory.

The three major components of the Fringe Reader shown in Figure 11-27 are described as follows:

1. Fringe Photograph Viewer:
- The photograph of the fringe pattern is placed under a clear glass retainer on the horizontal surface of the rotary positioning table, where it is viewed by the projection system objective.

- Illumination is provided by a high intensity, 100 watt microscope illuminator equipped with a heat absorbing glass filter and a beam focusing adjustment.

- The magnified image of the fringe pattern is projected onto the ground glass screen immediately in front of the operator of the system at either 10× or 20× magnification depending on the objective selected. An amici roof prism is included in the folded light path and produces an image having the same orientation as the photograph.

- The viewing screen is 7 × 7 inches square and consists of a ground glass plate overlaid with a thin plastic sheet prepared with mutually perpendicular inter-

Figure 11-26. Fringe Reader System

Experimental Methods—Full Field Techniques

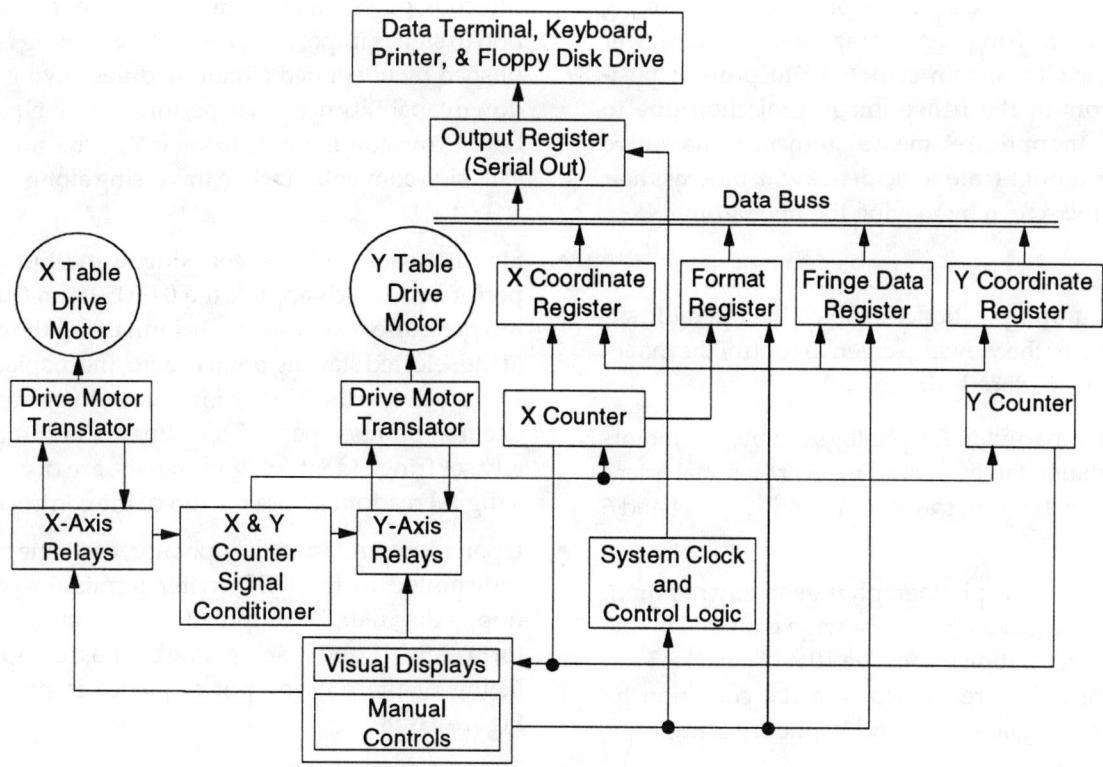

Figure 11-27. Fringe Reader Block Diagram

Figure 11-28. Fringe Reader Data Output Sequence

Order	Line Description (ASCII Code)		
1	Previous Data Acceptable? Yes = 2 (262)*, No = 1 (262)		
2	Carriage Return (215)	non-	
3	Line Feed (212)	printing	
4	Time Delay (Rubout)	control	
5	Time Delay (Rubout)	characters	
6	Sign + (352) or − (255)		
7	Most Significant Digit (260 Series)		
8	Decimal Point (256)	x axis	
9	1st Digit	coordinate	
10	2nd Digit (Right of Decimal Point)		
11	3rd Digit		
12	Least Significant Digit (260 Series)		
13	Comma Delimiter (254)		

(* Numbers are ASCII Code Equivalents)

Order	Line Description (ASCII Code)	
14	Sign + (253) or − (255)	
15	Most Significant Digit (260 Series)	
16	Decimal Point (256)	Y Axis
17	1st Digit	Coor-
18	2nd Digit (Right of Decimal Point 260 Series)	dinate
19	3rd Digit	
20	Least Significant Digit (260 Series)	
21	Comma, Delimiter	
22	Sign + (253) or − (255)	
23	Most Significant Digit (260 Series)	Fringe
24	1st Digit Left or Decimal Point (260 Series)	Order
25	Decimal Point (256)	
26	Least Significant Digit	
27	Comma, Delimiter (254)	
28	Most Significant Digit	Fringe
29	Least Significant Digit	Quality
30	Comma, Delimiter (254)	

secting lines to serve as target points for positioning the center of the fringes. All components of the viewer are designed to remain stationary to prevent positional errors in the fringe image projection due to changes in the optical element alignment. Consecutive fringes are brought into coincidence with the cross hair target on the screen by moving the photograph.

2. **X -Y Positioner:**
- The photograph positioner enables the operator sitting in front of the viewing screen to control the movement of the displaced fringes.

- Motion is imparted to the photograph by two motorized translation tables stacked in an orthogonal orientation with maximum travels of 10 and 15 cm. (4 and 6 in.).

  - Rotation of the photograph is generally required, and is provided by a rotary stage mounted on the upper translation stage. The 10×15 cm. (4×5 in.) photograph is restrained in a flat condition by means of a glass cover and boundary clamp.

  - Identical stepping motors (200 steps per revolution) are used to rotate the micrometer quality lead screws of the linear translation stages. The combination of 200 steps per revolution and 16 threads per cm (40 threads per inch) on the lead screws produces an advance of 0.00050 mm per pulse (0.0000197 in. per pulse) to the stepping motors.

  - The console is set up to provide three pulse rates in either be conducted to determine the ability of a typical operator to locate the centers of fifteen consecutive fringes.

- For the fringe reader described here, the optimum fringe density was found to be approximately 10 fringes per cm (25 fringes per in.). This density produced a range,location error of 0.0016 mm (0.000063 in.), and a worst case error of 0.0032 mm (0.000126 in.)

3. **Data Recorder:**
- Recording the fringe coordinates together with other pertinent data is made simultaneously on a DEC writer terminal and a floppy disk file are the data recording sections of the Fringe Reader system. Figure 11-27 shows a block diagram of the data recording system.

- The determination of fringe coordinates requires counting the number electrical pulses that are directed to the stepping motors on the translation stages while moving the fringe under the reference cross hairs on the viewing screen. Since both forward going and reverse direction pulses can be sent to the drive motors, attention to sign is important. The pulse counting is accomplished by integrated circuit modules having the up/down capability needed to perform the algebraic addition. Provision is made to maintain the normal X-Y axes sign convention when traversing along X = 0, or Y = 0.

- Simultaneous with the counting, a multiplication is performed which applies the 0.00050 mm (0.0000197 in.) per pulse relationship. By initializing the counters at the selected starting point to zero, the displacements serve to define the fringe coordinates, in inches, to four decimal points. The X-Y coordinates, and the manually set fringe identifying numerals, are displayed on a digital readout across the top of the control console.

- Upon command from the operator, each line of data is transmitted to the DEC writer terminal and to the floppy disk file. The output of the data buss is in serial form (ASCII), and is acceptable by most computers. A listing of the data output sequence is provided in Figure 11-28.

- Numerous safety features and conveniences have been built into the data recorder section of this device. Four of the more significant features are:

  - The data displayed on the control console is locked in a self-contained memory at the instant the record button is pressed. This permits the operator to advance to the next fringe without waiting for the data output to be completed.

  - While data is being transmitted, the record button is rendered inoperative. Thus there is no danger of interrupting the data transmission of the previous data point.

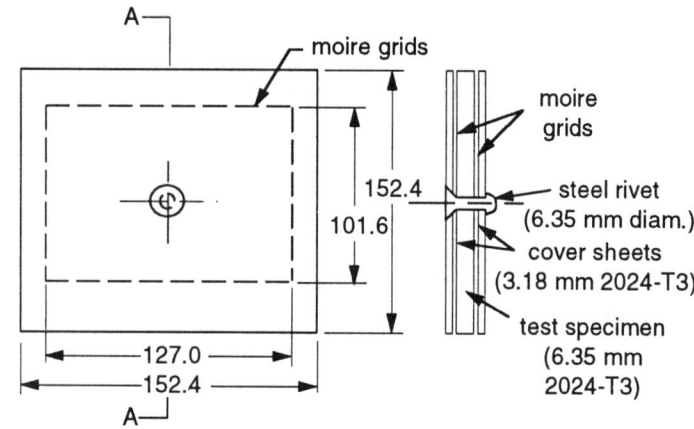

Figure 11-29. Rivet Specimen

- Should an error be made by the operator in the previously recorded line, there is a "PREVIOUS DATA ACCEPTABLE" switch on the control console. If an error has been noted by the operator, this switch is moved from it normal "YES" position to the "NO" position. This causes a numerical character to be changed in the data string. During subsequent processing of the data strings this character can be tested to eliminate erroneous lines of data automatically.

- Motion in both the X and Y directions can be controlled independently.

After completing the data recording process, the data must be converted into strains. This is accomplished by fitting a smooth curve to the fringe number/coordinate data, finding the slope of the curve at selected points, and applying the appropriate strain factor based on master grill pitch. A large number of curve fitting routines were tested at the time of construction of this device and found to be unacceptable. The program we are using was acquired from the AT&T Labs, Murray Hill, NJ. It computes polynomial splines and does a least squares fit to the tabulated data. The calculation of the splines is based on the method of Carl De Boar (Ref. 10). This program has a great deal of flexibility in that the order of the spline segments can be varied as well as locations and number of segments. Our experience with this program has led us to fix the order of the spline at 3, and to vary the number and location of the junctions to achieve a smooth closely fitting curve to the data. The slope of the curve (fringes per unit length—inches) multiplied by the pitch of the master grill (inches per line) provides an apparent strain value. As stated earlier, optimum accuracy in fringe position reading was obtained when the density was approximately 10 fringes per cm (25 fringes per inch). Over the years we have found that an optically induced tare of 2-7 fringes per cm (5-10 fringes per inch) provides the optimum fringe spacing in the test area for most cases. This density provides a reduced gage length for strain measurements, and also provides fringes in the test area with sufficient sharpness to minimize measurement error during the recording process. The actual strain is found by:

$$\varepsilon_a = \varepsilon_{ap} - \varepsilon_t \qquad \text{Eq. 11-39}$$

where ε_a = actual strain
 ε_{ap} = apparent strain
 ε_t = tare strain

11.3.2 Application to Determining Residual Strains in Riveting

The design of a current high performance aircraft requires a detailed analysis of fatigue critical areas of the structure to ensure adequate fatigue life. One area of continuing concern is the stress fields developed around various forms of fasteners. Numerous tests have shown the interference fit fastener to be superior to that of a loosely fitting bolt. One of the more recent developments in the methods of installing a fastener capable of introducing beneficial residual strain fields has been Grumman Corp.'s Stress Wave Riveter (SWR). This system, which causes the fastener to expand by a high energy stress wave, is being used in the manufacture of the F-14 aircraft. Cyclic fatigue tests have shown the system to have beneficial properties. To better understand the residual strain fields produced by the SWR system, Grumman's Research Department requested that several test specimens be manufactured, instrumented with moiré grids, riveted, disassembled, and experimentally analyzed. The typical specimen is illustrated in Figure 11-29. A sandwich specimen configuration was chosen to allow an analysis at the central interfaces of a riveted joint and to minimize the damage to the grid during the riveting process. The experimental data were to be used to aid and refine a mathematical analysis of the same problem. It was assumed that a radial strain field was the basic form, and therefore, there would be strain symmetry about a radial line. This simplified the analysis to determining radial and tangential strain components for four radial lines aligned with the axis of the moiré grid. The moiré strain recording camera is capable of adjusting the optically induced tare for either apparent compression or tension. The best results were obtained by taking two plates, one with a compressive tare and one with a tensile tare. These two plates were then used in the fringe multiplier to obtain the four strain photographs shown in Figure 11-30. The two compressive tare photographs (Figs. 11-30c and 30d) were used to obtain the compressive radial strain components, and the two tensile tare photographs (Figs. 11-30a and 30b) were used to determine the tensile tangential strain components. A typical plot of the resulting strain measurements is presented in Figure 11-31. It should be noted that this process was repeated for both sides of the two specimens for each of two different aluminum alloys. At the time this limited test was performed the theory of dynamic riveting was proven to be very useful in that it confirmed the theoretical analysis needed to compute the fatigue life of this type of fastener installation. It also demonstrated the value of moiré methods in experimentally determining the strain distribution in a complicated elastic/plastic strain field.

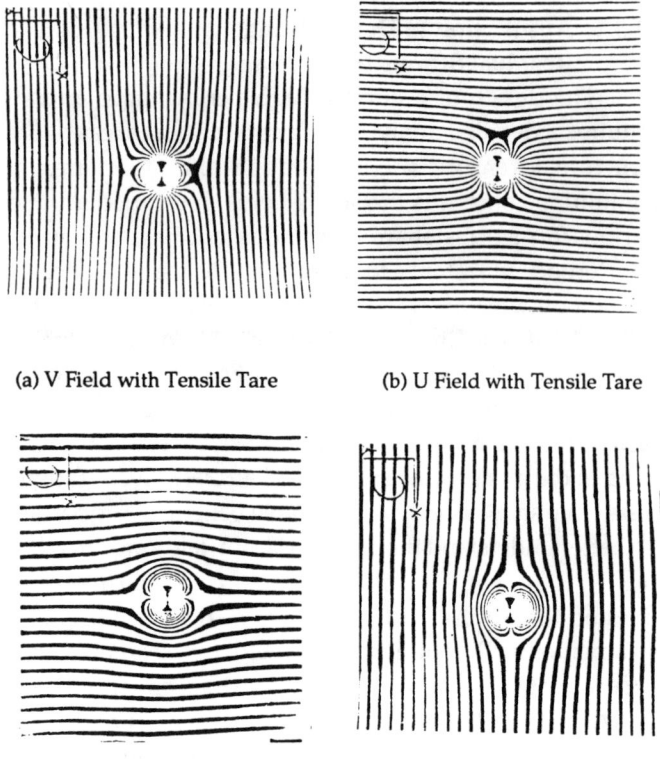

(a) V Field with Tensile Tare
(b) U Field with Tensile Tare
(c) U Field with Compression Tare
(d) V Field with Compressive Tare

Figure 11-30. Moiré Fringe Patterns

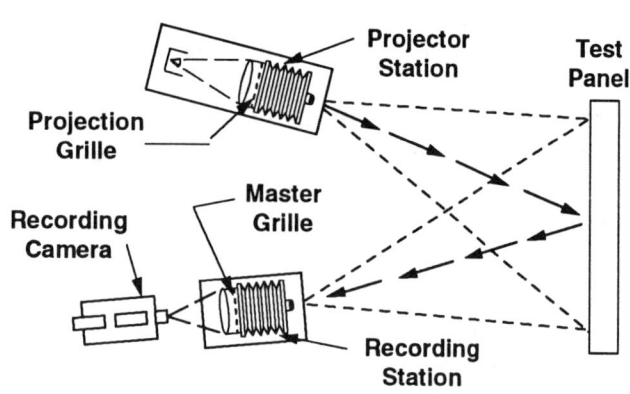

Figure 11-32. Projection Moiré Schematic

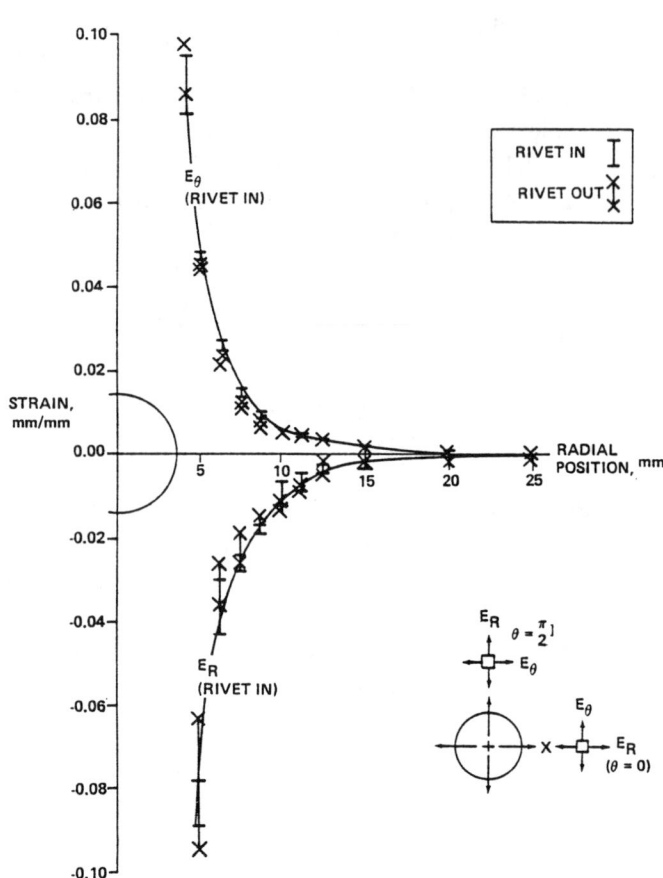

Figure 11-31. Strain Components for Stress Wave Driven Rivet-Determination of Out-of-Plane Displacements

In the Grumman's Experimental Mechanics Laboratory, the determination of out-of-plane displacements has also been accomplished by utilizing moiré interferometry. Two methods are used depending on the geometry of the test specimen and the desire for real time analysis. The two techniques are generally called "The Projection Moiré Method," and "The Shadow Moiré Method." Each method has some advantages and disadvantages. The information gained from both methods is identical; that is, full field contour maps of the deformed surface being viewed.

The use of projection and shadow moiré methods to measure out-of-plane displacements has several very important advantages. These two methods are full field, non-contacting, and have no inertia. This makes them ideal tools for measurements made under dynamic loading conditions, or for structural elements which are extremely light in weight which would preclude attaching some form of instrumentation with associated mass and inertia. The fact that these are optical methods also enables them to be used where the test element must be placed in some hostile environment. The limiting factor being that the surface must be capable of being viewed and illuminated (Ref. 11).

11.3.3 Determination of Out-of-Plane Displacements Projection Moiré

Figure 11-32 shows schematically the test set-up for the Projection Moiré Method. The principle of operation is quite simple. A line ruling is projected on the surface of the structure to be examined. Of necessity the surface must have a color and texture which allows the line ruling to be imaged on it. In practical terms the surface is painted with a flat white paint. The image of the line ruling on the test surface is then viewed by the lens of the recording station, and imaged to another line ruling located at the image plane of the recording station. If the image of the projected grill on the recording station image plane, and the recording station master grill have different pitches, moiré fringes will be produced. Thus, with the test surface in its undeformed state a projection grill is created which when viewed by the recording station has the same pitch as the master grill in the recording station. As the test surface deforms, the pitch of the projected grill on the surface changes following the laws of geometric optics. Equally the image of the changed projected grill at the recording station image plane is changed by the recording station lens. Thus as the test surface deflects away from its initial position, moiré fringes proportional to out-of-plane displacements are created at the recording station image plane. These fringes are then recorded by a camera for later use.

The advantage of the projected moiré method is that small irregularities, other deviations from flatness in the test surface can be "zeroed out" when the projection grill is made from the image of the master grill on the specimen surface. That is, even if the surface is not a perfect plane it is possible to start with a field without any residual fringes. The disadvantage of the projection method is that the displacement fringes are only visible on the image plane of the observation station, and if a recording camera is setup to record the displacement fringes as a function of applied load, then the fringes cannot be viewed easily in real time.

To minimize distortion in the image of the test article, the optic axis of the viewing station is set perpendicular to the test surface. Optical distortion is minimized by ensuring that the vertical axis of the projection grill is normal to the optic axis of the projector system. Also, that the normal to the surface to be measured, the optic axis of the projector, and the optic axis of the viewing system all lay in a plane.

Figure 11-33 shows the Projection Moiré system set up for room temperature test of a reinforced composite panel subjected to a shear loading.

Shadow Moiré

Figure 11-34 shows schematically the test set-up for the Shadow Moiré Method. The principle of the operation

Figure 11-33. Projection Moiré Test Set-up

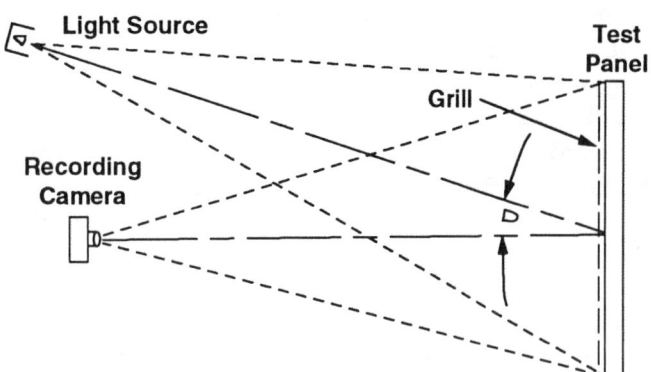

Figure 11-34. Shadow Moiré Schematic

is described as follows. A grill is prepared which is larger than the area of the test article to be measured. The grill is mounted in front of the test surface as close to it as possible (allowing that displacement during loading is going to occur) and adjusted to be parallel to the largest percentage of the surface. A camera is set up to view the test surface through the grill with its optic axis normal to the grill/test surface. A light source is positioned such that the line between it and the center of the test surface, and the optic axis of the camera form a plane, with the lines of the grill normal to this plane. This method relies on the shadows of the master grill lines interfering with the grill lines themselves to produce the moiré effect. Only one grating is required instead of the two required for the projection moiré method.

The advantage of this method is that all observers standing in the vicinity of the recording camera can see the displacement fringes in real time. There are two principal disadvantages to this method. The first is that large flat grills must be made through which the test surface will be viewed. The second is that most structural elements, such as those used in the aircraft industry, are not truly flat; therefore, it is not possible to zero out the entire field of view. It will be seen, however, that for many typical test specimens, the surfaces are flat enough that the residual fringe or fringes are usually quite wide and do not substantially affect the final photographs of displacement. Figure 11-35 shows the Shadow Moiré test setup for an elevated temperature test of a stiffened composite compression panel.

Sensitivity and Calibration

By applying the geometric constraints discussed in the preceding sections, the sensitivity S, of both methods can be simplified to:

$$\text{Sensitivity} = P/\tan a \text{ (length per fringe)}$$
Eq. 11-40

Figure 11-35. Shadow Moiré Test Set-up

where P = pitch of the ruling on the test surface
       a = incident angle of the light source on the surface

A theoretical analysis of the projection moiré phenomena employed here (Ref. 3) shows that in a general sense the absolute displacement between fringes in not a constant, but changes as a function of the absolute value of fringe order. however, the small included angles used by the optical elements in our tests have reduced this error to a negligible part of the whole. Thus, displacement w from the initial plane is given by:

$$w = N \times S \qquad \text{Eq. 11-41}$$

where: N = fringe number
        S = Sensitivity (Eq. 11-40)

Similarly, an analysis of the shadow moiré method used here (Ref. 12) shows that fringe formation is affected by the distance between the grill and the surface. However, if the ratios of the distance between the camera lens and the grill to the grill and the surface is very large, this effect also becomes negligible, and the same two relationships are found to govern the sensitivity of the process and the displacement.

While it is theoretically possible to determine the sensitivity of the set-up knowing the geometry and grill pitch, an actual calibration is recommended for each test (one picture is worth one thousand expert opinions). This can easily be accomplished by placing a ramp with a known slope into the field of view prior to beginning each test. A photograph of the calibration ramp is made and the actual sensitivity determined by a computer program which determines the slope of the calibration ramp (inches of rise per inch of inplane displacement) and the slope of the out-of-plane displacement fringes (fringes per inch of in-plane displacement). Dividing the fringe slope by the ramp slope produces the required value of sensitivity. Figure 11-36 illustrates this process.

It must be pointed out that the fringes produced by either of these methods are lines of constant out-of-plane displacement. There is no simple way of determining from the photographs whether the displacement shown has occurred in a direction towards or away from the recording camera. To provide this information some additional instrumentation must be included in the test setup. One or more calibrated electronic displacement transducers can be placed so that the recording camera will record their value at the instant it captures the displacement fringes. This allows a fringe order (positive or negative) to be assigned to the nearest fringe. Boundary conditions and the physical constraints from the laws of continuity will provide the remaining clues necessary to completely identify the displacements over the entire surface.

If the test is to be dynamic in nature or if a failure test is envisioned (the displacement of the load train would continue to increase until the specimen or structure fails), then a digital readout of the load can also be placed in the field of view of the recording camera.

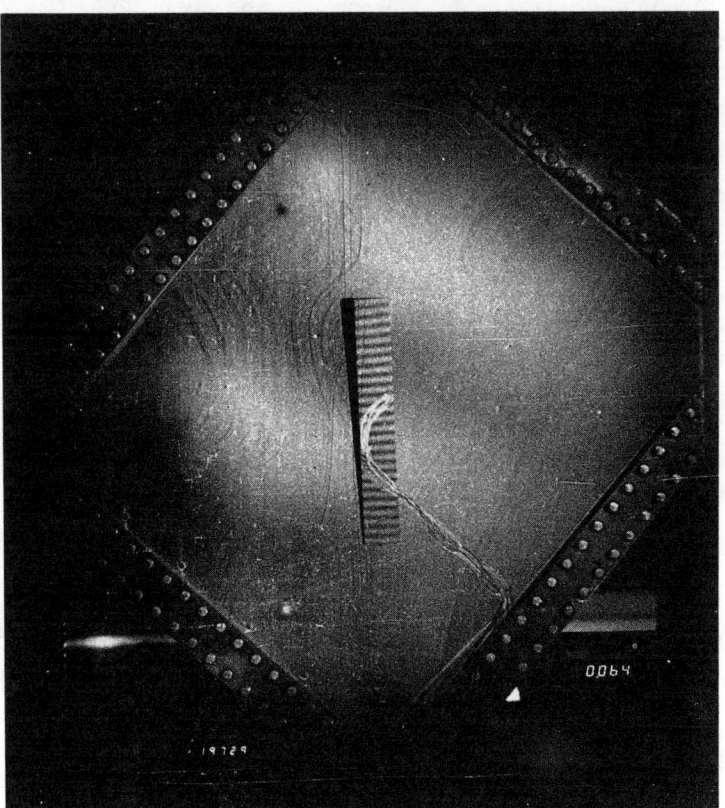

Figure 11-36. Out-of-Plane Calibration Test Set-up and Procedure

Calibration for Room Temp. Shear Panel (Frame No. 1-5)

Run Calib. Sav

	Photo Data	
X-Coord	Y-Coord	Fringe
0.0000	0.0000	0.0
0.0000	1.4667	0.0

Are photo scale end points in X or Y? Enter X or Y.
Y

Enter actual length of calib ramp - F10.4
0.125
1.125

	Photo		Actual	
X-Coord	Y-Coord	Fringe	X-Coord	Y-Coord
0.0000	0.0439	1.0	0.0000	0.2993
0.0000	0.0977	2.0	0.0000	0.6661
0.0000	0.1561	3.0	0.0000	1.0643
0.0000	0.2078	4.0	0.0000	1.4168
0.0000	0.2628	5.0	0.0000	1.7918
0.0000	0.3180	6.0	0.0000	2.1681
0.0000	0.3760	7.0	0.0000	2.5636
0.0000	0.4285	8.0	0.0000	2.9215
0.0000	0.4867	9.0	0.0000	3.3183
0.0000	0.5445	10.0	0.0000	3.7124
0.0000	0.6001	11.0	0.0000	4.0915
0.0000	0.6602	12.0	0.0000	4.5013
0.0000	0.7158	13.0	0.0000	4.8803
0.0000	0.7752	14.0	0.0000	5.2853
0.0000	0.8352	15.0	0.0000	5.6944
0.0000	0.8908	16.0	0.0000	6.0735
0.0000	0.9515	17.0	0.0000	6.4874
0.0000	1.0107	18.0	0.0000	6.8910
0.0000	1.0701	19.0	0.0000	7.2960
0.0000	1.1259	20.0	0.0000	7.6764
0.0000	1.1862	21.0	0.0000	8.0875
0.0000	1.2470	22.0	0.0000	8.5021
0.0000	1.3056	23.0	0.0000	8.9016
0.0000	1.3667	24.0	0.0000	9.3182
0.0000	1.4250	25.0	0.0000	9.7157

Total Absolute Error = 0.683 Percent
Out-of Plane Fringe Sensitivity = 0.0393 inches/fringe
Stop--

Projection Moiré —Test Set-up and Procedure

The first step in setting up the Projection Moiré System is to establish the line of the recording station optic axis. It is desirable that this be normal to the test surface, and passing through its geometric center. This can be achieved by using an aligning mirror placed on the test surface. The recording station is then moved until the laser beam is reflected back on itself, and the desired spacing between the recording station lens and the test surface is obtained.

The master grill on the recording station is a high quality, 500 lpin ruling, oriented with the lines vertical. For our system, the master grill is made on holographic grade photographic plate film by contact printing from a purchased master grill. This guarantees a high quality grill for making the projector grills and later for fringe formation at the recording station.

The projector and recording station lenses are process quality lenses (Schneider-Kreuznack Componon-S) with a 300 mm focal length.

If both stations are set 327 cm (129 in) from the center of the test surface and there is a 22 degree included angle formed by the optic axes of the two stations, a 10:1 magnification of the 500 1pin master grill is produced resulting in a grill on the test surface with a density of approximately 50 1pin. The combination of a projected grill with a pitch of 0.050 cm (0.020 in) and an included angle of 22 degrees will produce a sensitivity of approximately 0.13 cm (0.050 in) of displacement per fringe. The actual value of sensitivity for each test set-up should be obtained by calibration.

The projection grills must be made up new for each test set-up since the orientation and shape of the test specimen changes with each test. In each case, the light source is removed from the projection station and placed behind the recording station. The optics of the recording station are adjusted until a sharply focused image of the master grill is obtained on the test surface. It should be noted that this test method requires that there be no free body motion of the test surface once the projection grill is made. To ensure this, the test set-ups should include guides and restraints which would preclude rotation of the test specimens during loading. With the master grill projecting a sharp magnified grill on the test surface, the back plane of the projector station is carefully adjusted to obtain a well focused image of the grill on the test surface. When this has been accomplished, the clear glass back plane is removed and a photographic plate holder is installed in its place. An exact replica of the projected master grill is obtained. The photographic plate is removed, developed, and reinstalled carefully to match the projected grill, thereby clearing the field of fringes. Finally, the light source is removed from the viewing station and returned to the projector station and the set-up is completed.

Shadow Moiré Method—Test Set-up and Procedure

This test set-up procedure for implementing a shadow moiré test is much simpler than that for the projection moiré technique. The recording station consists of any standard camera. Our tests used a single lens reflex type with a Kodak 2475 type film for high contrast in a reduced light situation. To simplify the subsequent data reduction, the camera is placed at the same height as the test surface and normal to it. The added complexity of implementing the shadow moiré method comes in making and positioning the large grills required. This laboratory has chosen to produce 91 × 66 cm (36 × 26 in) grills with a density of 50 lpin on clear photo reproduction mylar. To maintain the flexible mylar in a plane, a piece of plate glass is mounted in an aluminum frame, the edges of the mylar are taped to the aluminum frame, then a vacuum is applied to the space between the mylar and the glass plate. The ambient air pressure pushes the mylar against the plate glass forming an inexpensive, rigid grill which can be positioned next to the test surface. Adjustment of the grill to a position parallel to the plane of the test surface, and as close as possible allowing for the anticipated displacement is made easier if a three point support system forming two axes is used. The test set-up is completed by positioning a point source of light a a sufficient distance from the grill to minimize distortions in fringe sensitivity. If an incident angle of illumination of approximately 35 degrees is used, a sensitivity of approximately 0.08 cm (0.030 in) of displacement per fringe is obtained. The actual sensitivity should be obtained by calibration.

Data Reduction

Usually the objective of these test programs is to obtain out-of-plane displacement profiles of the various structural elements. This requires that the locations of the displacement fringes on the surface be determined accurately from photographs. Accurate fringe sensitivity determination requires careful calculations of the slope of the calibration ramp fringes. It can be seen that the task of accurately determining the fringe coordinates from the displacement photographs is the essence of the experimentalist's task. The requirement to process a large quantity of data points leads to the use of a data processing computer. This laboratory has developed and built a Fringe Reader System allowing the X-Y coordinates of the fringes on the photographs to be determined to a resolution of 0.0032 mm (125 microinches). This Fringe Reader System is described in detail in the previous section on in-plane moiré interferometry.

The fringe/coordinate data can be processed in two ways by an in-house developed Fortran computer program. In the first or simpler option, the photo scale factor, fringe sensitivity, displacement values, and surface coordinates are determined for each data point and then printed out for plotting. The second option fits a smooth curve to the data points. This allows the surface to be divided into a predetermined number of data points, at which interpolated values of the displacement can be obtained from the fitted curve. This approach is often necessary if computerized surface plotting is to be utilized.

11.3.4 Application to a Typical Design Problem

Out-of-plane moiré was used to assist in the analysis of stiffened composite structural elements. There were two objectives to these tests. The first was to determine the onset of buckling. This was accomplished by examining the series of photographs taken as the load was increased. Buckling was defined as that load at which a wave pattern was observed to have formed across the full width or depth of the structural element. Localized waves were observed at lower loads, but these were considered to be caused by uneven loading and test fixture edge effects.

The second objective of these tests was to obtain out-of-plane displacement measurements which could be used to correlate and refine the structural analysis of these complex structural elements. This objective was met by selecting pictures showing the fully developed buckling pattern. Using these pictures, sections were selected for analysis. The fringe orders were identified and their positions determined using the Laboratory's Fringe Reader System.

The test chosen to illustrate the usefulness of moiré methods was performed using a shear panel at room temperature, pressure, and humidity conditions. The projection moiré method was used for this test. Figure 11-37 shows the displacement contours at the six most significant points during loading. In these photos, (a) the panel is seen at the beginning of the loading sequence, (b) at the point at which a fully developed buckling pattern exists, (c) the maximum load prior to a structural failure, (d) the point at which one stringer has partially separated from the skin, (e) the maximum load sustained by the panel, and finally (f) the panel after it has collapsed in diagonal tension. Figures 11-38 through 11-40 show the displacement profiles along sections of interest before the partial failure of the stringer. The fringe sensitivity for this test was determined to be 1.0 mm (0.0393 in) of displacement per fringe as shown in Figure 11-36.

11.3.5 References
11.3 GEOMETRICAL MOIRE, Robert C. Schwarz

1. A.J. Durelli and V.J. Parks, *Moiré Analysis of Strain*. Prentice-Hall, Englewood Cliffs, NJ, 1970.
2. A.S. Kobayashi, Ed., *Manual on Experimental Stress Analysis*, Chapter 4, "Moiré Analysis of Strain," Chapter 5 "Photoelasticity," Chapter 7, "Moiré Interferometry," SEM, Bethel, CT, 1970.
3. J. Butriago and A.J. Durelli, "On the Interpretation of Shadow Moiré Fringes," *Experimental Techniques*, June 1968, pp. 221-226.
4. F. Chiang, "A Shadow Moiré Method with Two Distinct Sensitivities," *Experimental Mechanics*. October 1975, pp. 382-385.
5. F. Chiang, "Production of High Density Moiré Grids," *Experimental Mechanics*. June 1969, pp. 286-287.
6. G.L. Holister, and A.R. Luxmoore, "The Production of High Density Moiré Grids," *Experimental Techniques*, May 1968, pp. 210-216.
7. C. Sciammmarella, "The Moiré Method—A Review," *Experimental Mechanics*, November 1982, pp. 418-433.
8. D. Post, "New Optical Methods of Moiré Fringe Multiplication," Proc. of SESA, Vol. XXV, pp. 63-68.
9. D. Post and T.F. McLaughlin, "Strain Analysis by Moiré Fringe Multiplication," *Experimental Mechanics*, Sept. 1971, pp. 408-413.
10. D.G. Berghaus and J.P. Cannon, "Obtaining Derivatives from Experimental Data Using Smooth Spline Functions," *Experimental Mechanics*, January 1973, pp. 38-42.
11. M. Idesawa, "Scanning Moiré Method and Automatic Measurement of 3-D Shapes," *Applied Optics*, August 1977, pp. 2152-2162.
12. J. Marasco, "Use of a Curved Grating in Shadow Moiré," *Experimental Mechanics*, December 1975, pp. 464-470.

11.4 PHOTOELASTICITY
C. E. Taylor, University of Florida (emeritus), Gainesville, FL, and R. E. Rowlands, University of Wisconsin, Madison, WI

11.4.1 Introduction

The need for structures that are lighter, stronger, and more economical than their predecessors has created an urgent need for research and development in experimental mechanics. Availability of new materials and the introduction of better instruments for sensing, recording, and processing data have greatly increased the capabilities of photoelasticity (Refs. 1 through 11). Laser illumination significantly facilitates application of photoelasticity to dynamic and scattered light analyses, although it is not necessary for two-dimensional static cases.

This section assumes that the reader is familiar with stress and strain transformations (Mohr's circle), but no prior knowledge of optics or photoelasticity is required. Although this discussion is limited to photoelasticity, some of that technique's intrinsic attributes are shared by virtually all optical methods of mechanics. These include:

- They are usually full field methods and thereby provide information on the stress distribution and locations of critical stresses.

- They are non-contacting.

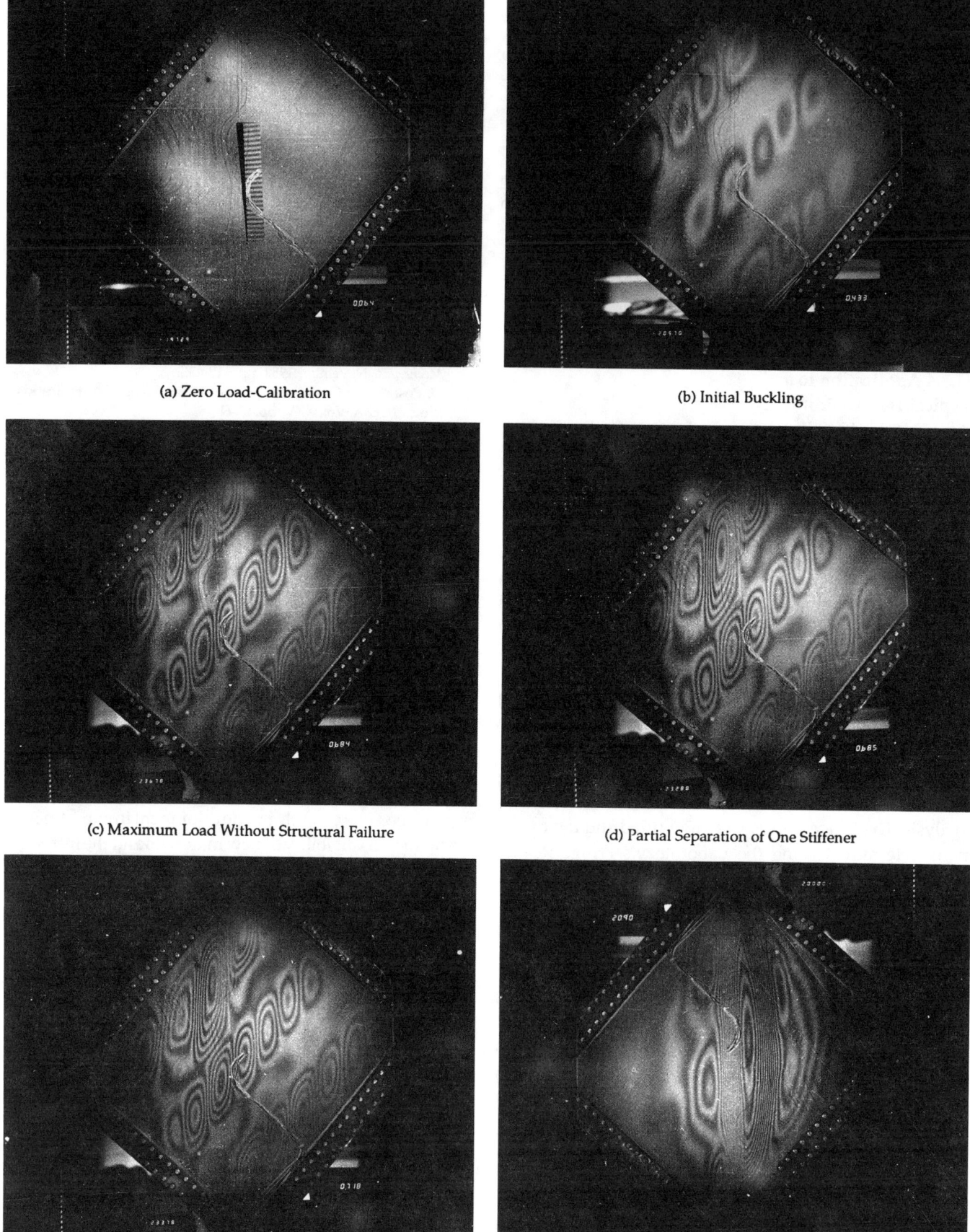

Figure 11-37. Displacement Fringes for the Shear Panel

EXPERIMENTAL METHODS—FULL FIELD TECHNIQUES

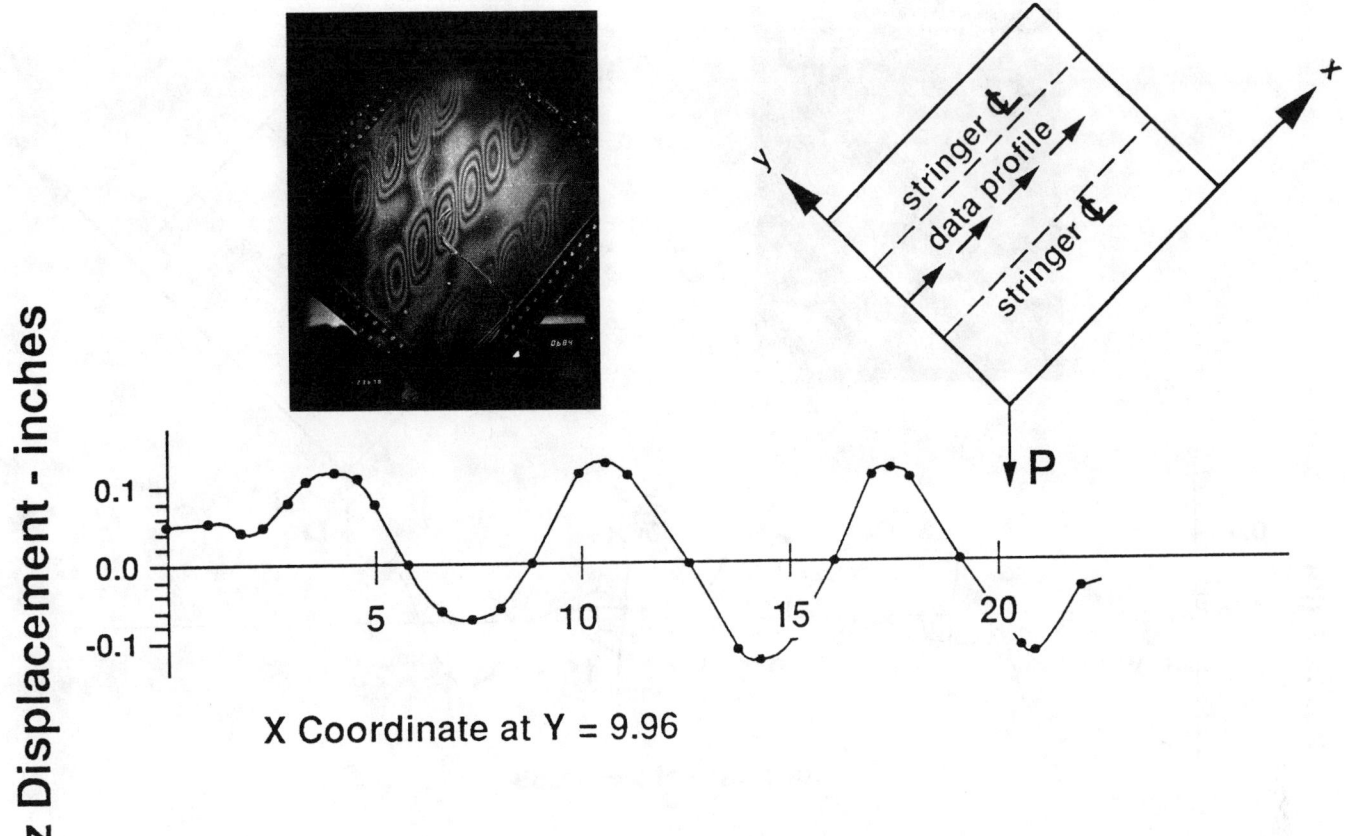

Figure 11-38. Displacements Between the Stringers

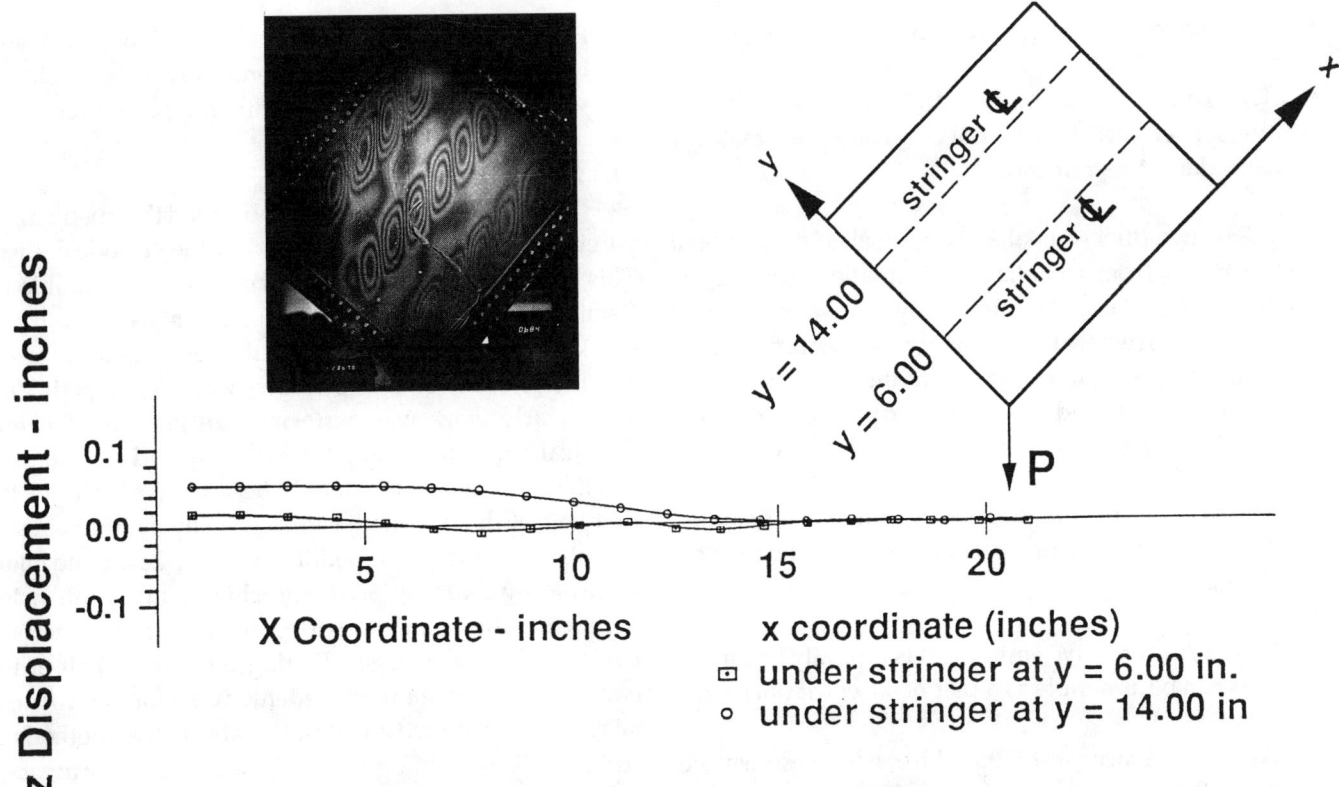

Figure 11-39. Displacements Under the Stringers

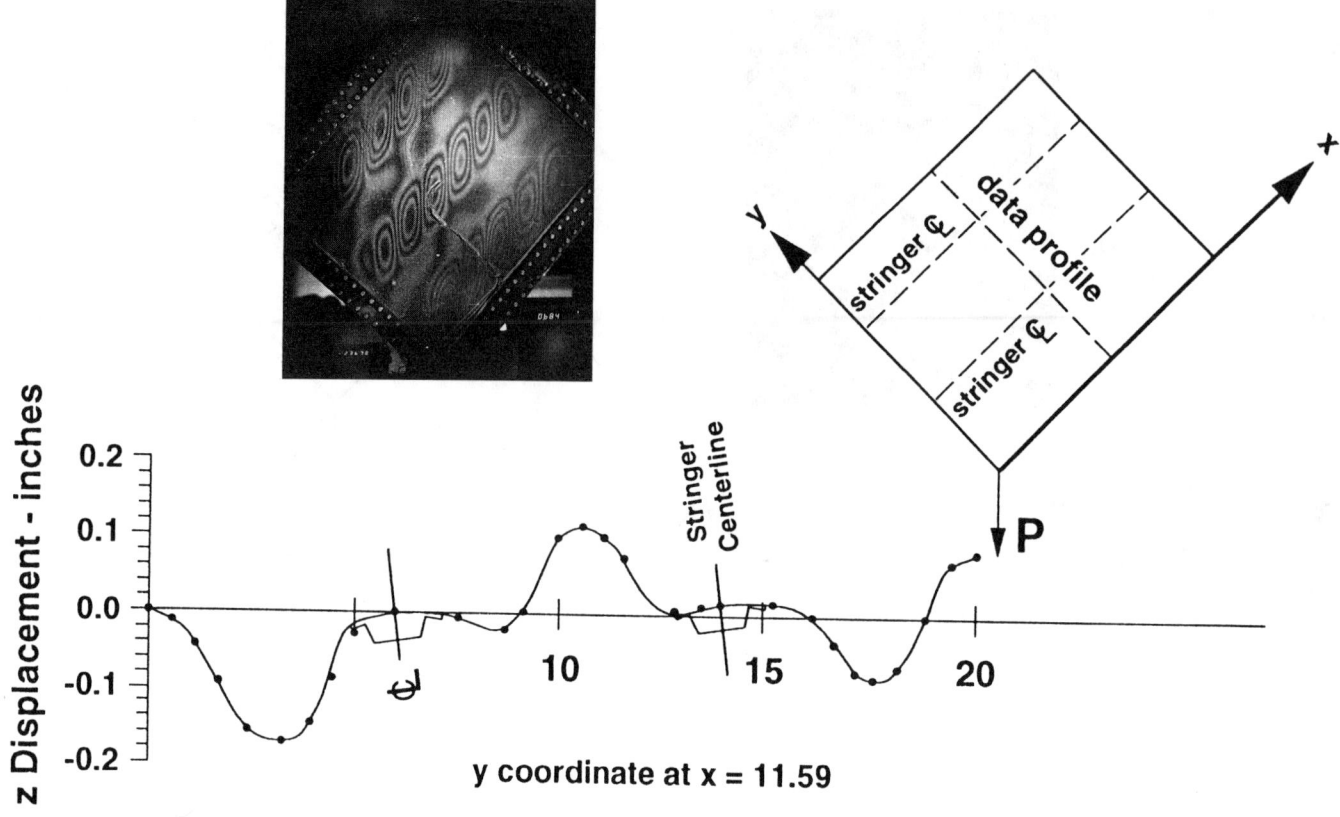

Figure 11-40. Displacements Normal to the Stringers

- Information is transmitted with the speed of light.

- They can frequently be used directly on the real structures in service. Thus it is not necessary to estimate the boundary constraints or loads.

Only two-dimensional, static photoelasticity is treated here. This far from trivializes the situation because the results apply to states of plane stress and plane strain, and all free surfaces (where the maximum stresses occur) are in a state of plane stress. In addition to the previously-cited advantages enjoyed by most optical methods, photoelasticity exhibits the following additional features:

- The most desired engineering information is obtained directly without differentiation or integration procedures.

- Simple, inexpensive equipment is typically adequate. This can be as simple as a pair of sheets of polaroid.

- Although space prohibits addressing these aspects here, the technique is also applicable to dynamic, inelastic, anisotropic, and thermal type problems.

Photoelasticity can often also be combined with analytical and/or numerical techniques to solve problems which cannot be solved by either approach alone.

11.4.2 Brief History

In 1816 Sir David Brewster observed that many transparent materials exhibited a load-induced optical effect. This optical effect, called "birefringence" or "double refraction," forms the basis for photoelasticity. In 1853, James Clark Maxwell related the optical phenomenon to stress and the so-called stress-optic laws were developed. Most of the early work was performed on glass models and practical applications of photoelasticity had to wait until suitable photoelastic materials became available. In the first half of the twentieth century, many synthetic resins and plastics were developed. The work of Coker and Filon, followed by contributions by Frocht and many others led to much enthusiasm and widespread use of photoelastic method of stress analysis. During the second half of the twentieth century the unprecedented development of digital computers offered an attractive alternative methods of stress analysis, although many believe this unfortunately detracted from the powerful array of experimental methods available.

11.4.3 Photoelastic Procedure
General Comments

Plane stress photoelasticity involves making a model, observing the laded model in a circular polariscope to determine the fringe order N at the critical area (or areas), and then computing the principal stress difference by Eq. 11-42.

$$\sigma_1 - \sigma_2 = Nf_\sigma/h \qquad \text{Eq. 11-42}$$

where σ_1 = maximum principal stress in the plane of the model
σ_2 = minimum principal stress in the plane of the model
N = the fringe order or relative retardation
f_σ = the material fringe value and is predetermined by calibration
h = the length of the light path within the model (usually the model thickness)

The model of an open-end wrench shown in Figure 11-41 illustrates the method. The maximum fringe order of Figure 11-41 is 7 at the location shown (the procedure for counting fringes will be described subsequently). Suppose that f_σ for the model material has been found by calibration to be 90 psi/fringe/inch, and the model thickness in 0.25 inches. By Eq. 11-42,

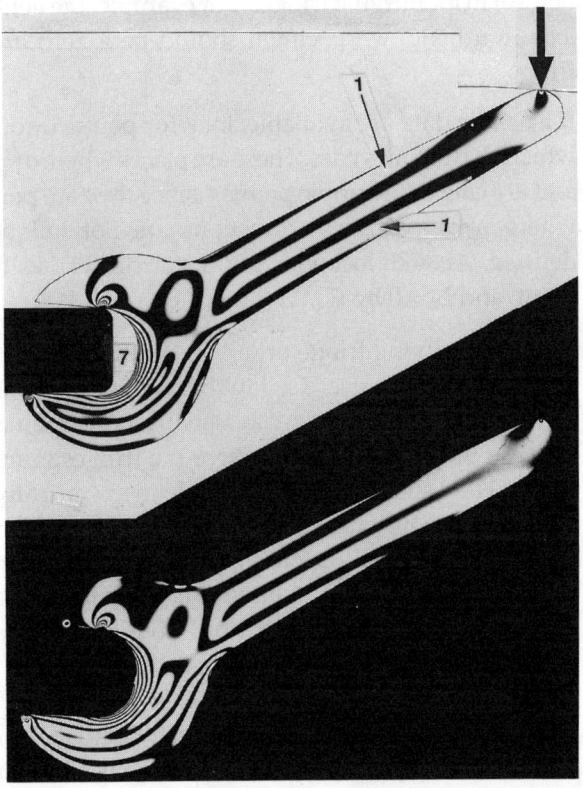

$$\sigma_1 - \sigma_2 = 7 \times 90 / 0.25 = 2520 \text{ psi}$$

Analysis of the forces acting on the wrench demonstrates that the inside surface is in tension and therefore

$$\sigma_1 = 2520 \text{ psi and } \sigma_2 = 0 \text{ psi}.$$

Such stress data is frequently all of the information required for the stress analyst.

Photoelastic Fringes

Most optical methods provide information in terms of interference patterns called fringe patterns. Photoelasticity gives *isochromatic fringes* and *isoclinic fringes*. Of these two types of fringes, the isochromatic fringes are by far the more useful and so they will be emphasized here. They are often called "photoelastic fringes" or just "fringes." A fringe is an area or band of constant color appearing on a model or on the image of a model. The analyst must interpret these fringes in terms of the desired information, usually the maximum stresses. Along a fringe the order N is constant, and therefore by Eq. 11-42, the principal stress difference ($\sigma_1 - \sigma_2$) is constant. If one finds the maximum value of N then the maximum value of $\sigma_1 - \sigma_2$ can be computed. As discussed previously, the maximum fringe order N_{max} almost always occurs at a free surface, where the shearing stress and normal stress are zero. Since the shearing stress is zero, the surface is by definition a principal plane and so the principal stress normal to the surface is zero. If one adopts the convention of $\sigma_1 > \sigma_2$, then in a region where the principal stress tangent at the boundary is zero and would be designated as σ_1. It follows that on an unloaded tensile boundary,

$$\sigma_1 = Nf_\sigma/h, \text{ and } \sigma_2 = 0 \qquad \text{Eq. 11-43}$$

whereas on an *unloaded boundary in compression*,

$$\sigma_1 = 0, \text{ and } \sigma_2 = -Nf_\sigma/h \qquad \text{Eq. 11-44}$$

It is therefore extremely easy to evaluate the principal stresses at a free surface and it merely requires determination of N at the critical point. Thus the most wanted information is readily obtainable. This fact explains one of the most significant advantages of the photoelastic method of stress analysis.

Counting Fringes

The chief task for the analyst is typically to find the location and magnitude of the maximum stress in the model. The amounts to determining the a maximum value of N, which may then be used in Eq. 11-42 to find the

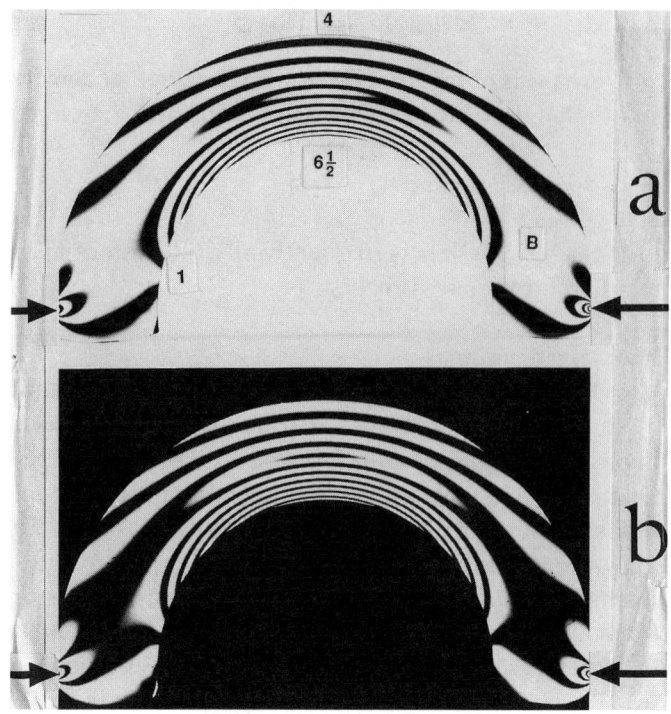

Figure 11-42. Isochromatic Fringe Patterns in a Horizontally Loaded Arch (a = light field, b = dark field)

maximum principal stress difference. Fringe orders occur in sequence. For example, next to a second order fringe must always be a first order, a third order, or another second order fringe (the latter occurs only in the neighborhood of a relative maximum or minimum). Fringe analysis usually begins by locating a zero order fringe, so that one may start counting from that point (e.g., the fringe adjacent to a zero order fringe is a first order fringe, followed by a second order, etc.). In the curved beam with opposed horizontal loads shown in Figure 11-42, the maximum fringe orders of interest occur at mid-span and are four (4) on the top surface and 6.5 on the bottom surface.

Identifying Fringe Orders with a Crossed Circular Polariscope

Methods of identifying the fringe order in a crossed circular polariscope include the following:

1. When using white light (or viewing a color photograph), any black isochromatic fringe is a zero order fringe.

2. An unloaded external corner is unstressed and therefore has N = 0.

3. An unloaded model should be completely covered by a large zero order fringe. A region that does not change color with load is a zero order fringe.

4. Start with an unloaded model as above and load the model very slowly, while observing where the first fringe order change in color occurs. This locates the position of maximum fringe order. Counting the number of color cycles (dark-to-light-to-dark) which occur while the load is gradually applied identifies the fringe order.

5. The reverse of the previous scheme may be used. Locate the point where the maximum fringe occurs. This is usually obvious when one views a fringe pattern. Gradually reducing the load to zero and counting the number of color cycle changes identifies the fringe order.

Identifying Fringe Orders in a Black and White Fringe Photograph

One can readily determine the fringe orders from a fringe photograph by the following steps:

1. First look at the filed (background area around the model). If the field is dark, the zero fringes will be dark. If the field is light, the zero order fringes will be light.

2. An unloaded external corner will be unstressed and will locate a zero order fringe.

3. Knowledge of elementary mechanics of solids is often helpful in identifying fringes. For example, the neutral axis in a bending specimen should be a zero order fringe.

4. If isoclinic data are available, look for points through which all isoclinics pass. These are places where $\sigma_1 = \sigma_2$ and are called "isotropic points," since they are points where principal stress directions are not uniquely defined. At such locations Mohr's circle shrinks to a point, and N = 0 by Eq. 11-42.

5. When identifying fringe orders in a photoelastic pattern, care must be exercised in the vicinity of maxima or minima. Those areas can usually be recognized because they are locations where the fringes branch. Figure 11-43 shows a curved beam loaded vertically at three points. The letter "B" on the fringe photograph in Figure 11-43 identify some fringes which branch and therefore deserve special attention. They usually are the fringes where one must stop "counting up" and start "counting down" or vice versa.

When white light is used, a crossed circular polariscope is almost always employed. For photographing fringe patterns, black and white film is most commonly used together with monochromatic light and a circular polariscope.

A light field (rather than a dark field) is often preferred so as to clearly define the edges of the model seen in Figure 11-41. In Figure 11-42b, where the tope of the visible fringe pattern at mid-span is white, it is not clear whether the area just above the white fringe is a dark fringe or is part of the background. If only Figure 11-42b were available, the fringe order could readily be reported incorrectly.

Figure 11-43 shows the curved beam illustrated in Figure 11-42, but with the loading direction changed. Whereas the beam of Figure 11-42 was loaded horizontally, that of Figure 11-43 is subjected to vertical three-point bending. The small "island" fringe in each model of these patterns in Figs. 11-42 and 11-43 is a zero order fringe loading the neutral axis. Other fringes are approximately parallel to the longitudinal axis of the beam. This is characteristic of members subjected to bending and is predicted by the flexural stress formula.

The model shown in Figure 11-44 illustrates an axially loaded horizontal member. Note that the "island" in the center of these patterns is <u>not</u> a zero order fringe; axially loaded members generally do not have a neutral axis. Free corners are more reliable locations of zero fringes that are "islands." The free corners are not visible in the dark field pattern of Figure 11-44a. This again illustrates why light filled fringe patterns are preferred by many experimentalists.

Fractional Fringe Orders

If dark fringes represent integral fringe orders with monochromatic light, then the light fringes represent the half orders. It is therefore easy to obtain the accuracy of at least a half fringe order and one can usually estimate to the nearest quarter fringe with difficulty. Associated optical methods have also been developed for measuring fractional fringe order. One such approach, proposed by Tardy, is popular because it does not require any equipment other than a circular polariscope. By the Tardy method, one can determine the fringe order within one fiftieth of a fringe using the human eye. Although the use of a photomultiplier tube can double this sensitivity, one fiftieth of a fringe is usually more than sufficient.

When one rotates the analyzer of an initially crossed circular polariscope 90°, the dark fringes move to the position formerly occupied by the light fringes, and vice versa. The Tardy method carries this idea further. Starting with a crossed circular polariscope and the polarizer parallel to one of the principal stresses, one rotates the analyzer until a dark fringe is move to the point under study. The fractional fringe order is then the angle of analyzer rotation (in degrees) divided by 180°.

Fringe Multiplication

It is difficult to obtain reliable results when inadequate fringes are available. A fringe multiplication technique developed by D. Post uses partial mirrors before and

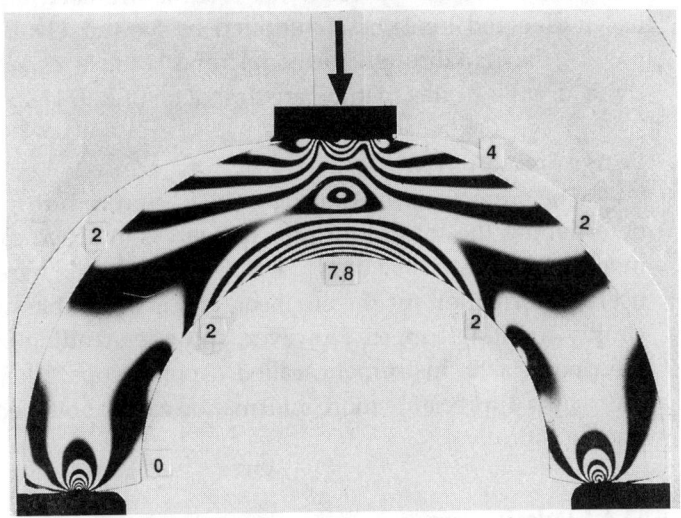

Figure 11-43. Isochromatic Frinte Pattern in an Arch Subject to Three point Loading Producing Vertical Bending (light field)

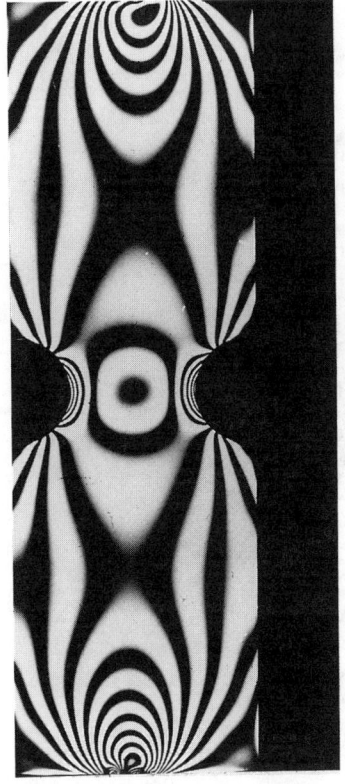
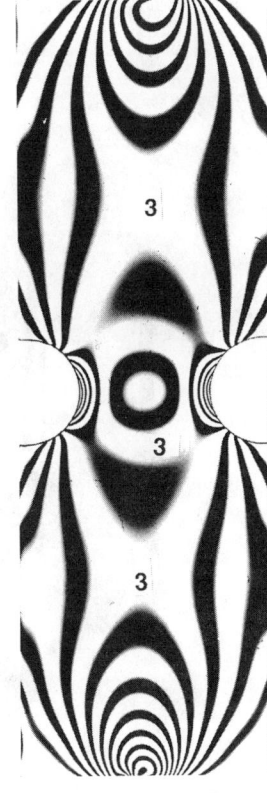

Figure 11-44. Isochromatic Fringes in a Vertically Loaded Member Containing Side Notches (a = dark field, b = light field)

after the model to pass the light through the photoelastic model several times. This has the effect of increasing the path length, and by Eq. 11-42 thereby increases the photoelastic effect. Fig. 11-45 illustrates this ability to increase the number of fringes. In this case the original fringe order (denoted by 1×) is multiplied by 5× and 11× by passing the light through the model 5 and 11 times respectively. Details of this method are described in Ref. 11-2.

Fringe Formation

The previous discussion describes the procedure for determining the maximum principal stress in a two-dimensional photoelastic mode. This is probably the most useful information for the engineer, and is attainable by simply counting fringes. However, an understanding of the photoelastic instrument, called a polariscope, illustrates how appreciably more information can be obtained photoelastically.

11.4.4 Polariscopes
General Comments

A polariscope includes linear polarizers and perhaps quarter-wave plates. A typical optical bench also includes a light source, a monochromator, lenses, a loading device, and a camera or viewing screen.

Light

Light is a periodic disturbance which travels through space with extremely high velocity. It will suffice here to think of light as a vector acting in the plane perpendicular to the propagation direction. For concreteness, we will designate the direction of propagation as the z-direction such that the *light vectors* act in the xy-plane. *Ordinary-light* contains many waves whose vectors act in various directions in the xy-plane.

Linear Polarizers

Linear polarizers pass one component of the light vector and absorb the other. For example, if the plane of polarization of the polarizer (this will be referred to simply as the plane of the polarizer) is the x-direction, then all of the light vectors would be resolved into their x- and y-components. The y-components would be absorbed and the x-components would be transmitted. The resultant light would contain only waves vibrating in the xz-plane, and would therefore be called *plane polarized light* (in the x-direction), or *linearly polarized light*.

Wave Plates

When a ray of plane polarized light enters a birefringent material, it is split into two rays. These two rays vibrate in mutually perpendicular planes, travel through the model with different velocities, and emerge from the model out of phase. The direction in which the faster wave vibrates is called the fast *axis* and the direction in which the slow wave vibrates is called the *slow axis*. The distance by which the slow wave lags the fast wave is the relative retardation δ (a distance).

Wave plates are birefringent materials. If the relative retardation of monochromatic light (single wavelength) of wavelength λ is $\delta = \lambda/4$, then the plate is called a *quarter-wave plate*, and if the relative retardation is $\lambda/2$, the plate is called a *half-wave plate*. The relative retardation Δ (in radians) is $\pi/2$, and π, for a quarter-wave and half-wave plate, respectively. It is not feasible to have a quarter-wave plate for white light, because may different wavelengths would be present and the δ (which is independent of wavelength) can be a quarter-wave of only one of them.

Quarter-wave plates are very useful in a photoelastic laboratory because they provide an inexpensive and convenient way to produce *circularly polarized light*.

Light becomes plane polarized when it is transmitted through a polarizer. When light is passed in succession through a polarizer and then through a quarter-wave plate, the existing light may be: (1) *plane polarized* (either the

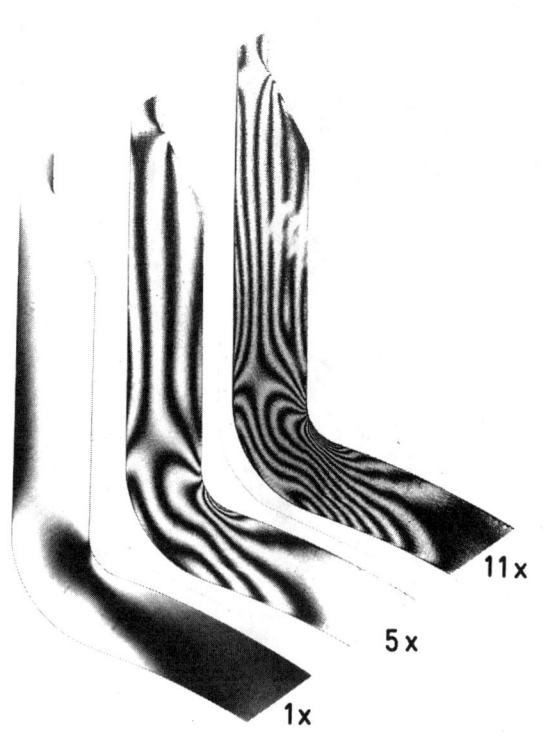

Figure 11-45. Application of Fringe Multiplication to a Slice Removed from a Three-Dimensional Photoelastic Model of a Pressure Vessel.

fast or slow axis of the quarter-wave plate is parallel to the plane of the polarizer), (2) *circularly polarized* (the fast and slow axes of the quarterwave plate make an angle of 45° with the plane of the polarizer), or (3) *elliptically polarized* (neither of the above conditions is satisfied).

Types of Polarizers

One common way to classify polariscopes is by the type of polarized light that they direct through a photoelastic model. *Plane polariscopes* pass plane polarized light through a model, and *circular polariscopes* provide circularly polarized light for the model. The optical systems are usually designed so that they can quickly be converted from a plane polariscope to a circular polariscope and vice versa.

Plane Polariscope

A typical plane polariscope consists of two identical linear polarizing elements as shown in Figure 11-46. The first element which is reached by the light is called the polarizer (P) and the last element is called the analyzer (A). The polarizer and analyzer are usually arranged in the "crossed" position so that the plane of the polarizer is perpendicular to the plane of the analyzer. M_1 and M_2 represent the components of light transmission through the photoelastic model.

Isoclinic fringes, called "isoclinics," as well as the isochromatic fringes can be viewed with a plane polariscope. If M_1 in Figure 11-46 is parallel to the polarizer, then the component M_2 is zero. The analyzer is perpendicular to M_1 and passes no light. The dark fringe which results is an isoclinic. It is the locus of points on the model where one of the principal stresses is parallel to the analyzer. For example, if the polarizer makes an angle of 30° with the x-axis (and the analyzer therefore makes an angle of 120°), the resulting fringe is called the 30° isoclinic, etc.

Circular Polariscopes

If the fast axis (F) of the first quarter-wave plate ($\lambda/4$) is parallel to the slow axis (S) of the second quarter-wave plate as shown in Figure 11-47, then this system is called a crossed circular polariscope because the polarizer (P) and analyzer (A) and the quarter-wave plates ($\lambda/4$) are crossed. Such a "crossed" position for quarter wave plates should always be used to minimize errors due to imperfections in the wave plates. This arrangement produces a dark field. Figure 11-48 shows a parallel circular polariscope (e.g., the polarizer (P) and analyzer (A) are parallel) (the quarter-wave plate remains crossed). Such an arrangement produces a light field.

Aligning a Polariscope

To align a polariscope, it is best to use a white light source. If the light is too bright, a piece of ground glass or a similar diffuser may be placed in front of the light source to render it more comfortable to the eyes. The procedure is as follows:

1. Place the polarizer and analyzer in the system While viewing the light source through both of these elements, rotate the analyzer until extinction of the light is obtained. This is a crossed position. Now try rotating the analyzer until maximum brightness is obtained. This is a parallel position. The human eye is much more sensitive to minimum brightness than to maxi-

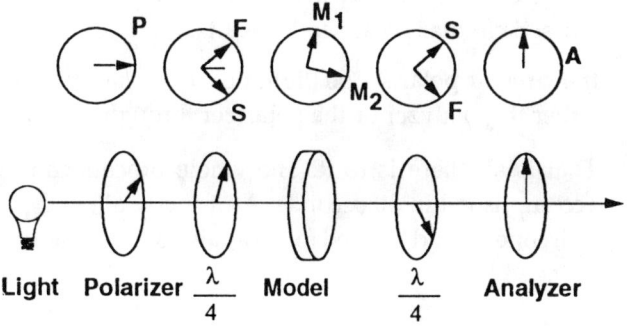

Figure 11-47. Crossed Circular Polariscopes

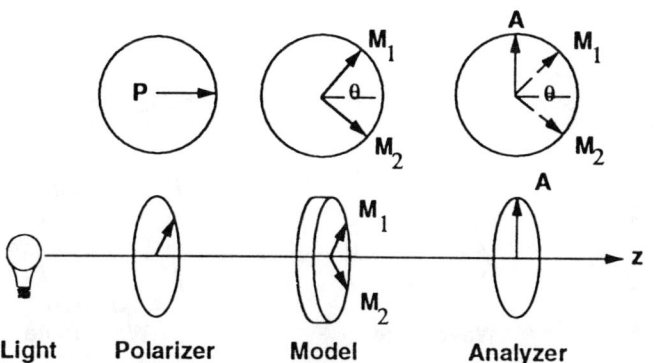

Figure 11-46. Plane Polariscope

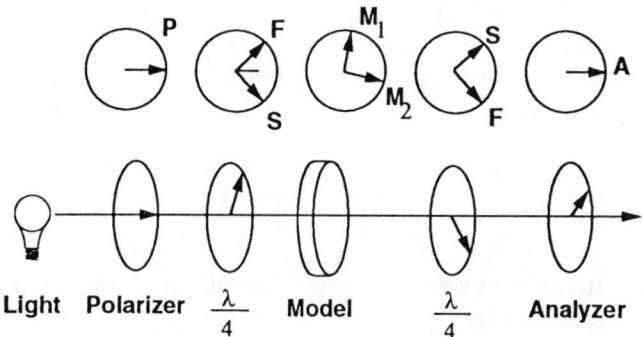

Figure 11-48. Parallel Circular Polariscope

mum brightness. Hence one should *always work with the extinction position rather than the position of maximum brightness.*

2. Extinction occurs any time that the polarizer and analyzer are crossed, but the angles they make with horizontal and vertical can be unknown. These angles can be determined by placing a model, whose principal directions are known, between the polarizer and analyzer. A simple tension specimen can be employed, but a circular disk is usually a more convenient model to use.

3. To align the quarter-wave plates, place one of them between a crossed polarizer and analyzer, and rotate the wave plate until extinction (or minimum intensity) occurs. Its fast axis is then parallel to the direction of transmission of either the polarizer or the analyzer. Leave this quarter-wave plate in place and put the second quarter-wave plate in the system. Rotate the second quarter-wave plate until extinction (or another relative minimum) occurs.

4. Next, rotate each quarter-wave plate through exactly 45° and observe the background. If it is dark, the system is properly set up as a crossed circular polariscope. If the field is bright, the quarterwave plates are parallel and not crossed as they should be. In this case, rotate either quarter-wave plate through 90° and a crossed circular polariscope results.

5. If a circular polariscope (light field) is desired, rotate either the analyzer or the polarizer through 90°.

6. That is all there is to it. The whole process can be accomplished in 30 seconds. A properly aligned polariscope should be used for all analysis and photography.

Lens Polariscopes

Polariscopes are also classified according to whether or not field lenses are employed. A lens polariscope typically includes a collimating lens, a field lens, and a focusing lens as shown in Figure 11-49. Parallel or collimated light passes through the model. The advantages of a lens polariscope include:

- It uses light efficiently. This becomes important for dynamic photoelasticity where short exposure times are frequently required.

- It can be used for fringe multiplication and fringe sharpening.

- A magnified image may be obtained.

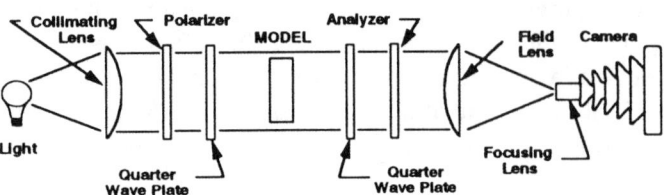

Figure 11-49. Lens Polariscope

Diffused Light Polariscope

Figure 11-50 illustrates a diffused light polariscope. In this case the light source is usually an array of fluorescent tubes, but it could be an ordinary incandescent light or a mercury source with ground glass diffusers placed in front of it. Photographs may be taken with a telephoto lens place several feet in front of the model. Light rays that are not traveling nearly parallel to the z-axis (as assumed in the theory) are simply not collected by the camera lens and are lost. This leads to inefficient use of the available light energy, and often long exposure times. The latter is not significant for static work but is important for dynamic analysis. The advantages of diffused light polariscopes include:

- They are inexpensive, since no lens are required.

- The are compact; the length of the optical bench may be as short as two feet. Thus automated systems are not essential since all elements of the system are within easy reach of the observer.

- Unlike lens polariscopes, they do not require a highly polished model.

Other Polariscope Components

In addition to the linear polarizers and quarter-wave plates, a polariscope typically includes a light source, lenses, a monochromator, a loading frame, and a camera or viewing screen.

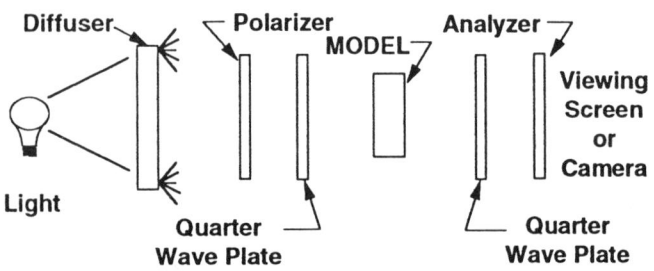

Figure 11-50. Diffused Light Polariscope

Light Sources

Mercury vapor light sources and lasers are used extensively for photoelasticity, but incandescent lights are adequate for static, two-dimensional studies. For a *lens polariscope*, the light source is placed at (or near) the focal point of the collimating lens so as to obtain collimated light. *Diffused light polariscopes* use extended light sources. An array of fluorescent tubes located behind a ground glass diffuser is commonly used for this purpose.

Lenses

With a lens polariscope, a *collimating lens* is usually placed so that the light source is at the focal point of the lens. Since light radiates out in all directions from an incandescent light source, the closer the lens is to the light source, the more light is collected. Short focal lengths usually means thick lenses, which can be expensive. High quality is not essential for collimator lenses, and strain-free lenses are not required as long as they are not located between the polarizer and analyzer. However, lenses should be of sufficient quality so that images of the model are not distorted. If plano-convex lenses are used for the collimating and the field lenses, the convex side of each such lens should be nearest to the model in order to minimize spherical aberration. *Focusing lenses* should be of high quality. For this, good camera lenses or enlarging lenses are satisfactory. Photography for a diffused light system is usually done using a telephoto (long focal length) lens.

Monochromators

Interference filters selectively pass wavelengths only in a narrow part of the spectrum and thereby transform white light into (sufficiently) monochromatic light. The wavelengths passed by the monochromator should match the wavelength for which that wave plates are quarter-wave plates. A monochromator can be placed anywhere in the light path, such as in front of a camera lens or in front of the eye when using a diffused light system. Care should be exercised not to place the monochromator too close to a strong light source, where it can be damaged by heat.

Magnification

With a lens polariscope the image is ordinarily observed on the ground glass of a view camera and the image size can therefore be controlled. Greater magnification may be obtained by: (1) moving the field lens closer to the model, (2) using a short focal length lens for the field lens, or (3) using a long focal length lens for the focusing (or camera) lens.

It is good experimental practice for the observer to be able to operate any controls that need to be adjusted. For example, in applying the Tardy method, the analyzer must be rotated until the center of a fringe is moved to the point under study. Yet in a simple lens polariscope, the observer may be several feet away from the model and analyzer. The difficulty is frequently overcome by providing a means for controlling the optical elements remotely from the observer's position. An inexpensive alternative is to use mirrors so that the image is formed on a screen close to the analyzer and the model. In such cases one should be certain that no mirrors are placed between the polarizer and the analyzer.

In a *diffused light polariscope*, the model is usually observed directly. Magnifying lenses may be used, but care should be exercised so that the direction of observation is parallel to the z-axis. Fringe orders formed by oblique rays are slightly different that those formed by normal rays. When photographing with a diffused light polariscope, telephoto lenses are usually used.

Loading Devices

Many loading devices are available. Forces may be applied and measured by dead loads, mechanical dynamometers, or electrical receivers. Electronic load measuring devices are excellent for a highly-active laboratory. Frequent calibration may be required but may well be worth the time an efforts because of the time saving they facilitate in the testing procedures. Mechanical dynamometers are a good compromise between the above mentioned other two approaches. Dynamometers are less expensive and require less frequent calibration than electronic systems, and are convenient to use. Dead loads, using weights and levers, are an inexpensive and reliable way to apply forces to a model. They generally do not need to be calibrated, even after long periods of inactivity.

Acknowledgments

Pat Grinyer proficiently typed this section. Robert Reese provided the line drawings and Dr. D. Post provided several of the fringe photographs.

11.4.5 References

1. E.G. Coker and L.N.G. Filon, *Treatise on Photoelasticity* Cambridge University Press, London, 1957.
2. M.M. Frocht, *Photoelasticity*, Vol. 1, J. Wiley and Sons, (5th Printing), New York, 1960.
3. M.M. Frocht, *Photoelasticity*, Vol. 2, J. Wiley and Sons, New York, 1948.
4. A.J. Durelli and W.F. Riley, *Introduction of Photomechanics*, Prentice-Hall, Englewood Cliffs, NJ, 1965.
5. A. Kuske and G. Robertson, *Photoelastic Stress Analysis*, J. Wiley and Sons, New York, 1974.
6. F. Zandman, A. Redner, and J.W. Dally, *Photoelastic Coatinas*, SEM (SESA) Monograph No. 3, Iowa State University, Ames, Iowa, 1977.
7. A.S. Kobayashi, Ed., *Handbook on Experimental Mechanics*, Prentice-Hall, Englewood Cliffs, NJ, 1987.
8. J.F. Doyle and J.W. Phillips, *Manual on Experimental Stress Analysis*, 5th Ed., SEM, 1989.,
9. J.W. Dally and W.F. Riley, *Experimental Stress Analysis*, 3rd. Ed., McGraw-Hill, New York, 1991.

11.5 PHOTOELASTIC COATINGS

Alex Redner, Ph.D., President - Strainoptic Technologies, North Wales, PA.

The finite element method and photoelastic modeling are effective methods of stress analysis, practical for solving problems where well-defined forces and boundary conditions can be applied to models on a simulated bases. These methods do not take into account such conditions as hidden material defects, assembly stresses, material behavior, and other parameters that are contributing factors to the structural integrity of a part or structure.

Structural tests using photoelastic coatings are conducted on real parts operating under actual service conditions. The coating reveals the true surface strains occurring on a part, even when the forces acting on the test item are not well known.

Photoelastic coatings are easy to apply and, with proper test planning, are very economical to use. Since a visible picture of the stress field is provided over the entire area coated, intelligent application of the coatings can save many hours of testing time and provide quick solutions to design or service/failure problems.

The principle of photoelastic coatings was first proposed by Mesnager in 1930 (Ref. 1). However, it was only after electrical-resistance strain gages were extensively used that photoelastic coatings generated interest, and the practical use followed (Refs 2, 3, and 4). Over the years, a substantial number of technical papers have been produced, establishing the technique as an important method for treatment of various strain-measuring problems. The references cited deal with the subject of accuracy of measurement and correction factors and methods of application and limitations. This reference list represents but a few of the volumes of work published on the technique which are available.

To obtain full economic and technical advantage, it is necessary to design a detailed test program that is best suited to accomplish the desired test objectives. The test program should include the following:

- Methods, materials, and procedures used to apply the coating to the test item.
- The type of data required to achieve the test objectives and give insight into the structural behavior so that problem can be solved.
- The loading sequence to be used.
- Provision to adjust the test procedures to meet the stated objectives.
- Analysis and presentation of the test data.

A poorly planned test program will usually generate technical difficulties, produce cost overruns, and will eventually fail in to achieve the test purposes.

The information provided in this handbook should help to select suitable techniques for a successful test.

11.5.1 Fundamental Relationships and Measuring Techniques

If ε_i is the principal strain on the surface of the structure, the principal strains in the coating are (see Figure 11-51):

$$(\varepsilon_i)_c = C_i (\varepsilon)_s \qquad \text{Eq. 11-45}$$

where C_i = a correction factor (Ref. 5) which is a function of of the coating thickness; this correction factor approaches 1.0 when the thickness of the coating is small.

The frequently used "bending correction" factors are shown in Figure 11-52, (Ref. 5)

As a result of strains in the coating, the index of refraction changes and the material becomes "birefringent," (e.g., light vibrating along minimum and maximum strains propagates at different velocities). The resulting retardation between light waves vibrating along ε_1 and ε_2 is related to strains and stresses in the coating by the Brewster Law.

Using a reflection polariscope arrangement shown in Figure 11-53, the optical retardation δ and the fringe order N are related to the principal strains ε_1 and ε_2 by Eq. 11-46:

$$\delta = 2tk(\varepsilon_1 - \varepsilon_2) = N\lambda \qquad \text{Eq. 11-46}$$

where λ = wavelength of light
  k = is the strainoptic material constant obtained by calibration
  t = thickness of photoelastic coating

The strains are then calculated by Eq. 11-47:

$$\varepsilon_1 - \varepsilon_2 = (\lambda/2tk) \cdot N = f \cdot N \qquad \text{Eq. 11-47}$$

where f = "fringe value" is a characteristic of the coating material

The light intensity I is modulated by the strains and their direction β by Eq. 11-48 (Ref. 3)

$$I/I_o = \sin^2 \pi\delta/\lambda \quad \text{(circular polariscope)} \qquad \text{Eq. 11-48}$$

where I = modified light intensity

I_o = original light intensity

$I/I_o = \sin^2 \frac{\pi\delta}{\lambda} \times \sin^2 2(\beta - \alpha)$ (plane polariscope)

Eq. 11-49

These expressions permit the interpretation of the observed black and color fringes.

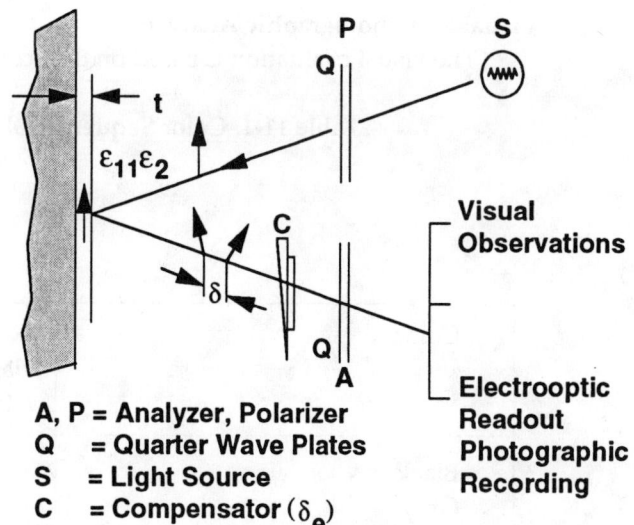

A, P = Analyzer, Polarizer
Q = Quarter Wave Plates
S = Light Source
C = Compensator (δ_e)

Figure 11-53. Reflection Polariscope Schematic

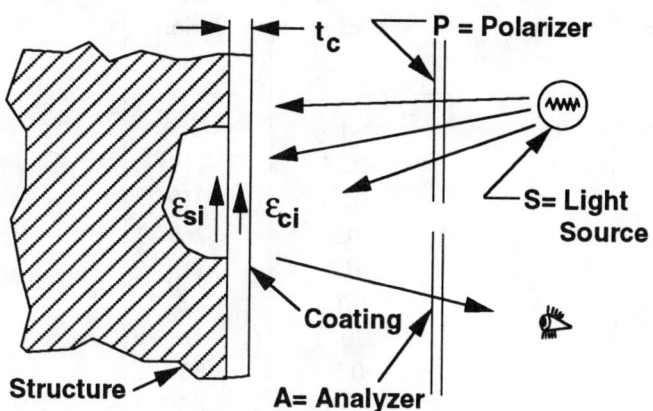

Figure 11-51. Principle Schematic for Photoelastic Coatings

Measuring Direction of Principal Stress

In a "plane" polariscope (quarter-wave plate removed), black "isoclinic" fringes are observed, whenever the direction of strains is parallel to the polarizers
($\beta = \alpha$, I = 0).
Common rotation of crossed polarizers moves these black fringes to any desired point making I = 0. The light intensity becomes zero when $\beta = \alpha$. Knowing the position of polarizers to a selected reference (vertical for example) quantifies the direction of principal strains (Fig. 11-54).

Measuring of Fringe Order N

The fringe order observed in a stressed coating can be evaluated visually, or measured using a suitable instrument. In a circular polariscope, the monochromatic light intensity becomes zero when the fringe order N is an integer (N = 0, 1, 2, ...). In white light colorful fringes are observed.

Reflection Polariscope - Compensators

The fringe order N can be measured using a compensator when it is introduced in the path of light. The compensator adds its own retardation δ_c to the measured δ (see Figure 11-53). When the total $\delta + \delta_c - 0$, a black fringe (I=0) is observed. Since the compensator is calibrated (e.g., δ_c is known), the value of δ is determined directly. The precision of the compensator is typical ±0.02 fringes.

Reflection Polariscope Analyzer-Rotation Methods

In a circular polariscope that includes quarter-wave plates, the fractional fringe order r (fraction of an order at a point located between fringes) can be measured using

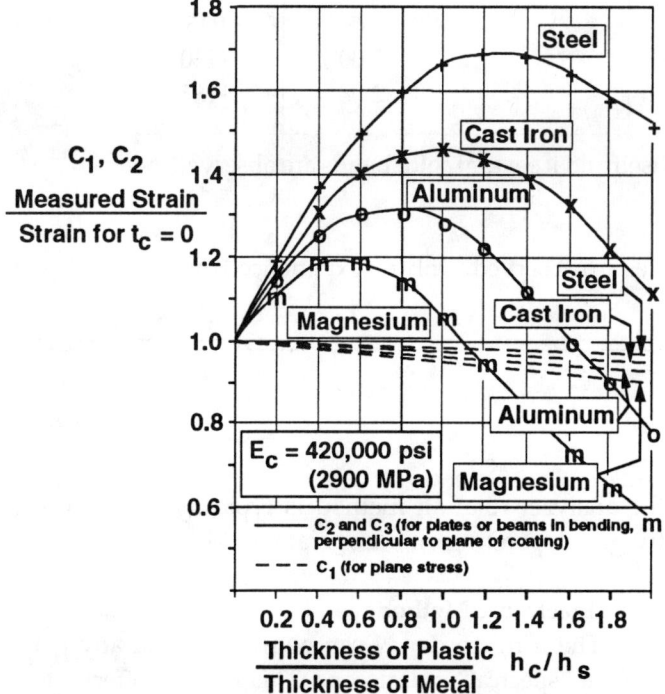

Figure 11-52. Plane-Stress and Bending Correction Factors C_1 and C_2

Visual and Photographic Analysis

The visual evaluation is based on the "color versus fringe order chart" given below:

Table 11-1. Color Sequence Observed versus Retardation Values

Color	Retardation λ (mm)	Fringe Order	Color Pattern Observed with Reflection Polariscope Color-Enhancing Plate	
			Fringe Order	Retardation λ (mm)
Black	0	0	-1	-585
Gray	150		-0.73	-415
White Yellow	250		-0.55	-315
Yellow	300	1/2	-0.5	-265
Orange (Dark Yellow)	450		-0.2	-115
Red	500		-0.11	-65
Indigo-Violet (1st Order Fringe)	570	1	0	0
Blue	600		0.5	35
Blue-Green	650		0.15	85
Green-Yellow	750		0.32	185
Yellow	850	1-1/2	0.50	285
Orange (Dark Yellow)	950		0.69	385
Red	1050		0.85	485
Indigo-Violet (2nd Order Fringe)	1100	2	1.00	565
Green	1300		1.29	735
Green-Yellow	1400		1.46	835
Pink	1500		1.64	935
Violet (3rd Order Fringe)	1695	3	2.00	1130
Green	1750		2.08	1185

The color sequence (yellow →red →green) repeats itself and if several color bands are observed, one must first identify fringe orders.

Results are obtainable only when several fringes are observed. Only an experienced operator should attempt to obtain quantitative data.

---

rotation of the analyzer method. If a point is located between fringe n and n + 1 and a rotation of γ brings the lower order n to a point; the fringe order N at this point is:

$$N = n + \gamma/180 = n + r \qquad \text{Eq. 11-50}$$

In order to use these methods (e.g., Tardy as shown in Figure 11-55), all elements must be carefully aligned, with respect to the direction of principal stress. The sensitivity of the analyzer-rotation method is typically ±0.01 to 0.02 fringes, depending on the operator's skill.

Opto-Electronic Methods

The fringe order N can be measured at any point using a "spectral" polariscope. This method offers a tenfold increase in sensitivity and eliminates the subjective judgmental factor of visual methods. The spectral polariscope (see Figure 11-56) uses a personal computer based

Experimental Methods—Full Field Techniques

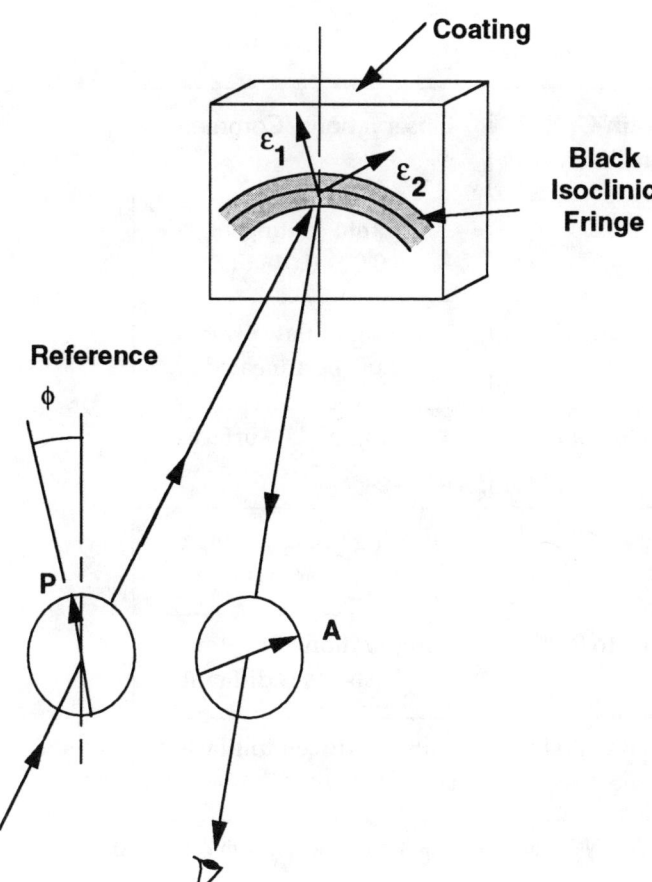

Figure 11-54. Formation of Isoclinic Fringes in a Photoelastic Coating

analysis of color and yields data independently of direction of principal stress. Table 11-2 summarizes the relative merits of various methods.

11.5.2 Application of Photoelastic Coatings to Structures

The selection of the coating material and its application to the surface of a structure is critically important in the photoelastic coating success. The coating materials should exhibit a good sensitivity, high "k" factor (strain optic material constant), good bondability to the surface, and capability of sustaining the prescribed strains without cracking. In coating plastic or rubber parts, low-modulus coatings must be selected to avoid reinforcing efforts.

Table 11-3 gives the typical material properties of coating materials.

Calibration

To ensure precision in the test measurements, the photoelastic materials should be calibrated. A simple calibration setup is shown in Figure 11-57. Here a small strip of the coating material is cemented to a cantilever beam clamped to a table or bench. The beam is loaded with dead weights. The stress in the beam (structure) is given by Eq. 11-51:

$$\sigma = 6 F L / (w\, t_s^2) \qquad \text{Eq. 11-51}$$

Table 11-2. Comparison of Measuring Methods

	Sensitivity Fringes	Coverage	Speed	Application	Limitations
Visual Observation	0.2	full field	visual	qualitative evaluation	at least one fringe must be visible
Photography & Video	0.2	full field possible	high speed	dynamic tests, full field	low resolution
Compensator	0.02	point	very slow	static test point data	slow, expertise required
Analyzer Rotation	0.02	point	slow	static test point data	slow, expertise required
Spectral Analyzer	0.002	point	millisecond	high precision static or dynamic conditions	point information expertise required

Table 11-3. Coating Materials*

Material	Typical Use	Strain-Optic Sensitivity K	Observations/Comments
Polyethermide	Flat Sheets	0.2	thin coatings 0.02 - 0.046 in.
Polycarbonate, Polysulfone	Flat Sheets	0.14 to 0.16	coating of flat surfaces must be annealed
Cast Epoxy Sheets	Flat Sheets	0.08 to 0.13	coating of flat surfaces
Polyurethane	Flat Sheets	0.01	coating of high elongation surfaces (tires)
Liquid Epoxy Resin	Coating of Curved Surfaces	0.08 to 0.11	preparation of contourable sheets is difficult
Prepolymerized Epoxy Sheets	Coating of Curved Surfaces	0.08 to 0.11	must be frozen until used

*Coating materials are available from Measurements Group (Raleigh, NC) and Strainoptic Technologies (North Wales, PA).

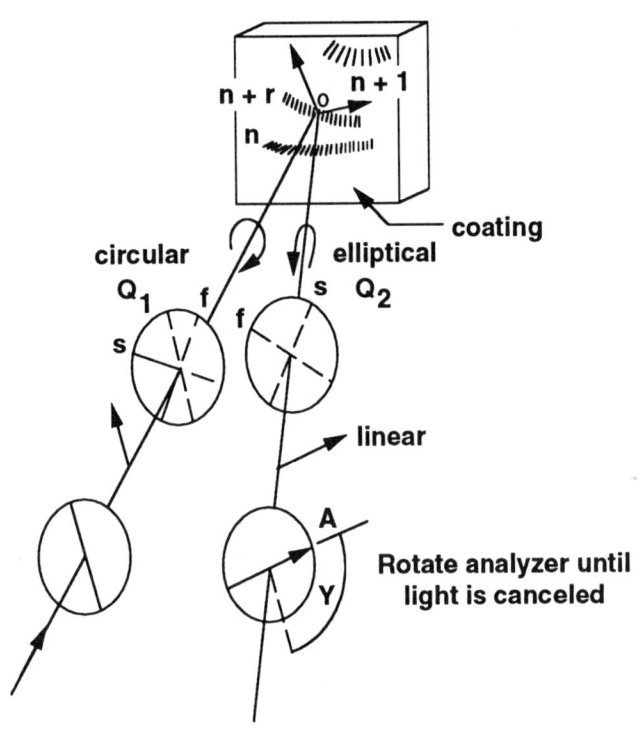

Figure 11-55. Analyzer Rotation in the Tardy Method to Adjust for γ Extinction

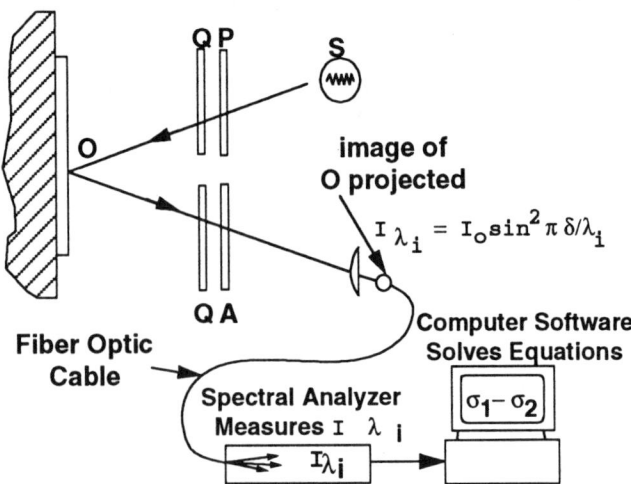

Figure 11-56. Spectral Photoelasticity Schematic

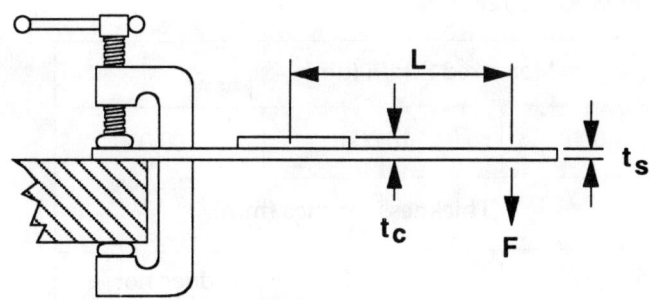

Figure 11-57. Calibration of Photoelastic Coating Material

where   F = applied force
L = distance from force location to center of photoelastic coating
w = width of the beam
t_s = thickness of the beam

and the difference in principal strains ($\varepsilon_1 = \varepsilon_2$) on the surface ($_s$) is:

$$(\varepsilon_1 - \varepsilon_2)_s = \sigma(1 + v) / E \qquad \text{Eq. 11-52}$$

where σ = from Eq. 11-51
v = Poisson's ratio for the beam
E = modulus of elasticity of the beam

The strains in the photoelastic coating must be computed using the correction factor C as given in Eq. 11-53.

$$(\varepsilon_1 - \varepsilon_2)_s = (\varepsilon_1 - \varepsilon_2)_s \times C = N\lambda / 2t_c k \qquad \text{Eq. 11-53}$$

where   N = measured fringe order
λ = wave length of light
t_c = thickness of coating
k = strainoptic material constant

Combining the above equations, the strainoptic material constant k is related to the measured fringe order N by:

$$k = E N \lambda \times w t_s^2 / [12 t_c (1+v) F \times L] \qquad \text{Eq. 11-54}$$

where E, N, λ, w, t_s, t_c, v, F, and L have been previously defined.

The force F should be applied in four or five increments, minimizing the uncertainty of individual measurements and offset due to an initial birefringence in the sample material.

Selection of Coating Thickness (Refs. 8 and 9)

The sensitivity of the coating is directly proportional to its thickness. Unfortunately, only thin coatings can conform to curved surfaces and produce accurate results, minimizing uncertainties due to the reinforcing effects and of the correction factor C. Logically, one should select the smallest thickness that provides sufficient sensitivity consistent with the test purposes and instrumentation employed. The minimum thickness is estimated, taking into consideration the measured strain, and the coating sensitivity k from Eq. 11-55:

$$t = \frac{\text{Desired minimum fringe order } N \times \lambda}{2k \text{ (nominal measured strain)}}$$

Eq. 11-55

The "desired" minimum fringe order is dictated by the sensitivity of the measuring instrument (see Table 11-2) and also by the precision requirements summarized in Table 11-4.

The visual and photographic methods can only provide qualitative information when thin coatings are used. The coating thickness needed when measuring using the reflection polariscope and point-measuring technique is shown in Table 11-5.

Table 11-4. Measuring Methods, Sensitivities, and Desired Fringe Orders

Measuring Method	Sensitivity of Measurement of N	Desired Fringe Order N for 1% Resolution
Visual and Photographic Evaluation	0.2	20
Compensator, Tardy, or Senarmont	0.02	2
Electro-optic Spectral	0.002	0.2

Table 11-5. Coating Thickness Requirements

Coating Type	Measuring Method	Desired Fringe Order	Measured Strain (μin)		
			1,000	10,000	100,000
			Thickness - inches (mm)		
Epoxy $k = 0.1$	Visual Electro-Optic	$N = 2$ $N = 0.2$	0.250 (6) 0.025 (0.6)	0.025 (6) 0.0025 (0.06)	does not apply
Polyurethane $k = 0.01$	Visual Electro-Optic	$N = 2$ $N = 0.2$	does not apply	0.250 (6) 0.025 (0.6)	0.025 (0.6) 0.002 (0.06)

As shown in Table 11-5, only high precision compensation methods, and spectral method offer sufficient resolution to accurately measure the fringe order N.

With the exception of those few cases where flat sheets are used, coatings are applied in most instances by the "contouring" method as shown in the photographs in Figure 11-58. These photographs show the following:

- Liquid resin is poured onto a flat plate
- The plastic form is lifted from the plate before fully hardened
- The plastic form is shaped to fit the contours of the test part and allowed to harden
- The plastic shell is cemented to the test part

While this method is very simple, some practical experience is required to become proficient in its application; it is not reasonable to expect a technician to achieve optimum results on their first attempt. Preparation and contouring of one sheet (typical size - 8 in. by 8 in. by 0.10 in. thick or 200 mm by 200 mm by 2 mm thick) can be accomplished in a few hours of work. On large structures, when several sheets are required, a carefully planned schedule must be outlined, not only for reasons of economy, but to ensure consistency of quality and calibration (Ref. 10).

Instead of the lengthy polymerization process, frozen sheets that are prepolymerized to the contouring stage are commercially available (Ref. 11). These sheets are preserved in a freezer and are ready for forming upon removal from the freezer and reheating to room temperature.

Cementing

Photoelastic coatings must be securely cemented to the structure, to ensure the proper strain transmission. In most applications, aluminum powder filled epoxy cement (Ref. 12) are most suitable, since addition to their cementing function, they also provide the needed reflectivity.

Room-temperature curing cements are most adequate, and typically require 12 to 24 hours to cure to reach sufficient strength and absence of creep. Fast curing and heat curable cements are not recommended, since they shrink during polymerization, developing stress in the coating material and initial fringes, that cannot be easily accounted for.

11.5.3 Separation of Principal Stresses

The photoelastic pattern observed in normal incidence of light (see Figure 11-51) reveals the following:

- The difference of principal stresses, $\sigma_1 - \sigma_2$, obtained by measuring the fringe order N.
- Direction of principal stresses, β, from observed isoclinic fringes.

When measuring uniaxial stresses, this information provides the complete solution of the following stress problems:

- Maximum strain: $\varepsilon = (Nf_\varepsilon) / (1 + \nu)$ (full field)
- Maximum stress: $\sigma = Nf_\sigma E_s / (1 + \nu)$ (full field)
- Direction: β = available directly in degrees at any desired point

where f_ε and f_σ are the material fringe values for strain and stress respectively

In biaxial stress field the two measured quantities (N, β) are revealing:

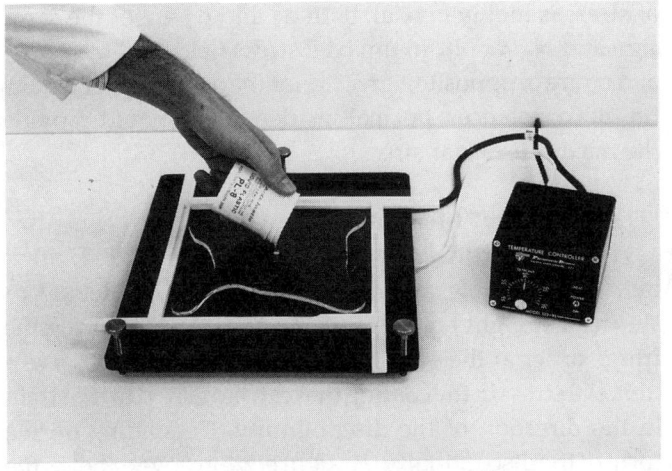
(a)

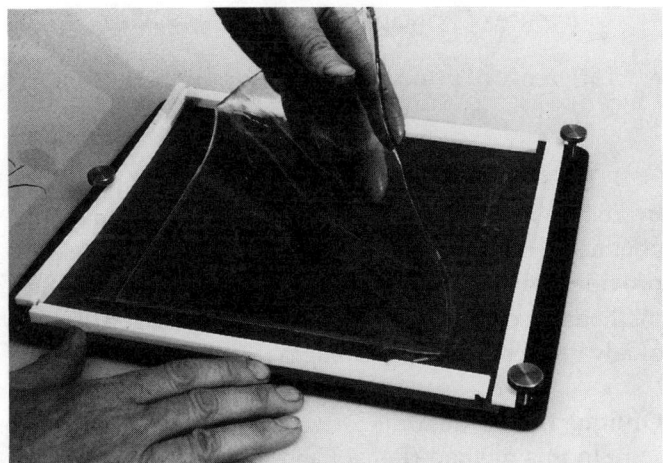
(b)

(c)

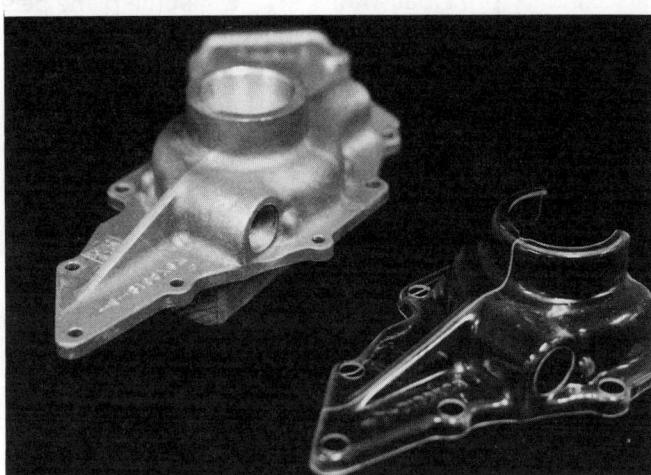
(d)

(e)

Figure 11-58. Calibration of Photoelastic Coating Material sequence of operations in applying a photoelastic ciating: (a) liquid resin is poured onto a flat plate; (b) plastic form is lifted from the flat plate before the form is fully hardened; (c) pastic form is shaped to the contours of the test item; (d) formed plastic prior to being cemented to the test item, and (e) photoelastic coating attached to test item schematic diagrams of oblique incidence polariscope.

- Difference of principal strains: $\varepsilon_1 - \varepsilon_2 = N \times f_\varepsilon$
- Difference of principal stresses: $\sigma_1 - \sigma_2 = N \times f_\sigma E / (1 + \nu)$

The stresses σ_1 and σ_2 can be "separated" providing σ_1 and σ_2 individually only if additional information is obtained. Several experimental procedures can be used to provide this additional (measured) information. The three methods described in this section are summarized in terms of advantages and disadvantages in Table 11-6.

Oblique Incidence

In this method (Ref. 13 and 14), light is directed to cross the coating at an angle D to the normal and the measured fringe order ND is proportional to the difference of secondary principal strains, $\varepsilon^1_x - \varepsilon^1_y$ in the plane perpendicular to the ray as shown in Figure 11-59. The fringe order is measured twice; one measurement is performed in the normal incidence (yielding N_n) and another measurement in the oblique incidence plane using an incidence angle θ (giving N_θ). From these two measurements, the principal strains ε_1 and ε_2 and stresses σ_1 and σ_2 are computed (Ref. 14) as given in Eq's. 11-56 and 11-57:

$$\varepsilon_1 = f_\varepsilon \left[N_\theta (1-\nu)\cos\theta - N_n (\cos^2\theta - \nu) \right] / 2t(1+\nu)\sin^2\theta$$
Eq. 11-56

$$\varepsilon_1 = f_\varepsilon \left[N_\theta (1-\nu)\cos\theta - N_n (1 - \cos^2\theta) \right] / 2t(1+\nu)\sin^2\theta$$
Eq. 11-57

This method yields satisfactory results only when the state of stress is highly biaxial, both σ_1 and σ_2 are of the same sign and $N_n \ll N_\theta$. In uniaxial stress field, and in case σ_1 and σ_2 are of opposite signs, the method yields poor results (in those instances, normal incidence N_n should provide the maximum shear stress).

"Discontinuity" Method

Here an artificial edge in the coating is created, either by drilling a hole in the coating (Ref. 15, O'Regan) or slitting a straight groove (Refs. 16 and 17). Measuring the fringe order at the edge of the groove, one can observe a uniaxial stress in the coating that can be related to the strain in the direction of the discontinuity. Depending on the form of the discontinuity (straight edge groove or hole), the geometry of the cut influences slightly the results and correction factors shown in references should be used if accurate results are needed (Ref. 17).

Strain Gages

In some applications, strain gages cemented to the photoelastic coating can be used to provide the information required for a complete solution at a point. The strain gage should be cemented along the principal axes and a low-current strain-gage indicator used, since the self-hearing becomes a serious problem when the gage is not in contact with the structure. Correction factors are also needed to account for the increased distance to the neutral plane and local reinforcing effect (Ref. 18).

11.5.4 Test Report Preparation Guidelines

The test report is an important document since it contains the information gained in testing. One should not

Table 11-6. Summary of Selection of Stress-Separation Methods

Method	Advantages	Disadvantages
Oblique Incidence	• Coating is not destroyed • Accurate in biaxial field • No addition or recalibration required	• Poor precision in uni-axial field • Additional instrument required • High precision measurement needed
Discontinuity	• No additional instrumentation needed • Low cost • Expedient	• Cutting skill needed • Coating destroyed locally
Strain Gage	• Electrical readout	• High cost • Strain gage readout required • Complex corrections required

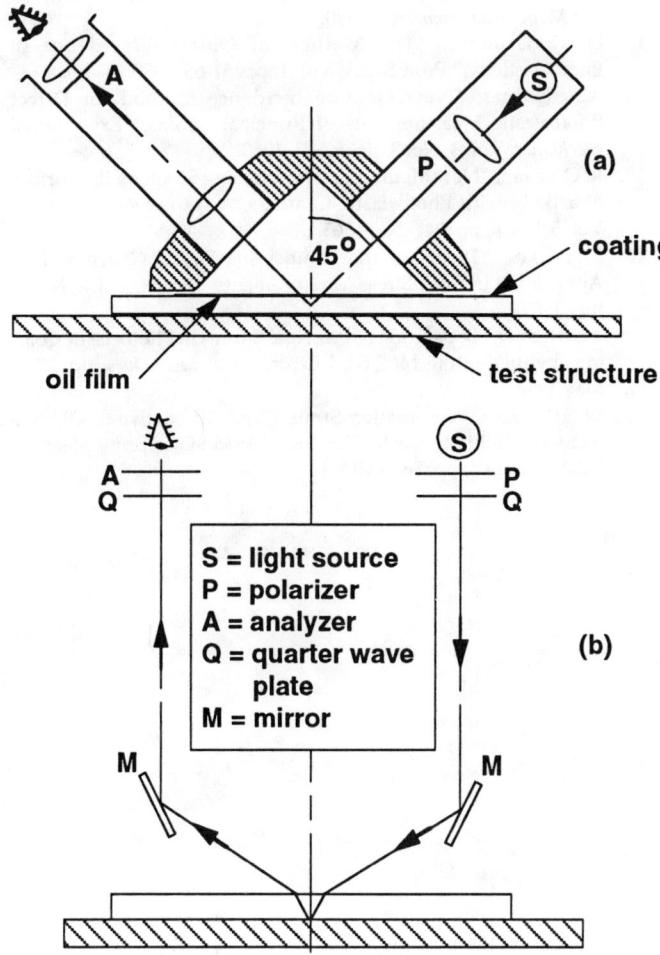

Figure 11-59. Schematic Diagrams of Oblique Incidence Polariscope

loose sight of this key fact, and should concentrate the test effort on tasks that will yield information that will be included in the test report. The report should contain the following information organized in a suitable chronological sequence.

1. Test Objectives: the purpose of the test, its expectations and objectives.

2. Description of the Test Item or Part: the structure should be described including drawings, test specifications, test plans, material property information, and related information.

3. Applications of Photoelastic Coating: the selection of coating material, application method, cementing procedure, and calibration results should be reported. Also, it is desirable to obtain photographs showing the coated areas and location of where data was obtained.

4. Test Program and Loading Sequence: the test dates, schedules, loading systems, and controls used should be discussed. Special tooling and fixtures should also be described. Loading equipment and sequences (e.g., rates, holds, cycles) should be documented.

5. Data: The test data (raw) should be reported as measured together with a sample data reduction calculation, formulas used, and stresses measured. All photographs of the fringe patterns observed should be carefully marked up with the loading conditions used to avoid mistakes in interpretation.

6. Conclusions: the conclusions should be based on the reported test results, visual observations not included in the photographs, findings not anticipated in the original test objectives, and suggestions based on the test results.

11.5.5 References

1. M. Mesnager, "Sur la Determination Optique des Tensions Interieure dens les Solides a Trais Deminsion, *Comptes Rendue*, Paris, Vol. 190, p. 1149, 1930.
2. F. Zandman, A.S. Redner, and J. Dally, *Photoelastic Coating*, SESA Monograph No. 3, Iowa Press, Ames, Iowa, 1977.
3. F. Zandman, A.S. Redner, and E.I. Reigner, "Reinforcing Effect of Birefringent Coatings," *Experimental Mechanics*, Vol 2, No. 2, pp. 55-64, 1962.
4. F. Zandman and M. Wood, "Photostress Analysis," *Product Engineering*, March 1959, pp. 167-178.
5. H.T. Jessep, "On the Tardy and Senarmont Methods of Measuring Fractional Relative Retardations," *British Journal of Applied Physics*, Vol. 4, pp. 138-141, May 1953.
6. W.E. Nickola, "Summation Strain-Gage Alternative to Oblique Incidence in Photo-elastic Coatings," 1985 SEM Spring Meeting, Las Vegas, NV, pp. 869-880.
7. A.S. Redner, "Photoelastic Measurements by Means of Computer-Assisted Spectral Contents Analyzer," *Experimental Mechanics*, Vol. 25, No. 2, pp. 148-153, 1985.
8. D. Post and F. Zandman, "Accuracy of Birefringent Coating Method for Coating of Arbitrary Thickness," *Experimental Mechanics*, Vol. 1, No. 1, pp. 21-32, 1961.
9. A.S. Redner, "How to Select Photoelastic Coatings," *Materials in Design Engineering*, Sept. 1964.
10. A.S. Redner, "Photoelastic Coating," *Experimental Mechanics*, Vol 20, No. 11, pp. 403-408, Nov. 1980.
11. _____, Instruction Manual, "Application, Forming, and Cementing Strainoptic™ Photoelastic Coating Frozen Sheets," Strainoptic Technologies, North Wales, PA.
12. _____, Instruction Manual, "Instructions for Bonding Flat and Contoured Photoelastic Sheets to Test Part Surfaces," Measurements Group, Raleigh, NC.

11.6 REFERENCES

1. E.G. Coker and L.N.G. Filon, Treatise on Photoelasticity. Cambridge University Press, London, 1957.
2. J.W. Dally and W.F. Riley, *Experimental Stress Analysis*, 3rd. Ed., McGraw-Hill, New York, 1991.

3. J.F. Doyle and J.W. Phillips, *Manual on Experimental Stress Analysis*, 5th Ed., SEM, 1989.
4. A.J. Durelli and V.J. Parks, *Moiré Analysis of Strain*. Prentice-Hall, Englewood Cliffs, NJ, 1970.
5. A.J. Durelli and W.F. Riley, *Introduction to Photomechanics*, Prentice-Hall, Englewood Cliffs, NJ, 1965.
6. M.M. Frocht, *Photoelasticity*, Vol. 1, John Wiley and Sons, (fifth printing), New York, 1974.
7. M.M. Frocht, *Photoelasticity*, Vol. 2, John Wiley and Sons, New York, 1948.
8. E. Hecht, *Optics*, Addison-Wesley, pp 112-113, Reading, MA, 1987.
9. A.S. Kobayashi, Ed., *Manual on Experimental Stress Analysis*, Chapter 4, "Moiré Analysis of Strain," Chapter 5 "Photoelasticity," Chapter 7, "Moiré Interferometry," SEM, Bethel, CT, 1970.
10. A.S. Kobayashi, Ed., *Handbook on Experimental Mechanics*, Prentice-Hall, Englewood Cliffs, NJ, 1987.
11. A. Kuske and G. Robertson, *Photoelastic Stress Analysis*, John Wiley and Sons, New York, 1974.
12. W.W. Stinchcomb, Ed., *Mechanics of Nondestructive Testing*, "Optical Interference for Deformation Measurements - Classical. Holoaraphic. and Moiré Interferometry," 1980.
13. D. C. Drucker, "The Method of Oblique Incidence in Photoelasticity," Proc. SESA, Vol. 1, pp. 51-66, 1950.
14. A.S. Redner, "New Oblique Incidence Method for Direct Photoelastic Measurements of Principal Strains," *Experimental Mechanics*, Vol. 3, No. 3, pp. 67-72, 1963.
15. R. O'Regan, "New Method for Determining Strain on the Surface of a Body with Photoelastic Coatings," *Experimental Mechanics*, Vol. 5, No. 8, pp. 241-246, 1965.
16. I. Hawkes, "Theory of the Photoelastic Biaxial Gauge and Its Application in Rock Stress Measurements," *Strain*, Vol. 3, No. 3, July 1967.
17. A.S. Redner, "Separation of Principal Strains in Photoelastic Coatings by the Slitting Method," *Experimental Techniques*, pp. 2932, May 1987.
18. W.E. Nickola, "Summation Strain-Gage Alternative to Oblique Incidence in Photo-elastic Coatings," 1985 SEM Spring Meeting, Las Vegas, NV, pp. 869-880.

Chapter 12
Experimental Methods—
Nondestructive Evaluation Techniques

12.1 INTRODUCTION

This chapter contains four main parts: acoustic emission, ultrasonic methods, impact-echo techniques, and real-time radiography. Nondestructive techniques are not limited to these four methods. The intent of this chapter is to present information useful to structural testing which is not available in other texts or references. The section on acoustic emission is a tutorial which gives significant background and basis for this method. This method is not available in other publications within the Society for Experimental Mechanics. The ultrasonic method is developed for direct application to structural testing. The impact-echo and real-time radiography techniques have direct application to structural testing as well.

12.2 ACOUSTIC EMISSION, PRINCIPLES AND INSTRUMENTATION

Alan G. Beattie, Ph.D., Nondestructive Technology, Electromagnetics, and Optics Department, Sandia National Laboratories, Albuquerque, New Mexico. Reprinted by permission from the Journal of Acoustic Emission, V.1.2, No. 1/2

12.2.2 Introduction

Acoustic emission (AE) can be defined as acoustic waves generated by a material when subjected to an external stimulus. The current technological uses of the phenomena include flaw detection and early warning of structural failure and experimental studies of materials in many areas. However, as a useful tool, acoustic emission predates the human race. Most land animals are aware of the implications of the sounds of breaking wood. The first acoustic emission instrument then is the ear, human or animal. Even in this day of electronics, the ear is still a primary instrument and many experienced investigators will include an audio channel on acoustic emission tests.

Despite the convenience and sensitivity of the natural instrumentation, there is always a desire to extend the range of instrumentation and quantify the results. In acoustic emission, as in other fields, this has led to a large number of new measurement techniques as well as to claims and counter claims as to the effectiveness of these techniques. This is normal in any growing field but tends to leave the newcomer to the field at a disadvantage. The primary source of information is often the advice in brochures of the instrument manufacturers. These sources, which are generally quite honest for advertising, tend to stress the capabilities of the individual instrument and say very little about what should be or is being measured. Thus, the newcomer, with good faith on all sides, often ends up with a beautiful piece of instrumentation which is not particularly applicable to his needs.

To intelligently use instrumentation, one must first understand what is being measured. Therefore, the first half of this chapter considers what acoustic emission is, how it is produced and how the sample and the detection effect the final signal. The second half describes most of the currently used acoustic emission instrumentation techniques. Emphasis is placed upon what parameter of the acoustic emission signal is being measured and how it relates to the acoustic emission. Models rather than mathematics are used throughout with equations included primarily to illustrate the various points.

To keep this chapter to a reasonable length, it has been limited to analysis techniques used for signals from a single sensor. The vast topic of location of acoustic emission sources has been left out. It is hoped, despite this limitation, that the information presented here will allow the reader to make an informed decision as to what instrumentation or analysis technique to use in a given situation.

The information presented in this chapter has been developed and checked by the author. No attempt has been made to extensively document the development or usage of the different techniques. Instead, the researcher is referred to the following books and articles. An excellent history of the field is contained in the introduction to Drouillard's Bibliography (1979). The reviews of Liptai, et al. (1979), Spanner (1974) and Lord (1975) cover much of the early development work and applications. Many applications to structures are covered in the books by Williams (1980) and Nichols (1976) and in three ASTM Special Technical Publications (1972, 1975, 1979). Relatively current work in the field is covered by papers in two conference proceedings (Dunegan and Hartman, 1981; Nicoll, 1980). Several proceedings of biennial symposia held in Japan (1974, 1976, 1978, 1980, 1982) are also helpful in many areas.

12.2.2 Principles of Acoustic Emission

The terms "acoustic emission" and "acoustic emission signal" are often used interchangeably. Strictly speaking, an acoustic emission is an acoustic wave generated by a material and an acoustic emission signal is the electrical signal produced by a sensor in response to this wave. Acoustic emission instrumentation analyzes the signal and from this analysis, attempts to learn about the material which generated the emission. The characteristics of the signal are determined by the mechanism which generated the emission, the means by which it travels through the material and the sensor which transforms the emission into the signal. In this section, generating mechanisms, wave propagation and sensor characteristics shall be discussed before the electrical signal is studied.

Generating Mechanisms

The fundamental difference between acoustic emission and the field known as ultrasonics (Hueter and Bolt, 1955; Blitz, 1967), is that acoustic emission is generated by the material itself, while in ultrasonics the acoustic wave is generated by an external source and introduced into the material. In acoustic emissions, the investigator has no control over the sound generation mechanism but can only subject a material to conditions which will cause the production of acoustic emission. Typically, these conditions produce a static imbalance in the material. They may be a high stress level or a temperature which produces an instability in the phase of the material. In all cases, acoustic emission is generated by the dynamic processes by which the material attempts to return to a state of equilibrium.

Types of Mechanisms

Acoustic emission can be generated by a large number of mechanisms, Some of these are: fracture of crystallites; crack nucleation and growth; several mechanisms involving dislocations; fracture of inclusions in a material; phase transformations in solids; boiling; and electrical discharges. All of these mechanisms are characterized by a rapid collective motion of a group of atoms. To generate a propagating acoustic wave, the motion must occur at a velocity close to an acoustic velocity of the material. While the minimum necessary number of atoms involved in the collective motion cannot be defined, it can be stated that motions of individual atoms, such as those occurring in a diffusive phase transition, will not produce acoustic emission.

Real Materials—Inhomogenity and Randomness

All real materials are inhomogeneous on a microscopic scale. Even a single crystal will have uneven distribution of dislocations. A real metal is composed of crystallites with random orientations and a range of sizes. These crystallites may also be of various metallic phases and compositions. To see the effect of such inhomogeneity on the generation of acoustic emission, take as a model a metal with a highly anisotropic crystal structure. Crystals of this metal will cleave easily on one plane and with difficulty on all others. Now apply a tensile force to each end of a polycrystalline bar of this metal. Inside the bar, some of the randomly aligned crystallites have an orientation in the local stress field which minimizes the stress necessary to cleave them. The local stress fields will also vary on a microscopic scale because of the anisotropy in the elastic properties of the crystallites. As the force is increased, the stress will reach a level where the most favorably oriented crystallite fractures. The fracture will produce an abrupt change in the local stress field surrounding the crystallites. This change will propagate into the metal away from the crystallite as an acoustic wave. It may also cause other nearby crystallites to cleave almost simultaneously with the original one. The amplitude, directionality and polarization of the resulting acoustic wave will depend both upon the size and orientations of its neighbors. Thus, the exact time of occurrence of the acoustic wave and its characteristics are functions of the local environment of the fracturing crystallite. The random variations in these environments produce an uncertainty in the times of occurrence of the acoustic waves. Another important point in this model is that the acoustic emission is produced by the fracture of individual crystallites and therefore occurs in the form of discrete packets of acoustic waves. These two characteristics, discrete packets of acoustic energy with no correlation either in time or in the characteristics of the packets, are the fundamental nature of acoustic emission. As such they determine what kind of instrumentation can be used and are the reason for the primarily statistical approach of acoustic emission analysis.

Burst and Continuous Emission

In the model, packets of acoustic waves were produced when the stress in the material exceeded the level necessary to fracture the crystallites. There was no other correlation in their time of occurrence. These discrete packets of acoustic energy are known as burst emission and their lack of correlation in time precludes the concept of a repetition rate. However, it is possible to define an average repetition rate as the average rate of occurrence of the bursts over some period of time. These average repetition rates differ widely for different systems. The lowest measured rates are for earthquakes which may occur a century apart on a given fault. The highest measurable rate is around 50 kHz. Above 50 kHz, the length of the bursts usually exceed the time interval between them. The superposition of the bursts produces continuously occurring acoustic waves which are called continuous emission. The only difference between burst emission is the average repetition rate. Both are generated by discrete processes. Figure 12-1 shows oscilloscope traces of burst emission and continuous emission signals.

The amplitude of continuous emission is usually much lower than that of burst emission. Generally, the amplitude of an acoustic emission burst will have some correlation with the volume of the region producing it. Since continuous emission is a superposition of a great many bursts, the volume of each region producing a burst is necessarily small. Therefore, the amplitude of each burst and thus the continuous emission is low.

It should be noted that an acoustic signal, similar to continuous emission, is produced by a gas or a fluid passing through a small opening. Such signals are the basis of acoustic emission leak detection. It is often impossible to distinguish a continuous emission signal from a leakage signal even though their origins are quite different.

12.2.3 Acoustic Waves

The link between the source and the acoustic emission signal produced by the sensor is the acoustic wave. Much of the complexity seen in acoustic emission signals is generated as the wave travels through the medium. Any understanding of acoustic emission signals requires a knowledge of the characteristics of acoustic waves (Kolsy, 1963). The place to start is the properties of the medium through which the wave travels.

All materials are collections of atoms. The atoms are held together by attractive forces but also prevented from approaching each other too closely by short range repulsive forces. The superposition of these forces results in an equilibrium position for the atom where it has its lowest energy and therefore the most stable configuration of the material. However, only at a temperature of absolute zero are the atoms actually at rest in these equilibrium positions. As the temperature is increased, the atoms gain some energy and perform pseudo-oscillatory or orbital motion about their equilibrium positions. The kinetic energy of the

Figure 12-1. (a) Oscilloscope Trace of Preamplified Noise Level; (b) Burst Emission; (c) Burst Emission Mixed with Continuous Emission; (d) Continuous Emission.

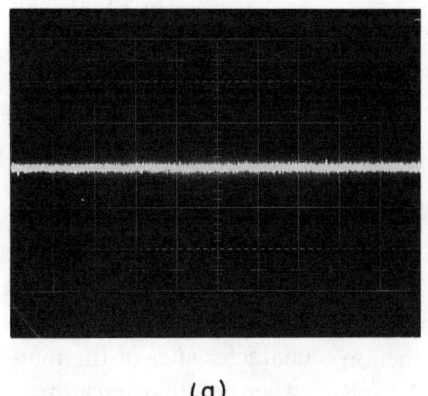

(a)

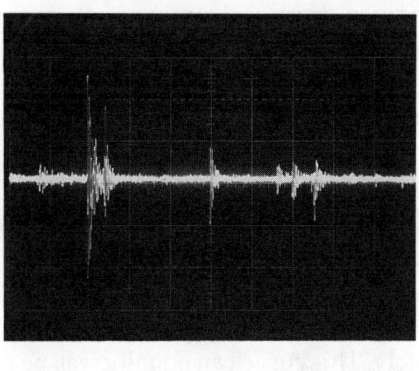

(b)

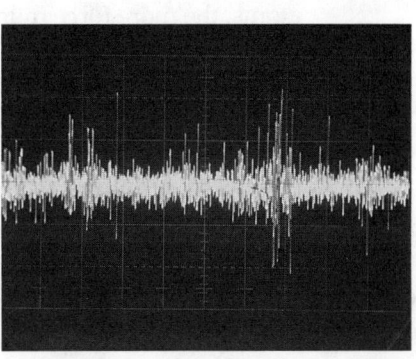

(c)

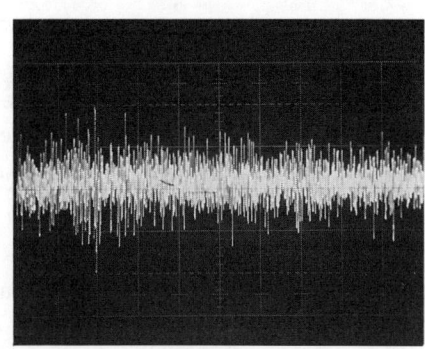

(d)

atoms is the thermal energy. The relationship between the magnitude of the thermal energy and the strength of the attractive forces determine whether the material is a solid, liquid or gas. In solids, the equilibrium positions retain a fixed geometrical relationship among themselves. The geometry of this relationship determines the crystal structure of the solid.

Because of the forces between the atoms the motions of adjoining atoms must be coupled in some fashion. This coupling allows the transfer of kinetic energy between atoms. For the higher energy (high frequency) components of the atomic motion, this coupling is weak and the correlations between the atomic motions may only extend over near neighbors. This type of short range correlated motion results in a basically diffusive transfer of energy in the material and is known as thermal conduction. However for very low frequency atomic motions the coupling may extend over a great many atoms. This long range correlated motion of atoms is an acoustic wave.

The long range correlations in an acoustic wave result in many atoms in a small region being displaced in the same direction from their equilibrium positions. This displacement is a local strain in the crystal. The strain is a dynamic one as its direction and magnitude are constantly changing as the atoms move. Since the atomic motion is pseudo-oscillatory, so is the dynamic strain. Thus an acoustic wave is an oscillating strain moving through a material. Because stress and strain are always directly related in a material there is also an oscillating stress field. Therefore an acoustic wave can be described as either a dynamic stress field or strain field in a material.

Wave Motion

The most familiar depiction of a wave is a sinusoidal curve such as shown in Figure 12-2(a). The amplitude oscillates between positive and negative limits at a fixed rate, known as the frequency, and the curve extends indefinitely. This curve can equally well be plotted as a function of space or time. In a medium, a wave has both a spatial and a time component. An equation for such a curve is

$$Y = A_0 \sin(\omega t - kx) \qquad \text{Eq. 12-1}$$

where A_0 is the amplitude, ω is 2π times the frequency, υ, and k is 2π over the wavelength, λ. The frequency, wavelength and wave velocity, v, are related by

$$v = \lambda \upsilon. \qquad \text{Eq. 12-2}$$

The wave described in equation 1 propagates in one direction only. in three dimensions, the wave front, which is a surface of constant phase for the wave, is a plane perpendicular to the x axis. Such a wave is known as a plane wave.

Most waves originating at a point in an extended medium initially have a spherical wave front. However, at some distance from the point or origin, the spherical surface will approximate a plane over a small area. For simplicity, in the rest of this discussion, plane waves will be assumed.

If two waves exist in a medium simultaneously, their amplitudes will add algebraically. Figure 12-2(b) shows the sum of the two waves

$$Y = A_1 \sin \omega_1 t + A_2 \sin \omega_2 t \qquad \text{Eq. 12-3}$$

where only the time component is plotted for clarity. One can, in this fashion, represent a complex waveform as a sum of simple waves. It has been long known that any arbitrary function which does not contain a discontinuity can be represented by an infinite sum of sinusoidal curves known as a Fourier series. One form of such a series can be written (Stearns, 1975)

$$f(t) = A_o/2 + \sum_{n=1}^{N} A_n \sin(n \omega_o t + \alpha_n) \qquad \text{Eq. 12-4}$$

where the A's are the amplitudes of the sine curves and the α_n are the phases. There is no restriction that the curve or waveform represented by the Fourier series be continuous. A transient wave such as that shown in Figure 12-2(c) may be described by a Fourier series. One very useful method of analyzing a wave is to look at its frequency components. A plot of the amplitude of the frequency components, A_n in Equation 12-4, against the frequency υ is known as the frequency spectrum of the wave. Figure 12-2(d) shows the spectrum for the wave shown in Figure 12-2(c).

Characteristics of the Materials

An acoustic wave can exist in any material, i.e., a solid, a liquid or a gas. Its velocity is determined by the characteristics of the material. The stronger the force between neighboring atoms, the more closely coupled will be their motion On the other hand, the larger the mass of the atoms, the more force must be applied for the same acceleration. Because a wave is a synchronized movement of a large number of atoms, it is actually the density of the material, ρ, rather than the mass of the individual atoms which governs wave motion. Thus the wave velocity should be directly proportional to the atomic restoring force and inversely proportional to the density. The actual relationship is

$$v_i = \sqrt{C_i/\rho} \qquad \text{Eq. 12-5}$$

where v_i is the velocity for the particular type of wave and

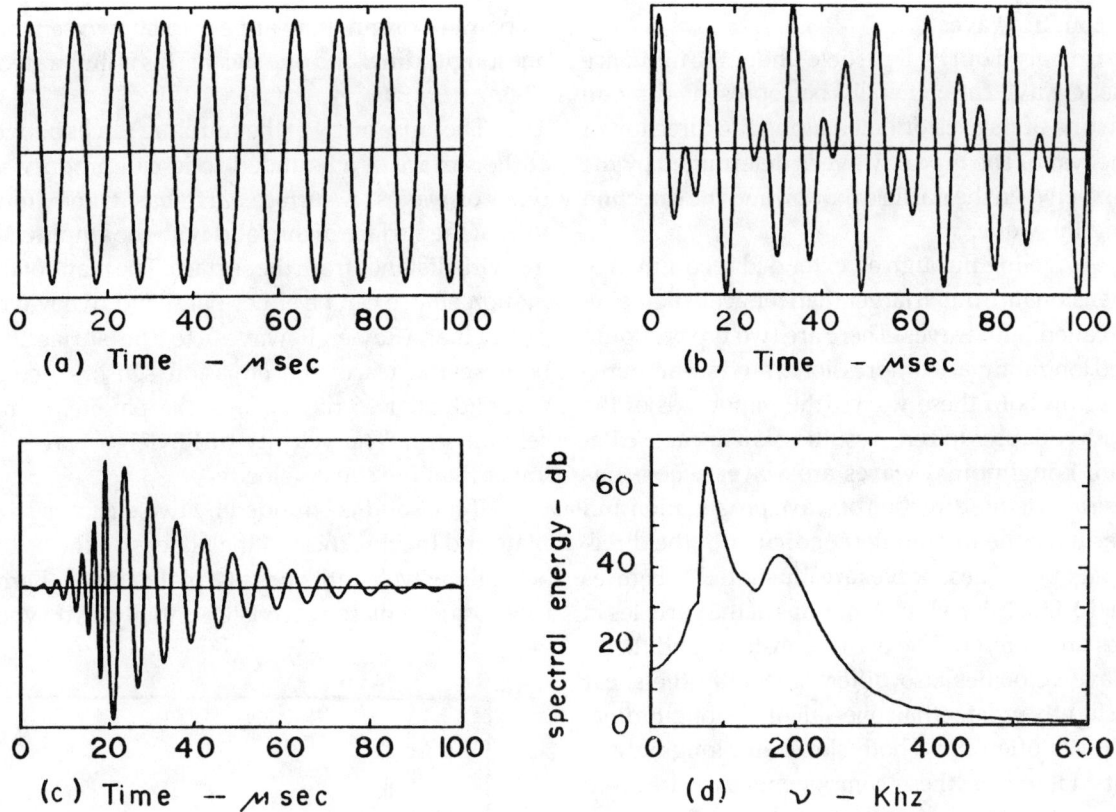

Figure 12-2. (a) Sine Wave; (b) Sum of Two Sine Waves; (c) Transient Wave; (d) Spectrum of Transient Wave (The Amplitude is Plotted on a Logarithmic Scale).

C_i is known as the elastic constant for that type of wave. The elastic constant is a measure of the strength of the coupling between atoms for that particular kind of motion. Different relative motions of the atoms will have different values of the elastic constant.

Another property of the material is the characteristic impedance. This is defined as

$$Z_i = \rho v_i = \sqrt{\rho C_i}. \qquad \text{Eq. 12-6}$$

The reflection and transmission of acoustic wave at an interface depend upon the characteristic acoustic impedances of the two materials.

Acoustic velocities, acoustic impedances and densities for some materials often seen in acoustic emission tests are given in Table 12-1.

Table 12-1. Acoustic Velocities and Impedances for Longitudinal, Shear and Rayleigh Waves in Several Materials

Material	V_l	V_s	V_R	ρ	Z_l	Z_s	Z_R
Aluminum	6.42	3.04	2.87	2.70	17.30	8.20	7.70
Brass	4.70	2.11	1.99	8.60	40.60	18.30	17.10
Steel	5.94	3.25	3.03	7.80	46.50	25.40	23.60
Nylon	2.62	1.07	1.01	1.11	2.86	1.18	1.12
Lucite	2.68	1.10	1.04	1.18	3.16	1.30	1.23
Water	1.50	—	—	1.00	1.50	—	—
Air	0.33	—	—	0.00123	0.0004	—	—

Velocities are in units of 10^3 μs
Densities are in units of gm/cm^3
Impedances are in units of 10^6 $kgm/\mu s^2$

Types of Acoustic Waves

The path traced out by a particle under the influence of an acoustic wave can generally be represented by an ellipse with one of its axes oriented along the direction of travel of the wave. The type of wave is determined by the relationship between the particle motion and the direction of travel and the wave.

Waves traveling through an extended medium (one whose dimensions are much larger than the acoustic wavelength) are called bulk waves. There are two types of pure waves called longitudinal (compressional) and shear (transverse) waves. In both these waves, the minor axis of the elliptical path collapses to zero, resulting is a linear oscillatory motion. Longitudinal waves are waves where this motion is parallel to the direction of wave propagation and shear waves have the motion perpendicular to the direction of propagation. These waves are illustrated in Figures 12-3(a) and (b). Since the relative motions of the particles in these waves are different, the elastic constants and therefore the wave velocities also differ. Generally the shear velocity is slightly greater than one half of the longitudinal velocity. Waves often have both shear and longitudinal components. However, these components each travel at their own velocity. In a nonattenuating, nondispersive medium, a transient wave, sampled at some distance from its point of origin, may appear to be two separate waves, one longitudinal and one shear. This is illustrated in Figure 12-4.

The anisotropy of the coupling forces between atoms at the surface of a bounded solid can produce additional types of waves. A surface wave has its maximum amplitude at the surface of the solid with the amplitude decreasing with distance from the surface. The plane of the particle motion ellipse can be either parallel (Love waves) or perpendicular (Rayleigh waves) to the surface. However, because most acoustic emission sensors detect motion perpendicular to the surface, the parallel component is seldom seen. The velocity of Rayleigh waves is slightly lower than the shear velocity.

If the solid is bounded by two surfaces so that it is a plate and the thickness of the plate is on the order of a few acoustic wavelengths or less, plate waves (Lamb waves) can occur. A plate wave is essentially two synchronized

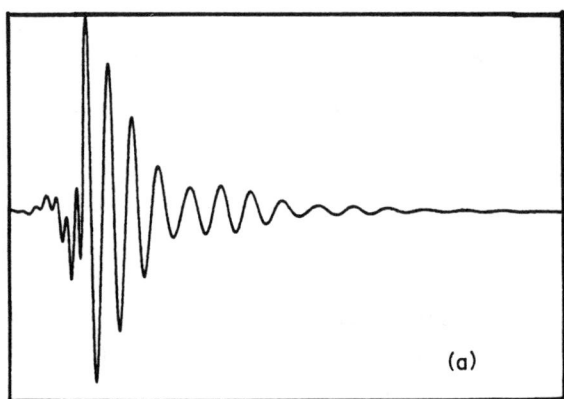

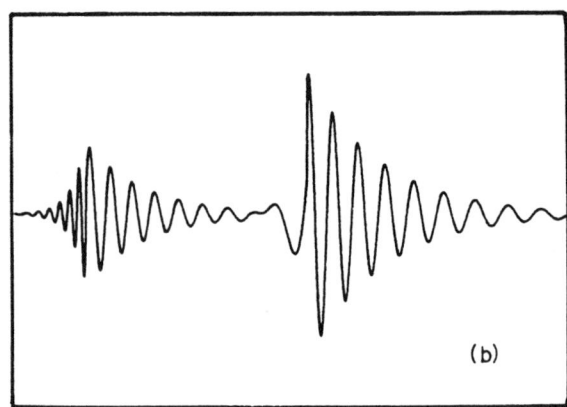

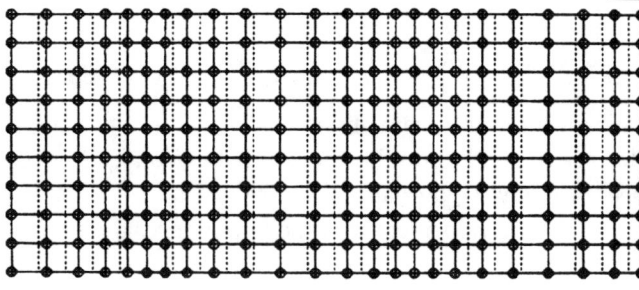

(a)

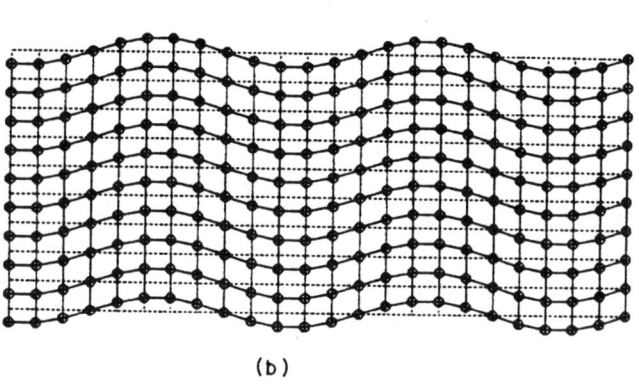

(b)

Figure 12-3. Particle Displacements for Bulk Acoustic Plane Waves Frozen at an Instant of Time. (a) Longitudinal Waves; (b) Transverse Waves.

Figure 12-4. The Effect of Different Acoustic Velocities on the Waveform. (a) Longitudinal and Shear Waves Detected near Point of Origin; (b) Longitudinal and Shear Waves Detected Some Distance from the Point of Origin. The Effects of Attenuation and Dispersion have been Ignored.

surface waves which can be synchronized either symmetrically or antisymmetrically. Particle motions in Rayleigh waves and plate waves are shown in Figure 12-5.

Bulk waves, surface waves and plate waves are the most important types of waves seen in the field of acoustic emission. However, there are not the only types of waves which can be found in solids. In general, bounded solids of moderately symmetrical geometry can support unique types of waves.

Dispersion and Group Velocity

The wave velocity defined in Equation 12-2 is the phase velocity. For unbounded media and surface waves on a single surface, this velocity is the only wave velocity and it is independent of frequency. However, all waves traveling in bounded media where the physical dimensions are within an order of magnitude of the acoustic wavelength are dispersive; that is, the phase velocity is a function of frequency. This is illustrated in Figure 12-6 where the frequency dependences of the velocities for symmetric and antisymmetric plate waves are shown.

Dispersion would have little effect on continuous waves. However, acoustic emissions are packets of waves which can be thought of as a superposition of continuous waves, as shown in Equation 12-4. If each wave train making up the packet travels at a different velocity, the wave packet will change shape as it travels through the medium. The result is that the same acoustic emission may look quite different when detected by the same sensor at different positions.

Energy in a wave packet does not travel at the phase velocity, but at the group velocity. The phase velocity can be defined by rewriting Equation 12-2 as

$$v_p = \omega/k \qquad \text{Eq. 12-7}$$

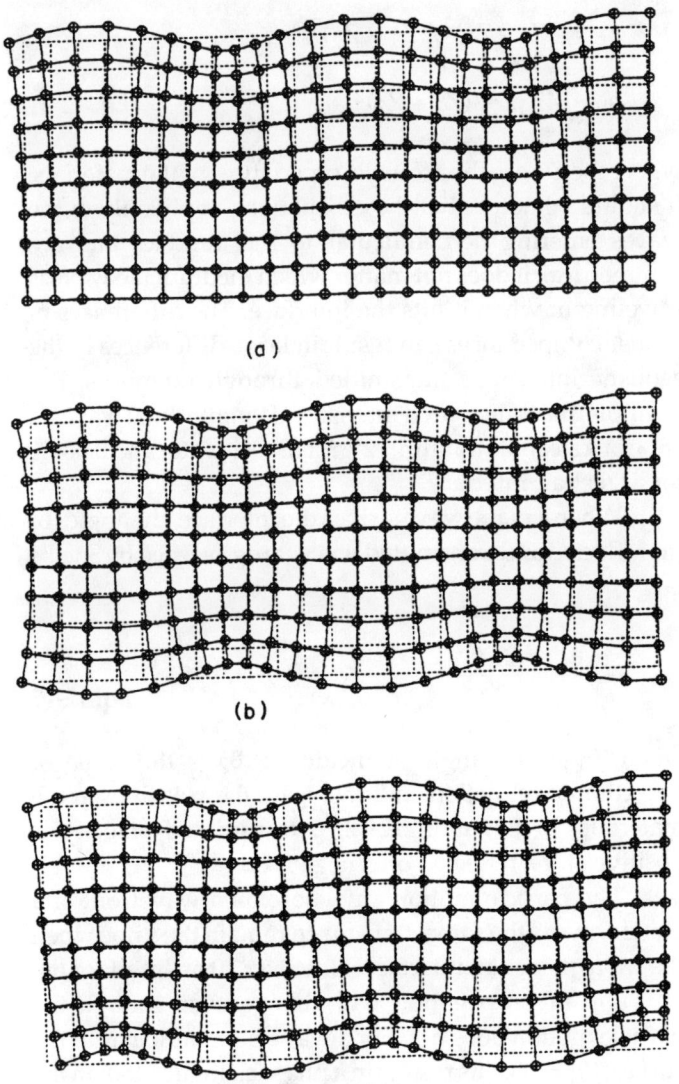

Figure 12-5. Particle Displacements for Acoustic Waves Frozen in Time. (a) Rayleigh Wave; (b) Plate Wave, First Symmetric Mode; (c) Plate Wave, First Antisymmetric Mode.

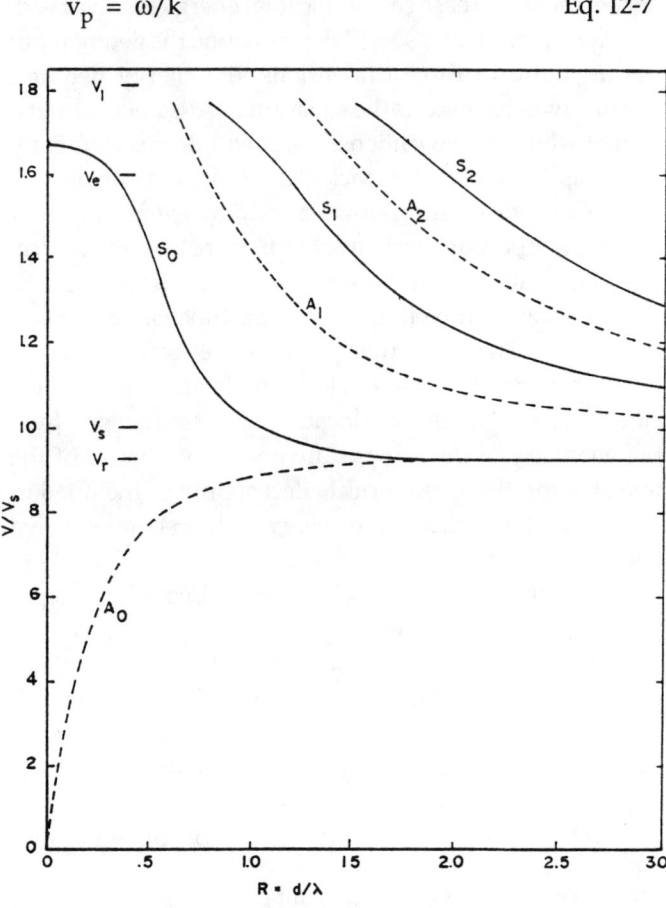

Figure 12-6. Phase Velocities for Different Plate Wave Modes Plotted as a Function of the Ratio of the Plate Thickness to the Acoustic Wave Length. This Figure is Drawn for Steel with a Poisson's Ratio of 0.28. The Longitudinal, Expansional, Shear and Rayleigh Wave Velocities are Shown.

while the group velocity is defined as

$$v_g = d\omega/dk. \quad \text{Eq. 12-8}$$

In the absence of dispersion, these are the same velocity, but in most bounded solids the group velocity will be less than the phase velocity. This can have real effects in acoustic emission when one is attempting source location by measuring the differences of arrival times at two or more sensors. One can also get strange results when trying to compute the locations of acoustic emission sources by using a phase velocity for a thin sheet of metal.

Attenuation

A wave packet is generated with a well defined energy. As the packet propagates away from its source, the energy content will remain constant in the absence of any dissipative mechanisms. However, if the wavefront of the packet is expanding, the energy per unit volume of the packet must decrease so that the total energy is conserved. The rate of this decrease will depend upon the geometry of the medium. In three dimensions, the energy per unit volume will decrease as the square of the distance from the source while in two dimensions, it will decrease linearly with this distance. If the packet is confined to one dimension, as in propagation down a rod, the energy per unit volume will be independent of the distance from the source.

Normally, when thinking about acoustic waves, one assumes a wave traveling in only one dimension; that is, a plane wave. Therefore, this geometrical effect on the wave packet's energy is ignored. However, in an acoustic emission test, where neither the location of the source nor, often, the geometry of the sample, are under the control of the investigator, this geometrical effect should be included in any attempt to measure the energy of the generated wave packet.

The attenuation of a plane wave arises from dissipative mechanisms or scattering as the wave propagates. In a homogeneous medium, these losses usually occur as a fixed percentage of the wave packet energy per unit length of travel. Mathematically this is an exponential decrease in the wave amplitude with distance and can be expressed as

$$A = A_o \exp(-ax) \text{ or } A = A_o \exp(-\beta t) \quad \text{Eq. 12-9}$$

where a is an attenuation constant per unit length and β is an attenuation constant per unit time. The two constants are related by the acoustic velocity

$$\beta = \alpha v \quad \text{Eq. 12-10}$$

Both forms of the attenuation constant are seen in the literature.

There are many types of acoustic attenuation mechanisms are most have some form of frequency dependent. Fortunately, in the normal acoustic emission frequency range of 50 kHz to 1 MHz, both the frequency dependence and the magnitude of many of these attenuation mechanisms is small in structural materials. Only in composites and in geological materials is attenuation a severely limiting factor in acoustic emission tests.

Interfaces: Reflection, Transmission and Mode Conversion

If a plane wave strikes a surface between two materials with different acoustic impedances, part of the wave will be reflected and part transmitted. The intensities of the reflected and transmitted components are given by:

$$a_r = (Z_1 - Z_2)^2/(Z_1 + Z_2)^2$$

$$a_t = 4Z_1Z_2/(Z_1 + Z_2)^2 \quad \text{Eq. 12-11}$$

where the Z's are the acoustic impedances of the materials. It should be noted that these equations apply only to the waves entering perpendicular to the interface and are symmetrical; it does not matter which medium the wave is traveling in when it hits the interface. The differences in acoustic impedances can result in large differences in the acoustic intensities transmitted through interfaces. For example, the transmitted intensity of longitudinal waves is 78% for a steel-aluminum, 12% for a steel-water and 0.004% for a steel-air interface.

When a plane wave strikes the interface, the angles of the reflected and transmitted waves are governed by Snell's law

$$\frac{\sin\theta_1}{v_1} = \frac{\sin\theta_2}{v_2} \quad \text{Eq. 12-12}$$

where θ_1 is the angle of incidence, θ_2 is the angle of reflection of refraction, and the v_i are the velocities in the materials. In Equation 12-12, a transmitted velocity is positive and a reflected one, negative.

The particle motion anywhere on a wavefront of a plane wave is the same. It wants to remain the same when the wave passes an interface. However, at an interface the direction of propagation will change even though the particle motion does not. For a wave perpendicular to the surface ($\theta_1 = 0°$), this results in a phase change of 180° in the relative motion of the particle to the wave direction. This does not change the character of the wave. For nonperpendicular angles of incidence, the reflected and transmitted waves will have both longitudinal and shear components (unless the particle motion is parallel to the interface) because of the change in the angle between the particle motion and the propagation direction. This is illustrated in Figure 12-7. The process of generating both modes of bulk waves upon reflection or refraction is known as mode conversion. In acoustic emission, where there is no control of the wavepath, the almost inevitable result of mode conversion is that the wave reaching the sensor is composed of both longitudinal and shear components, no matter what its original polarization. In most situations, surface waves are also present. Since mode conversion occurs at almost every reflection, it is an almost continuous process as the wave propagates in a bounded medium. Because of this continuous transformation between modes traveling at different wave velocities, the transient waveform may lengthen in time as it travels instead of dividing into separate longitudinal and shear components as shown in Figure 12-4.

12.2.3. Sensors

A sensor is a device which generates an electrical signal when it is stimulated by an acoustic wave. The exact relationship between the characteristics of the wave and those of the signal will depend upon both the sensor and the wave. An ideal sensor would produce a voltage-time curve identical to the amplitude-time curve of the wave at the point where the sensor is located. No sensor approaches this ideal although for certain types of acoustic waves, laser interferometry may come close. Many available sensors operate quite well for specific types of waves over limited ranges of parameters. Because of the wide range of frequencies and different acoustic modes contained in most acoustic emission signals, the choice of a sensor is usually not critical to detect acoustic emission. However, the optimal choice of a sensor will always improve the data and may be the difference between a successful or unsuccessful test or experiment where low amplitude emissions are involved.

Acoustic emission sensors can be based upon several physical principles. The signals can be generated by electromagnetic devices such as phonograph pickups, by capacitative microphones, by magnetostrictive devices, by piezoelectric devices and by the use of laser interferometers to detect the surface motion of the sample. The large majority of acoustic emission sensors are piezoelectric and the rest of this discussion will be limited to such devices.

Piezoelectricity

Piezoelectricity (Cady, 1984, Jaffe, 1971, Mason, 1950, 1958) is the name given to the coupling between strain and electric polarization which occurs in many crystals. It is a geometrical effect and occurs only in materials which lack a center of crystal symmetry. This is not a severe restriction as 21 of the 32 classes of crystal structure lack a center of symmetry. In such crystals, a strain will shift the centers of positive and negative charge distribution so that they no longer coincide. This produces an electric dipole moment throughout the crystal. The polarization of a crystal is defined as the dipole moment per unit volume. When a polarization exists in a nonconducting crystal, electric charge will appear on certain surfaces. Conducting electrodes on these surfaces allow the measurement of the charge which is directly proportional to the strain. The effect is symmetric in that application of charges to the electrodes changes the strain. Since the stress field and the strain field in a material are directly related, the piezoelectric effect can equally be defined as the coupling between the stress and the polarization in a crystal. The exact stress or strain which is measured (or generated) can be chosen by a careful selection of the crystal surfaces to be electroded.

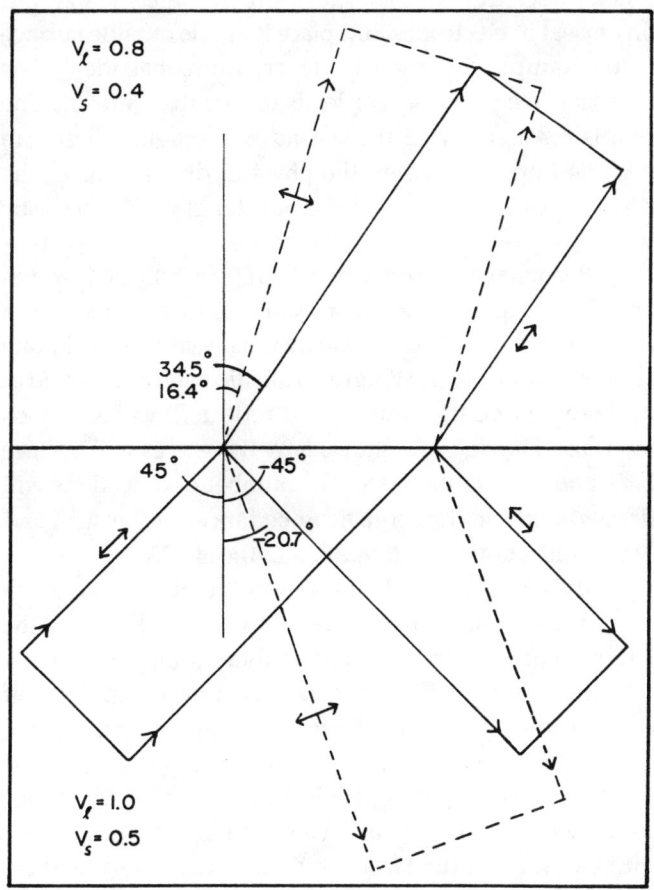

Figure 12-7. Reflected and Transmitted Waves Across an Interface. The Incident Wave is a Longitudinal Wave with an Angle of Incidence of 45°. The Velocities in the First Medium are $v_l = 1.0$ and $v_S = 0.5$ while in the Second Medium $v_l = 0.8$ and $v_S = 0.4$. The Double Arrows Show the Direction of the Component of Particle Motion Associated with Each Wave.

Initially all piezoelectric devices were made from single crystals. The most useful of these were quartz, Rochelle salt and ammonium dihydrogen phosphate. Later a class of materials known as ferroelectrics (piezoelectric materials which have a polarization even in the absence of a strain) were investigated and methods were discovered in which ceramics made of ferroelectrics could be given a uniform direction of polarization similar to that found in a piezoelectric crystal. It became possible to produce ferroelectric ceramics with many properties superior to piezoelectric single crystals. The result has been that almost all acoustic emission sensors are made from ferroelectric ceramics of various types.

Size Effects

An infinitesimal piece of piezoelectric material with many different sets of electrodes totally embedded in a sample would come close to the ideal sensor. However, when we scale up the piezoelectric to manageable size, put on one set of electrodes and place it on the outside surface of the sample, we rapidly depart from that ideal. The physical size of the sensor leads to two main effects. The first is resonance and the second is averaging. Both can become important when the physical dimensions of the sensor approach or exceed the wavelength of the acoustic wave.

Resonance occurs when there are reflected waves and regular geometries. For example, place a plate of a material between two other materials such as a steel plate submerged in water. When an acoustic wave is directed at it, there will be transmitted and reflected waves at each interface. Figure 12-8a shows how the waves will bounce back and forth in the plate. The number of reflections will depend both on the acoustic impedances of the steel and water and upon the attenuation in the steel.

If the plate is one-half wavelength thick, as shown in Figure 12-8b, each reflected wave will be in phase and the strains will algebraically add. If there are a great many reflections, the peak strain can reach a very high level. If there are only a few reflections, the frequency of this wave need not be exactly that of a half wavelength to get some reinforcement, but the greater the number of reflections, the narrower will be the allowed frequency-range at maximum strain and the larger this maximum strain will be. This is illustrated in Figure 12-c. The sharp high amplitude peak is said to have a high Q where Q is defined as the ratio of energy stored per cycle to the energy dissipated per cycle. This increase of the stress or strain level in a material at a half wave thickness is known as a resonance and v_0 is the resonant frequency. From Figure 12-8b, we see that at resonance, the average strain throughout the plate is a maximum. Since the output of a piezoelectric crystal is proportional to the strain, and to the average strain for a crystal of finite dimensions, the maximum output of a sensor is at its resonant frequencies. If the frequency is increased until there is one full wavelength in the crystal, there will again be strain reinforcement due to the reflected waves. However, we can see in Figure 12-8d that while the strain level may be very great at this frequency, $2v_0$, the average strain over the crystal exactly cancels so that the output of the sensor is zero. Increasing the frequency to $3v_0$, we see 1.5 wavelength in the crystal, reinforcement again occurring, and the average strain over two-thirds of the crystal canceling but the average strain over that last third being a maximum. The result is that a piezoelectric sensor will have a maximum output whenever the thickness, d, is

$$d = (2n-1)\lambda/2 \qquad \text{Eq. 12-13}$$

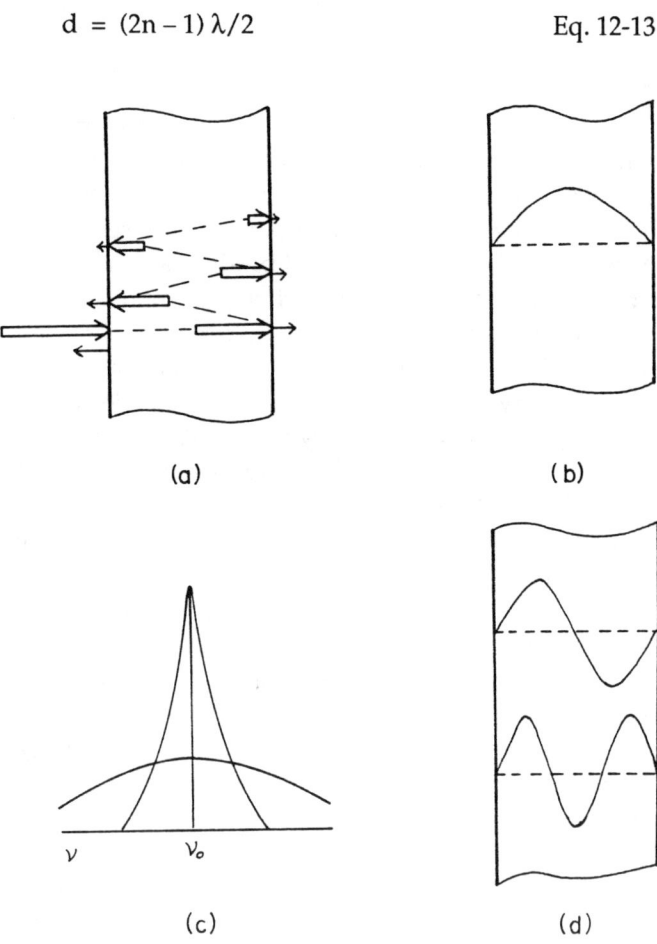

Figure 12-8. (a) Representation of Reflected and Transmitted Waves Inside a Plate Immersed in Water. Successive Reflections have been Displaced Upward for Clarity; (b) Representation of Strain in a Plate 1/2 Wavelength Thick; (c) Amplitude as a Function of Frequency in a Bounded Solid for High and Low Q Materials; (d) Representation of Strain in a Plate One Wavelength and One and One Half Wavelength Thick.

and no output when

$$d = n\lambda \qquad \text{Eq. 12-14}$$

Thus, a sensor can be operated either at its fundamental frequency, υ, or its harmonic frequencies $n'\upsilon_0$, where n' is odd. The Q of the transducer depends only on the number of reflections in the sensor and therefore the Q is independent of the harmonic at which the sensor is operating as long as the material of the sensor does not show a frequency dependent attenuation. Strictly, the sensor is resonant at the frequencies of $n\lambda/2$ but because of strain averaging across the transducer the output is zero when n is even.

The sensor will always have an output at frequencies below the fundamental frequency, υ_0. At frequencies less than about $3\upsilon_0/4$, the resonance will have no effect and the output will be essentially independent of frequency.

In reality a material cannot be strained in one dimension without producing strains in other directions as shown in Figure 12-9a. Piezoelectricity is quite complex and a large part of its history in acoustics has been a search for particular crystal orientations where only certain strains will generate charge on a pair of electrodes. Modern acoustic emission sensors, made of polarized ferroelectric ceramics, also suffer from multiple resonances. However, because the ceramics have an intrinsically low Q neither the peaks nor valleys are especially sharp. Thus, most sensors have a broad but highly "colored" spectral response curve as shown in Figure 12-9b.

There is another important aspect of strain averaging by a sensor. Figure 12-10a shows a block with a sensor mounted on it. If the sensor is excited with a compressional wave moving perpendicular to its surface, the entire sensor face will move in phase and, excluding resonances, the average strain in the sensor will be independent of frequency. Next look at a Rayleigh wave, again with the particle motion perpendicular to the sensor face, traveling parallel to the face. In this case, the strain distribution in the transducer will vary as a function of distance along the wave. Figure 12-10b shows the strain variation where the diameter of the sensor is less than $\lambda/2$. Here, the output is still proportional to the amplitude of the wave. In Figure 12-10c, we see the case where the diameter of the sensor is larger than the wavelength. In this case, for every complete wavelength under the sensor, the strain averages to zero. Only the extra fraction of the wavelength under the sensor contributes to its output. This averaging essentially reduces the effective area of the sensor and the higher the frequency, the greater the reduction. Another effect is that at certain frequencies, dependent upon the shape of the sensor and the acoustic velocity of the sample, the total strain averages to zero. These effects are illustrated in Figure 12-10d where the response for this type of surface wave is plotted for a sensor with a flat response to compressional waves perpendicular to it face. It is obvious that the high frequency response of such a sensor is going to depend drastically upon the angle of incidence with which the wave strikes the sensor. This averaging effect depends upon the acoustic wavelength <u>in the material</u>. Therefore, the sensor response not only is going to vary with frequency and angle of incident, it is going to vary when used with different materials. The best answer to this problem is to make the sensor physically small. For steel, a 3mm diameter sensor should work reasonably well below 500 kHz. The inevitable tradeoff is that the smaller sensor has a lower capacitance and thus, as will be discussed later, a reduced effective sensitivity. Another approach is a more complicated polarization geometry than a simple disk polarized between its faces. Some sensors of this type are commercially available and the reader should contact the various manufacturers for details and specifications (the

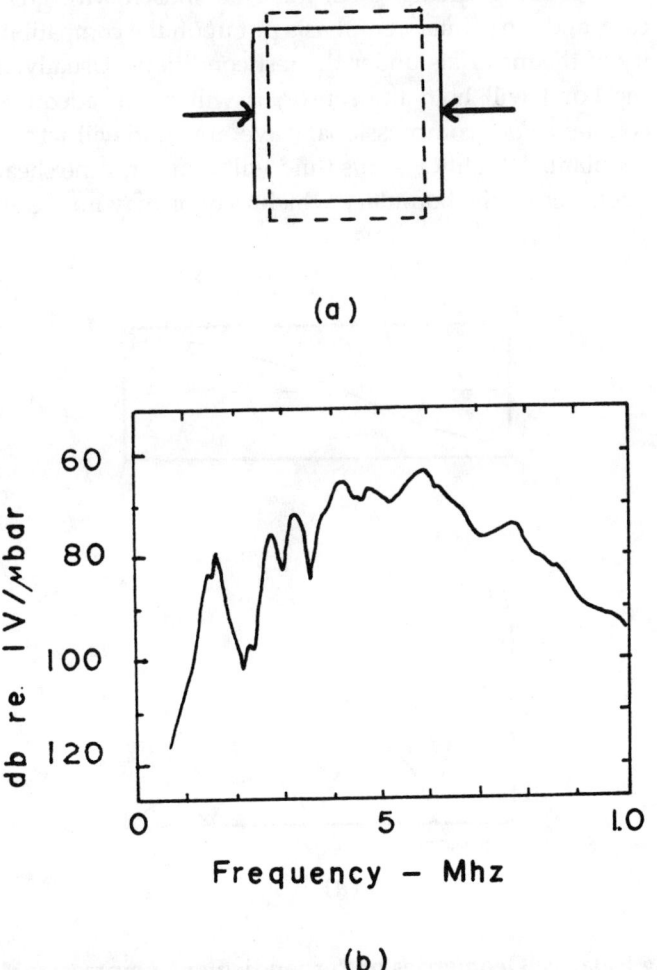

Figure 12-9. (a) Deformation of a Material Showing Multiaxial Strain Resulting from a Uniaxial Force; (b) Spectral Response of an Acoustic Emission Sensor.

geometry and materials in such sensors are often treated as trade secrets).

Couplants

To this point, it has been assumed that the sensor has simply been placed on the surface of the material containing the acoustic wave. When this is tried, it is found that the sensor produces a very weak signal. If a thin layer of a fluid is placed between the sensor and the surface, a much larger signal is obtained. The use of some type of couplant is almost essential for the detection of low level acoustic signals. Physically, this can be explained by looking at the acoustic wave as a pressure wave transmitted across two surfaces in contact. On a microscopic scale, the surfaces of the sensor and the material are quite rough so that only a few spots actually touch when they are in contact. Stress is force per unit area and the actual area transmitting a force is very small. If the microscopic gaps are filled with a fluid, the pressure will be uniformly transferred between the surfaces. For a shear wave with a variable strain component parallel to the surfaces, again very little strain will be transferred between the surfaces because of the few points in actual content. In this case, filling the gaps with a low viscosity liquid will not help much since a liquid will not support a shear stress. However, a high viscosity liquid or a solid will help transmit the parallel strain between surfaces. The purpose of a couplant then, is to ensure good contact between two surfaces on a microscopic level.

Much has been written about couplants for both ultrasonics and acoustic emission and much of this is either wrong or misleading. One basic problem is in using the terms bond and couplant interchangeable, a practice of which many people, including the author, have been guilty. Strictly, a couplant is any material which aids the transmittal of acoustic waves between two surfaces while a bond is a couplant which physically holds the sensor to the surface. Water is a couplant and cured epoxy resin is a bond. Many problems have come about from using a bond in an inappropriate way. If a rigid bond is used to attach a sensor to a sample which elastically deforms during the test, the normal result is a broken bond and poor or no sensitivity to the acoustic wave. Similarly, in an experiment where the temperature is changed appreciably, the use of a rigid bonding material can lead to broken bonds due to differential thermal expansion between the sensor and the sample.

Bonding agents, then, must be chosen with great care, and the primary emphasis put upon the compatibility of the materials under the test conditions. Usually, if the bond will hold the sensor, it will be an adequate couplant. For a compressional wave, any fluid will act as a couplant. A highly viscous fluid will transfer some shear stress across the boundary which may or may not be an

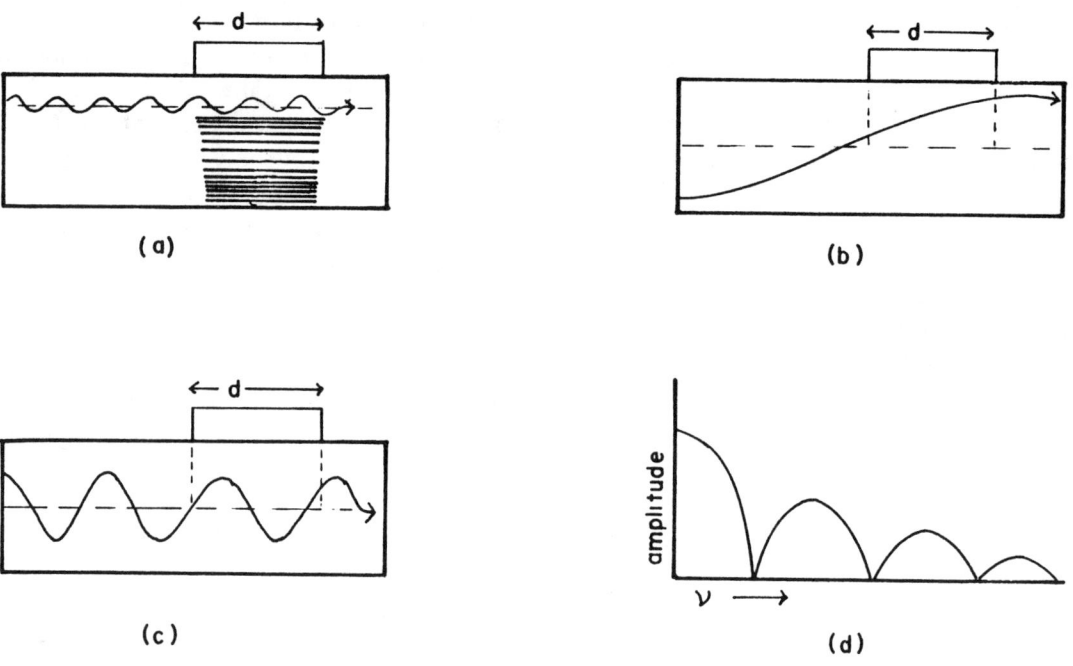

Figure 12-10. (a) Block with Sensor Mounted on it Showing Relative Geometries of Perpendicular Compressional Wave and Parallel Rayleigh Wave; (b) Representation of Strain Produced by Rayleigh Wave Under Sensor for $\lambda>d$; (c) Representation of Strain Produced Under Sensor for $\lambda<d$; (d) Response of Finite Diameter Sensor to Rayleigh Waves as a Function of Frequency. The Response of the Sensor to Perpendicular Compressional Waves in Independent of Wavelength.

advantage. In one study, the author tested a large number of couplants with compressional waves. Almost all couplant showed an increase of 30 ± 2 dB in the signal strength over no couplant. The variation was little more than the uncertainty of the measurement. The most practical rule is to use as a couplant, a thin layer of any viscous fluid which wets both surfaces. The sensor should be held against the surface with some pressure furnished by magnets, springs, tape, rubber bands, etc. The secret is to use as thin a layer as possible. If a rigid bond is used, there must be no differential expansion between the two surfaces. While flexible bonds have been tried, I have never found one which I would trust in a critical situation. In Table 12-2, a few commonly used couplants are listed along with the temperature range where they can be used.

Temperature Effects

The temperature dependence of piezoelectricity is complicated and detailed discussion of its effect on sensors is beyond the scope of this chapter (Cady 1964 and Mason 1950). However, there are certain effects which can lead to problems when a sensor is used at different temperatures. First, ferroelectric materials (Jaffe, 1971), such as the PZT ceramics, have a temperature (the Curie temperature) above which the material transforms to another, and usually nonferroelectric, phase. Taking a ferroelectric sensor through the Curie temperature will remove the polarization, destroying the piezoelectric properties of the sensor, and may shatter the ceramic as well. Ferroelectric sensors will usually work well up to temperatures within 50°C of the Curie temperature, if the other materials in the sensor can stand the temperature. The Curie Temperatures of PZT ceramics lie between 300 and 400°C.

Ferroelectric ceramics and polycrystallite. Each crystallite may have one or more ferroelectric domains in it. These domains are regions where the spontaneous polarization is all in one direction. This polarization can be only along certain directions in the crystal structure. When the ceramic is poled, these domains are aligned as closely as the crystal orientation allows to the direction of polarization. Because of the random orientation of the crystallites, there are always going to be a fair number of domains which are almost equally close to the direction of polarization in the ceramic. Small strains may be enough to cause the domain to change orientation. Such a flip of a domain will cause a very small change in the polarization of the sensor. However, this change is the same order of magnitude as the change caused by an acoustic wave. In general, it is impossible to distinguish an electric signal caused by a flipped domain from one caused by acoustic emission. The strain necessary to produce a domain flip can come from changing the temperature of a sensor. Thus, changing the temperature of a ceramic sensor can produce signals, indistinguishable from acoustic emission, which arise in the sensor and not the sample. Empirically, the temperature change necessary to produce appreciable amounts of these domain flip signals is around ±100°C in PZT. This effect does not prevent ferroelectric ceramic sensors from being used at different temperatures but thermal equilibrium should be achieved before data are taken. If one wishes to measure acoustic emission while changing temperature, it is recommended that ferroelectric ceramics be used only for small temperature deviations. To measure acoustic emission over large temperature deviations, a sensor made of a single crystal, such as quartz, should be used. Another approach is to use an acoustic waveguide which isolates the sensors from the temperature fluctuations of the sample.

Table 12-2
Some Common Acoustic Emission Couplants and the Approximate Temperature Range where They Can Be Used.

Dow Corning V-9 resin	~ –40°C to 100°C
High vacuum stop cock grease	~ –40°C to 200°C
Ultrasonic couplants	Room temperature
Petroleum grease	Room temperature
Water	1°C to 99°C
Dow Corning 200 fluid	–273°C to –70°C and –30°C to 200°C
Salol	~ –40°C to 40°C
Nonaq stop cock grease	–273°C to 100°C
Dental cement	~ 0°C to 50°C
50% Indium-50%Calium mixture	20°C to 700°C

Sensor Sensitivity—Effects of Cables

The sensitivity of a sensor is governed by the intrinsic sensitivity of the piezoelectric material, the dimensions of the piezoelectric element and the design and materials used in its case. Practically, one receives from a manufacturer a measured response curve to a standard signal and the capacitance of the sensor. This curve is often presented as if it were independent of the measurement technique. However, the sensitivity of a sensor will always depend, in part, upon the equipment with which it is used.

The open circuit voltage produced by a sensor is a property of the piezoelectric element and is

$$V_o(S) = Q(S)/C_o \qquad \text{Eq. 12-15}$$

where Q is the charge produced by a strain S and C_o is the capacitance of the sensor. When connected to a preamplifier, the actual voltage across the input resistor of the preamplifier (if the input resistance is large) is

$$V(S) = Q(S)/(C_o + C_c + C_I) \qquad \text{Eq. 12-16}$$

where C_c is the cable capacitance and C_I is the input capacitance of the preamplifier. Acoustic emission sensors usually have a capacitance that falls between 100 and 1500 pF. Preamplifier input capacitances range between 20 and 40 pF. The capacitance per unit length of coaxial cables depends upon the cables' impedance with approximate values of 100 pF/m (30 pF/ft) for 50 ohm cable, 67 pF/m (20 pF/ft) for 75 ohm cable and 50 pF/m (15 pf/ft) for 97 ohm cable. To see the range of this reduction in sensitivity, assume an input capacitance of 30 pF and 1.5 m of 50 ohm cable. For a sensor with a capacitance of 100 pF, the output voltage would be V = $0.36V_o$ or a decrease in sensitivity of 9 dB. A sensor of 1600 pF capacitance would have V = $0.9V_o$, a decrease in sensitivity of 1 dB. Thus, the loss in sensitivity can be appreciable for low capacitance sensors. Since most of the extra capacitance comes from the cable, it is a good practice to keep the length of the cable between the sensor and preamplifier as short as possible.

Sensor Sensitivity—Effect of Preamplifier Noise

The spectral response curve of a sensor-cable-preamplifier combination to an acoustic signal is shown in Figure 12-11. Also shown are three response curves of the preamplifier without an AE signal. The exact shape of the spectral response of a sensor to an AE signal will depend to some extent on the preamplifier input characteristics and on the cable capacitance. Too much cable capacitance will tend to short out the high frequencies relative to the low frequencies. Different preamplifiers will give different response curves even with the same cable-sensor combination, although the difference is small between most commercial AE preamplifiers. The use of preamplifiers not designed for acoustic emission can have an appreciable effect on a sensor's spectral response characteristics.

The three preamplifier spectral response curves in Figure 12-11 are, from top down, the open circuit response, the response with cable and sensor in the absence of an AE signal and the response of the preamplifier with the input shorted with a 50 ohm resistor. The noise in a shorted preamplifier is generated by current fluctuations in the first amplification device. If the input is not shortened, the noise can be regarded as being generated by current fluctuations in the input resistor. The rms noise voltage (Bell, 1967), for such a resistor is given by

$$V^2 = 4kTR \Delta \upsilon \qquad \text{Eq. 12-17}$$

where k is Boltzmann's constant, T is the absolute temperature, R is the resistance, and $\Delta \upsilon$ is the bandwidth of the preamplifier. If the input stage of the preamplifier is open, R is the input resistance of that stage, but if there is a sensor connected, R is some equivalent resistance. The capacitance of the sensor and cable will tend to short out the higher noise frequencies. The peaks in the sensor curve in Figure 12-11 are caused by the mechanical resonances of the sensor raising the impedance of the sensor-cable combination at the resonant frequencies.

The noise level of a preamplifier is commonly specified by the manufacturers as the output rms voltage di-

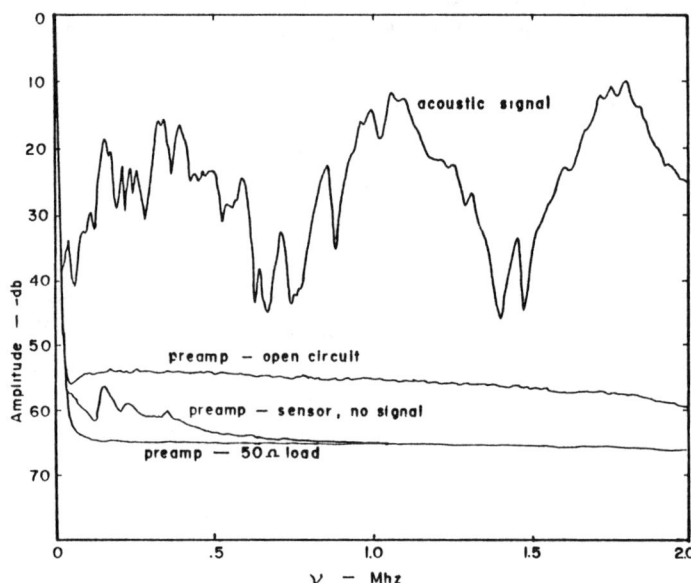

Figure 12-11. Frequency Response of an Acoustic Emission Sensor-Preamplifier Combination to an AE Signal. The Response of this Combination is also Shown for No Signal. The Responses of the Preamplifier with an Open Input Circuit and with a Load are Also Shown.

vided by the gain when the input is shorted.

$$V_s = V_o/G \qquad \text{Eq. 12-18}$$

The noise contributed by the sensor-cable combination can be estimated from Equation 12-17. A value of 295 K may be used as the temperature and the output impedance of the sensor-cable combination used as the equivalent resistance, R. This output impedance can be measured by a simple circuit. A variable resistance in parallel with the preamplifier input is reduced until the preamplifier output is one-half the level with no parallel resistor. The output impedance of the sensor-cable combination is then equal to the output impedance of the preamplifier which is essentially the value of the parallel resistor network formed by the preamplifier input resistance and the value of the variable resistor. For most acoustic emission sensors this impedance will be between 50 and 1000 ohms. Equation 12-17 can be rewritten

$$V_n = 0.004\sqrt{Z_o \Delta\upsilon} \qquad \text{Eq. 12-19}$$

where the sensor impedance, Z_o is in kilo ohms and $\Delta\upsilon$ is in hertz.

The approximate noise voltage of the preamplifier-sensor combination will be

$$V_{noise} = \sqrt{V_s^2 + V_n^2}. \qquad \text{Eq. 12-20}$$

This noise voltage is measured by an rms voltmeter which measured heating power (see the section on rms measurement). It assumes that the noise voltage is flat over the bandwidth, $\Delta\upsilon$. In Figure 12-11, we saw that this is not strictly the case and that the noise level at certain frequencies may be six to eight dB higher than the rest of the bandwidth. As long as the acoustic emission signals contain a large range of frequencies, this should not matter but if they are confined to one narrow frequency band and are relatively low level, these noise peaks may present problems.

Sensor Calibration

All manufacturers furnish calibration curves with their sensors. The currently accepted units are dB referenced to 1 V/m/s and dB referenced to 1 V/μ bar. The AE signal used to calibrate the sensors may be either a surface wave or a compression wave. A typical calibration curve is shown in Figure 12-11. Calibration curves for the same sensor will not be the same for calibrations in velocity (m/s) or in pressure (μ bar). The reason is that pressure is proportional to strain which is related to the surface displacement while velocity is the time derivative of surface displacement. For a constant frequency, the relationships are

$$D = \sin\omega t \qquad \text{Eq. 12-21}$$

$$v_s = \frac{dD}{dt} = \omega\cos\omega t$$

where D is the surface displacement and v_s is the surface velocity. Thus, the velocity calibration curve is approximately the pressure calibration curve multiplied by the frequency. This tends to make the velocity curve seem flatter since it does not fall off as fast at higher frequencies.

In selecting a sensor, one should have some idea of both the frequency range and the type of waves to be expected. Calibration curves can then be compared provided that they are for the same type of source and the same units. Recently, the National Bureau of Standards (now NIST) (Breckenridge, 1982) has developed methods for the absolute calibration of sensors for Rayleigh waves and longitudinal waves. Manufacturers' calibration curves which are referenced to an NIST calibrated sensor can be directly compared regardless of type of source. However, a surface wave calibration should not be compared to a longitudinal wave calibration.

12.2.4 Acoustic Emission Signals

An AE signal is defined as the electrical output of a sensor responding to an AE. The separation of AE into continuous and burst emission is a natural result of the appearance of the signals on an oscilloscope. Continuous emission has the appearance of an increase in the preamplifier noise level while burst emission appears as well defined transient events as shown in Figure 12-1. The only distinguishing characteristics a continuous emission signal has are its amplitude and its frequency content. Burst emission signals, appearing as individual events, offer more parameters to analyze.

Signal Characteristics

Acoustic emission signals have the same random character as the AE from which they are derived. The correlations between the timing and characteristics of the bursts are strictly statistical. While similarities between burst signals exist, especially for a given test, seldom, if ever, are two signals identical in all their characteristics. For signal analysis, the most important features are the average repetition rate of the signals, the individual burst amplitudes, the energies of the bursts, and the frequency content, both of the individual bursts and the average content. Other important characteristics are the burst signal rise time, delay constant and signal length.

A voltage-time curve for an AE signal is shown in Figure 12-12a with the frequency spectrum for this curve displayed in Figure 12-12b. The curve is moderately complex and this complexity is typical of signals seen in AE experiments. The damped sine wave type of signal, which is discussed in the section on mathematical models of signals, is conspicuous by its rarity in real signals.

Figure 12-12a shows first the preamplifier noise, then the start of the signal at about 20 μs, displaying a rapid rise time and reaching a peak amplitude followed by rapid decay (35 to 60 μs). At this point, a larger amplitude component occurs. This exhibits a higher peak amplitude accompanied by a shift to lower frequencies. The signal then decays, at a slower rate, until it starts to disappear back into the preamplifier noise level. This signal was chosen to show that the concepts of rise time, peak amplitude, decay time and signal length, while valuable, may be somewhat ambiguous for any given burst signal.

The signal rise time can be defined as either the time from which the signal first occurs until the first maximum signal excursion or until the peak amplitude occurs. In Figure 12-12a, these are quite different.

The signal length is impossible to rigorously define for the signal slowly loses itself in the preamplifier noise. Figure 12-12a appears to have two different decay rates for the first and second parts of the signal. These decays may or may not be exponential.

The peak amplitude in Figure 12-12a appears on the positive cycle near 65 ms. It is different, however, from the first maximum amplitude which appear at 28 μs. The peak amplitude can as easily have been a negative going cycle. In Figure 12-12a, positive and negative cycles have similar amplitude although differences between positive and negative maximum amplitudes of over 5 dB are commonly found.

The variations in the magnitude of these parameters for burst signals in a single experiment can be quite large. Peak amplitudes can easily range over 60 dB for a typical experiment and an 80 dB range is common. Rise times can be effected by dispersion of the AE wave and so may depend upon the distance the emission travels from source to sensor. The signal length is dependent upon both the signal amplitude and the decay rate. Low amplitude, high frequency signals as short as 10 ms can be seen in some experiments while in structural tests, signal lengths can exceed 50 ms. The decay rate is usually fairly constant for a given specimen although it may be higher frequency dependent, especially in composite and other heterogeneous materials.

Figure 12-12a shows a second maximum which may be the result of the emission traveling different paths to reach the sensor. Such multiple paths can lengthen and distort the signal as well as completely distorting the rise time. Such a signal is illustrated in Figure 12-13 where an artificial signal is taken through six different paths to the sensor. The resulting signal shown in Figure 12-13c shows the distortions.

Frequency Spectra

The frequency spectrum of an AE signal will be determined by the bandpass of the amplification system, the frequency response of the sensor, the geometry and AE characteristics of the sample and the characteristics of the source. The norm for AE is that no two signals have exactly the same frequency content. The variation may be small for a localized source in a well defined geometry and homoge-

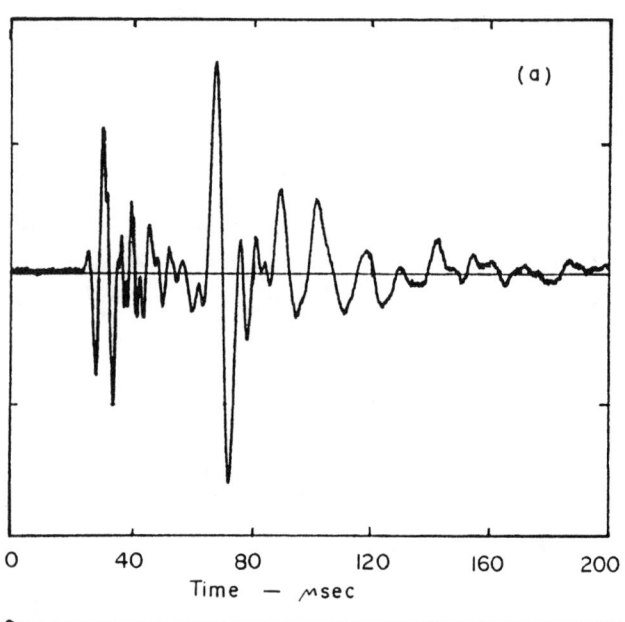

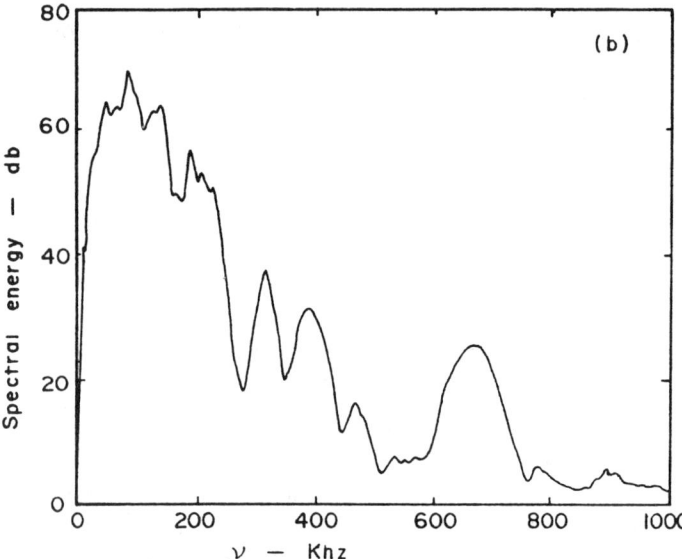

Figure 12-12. (a) The Voltage Time Curve for a Real Acoustic Emission Signal; (b) The Spectrum of the Signal Shown in (a).

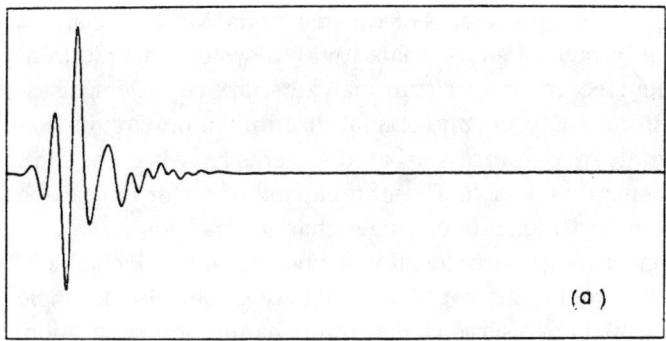

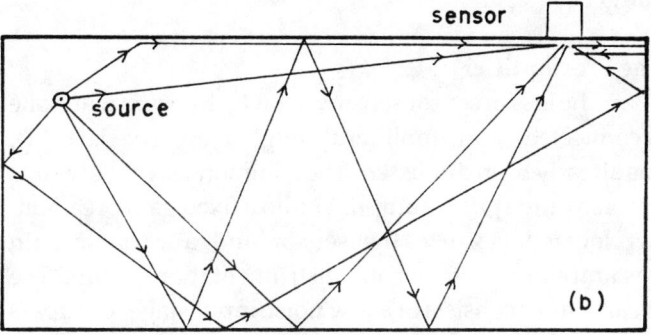

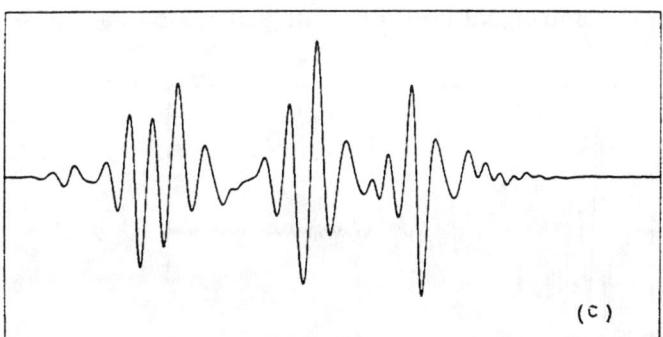

Figure 12-13. The Effect of Multiple paths on the Waveform of an Acoustic Emission Signal. (a) The Initial Waveform; (b) The Different Paths Traveled to Reach the Sensor; (c) The Waveform at the Sensor.

neous material. Very large variations will occur in heterogeneous materials such as composites and in widely scattered sources in complex geometries.

Generally, the spectrum of an AE burst will contain numerous peaks at different frequencies as can be seen in Figure 12-12. A short impulsive source will generate an emission with a broad frequency content. The spectrum of a delta function (a large pulse with zero width) will contain all frequencies equally. As this pulse is modified by transmission through the medium, and by detection and amplification devices, some of the frequencies are filtered out. The resulting signal contains many frequencies and is far from the idealized damped sine wave. The frequency spectra of a signal can be looked at as a complex record of everything that has happened to that signal from generation to display but at this point in time, it is not clear that this record can be read for anything but the crudest details.

Signal Energy

The one characteristic of an AE signal that is defined is its energy (Beattie, 1976). This energy is

$$E = \frac{1}{R} \int_o^\infty V(t)^2 \, dt \qquad \text{Eq. 12-22}$$

where R is the resistive load for the sensor and V(t) is the time dependent voltage output of the sensor. In this definition, it is assumed that there is no noise voltage; otherwise, the integral will be infinite. Practically, the limits of integration are the pulse length.

The energies of AE signals can be extremely small. An easily detectable signal would have an amplitude of $10\,\mu V$ and a length of $10\,\mu s$. Assuming a square nonoscillatory pulse for simplicity and a preamplifier input resistance of 20 kilo ohms, the signal energy if 5×10^{-13} ergs. This is the energy of the signal, not that of the AE. Sensor coupling efficiencies, sample attenuation and all the problems of partitioning energy between acoustic modes make it difficult to accurately estimate the energy in an AE burst even under highly controlled laboratory conditions. Under normal conditions, the energy in the smallest AE burst will be several orders of magnitude larger than the above figure. Conservatively, one should be able to detect an AE burst with an energy of 10^{-9} ergs. To illustrate how small this energy is, a one mm piece of #20 copper wire (about 0.8 mm diameter) will acquire an energy of about 5 ergs in falling 1 cm.

Mathematical Models of Signals

Acoustic emission signals are often modeled in the literature (Harris and Bell, 1977) by a decaying sine wave of the form

$$Y(t) \quad A_o \exp(-\alpha t) \sin \omega t \qquad \text{Eq. 12-23}$$

where $\omega = 2\pi \upsilon$, A_o is the amplitude and a is the damping factor. The graph of this curve is shown in Figure 12-14a. This model is a quite crude approximation of an AE signal as shown in Figure 12-12a. The rise time and complex frequency content are missing. However, it can be useful to show the approximate dependence of various signal analysis parameters on the signal characteristics. For example, the energy of such a signal is

$$E = \frac{A_o^2}{4a} \frac{1}{1 + a^2/\omega^2}$$ Eq. 12-24

Thus, the signal energy will depend directly upon the damping factor and the amplitude but is essentially independent of the frequency as long as $a \ll \omega$, which is the case for most AE signals.

To show the differences between the model signal in Equation 12-23 and a more realistic signal, a complex artificial signal was constructed on a computer. This signal is shown in Figure 12-14b. The spectra of the two artificial signals are shown in Figures 12-14c and 12-14b. It should be noted that while the signal in Figure 12-14b is much closer to a real signal such as seen in Figure 12-12a, it is much more difficult to manipulate mathematically than is Equation 12-23. However, if numerical methods are used on a computer, manipulation of artificial signals such as that in Figure 12-14b will give a more accurate estimate of instrumentation performance.

12.2.5 Acoustic Emission Instrumentation
The Acoustic Emission System

An AE system will always have a sensor and a preamplifier. A system usually includes postamplifiers and signal processors of various kinds. More specialized equipment often associated with a system include transient recorders, spectrum analyzers, tape recorders, distribution analyzers and spatial discrimination circuits. Recently, microprocessor-based systems have become commercially available. These are capable of performing all the standard methods of single channel analysis as well as performing source location for two to eight AE channels. The larger minicomputer based systems can also do single channel analysis as well as multichannel source location. Acoustic emission systems can be as simple or complex as one wishes, depending both on the experiment and the budget.

The Preamplifier

The loss in sensor sensitivity, which occurs when one is connected to an amplifier through a long coaxial cable, has already been discussed. The common way to solve this problem is to split the amplifier into a fixed gain preamplifier located close to the sensor and a variable gain postamplifier in the main instrumentation group. The preamplifier consists of a low noise input stage, bandpass filters and a low impedance output stage capable of driving a 50 ohm cable. A preamplifier is normally powered from the main instrumentation group, either by separate power and signal leads or by dc voltages on the signal

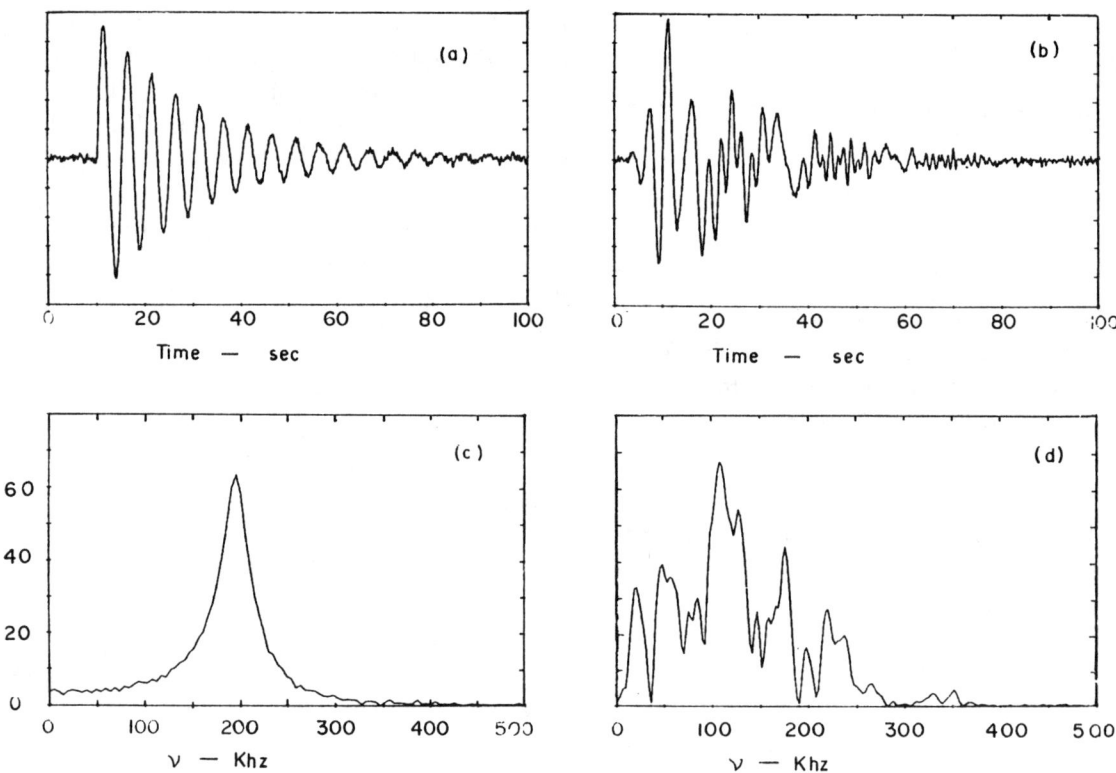

Figure 12-14. (a) Damped Sine Wave with Simulated Amplified Noise; (b) A Realistic Simulated Acoustic Emission Signal; (c) The Spectrum of Figure 12-14a; (d) The Spectrum of Figure 12-14b.

cable. Acoustic emission preamplifiers are designed to have a relatively flat frequency response between about 20 kHz and 2 MHz, without the bandpass filters. They almost always have a fixed gain of either 40 or 60 dB.

One feature common to all commercial AE preamplifiers is over-voltage protection in the input circuit. This protection usually takes the form of crossed semiconductor diodes which do not conduct until a few tenths of a volt have been reached, therefore limiting input signal amplitude to levels less than this. Such protection is necessary for semiconductor amplifying devices since piezoelectric sensors are capable of generating signals with amplitudes of hundred of volts if excited strongly enough. Besides the possibility of destroying the input stage, large signals can also saturate the preamplifier causing it to be inoperable for milliseconds or longer.

The input noise of the preamplifier has been discussed in the section on sensors. Here, it should be stressed that all other parameters being equal, the input noise level is proportional to the square root of the bandwidth. The input noise level will be reduced by 20 dB if the bandwidth of the preamplifier is reduced from 2 MHz to 20 kHz. This is one of several reasons why bandpass filters are used in preamplifiers. If nothing is known about the frequency content of AE in an experiment, a wideband filter should be used. However, if the frequency content is approximately known, a narrow bandwidth will lower the noise level.

Acoustic emission preamplifiers are divided between those with fixed bandwidths and those with plug-in variable bandwidth filters. In general, plug-in filters are preferable as they allow greater flexibility and require stocking only a variety of filters instead of preamplifiers. The filters in commercial preamplifiers do not all have the same roll-off characteristics. The roll-offs range from 12 dB/octave to 48 dB/octave (an octave is a range in frequencies of a factor of two; i.e., 125 kHz to 250 kHz). The signal amplitudes as a function of frequency for several roll-offs are shown in Figure 12-15.

It should be noted in comparing roll-off figures that several AE systems contain bandpass filters in both the preamplifier and the postamplifier. The roll-off figures of the two filters in dB should be added to obtain the characteristics of the system.

When a filter bandpass is specified, the upper and lower frequencies refer to the half power or 3 dB points; i.e., the output voltage is reduced to 0.707 of the output in the mid-range. This is the standard definition of bandpass and should be used in noise calculations. However, another function of the filter in AE is to remove unwanted noise signals (either acoustic or electromagnetic in origin) at low and high frequencies. Since such signals often have appreciable energies a useful definition of bandpass in AE would

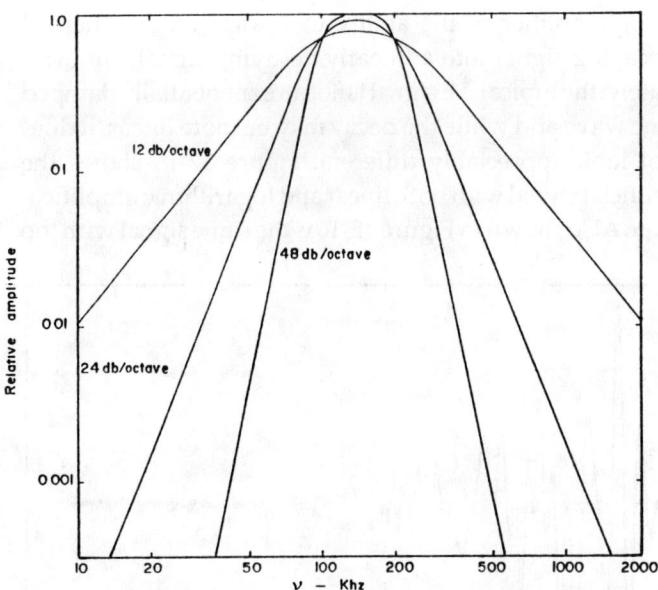

Figure 12-15. Signal Amplitudes as a Function of Frequency for 100 to 200 kHz Band Pass Filters with Three Different Filter Slopes. The 40 dB Band Widths for These Filters are 56-360 kHz for the 48 dB per Octave Filter, 32-630 kHz for the 24 dB per Octave Filter and 10-2000 kHz

be the 40 dB points. Such 40 dB bandpasses are shown for the different roll offs in Figure 12-15. It is obvious that the steeper the roll off, the better the suppression of unwanted signals. This can be highly important in an industrial plant where there are many sources of electromagnetic noise. For example, many power lines seem to have very short (~5 μs) large amplitude signals around 1.5 MHz while most room noise is usually below 20 kHz.

Another parameter that can be important in a preamplifier is the dynamic range. This is roughly defined as the ratio of the maximum output voltage the preamplifier can deliver without distortion and the amplified input noise voltage. The dynamic range of commercial preamplifiers normally falls between 60 and 80 dB. The 40 dB gain preamplifiers have a somewhat higher dynamic range. When using a signal processing technique, such as amplitude distribution, a large dynamic range is desirable. It should be realized, however, that 100 dB of dynamic range with 60 dB of gain would require an input noise level of 1 μV and 100 V of undistorted output.

One method of handling signals with a large dynamic range is to use a logarithmic amplifier. Some commercial systems are now using such amplifiers. A true long-amplifier is not possible for bipolar (positive and negative going; see Figure 12-16) signals. The actual gain curves for bipolar log-amplifiers are linear in the small signal region, logarithmic in the large signal regions and neither is the transition region. One of the touted virtues of a log-amplifier is the ability to change an exponential decaying signal into a linearly decaying signal. Unfortunately, the typical AE signal is not an exponentially damped sine wave and while the decay may be more linear, it does not look appreciably different. Figure 12-16 shows the artificial signal with both linear and logarithmic amplification. Also shown in Figure 12-16 is the same signal with too much gain in the log-amplifier so that the noise is accentuated at the expense of the signal. Log-amplifiers, when properly used, will not effect the accuracy of the count or peak amplitude data. They may slightly increase the ease of measuring signal lengths but will complicate the electronic measurement of signal energy.

There are a few sensors available which have the preamplifier built in the same package as the sensor. This has several advantages such as no cable capacity effect and having the preamplifier characteristic tailored to match the sensor. The disadvantages of such units, beside their higher cost, are that they are restricted to temperatures near room temperature (the preamplifier will not work properly at high or low temperatures) and that a separate preamplifier has to be purchased for each sensor.

Postamplifiers and Signal Processors

Many systems have variable gain postamplifiers. This allows the use of signal processors with fixed input ranges or thresholds in conjunction with fixed gain preamplifiers. The adjustable gain can either be linear by means of a potentiometer or logarithmic in one dB steps. The logarithmic gain is certainly easier to write down (i.e., 94 dB referenced to a one-volt threshold), but otherwise there is not much difference. The total gain of the system is, of course, the sum of the preamplifier and postamplifier gains (when expressed in dB).

As was mentioned, many postamplifiers also have bandpass filters which give an additional noise rejection capability to the system.

Many systems contain signal processing capabilities in addition to the analysis technique discussed in later sections. Some of those are voltage controlled gates which allow data to be collected only on certain portions of a load cycle, envelope processors which attempt to filter out the high frequencies leaving only the signal envelope to be counted, logarithmic converters which allow the output of the signal analyzer electronics to be plotted in logarithmic form and a unit which allows combining the output of several preamplifiers so that several sensors can be monitored by one channel of electronics.

Transient Recorders

One of the most versatile instruments for studying individual AE burst signals is the transient recorder. These instruments digitize a signal in real time and store the digitized data in a memory. The signal can be reconstructed at a slower speed from the memory and played back repetitively into an oscilloscope or spectrum analyzer. Or the contents of the memory can be transferred to a computer, thus using the transient analyzer as a signal digitizer.

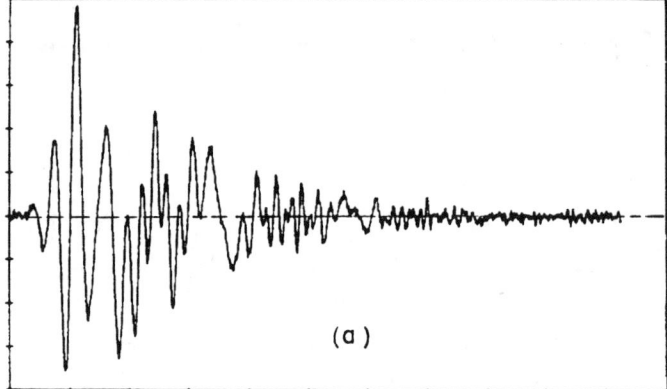

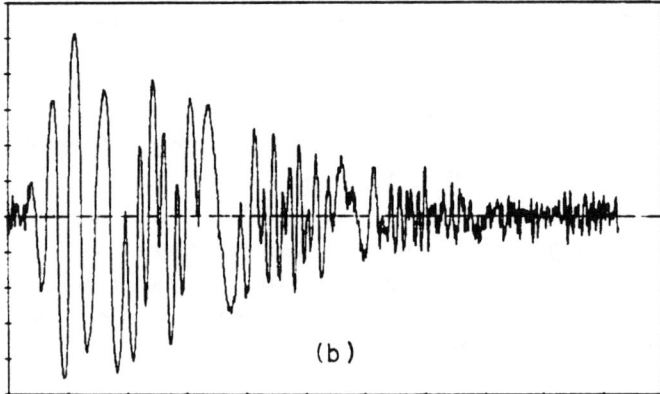

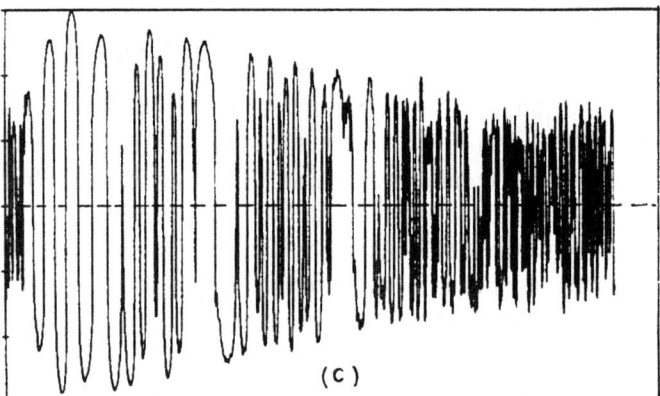

Figure 12-16. Effect of Logarithmic Amplification on Simulated Acoustic Emission Signal. (a) Original Signal; (b) Logarithmic Amplification; (c) Logarithmic Amplification with Excessive Gain Emphasizing Preamplifier Noise.

Transient recorders usually have a wide range of digitizing rates on the same instrument. The fastest rate is the limiting rate with some instruments sampling up to 1 word/ns. The slower rates are for convenience and may extend down to one sample per 10 seconds. These recorders sample the signal, and do not average it over the sample period. If the maximum sampling rate is one sample per 10 ns, the sample length will be somewhat shorter than 10 ns long. This is the sample width no matter that the sample rate is. In Figure 12-17, two sampled signals are shown. It is obvious that if one samples at two slow a rate for the signal, information is lost. The precise formulation of this principle is the Nyquist theorem (Stearns, 1975), which states that all the frequency information in a signal will be retained if the signal is sampled at a rate greater than two samples per period of the highest frequency component present. Thus, a signal containing frequencies up to 1.0 MHz should be sampled at more than two samples per µs. If the Nyquist criteria are followed, the signal can be accurately analyzed using Fourier analysis. However, if an accurate representation of the waveform is desired, a better rule of thumb is to take around 10 samples per second of the highest frequency component. This is illustrated in Figure 12-17a.

Sampling too slowly produces a distortion called aliasing (Stearns, 1975), an extreme case of which is illustrated in Figure 12-17b. The presence of frequency components higher than the Nyquist frequency give rise to image frequency components at lower frequencies. Once aliasing occurs, there is no way to recover the original waveform. To prevent aliasing, one should always use a low pass filter in the amplifier system before digitizing a signal. This filter should attenuate the signal by 20 dB at the Nyquist frequency. Failure to use an antialiasing filter may lead to significant <u>unrecognized</u> errors when analyzing digitized signals.

A simple solution to the aliasing problem would be to sample all AE signals at 10 to 20 samples per µs. However, there are two problems. The first is that the faster one samples in real time, the smaller is the digital word which can be used. A rough estimate is that one can use a 10-bit word at 20-60 samples per µs, an 8-bit word at up to 100 samples per µs and a 6-bit word at up to one sample per ns. The dynamic range of a bipolar 10-bit word is 54 dB, that of an 8-bit word, 42 dB and 30 dB for a 6-bit word. Thus, there is a tradeoff between dynamic range and sampling speed. Furthermore, the word size is fixed for a particular instrument and does not increase when the sampling speed is slowed.

The second problem with a high sampling speed is finite memory size. Transient recorder memories typically range between 512 and 16384 words, in powers of 2. If one samples at 20 samples pere µs with a 4096-word memory, only 205 µs of data can e stored. For long AE signals, the end of the pulse will be lost. Thus, the sampling rate should be as slow as possible. One to five samples per µs, depending on the frequencies present, are the most useful rates for AE signals. While these rates may still drop off the end of long signals, they appear to be the best compromise with present instruments.

This discussion applies to transient recorders which electronically digitize the signal. There is another kind where the signal is stored on the face of a fast storage oscilloscope. The face is then scanned and the signal digitized. This allows higher resolution at the expense of complexity and a longer waiting period before the next signal can be accepted.

One advantage of transient recorders is an additional mode of triggering, usually called the pretrigger mode. In this mode, the input signal is continuously being digitized and the data fed into the memory. When the memory is full, the earliest word in the memory is lost at one end while the latest word to be digitized is fed into the other end. This results in the memory always containing a digitized "pic-

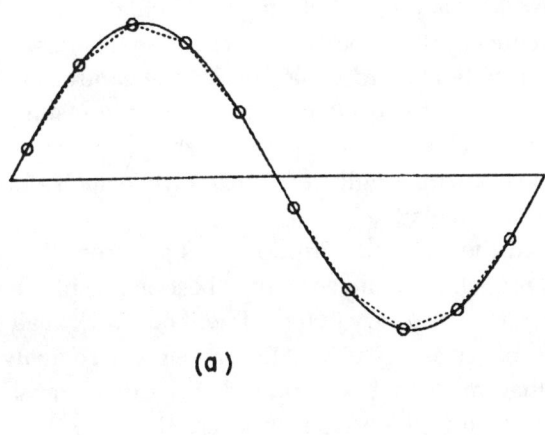

(a)

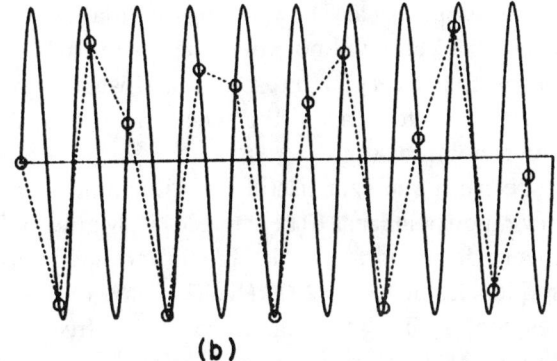

(b)

Figure 12-17. Effect of Sampling Rate on Digitization of Signals. (a) 10 Samples per Period; (b) Less than Two Samples per Period Showing Aliasing.

ture" of the signal, extending from the current instant of time back to the length of the memory. When a trigger signal comes (either internal or external), it can be set to stop this memory cycle with the trigger word being the first word of the memory, the last word, or any word in between. Thus, one can examine the signal that occurs either after or before the trigger. If the trigger is set to stop the memory with it at 10 to 15% of the memory (say word 400 to 600 for a 4096-word memory), the recorder will capture the entire signal including the low level portion occurring before the trigger level was reached.

Some of the more elaborate recorders allow the recording of two or more signals simultaneously. This allows the direct comparison of the time of occurrence of two (or more) AE signals or AE signal and some other signal of interest. An example is shown in Figure 12-18. This can also be done with two transient recorders, both using the same trigger.

A transient recorder is an extremely useful instrument, whether used as a digitizer with a computer or simply to examine AE waveforms. For most laboratory studies, a machine with a 10-bit word and a maximum sampling rate of ten samples per μs is recommended. The main justification for a more expensive machine is a need to record two data channels simultaneously.

Spectrum Analyzers

Spectrum analysis can be done either directly with analog electronics or digitally. This section will discuss only analog spectrum analyzers, reserving digital spectrum analysis for discussion in a later section.

The basic spectrum analyzer is simple in concept. A local oscillator signal is mixed with the input signal and the best frequency passed through a chain of intermediate frequency (I.F.) amplifiers, after which it is measured by a voltmeter. The oscillator is swept in frequency so that the frequency components of the signal selected by the I.F. amplifier are continuously changing. The output of the voltmeter is plotted on the vertical axis of an oscilloscope while the horizontal axis is driven by a signal synchronized to the local oscillator frequency. Thus, one obtains a plot of signal strength vs. frequency. If a logarithmic display is wanted, the vertical signal is passed through a log amplifier. The width and speed of the local oscillator scan, the center frequency of the local oscillator and the sharpness of the I.F. amplifier filters are all under the control of the operator. While it is quite easy to distort the results by inappropriate choice of these parameters, the better instruments have warning lights to indicate when such a poor choice has been made.

From the above description, it is apparent that an analogue spectrum analyzer works best on a continuous signal that does not vary in time. It will not work at all on a short transient such as an AE burst signal. To analyze bursts, they must first be captured either on a transient recorder or on a video tape recorder. They can then be played back repetitively onto the spectrum analyzer, simulating a continuous signal. The resulting modulation (typically, 60 Hz to 1 kHz) will not effect the spectrum unless one attempts to get very high resolution. Most AE bursts have rather broad features in the spectrum so that high resolution is not necessary.

A spectrum analyzer useful for AE should have a frequency range of at least 10 kHz to 2 MHz. Many on the market were designed for audio signals and have maximum frequencies of 100 or 200 kHz. These are of limited usefulness for AE since most signals have some frequency content up to 500 kHz.

Spectrum analyzers with both horizontal and vertical signal outputs are the most useful for AE. This allows the signal to be plotted on an X-Y recorder which gives a more useful and detailed permanent record than an oscil-

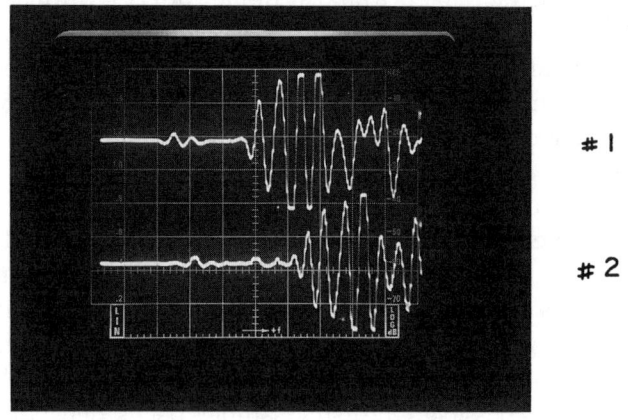

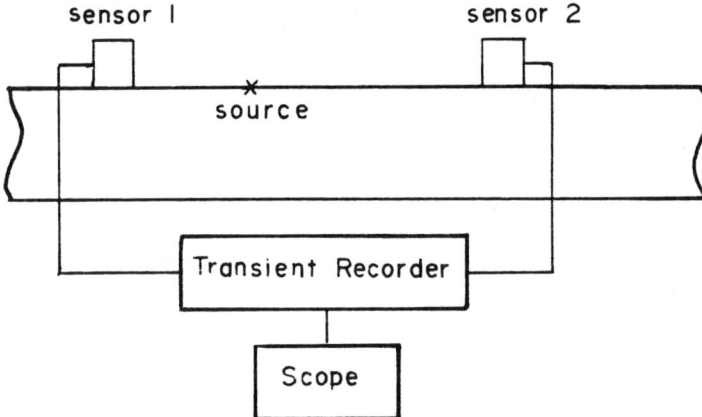

Figure 12-18. Transient Recording of Two Channels Simultaneously Showing the Time Relationship of the Signals. (a) Signals; (b) Experimental Setup.

loscope picture. Figure 12-11 shows an X-Y plot of the output of a spectrum analyzer.

Data Storage

The intrinsic irreversibility of most AE sources often makes the storage of the raw data desirable. This will allow the direct comparison of experiments when data analysis parameters are changed or new techniques of data analysis are used on later experiments. When testing in remote areas or performing tests which last for only a short time, storage of the raw data will allow extensive post test analysis in the laboratory at the leisure of the investigator.

At present, the only practical method for storing raw AE data is on magnetic tape. The frequency range of interest for most AE data, 50 kHz to 1 MHz, is too high for audio or small instrumentation recorders. Only large instrumentation recorders or video tape recorders have the necessary frequency capabilities. To record these high frequencies, a large relative velocity between the tape and the record and reproduce heads is necessary. Instrumentation recorders and video recorders differ in how they achieve this relative velocity as well as in other details. For this reason the two types of tape recorders will be discussed separately.

Instrumentation Recorder

In an instrumentation recorder, the record and reproduce heads are fixed and the tape is moved over them at a high velocity. Tape velocities range from a high of 3.05 m/s (120 ips) down to 2.36 cm/s (0.9275 ips). The velocities are usually stepped by factors of 2. The tape heads for individual channels are quite narrow and record on a track parallel to the tape. Tape widths are either 12.7 mm (1/2") or 25.4 mm (1"). Half-inch tape has 7 data channels while 1-inch tape may have either 14 or 28 channels. Both widths also have an audio channel.

Data can be recorded on the tape either as an amplitude or frequency modulated signal. The amplitude modulated (or direct) record electronics have a maximum frequency pass-band of 400 Hz to 2 MHz at a tape velocity of 3.05 m/s. The signal-to-noise ratio is about 24 dB for this pass-band. The frequency modulated (FM) electronics have a maximum pass-band from 0 to 500 kHz at 3.05 m/s with a signal-to-noise ratio around 34 dB. The maximum frequency is proportional to the tape speed. For example, the pass-band of the same FM electronics would be 0 to 62.5 kHz at a tape speed of 38.1 cm/s.

Tape comes in several size reels, the largest holding about 2800 m of tape. While this seems like a large amount of tape, at 3.05 m/s it will last only 15 minutes. This means that for instrumentation recorders, one is always faced with a tradeoff between high frequency response and the length of a test which can be recorded on one reel of tape.

The playback speed on an instrumentation recorder is usually independent of the speed at which the tape was recorded. This allows great flexibility in the ability to examine short records in detail. Playing the tape at a slower speed also down shifts the frequencies of the data by the ratio between the tape record and reproduction speeds. One problem in older machines in playing back the tape at a slower speed is that it may be tedious to find the few meters of tape which are of interest. At 2.4 cm/s, a 2800 m reel will play for 32 hours. Searching at 3.05 m/s and then slowing the tape down at the desired spot may not be easy since the inertia in the reels often carries the tape many feet past the data of interest. Newer, microprocessor controlled machines allow one to find the region of interest and then to shuttle the tape within this region back and forth at any speed as well as to easily return to that portion of the tape. To anyone who has tried to relocate 3 m of tape in a 3-km-long reel, this is a major advantage.

Instrumentation recorders have long been bulky, heavy machines that are awkward to transport between buildings or sites. The newer machines can be much smaller and lighter. While they still weigh 50 kg or more, this is almost weightless compared to the half ton of their predecessors. The older recorders are complex machines and do not have a reputation for high reliability, although a well maintained machine can be relatively trouble free. One can only hope that the size and weight reduction of the newer machines does not reduce their reliability. Another major disadvantage has been the high cost of these recorders. There seems to be little hope of any substantial cost reduction.

Video Tape Recorders

A video tape recorder differs from an instrumentation recorder in the manner in which the relative velocity between the head and the tape is achieved. In a video recorder, the tape moves at slow speed past a rotating head. This geometry is shown in Figure 12-19. The data tracks are parallel to each other but diagonal to the tape as shown in the figure. This configuration limits a video recorder to a single data channel with possibly two audio channels, one on each edge of the tape. There are generally two heads as shown in the figure. These heads read or write on the tape sequentially which requires switching the signal between them every 180° of rotation. The speed of the head rotation is often set so that each record is 16-2/3 ms long to conform with television transmission characteristics. For television recording, there is a small gap between each record but for AE use, the recorder is modified to eliminate this gap. This modification can leave small transients where the heads are switched. The transients may be

20 or more dB above the recorder noise level and if played into a counter may give 1 to 10 more counts, 60 times a second. For a test with large amounts of high energy AE, the transients can be ignored, but for quiet tests it may be necessary to use a gating circuit to remove them (a word of caution—poor gating circuits can add more transient than they remove). Since the transients are on the order of 0.2 to 0.4 ms long, at most only 2-1/2% of the data is lost by such gating. It is possible to make transient free video recorders and such video instrumentation recorders are available but more than an order of magnitude difference in their price over that of the small TV units does not make them very attractive for AE. The new video cassette recorders are smaller and cheaper than the real units currently used for AE and they should be able to be modified for this use.

Typical video tape recorders have bandwidths from 0 to 3.5 MHz or higher. Signal-to-noise ratios are between 30 and 40 dB. Tape speeds are on the order of 19 cm/s (7.5 ips) which will give an hour or recording time for a 700 m tape. The reel-to-reel recorders all weigh between 20 to 32 kg (45 and 70 pounds) which make them somewhat portable.

Since the necessary velocity between the head and the tape is furnished primarily by the rotation of the head, the tape can be stopped and one record played repetitively.

This allows examination of single AE bursts. Commercially modified units are available with trigger and gating circuits which allow one single AE signal to be selected and played into a spectrum analyzer. Thus, burst-by-burst analysis of an AE test is possible with such recorders. Another available modification is voltage-to-frequency and frequency-to-voltage converters which allow the recording of DV voltages on one of the audio channels.

The limitations of the video recorder are the single data channel, the presence of head switching transients and the limited dynamic range. The advantages are low cost, portability and the stop action feature.

Spatial Discrimination

Spatial discrimination (Nakamura, 1971) is a technique for accepting AE signals which originate only within a region of interest. Other names for spatial discrimination seen in the literature are master-slave transducers and guard transducers. While spatial discrimination circuits use several transducers, there is usually only one data channel and no provisions for source location.

Figure 12-20 shows a spatial discrimination setup as it would be used in a tensile test. The purpose of this setup is to eliminate AE coming from grip region and accept signals only from the gage region. The circuit is quite simple. If a signal from a guard sensor reaches the discriminator before a signal from the data sensor, it inhibits the data analyzing device—usually an electronic counter. If the signal from the data sensor reaches the circuit first, the guard sensor signal has no effect. This circuit effectively draws a line half way between the data sensor and the guard sensor. Emission bursts originating closer to the guard sensors are ignored. By paralleling several guard sensor-preamplifiers on the same input or several guard sensors on the same preamplifier, a two-dimensional acceptance region can be set up. If a guard sensor is acousti-

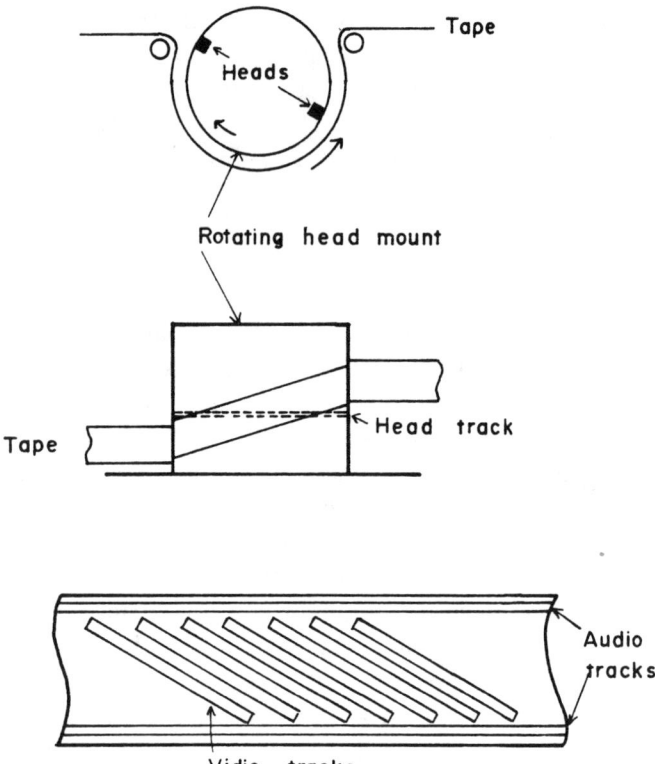

Figure 12-19. Geometry of Video Tape Recorder. (a) Top View of Tape and Head; (b) Side View of Tape and Head; (c) Position of Recorder Tracks on Tape.

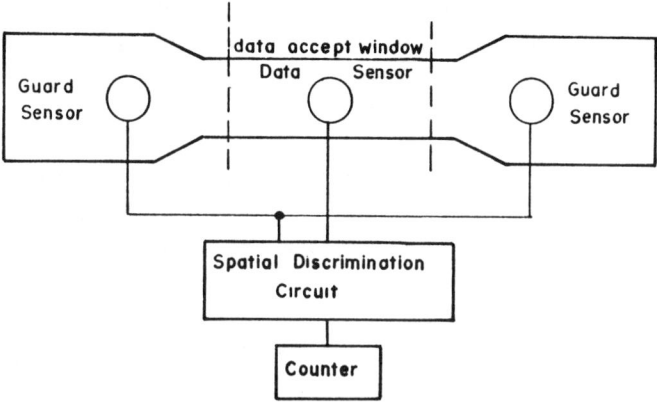

Figure 12-20. Setup of Spatial Discrimination Circuit on Tensile Specimen.

cally isolated from the experiment and a small time lag introduced between the data sensor and the discrimination circuit, one can eliminate much of the electrical noise that may be present in a test area.

12.2.6 Signal Processing Methods

Most AE processing methods measure the characteristic of the individual burst emission signals. An analog system will usually present these characteristics either as a running sum or as a sum per unit time. A microprocessor based system can store the measurements in memory for future analysis as well as presenting these same sums in real time. These sums are a form of statistical analysis since the summing process merges the individual characteristics of each signal into a somewhat smoother function of time. Seldom are the characteristics of each signal analyzed individually since the large number of burst signals and the statistical distribution of the bursts make such an analysis both cumbersome and not very meaningful for the average AE test.

Processing methods where the signal characteristics are presented individually are audio and visual analysis. In these methods, the averaging procedure is left to the human brain. The rms voltage measurement also differs in that the signals are averaged in the electronics before the desired characteristic is measured.

All of the signal processing methods discussed in this section can give useful information in an AE experiment. The two most useful, especially for NDT, are visual analysis and the AE count. The running display of the raw signals on an oscilloscope give the experienced investigator a good feel for what is happening. The plot of the total count as a function of time, pressure or temperature gives a quantitative record which is easy to read and interpret. An audio channel allows continuous monitoring of the test even when reading the plot of the count and allows a qualitative interpretation for some characteristics of the longer signals.

Audio Conversion

Many early AE experiments were restricted to the audio (~ 20 Hz to 15 kHz) frequency range. It was natural to amplify these signals and feed them to a loud speaker. An experienced investigator could learn much about the experiment from listening to such signals. When most AE experiments went to higher frequencies to eliminate interference from the background noise in the room, the AE signals were down converted in frequency to allow the continued use of audio monitoring. This down conversion is done by mixing the AE signal with the output of an oscillator. For two continuous waves we can write

$$\cos(\omega_1 t) \cos(\omega_2 t) \qquad \text{Eq. 12-25}$$
$$= \left((\omega_1 + \omega_2) t + \cos(\omega_1 - \omega_2) t \right)$$

The frequency $\omega_1 - \omega_2$ is called a beat frequency and can be made to be in the audio range by a correct choice of oscillator frequency.

There are two problems with the use of such a simple analysis on AE signals. The first is that they are transients containing a wide range of frequency components. Thus, it becomes difficult to pick a good oscillator frequency. This can be solved by using a square wave generator instead of a sine wave generator. A square wave has a large harmonic content and its Fourier series representation is

$$f(\omega t) = \sum_{n=1}^{N} \sin(2n-1)\omega t/(2n-1). \qquad \text{Eq. 12-26}$$

Notice that only odd harmonics are present and the amplitudes of the higher harmonics drop off relatively slowly. Each frequency component in the AE signal that is within about 10 kHz of one of the square wave's harmonics will produce a useful audio signal. With a variable frequency square wave oscillator covering the range of 50 kHz to 200 kHz, the entire AE frequency spectrum can be covered. The oscillator frequency is set by ear to within 10 kHz of a dominant AE signal frequency for the test. The audio output will consist of a mixture of all signal frequencies near the oscillator frequency of any of its harmonics.

The other problem with audio conversion is that often AE signals are quite short. A 100-μs-long signal can give only one cycle of a 10 kHz signal no matter how it is processed. Therefore, the audio signal is often a short, high-frequency burst that sounds like static. Definite frequencies are apparent only for the longer signals. Even so, audio conversion is almost always useful as an alarm that AE is occurring. For structures, where longer signals often occur, the change in tone or "quality" of the signals can be the first sign of the beginning of some type of failure. The discriminating ability of a trained ear is extremely good and most experienced practitioners of AE testing will include an audio channel on structural tests.

Visual Analysis

Generations of electrical engineers have watched electrical signals on oscilloscopes. It was natural to look at AE signals on oscilloscopes even though they are transients. While it might seem that the transient bursts flash by too quickly to allow much of an impression, any change in the overall character of the signal is readily apparent. Event rate, signal amplitude, signal duration, ratios of different types of signals all give a definite if subtle visual character to the signals. A change in this character is usually easy to spot although it may be hard to define. Visual analysis should be used for all but the most routine AE experiments.

The combination of preamplifier and oscilloscope gain is usually set so that the preamplifier noise is visible on the trace. The best sweep speed for most experiments of 2 ms/division. Slower speeds may flicker annoyingly and faster speeds may not allow the display of several bursts on the same trace which is useful when a change in the average repetition rate occurs.

RMS Voltage or Signal Level

The best measure of continuous emission is the average signal amplitude. Two common instruments for measuring this amplitude are the rms (root mean square) voltmeter and the ac voltmeter. The rms voltage is defined such that the signal power is proportional to the square of the rms voltage.

$$\bar{P} \approx (\bar{V}_{rms})^2 \qquad \text{Eq. 12-27}$$

The ac voltage is the average value of the rectified signal. In both measurements, the signal is averaged over a period of time. Generally, the rms voltage will be equal to the ac voltage only when the signal is a continuous wave with fixed frequency components. For AE signals, with their wide range of amplitudes, average repetition rates and frequencies, the two measurements can give quite different values although usually the difference will not be significant.

Rms voltages are measured in an instrument such as the Hewlett Packard 3400A voltmeter by measuring the heating power of the signal. The signal is amplified and applied to one resistor while a dc voltage is applied to an identical resistor and varied so as to keep the temperatures of both resistors the same. The dc voltage is then measured. Because this involves a thermal measurement, the response is relatively slow. A measured response time for the HP 3400A is 200 to 300 ms. Thus, an rms voltmeter cannot follow the envelope of AE bursts. This slow response effectively averages the incoming signals. Bursts with a high average repetition rate and a fairly small range of amplitudes will appear as a constant output voltage. High amplitude pulses that occur at large intervals will show up as a spike for each burst. While the long time constant changes the appearance of the signal record, it does not effect the accuracy. The energy measured by the rms voltmeter is the energy of the signals. That is

$$E = \frac{1}{R} \int_0^T V(t)^2 = \frac{1}{R} \int_0^T V_{rms}^2(t) dt \qquad \text{Eq. 12-28}$$

as long as the integration time, T, significantly exceeds the time constant of the rms meter.

An rms voltmeter measures not only the AE signal voltage, but also the noise voltage of the preamplifier. If we assume that there is no correlation between the preamplifier noise voltage and the AE signal, then we can write

$$V_{rms} = \sqrt{\gamma_{rms}^2 - V_n^2} \qquad \text{Eq. 12-29}$$

where V_{rms} is the actual signal voltage, γ_{rms} is the meter measurement, and V_n is the rms noise voltage of the amplifier chain.

AC voltmeters are much simpler, electrically, than rms voltmeters. They are basically just a rectifier followed by a filter. The time constant will depend upon the frequency range that they are designed to measure. For meters designed to measure frequencies above 10 kHz, this time constant can be less than a millisecond. Most microprocessor based systems and many of the newer analog systems incorporate an ac voltmeter in them to give a measure of the signal level. For most nondestructive testing applications rms voltmeters and ac voltmeters can be used interchangeably. It is only in the laboratory, where the value of the signal energy or power may be important, that rms measurements are superior. The reader is referred to the work of Hamstad (1974) for more information about rms voltage measurement and its uses and to an experiment by the author for a use of signal level measurement (Beattie, 1982).

Acoustic Emission Count and Count Rate

The measurement of the AE count is one of the easiest and most useful methods of analyzing AE. In practice the signal is amplified and fed into an electronic counter. When the counter runs continuously without being reset, the output is the total count. If the counter is reset after a period of time (usually between 0.1 s and 1 hour), the maximum count for each time period is the count rate. In both measurements, the counter output is often transformed by a digital-to-analog converter so that it can be plotted on an Z-Y recorder.

The AE count is an excellent qualitative and a very poor quantitative measure of AE. This can be seen by an examination of the method by which the count is measured. Figure 12-21 shows the artificial AE signal first seen in Figure 12-14b. Several different parameters are shown on the drawing. The electronic counter will register one count any time the threshold voltage, V, is crossed in a given direction (in this case, positive to negative). The combination of the gain and the threshold voltage is usually set so that the threshold level is near the noise level. Some practitioners prefer to set the threshold level so that

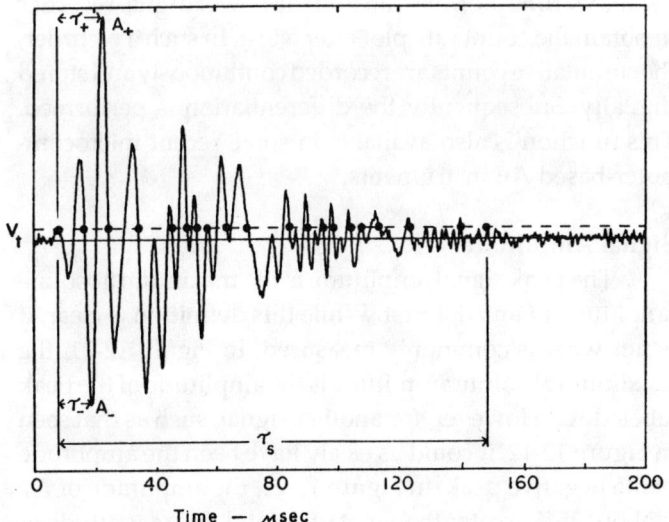

Figure 12-21. Simulated Acoustic Emission Signal Showing the Triggering Points for the Acoustic Emission County, the Peak Amplitudes, the Rise Times and the Signal Length.

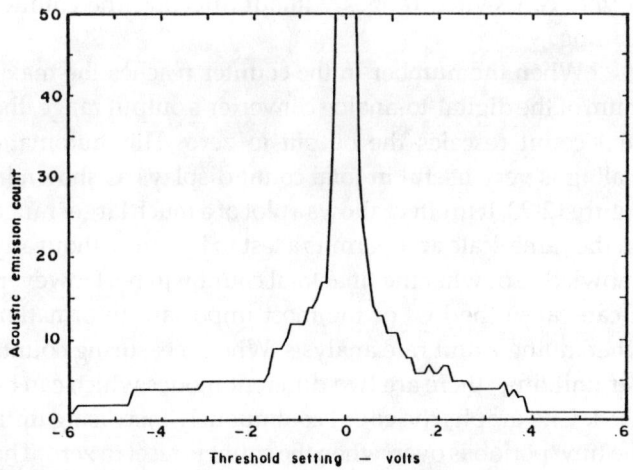

Figure 12-22. Effect of Counter Threshold Setting on Measured Count for Simulated Acoustic Emission Signal.

a few of the largest noise peaks are counted. That way they know that the system is functioning and that any signal above the noise will be counted. Others prefer to set the level just above the largest noise peaks so that only the AE signals are counted.

It is obvious that the number of counts registered for an AE signal will depend upon the gain and threshold settings (some commercial systems have a fixed threshold, usually one volt, so that only the gain can be varied). In Figure 12-22, the count for the signal in Figure 12-21 is plotted as a function of threshold voltage. This curve is neither smooth nor symmetric. The occasional decrease of one count as the absolute value of the threshold voltage is decreased results from the small noise peaks superimposed on the signal peaks. The central maximum is produced by the noise background which in this case has a maximum value of ±0.5V. Thus, the value of the count from a given signal depends not only on the gain-threshold setting, but also on the symmetry of the signal.

The dependence of the count upon the threshold setting can lead to problems when testing pressure vessels. The increase in the background noise level produced by a leak (even a very small one) can saturate the counter so that all meaningful AE signals are lost in the noise. To allow count data to be collected despite such leaks, floating threshold circuitry has been devised. This circuitry measures the background noise level and sets the threshold a fixed amount above it. One of these circuits sets the threshold a fixed voltage above the background and another sets it a fixed dB level above the background. Both methods seem equally effective. Such circuitry should be available whenever a pressurized system is tested.

Figure 12-22 graphically illustrates that while the count can give some qualitative estimate of AE activity, there is no fixed mathematical relationship between the count and any other signal parameter for real AE signals. The best that can be said is that if the AE signals all have similar amplitudes, decay times and frequency content, then the AE count can be shown to be related to the signal energy.

The AE count is an indicator that AE is occurring and gives a rough estimate of the rate and amount of the emission. In a single channel test on a structure, this is really all the information that is essential. The two types of displays, total count and count rate, are shown for the same test in Figure 12-23. In this case, both are displayed as a function of time instead of an external parameter such as pressure. Of the two displays, I prefer the total count for the following reasons. First, it appears easier for the eye to differentiate a curve than to integrate it. An increasing AE count rate is more obvious on the total count curve. Second, data cannot be lost on a total count curve as it can on a count rate curve. This possible loss is described below.

Electronic counters are digital devices and a digital-to-analog converter is needed to produce graphs like Figure 12-23. These converters take three decades of input and convert them into a thousand different voltage levels. Which three decades of a number are converted is under control of the operator. One can choose to plot the counts from 1 to 1,000, 10 to 10,000, etc. In most commercial AE equipment, when the plot range is switched, the number in the counter is lost. This is usually only a minor annoyance but it is not necessary. Laboratory digital-to-analog converters can switch ranges without affecting the counter reading.

When the number in the counter reaches the maximum of the digital-to-analog converter's output range, the next count rescales the output to zero. This automatic scaling is very useful in total count displays as shown in Figure 12-23. It in effect allows a plot of a much larger range on the same scale and permits a test to be run without any knowledge of what the final total count will be. However, it can cause the loss of the most important information when doing count rate analysis. When measuring counts per unit time, there are two different modes which can be used. One graphs the count continuously increasing until the time period is over, when the count is reset to zero. The other displays the final count for the previous time period as the count for the current time period is being measured. This latter method makes it much easier to read the graphs but can cause the loss of large bursts. If the digital-to-analog converter is set to a maximum of 1,000 counts, a count of 1,050 in that second will reset the output of 50 counts and give no printed record of the 1,000 counts. Thus, one can lose all record of the important large bursts in this mode. If the summing mode is used and several zeroings occur during the time period, the data will not be lost but may be hard to interpret.

Since the rate of AE counts is the derivative of an AE counts vs. time plot, one can utilize a newer digital recorder to obtain the count rate plot after a test. In such a recorder, the cumulative counts are recorded continuously and stored digitally. Subsequently the differentiation is performed. This function is also available in some recent microcomputer-based AE instruments.

Signal Amplitude

The peak signal amplitude is the maximum absolute amplitude of an AE burst. While this definition is clear, it is not what is commonly measured. In Figure 12-21, the maximum absolute amplitude is the amplitude of the peak labeled A_+. However, for another signal, such as that seen in Figure 12-12, it could as easily have been the amplitude of the negative peak. In Figure 12-21, the amplitude of A_+ is about 25% greater than that of A_- and this magnitude of disparity between positive and negative peak amplitudes is typical of AE signals. Most of the signal amplitude measuring circuits measure only the positive (or negative) peak amplitude. Therefore, there is a built-in uncertainty of around 3 dB in each measurement.

One method of measuring the signal amplitude is to measure the amplitude of each cycle and to replace the previous value in memory by each higher value. How long this process is continued depends upon the circuit used. It can either be terminated after a fixed time or terminated

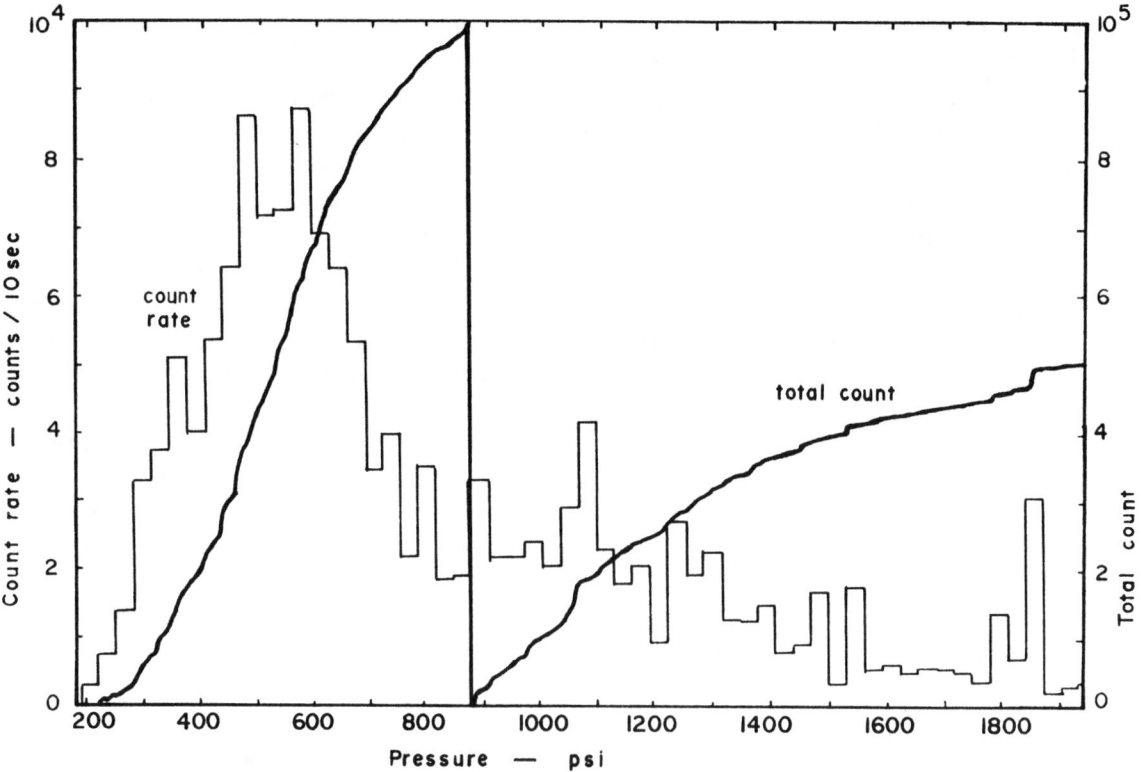

Figure 12-23. Plot of Total Count and Count Rate Curves as a Function of Pressure for an Acoustic Emission Test. Note the Rescaling of the Counter in the Middle of the Total Count Curve.

when the amplitudes of succeeding cycles start decreasing. Either way, a large peak occurring late in the signal as in Figure 12-13c can be missed.

Such uncertainties in the measurement are matched by the uncertainties in the origin of the signal. In Fig. 12-13 we saw that the peak amplitude may be caused by acoustic wave packets traveling different paths. In fact, the value of the peak amplitude may be due to a fortuitous superposition of several wave packets and have little relationship to the peak value of any of the individual packets. Acoustic attenuation of the emission also adds to the uncertainty.

The result is that the peak amplitude of an individual signal may mean very little. However, when some type of statistical average is taken, it is a very significant parameter. The digital values of peak amplitudes are sometimes put into electronic counters and plotted as are AE counts. By themselves, these data are no more significant than the count and much harder to obtain. However, a microprocessor system which can plot the average amplitude per burst as a function of parameter may give a significant warning of increasing flaw growth rate.

In measuring the average amplitude per burst, different averages will be obtained depending upon whether a linear or logarithmic preamplifier is used. If a linear preamplifier is used, the average amplitude will be

$$\bar{A} = \frac{1}{n} \sum_{i=1}^{n} A_i \qquad \text{Eq. 12-30}$$

where n is the number of bursts in the average. If a logarithmic preamplifier is used, the average amplitude will be

$$\bar{A} = \left[\prod_{i=1}^{n} A_i \right]^{1/n} \qquad \text{Eq. 12-31}$$

where π indicates a product. These are both legitimate averages but they are not equal. For NDT applications, the difference may not be significant, but in a laboratory experiment where one is trying to get quantitative data, the difference should be recognized.

The other statistical method of using peak amplitudes is an amplitude distribution. This will be discussed in the section on statistical distribution. Its usefulness is based in part upon the empirical observation that different AE source mechanisms can produce different statistical distributions of signal amplitudes.

Signal Rise Time

The signal rise time can be defined as the time period between the time when the AE signal is first detectable above the noise level and the time when the peak amplitude occurs. The rise time is labeled τ in Figure 12-21. Because only the positive (or negative) peak amplitude is usually measured, the electronics will measure only τ_+ (or τ_-) no matter which is the actual rise time. There is another ambiguity in the rise time measurement which is the determination of the first detectable signal. This will depend upon the threshold voltage and the noise level. In thin plates where the primary propagation is by plate waves, the electronics may detect the longitudinal wave generated by sources near the sensor but only the plant wave for sources far from the sensor. The peak amplitude will be that of the plate wave. Dispersion will produce increases in the rise times that is a function of the distance the wave travels. Thus, the measured rise times may depend in several ways upon the distance of the source from the sensor.

Rise time measurements are a relatively new addition to AE analysis techniques, although some attempts were made to use its value as an indicator of the source-to-sensor distance. The above discussion indicates that a plot of the average rise time could be useful but at this date there is little experimental data available.

Signal Duration

The signal duration is defined as the length of time that the burst emission signal is detectable. This is a very ambiguous definition since it depends upon the preamplifier noise level and the method of detecting the signal. A typical detection method uses trigger circuits similar to those in electronic counters (often the output pulses from the counter registering AE counts are used). The trigger pulses start and stop a separate counter that counts clock pulses. Thus, the signal duration measurement, unlike the count, is independent of the frequency content of the burst signal. Figure 12-21 shows the signal duration, τ_s, for the artificial signal and for that particular threshold setting.

It is obvious that absolute values of signal duration cannot be measured. Their values depend not only on the electronics, but also on the sensor sensitivity and the efficiency of the couplant. However, the duration can be a useful measurement when measuring the relative durations of signals from the same test. These durations will usually be between 10 μs and 50 ms. They depend upon the signal amplitudes and decay constants. A change in either the average signal duration or the distribution of durations can indicate either a change in the signal path to the sensor or a change in the generating mechanism. Both can be important in structural tests. For example, in a glassfiber composite, matrix crazing generally produces short durations while propagating cracks produce long signals. Another use is that electrical noise is often quite short, <10 μs, while AE signals are usually longer. The signal duration can be used as a means of discriminating between the two signals.

Signal Energy

The energy is the one characteristic of an AE signal that can be unambiguously defined. The definition is given by Equation 12-32. This definition assumes a large signal-to-noise ratio. A definition in the presence of noise is

$$E = \frac{1}{R} \int_0^T V^2(t)\,dt - \frac{1}{R} \bar{V}_n^2 T \qquad \text{Eq. 12-32}$$

where $\bar{V}_n$ is the average rms noise voltage and time T completely contains the signal. The problem with energy measurement is not in the definition but in the practice.

The signal energy can be measured (Beattie, 1974) by a circuit similar to the block diagram shown in Figure 12-24. Basic to such a circuit is the squaring module. This module must operate at frequencies up to 2 MHz and have a large dynamic range. The dynamic range is the major problem in energy measurement. Acoustic emission experiments can easily have a range of signal amplitudes exceeding 60 dB. If these signals are squared, the output amplitude range of the squaring module will be 120 dB. Currently available modules have an output amplitude range of slightly over 80 dB, which will limit the input range to about 40 dB. This is still a better dynamic range than some commercially available circuits or published circuits which are linear in energy over about 25 dB.

One solution to the problem of dynamic range in energy measurements is to use several parallel amplifiers with different gains, each feeding into a different energy circuit. A test for signal clipping could let a microprocessor determine which reading to keep and what scale factor to multiply by. Such a circuit is complex but it is certainly possible with existing electronics and would permit energy measurements over the complete signal dynamic range.

Another proposed method for measuring the signal energy is to multiply the square of the peak amplitude by the signal duration. Computer modeling shows that for similar signals varying only in amplitude, there is an approximate linear relationship between this product and the energy. However, as the damping factor or other parameters related to the signal shape change, the relationship between this product and the energy is no longer linear. When this is added to the inaccuracies in the measurement of the peak amplitude and the signal duration, it is doubtful that this product has any closer relationship to the signal energies than does the AE count.

Outside of a laboratory experiment, where the AE energy is probably the best parameter to use, the main reason to use energy analysis is to accentuate the signal with either abnormally large amplitudes or durations. If the expected failure mechanisms produce such signals, then energy analysis can be valuable. Otherwise the added complexity and reduced dynamic range of energy analysis circuitry make the AE count a more useful analysis method.

Event Count

An AE event is defined as a detected AE burst. Thus, an event describes the AE signal and not necessarily the AE. The propagation of the AE by several different paths may result in one disturbance producing several AE events. An example of this is a hydrostatic test of a pressure vessel where the AE traveling the steel surface and that going through the water may arrive at the sensor at sufficiently different times to be classified as two events.

Nevertheless, the concept of the AE event is essential when using a multichannel source location system and useful in single channel tests. For some laboratory tests on systems where the generating mechanisms are thought to produce discrete signals (breaking fibers in a composite or transgranular cracking in a large grain metal), the event count is the most desirable analysis method.

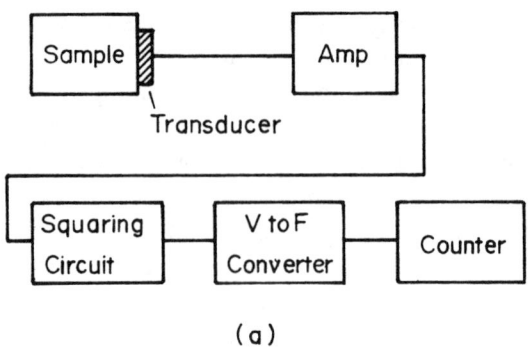

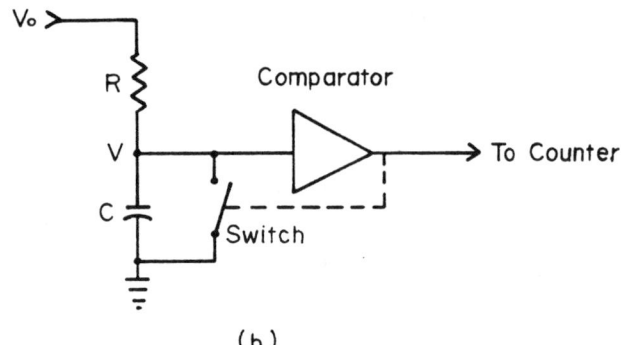

Figure 12-24. (a) Block Diagram of the Energy Measuring Circuit; (b) Block Diagram of a Simple Voltage-to-Frequency Converter.

The measurement of events is usually done by demodulating the signal so that the only burst envelopes are left and then counting the envelopes. Such demodulation involves a time constant in the circuitry. This in turn implies that the event count will be correct only when all the AE signals have about the same decay constants. A mixture of decay constants will often confuse the event counting circuitry. The circuitry will also be confused if the events occur rapidly enough so that some start to overlap. The first problem could be overcome by measuring the duration of each signal and locking out the event counter until that signal was finished. Once the events are processed by an electronic counter, the displays can be handled the same as the AE count.

Count or Energy per Event

Often, as a sample approaches failure, the AE bursts seen on the oscilloscope appear to get larger. This increase, if real, can be measured by a running calculation of the counts or energy per event. Such a calculation is most easily done with a microprocessor-based system and can also be done with a two-channel counter which can take the ratio of two frequencies. For most experiments, either the count or the energy per event will give about the same results. It is only when there is a change in either the damping factor of the frequency that energy per event is the better parameter. Considering the dynamic range problems and complexity of energy measuring circuits, the count per event measurement seems preferable.

12.2.7 Advanced Analysis Methods
Distribution Analysis

The randomness inherent in the generation of AE, the uncertainties in the path and wave modes during the transmission from source to sensor, and the instrumentation errors in quantifying the signal parameters all argue for a statistical analysis of AE signals. One type of statistical analysis, distribution analysis, has come into wide usage in recent years.

In distribution analysis, some parameter of a burst emission signal is measured. This parameter is divided into a range of values (on either a linear or logarithmic scale). As each signal is measured, a counter assigned to that particular range of values for the parameter is increased by one unit. For example, if the parameter is peak amplitude, the amplitude may be divided into one dB bins and a signal would increase the counter assigned to, say, the 43 dB bin by one unit. After many burst signals have been measured, the resulting counts may be plotted as a histogram or differential curve such as that shown in Figure 12-25. Another way of representing this data is to plot each point as total number of all signals with that value

or a higher value of the parameter. This summation curve is also shown in Figure 12-25 for the same data. In this figure both the parameter values and the number of events are plotted logarithmically. Both axes can also be plotted linearly.

The most common parameter used in distribution analysis is the peak amplitude although signal energy, count and duration are also used. Any signal parameter which can be measured can be used in distribution analysis. While the idea of distribution analysis is relatively simple, the electronics used to perform such analysis are complex and moderately expensive. However, all modern micro or minicomputer based systems can perform this type of analysis and several stand-alone systems are also available. The main problem in the use of distribution analysis is not in the acquisition of the data but in its interpretation. The simplest type of analysis is to take distributions, such as that shown in Figure 12-25, during the course of an experiment and to look for the appearance or disappearance of features, such as the peaks, as the experiment progresses. This can be a very useful type of analysis. For example, the appearance of a number of high amplitude or energy events in a test of a fiber composite may signify a growing crack. However, all too often, the examination of a sequence of such curves taken during an experiment will lead to as many scenarios as there are investigators.

To help remove the problem of the interpretation being based primarily on the basis of the investigator,

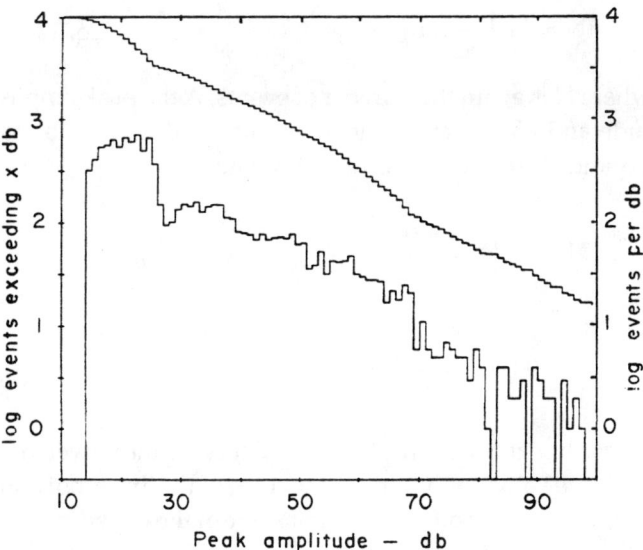

Figure 12-25. Two Methods of Plotting Peak Amplitude Distributions. The Bottom Curve is a Differential Plot. The Top Curve is a Plot of the Total Number of Events with Amplitudes Greater than the Amplitude at that Point.

attempts have been made to quantify the interpretation of distribution analysis data. In this discussion we shall restrict ourselves to peak amplitude data. However, the mathematics can be applied to any signal parameter.

The first mathematical approach was borrowed from seismology. It had been discovered that when the amplitude of the seismic waves was plotted on a log-log summation curve, such as that in Figure 12-25, the higher amplitude data could be fit with a straight line. This line is a solution of the equation (Mogi, 1962; Pollock, 1973)

$$N = CA^{-B} \qquad \text{Eq. 12-33}$$

which can be written

$$N = N_o \left(\frac{A}{A_o}\right)^{-B} \qquad \text{Eq. 12-34}$$

where N is the number of events with amplitude A or greater, B is a constant and is the slope of the straight line on the log-log plot, and $C - N_o/A_o^{-B}$ is another constant. Equations 12-33 and 12-34 are known as the power law distribution function. The major problem with this distribution is that as A gets small, N gets large without bounds, a situation which is physically unrealistic.

Physically, there has to be a minimum value for the amplitude of an AE burst as well as a minimum detectable amplitude. To account for this, another distribution function (Graham, 1977), called the extreme value function, was introduced. This function is

$$N = N_o \left[1 - \exp(A/A_o)^{-B}\right] \qquad \text{Eq. 12-35}$$

where N is again the number of events, A the peak amplitude and N_o A_o and B are constants. If this equation is expanded in a power series, it becomes

$$N = N_o \left(\frac{A}{A_o}\right)^{-B}$$
$$\left[1 - \frac{1}{2!}\left(\frac{A}{A_o}\right)^{-B} + \frac{1}{3!}\left(\frac{A}{A_o}\right)^{-2B} \cdots \right] \qquad \text{Eq. 12-36}$$

Thus, the extreme value function reduces to the power law for large values of A with, B, the slope of the linear portion of the curve in both distributions. From the experimental viewpoint, they can be viewed as the same distribution most of the time and one only has to determine the slope of the linear portion of the data plot.

These distribution functions agree fairly well with data from many experiments. However, they are basically empirical with little physical justification. A third distribution has been proposed, based upon the following reasoning. For AE bursts produced by breaking inclusions, it is reasonable to expect the amplitude of the burst to be proportional to the size of the inclusion. Experimental measurements of inclusion sizes in some materials shows that the size distribution fit a Weibull distribution. From this reasoning, the AE peak amplitude data would be expected to fit a Weibull distribution (Ono, et al., 1978). This distribution function can be written

$$N = N_o e^{-CA^B} \qquad \text{Eq. 12-37}$$

where N_o, C and B are constants, The best fits of these three distribution functions to data produced by the formation of Niobium Hydride are shown in Figure 26. Note that the curvature at the low amplitude end is produced by the electronic threshold which prevents the preamplifier noise from saturating the circuits.

The curves in Figure 12-26 were computer fits to the data using a nonlinear least squares fitting program. Such a program is not necessary for a power law distribution where the slope of the straight line is the important parameter and can be fit by hand. However, the extra parameter in the Weibull distribution makes a computer fit almost essential for an accurate determination of the constants. The interaction between the constants B and C in the Weibull distribution, Equation 12-37, allows various combinations to give approximate fits, especially if the data do not fit the distribution very well or have a low number of events so that there is a large statistical fluctuation in the values of the data points. The weighting of the data points must also be carefully considered in using a computer fitting routine. Where there is a large variation in the magnitude of each data point, as is the case in almost all distribution functions, it is often better to minimize the sum of the squares of the percentage deviation rather than the sum of the squares of the deviation. That is, the computer program should minimize P and D in the equations

$$P = \sum_{i=1}^{N} \left(\frac{A_i - T_i}{A_i}\right)^2 \quad D = \sum_{i=1}^{N} (A_i - T_i)^2 \qquad \text{Eq. 12-33}$$

where the A_i are the data points and the T_i are the calculated points.

In Figure 12-26, the data best fits a Weibull distribution. This often appears to be the best distribution for processes where the AE is generated randomly from many sites in the sample. However, for both earthquakes and crack propagation, the power law appears to be the best fit. This is shown in Figure 12-27 where the best fit for both a Weibull and a power distribution are shown for data from stress corrosion cracking.

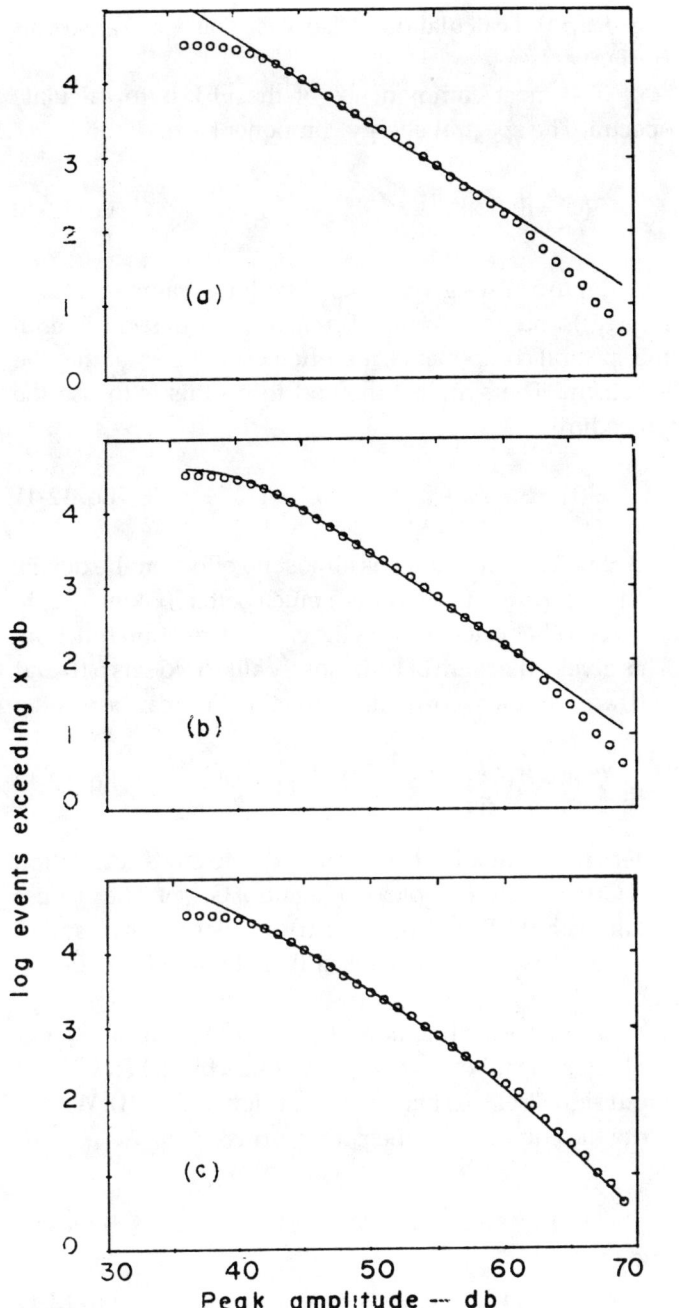

Figure. 13-26. Computer Bits of Three Analysis Functions to Data Obtained During the Formation of Niobium Hydride. (a) Power Law; (b) Extreme Value Function; (c) Weibull Distribution Function.

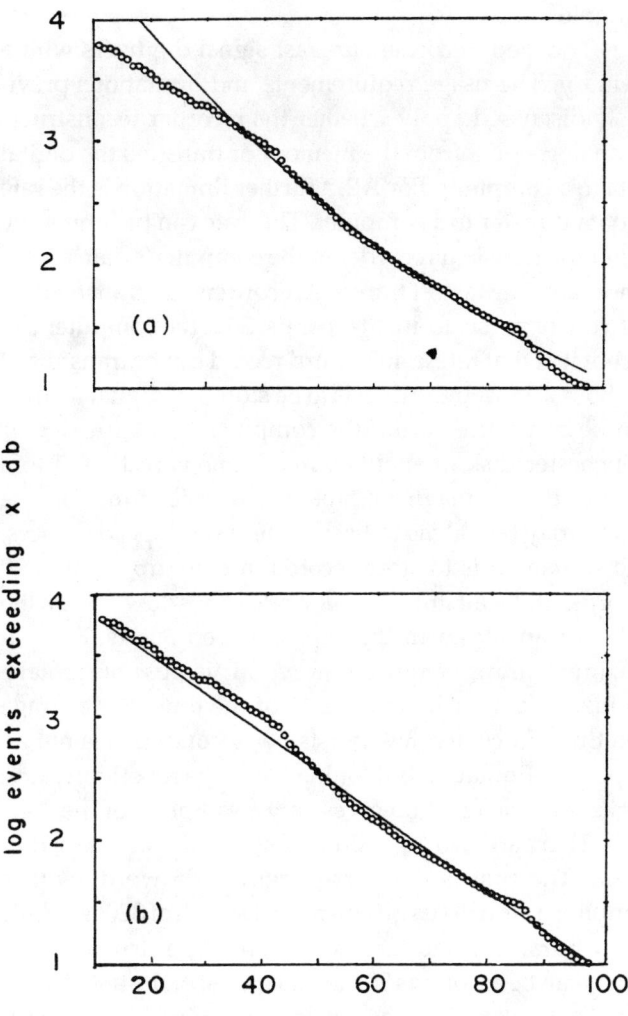

Figure 13-27. Computer Bits of Data from a Stress Corrosion Cracking Experiment. (a) Weibull Distribution; (b) Power Law.

Distribution analysis can be very useful in a real time test by looking for changes in the differential curve which can indicate a change in the source mechanism. However, the real future would appear to be with computer based systems. Programs can be developed which will track changes in the distribution constants during a test. The indications that the peak amplitudes may fit different distributions depending on the mechanisms generating the emission could add significantly to the ability of large computer based systems to grade sources. With further research, amplitude distributions and other distributions may be the key to determining flaw severity with AE.

Digital Analysis

In recent years, there has been an increase in both the number of AE systems incorporating digital computers and the capabilities of the digital computers. There has also been an increase in the availability and in the ease of interfacing of fast signal digitizers such as transient recorders. This has changed digital signal analysis from a tedious mathematical curiosity to an easily possible and perhaps practical signal analysis technique. In this section, some of the fundamental ideas will be briefly covered with the reader referred to a previous article (Beattie, 1979) for discussion in somewhat more depth and for experimental examples.

Transient recorders are fast signal digitizers with a memory. The usage requirements and limitations previously discussed apply whether the recorder reconstructs an analog signal from the memory or transfers the digital data to a computer. For AE, a further limitation is the rate of data transfer to a computer. This rate can be limited by either the transient recorder or the computer's data acceptance rate. The fastest transient recorders can transfer data at two words (six to 10 bits) per µs. So if the computer can accept it at that rate, a 4096 word record can be transferred in about 2.1 ms. If the data is to be stored in digital form, it can be transferred from the computer to a cartridge or Winchester disk in about 10 ms. If one is restricted to a floppy disk or magnetic tape, the transfer time for one record may be 200 ms to 1 s. For these slow speed devices, one stratagem is to store records in core (up to 10 to 20 records) and wait until the burst rate slows down. Then the data is transferred in the gaps between bursts. Even at maximum transfer rates, however, it is possible to take digitized data for less than 20% of the time. For average repetition rates of a few bursts per second, this is not an important limitation but for high average repetition rates, digital data acquisition gives only a sampling of the AE.

There are two other limitations on digital data acquisition. The first is the burst length. 4096 words with a sampling rate of 5/µs gives a record length of 820µs. While many bursts are shorter than this, in large structures the bursts can be as long as 50 ms. Since most of the tail of these long bursts contains frequencies determined by the resonant frequencies of the structure, this may not be an important limitation for spectral analysis. However, it has a drastic effect on energy, pulse length or envelope measurements. The other limitation is digital storage space. With maximum packing density, less than 2000 records containing 4096 10-bit-words can be placed on a 10-M byte disk. For a large test this is a serious limitation. Not only are 50-M byte disks expensive but processing that much data can take considerable time, even on a large computer.

Once the data are acquired and stored, analysis can be performed. The basic calculation for several types of analysis is the discrete Fourier transform (Stearns, 1975). This process transforms a sample set $f_i(t)$, in the time domain into a sample set $F_i(j\omega)$ in the complex frequency domain. This can be indicated mathematically by

$$F_i(j\omega) = T(f_{i(t)}) f_i(t) = T^{-1}(F_i(j\omega)) \quad \text{Eq. 13-39}$$

where T indicates the discrete Fourier transform and T^{-1} is the inverse discrete Fourier transform. Standard programs are available for most computers to take discrete Fourier transforms. They almost universally use algorithms to speed up the calculation called Fast Fourier Transforms (FFT).

The most common use of the FFT is to calculate spectra. The spectral energy component are

$$S_i(\omega) = F_i(j\omega)^2. \quad \text{Eq. 12-40}$$

Figure 12-14 gives plots of both the value of a function, $f_i(t)$, and its spectrum, $S_i(\omega)$. Because of some jitter in the spectral components, it is often useful to smoother the spectrum. The simplest method to do this is to use the procedure

$$S'(I) = S(I)/2 + (S(I-1) + S(I+1))/4 \quad \text{Eq. 12-41}$$

Such a procedure has almost no effect on the details of the spectrum but presents a much better looking graph.

Another calculation is the cross correlation function. This gives a measure of how closely alike two sets $f_i(t)$ and $g_i(t)$ are. The cross correlation function $C_i(t)$ can be written

$$C_i(t) = f_i(t) * g_i(t) = \supset^{-1}\left(F_i(j\omega)*G_i(j\omega)^*\right) \quad \text{Eq. 12-42}$$

where the symbol * is used to indicate cross correlation and $G_i(j\omega)^*$ is the complex conjugate of $G_i(j\omega)$. The process of taking the FFT's of $f_i(t)$ and $g_i(t)$ and then the inverse FFT of $F_i(j\omega) \cdot G_i(j\omega)^*$ is much faster than the standard calculation.

The transfer function of a linear system can be described in the following way. Assume a black box with an input signal set $f_i(t)$ and an output signal set $h_i(t)$. We can write the equation for their transforms

$$H_i(j\omega) = T_i(j\omega) \cdot F_i(j\omega) \quad \text{Eq. 12-43}$$

and thus

$$T_i(j\omega) = H_i(j\omega)/F_i(j\omega) \quad \text{Eq. 12-44}$$

where $T_i(j\omega)$ is called the transfer function and is independent of the input signal for a linear system. Thus, by measuring the input and output signals of a linear system we can calculate the transfer function. Once it is known, we can calculate the input signal, $d_i(t)$, to the system for any output signal $g_i(t)$ by

$$d_i(t) = T^{-1}\left(G_i(j\omega)/T_i(j\omega)\right). \quad \text{Eq. 12-45}$$

$$T_i(j\omega) = TA_i(j\omega) \cdot TB_i(j\omega). \quad \text{Eq. 12-46}$$

To indicate how this can be used in AE, if one can measure the transfer function of a sample and of the AE sensor, then one can calculate the waveform generated by an AE source from the AE signal. From this one might be able to identify the type of the source.

The signal energy can be calculated by

$$E = \sum_{i=1}^{N} f_i(t)^2 \quad \text{Eq. 12-47}$$

or

$$E = \frac{2}{N} \sum_{i=1}^{N} S_i(\omega). \quad \text{Eq. 12-48}$$

Thus, it is easy to calculate the energy of the signal at the same time that you calculate its spectrum. To get the actual energy of the signal, one should subtract the digitized preamplifier noise signal if it is present. This is

$$E_n = (V_{rms})^2 N \Delta t \quad \text{Eq. 12-49}$$

where V_{rms} is the rms noise amplitude, N is the number of samples in the record and Δt is the sample interval.

Only a few of the more useful types of calculations have been presented here. Once the signal has been digitized, the types of calculations are limited mainly by the imagination and knowledge of the investigator. However, it should be pointed out that with a large number of signal records, each with several thousand points, one can easily strain the capacities of even the largest computers if one is not careful. It is not difficult to acquire data at rates ten to a hundred times faster than the calculations can be performed.

12.2.8 Conclusions

It is hoped that the information contained in this paper will help the new practitioner conduct successful acoustic emission tests. An understanding of the principles presented should aid in the design of the test or experiment. The instrumentation and analysis techniques discussed are basic to the current practice of acoustic emission. In no sense, however can the discussion be considered comprehensive since, with the proliferation of microcomputer based systems, new analysis techniques and parameters are constantly being devised. Such parameters may prove extremely useful. However, it should always be remembered that the most important point in acoustic emission analysis is to understand the parameter that is being used. There is no one right technique or parameter for any test. All methods can give valid information if the method is understood. As is the case in many young technologies, the most important parameter in the test is the knowledge, experience and ingenuity of the testing engineer.

12.2.9. References

12.1 ASTM (1972), *Acoustic Emission*, ASTM STP 505, American Society for Testing and Materials, Philadelphia. [D-1907]

12.2 ASTM (1975),*Monitoring Structural Integrity by Acoustic Emission*, ASTM STP 571, American Society for Testing and Materials, Philadelphia. [D-1948]

12.3 ASTM (1979), *Acoustic Emission Monitoring of Pressurized Systems*, ASTM STP 697, American Society for Testing and Materials, Philadelphia.

12.4 A.G. Beattie and R.A. Jaramillo (1974), "The Measurement of Energy in Acoustic Emission," Rev. Sci. Instrum. 45:352. [D-121]

12.5 A.G. Beattie (1979), "Studies in the Digital Analysis of Acoustic Emission Signals," in *Fundamentals of Acoustic Emission*, ed. K. Ono, Materials Department, UCLA, Los Angeles, pp. 17-47.

12.6 A.G. Beattie (1976), "Energy Analysis in Acoustic Emission," Materials Evaluation, 34:73. [D-120]

12.7 A.G. Beattie (1982), "An Acoustic Measurement of Boiling Instabilities in a Solar Receiver," Journal of Acoustic Emission, 1:21.

12.8 D.A. Bell (1967), "Noise," in *Handbook of Electronic Engineering*, eds. L.E.C. Hughes and F.W. Holland, Chemical Rubber Company, Cleveland, Ohio.

12.9 J. Blitz (1967), *Fundamentals of Ultrasonics*, 2nd edition, Plenum Press, New York.

12.10 F.R. Breckenridge (1982), "Acoustic Emission Transducer Calibration by Means of the Seismic Surface Pulse," Journal of Acoustic Emission, 1:87.

12.11 W.G. Cady (1964), *Piezoelectricity*, Dover Publications, New York.

12.12 Deutsche Geellschaft fur Metallkunde (DGM) (1980), *Acoustic Emission*, trans. by A.R. Nicoll, DGM, Oberusel, Germany.

12.13 T.F. Drouillard (1979), *Acoustic Emission, A Bibliography with Abstracts*, IFI/Plenum Data Co., New York.

12.14 H.L. Dunegan and W.F. Hartman eds (1981), *Advances in Acoustic Emission*, Dunhart Publishing, Knoxville, Tennessee.

12.15 L.J. Graham (1977), "Characterization of Acoustic Emission signals and the Application to Composite Structures Monitoring." *Interdisciplinary Program for Quantitative Flaw Definition*, Special Report, Third Year ARPA-AFML Contract No. F33615-74-C-5180, Rockwell International Science Center, Thousand Oaks, CA.

12.16 M.A. Hamstad (1974), "On Energy Measurement of Continuous Acoustic Emission," UCRL Report #76283. [D-670]

12.17 D.O. Harris and R.L. Bell (1977), "The Measurement and Significance of Energy in Acoustic Emission Testing," Experimental Mechanics, 17:347. [D-711]

12.18 T.F. Hueter and R.H. Bolt (1955), *Sonics*, John Wiley and Sons, New York.

12.19 H. Jaffe (1971), "Piezoelectricity," in The Encyclopedia Britannica.

12.20 Japan, Comm. on AE (1974) *The Second AE Symposium*, Japan Ind. Plann. Assoc., Tokyo. [D-1984]

12.21 Japan, Comm. on AE (1976), *The Third AE Symposium*, Japan Ind. Plann, Assoc., Tokyo. [D-1985]

12.22 Japan Comm. on AE (1978), *The Fourth AE Symposium*, Japan Ind. Plann. Assoc., Tokyo.

12.23 Japan. Soc. NDI (1980), *The Fifth International AE Symposium*, Japan Society, NDI, Tokyo.

12.24 Japan. Soc. NDI (1982), *Progress in AE, Proc. 6th Int. AE Symp.*, eds. M. Once, K. Yamaguchi and T. Kishi, Japan Society NDI, Tokyo.

12.25 H. Kilsky (1963), *Stress Waves in Solids*, Dover Publications, New York.

12.26 R.G. Liptai, D.O. Harris, R.B. Engle and C.A. Tatro (1971), "Acoustic Emission Techniques in Materials Research," Int. J. NDT, 3;215-275. [D-1085]

12.27 A.E. Lord, (1975), "Acoustic Emission," *Physical Acoustics*, vol. 11, eds. W.P. Mason and R.N. Thurston, Academic Press, New York. [D-1094]

12.28 W.P. Mason (1950), *Piezoelectric Crystals and Their Applications to Ultrasonics*, Van Nostrand, Princeton, New Jersey.

12.29 K. Mogi (1982), Bull. Earthquake Res. Inst., 40:831-853. [D-1204]

12.30 W.P. Mason (1958), *Physical Acoustics and the Properties of Solids*, Van Nostrand, Princeton, New Jersey.
12.31 Y. Nakamura (1971), "Acoustic Emission Monitoring System for Detection of Cracks in a Complex Structure," Materials Evaluation, 29:8. [D-1239]
12.32 R.W. Nichols (1976), Acoustic Emission, Applied Science Pub. Ltd., London, GB. [D-1274]
12.33 K. Ono. R. Landy and C. Oudhi (1978), "On the Amplitude Distribution of Burst Emission Due to MnS Inclusions in HSLA Steels," *Proc. 4th AE Symp.*, Japan Ind. Plann. Assoc., Tokyo, VI-33-45.
12.34 A.A. Pollock (1973), "Acoustic Emission Amplitudes," Nondestructive Testing, 6:264. [D-1404].
12.35 J.C. Spanner (1974), *Acoustic Emission, Techniques and Applications*, Intex Publishing, Evanston, Illinois. [D-1660]
12.36 S.D. Stearns (1975), *Digital Signal Analysis*, Hayden Book Co., Rochelle Park, New Jersey.
12.37 R.V. Williams (1980), *Acoustic Emission*, Adam Hilger Ltd., Bristol, GB.

12.3 ULTRASONIC METHODS
Dr. Wei-yang Lu, Engineering Mechanics Dept., University of Kentucky, Lexington, KY.

12.3.1 Introduction

In a typical ultrasonic test, a known ultrasonic stress wave is generated by a transducer at some surface area of a structural component, the wave propagates into the interior of the body and is then received by a transducer at the surface. The received signal carries rich information about its path and the portion of the material it traveled through. By analyzing the received signals, the characteristics of the component can be determined.

For the purpose of nondestructive evaluation, the waves should not alter the state of the structure. The propagating stress waves usually have very small amplitude and are in the form of pulses. The frequency of the wave is generally in the ultrasonic range (> 20 kHz). Acoustic emission technique (previous section) also utilizes such stress waves; however, the waves are not generated by transducers but by the structure itself such as crack propagation and dislocation motion. Ultrasonic method discussed in this section is considered as an active technique while acoustic emission is a passive method.

12.3.2 Applications

The primary application of ultrasonic method is flaw detection and characterization. Ultrasonic technique has great sensitivity to detect and locate flaws deep within the structural element. It has become an important method for inspecting structurally critical areas such as weldments, joints, etc. Ultrasonic equipment is usually portable and nonhazardous which makes inspection safe and convenient.

Ultrasonic thickness measurement is another application which has been used extensively to determine the wall thinning of structural elements caused by wear, corrosion, or erosion. It only requires that one side of the part is accessible. Examples include in-service and maintenance inspection of piping components, boil tubes and storage tank walls of chemical and nuclear plants.

The applications of ultrasonic method in material characterization have increased rapidly in recent years. In addition to the determination of material constants such as Young's modulus, shear modulus and Poisson's ratio, the technique has been used to measure grain size, porosity, damage, texture, adhesive joint strength, and to monitor heat treatment, environment degradation, hydrogen attack, etc.

Ultrasonic waves are also used for nondestructive stress measurement. In most stress measuring techniques, strains are measured and the stress-strain relation of the material is needed to determine the stresses. The techniques based on strain measurement are not reliable when the initial stress state is unknown or plastic deformation is involved. Ultrasonic method has been applied to measure the tensile stress in a bolt and the residual stresses near weldments.

12.3.3 Basic Principles

The response to an ultrasonic wave varies with different materials. It is known that a real material always has certain degrees of anisotropy and damping; however, an idealized solid, which is isotropic and lossless, will be discussed first. The idealized material model often offers excellent approximations in many ultrasonic inspection problems. The effects of material damping and stresses will be discussed later.

There are two different types of stress waves, longitudinal and shear, may propagate in the interior of an isotropic solid. In longitudinal waves, often called compressive waves or P waves, the particle motion of the material is essentially in the same direction as the wave propagation. In shear waves, often called transverse waves, the particle motion is perpendicular to the direction of propagation. The phase velocities of these waves V_l and V_s depend on the density ρ and the elastic constants of the solid

$$V_l = \sqrt{\frac{(1-\upsilon)E}{\rho(1+\upsilon)(1-2\upsilon)}} \qquad \text{Eq. 12-50}$$

and

$$V_s = \sqrt{\frac{\mu}{\rho}} = \sqrt{\frac{E}{2\rho(1+\upsilon)}} \qquad \text{Eq. 12-51}$$

where μ is the shear modulus, E is Young's modulus, and υ is Poisson's ratio; subscript l and s indicate longitudinal wave and shear wave, respectively.

Longitudinal waves are always faster than shear waves. Wave velocities of some materials are listed in Table 12.3. The following equations can be obtained from Eqs. 12-50 and 12-51.

$$E = \frac{\rho V_s^2 (3V_l^2 - 4V_s^2)}{(V_l^2 - V_s^2)} \quad \text{Eq. 12-52}$$

$$\mu = \rho V_s^2 \quad \text{Eq. 12-53}$$

$$\upsilon = \frac{1 - 2(V_s/V_l)^2}{2[1 - (V_s/V_l)^2]} \quad \text{Eq. 12-54}$$

The elastic properties of a material can be determined from ultrasonic wave velocities.

Stress waves will scatter at a boundary between two different isotropic media. For the case of shear waves, it is desirable to resolve the particle motion into two components perpendicular to each other. By using the boundary as a reference, the component of particle displacement parallel to the boundary is termed the horizontally polarized shear (SH) wave, and the other is called the vertically polarized shear (SV) wave.

The scattering of SH waves at a boundary is analogous to the scattering of light. An incident SH wave excites one reflected SH wave and one refracted SH wave, Fig. 12-28(a). For the case of an incident SV wave or P wave, the scattering is more complicated. In order to satisfy the boundary condition, both SV and P waves may be generated in the reflected waves and in the refracted waves as shown in Fig. 12-28(b & c). Propagation directions are given by acoustic Snell's Law as follows:

$$\frac{\sin\theta_s}{V_s} = \frac{\sin\theta_l}{V_l} = \frac{\sin\theta_s'}{V_s'} = \frac{\sin\theta_l'}{V_l'} \quad \text{Eq. 12-55}$$

where prime indicates transmitted waves. The occurrence of refracted waves is dependent on the incident angle. When $\sin\theta_l < V_l/V'_l$ or $\sin\theta_s < V_s/V'_l$, both SV and P waves are transmitted; when $\sin\theta_l < V_l/V'_l < V_l/V'_s$ or $\sin\theta_s < V_s/V'_l < V_s/V'_s$ only SV wave is transmitted; when $\sin\theta_s < V_s/$

Table 12.3 Wave Velocities of Various Materials

Material	E (GPa)	μ (GPa)	ν	ρ (Kg/m³)	V_t (m/s)	V_s (m/s)	V_g (m/s)
Steel	207	81	0.29	7.7×10³	5940	3230	2990
Copper	110	45	0.34	8.9×10³	4360	2250	2010
Aluminum	72	26	0.34	2.7×10³	6410	3100	2900

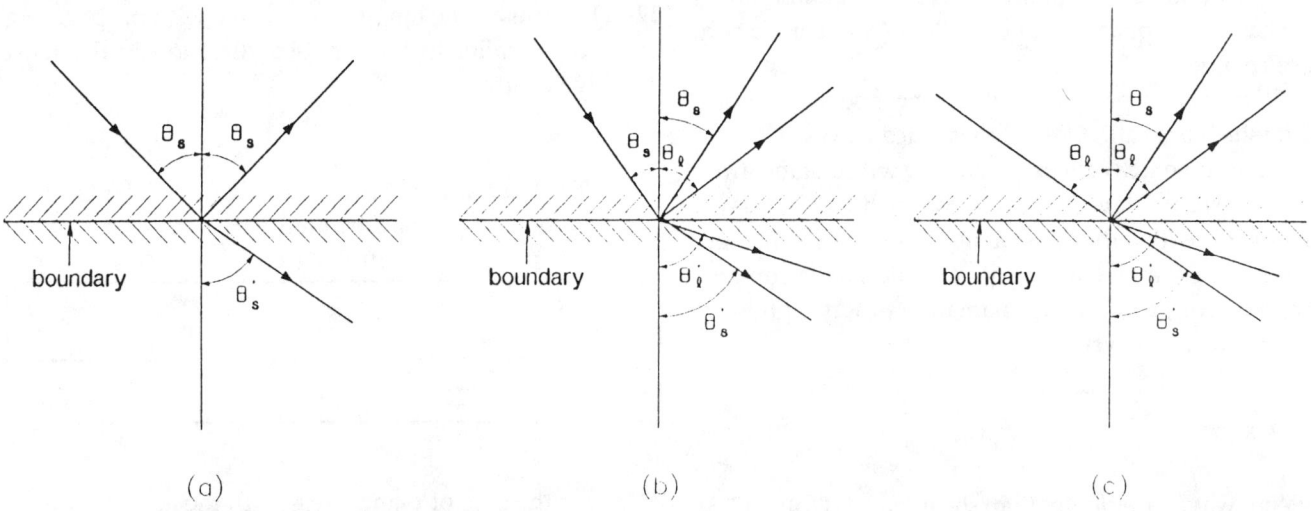

Figure 12-28. Reflection and refraction of (a) SH wave, (b) SV wave and (c) P wave at a boundary between isotropic media.

V'_l or $\sin\theta_s < V_s/V'_l$ no wave is transmitted, the ultrasonic energy propagates on the interface. The wave propagating at the interface of two solids is called a Stonely wave.

An important practical problem is the reflection at a free boundary. The propagation directions of the reflected waves are the same as described above, but there is no refracted wave. An SH wave is totally reflected at a free boundary. The amplitude of the reflected wave is equal to that of the incident wave with zero phase change. For an incident SV wave or P wave, the reflection coefficients are

$$\begin{bmatrix} \Gamma_{ll} & \Gamma_{sl} \\ \Gamma_{ls} & \Gamma_{ss} \end{bmatrix} = \frac{1}{K}$$

$$\begin{bmatrix} \sin2\theta_l \cos2\theta_s - (V_l/V_s)^2 \cos^2 2\theta_s & 2(V_l/V_s)\sin2\theta_s \sin2\theta_l \\ 2(V_l/V_s)\sin2\theta_s \cos2\theta_s & -\sin2\theta_s \sin2\theta_l + (V_l/V_s)^2 \cos^2 2\theta_s \end{bmatrix}$$

Eq. 12-56

where $K = \sin2\theta_s \sin2\theta_l + (V_l/V_s)^2 \cos^2 2\theta_s$, and Γ_{sl} = (reflected SV wave amplitude)/(incident P wave amplitude) with analogous expressions for $\Gamma_{ll}, \Gamma_{ls}, \Gamma_{ss}$. When the incident SV wave or P wave is normal to the boundary, i.e., $\theta_l = \theta_s = 0$. Eq. 12-56 becomes

$$\begin{bmatrix} \Gamma_{ll} & \Gamma_{sl} \\ \Gamma_{ls} & \Gamma_{ss} \end{bmatrix} = \left(\frac{V_s}{V_l}\right)^2 \begin{bmatrix} -(V_l/V_s)^2 & 0 \\ 0 & (V_l/V_s)^2 \end{bmatrix}$$

$$= \begin{bmatrix} -1 & 0 \\ 0 & 1 \end{bmatrix} \qquad \text{Eq. 12-57}$$

The waves are totally reflected and no other wave types are generated at normal incidence. The reflected longitudinal wave has a 180° phase change, while SV wave has a zero phase change.

Waves may propagate along the free surface of a body, which are called Rayleigh surface waves. The particle displacement decays exponentially with distance from the free surface. Wave motion is confined to a thin layer near surface with a thickness about twice the wavelength of the surface wave. The velocity equation has a complex form; however, a good approximation of Rayleigh wave velocity V_R can be written as

$$V_R = \frac{0.862 + 1.14\upsilon}{1+\upsilon} V_s . \qquad \text{Eq. 12-58}$$

Rayleigh waves are slower than shear waves, (Table 12-3). When propagating around a curved surface, Rayleigh wave speed changes slightly and reflection may occur. The speed increases as the radius of curvature decreases. The Rayleigh wave is partially reflected when the ratio of radius of curvature to wave length is less than two; when the ratio is greater than two, reflection is almost zero.

Acoustic losses in many materials may be modeled by a linear viscous damping element as shown in Fig. 12-29. The damping results in two effects: a small increase in phase velocity and a decay of wave amplitude in the direction of propagation. Assume a x-propagating wave, the amplitude varies according to the exponential function $e^{-\alpha x}$, where α is called attenuation factor. Figure 12-30 shows multiple backface echoes fitted by an exponential curve. The model shows the attenuation factor is proportional to the square of the frequency f, i.e.

$$\alpha = cf^2 \qquad \text{Eq. 12-59}$$

where c is a constant. The frequency dependence of attenuation, which may deviate from the equation for real materials, is used for characterizing material properties.

Ultrasonic wave propagation in a solid is also affected by stresses. This is known as the acoustoelastic effect. For an isotropic elastic material in a stress free state, the velocity dependence on uniaxial stress σ_l can be expressed by

$$\Delta\upsilon_{ij} = \beta_{ij} \sigma_l \qquad \text{Eq. 12-60}$$

where β_{ij} is the acoustoelastic coefficient, $\beta\upsilon_{ij}$ is the relative change in speed of the wave propagating in the i-direction and particle motion in the j-direction. For example, $\Delta\upsilon_{12} = (V_{12} - V°_{12})/V°_{12}$, where superscript "°" indicates the stress free state. (Note that $V°_{11} ... = V_l$ and $V°_{12} = ... = V_s$.) The acoustoelastic effects of carbon steel are shown in Fig. 12-31 and the coefficient β_{ij} is listed in Table 12-4. In general, the longitudinal wave with propagating direction parallel to the stressed direction is the most sensitive to stress.

Table 12-4. The Acoustoelastic Coefficients for Carbon Steel ($\times 10^{-5}$ MPa^{-1})

β_{11}	β_{11}	β_{11}	β_{11}	β_{11}
−1.26	−0.24	0.21	0.07	−0.72

For the case of plane stress with the acoustic waves propagating normal to the plane of stress, the following equations can be derived:

$$\frac{V_{11} - V_l}{V_l} = p\,(\sigma_2 + \sigma_3) \qquad \text{Eq. 12-61}$$

$$\frac{V_{12} - V_{13}}{V_s} = q\,(\sigma_2 - \sigma_3) \qquad \text{Eq. 12-62}$$

where p and q are functions of density and elastic constants. As shown in Fig. 12-32, Eq. 12-62 is analogous to the photoelastic birefringence method. One practical difficulty is that the influence of stress on wave velocities is small (it is a second order effect). A small degree of material anisotropy (texture) or microstructure variation may change elastic constants and cause a large error in stress evaluation. In this case, additional information on texture or microstructure is necessary.

One way to bypass this difficulty is to use the universal relations for orthotropic media, Fig. 12-33

$$\sigma_3 - \sigma_1 = \rho\,\frac{\{v_a^2(\theta_2) - v_b^2(\theta_2)\}\cos^2\theta_1 - \{v_a^2(\theta_1) - v_b^2(\theta_1)\}\cos^2\theta_2}{\cos^2\theta_1\sin^2\theta_2 - \cos^2\theta_2\sin^2\theta_1};$$

$$\text{Eq. 12-63}$$

here the coordinate planes are parallel to the planes of symmetry; σ_1 and σ_3 are principal stresses; $v_a(\theta)$ and $v_b(\theta)$ are the phase velocities of shear waves with propagation directions $(0, \pm\cos\theta, \pm\sin\theta)$ and $(\pm\sin\theta, \pm\cos\theta, 0)$, directions of motion $(\pm 1, 0, 0)$ and $(0, 0, \pm 1)$, respectively. Universal relations directly relate wave speeds to stress and density and require no other constant to be determined.

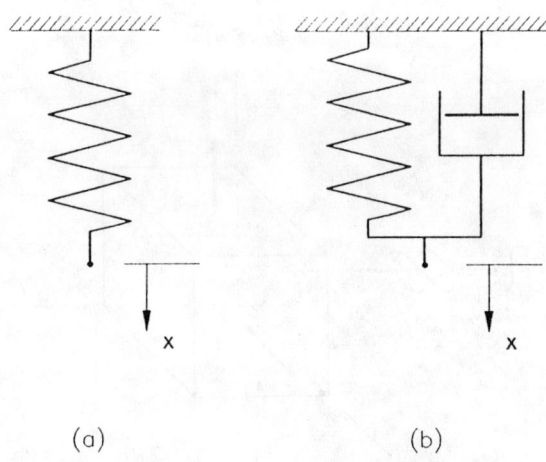

Figure 12-29. Models of (a) ideal and (b) damped elastic material.

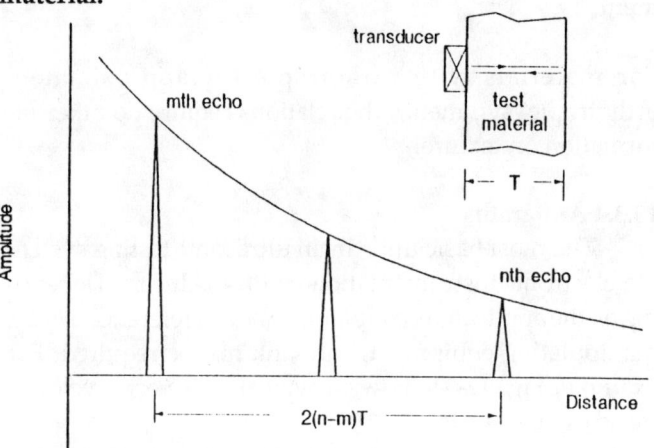

Figure 12-30. Multiple backface echoes fitted by an exponential curve.

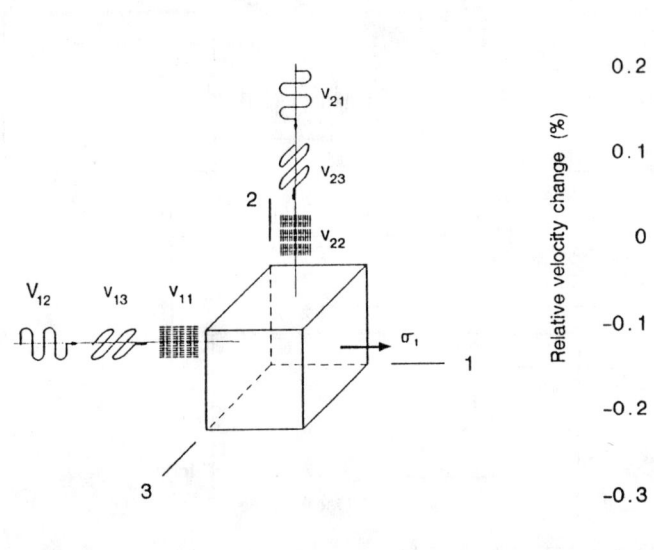

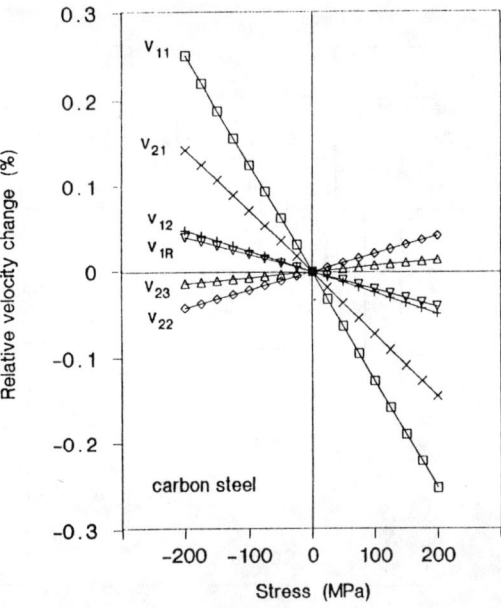

Figure 12-31. Velocity dependence on stress.

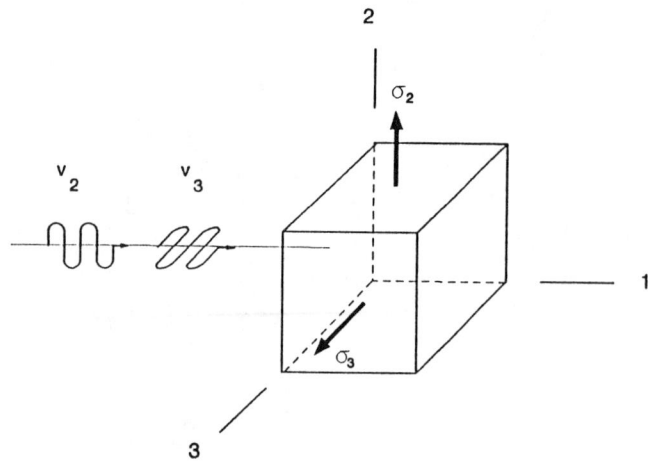

Figure 12-32. Ultrasonic birefringence stress measurement.

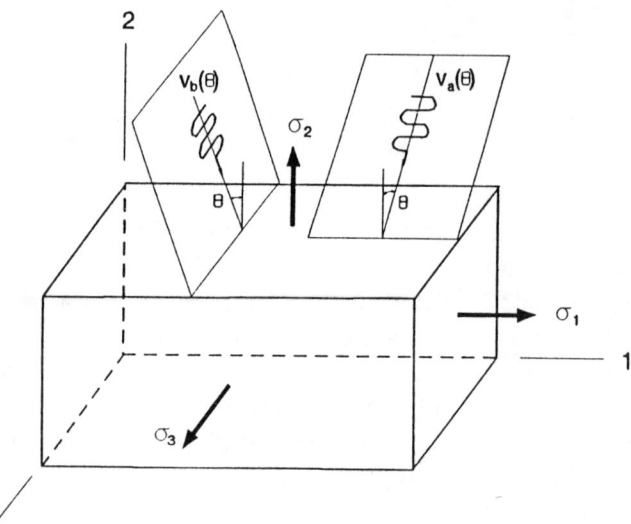

Figure 12-33. Ultrasonic stress measurement based on universal relations.

For materials with orthotropic (or approximately orthotropic) symmetry, the relations require no other information on texture.

13.3.4 Apparatus

The most basic units in an ultrasonic testing system are electronic instrumentation and transducers. Depending on the application, couplant, standard reference blocks, manipulating equipment, and tank maybe required. For example, Fig. 12-34 shows a typical pulse-echo velocity measurement system.

Instrumentation

The main functions of ultrasonic instrumentation are generating electrical pulses to excite transducers, receiving and amplifying ultrasonic signals detected by transducers, and displaying test results. The typical components corresponding to these functions are pulsers, receivers and an oscilloscope. Many commercial flaw detectors integrate all three components into one compact unit.

The selection of ultrasonic instruments is based on the ultrasonic frequencies and energy levels required to perform the test. General considerations are the attenuation factor of the material to be tested, material thickness, and the minimum discontinuity to be detected. Theoretically, the higher the frequency the smaller the discontinuity that can be detected. However, the attenuation factor is proportional to the square of the frequency for viscously

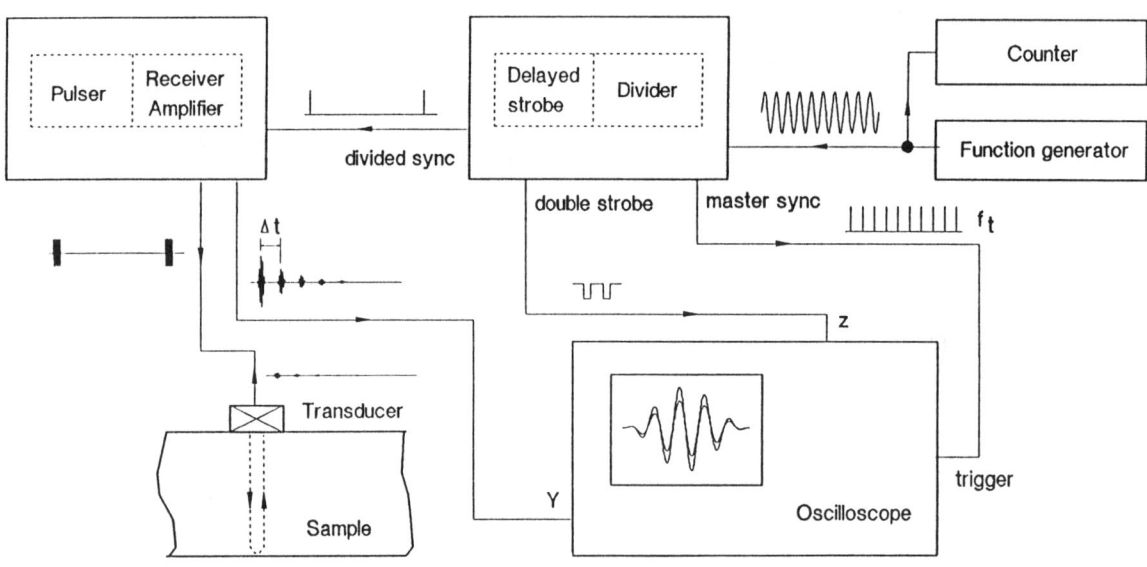

Figure 12-34. Pulse-echo-overlap velocity measurement system.

damped materials. Lower frequency and higher energy levels provide better penetration.

Some advanced applications demand further conditioning and processing of received ultrasonic signals. A gate unit is required when a specific region of interest of an ultrasonic signal needs to be selected for analysis. Several special purpose units in ultrasonic testing are also available commercially, such as peak detector for peak detection and pulse comparator for attenuation measurement.

Recently developed computer-controlled ultrasonic testing instruments add great flexibility in system configuration, test control, data analysis, and result display. In addition to pulsers and receivers, a computer and waveform digitizers are essential. The received ultrasonic signals are digitized at a high rate. All signal processing and analyzing can be done by computer. Software programs could replace those special purpose units mentioned in the last paragraph. However, the sampling rate limits the applied frequency of the ultrasonic wave. At the present time, a PC plug-in Analog-to-Digital (A/D) converter is capable of sampling at 100 MHz for a single shot signal or 800 MHz for repeated signals. The sampling rate is expected to be higher in the future. This type of instrument offers better accuracy, faster speed and simpler operation.

Transducers

Transducers transmit and receive ultrasound in the material. The transmitter converts the electrical pulses to stress waves, the receiver senses the motion of the particle and converts it to electrical signals. In many cases the same transducer is used as both transmitter and receiver. Several types of transducers are available today, they are piezoelectric transducers, electromagnetic acoustic transducers (EMATs), and laser.

Piezoelectric transducers are used more frequently than any other transducers in NDE. They are simple and easy to use. Couplant is required between the surface of the transducer and the testing object. Typical transducer sizes range from 0.125 in. (3.2 mm) to 1.125 in. (28.6 mm) in diameter. Longitudinal transducers are the most popular type. Figure 12-35 shows a typical ultrasonic beam and the pressure fluctuation patterns of an unfocused longitudinal transducer. A ultrasonic wave radiates away from a transducer and diverges. The beam can be divided into two regions with different characteristics. In the near field, interference effects cause strong pressure fluctuations; in the far field, the pressure decreases with increasing distance. The near field effects may cause sensitivity inconsistence; therefore, inspection for flaws within the near field should be avoided. The transition between near field and far field is approximated by

$$N = \frac{d^2}{4\lambda}$$ Eq. 12-64

where d is the transducer diameter and λ is the wavelength ($\lambda = V/f$). The beam spread is given by

$$\psi = \sin^{-1}\left[1.2 \times 10^{-3} \frac{V_l}{fd}\right]$$ Eq. 12-65

where f is the frequency (MHz).

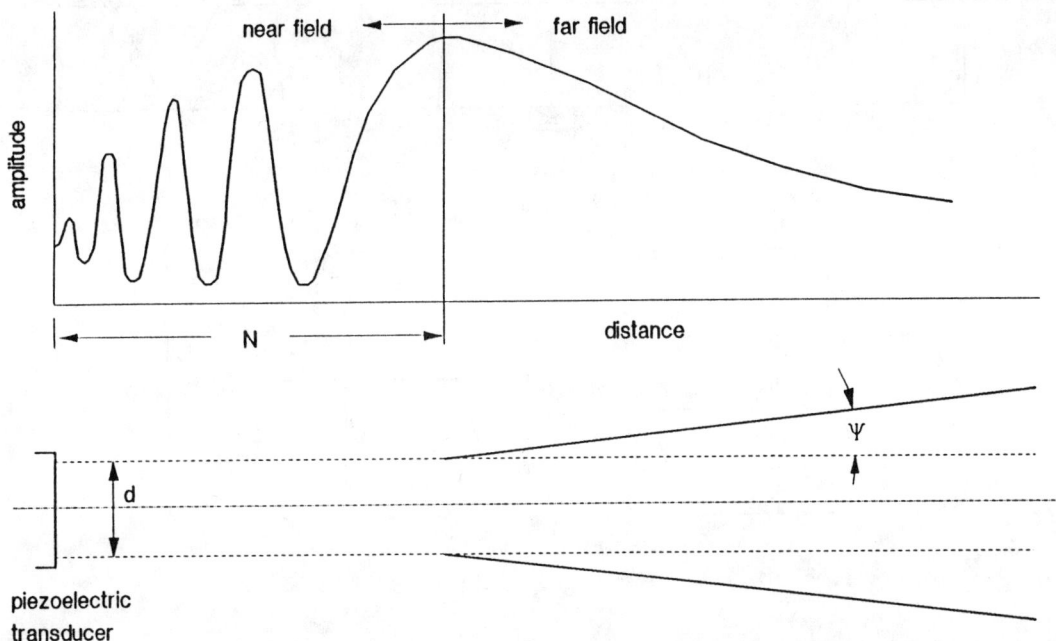

Figure 12-35. Ultrasound field pattern and beam spread.

Similar to optical lenses, there are focussed transducers which can focus the ultrasonic beam at localized regions. Focussed transducers contract the near field length and improve sensitivity at desired specific depth in the material being tested.

Bandwidth, frequency and size affect the performance of a transducer. Narrowband tuned transducers provide excellent sensitivity and penetration. However, this type of transducer usually has a long initial oscillation (i.e. ring-down) which may mask early received signals and cause poor near surface resolution. Broadband transducers have better resolution.

High frequency transducers provide a more direct ultrasonic beam with less beam spread and better depth and lateral resolution. The ultrasonic beam spreads wider with low frequency, therefore, low frequency transducers provide better detection of misaligned discontinuities. Lower frequencies also offer better penetration.

The diameter of an ultrasonic beam is proportional to the diameter of the transducer. Larger diameter transducers give a longer near field zone but less spread beams.

Figure 12-36 shows some common transducer arrangements. Pulse echo is the most popular arrangement where relatively smooth and parallel surfaces are required. Through transmission is the most efficient arrangement. It is used in composite and other highly damped materials. To avoid near field operation, a buffer rod is sometime used to provide a time delay between the transmitter pulse and the first received signal. The buffer rod arrangement is also very useful for detecting flaws near the front surface. Pitch and catch technique requires a separate transmitter and receiver and may be used when the ring-down interferes with the test or when the surfaces are not parallel. Angle beam transducers are usually used in this arrangement, Fig. 12-36(c). They can be used in other arrangements as well. As discussed earlier, the reflection and refraction of stress waves at interface boundaries are complex for oblique incident waves. Different wave types may be generated. In many practical inspection situations, it is desirable to excite only one wave type so the received signals can be clearly interpreted. The incident wave type and angle need to be selected properly.

The arrangements shown in Figs. 12-36(a)-12-36(d), where transducers are in contact with the testing object through a thin layer of couplant, are called contact method. The other common method is immersion method. Figure 12-36(e) shows an immersed pulse echo arrangement, where transducer and testing object are separated by water. Contact method and immersion method are usually applied in manual scanning and automatic scanning, respectively.

The requirement of couplant somehow limits the application of piezoelectric transducers. Since EMATs do not need couplant, they have the advantages of high speed operation and high temperature applications. Another advantage is that EMATs can generate SH waves more easily than fluid coupled piezoelectric transducers. The use of SH waves in NDT is preferred in many situations since the results are much easier to interpret than with

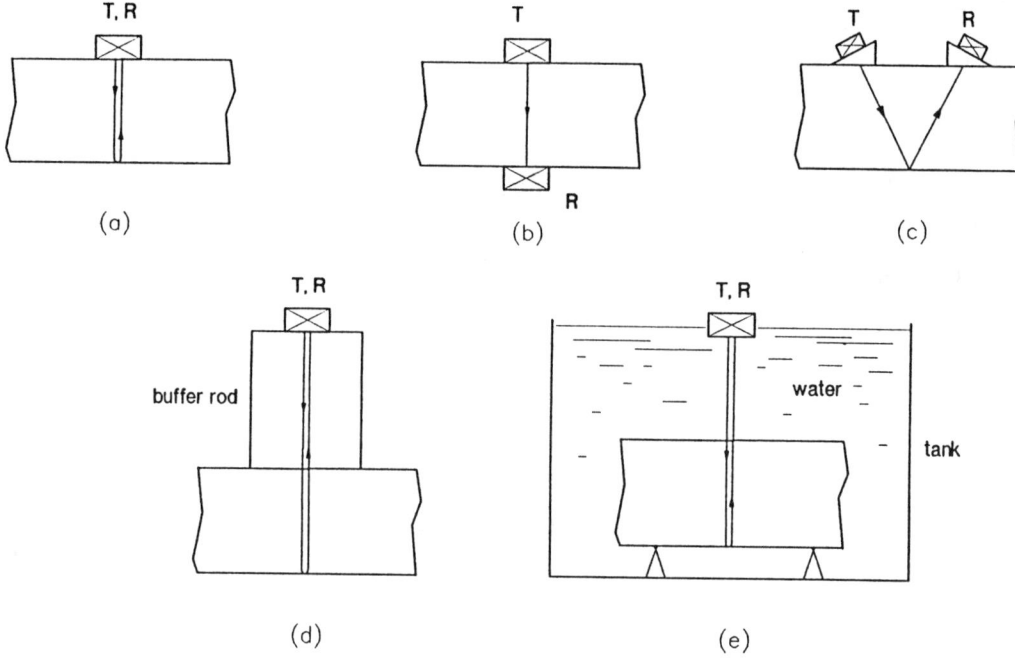

Figure 12-36. Transducer arrangement: (a) pulse echo, (b) through transmission, (c) pitch and catch, (d) buffer rod, and (e) immersion method.

other waves. In comparison to piezoelectric transducers, EMATs have the disadvantage of lower efficiency which leads to lower signal-to-noise ratio. EMATs are usually larger and operate at lower frequencies.

The use of laser to excite and receive ultrasonic waves also has the advantage of noncontact operation. Laser generates all types of waves over a wide frequency range by rapid localized heating of the surface of a material. The area of the laser beam could be very small, on the order of millimeters, which is equivalent to a very small size transducer. Laser method may have many important applications in the future.

Couplant

A couplant is required between the face of the piezoelectric transducer and the examination surface to transmit ultrasound. Typical couplants include water, cellulose gel, oil and grease. During ultrasonic examination, the couplant layer must be maintained to adequate thickness and the contact area held constant. Couplant variations will change examination sensitivity. High viscosity couplant should be used for shear wave transmission. The selection of couplant also depends on the roughness of the examination surface. Rougher surface requires higher viscosity couplant.

Reference Standards

In ultrasonic testing, a standard reference usually is necessary to establish the standard of the test equipment and the basis for signal evaluation. Reference blocks with known discontinuities and configurations may be fabricated. The surface roughness, velocity and attenuation of the reference block should be similar to the examined object.

12.3.5 Testing Techniques

In spite of wide applications of ultrasonic method, the testing techniques, in general, fall into three basic categories: discontinuity (flaw) detection and evaluation, velocity measurement, and attenuation measurement.

Flaw detection and evaluation

Ultrasonic waves reflect at a boundary. Thus, the echo appearing between the front and back echoes indicates the presence of a discontinuity. The location of the discontinuity is determined from the wave velocity and the transit time. Figure 12-37 shows the screen appearance of a typical flaw detector. The horizontal and vertical axes indicate time (or distance) and amplitude, respectively. This is the A-scan presentation. To evaluate the discontinuity quantitatively, reference blocks with different size reflectors at the proper metal travel distance (i.e. close to the discontinuity depth) are usually required. However, this

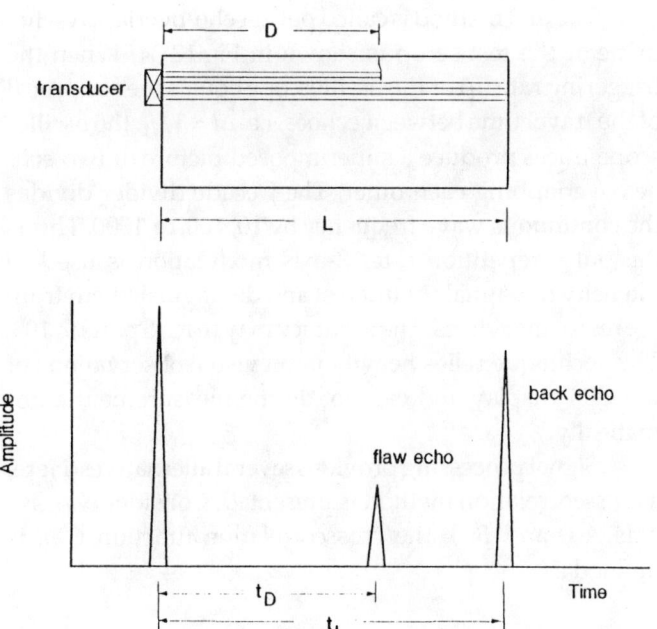

Figure 12-37. Pulse-echo flaw detection and location.

only gives the indication of an equivalent response, not the exact size, shape, or orientation.

For large discontinuities, the 6 dB down method is widely used. The echo amplitude reduced to the one half the amplitude of the indication from the reference reflector at the equivalent depth, marks the boundary of the discontinuity. The distance between two half-amplitude positions approximates the discontinuity length. When automatic scanning is used, the image of discontinuities can be displayed. B-scan displays a cross section of the specimen indicating the approximate length of the discontinuities and their relative positions. C-scan displays a plane view of the test object and discontinuities.

Velocity measurement

To determine the absolute velocity of a wave,
$$V = L/t,$$
one must measure the wave path length L and the corresponding transit time t. Wave path length can be measured by using a scale or caliper; transit time can be obtained from the A-scan presentation. Between two consecutive echoes the wave travels a distance twice the thickness T of the specimen. From Fig. 12-30 the wave velocity is

$$V = 2T(n-m)/\Delta t \qquad \text{Eq. 12-66}$$

The measurement described above does not provide the accuracy required by modern applications. Several techniques have been introduced for the precise measurement of transit time. Here, two techniques are briefly described.

The first method is called pulse-echo-overlap. A schematic of the test setup is shown in Fig.12-34. When the triggering rate (f_t) of the oscilloscope equals the reciprocal of the travel time between echoes, i.e. $\Delta t = 1/f_t$, the oscilloscope traces produce a superimposed picture of two echoes overlapping each other. The Decade divider divides the continuous wave frequency by 10, 100, or 1000. This is the pulse repetition rate. Z-axis modulation is used to intensify the signals of interest and distinguish them from the rest of the echoes. The accuracy may reach 5 parts in 10^6. This technique relies heavily upon visual observations of the CRT display and can not do the measurement automatically.

Signal processing provides several alternatives. Here, a crosscorrelation method is presented. Consider two signals $A(t)$ and $B(t)$, the crosscorrelation function C_{AB} is defined as

$$C_{AB}(\tau) = \int_{-\infty}^{\infty} B(t)A(t-\tau)dt .\qquad \text{Eq. 12-67}$$

The value of $C_{AB}(\tau)$ gives an indication of similarity between signals $B(t)$ and $A(t-\tau)$.. If the maximum value of C_{AB} occurs at $\tau = \tau_0$, then the signals $B(t$ and $A(t-\tau_0)$ have the greatest similarity. Now consider $A(t)$ and $B(t)$ as two ultrasonic echoes. Since $B(t)$ is similar to $A(t)$ with a time delay, the value τ_0 is exactly the delay time between these two echoes. During application, signals are digitized and digital cross correlation is performed. The correlation function is then approximated by a least-squares curve fitting of the calculated discrete values. The delay time corresponds to the maximum of the interpolation function. The time resolution using this method is better than that given by the digitizing frequency alone. It has been demonstrated that with signals of 400 kHz center frequency, digitized at 20 MHz, the resolution of transit time is about 1 ns.

Attenuation measurement

Figure 12-30 is the typical A-scan presentation showing multiple back reflection signals. The amplitude of these pulses can be fit by an exponential curve. The apparent attenuation α_A (i.e. the observed decay of ultrasonic signals) is calculated in terms of decibels per unit length as

$$\alpha_A = \frac{20 \log_{10}(A_m/A_n)}{2(n-m)T} \qquad \text{Eq. 12-68}$$

where A_m and A_n are amplitudes of the m th and n th back reflections ($n > m$). These back reflections are necessary in the far field of the ultrasound beam.

The apparent attenuation as shown in the figure includes energy losses attributable to testing material, geometry, instrumentation, beam divergence, interface reflections, measurement procedures, etc. The portion of the observed ultrasonic energy loss which is attributable to the testing material is the absolute (true) attenuation. For true attenuation measurement, corrections for beam divergence, geometry, and other factors are necessary. Fortunately, the relative apparent attenuation between materials gives enough information in many practical applications. When comparing the apparent attenuation of different materials or components, similar specimen geometry and same test conditions (i.e. instrumentation, transducer, procedures, etc.) are necessary.

References

1. D.E. Bray and R.K. Stanley, *Nondestructive Evaluation*, McGraw-Hill, Inc., New York, 1989.
2. 1990 *Annual Book of ASTM Standards, Nondestructive Testing*, vol. 3.03, American Society for Testing and Materials, 1990.
3. B.A. Auld, *Acoustic Fields and Waves in Solids*, Robert E. Krieger Publishing Company, Malabar, FL, 1990.
4. H. Kolsky, *Stress Waves in Solids*, Dover Publications, Inc., New York, 1963.
5. Analytical Ultrasonic in Materials Research and Testing, NASA Conference Publication 2383, 1986.
6. E.P. Papadakis, "Ultrasonic Velocity and Attenuation: Measurement Methods with Scientific and Industrial Applications," Physical Acoustics XII, eds. W.P. Mason and R.N. Thurston, pp. 277-374, 1976.
7. N.N. Hsu, "Acoustical Birefringence and the Use of Ultrasonic Waves for Experimental Stress Analysis," Exp. Mech., vol. 14, pp. 169-176, 1974.
8. R.B. Thompson and D.O. Thompson, "Ultrasonics in Nondestructive Evaluation," Proc. of IEEE, vol. 73, pp. 1716-1755, 1985.
9. C.-S. Man and W.Y. Lu, "Towards an Acoustoelastic Theory for Measurement of Residual Stress," J. of Elasticity, vol. 17, pp. 159-182, 1987.
10. W.Y. Lu, B.W. Maxfield and A Kuramoto, "Ultrasonic Velocity Measurement by Correlation Method," 1990 Spring Conference on Experimental Mechanics, pp. 279-284, 1990.

12.4 THE IMPACT-ECHO TECHNIQUE
A.K. Maji and M.L. Wang
Assistant and Associate Professors of Civil Engineering
University of New Mexico, Albuquerque, NM 87131

12.4.1 Introduction

In spite of recent advances in concrete technology, there are no satisfactory methods for detecting some of the internal defects in hardened concrete. Various destructive and nondestructive techniques have been developed in recent times to aid the construction industry in quality control and determination of concrete strength [1]. The routine method for verifying the existence of defects within a concrete structure is the taking of drilled core samples. This approach has the disadvantages of being expensive, destructive, and at best, only a few samples can be tested at arbitrarily selected locations. It is possible to look through concrete by using X-ray radiography and Penetrating Radar [2], such methods are expensive, cumbersome and need considerable expertise in data interpretation. Some new methods such as Infrared thermography [3] are finding more applications, but are somewhat dependent on the environment, weather etc. Impulse radar and thermography [4] are evolving as possible candidates for subsurface inspection in concrete. These later method shows promise for rapid scanning of an entire road section or structure. The Ultrasonic testing technique that has been enormously successful in other materials has found limited application in concrete because of its heterogeneous nature which leads to a wide scatter in results [4]. However, the penetrating ability of ultrasonics in concrete is limited and the experiment is time consuming. Therefore, the need for a fast and inexpensive nondestructive technique capable of determining the quality of concrete construction continues to exist. A recent conference on nondestructive testing of structures [S] has summarized many innovative nondestructive testing techniques; some at their infancy, others applied for practical engineering problem solving.

In recent years the Impact-Echo (IE) technique has become a useful technique for inspecting concrete structures, asphalt overlays, bridge decks, and other structures. Striking an object with a hammer and listening to the ringing sound is a common qualitative way of detecting the presence of internal voids, cracks or delaminations in any material. This technique is essentially a modification of the ultrasonic testing techniques. Since the 1930s, stress waves have been used for examining the integrity of materials. World War II saw an increase in the utility of this method for various industrial inspections. The ultrasonic pulse-echo technique uses the time taken by a stress wave to return to a transducer after being reflected from an internal flaw. The major problem in using this technique in concrete is the amount of scattering of the stress waves in inhomogeneous concrete, which greatly inhibit its penetration capability.

The Impact-Echo technique therefore developed as an offshoot of the ultrasonic pulse-echo technique that could be used for concrete. It is a quantitative non-destructive test technique to detect flaws in concrete such as honeycombing, delaminations (within the concrete or between concrete and asphalt overlays), cracks and voids [6-9]. The strength of the technique is its simplicity and ease of application. In this method, an impulse force is applied to the object and its response is monitored by a transducer. Both the source of the impulse and the receiving transducer are mounted on the same surface of the specimen. This has the additional advantage that the technician does not have to have access to both the surfaces of the structure that is being inspected.

Once the impact is done, the waves introduced travel in the material and are reflected back and forth whenever it meets a new solid material, or air. The transducer monitors the wave as it is reflected from internal defects, delaminations or external boundaries of the structure. The response of the transducer has to be monitored by a data acquisition system. Depending on the type of transducer used, the data acquisition system records the displacement, or acceleration at the location of the transducer. The transducer response is dominated by frequencies that are caused by the back and forth bounding of the impact waves in between the structure's boundaries, or flaws. Hence, if the velocity of the sound in the material, concrete, or asphalt is known, the distance to any internal defect or an interface can be calculated. The thickness of a wall or slab can be similarly determined if one has access to only one of its sides.

Since the technique discussed so far is similar to what is used in ultrasonic inspection, it is important to point out the differences. Ultrasonic typically refers to frequencies of sound above the audible range Typical frequencies of the sound generated in ultrasonics is 25 kHz or more, often in the range of 100s of kHz. This is necessary, because the wavelength of the sound is inversely proportional to the frequency. Also, if the wavelength is much larger than the flaw that is to be detected, then it is not effected by the flaw, which goes undetected. The high frequencies used in ultrasonics tend to be scattered very easily in a heterogeneous material like concrete. This limits the depth of concrete that can be inspected with the ultrasonic technique (to about 30 cm). It is also necessary to have access to both the surfaces of the structure being inspected to use the ultrasonic through-transmission technique. Increasing the power of the ultrasonic waves to penetrate larger thicknesses of concrete becomes cost prohibitive. Therefore, to improve the resolution of the ultrasonic technique small wavelengths are necessary, which is why high frequencies are necessary. The impact-echo technique is geared towards detecting larger flaws and defects which are typically detrimental to concrete and asphalt structures. Since the flaws to be detected are larger, one can afford to used smaller frequencies (0.5-25 kHz). The advantage of this is that the sound waves can be generated by impact (typically hitting the surface with a hammer). Also, the transducer and the data acquisition system required for this test are simpler, and consequently cheaper.

12.4.2 Theoretical Background

The Impact-Echo technique involves introducing mechanical energy in the form of short pulse into a structure. As the resulting stress wave propagates through the medium, it is reflected by free boundaries and by internal defects of sufficient size. A transducer mounted on the surface receives the reflected pulses or echoes from the discontinuities in the material and converts them to electrical signals which can be digitized, stored and analyzed.

There are three primary modes of stress wave propagation through an isotropic, elastic media: dilatational, distortional and Rayleigh waves. Dilatational and distortional waves commonly referred to as compression and shear waves, or P- and S-waves and are characterized by the direction of particle motion with respect to the direction of wave propagation. In infinite elastic solids, the compression wave velocity, Cp is a function of Young's modulus of elasticity, E, the mass density ρ and Poisson's ratio ν. The compression wave velocity Cp, in an infinite solid is given by the following equation:

$$C_p = \sqrt{\frac{E(1-\nu)}{\rho(1+\nu)(1-2\nu)}} \qquad \text{Eq. 12-69}$$

For the concrete, E and ν can be determined from a uniaxial compressive test. The density ρ can also be measured easily. Alternately, the velocity has to be measured at the site, *in situ*, on the concrete. This can be done by mounting two transducers on the concrete and impacting near one of the transducers. The response of the two transducers is analyzed by the data acquisition system. The time taken by the wave to reach the two transducers is different, and this difference in time can be used to find the velocity of the sound wave. This experimental measurement of the wave speed requires that the data acquisition system has at least two channels for the two transducers, which can be triggered simultaneously. Fortunately, most available systems have at least two channels. Alternatively one can use the impact echo technique itself to determine the wave velocity on a portion of the structure that is known to have no defects. When this is done a single channel system will suffice. The shear wave velocity in concrete is about 61% of the compression wave velocity and the surface wave velocity is 56% of the compression wave velocity.

12.4.3 Performing the Test

An Impact-Echo test on a slab is illustrated in principle in Figure 12-38. The impact on the surface generates a stress pulse, which is reflected back and forth between the top and bottom surfaces of the slab. This impulse is generated by hitting the surface with a light hammer. Different hammer heads are available for imparting waves of different frequencies, as may be necessary for structures of different dimensions. Hence, a transient resonance condition is created and the waveform exhibits a characteristic periodic pattern [6-8]. Due to the nature of the impact (vertical drop of a steel ball) and the proximity of the transducer to the impact point, the P-wave portion of the pulse dominates the displacement waveform. The time δt required for the wave to return from the opposite surface of the slab for each successive reflection is:

$$\delta t = 2T/C_p \qquad \text{Eq. 12-70}$$

where T is the slab thickness, and C_p is the P-wave speed. The frequency with which the wave returns to the top surface after being reflected from the bottom surface is equal to the inverse of the time it takes to travel back and forth:

$$f = 1/\delta t = c_p/2T \qquad \text{Eq. 12-71}$$

The thickness of the slab is therefore:

$$T = C_p/2f \qquad \text{Eq. 12-72}$$

The frequency content of the waveforms detected by the transducer is determined by the 'Fourier Transform' and reflection frequency 'f' is determined as the peak in the frequency content of the signal. Many data acquisition systems can be used as 'frequency (FFT) analyzers' which automatically calculate the frequencies and report the peak frequency. Alternately, the transducer response data is Fourier transformed using programs that can be used on a personal computer, or almost always, directly on the data acquisition system itself. Figure 12-39 shows some typical IE results in terms of the frequency of the resulting waveform [6]. It may be observed that as the flaws are closer to

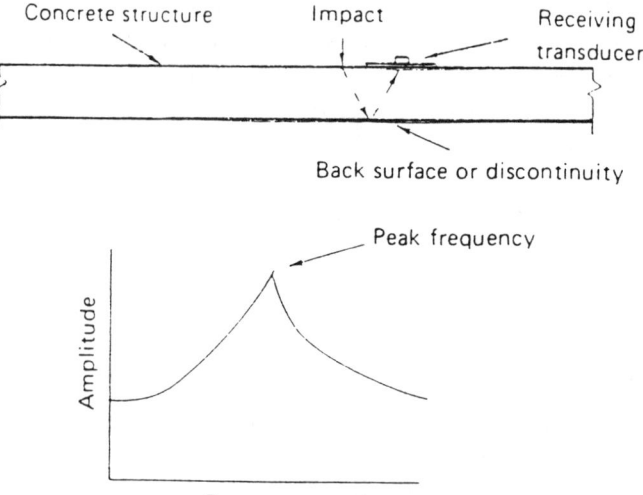

Figure 12-38. Conceptual Representation of the Impact-Echo Technique.

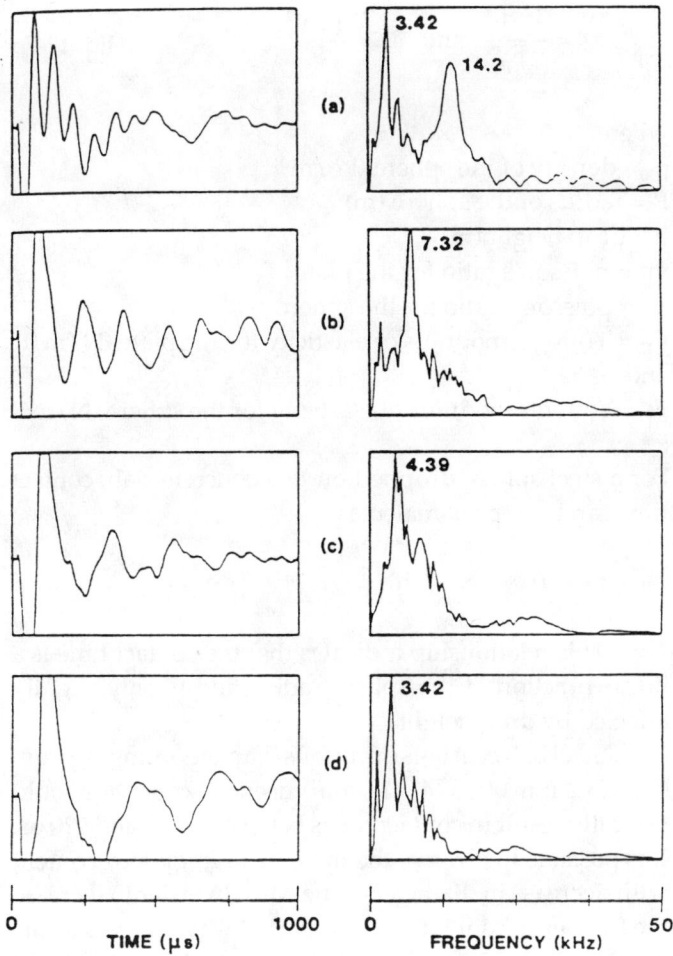

Figure 12-39. Waveforms and spectra for flaws at depths of (a) 127 mm, (b) 258 mm, and (c) 380 mm; and unflawed 500 mm thick plate. *(Sasalone and Carino, 1989)*

the surface, the peak frequency shifts to higher frequencies, as indicated by Equation 12-71.

Determining the location of an acoustically denser materials, such as determining the location of rebars in concrete is done in a similar manner. The only difference is that during a reflection from an acoustically softer media (waves in concrete reflecting from air at a free surface boundary) the returning wave has a reversal of phase. However, an acoustically denser material (steel, as compared to concrete) reflects the impinging waves without any phase reversal [9]. This results in the echo frequency for reflections from steel being half of what it should be using Equation 12-71. Therefore the correct distance from the transducer to the steel is given by:

$$T = C_p/4f \qquad \text{Eq. 12-73}$$

Another means of applying the IE technique is to compare the wave propagation in the vicinity of the flaw with the wave propagation in a region without a flaw. This requires comparing the two signals by performing a cross-correlation [10]. The main advantage of this extension of the IE technique is that the transducer and the impact do not have to be at the same location, consequently it is not necessary to inspect the structure point by point. On the other hand, one seldom encounters a structure that has the same geometric constraints (flaws, material property, distance from boundaries etc.) at different locations.

The three basic steps in performing the IE inspection are therefore:

i) Determine the wave speed in the material
ii) Mount the transducer and connect it to the data acquisition system.
iii) Use an impact source near the transducer, and determine the dominant frequency

12.4.4 Instrumentation

Three types of impact source have been used: a rebound hammer, steel ball bearing balls of various diameters [6, 7] and a Ball Bearing gun using 4.5 mm steel balls [10]. The receiving transducer used was a specially designed, commercially available broad band, high frequency, normal displacement transducer from Industrial Quality Inc. (model 501). The design of the transducer is based on the work done by Proctor [11]. This consists of a small, conically-shaped, piezoelectric element cemented to a large brass backing. A thin sheet of aluminum foil is used between the conical element and the concrete surface to complete grounding of the transducer circuit. Figure 12-40 shows IE being conducted using the displacement transducer, and ball bearing drops as the impact source. The impact was conducted directly on the concrete surface and not on the aluminum foil. This type of transducer has a

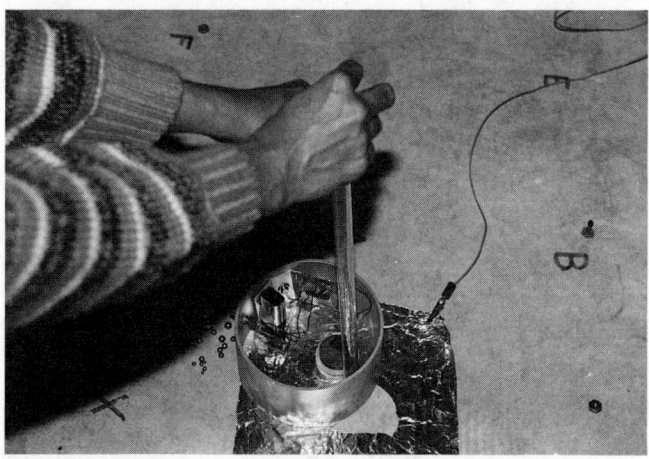

Figure 12-40. Dropping Steel Balls for Performing Impact-Echo

frequency response limit of about 2 MHz. Capacitance type transducers [12] and displacement monitoring based on laser interferometry have also been used for obtaining the displacement vs. time history. The cheapest alternative which seems to suffice for most low frequency impact-echo applications is to use accelerometers that are typically used for dynamic testing of structures; most laboratories already have such transducers for conventional structural testing. It is necessary to mount the accelerometer to the specimen using mounts that are attached with strong adhesives. This means that the accelerometer cannot be moved around at will to inspect different areas which is possible with the displacement transducer. Table A (following references) lists the sources of some of the instruments that are being discussed in this section.

Fourier Transform analysis can be performed directly using a F.F.T Analyzer (Zonic A& D, Model AD 3524/25) with a frequency range up to 100 kHz. It handles a transform size of 2048 points. A Nicolet Digital Oscilloscope (Model 4094 and accessories) can be used to find the wave speed, and to perform the signal processing algorithms such as frequency domain filtering, cross-correlation etc. Many FFT analyzers can be used in the averaging mode which enables one to find the average frequency content from a number of impacts at the same location. This added feature helps to average out some of the spurious frequencies and improve the results.

12.4.5 Other Concerns
Nature of the Impact Source

The impact source influences the energy and frequency content of the propagating pulse. The frequency content of the stress waves produced by impact depends upon the contact time of the impact. A shorter contact time produces a broader range of frequencies in the waves; however the amplitude of each component frequency is lower. The amplitude can be increased by using a larger size impact source, or imparting higher velocities of impact.

Three types of impact sources can be used: i) a rebound hammer, ii) steel ball bearing dropped on to the surface, and iii) shooting 4.5 mm diameter steel balls using a Ball Bearing gun.

The contact time of an impact produced by dropping a steel sphere on a concrete slab can be approximated by the Hertz elastic solution for a sphere dropped on to a thick plate [7]:

$$t_c = 5.97 \left[\rho_s \left(\delta_s + \delta_p\right)\right]^{0.4} \frac{R}{(h)^{0.1}} \qquad \text{Eq. 12-74}$$

$$\delta_p = \frac{1-v_p^2}{E_p} \quad \text{and} \quad \delta_s = \frac{1-v_s^2}{E_s} \qquad \text{Eq. 12-75}$$

where
ρ_s = density of the sphere (kg/m^3)
R = radius of the sphere (m)
h = drop height (m)
v_p = poisson's ratio for the plate
V_s = poisson's ratio for the sphere
E_p = Young's modulus of elasticity for the plate (N/m^2)
and
Es = Young's modulus of elasticity for the sphere (N/m^2)

For a steel sphere dropped on to a concrete slab, contact time can be approximated as:

$$t_c = 0.00858 \, R/(h)^{0.1}$$

This relationship indicates that the contact time is a linear function of the sphere radius and is only slightly affected by drop height.

Steel ball bearings of various diameters ranging from 1/8" (3.2 mm) to 3/4" (19 mm) used on concrete would typically result in contact times between 20 µs and 120 µs. It is possible to observe the increase in frequency content with decrease in diameter of the balls. In order to increase the frequency content of the imparted wave, as well as the penetrating ability (amplitude or energy) of the impact, a Ball Bearing gun can be used [10]. These use 4.5 mm diameter steel balls that are fired on the surface. The higher frequencies generated enable the detection of voids as small as 5 cm in diameter. Figure 12-41 shows the frequency content of the signal generated at a transducer when different types of impact sources are used. The use of extremely high impact, may cause aesthetically unacceptable damage to the concrete surface, and contribute to the high frequency waves generated. However, the ball-bearing impact used leaves only a small (about 1 mm. in diameter) scratch on the concrete surface, and does not pose any safety hazards. Using the ball bearing gun, it is possible to detect flaws as small as 5 cm in diameter in a concrete slab [10]. Due to the presence of the void, the sound wave has to travel a greater distance to reach the other side of the slab, and return to the transducer. This increases the travel time and consequently decreases the peak frequency from 18.5 kHz to 16.1 kHz. at the location of the void. Figure 12-42 shows the frequency content of the signal generated at the location of a 5 cm void using this technique. The peak frequency is at 16.1 kHz. A second weak peak of 35.25 kHz. is caused by the reflection of the waves from the top surface of the void. For the determina-

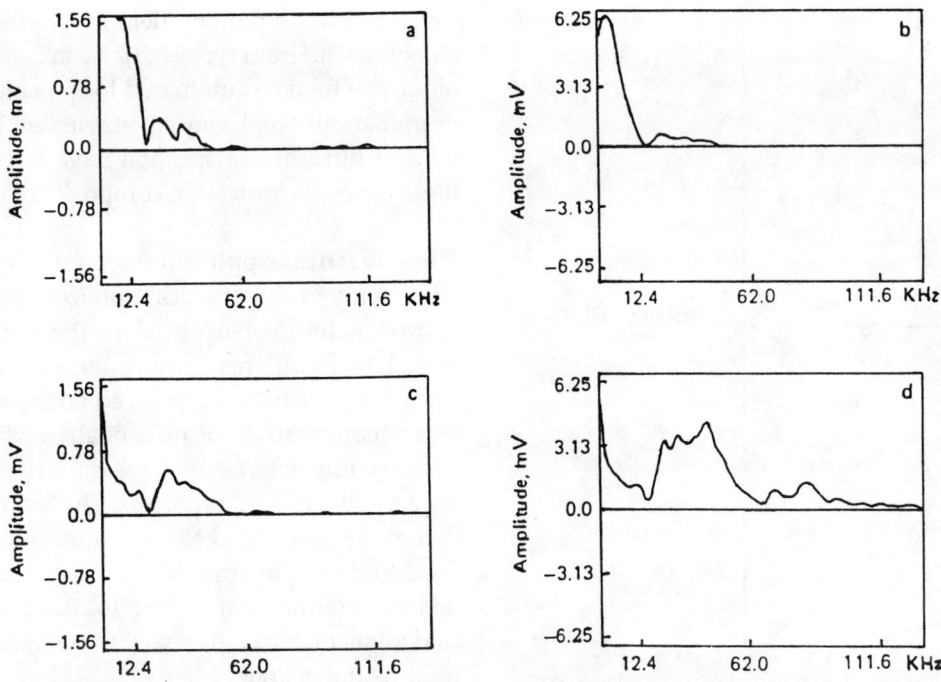

Figure 12-41. Frequency content response by dropping steel ball: (a) 1/2 in. (13mm); (b) 3/4 in (19 mm); and (c) 1/4 in (6.4 mm). (d) Frequency content of response due to a B.B. gun (4.5 mm).

tion of large flaws and delaminations (about 20 cm in size, or larger), however, an impact hammer, or steel balls are adequate.

Data Acquisition System

Since this is the most expensive part of the necessary arsenal, it requires additional thought. Presumably, the same system is used for numerous other experiments, besides IE, and the required specifications are dictated by the overall need of the laboratory or the company. Specifically for IE applications, one has to have an idea regarding the size of the flaws that will be tested for and determine the frequencies of interest from the size of the structure being inspected. The data acquisition system should then have a digitization rate fast enough to capture the relevant frequencies, and a time window (duration of acquisition and storage) large enough to capture the response for a sufficiently long time to monitor a few reflections. A frequency range of up to 100 kHz and a window of 1000 points will suffice for practically all applications, and are available in most FFT analyzers or data acquisition systems. Also for this reason, it is usually unnecessary to purchase a separate system for the sole purpose of doing the IE tests.

Possible Extensions of the Method

One of the shortcomings of the IE technique is the fact that it is a point by point inspection technique. This comes from the fact that the transducer and the impact both have to be located right above the void in order to detect it. This makes it cumbersome to use the technique for large sections of highways or for automated inspection. Maji et. al. [10] have proposed the cross-correlation technique to overcome this issue. Figure 12-43 shows a schematic of this technique. In this picture a 5 cm void was located at Z', in a 10 cm deep slab. A ball bearing gun impact was made at point A and the response of the slab was recorded at the locations X, Y, and Z. The cross correlation technique was then used to compare the response at X in turn with the responses at Y and Z.

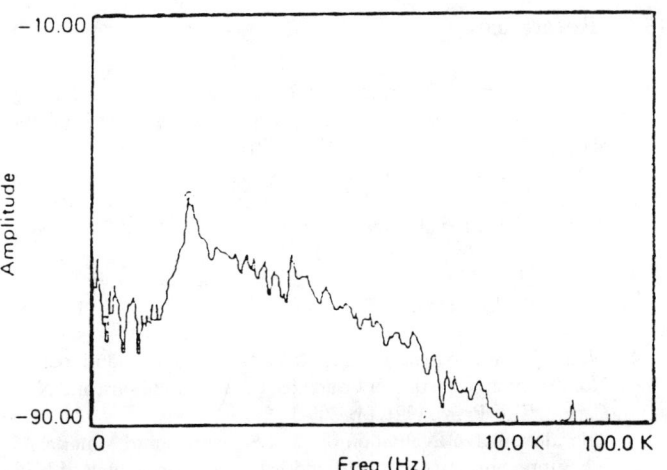

Figure 12-42. Frequency Spectrum over a 5 cm void. *(Maji et. al. 1990)*

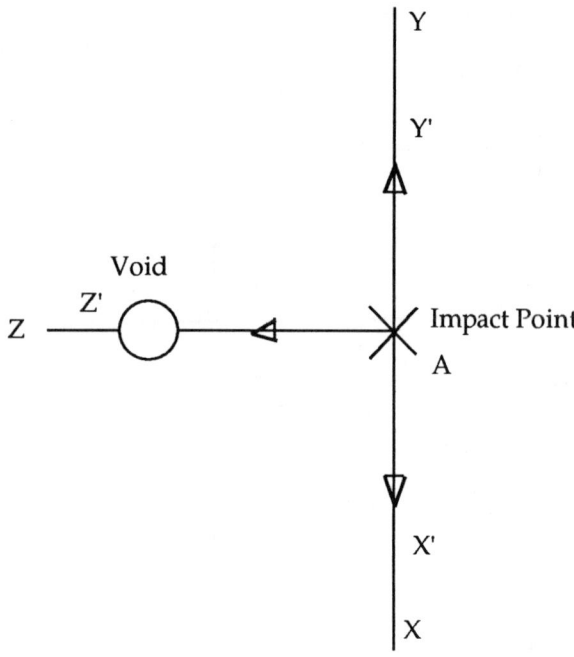

Figure 12-43. Transducer and Impact Points for the Cross-Correlation Method

The cross correlation technique uses the following equation to compare two signals A(i) and B(i), and reports the similarity between the two signals:

$$CCf(t) = \sum_{i=0}^{N-1} A(i) \cdot B(i+t) \qquad \text{Eq. 12-76}$$

where N is the number of digitized data points being compared, and t is the time variable. As a result of this calculation, the signals at X and Y show a greater correlation coefficient (0.82) as opposed the signals X and Z (0.73), because of the proximity of Z to the void. If the transducer location is moved closer to the void at Z', and the other points are correspondingly moved to X' and Y', the correlation coefficients comparing the responses at X' and Y' remain high (0.83). However, the correlation coefficients comparing the responses at X' and Z' decrease further (0.54) because of the proximity of the transducer at Z' to the void. This technique illustrates a possible extension of the IE technique where it is not essential to place the transducer and the impact at the location of the void, and therefore it is not essential to inspect a slab on a point by point basis. This also means that unlike the point by point measurement technique which utilizes only the compression wave (P-wave), the correlation technique is influenced by the other types of wave propagation as well. The cross correlation software is readily available from most manufacturers of data acquisition systems, or personal computers. An additional extension of the techniques to use the newly emerging field of 'Neural Networking' to recognize different types of flaws in concrete. The IE technique can then be automated for repeatable tasks such as examination of high way pavements etc. Researchers in the field of ultrasonic inspection have already made use of these concepts for testing composite materials [13].

Some Practical Applications

Many of the practical applications have been highlighted by the developers of the IE technique; Carino and Sansalone [6-8]. These applications have included detection of large artificially induced honeycombs in concrete, delaminations underneath asphalt overlays etc. Some well known companies have adopted the IE technique for solving a variety of field problems. The technique was used to determine voids in shear wall to ramp beam joints; was used to determine the thickness of slabs poured in residences to within half an inch;to check for the presence of cold joints,to detect internal flaws in massive foundations used as boiler foundations [14, 15]. In the inspection of concrete arch bridges, the slabs, spandrel walls and abutment walls are accessible only from one side. The IE technique is particularly suitable for inspecting such members for internal corrosion induced damage and delaminations [16]. It has however, been suggested that these tests be supplemented wherever possible with some means of independent verification, such as examination of cores. The nature and location of delaminations in a 30 year old parking structure were determined by inspecting the slab from underneath using IE [17]. Delamination at the steel or asphalt interfaces could be specified. Honeycombing in the diaphragm of box-girder bridges could be determined and successfully verified by chipping out concrete from the suspected areas. This led to recommendations for repairing with hotpatch epoxy fillers [17].

13.4.6 References

1. Green Robert Jr. "Current and Emerging Techniques for the Nondestructive Evaluation of Civil Structures," Proc of Int. Workshop on NDE for Performance of Civil Structures, editors: Agbabian M.S. and Masri S.F., Feb. 1988, p. 81-100.
2. Cantor T.R. "Review of Penetrating Radar as Applied to Nondestructive Evaluation of Concrete," ACI-SP 82, 1984, pp. 581-602.
3. Teodoru G.V. "The Use of Simultaneous Nondestructive Tests to Predict the Compressive Strength of Concrete," ACI-SP 112, pp. 137-152.
4. Weil G.J. "Infrared Thermographic Techniques," in 'Handbook of Nondestructive Testing of Concrete', ed. V.M. Malhotra and N. J. Carino, CRC Press, 1991, pp. 305-3 16.
5. "Nondestructive Evaluation of Civil Structures and Materials", ed: Suprenant, Sture, Noland and Schuller, Proceedings of NSF conference at Boulder, CO, PIP press, CO, October 1990,
6. Sansalone M. and Carino N.J. "Laboratory and Field Studies of the Impact-Echo Method for Flaw Detection in Concrete," ACI-SP 112,

1989, pp. 1-20.
7. Carino N.J., Sansalone M. and Hsu N.N. "A Point Source, Point Receiver, Pulse-Echo Technique for Flaw Detection in Concrete," J. of ACI, V 83, No. 2, 1986, pp. 199-208.
8. Sansalone M. and Carino N "Impact-Echo Method," Concrete International, April 1988, pp. 38-46.
9. Sansalone M. and Carino N.J. "Impact-Echo: A Method for Flaw Detection in Concrete Using Transient Stress Waves," NBSIR 86-3452, National Bureau of Standards, Sept. 1986, pp. 222 (NTIS PB # 87-104444/AS)
10. Maji A.K., Paul S. and Wang M.L. "Improved Impact-Echo Technique by Signal Processing," Research in Nondestructive Evaluation, V 2, No. 1, 1990, pp. 45-56.
11. Proctor T.M. Jr. "An Improved Piezoelectric Acoustic Emission Transducer," J. of Acoustic Soc. of America, 71(5), May 1982, pp. 1163-1168.
12. Boler F.M, Spetzler H.S., and Getting I.C. "Capacitance Transducer with a Point-like Probe for Receiving Acoustic Emissions," Rev. Sci. Instrum., 55, (8), August 1984, pp. 1293-1297.
13. Thomsen J.J. and Lund K. "Quality Control of Composite Materials by Neural Network Analysis of Ultrasonic Power Spectra," in Materials Evaluation, V 49, No. 5, 1991, 594-600
14. Limaye H.S. "Laboratory and Field Testing of Concrete Using Impact-Echo Technique and an FFT Analyzer," in Proc. of Int. Workshop on NDE for Performance of Civil Structures, USC, Los Angeles, CA, Feb.1988, pp. 282-291.
15. Limaye H.S. "Applications of Impact-Echo Technique," Proc.of SEM Spring Conference, May-June 1989, pp. 683-688.
16. Limaye H.S. and Klein G.J. "Investigation of Concrete Arch Bridges Using the Impact-Echo Method," in 'Nondestructive Evaluation of Civil Structures and Materials', Proceedings of NSF conference at Boulder, CO, PIP press, CO, October 1990, pp. 221-232.
17. Olson L. "Nondestructive Evaluation of Structural Concrete with Stress Waves," in 'Nondestructive Evaluation of Civil Structures and Materials', Proceedings of NSF conference at Boulder, CO, PIP press, CO, October 1990, pp. 61-70.

Table A

Item	Source
Impulse Hammer (model 9728)	Kistler Instrument Corp., Amherst, NY
FFT Analyzers	Keithley Instruments, Inc., Cleveland, OH; Bruel & Kjaer Instruments, Inc., Marlborough, MA
Data Acquisition Systems	Nicolet Instruments Corp., Madison, WI; Lecroy, Spring Valley, NY
Displacement Transducer	Industrial Quality Inc., Gaithersburg, MD; Physical Acoustics Corp., Princeton, NJ

12.5 REAL-TIME RADIOGRAPHY
Ming L. Wang
Department of Civil Engineering, The University of New Mexico

12.5.1 Introduction

Real-time radiography is a nondestructive testing technique that uses penetrating radiation to produce X-ray images. The arrangement of the radiation source, object and image plane is similar to conventional radiography. Real-time radiography allows radiographic interpretation to be performed simultaneously with the progress of the event. The most important process in real time radiography is the conversion of radiation to light by means of a fluorescent screen or an image intensifier. The light signal can then be observed directly, amplified and converted to a video signal for presentation on a television monitor and recording in a video recorder. This technique is often applied to samples on assembly lines for rapid inspection. Remote adjustment of the specimen position allows inspectors to freely view the details of interest or to move to another location of interest without the delay and expense of film development. For dynamic events, real-time radiography is typically operated in the range of 30 frames per second. With the use of image processing techniques, real-time radiography has gained wide acceptance for various industry applications.

In general, radiography is most useful for finding internal, nonplanar defects such as porosity and voids. It is also suitable for detecting changes in material composition, for thickness measurement, for detecting corrosion of a structure component and for locating unwanted or defective components hidden from view in assembled parts. Detection and identification of the type of weld flaw has been applied in the industry for increasing the inspection speed, and for reducing operating costs. Using scattered x-ray radiation, density measurements of liquids, solids and gases, both organic or inorganic compounds can be achieved.

The material presented herein is based on references [1-8] and other specialized works [9-12]. It is intended only to provide an overview of the real-time radiography technique. For more in-depth understanding of the theory of x-ray radiography and image processing technique, the reader should refer to the references.

12.5.2 Basic Principles

X-rays are a form of electromagnetic radiation, as is light. Their distinguishing feature is their extremely short wavelength about 1/10,000 that of light, or even less. This characteristic is responsible for the ability of x-rays to penetrate materials. The mechanism of x-ray inspection is the propagation of energy from a source through an object and the evaluation of the energy pattern as received. X-rays are produced when electrons, traveling at high speed, collide with matter or change direction. The radiation source emits energy that travels in straight lines and penetrates the object under investigation. Once the radiation energy has passed through the item, an image is produced on a recording plane opposite to the source which is used to evaluate the condition of the part being inspected. Film is most frequently used as the recording medium. However, the use of television cameras, microcomputers with digital image processing and storage and an image intensifier to convert the x-ray image to light have produced much improvement in the speed and performance of x-ray inspection systems.

The video signal from a TV camera or CCD camera focused on an image intensifier must be digitized through an analog-to-digital converter and stored in a computer. The digitized image is quantified in space and intensity. Any image can be regarded as a series of very small elements (pixel) and the position and brightness of each pixel in the image area are determined. Commonly an image is taken as 512 lines across the image, with 512 pixels on each line. This gives a square array of 262,144 picture elements (pixels). The brightness of each pixel is measured on an 8-bit scale (256 intensity levels) or a 9-bit scale (512 intensity levels). The basic quantization method gives 256 or 512 contrast steps when going from black (lowest intensity) to white (highest intensity).

12.5.3 Basic Equipment

Basic equipment is shown in Figure 12-44. A conventional x-ray image intensifier tube (6 in) is used to convert the x-ray image to light, and a CCD (charge couple device) camera is focused on the intensifier tube output screen. The intensifier and camera can be placed in the direct line of the radiation for x-ray energy below 300 kV without any damage to their electronic and optic components. The output from the camera is amplified, digitized, stored, enhanced and displayed on the monitor. A video recorder is often used to provide a permanent record of the inspection. The original and processed images can then be stored in a laser disk system. Image recording with a laser disk digital recording system appears to be a standard practice for long-term recording. A 300 mm disk can carry 8000 frame images.

In a conventional x-ray tube, a filament supplies the electrons and forms the cathode or negative electrode of the tube. A high voltage applied to the tube generates the electrons to the anode (target). Once the moving electrons strike the target, x-rays are generated. The high melting point of tungsten makes it a very suitable material for the target of an x-ray tube. The higher the voltage, the greater the speed of electrons striking the focal spot. The result is a decrease in the wavelength of the x-rays emitted and an increase in their penetrating ability and intensity. Typical x-ray machines with 150 kV will penetrate 5" thick aluminum or 1" thick steel.

12.5.4 Effects of Scattered Radiation

Scattered radiation affects contrast of the image by raising the background brightness level. Scattered radiation comes from many sources in a real time imaging setup. It is also called secondary radiation as it results from deflected photons. There is room scatter, objective scatter, fixture scatter, air scatter in the primary radiation beam, and scatter from any and all objects placed in the path of the radiation beam. The control of scatter is a very important issue in the application of real-time radiography. Specific techniques to reduce scatter are discussed.

1. Collimate the primary beam to the minimum viewing area necessary. Photon streams striking the object at normal incidence would be less prone to excite scatter than streams entering at an oblique angle. Greater distances from the source to the object would increase the sharpness of the image. At larger distances, the width of the source becomes smaller compared to the source-to-object distance and the size of the shadow of the edge is decreased.

2. Filter the primary beam to remove the low energy portion of the spectrum, which is more susceptible to scatter. The lower energy x-rays, (longer wavelengths) are seen to produce much greater scattering than those having higher energy contents.

3. Filter the radiation beam between the object and the image intensifier, because many filters preferentially remove the lower energy scattered radiation.

4. Shield the setup to reduce room scatter from walls, ceiling and floors.

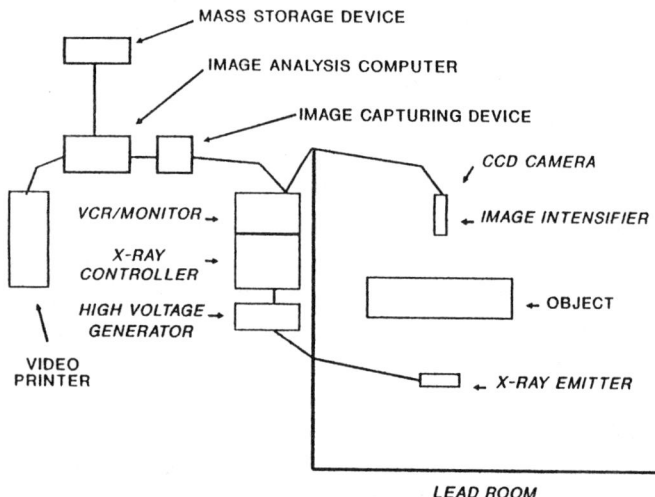

Figure 12-44. Basic Real-Time X-Ray System

Several devices are available to minimize the effect of the scatter. Some of the frequently used items are the various filters and screens which have been used by others [2, 4, 5]. Filters are placed near to the source, while screens may be placed between the image intensifier and object. The purpose of the filter near to the x-ray source is to remove those energy levels that are more likely to cause scatter in the test specimen.

Longer wavelength (lower energy) x-rays are less penetrating and producing greater scatter. Lead foil screens are placed adjacent to the image intensifier to reduce scatter from all source. Lead screens can also serve as an intensifier. Moores (13) has described recent applications of screen for image quality improvement.

12.5.5 Image-Processing Techniques

Digital image processing methods provide a mathematically based approach to improve the interpretation of data from x-ray images. In practice, the purpose of image enhancement in nondestructive testing is to display low contrast detail relating to cracks, boundaries, corrosion, part orientation, voids and inclusions. In some cases the objective is to locate edges more precisely in order to improve measurement accuracy. Real-time radiographic systems compare unfavorably with film-based radiography in two major respects, low contrast sensitivity and limited resolution. Image enhancement techniques becomes necessary in order to compensate these limitations. Real-time radiographic images are analog or digital signals, generated by television cameras or solid state devices (CCD Camera).

While the emphasis of the following presentation is on real-time radiography, it should be understood that many of these techniques are also useful in film based applications as well as in operations for videotape recorded signals. Only enhancements in the spatial domain are presented; manipulation in the frequency domain has not been addressed, although some of the operations to be described have a frequency filtering effect.

Real-time image enhancement can be grouped into two major categories, namely, point processing operations and neighborhood processing operations. Point processing performs the same operation on each pixel in the image independent of the values of neighboring pixels. Point processing includes summing, averaging, subtracting, gray level mapping and histogram equalizing. Neighborhood operations are performed on each pixel based on the values of neighborhood pixels. Neighborhood processing includes differentiation, gradient operations edge detection and spatial filtering. All these operations are available in most commercial digital image processing software. The video output of a real-time radiography system is a mixture of signals from the object being examined, systematic noise and random noise. Systematic noise includes camera target burns, light leaks, power line interferences. The random noise component is usually a mixture of signals induced noise caused by the statistical nature of the quantum processes involved plus an additive component due to thermal and bias current noise originating in the video amplifiers. Random noise may be reduced by summing or averaging successive frames while some types of fixed pattern noise may be reduced using subtraction techniques.

12.5.6 Example 1—Flaw Detection of Composite

To demonstrate the utility of using X-rays to detect flaws in a carbon-carbon composite, a sample of a carbon-carbon composite was x-rayed. A radiograph was prepared using an IRT 160/3200 real-time x-ray system. An 6" Image Intensifier and a CCD camera were used to record the images. An Imaging Technology FG-100-1024 Frame Grabber was used to convert analog data to digital data. The layers of the carbon fabric are parallel to the x-rays. Figure 12-45 shows a typical view showing flaws in the binder. This image is an average of 32 frames added together to reduce the random noise and to improve the signal to noise ratio. Not all the flaws are apparently detected as will be shown in later figures. The images produced by the x-ray often hide some of the details.

By applying some simple image analysis technique the features of the images can be enhanced. Figure 12-46 shows the same image after it has been added to itself several times. Each time the image was added to itself, a value equal to 70% of the lowest grey level was uniformly subtracted from the resulting image. By repeating this

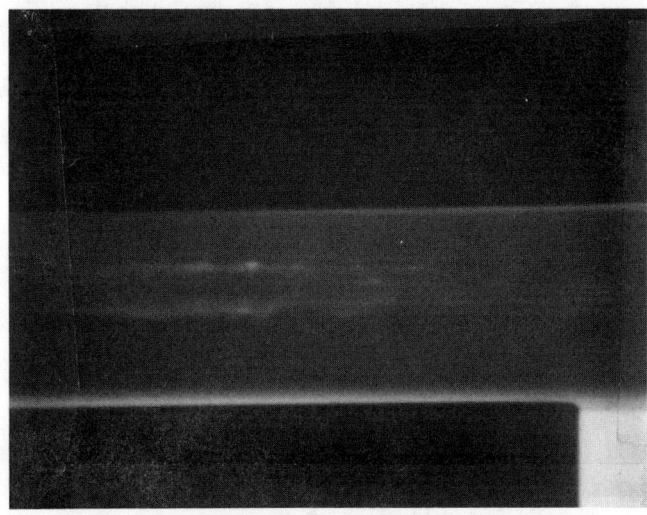

Figure 12-45. Real-Time X-ray Image of a Carbon-Carbon Composite under 100 kV Energy.

process the contrast of the image is increased, which shows some additional detail that was not present in the original image. It brings out some additional flaws and delamination details that were not apparent in the original image. By using a simple horizontal edge filter, the delaminations are seen more clearly as shown in Figure 12-47.

By utilizing a general convolution filter the image from Figure 12-45 can be enhanced in a different manner. Figure 12-48 shows that the relative magnitude (size) of the flaws are enhanced by using one type of convolution while Figure 12-49 shows that the locations of any flaws above a certain size. Note, a dimple is shown in the lower right hand side of Figure 12-48. This is due to a small hole that was burned into the specimen. The bright horizontal region on top of the specimen depicts a groove burned on the top of the specimen. This particular feature is almost invisible in the other figures. Thus proper image analysis can uncover many details that are not readily apparent to the unaided eye.

12.5.7 Example 2— Internal Strain Measurement for Concrete

Recently, an Instron machine has been acquired for testing concrete specimens, both torsionally and axially, to failure. As one method for measuring deformations, it is proposed that a lead-wire grid be embedded in the specimens and X-rays be used to obtain images of the wire mesh as the load is applied. Specimens are loaded axially. The X-rays are taken continuously during the loading process. The deformation of the lead grid provides a method for

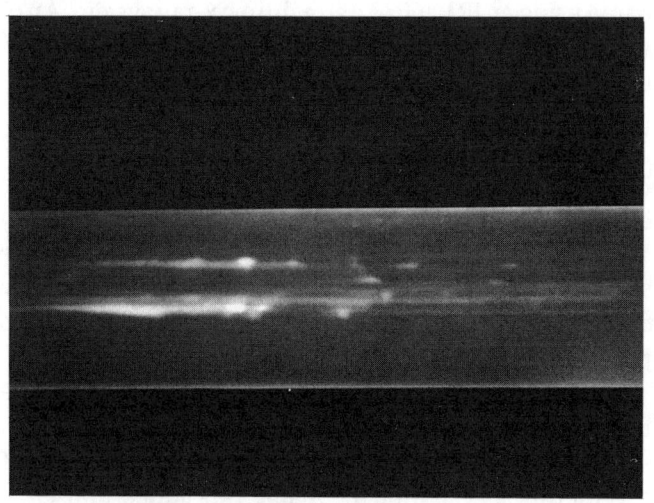

Figure 12-46. Processed Image of Figure 12-45 after Using Image Addition Operation.

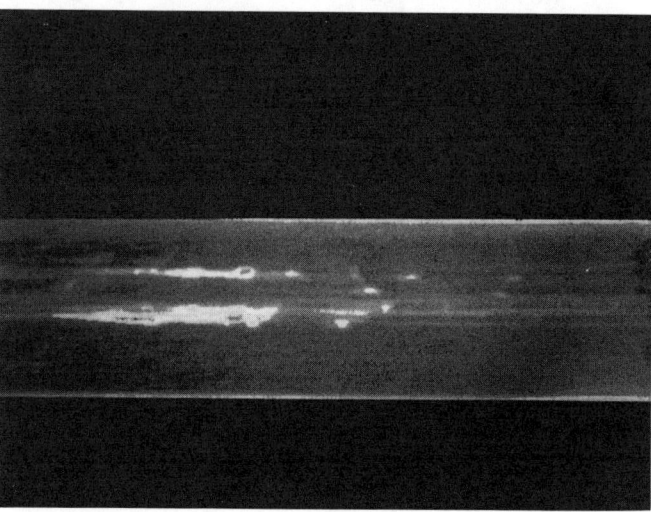

Figure 12-48. Processed Image of Figure 12-45 after Using a Convolution Filter.

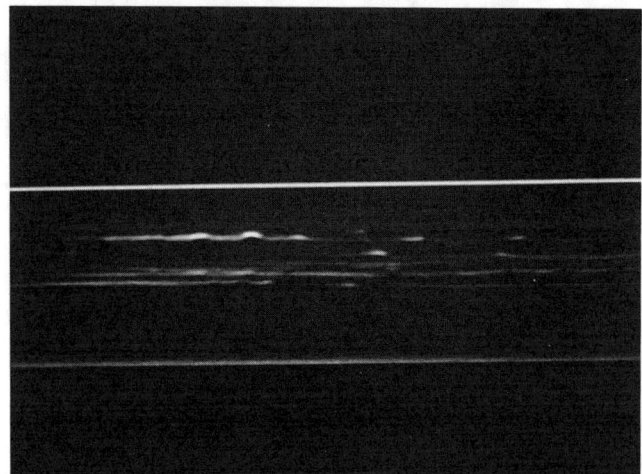

Figure 12-47. Processed Image of Figure 12-45 after Using Horizontal Edge Filter Operation.

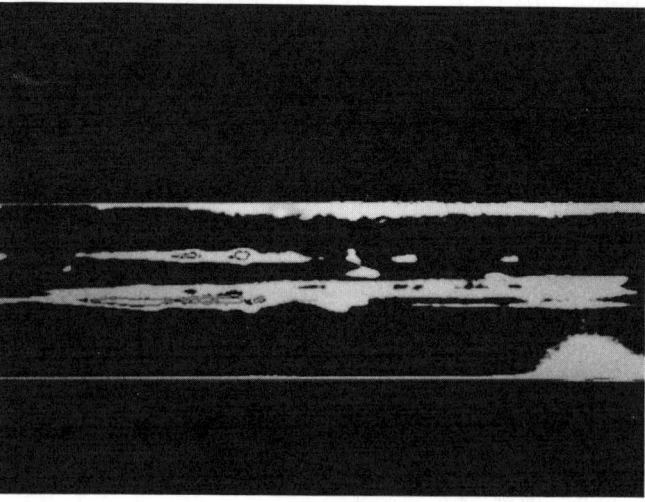

Figure 12-49. Processed Image of Figure 12-45 after Using a Convolution Filter with Different Kernel Values.

determining displacements, and consequently strains, as a function of time within the specimen. Furthermore, the changing gray levels on the radiograph are a measure of variation in density which can be enhanced with color image analysis. Experiments indicate the real-time X-ray and image analysis is a viable tool for providing internal strain measurements. The density map of the failure process zone may also be related to the internal mean stress distribution.

Equipment

An Instron 1323 Biaxial Loading System was used to test the specimens. This system has a 110 kip capacity in tension and in compression and a 50 in-kip capacity in torsion. Furthermore, the system can be controlled through load, displacement or strain. Experimental data are converted with an Isaac 2000 (12 bit) analog to digital converter. The data are stored and processed on an IBM AT personal computer.

During the tests the samples are X-rayed continuously using an IRT IXRS 160/3200 Industrial X-ray System. The X-ray has a 160 kV and a 3200 Watt maximum output. Using a 6" image intensifier and a CCD camera the continuous fluoroscopy of the samples during loading is recorded. The images are stored on a video recorder for later analysis. Images from the test are digitized using an IMAGING Technologies, Inc. FG100-1024 video capturing board (image digitizer) in an IBM AT computer system. The images are stored and processed on the IBM AT. Digital image enhancement software is available. Figure 12-50 provides a graphical layout of the equipment.

Specimens

Cement paste samples were prepared using the following mix design:
2 lbs sand (#16-#30 sieve), 4 lbs sand (#30-#50 sieve), 3 lbs type 1 cement
1.1 lbs water, 14 grams superplasticizer

Samples were measured 1.5 inches by 1.5 inches by 3 inches. Prior to casting, a lead-wire mesh was placed in each of the samples parallel to the longest lengths and the bottom of the mold. The lead wire was approximately 0.25 mm in diameter. The mesh in the samples measured 0.375 inches by 0.425 inches. The samples were cured in a lime bath at 110 degrees F for five days and tested four days after being removed from the bath. The average compressive strength of these specimens was about 5000 psi.

Test Procedure

The samples were tested in uniaxial compression with the mesh parallel to the direction of loading. The loading

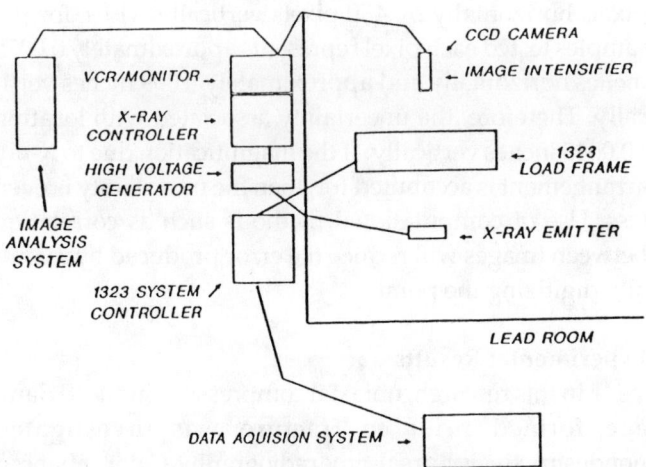

Figure 12-50. Test Setup

system was displacement controlled with the displacement rate set to approximately 0.0002 inches per second. The samples were loaded to failure and continuously X-rayed during the loading process. Load-displacement data were collected.

Image Analysis

Several images were taken from each test. These images were analyzed using two techniques. In the first technique (Digital Tracking), the displacement of discrete points in the sample from image to image is tracked from the beginning of the load to failure. A lead wire mesh included in the specimens during casting provided these points. A plot of the displacements of these points provides a representation of the displacement field within the sample at the location of the mesh.

In the second method (Color Enhancement), image processing filters are used to add color to the images. Each digitized image is made up of pixels which are small discrete blocks. The value (0255) of the blocks are determined by the grey level (shade of grey) of the corresponding portion of the original image. The grey levels vary according to the densities of the materials in the path of the X-rays. By utilizing the color enhancement filtering technique changes in grey levels (densities) can be accentuated within an image and between images and, thus, a way to qualitatively examine the evolution of the failure zone is obtained.

By comparing the results of these two techniques to the force/displacement data, an improved method is provided for describing the evolution of failure in concrete. These data are also valuable for evaluating constitutive models of concrete.

Uncertainty Analysis

When images are digitized, each image measures 512

pixels horizontally by 480 pixels vertically. Thus, for the samples tested each pixel represents approximately 0.0075 inches horizontally and approximately 0.009 inches vertically. Therefore, the uncertainty associated with locating ±0.0045 inches vertically. If the magnification due to X-ray arrangement is accounted for, then the uncertainty is even less. Use of computational methods such as correlation between images will reduce the error produced by manually digitizing the points.

Experimental Results

In this research, uniaxial compression-induced damage formed prior to fracture was investigated nondestructively by real-time radiography. Color enhanced X-ray damage evaluation was used to establish load levels at which damage initiates and to document the progression of damage as the load was increased to failure.

Experiments have demonstrated that real-time radiography with image analysis is a powerful tool for providing internal strain measurements and a density map of the failure process zone. As an example, consider a rectangular prism specimen loaded axially. Within the specimen a grid of lead wire (0.25 mm diameter) was embedded. The embedded lead mesh can be seen using real time X-ray as shown in Figure 12-51. If X-rays are taken continuously during the loading process in real-time, then the deformation of the lead grid provides a method for determining displacements as a function of time. Consequently, numerical differentiation yields strains within the specimen. Furthermore, the various levels of grey on the radiography are measures of variation in density which can be enhanced with color.

Figure 12-52a and b show the deformed and original grids at 40% of peak load and after the peak load, respectively, for end-confinement conditions caused by direct contact of the specimen with the loading platens. The grids are isolated from the radiograph through image analysis software. Since the images are stored, internal displacements can be obtained by comparing the deformed grid with the undeformed one. As shown in Figures 12-53a through 12-53d, the displacement field can be displayed and quantitative values can be determined. Figures 12-53a through 12-53d show the localized aspect of the deformation field that is typical of the failure process. The failure zones in most cases emanate from corners at an inclination of about 45 degree with an interior domain showing a higher angle of inclination with respect to the end face.

Figures 12-54 a to 12-54d provide a sequence of color images which show changes in density patterns as the load is applied to the specimen. Initially, the existence of blue splotches as originally shown in color indicates that the specimen is manufactured with some density variation which is expected for cement mortar. As the load is applied the density variation reduces as existing voids and microcracks are closed. Then as the load is increased fur-

Figure 12-51. Lead Mesh Embedded Inside the Cement Paste

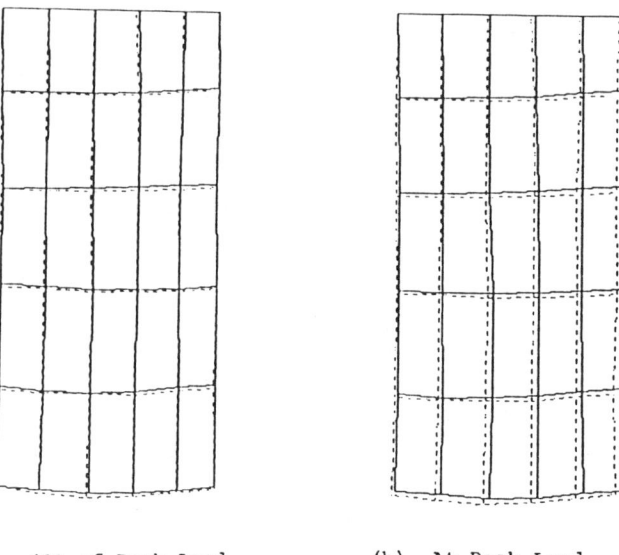

(a)   40% of Peak Load   (b)   At Peak Load

Figure 12-52. Comparison between Original and Deformed Meshes

- - - - - - - - Undeformed   ——————— Deformed

Experimental Methods—Nondestructive Evaluation Techniques

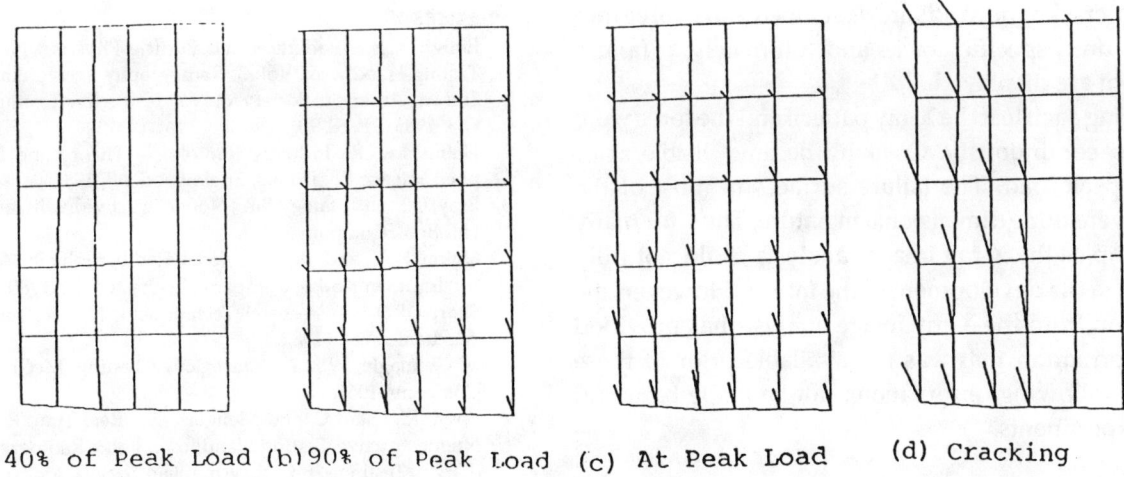

(a) 40% of Peak Load (b) 90% of Peak Load (c) At Peak Load (d) Cracking

Figure 12-53. Displacement Vectors as Loading Progresses for Minor Confinement.

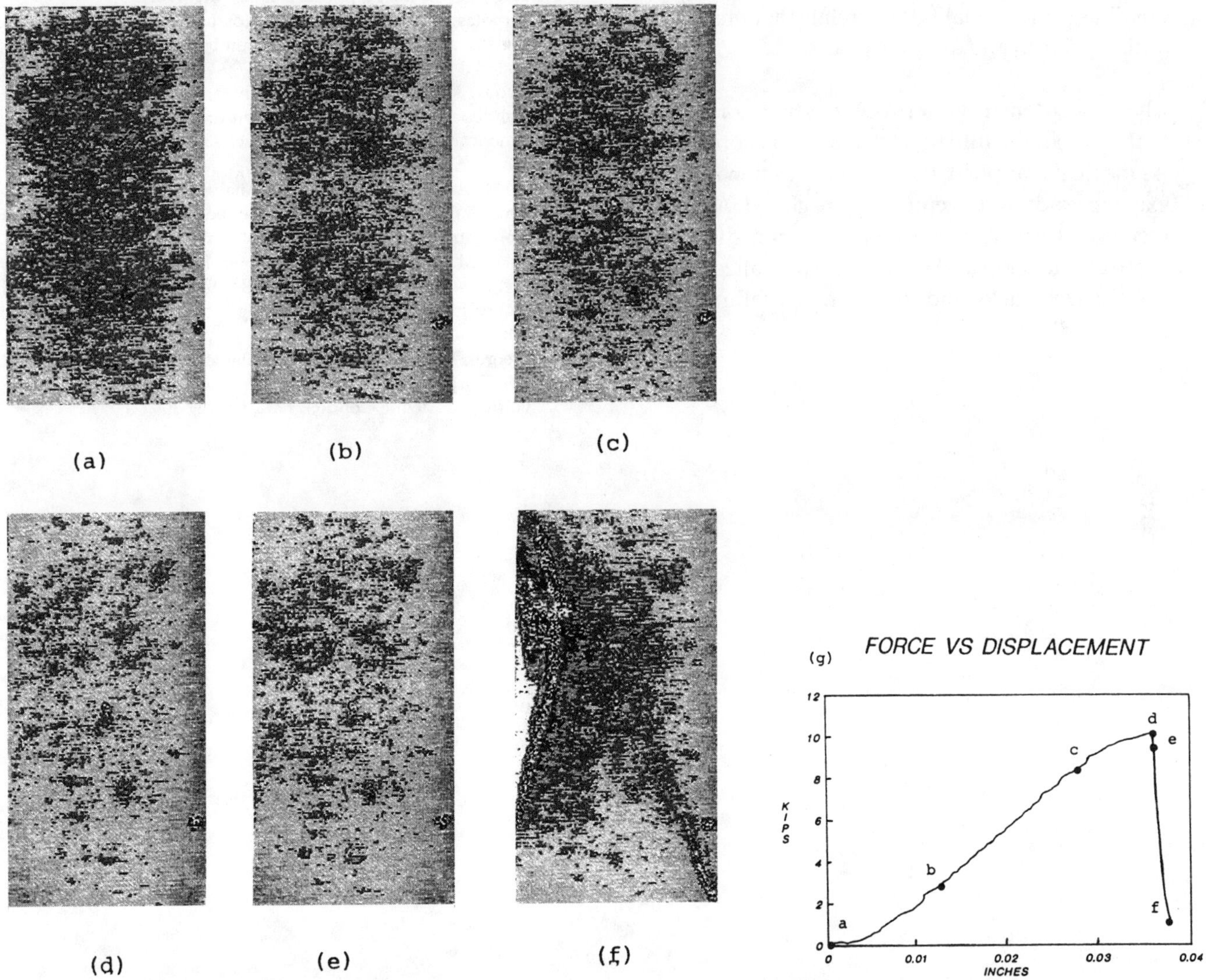

Figure 12-54. Density Patterns and Corresponding Loading Level

ther, microcracks and voids are developed and evolve into a pattern until specific cracks and, ultimately, a failure mechanism are displayed.

During each test the X-ray pattern and the force were monitored continuously. Cracking became visible at or after the peak load. The failure surfaces in most of the samples were three dimensional in nature, Thus, for many of the samples the X-ray images analysis could not fully characterize the development of the failure. However, the information from the X-ray image analysis has provided much information that was not available prior to these tests. The following observations can be made based on several experiments.

1. The X-ray real-time radiograph and image analysis is an effective approach to measure the quantitative internal strain values and evolution and provides a possible guide to qualitatively relate the density distribution prior to and after the peak.

2. Changes in density pattern occur as the load is applied to the specimen. Initially, the density is non-uniform. As the load is applied, the density variation reduces as existing voids and microcracks are closed. As the load increases further, a non-uniform strain distribution begins to develop and evolve into a localized pattern until major cracks and, ultimately, a failure mechanism are displayed.

References

1. Emigh, C.R., "Radiation and Particle Physics," Nondestructive Testing Handbook, Vol. 3, Radiography and Radiation Testing, 2nd ed., American Society for Nondestructive Testing, Columbus, OH, 1985, PP. 62-90.
2. Halmshaw, R., Industry Radiology—Theory and Practice, Applied Science Publishers, Englewood, NJ, 1981.
3. Bray, D.E. and Stanley R.K., Nondestructive Evaluation, McGraw-Hill Book Company, 1989.
4. Quinn, R.A., and Sigl, C.C. (eds), Radiography in Modem Industry, Eastman Kodak Company, Rochester, NY 1980.
5. Sharpe, R.S. (ed), Research Techniques in Nondestructive Testing, Academic Press, 1982.
6. McGonnagle, W.J., Nondestructive Testing, McGraw Hill Book Company, 1961.
7. Bossi, R., Oien, C. and Mengers, P., "Real Time Radiography," Nondestructive Testing Handbook, Vol.3, Radiography and Radiation Testing, 2nd ed., American Society for Nondestructive Testing, Columbus OH, 1985, SEc. 14, pp.593-640.
8. Halmshaw, R. "An Analysis of the Performance of X-Ray Television-Fluoroscopic Equipment in Weld Inspection," Material Evaluation, Vol 45, no. 11, pp. 1298-1302, Nov. 1987.
9. Berodias, M.G. and Peix, M.G., "Nondestructive Measurement of Density and Effective Atomic Number by Photon Scattering," Material Evaluation vol. 46, pp. 1209-1213, Aug. 1988.
10. Jacoby, M.H., "Image Data Analysis," Nondestructive Testing Handbook, vol. 3, Radiography and radiation Testing, 2nd ed., American Society for Nondestructive Testing, Columbus OH, 1985, sec. 15, PP. 641-673.
11. De Meester, P., and Aerts, W., "Analysis of Radiographic Image Recording Systems," R. S. Shape, ed., Research Technique in Nondestructive Testing, vol. 5, Academic, London, 1982, pp. 1-52.
12. Paker, M.E., "The Use of Digital Image Processing in the Detection of corrosion by Radiography," in R.S. Sharpe (ed), Research Techniques in Nondestructive Testing, vol. V, Academic, London, 1982, pp. 53-74.
13. Moores, B.M., "Screen/Film Combinations for Radiography," in R.S. Sharpe (ed), Research Techniques in Nondestructive Testing, vol. VI, Academic, London, 1982, pp. 289-308.

CHAPTER 13

TEST ORGANIZATION AND SAFETY, HEALTH, AND ENVIRONMENTAL ISSUES

Robert T. Reese, Sandia National Laboratories,
Albuquerque, New Mexico

13.1 INTRODUCTION

Structural tests are performed because insights are needed. There may be legal or regulatory requirements which must be fulfilled, apprehensions regarding structural integrity which require proof or load tests, or experimental evidence needed to verify analytical models. When tests are necessary, organizing them in the best manner possible will ensure quality and obtain the desired results. The basic method in organizing test activities is to proceed as though an experimental investigation is in progress and that all facts and details will be considered. The planning and organization efforts guide the investigation to a successful conclusion. The underlying themes in organization efforts are to obtain the essential information and insights, limit expensive activities so that only necessary tests are performed, and eliminate the extraneous data or invalid responses.

This chapter explains in general terms and with some specific details the organization of activities and operations covering the wide ranges of possible structural tests and associated equipment. Also included is a description of the safety, health, and environmental (S/H/E) issues associated with structural testing. These issues are organized and presented in this manner because this is the order of their importance in structural testing. A methodology is presented for the organization, evaluation, and how compliance on S/H/E concerns and requirements can be achieved in performing structural tests.

The first item considered in organizing a test is the structure or item to be tested. It is the reference for planning and organizing efforts. Virtually all test discussions center around what the structure is, how it will be loaded, how it will respond, what measurands will be used to describe its behavior, and what hazards could be present prior to, during, and after completion of testing operations. These topics vary in importance and each must be considered and evaluated on a case-by-case basis. For example, if a structure or assembly presents significant hazards (e.g., sudden release of stored energy) during test operations, then the potential hazards involved become a principal part of planning and organizing a test. However, if a structure or an item has hazards which are routinely handled in testing operations, then the concerns for potential hazards are relegated to a less important role in organizing a test (Ref. 13-13).

The second item in organizing a test are the reasons for the test to be performed. The reasons may be defined as a test objective (e.g., proof test), in terms of a sequence of tests in a preliminary test plan (components subjected to static loads, vibration, and shock conditions), or as acceptance or certification requirements (e.g., material properties). These reasons were explained in the types of tests described in Chapter 4. The structure and the reasons for the tests form the major considerations in planning.

This chapter is written to cover the wide range of organizations and personnel who could perform tests in laboratory, field, or other locations. This chapter consists of four basic sections: (1) basic test planning which includes a discussion of how the structure and test items influence test planning, (2) test preparation which consists of carrying out the requirements of test plans and following through on the many details in getting ready to perform including data acquisition, analysis, and presentation, (3) explanation of the safety, health, and environmental concerns and issues associated with structural testing, and (4) techniques and methodologies for performing tests which protect the personnel involved, mitigate or eliminate the potential hazards, and protect and preserve the environment. There is a limited discussion on hazards and their reduction or mitigation in Sections 13.2 and 13.3. However, most of the discussion on safety, health, and environmental concerns is in found in Sections 13.4 through 13.6.

Experienced test personnel are familiar with how structures will respond to loading conditions. However, there can be some surprises (e.g., unanticipated structural behavior, potential hazards, malfunction of test equipment or test item, or emergencies such as an electrical power failure) as described in examples in Chapter 15. Personnel who are less experienced in testing are not as likely to be aware of the potential hazards involved in structural testing. This broad approach gives the reader a coverage of the essential parts of organizing structural tests. This coverage ranges from the more familiar aspects of the five basic phases of structural engineering with obvious emphasis on testing to the less known constraints in attempting to comply with the evolving federal and state rules and regulations for control of hazards in the work place, company policies, and other requirements which can affect how structural tests are actually accomplished.

13.2 BASIC TEST PLANNING

Test planning and preparation follows basic rules and common sense procedures. Test planning outlines and describes what is to be accomplished and adequate test preparation ensures that the test plan can be completed. The structure to be tested and the reasons for performing the tests have more influence on test planning than any other factor. In some organizations where subject to strict cost controls and accounting procedures are enforced, then test costs and schedules become as important as the reasons for performing the tests. Planning structural tests needs to include the following details:

1. The structure or assembly to be tested along with the reasons for performing structural tests (also include whether the test is a one time event or that many similar items will be tested to develop a statistically sound data base)
2. A statement of work or a test plan
3. The equipment for applying forces and loading conditions to perform tests
4. Handling and rigging equipment for lifting and positioning test items, fixtures, and loading systems
5. The types of fixtures and interfaces for attachment of test items and for loading purposes and boundary conditions
6. The anticipated responses of the structure as reflected in the types of instrumentation needed
7. How the data are to be acquired, analyzed, and interpreted
8. An understanding of the potential hazards and methods for controlling them
9. Definitions of the independent and dependent variables
10. The funding support (e.g., the compromises between available test money and the tests needed to obtain the insights desired).

Any of these basic efforts described above in planning can become the controlling factor in completing structural tests. Implied in this planning effort are the abilities and cooperative efforts of test personnel, project engineering groups, those overseeing their work, and others associated with the testing activities to develop and implement test plans which can accomplish the necessary tests in a cost effective, safe, and environmentally sound approach.

A sample static test request form (Figure 13-1) outlines the essential information requested for a laboratory performing test services. This sample form summarizes the information just described into a one-page summary and is used here to simplify this discussion. As the structure or assembly becomes more complicated obviously there will be more detailed information needed on a test request. Some test plans require many pages of explanations, descriptions, and approvals. (A similar form is used in many testing laboratories which offer services to other organizations in research and development work.) This form consists of six parts:

1. administrative
2. test purposes
3. test description
4. potential hazards
5. instrumentation needs
6. requested records and reports

The intent is to organize the information describing the structure, to define the purposes of the tests (insights desired), and to develop the instrumentation and data requirements. This form is consistent with the underlying theme of defining the essential information to perform a quality test in which the correct tests were performed and the data and results are valid. Test requests do not need to be complicated, but they do need to be complete. It is typical for most organizations to require some type of form to be completed prior to initiating a test and have it maintained as a part of a test record. Project engineering groups are usually required to complete various types of reports which would include details of tests performed on prototype or actual structures.

Test plans usually include a statement or description of the work to be performed. A test plan may be developed by a project organization with requests to furnish projected costs, schedules, and other items to be delivered as part of

Figure 13-1. Sample Test Request

XYZ STRUCTURAL TESTING LABORATORY	ANY TOWN, USA		909-222-6789

Requester:	Phone:	Address:	Date:
Funding Source:		Expected Costs	

Test Purposes: (type of test)

_____ Acceptance/QA	_____ Failure Loads/Modes	_____ Service Life
_____ Calibration	_____ Full Scale/Model	Extension
_____ Connections/	Inertial	_____ Stiffness/
Joints	_____ Mass Properties	Compliance
_____ Development of	_____ Materials	_____ Structure-Control
Test Units	Characterization	Interaction
_____ Energy Absor-	_____ Modal Analyses	_____ Time Dependent
bed (Crush)	_____ Proof/Load	Creep
_____ Environmental	_____ Residual Stresses	_____ Fatigue
/Materials Compatibility		_____ Wind Tunnel

Test Description (include sketches of proposed tests):

Hazards: _____ Anoxic _____ Corrosive _____ Electrical _____ Explosive _____ Other
_____ Pressure _____ Shrapnel _____ Temperature _____ Toxic _____ Radioactivity

Instrumentation Requirements (Measurements to be Made:)

_____ Acceleration _____ Displacement _____ Extensometer _____ Force _____ Other
_____ Strain _____ Temperature _____ Velocity _____ Nondestructive Techniques

Instrumentation Used: (serial numbers, calibration dates, gages, gage factors, attachment/adhesives, cables, transducers, etc.)

Number of Channels:

Responsible Engineer:	Technician(s):
Other Personnel:	

Comments (Engineer/Technician):	Type of Report Requested
	_____ None (Data Only)
	_____ Data/Plots/Listings
	_____ Work Statement and Data
Disposition of Test Hardware:	_____ Complete w/Photos/Video

the test. A test plan can also be developed by a test laboratory in response to a request for test services. This second type of test plan would require some type of approval by the organization funding the test. Many test plans are worked out in cooperation among the organizations involved. The bigger the project and more complicated test requirements necessitate better communication and planning efforts.

Test Structure or Component

The physical description (e.g., size, shape, mass) of the structure or assembly to be tested along with the desired loading conditions defines the requirements and capabilities for the equipment needed to perform the tests. Most testing systems, equipment, and facilities have limits on their capabilities: widths, heights, and depths that can be accommodated, load capacities and rates, pressure capabilities, temperature ranges, frequencies and accelerations, alignments and orientations. Testing systems also have limits on how the inertial properties and thermal characteristics of test structures can influence the choice of available equipment. Having the structure tested using the appropriate equipment is essential in performing quality tests.

The physical constraints in terms of test item size, shape, and other physical parameters typically determine the equipment needed for testing in a laboratory or field location. For example, most vibration shakers and shock testing machines can deliver certain acceleration-time (g-time) histories to certain limiting sizes of test items. Most test equipment have been designed to accommodate structures which are typically tested. For example, automobiles and trucks (chassis, suspensions systems) require the application of forces at four or more tire pad locations or through axle spindles. Aircraft structural sections (wings, elevators, tail sections, landing gears) or entire prototype or actual aircraft can require multiple applications of forces simulating taxiing, takeoff, climb, cruise, descent, landing as well as internal pressure changes, shifting of fuel, and various passenger and cargo loadings. Small structures such as electrical lead wires, printed circuit boards, and other miniature hardware can require the application of very small forces or temperature gradients on contacts, printed circuit boards, and related hardware. Combined with the requirements for applied forces, temperatures, accelerations, etc., these physical conditions of size, shape, and mass usually dictate the kind of equipment needed for performing the tests.

Extensive experience in structural testing identifies two basic problems in testing prototype or actual structures or components: (1) test items are rarely designed with testing in mind and (2) designers are often reluctant to include any features which would simplify structural testing. For example, a routine structural test is to perform static load tests on missile flight hardware. These tests verify structural integrity and give evidence that the structure can withstand static load equivalents of in-flight accelerations simulating steady state launch accelerations, maneuvering, and shocks from separation of stages while in flight. These loads are typically specified on the three principal axes of the vehicle. Test requirements usually specify that the loads are to be applied at the center of gravity of the missile. Missile structures are typically hollow allowing room for fuel, payloads, etc. The missile shells are usually tested separately from the contents as the shell structure is the principal load carrying member. The center of gravity is typically located along a longitudinal axis in the empty section. Often there is no physical structure at the center of gravity. There are no magic sky hooks or other ways of applying forces to this missile shell when no connecting structure exists. Three potential solutions exist for this problem: (1) fixtures are made so that a point loading condition can be distributed to the missile shell structure, (2) high strength straps can be attached around the missile shell and used for lateral loadings, and (3) the test structure can be stiffened locally where point or line loads are applied. The missile shell is a often a thin skin and there is danger of it buckling when loaded laterally, particularly because the full effect of the applied force is concentrated on a line around a portion of the structure. The point in this discussion is that it is often easy to define a loading condition in a test plan but it can be much more difficult to develop an adequate simulation in laboratory or field conditions. Prevention of undesired structural behavior (e.g., buckling of a missile shell) may be an important part of test planning. Methods of applying realistic forces to sections of thin, flexible hardware (e.g., missile shell structures, fins used for guidance, printed circuit boards) must be considered in planning.

The locations and methods used to apply loading conditions are also important in planning. Simulation of gravity or acceleration loads with discrete forces is often required. The use of water as a pressure medium rather than gases is often preferred because of the reduced potential energy (see Chapter 5, Table 5-3). However, a structure pressure tested using liquids must be sufficiently strong to withstand its added weight. The application of torques using forces and lever arms rather than torque motors can simplify test equipment requirements. A typical approach is to determine what is actually needed in terms of loading conditions and then to use simpler methods to apply forces and other conditions. For example, it is often easier in test equipment to apply forces in mutually orthogonal axes rather than trying to develop fixtures and adapters to perform static or dynamic tests on a structure oriented at some oblique angles or mounted upside down.

Some types of fixtures and interface hardware are needed for almost all types of test equipment. For example, the typical threads on tensile test equipment are 1-14 UNF which are finer than the regular fine threads (1-12 UNF). Adapters and hardware are often required to attach test items to test equipment. Vibration test equipment almost always has some fixtures used to attach and orient the test items on the shaker tables. Shock test machines need to have some type of pads (e.g., typically called programmers) to define and shape the shock pulse imparted so that required acceleration time histories can be delivered. These fixtures and interfaces are essential in performing structural tests. Existing test laboratories often have a large inventory of used fixtures. Usually some standard types of fixtures are available and can be modified for use. The test equipment can also be modified or additional hardware can be developed for test purposes as shown in Fig. 13-2.

The anticipated responses (the insights needed) usually define the types of instrumentation required. These responses are often located at key points within a structure or may be for the action of the entire structure. The key locations often relate to three typical test requests: (1) correlation of analytical models with test results, (2) concerns for structural integrity (e.g., around joints, connections, or other areas), or (3) overall insight into structural behavior (e.g., proof of performance). The loading points,

axes where accelerations or shocks are applied, or the orientation of a structure in a test can influence test planning. For example, a shipping container for computer parts may require tests involving three axes of vibration testing, three axes of shock testing, and a failure test consisting of dropping the container on a side or corner. Puncture tests on the container skin or covering may also be required. The weight, size, and orientation of the container and contents will define the basic test parameters. On the other hand, suppose that no requirements exist for a new product and the development tests that go along with it. Then test planners will need to consider the various types of tests needed (see Chapter 4) to determine structural responses obtained from prototype testing, laboratory simulations, and other approaches. They will need to develop a test plan to adequately investigate the structural behavior of the system or assembly.

A static load test may require the measurements of pressures or forces only while a vibration test typically requires measurements of accelerations. Some other types of tests may need strain and temperature measurements. Planning the test may require the installation of the gages and sensors in the structure while it is being manufactured or assembled. For example, the static or fatigue tests on an aircraft wing usually require the installation of strain gages and other sensors which are often installed on internal surfaces of critical wing structural sections while the wing is under construction and prior to the time when the covering metal or composite skin is installed. One approach used is to install two gages at every critical point so that if one gage proves to be faulty, then the backup gage is used. Suppose that a critical strain gage installation or an acoustic emission transducer proves defective after the wing has been assembled, then some difficult decisions (e.g., forego obtaining the information versus the efforts and costs to replace the sensors) are required before a test is performed. Developing a manufacturing schedule which includes time windows for installation of gages and sensors can be an essential part of a successful test program.

The data to be acquired also dictate the types of equipment needed. For example, static testing is thought to be independent of time. However, if a load-deflection curve to failure is needed in a static or low-rate crush test on an automobile shock absorbing bumper, that curve must be obtained in real time in order to capture the detailed shape of the curve. Crushing characteristics of most structures show a repeating cycle of load buildup, reduction in force when buckling or other deformation occurs and then an increase in applied force and then a relaxation of force while another deformation occurs. If the desired insight in the test is to have a detailed understanding of the energy absorption capabilities, then discrete

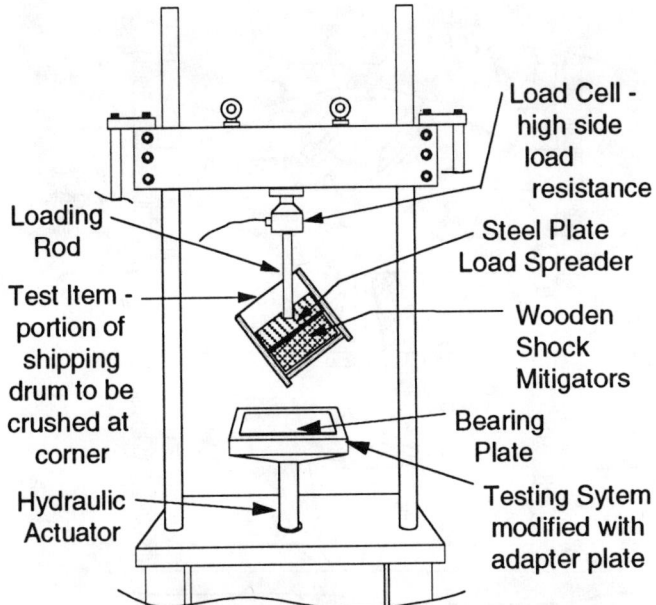

Figure 13-2. Modification of Test System to Include a Test Plate Used for a Corner Crush Test of a Shipping Container

digital sampling of the data will yield an accurate load-deflection curve. However, if only 50 to 100 data points can be sampled throughout the test, then it is very unlikely that the detailed curve can be defined. As the complexities of shock and vibration tests continue to increase, the data to be acquired can have significant effect on the types of equipment needed.

The estimates of applied forces and loading conditions along with structural responses are invaluable in determining how a test is to be planned and completed. Testing machines and equipment have limitations on their capabilities—forces, pressures, displacements, frequencies, masses, temperatures, etc. Testing machines can be operated to their useful limits although that is not always in their best interest. It is common for test equipment to have practical limitations on forces, pressures, shock levels, and test item sizes below suggested capabilities. Planners need to know what limiting conditions exist for equipment operation. While this knowledge appears obvious, test plans often have requests for equipment operation above its capabilities.

With adequate understanding of the structure and how it is expected to respond along with limits in terms of expected worst case behavior, tests can be planned and safely completed. For example, a proof test on a structure can result in the structure exhibiting no structural degradation during the test or can have the structure fail in the worst possible way (e.g., formation of shrapnel and rapid release of stored energies.) Both events need to be anticipated and the necessary plans made.

Other Concerns in Test Planning

Most laboratory or field test locations have various types of lifting and rigging equipment—cranes, hoists, cables, straps, eyebolts, clevises, hooks, etc. This equipment along with experienced operators is needed to lift and position the structure to be tested, to attach test equipment (e.g., vibration shakers), to install fixtures and other associated instrumentation and test equipment so that tests can be performed (Ref. 13-6)

Another concern in planning is the ability to compromise. For example, a test request was received for a laboratory to perform external pressure tests on sealed diving equipment to 1,100 psi. The laboratory had only one pressure vessel large enough to accommodate the diving equipment but the vessel had a maximum allowable working pressure (MAWP) of 1,000 psi. The test requesters could see no harm in operating the pressure system and vessel to 1,100 psi. However, operation of this equipment above its rated capacity could violate basic company rules and policies as well as changing the mode of operation (from manned to remote). The requesters needed a test performed at pressure levels greater than the operational range of the vessel. A solution to this potential problem was reached by conducting a required proof test for the vessel and system in conjunction with the requested tests. Sometimes tests costs are increased substantially because of inabilities to seek solutions to problems.

The modes of failure are also important in test planning. Equipment and procedures are needed to ensure that test items and fixtures are "caught" before they can become too active. For example, a test request was received to determine the strengths of some long, large diameter threaded rods. The expected strength of these rods exceeded 300,000 lbs. No conventional fixtures were available for testing these rods. The tests were conducted as shown in Fig. 13-3 in which two rods were connected to a coupler. The rods were supported by a plate on the stationary crosshead on the universal testing machine. The upper portion of the rod was connected to the testing machine by a series of ropes. The holes through the testing machine were covered with plates which were secured to the testing machine. The plates had holes which were larger than the threaded rod but smaller than the threaded coupler. The lower end of the assembly had a catch bucket filled with rags. When either rod broke it was captured by the ropes and the bucket while the coupler was captured within the crossheads.

Another typical safety device used is an interlock. This device may be mechanical in nature such as a padlock positioned on test equipment so that power cannot be

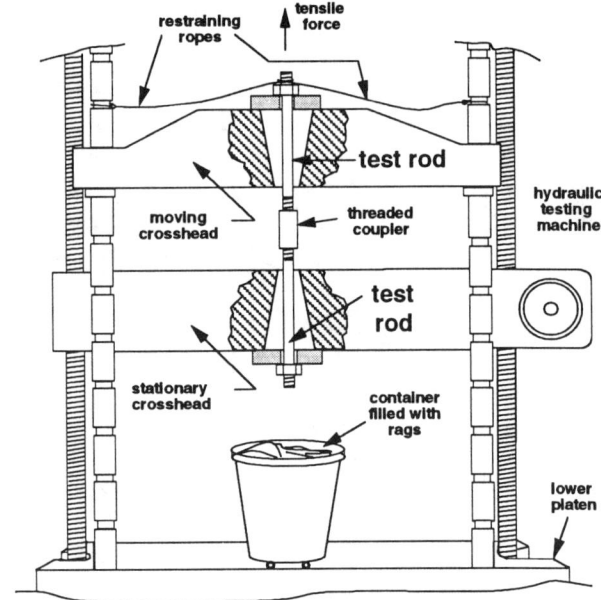

Figure 13-3. Safety Restraints for Tensile Tests on High Strength Rods; Safety Restraints Capture the Portions of the Fractured Rods

turned on until the lock is removed. This device may also be electrical in nature such as sequential operation of switches in order to arm detonators for explosive testing. Another example is the use of an interlock to prevent unauthorized access. If an access door to a test system is opened (the interlock is violated), then the test system is rendered inoperable until the door is closed. A variation on an interlock is the use of lockout/tagout procedures. For example, if test personnel require the electrical power supply to be turned off while they are maintaining some test equipment, then the electrical power circuit can be locked out while the work is being performed. A lockout is a padlock which has only a single key. This lock and key belongs only to one person. The lock is to be attached to the power switch (typically through metal eyes) so the switch cannot be activated. The tagout indicates the owner of the lock and often the type of work underway (Refs. 13-1, 13-3, and 13-7).

The potential hazards associated with conducting structural tests can be identified in many ways. A preliminary categorization of hazards with a few typical examples includes the following: (1) electrical (shocks), (2) gravity (lifting, rigging, obstacles), (3) hazardous materials (flammable, corrosive, toxic, reactive, asphyxiants,), (4) mechanical (power tools, chips, shrapnel, stored mechanical energies), (5) pressure (pressure vessels and systems, noise, compressed gases), (6) radiant energy sources (lasers, light, magnetic fields, ionizing radiation), (7) thermal (steam, flames, heaters, cryogenic materials), and (8) other (confined working spaces, weather phenomena, heights, snakes, spiders, and poor housekeeping). Details of the potential hazards and how to plan for their presence in structural tests are described in later sections in this chapter (Refs. 13-1 through 13-8, 13-11, and 13-12).

Test Insights and Understanding Desired Versus Measurements Needed

Test plans contain specifications for the measurements needed. The test results give the basis for determining how the structure responds which, in turn, gives the insights needed from the tests. The insights needed depend on what measurements are made.

The first question addressed regarding measurements is what measurements are needed. The first answer given is that most measurements including vast numbers in very complex conditions can be made given enough time, money, and equipment. However, it is typical that realistic constraints (e.g., funding, limited personnel, and test equipment) focus the tests to more limited requirements for measurements and corresponding insights. These constraints require compromises between what insights are actually needed and what measurements are needed to determine how the structure or test item responds. For those laboratories providing testing services, a typical question asked in test planning is how many data channels are available. Test personnel should be alerted to be somewhat skeptical of those requests which need every available channel. The point here is that a test may be performed which will gain the insight needed with only a minimum of channels. Test dollars are often scarce and the unnecessary embellishment of any part of the test is often not necessary. Adding excess channels is one of the most common ways to increase test costs. If the data obtained on each channel is needed to determine how the structure responds, then obviously it should be included. However, many times considerable effort is expended in installing gages and sensors and obtaining data which receives only a cursory analysis or is ignored.

A structural test laboratory, a field test assembly and equipment, or some other test configuration will likely be equipped with some these capabilities, but it is very unlikely they will be able to perform tests requiring the test and measurement capabilities described in Chapters 6 through 12. It is sufficient at this point in the handbook to suggest the following items to be considered based on observations of experienced test personnel and organizations:

1. It is typical for those familiar with or new to testing to request considerably more measurements than are actually needed in any particular test.

2. The inexperienced test requester usually encounters the following difficulties: determining exactly what measurements are needed, matching the needs versus the capabilities available for a given test, and evaluating the test data and effectiveness of the test effort.

3. The amount of test data actually studied and reviewed following a test is usually less than half of the data obtained, even for experienced test requesters.

4. Often there is little attention paid to the uncertainties associated with any data obtained. (Uncertainties are described in Chapter 14.

5. There is almost always a period of time needed for "data digestion" so there can be reconciliation between predictions of structural behavior and measured responses. This needed time period encourages a careful review and analysis of the data. This time period often prevents a period of "test results indigestion" in which a hurried judgment of the data suggests inadequate agreement between theory and test results.

6. Test planning is often inversely proportional to the

time in which structural response will actually occur. That is, a closed loop test conducted in a dynamic loading condition where structural response will occur and then be damped out in milliseconds usually requires significantly more planning and attention to detail than a test conducted statically. However, both types of tests may be difficult to perform. One subtlety here is that requesters often overlook the complexities of static tests and leave out important details.

7. Compromises are inevitable in structural testing.

8. Simulations are important in structural testing.

9. The type of data summaries, organizations, and presentations including real time charts, approximately real time plots, listings, sample rates, and other necessary descriptions of the data.

Facilities and Laboratory Evaluation

For those organizations which will be requesting tests to be conducted by others in their laboratories or facilities, it is important to have some criteria to evaluate their capabilities. The main point in this section is to suggest that a careful evaluation of the facilities and equipment to be supplied or used for testing can be necessary to completing a successful test or test program. The test equipment, facilities, instrumentation, and data acquisition systems should provide all or at least most of the capabilities needed to determine how a structure responds. Some suggested ways of evaluating the capabilities of and the corresponding insights needed versus the measurement requirements are given in Table 13-1. This table contains a summary of the types of tests (Chapter 4) versus the types of measurements usually made in structural tests.

A second approach is given in Table 13-2. This table summarizes the individual capabilities (e.g., measurements, test equipment, and services) in a listing that will be the starting point for evaluating a laboratory or facility. With so many potential details associated with testing, these lists will help in reminding the reader about these details. There are many more aspects of testing than are listed. The evaluation of a laboratory or facility often hinges on some small details.

Test Preparation

Test preparation is often thought of in terms of project engineering planning activities. Test preparation consists of organizing the many diverse activities associated with experimental investigations. While the details of project planning are beyond the scope of this handbook, the description of a few basic project planning fundamentals will significantly aid the discussion of this important aspect of test preparation. The basis for project planning consists of satisfying a triple constraint of (1) performance expectations (why the test is to be conducted), (2) time schedule, and (3) cost controls (Refs 13-14 and 13-15). If these constraints are properly balanced for the specific test to be conducted, then all three constraints will be satisfied and the test will very likely be considered a success.

Planning a test on the basis of satisfying the three constraints described above while including the details described below will ensure that preparations for a specific test will be thorough and adequate. Prototype or actual hardware must be designed or readied for testing, instrumentation methods and procedures must be defined, data acquisition systems need to be functional, and computer related equipment and systems need to be ready for testing. Among other items and activities that need to be considered in preparing for a structural test are:

1. Statement of work identifying the structure to be tested, preliminary schedules, funding sources, basic test purposes, instrumentation, and data to be obtained.

2. Test plan describing the work to be completed including loading conditions and sequences and related information.

3. Acquisition or delivery of test hardware and associated components which are a part of the items to be tested. Experience shows that it is often necessary to have the hardware in hand before test preparations can actually begin.

4. Definition of the detailed purposes for the tests. The purposes are also essential for test preparation. Test purposes are subject to change as test activities progress.

5. Evaluation of the potential hazards associated with the test items and their response to structural and other tests; this evaluation needs to include procedures for protecting test personnel and equipment from possible harmful effects.

6. Description of the physical parameters used to exercise the structure—loading conditions, directions, rates, controls, and other details for exercising the structure or test item.

7. Software—many of the tests are performed using computer controlled equipment for testing as well as data acquisition and analysis. The development, utilization, and modifications to software (in-house or

Table 13-1. Types of Tests/Insight Desired and Measurements Needed

Types of Tests/ Insights Desired	Typical Measurements Needed						
	Accelerations	Displacements	Forces or Pressures	Strains	Temperatures	Velocities	NDT
Acceptance/ Assurance	often	some	often	rare	often	rare	often
Calibration	often	rare	often	rare	often	rare	rare
Connection and Joint Behavior	often	often	often	often	often	rare	often
Energy Absorption	often	often	often	some	some	often	some
Environmental Compatibility	rare	some	some	some	some	rare	often
Failure Conditions	some	some	often	often	some	some	often
Fracture	some	often	often	often	often	some	often
Inertial Behavior	often	some	some	some	some	some	some
Loads	often	some	often	often	some	rare	rare
Mass Properties	often	rare	some	rare	rare	often	rare
Material Properties	some	often	often	often	some	some	rare
Modal Analyses	often	some	often	some	rare	some	rare
Model Behavior	some	some	often	often	often	some	some
Proof/ Operational	often	some	often	often	some	rare	rare
Residual Stresses	rare	some	some	often	some	rare	rare
Service Life Extension	some	often	often	often	some	rare	often
Shock	often	often	some	some	some	often	rare
Stiffness/ Compliance	rare	often	often	often	some	rare	rare
Structure Control Interaction	often	often	some	often	some	some	rare
Time Dependent							
Creep	rare	often	often	often	some	rare	some
Fatigue	rare	often	often	often	often	rare	some
Vibration	often	rare	rare	some	some	some	rare

commercial) for test purposes needs to be completed before testing commences. The time required to develop and checkout software for use in testing can be very long and may need to be considered before any other activity.

8. Instrumentation—review of the independent variables, the acquisition and installation of gages, transducers, and sensors are needed prior to starting testing.

Table 13-2. Test Systems and Services to Consider When Evaluating Capabilities of Test Laboratories and Facilities

Measurement Capabilities	• accelerations (g's, frequencies and frequency responses, ranges • displacements (contact, non-contact, in-plane of structure, out-of-plane, frequency responses, ranges) • forces (load cells—strain gage, piezoresistive, piezoelectric, frequency response, ultrasonic, ranges) • strains (contact, non-contact, gage installation, point/whole field, in-plane, out-of-plane, ranges) • pressures (types and ranges of transducers) • temperatures (contact, non-contact, ranges) • velocities (contact, non-contact, ranges) • nondestructive evaluation methods and techniques
Equipment Capabilities	• Loading Equipment (capacities, sizes, strokes, instrumentation) cycles, velocities, modes) • Data Acquisition (strain, acceleration, displacement, sample rates, storage/retrieval, number of channels, plotters, printers) • Interfaces (adapters, hole patterns, sizes)/Available fixtures • Lifting and handling capacities (fixed and portable cranes, slings), forklifts, straps, eyes, shackles, clevises) • Shock Machines (drop tables, sizes, ranges, accelerations, programmers • Vibration actuators (g-force, axes, sinusoidal, random, orientations) • Centrifuges (rates, test item size limits and weights, environments, instrumentation, slip rings) • Spin test facilities (rates, test item sizes, instrumentation, slip rings) • Climatic chambers (altitude simulation, temperature ranges, sizes, humidity, other environments salt fog) • Temperature chambers (ranges, sizes, instrumentation) • Pressure equipment (gases, fluids, rates, instrumentation) • Pressure vessels (sizes, openings, instrumentation passthroughs) • Threaded fastener testing (fixtures, load washers, ultrasonic) • Repetitive test capabilities • Computer controlled testing systems • Portable test equipment (data acquisition systems, loading systems, force measurements) • Nondestructive evaluation techniques—acoustic emission, radiography, photometrics, ultrasonics, penetrants • Thermal capabilities—temperatures ranges, rates, thermography
Testing Services	• Calibrations and measurement traceability • Computer support (available software and hardware, programming) • Can facility handle test items which contain explosives, hazardous materials, radioactive sources, etc. • Adherence to government regulations and rules • Consulting services (engineering, instrumentation, test planning, equipment specification and acquisition)

9. Data acquisition—rates, storage, plots and curves needed, real-time information, videotapes, photometric coverage, data listings, etc.).

10. Fixtures and interfaces—Definition and review of the test set-up, development of appropriate boundary conditions, fabrication of fixtures and needed test equipment, ordering multiple tests on the same test item (e.g., coordination of activities and test items through various test facilities).

11. Development of test schedules including time peri-

ods for prototype design and fabrication, fixture design and construction, assembly of the prototypes, installation of gages and transducers, transportation, and other related aspects of the tests such as delays caused by bad weather, by required meetings and training, and, at times, by management intervention.

12. Suggestion and details of data reduction, data analyses, and evaluation of the tests performed along with reports needed.

Test purposes help guide planning efforts and a review of them is needed prior to beginning tests. Sometimes the reviews lead to reaffirming decisions already made. Sometimes the reviews help test requesters understand better why they are performing tests. These reviews also help limit the scope and interpretation of the test results. For example, an initial request to perform a pressure proof test on a pressure vessel could contain the requirement to certify that the vessel is ready to be placed in service. However, the results of a successful pressure test do not provide all the information needed to place the vessel in service. Other information is also needed before the vessel can be placed in service including material properties and their certification, structural analysis and verification of meeting applicable code requirements, design parameters which are reasonable for operation of the pressure system, and the use of the proper fittings and piping systems used to operate the vessel. The purpose in performing this pressure test would be to demonstrate pressure capability and to offer some of the information needed for certification.

Test preparation is basically following through on the details described and suggested in the test plan. Essentially, each aspect of testing must be considered and watched over to ensure that the test item, fixtures, instrumentation, data acquisition systems, and other items associated with a particular test are available and will provide their necessary functions during testing. On projects involving significant amount of structural testing it is customary to have weekly meetings to evaluate schedules, delivery of hardware, instrumentation, procurement problems, and all the details that must be completed to have a successful test.

Assembly and Checkout

Test items, test assemblies, and prototype test units come in many different configurations and degrees of assembly. In most tests on components and structures, there will be some assembly of the test items required so tests can be performed. The proper assembly of test hardware and fixtures is an essential part in performing a quality test. It is not uncommon to be ready for testing only to find that the test item does not fit into the test fixture, that some part of the data system is not operating properly, or any one of seemingly countless problems which occur. Performing trial assemblies, fit checks, and checking out the instrumentation systems (i.e., gages and transducers, data systems, recording devices, plotters, calibrations, etc.) prior to performing a test is a mandatory requirement in conducting larger tests. While it is difficult to define details of assemblies because there are so many different possible structural tests and test items, it is still possible to develop some general guidelines for assembly procedures which would include:

1. There should be instructions and procedures given to test personnel on the protection of the test item in transportation, in handling and installation in test fixtures, and protection from elements and the environments (e.g., extreme temperatures)

2. Assembly of test items should be left to those familiar with required procedures (e.g., fastener patterns, installation torque requirements, electrical connections, seals for vacuums and pressures)

3. Fit checks of test hardware and test items should be performed prior to testing.

4. Some type of review of the test set-up should be performed—an independent assessment by experienced test personnel not directly associated with the test can be very helpful. Merely explaining a test to others knowledgeable in test activities can help verify the simulations used, possible structural responses, and how the data will be obtained and interpreted.

5. There should be some type of test request or other instructions completed, followed, and then reviewed prior to a test.

6. Pretest meetings and reviews will help reduce or eliminate points of confusion. It is typical, even for the experienced tester, to have missed or not clarified some critical details associated with assembly of and installing the test item, test fixtures, loading systems, instrumentation systems, data acquisition and recording systems and other necessary parts associated with the tests.

Test Setup

Setting up a test involves gathering together the test items, test equipment, data systems, power supplies, and other equipment associated with the planned tests. Test setup also includes checking out computer software for test controls and data acquisition. Setup means the physical installation of test items or hardware into the test equipment (e.g., load frames) or the attachment of test equipment to the test item (e.g., a vibration shaker attached to a bridge or building). Also included is the installation of gages and transducers, signal cables, and related equipment. Often a check list becomes an important part of tracking the details.

One important phase of the setup is a trial or checkout procedure. This activity is used for instrumentation, test controls, and to ensure that the mating parts fit and will not introduce some undesired effects in the tests. One potential problem is alignment of fixtures in a testing machine. For example, failure tests on threaded fasteners requires the installation of many fixtures. A single shear test on a bolt can require some precise alignment so that bending forces are eliminated. Most testing machines maintain an alignment to a small dimension (e.g., 0.1 mm or 0.004 in.). However with a series of test fixtures installed, this misalignment will increase. The misalignment becomes a problem when the test results are affected. The difficulty comes in determining whether the results have been adversely affected. The potential misalignment is but one of many possible problems.

Experience shows that considerable time is required for setting up a test. The sooner the test item or hardware is ready, it should be made available for testing. Once the test item is available and positioned in the test equipment, then many other activities (e.g., gage installation, load trains installed, data acquisition systems hooked up) of testing can be completed. without the test item, many of these activities cannot begin. It is not uncommon for some interface problems to occur even as the test items are ready for assembly or as they are actually being assembled. One potential cause is the abuse of test equipment after repeated use in terms of deformed parts, damaged fixtures and equipment, elongated holes, and other damage from its original condition. It is also possible for the test items and fixtures to be made in patterns other than those that interface with the test equipment. Test laboratories usually have some portable equipment (e.g., grinders, drills, etc.) needed for some minor but essential modifications to test items and equipment before assembly is completed. For tests conducted in the field, this type of portable equipment can be indispensable in completing tests.

Another example of properly setting up a test is to double check the instrument settings for each gage and transducer used and to make a record of them. One effective way to double check is to have two different people record the information. While this procedure may appear obvious, it is not uncommon for these settings to be incorrect.

Another recommendation in setting up a test is to make a dry run which can consist of applying a portion of the load, vibration, or shock level which will maintain the structure well within the elastic range. This dry run gives the opportunity to check out each part of the test setup. For those dynamic conditions where there is only one opportunity to perform a test, then these checkouts and check lists become even more important.

Another choice in setting up a test is which way loading conditions will be applied. For example, a load can applied to a structure by pushing on one side of a beam or pulling from the other side. The effect of having a load applied at a point will very likely be the same. However, there can be a significant difference how the load train may be configured. A smooth structure (e.g., a wing or fin with smoothness required for aerodynamic purposes) with no places to attach a load may require adhesively attached load pads capable of supporting tension and compression forces. A compressive load against a flexible structure could result in having the load applied in different directions than originally planned. There may be additional dangers by having the loading equipment be ejected from the load train because of excessive deflection causing the load train to buckle under load.

There are standard fixtures used (e.g., a compression cage for application of tensile forces while maintaining test hardware in compression) to reduce or eliminate potential problems associated with difficult test set-ups. References 13-8, 13-9, 13-13, and ASTM have a variety of suggestions for test fixtures and set-ups.

13.4 SAFETY, HEALTH, AND ENVIRONMENTAL CONSIDERATIONS

The purposes of this section are to describe the safety, health, and environmental standards, requirements, and concerns associated with structural testing (Refs. 13-1 through 13-8, 13-11, 13-12, and the standards organizations listed in Section 13-7). These standards and requirements governing these considerations control hazards in the workplace. Structural testing is hazardous because in any test a structural system, mechanical assembly, or electrical component can fail or respond in ways which pose certain dangers to the people participating in the test, to the equipment used, and to the environment. Structural testing has many potential safety, health, and environmental concerns because the tests are often destructive or potentially destructive and they involve activities and materials which are or are perceived as hazardous because they are not routinely found in industrial applications nor are they familiar to the general public (Refs. 13-1, 13-6, and 13-10). The potential hazards need to be understood and controlled prior to planning, while organizing, and when conducting a structural test. These controls are developed through a detailed understanding of the hazards involved, adequate personnel training and experience, and use of safe and correct testing methods and procedures. It is also necessary to have a clear focus and direction in terms of test objectives which guide activities prior to, during, and after a test is completed. These efforts are essential in performing tests which are safe for the people involved, preserve the environment, protect the equipment and facilities used, and obtain the insights and understanding of structural behavior. The safety, health, and environmental considerations are discussed because some of these aspects are found in every test and because they can require more lead time to accomplish if permits are required. These three aspects will also have increasing emphasis and they will form many of the requirements and constraints in future tests.

In years past the performing of structural tests usually involved only the test personnel and observers. Currently, structural tests involve others as well. Some additional scrutiny has resulted from new requirements for performing structural tests as new laws are passed, regulatory agencies expand or modify their authority, and as problems have occurred. Today considerably more emphasis has been placed on health, safety, and environmental concerns as well as protecting the equipment and facilities involved. A methodology used to comply with the fundamental health, safety, and environmental requirements will be described initially along with approaches to prove or demonstrate compliance to governing requirements. The basics of how to work with health, safety, and environmental requirements are described in this chapter.

Structural testing may appear to be far afield from the use of hazardous materials and the generation of hazardous wastes. However, chemicals and hazardous materials are used in the installation of strain gages, painting and protection of equipment, the use and disposal of hydraulic oils, and the testing of structures which may contain hazardous materials and wastes. The first law protecting the environment was passed in 1899. Since then many laws have been passed along with creation of regulatory agencies. The primary law regulating hazardous wastes (Resource Conservation and Recovery Act [RCRA]) was passed in 1976 and amended in 1984. The goals of RCRA are to protect human health, the environment, and animal and plant life by regulating wastes from the moment they are created until they are rendered inert or they are not hazardous. This cycle is commonly called the "cradle to grave" concept for controlling these materials. These laws are strictly enforced and have significant liabilities associated with their violations.

The basis for safety and health standards in the work place are found in the regulations and requirements promulgated by the Occupational Safety and Health Administration (OSHA) (Refs. 13-1 through 13-3 and 13-7). These standards govern most of the activities associated with structural testing. These standards along with state, local, and industrial or university requirements will form the operational constraints and be the basis for how structural tests are to be performed in the laboratory or in the field.

Federal Regulations Overview

There are many possible governmental regulations that could apply to structural testing. These regulations are based in laws that have been enacted, by amendments to those laws, and by modifications and extensions of current regulations. Examples of the possible regulations that can affect and direct structural testing activities are:

1. OSHA rules requirements for safety in the workplace.

2. The standards associated with protecting the environment as defined in the National Environmental Policy Act (NEPA) through the Environmental Protection Agency along with the rules from the Hazardous and Solid Waste Amendments

3. Some of the rules associated with the Resource Conservation and Recovery Act (RCRA)

4. Provisions of the Clean Air and Clean Water Acts.

5. Rules and requirements associated with the Toxic Substances Control Act.

6. The rules of the Safe Drinking Water Act.

7. Some of the provisions of the Comprehensive Environment Response, Compensation, and Liability Act.

8. The rules of the Superfund Amendments and Reauthorization Act.

9. Some of the requirements of the Hazardous Materials Transportation Act.

In addition to these requirements, many states and companies have enacted laws that also require compliance initiatives. These rules and requirements must also be studied and the provisions adopted in performing structural tests. One observation is that the these additional rules and requirements sometimes tend to make compliance with their requirements a difficult thing to achieve (e.g., OSHA 1910.131 which states that no employee shall be allowed to consume food or beverage in any area exposed to a toxic material which could be interpreted to require no food or drink in any laboratory facility).

In evaluating these rules and requirements, it is apparent that the large portion of the hazards in structural testing are associated with safety and health in the workplace. There are other rules and requirements that must be followed; however, these typically have only a minor effect on structural testing with an occasional test which may be required to meet requirements of a specific act (e.g., clean water, safe drinking water). The reader is cautioned to ensure that the laboratory facilities, test equipment, field testing (which can be subject to NEPA requirements), and related activities comply with these other requirements. We will explain in some detail the rules and requirements for safety in the workplace in this section along with some descriptions of health and environmental concerns. Large organizations will need specialists in safety, health, and environmental areas to know the rules and properly apply them.

OSHA Requirements

The standards for occupational safety and health have been developed and implemented by the Occupational Safety and Health Administration and are found in the Code of Federal Regulations (CFR) in Title 29 CFR 1910 and 29 CFR 1990. Reference 13-1 contains the standards issued in 1990. The table of contents in Ref. 13-1 is found on p. 67 and the topical index begins on p. 759. Reference 13-2 contains a detailed compilation of the general standards used in industry. Both references have topical indexes needed for locating the many details found in these standards.

The basic purposes of these standards:

1. Recognize and identify the hazards associated with industrial activities, with particular emphasis on those found in your own workplace.

2. Determine how to take appropriate abatement actions.

3. Determine how to achieve and maintain compliance to federal requirements.

4. Explain how inspections are performed.

5. Explain how penalties are imposed and what employers can do about them (i.e., safe and proper industrial practices, good faith efforts,

6. Explain what the responsibilities are of the employers and employees in providing a safe and healthy work place.

To help introduce these standards, the following overview is given consisting of brief descriptions of 29 CFR 1910 subparts A through Z. This overview is given because readers are not as likely to be familiar with these regulations and how they can affect structural testing. The first three subparts define general requirements for these standards, adoption and extension of established federal standards, and general safety and health provisions for workplace activities. Subparts D through Z contain specific requirements. Table 13-3 contains a summary of subparts D through Z with general and specific references.

Safety Requirements

Safety is often considered as a set of straightforward and easily understood procedures and practices. However, safety can be difficult to define in a testing operation. One aspect of safety consists of very explicit requirements and regulations. Another aspect of safety consists of a wide variety of perceptions on what is safe, what is unsafe, and the reasons supporting either contention (Ref. 13-10). Safety requirements are primarily aimed at protection of the personnel directly involved in the test activities. Safety was once viewed as the absence of accidents involving injuries to personnel and damage to equipment. As greater understanding of the detrimental effects of hazardous conditions and materials have on personnel, the environment, and equipment, on-the-job (testing) safety has assumed new roles and definitions. Items such as near-misses are important because they represent possible adverse outcomes. Common safety items are personal protective equipment (e.g., safety glasses, safety shoes, protective clothing such as gloves and lab coats). Safety requirements go far beyond these personal items. A typical statement found in some testing laboratories is:

*"No job is more important than your safety,
your health, and
the protection of the environment."*

Safety sometimes consists of satisfying two apparent conflicts. For example, the OSHA rules require specific railings around working surfaces. If the working surface happens to be the wing of an aircraft to be tested, then no permanent railings can be installed because of damage to the wing structure. There is a need to protect personnel and a need to keep the aircraft undamaged and ready for testing. The solution to this problem was to erect a scaffold bridging over the wing and provide restraint systems to personnel so they would be kept from falling off the wing.

Safety of personnel and equipment is implied but the first major emphasis is the protection of people. (The protection of the environment will be discussed later in this chapter.) Most companies, laboratories, and universities have some type of organization responsible for setting safety policies. In addition, these organizations may also be responsible for some types of inspections and audits of procedures, practices, and reporting of incidents and problems as well as responsibility for equipment, the environ-

Table 13-3. Summary of OSHA Requirements—29CFR 1910 Subparts D Through Z

Subpart	Description—Standards for
D	Walking-working surfaces (1910.23), Fixed industrial stairs (1910.24), Portable wood ladders (1910.25), Portable metal ladders (1910.26), Fixed ladders (1910.27), Safety requirements for scaffolding (1910.28), Manually propelled ladders, stands, and scaffolds or towers (1910.29), and other working surfaces—dock boards, areas around equipment (1910.30). towers, other working surfaces, dock boards
E	Egress requirements, emergency action plans for fire and other emergencies, emergency escape routes are required
F	Standards for powered platforms, manlifts, and vehicle mounted work platforms
G	Occupational standards for health and environment (ventilation, noise, and radiation—both ionizing and non-ionizing)
H	Hazardous materials (compressed gases, flammable and combustible materials, explosives and blasting agents, storage and handling of liquefied gases and anhydrous ammonia)
I	Personal protective equipment for eyes, face, respiratory systems, head, feet, and electrical protective devices
J	Sanitation, potable water supplies, housekeeping, vermin control, and food and drink consumption in the work place, accident prevention signs and tags, lockout/tagout procedures
K	First aid and medical services
L	Fire protection
M	Compressed gases and compressed air equipment
N	Materials handling and storage (cranes, slings, lifting devices)
O	Machinery and machine guarding
P	Hand and portable powered tools and other hand-held equipment
Q	Welding, cutting, and brazing operations, gas manifold systems
R	Special industries (paper, textiles)
S	Electrical systems
T	Commercial diving operations
Z	Toxic and hazardous substances
1990.1	Carcinogen policy and model standards

ment, and the public interest. These safety organizations are likely to establish requirements meeting federal, state, local, and company requirements.

For example, specific "safe operating procedures" would likely be required for certain known hazardous operations such as the use of ionizing radiation equipment, the use of lasers in laboratory or field tests, the use of flammable, corrosive, or highly reactive materials, or the use of high energy electrical systems (Ref. 13-8). It is very likely that all companies, universities, and other organizations conducting structural tests will be required to have some types of formal procedures to provide guidance to personnel, protection from possible litigation, and perhaps, to prove to outsiders that their operations are conducted in a safe and prudent manner. Some organizations require safe operating procedures to be expanded to include additional requirements and concerns such as safe handling of explosives, chemicals, and radiation sources.

The following lists indicate both the basic information and additional materials that are typically required in these safe or standard operating procedures. In reviewing these lists the reader may find that there are many more things to be considered than may be required. However, experience shows that many organizations consider these items as routine.

1. Descriptions of the types of tests to be performed.

2. Identification and analyses of the potential hazards involved.

3. Reduction and/or elimination of the potential hazards through procedures, safety equipment, test operations, and related equipment and requirements.

4. Requirements for personnel (personal safety equipment, reading and understanding of the safe operating procedures, manufacturer's manuals).

5. General testing requirements (those that are found in all types of tests)

6. Specific requirements for tests involving certain loading conditions (e.g., forces, pressures, temperatures) or certain equipment (e.g., shock testers, vibrations shakers) or for tests conducted in certain locations (e.g., flight tests, field tests).

7. Certification of test equipment used (appropriate safety reviews, proof tests, tags, and other indications of serviceability)

8. A list of authorized personnel which is usually accompanied by a signature page indicating that they have read, understood, and will comply with the provisions of the safe operating procedures.

9. A sample test request form (an example is given Fig. 13-1).

10. The Material Safety Data Sheets (MSDS) required for virtually all materials with particular emphasis on those that have potentially hazardous characteristics.

In addition to these basic requirements, the following are also often included as necessary parts for performing tests:

1. Procedures for stopping operations which typically need to be coordinated through a test director.

2. Requirements for independent inspections, audits, and reviews of activities (e.g., new facilities, modifications to existing facilities, and test operations)

3. Requirements for automatic review and renewal of the procedures

4. Adherence to specific documents (e.g., personnel codes of conduct, company safety manuals, and governmental requirements)

5. Requirements for periodic safety meetings, training, and reviews along with required agenda items for each meeting

6. Requirements for employee training—courses, schools, manufacturer's instructions, and examinations

7. Requirements for handling and reporting of personnel injuries, illnesses, and emergencies

8. Additional safety systems required for personnel assigned to greater than average risk endeavors (e.g., remote or field locations, above ground work locations, exposure to hazardous materials or conditions)

9. Logs or records of equipment use (e.g., number of cycles on pressure vessels and systems, use of lifting and rigging equipment)

10. Records of equipment maintenance (e.g., testing machines, cranes, lifting and rigging slings and straps, test fixtures)

11. Safety analyses and proof tests of pressure vessels, load frames, test beams, and other equipment used in test activities

12. Requirements for adherence to specific regulatory and company policies (e.g., pressure safety practices, ASME codes, nuclear power industry and federal regulatory requirements).

13. Noise environments which may require periodic hearing tests.

14. Training and procedures required for entering confined work spaces

To apply these conditions to laboratory, field, or flight tests, some additional suggestions may be helpful:

1. Procedures for safe operation of testing equipment (e.g., machines, frames, pressure systems, temperature chambers)

2. Use of protective systems (e.g., barriers and protective shields)

3. Use of personal protective equipment (e.g., head and face protection in certain activities such as soldering and grinding)

4. Use of proven procedures for testing (e.g., follow manufacturer's operating manuals for equipment, development and approval of sound internal operating procedures for unique equipment)

Please note on the test request form (Fig. 13-1) that much of the form is directed to defining the test objectives and the hazards associated with performing each request. The first clue of possible problems is in the type of test(s) to be performed. If requested tests involve certain activities (e.g., crush testing, failure loads and modes, burst tests, materials testing, and evaluation of connections and joints), then the sudden release of stored energy is very likely because the test items are expected to fail. The second indicator of potential problems is found in the section where the proposed test is described. Unusual requests or actions that appear to be hazardous probably are. For example, a test request to map the growth of cracks in a concrete pressure vessel could pose significant hazards to personnel examining the cracks while the vessel is pressurized.

The third indicator of possible hazards is in the listing of potential hazards themselves. Those known hazards which could cause problems in a laboratory or field condition deserve special scrutiny. For example, tests involving the release of hazardous materials could be accomplished only under very controlled conditions and may even require permits from regulatory agencies before they can even be attempted. In spite of what appears to be a myriad of requirements and reasons, there is no reason to avoid understanding and complying with as many safety considerations as possible.

Health Requirements

The potential hazards to the health of test personnel need to be identified with a partial list given below. This listing is not inclusive and other hazards exist which will need to be assessed in order to conduct a safe test. If not specifically mentioned, it is implied that some types of protection are needed for personnel and the public.

1. Anoxic: does the structure or test facility have or contain materials that would deny oxygen which could affect the test item, data systems, and personnel conducting the tests? (e.g., gases like nitrogen, carbon dioxide, or argon).

2. Corrosive: are these materials in the test item or hardware and could they come into contact with them during a field or laboratory test? (e.g., acids, or materials that can be combined to make acids).

3. Electrical shock: does the structure or test item contain electrical power supplies or components which could short or discharge during a test? Are there competent service personnel to ensure that power supplies and wiring are properly installed?

4. Explosive: does the structure or test item contain explosives or other materials which could burn or detonate if exposed to the necessary conditions? Explosives come in various categories (e.g., self-contained) and quantities which need to be considered in terms of potential hazards.

5. Flammable materials: will the structure release flammable gases or liquids during a test; will the structure or test item come into contact with flammable materials; how will the flammable materials be handled?

6. Forces: will the forces required be greater than or near the capabilities of the testing machines, test frames, test fixtures, rigging, etc.?

7. Fragments and Shrapnel: can the test item release shrapnel or fragments when tested? (e.g., bolts, rivet heads, nuts, or parts of the test items).

8. Hazardous materials and wastes: would personnel be exposed to these materials in preparing for, conducting, or in posttest activities such as cleaning the test equipment?

9. Noise: will the test and equipment operation have accompanying noise levels in excess of recommended standards?

10. Pressures and Vacuums: will pressure or vacuum systems be used for the loading, will a test item or structure be pressurized or evacuated as part of the test, how are pressure or vacuum to be removed from the source and the test item, what pressurizing medium (fluids or gases) will be used and what energies are involved with each?

11. Radiant Energy: will the component or structure release radioactivity, microwaves, and other similar energy forms as a result of the tests, how will the test unit be isolated from personnel prior to, during, and after a test, how will posttest access be protected, how will the items be handled and disposed of (at least removed from the test scene) after the tests, etc.?

12. Temperatures: will the test item or structure be subjected to extremes of temperature?

Additional sources of health hazards can include: gravity type hazards (lifting, rigging, transporting equipment and test items), pinch points found in the use of tools (e.g., places where hands and fingers can be trapped), machines, and test fixtures, problems in working in confined or restricted spaces, the presence of trip hazards, etc.

Environmental Concerns

The Code of Federal Regulations (40 CFR) describes the applicable federal regulations. In addition, many states and communities have enacted their own requirements. Further, many companies have also developed their own requirements. The CFR are typically minimum requirements with those from other sources being more stringent. The Environmental Protection Agency (EPA) defines hazardous wastes as any solid wastes that meets any of the following criteria:

1. It is listed as a hazardous wastes in 40 CFR 261, or
2. It is listed as an acute hazardous waste in 40 CFR 261, or
3. It exhibits any characteristics of a hazardous waste, or
4. It is a mixture of any of the above.

The EPA has also defined four characteristics for hazardous wastes—ignitability, corrosivity, reactivity, and toxicity. Ignition includes a wide variety of possible combinations of material forms, pressures, and temperatures. Corrosivity is defined as those aqueous solutions with pH less than or equal to 2 or greater than or equal to 12.5, or a liquid that corrodes steel at a rate greater than 6.35 mm per year at 55° C (130° F). Reactivity also has a wide variety of actions—stability, violent reactions, formation of explosive or flammable mixtures, forms toxic gases, vapors or fumes, and is capable of detonation or explosive reaction. Toxicity is defined as containing specific metals or pesticides at specific concentrations as described in Table 1 of 40 CFR 261.

There are four basic phases to the presence of hazardous materials and wastes in structural testing: (1) acquisition, (2) storage, (3) use, and (4) disposal. The acquisition of any material must have a Material Safety Data Sheet (MSDS) included with delivery of the material. The MSDS contains considerable information on the material (e.g., trade names and identification, components and contaminants, physical data—boiling, melting, flash points, solubility, vapor density and pressure, specific gravity, and typical uses, how to fight fires involving the material, the requirements for transportation (49 CFR and U.S. Department of Transportation), toxicity data, first aid for health effects from inhalation, ingestion, or other exposure (eye contact), reactivity, conditions to avoid, procedures for handling spills and leaks, and protective equipment needed.)

The storage and use of hazardous materials can be done only in approved locations and with the appropriate containers. The containers should be in good condition without pin holes, no signs of rusting, or structural defects. Stored materials often have secondary containment requirements—either wrapped in a separate bag and placed in an approved storage container or have catch pans under the materials to contain leaks. Materials should not be stored with other incompatible materials or wastes and they should be segregated along specific lines so that leaks and spilled materials will not mix. Flammable and combustible materials are to be stored separately in National Fire Protection Association (NFPA) approved containers (Ref. 13-9).

These materials should be used under carefully controlled conditions. The requirements for a well ventilated work area, the use of face shields or safety glasses, gloves, and other protective equipment are customary. Personnel need to be trained in how to protect themselves from potential health hazards when they are in contact with these materials.

The disposal of hazardous materials and wastes is accomplished through carefully controlled procedures. Some materials (e.g., those with the right range of pH such as ketones, short chain alcohols, and dilute inorganic acids) could be candidates for disposal in sanitary sewers. Since the waste streams in sanitary sewers are often sampled and their origins traced, companies do not want to have their waste streams containing any materials that could cause harm to the environment much less the embarrassment. Even legal discharges of potentially hazardous materials can result in possible investigations, may require public explanations, and a host of other problems. The usual practice is that any hazardous material or waste is disposed of through approved procedures which include the following:

1. Accumulation requirements (volumes and time limits)—typically 60 days from the start date of the current waste cycle, when 55 gallons of typical wastes or when 1 quart of acutely hazardous wastes have been gathered

2. Procedures—forms, containers, labels, bagging requirements, notification, and collection

3. Requirements for empty containers—compressed gas cylinders when they have reached 25 psig and/or contain no more than 3 percent by weight of material remaining in the container

4. Collection—by company or contract personnel who uses approved disposal techniques (the source of these hazardous materials is liable if the disposing organization does not follow regulatory requirements)

Determining the procedures for orderly disposal of hazardous materials and wastes which comply with federal, state, and company requirements is not an easy task. As people become more accustomed to the procedures and requirements, disposal will become better understood and will be easier to comply with.

For all activities related to the health and environmental concerns, logs or records are required on the materials ordered, stored, used, and disposed. These logs typically have the name of the material, an identification code, the amounts of materials stored by location, how much material was used, and when the material was disposed of.

13.5 METHODOLOGY FOR EVALUATING HAZARDS IN STRUCTURAL TESTING

The basic concerns for health and protection of test personnel, the surrounding environment, and care for the test equipment require an increasingly formal approach in addressing each concern. A basic approach consists of defining the following items to a hazard assessment approach:

1. The description of the laboratory, facility, or project activity
2. The identification of the hazards involved
3. The techniques used to perform evaluations of the hazards to determine compliance to the requirements
4. The approaches used to maintain compliance to requirements

Each of the items listed above is described in a subsequent section.

Facility or Test Project Description

The first thing needed in the evaluation of hazards is a detailed description of the testing laboratory, facility, or site where structural tests or the project activity requiring structural tests are performed. This description provides the context in which the hazards will be identified and how they will be evaluated. This facility description needs to include the following: the equipment used, the location of testing facilities, the types of structural tests to be performed or the structural system to be tested, the adjacent activities and projects which could affect or be affected by structural tests, the personnel to be involved in testing including the requirements for their training, the documents and requirements that govern the activities of the laboratory or facility, where chemicals are stored and used along with locations of waste disposal containers, and other related information.

Identification of Hazards

Before the hazards can be evaluated they must be identified and categorized. Categories of potential hazards consists of eight basic types: (1) electrical, (2) gravitational, (3) hazardous materials and wastes, (4) mechanical, (5) pressure and stored strain energy, (6) radiant energy, (7) thermal energies, and (8) other types ranging from restricted spaces to natural elements and conditions to poor housekeeping. Some brief examples of each are given below:

1. Electrical—electric arcs, electrostatic discharges and shocks, and any unplanned contact with live sources above 50 volts and 50 milliamps
2. Gravity—lifting, rigging, transporting test items and hardware
3. Hazardous Materials and Wastes—explosives, flammable, solvents, and vapors
4. Mechanical—power tools and equipment, chips and shrapnel, pinch points, lifting and rigging
5. Pressure and Energy—the sudden release of stored potential and strain energies
6. Radiant Energy—intense light, lasers, ultraviolet and infrared light, magnetic fields, microwave and radio frequency (RF) fields
7. Thermal—heaters, furnaces, ovens, steam, welders, and other temperature extremes
8. Other—confined spaces, poor housekeeping, access restrictions, effects of other facilities (noise, natural elements (weather effects), animals and insects).

Evaluation of Hazards

The evaluation of hazards needs to include the following: (1) extent or magnitude, (2) potential severity, (3) potential impact (personnel, environment, equipment), (4) presence of hazard mitigation systems and devices (built-in such as barriers, personnel with safety glasses), (5) presence of operating procedures including emergency conditions, (6) status of training for personnel versus the job requirements, (7) qualifications of workers, (8) presence of warning and other types of signs, (9) acquisition of required permits and documents, and (10) proof of compliance including documentation, records, and guidelines.

The extent or magnitude describes the relative size or amount of potential hazards present in a test (e.g., how many bolt heads may be broken when a connection fails or the energy equivalent of contained pressure). The potential severity indicates the relative occurrence of the event (e.g., routine or not routine when compared with other better understood hazards). The potential impact indicates who and what could be affected in a structural test. The presence of hazard reduction equipment and systems addresses personnel safety as well as protection of the environment and equipment. Operating procedures indicating possible outcomes of a test prior to performing the test. If emergency procedures are included in these operating procedures then test personnel are aware of most of the possible problems. (It is very difficult to plan and be trained for emergencies when one is occurring; foresight is not always clear but hindsight always seems perfect. The paperwork and meetings needed to explain the lack of adequate emergency response procedures will very likely be much greater than the corresponding effort spent prior to the start of a test to reduce or eliminate potential emergencies.)

The evaluation of personnel training versus job requirements will show where additional training is needed and can avoid some unsafe conditions as well as potentially embarrassing moments. The presence of warning and other types of signs can help control hazards by reducing personnel at the test site, indicate the presence and description of potential hazards, and provide emergency instructions. Permits may be required for some activities (e.g., disposal of large quantities of hydraulic oil). These permits are often required for hazardous materials and wastes. The proof of compliance is a goal of the hazard identification, evaluation, and development of mitigation and elimination techniques and methods. Documentation of these efforts is needed because proof of compliance usually is based upon three factors: (1) documentation, (2) familiarity of test personnel with the requirements along with the signed corresponding procedures, and (3) adequate responses to questions as inspections are performed.

Reduction and Elimination of Hazards

With the hazards evaluated, plans can be developed and implemented to ensure that safe tests can be performed. That is, the potential hazards can be reduced or eliminated. Because many structural tests are naturally destructive, they leave the perception that these tests are dangerous (Ref. 13-10).. Proving that the dangers and hazards are controlled is not straightforward. The operating procedures or test plans need to address the actual hazards (and maybe even the perceived hazards) as well as how the hazards are to be reduced or eliminated.

Some common methods for control of mechanical hazards are:

1. Personnel barriers, labs often use commercially available plastic barriers which permit sight access but control shrapnel.

2. Use of proof tested fixtures, pressure vessels and systems including whip checks on pressure lines or hard mounts securing lines to structures.

3. Restriction of personnel access to test areas—only those people directly associated with the tests are permitted to participate in tests.

4. Use of remote test facilities which separate test items from personnel and controlling equipment, use of bunkers, reinforced control centers are common.

5. Personnel training for use of testing machines, drop tables, gas guns, and pressure systems.

6. Personnel training for use of auxiliary equipment such as fork lifts, cranes, ladders, scaffolds, hand and machine tools.

7. Use of safe operating procedures which define hazards and limit personnel actions to those considered or proven to be safe.

8. Use of operating procedures for equipment to ensure their safe use.

9. Use of only certified and properly calibrated test equipment.

10. Approved procedures for handling spills and disposal of hazardous materials and wastes along with necessary personnel training.

11. The acquisition of permits for certain operations.

12. Hazardous tests should have procedures which are reviewed and approved by safety organizations and/or personnel.

The plans and equipment may also need to be reviewed in terms of a critical self-evaluation or by outside groups. Plans are needed for some types of audits ranging from equipment evaluation and certification to assessment of personnel training to compliance to regulatory requirements for disposal of hazardous materials and wastes. Four observations about reviews are:

1. The most critical reviewers are typically those involved in conducting the tests, particularly if the people, the environment, or equipment is placed in some types of dangers as they have the most at stake and are the most knowledgeable about potential problems.

2. Outside groups are often more knowledgeable about the evolving federal and company requirements. There is a potential feeling among test personnel that these outside groups are "out to get you." However, a more reasonable attitude is that they also have information and understanding of requirements that test personnel may not be as familiar with and they can help the test personnel or project in their efforts to achieve compliance.

3. When a facility or laboratory is reviewed it is likely that there will some deficiencies found. Lab personnel should find and correct any major problems prior to having a review.

4. Cooperation with review personnel is necessary. Soliciting help and suggestions from review personnel can be helpful in improving safety in a testing facility.

Achieving and Maintaining Compliance

Continuing work is needed to achieve and maintain compliance with rules and regulations. Compliance to regulatory requirements and company rules is typically achieved by evaluations performed by people other than test personnel. There are six basic requirements to achieve compliance:

1. Completion of the necessary documentation
2. Training of personnel (knowing the rules and regulations and how to apply them)
3. Approved procedures in place and followed
4. Consistency within a framework provided by management
5. The abilities of test personnel to answer questions or to know where the answers can be found when inspections or audits occur.
6. Management responsibility, support, and involvement.

The safety, health, and environmental issues and problems are being addressed on an increasingly wider scale and to greater depths than ever before. Three approaches can be taken by test personnel: (1) ignore the increasing emphasis on these requirements and issues, (2) participate on a reluctant basis and respond only when absolutely required, or (3) get involved early and make a best effort to understand and comply with an evolving set of conditions and requirements. Even with the myriad of requirements (some of which may conflict) and with regulatory agencies seeming to compete for control of similar areas, compliance to rules and regulations is needed to continue to perform structural tests. To achieve compliance will likely require changing the ways things are accomplished and the procedures that go with them. These changes are not always comfortable or easy. If this process is delayed too long, then making the necessary changes is even more difficult.

It can be very helpful if competent safety and health professionals are available for planning and reviewing test activities. They are typically responsible for keeping up with the rules and implementing them on a daily basis. They will be the best source in helping with compliance requirements.

Information needed for compliance can be gathered in many ways including keeping up to date on the federal, state, and local requirements affecting structural testing, subscribing to various services that can aid businesses in responding to environmental actions, continuing training of personnel from a variety of sources, and from management. As the requirements for documentation continue to increase, records or logs of activities and work are useful in providing information for compliance. Examples are equipment usage, types of tests performed, hazardous materials and wastes disposed of, training, safety meetings held, test planning meetings particularly where safety and environmental concerns are discussed and protection systems planned (Refs. 13-1 through 13-3 and 13-7).

Compliance can be influenced by impressions of the attitudes of test personnel. For example, a facility which is neat and orderly reflects people who care about their work. A facility showing poor housekeeping can indicate the lack of attention to other aspects which can be more important in terms of possible safety, health, and environmental concerns. The overall approach to the assessment and evaluation of hazards can show evaluators how compliance to requirements is being met.

For example, in a project involving the failure testing of a prototype pressure vessel, various safety considerations can be identified. One possible failure mode would be for the vessel to leak prior to rupture. The leakage may be large enough that the vessel could no longer be pressurized. This possible leakage could cause the following hazardous situations: contact with high pressure gases or liquids, exposure to large noises, and acceleration of the vessel according to Newton's third law. Another possible failure mode would be for the vessel to rupture which could lead to the following hazardous situations: personnel and equipment contact with shrapnel and debris created when the vessel fails and other hazards similar to those created by leakage. In planning a test on this pressure vessel where leakage or rupture is possible, then the worst event (rupture) should be anticipated. The prudent course of action would be to keep test personnel well outside a zone where shrapnel is present. This stay-out zone would likely reduce or eliminate the exposure to the other hazards as well.

As a general statement in test planning and related safety concerns, one should consider the worst possible action of the test item in terms of the potential hazards to personnel and equipment. The resulting safe operating procedure will very likely ensure the safety of personnel, test facilities, and equipment. As a corollary, "Murphy's Law" indicates that if the worst event is not planned for then it is likely to happen. Safety is one of the areas of structural testing where compromises are not in order.

Stopping a test in progress can be difficult but necessary. Procedures are needed for an orderly stoppage and a resolution of the problems causing the stoppage. If the problems are caused by safety, health, and environmental concerns, then they should be brought immediately to the test director. These problems must be brought to a satisfactory conclusion before testing can continue. For order to be preserved, a test director must be given authority to act. The director will be the best person to act in the best benefit of all concerned. A prudent test director should not proceed with a test until disagreements regarding safety, health, and environmental concerns are resolved.

The question of hazards is a complex set of conditions, many of which can be only partially specified. Care taken in identifying and examining this list and adding other items for a specific test will help protect the test personnel, test hardware, facilities, and the environment is effective only if this process is translated into actions. That is, development of a safety program is not developing lists but having competent people capable of identifying, responding to, and solving hazardous situations. Check lists are helpful but cannot be considered to be a comprehensive list of all potentially hazardous situations.

Safe Operating Procedures

Procedures are needed to ensure the safety of the people performing the tests, to protect the facilities and capabilities of a laboratory or field test organization, to protect the test assembly or test item, and to protect the environment (Ref. 13-8). However, many aspects of safety are difficult to define in quantitative terms. Safety consists of three basic parts: (1) the real or actual hazards involved, (2) the perception and understanding of these hazards, and (3) the procedures, training, equipment, and response capabilities available to minimize the effects of the hazards should an accident or some other unplanned event occur (e.g., power outage). Structural testing is usually considered dangerous because failures do occur, there are often large amounts of stored energy in the structures, and many other potential hazards can exist in a test (e.g., loading conditions, fixture design). The safety procedures developed for a test laboratory or for a single test usually reflect both the actual assessment of the hazards involved and they also address the perceptions of possible hazards involved.

A safe operating procedure is a document typically developed by test personnel or those planning to conduct a test. Safe operating procedures can also be developed by a project engineering group whose aim is to protect the prototype hardware and instrumentation systems. The purposes of these document are to review the safety concerns, outline plans of actions, and indicate what should be done in the event of possible problems. Some of these documents are written for the operation of each type of equipment in a laboratory. Others may consolidate the operations of the laboratory equipment into a set of general guidelines and procedures. Whichever approach is used, the laboratory personnel are usually required to review and update these procedures on a periodic basis.

The following summarizes the contents of a safe operating procedure for a structural test laboratory or for tests conducted in the field:

1. The locations where structural tests would be performed.

2. The sources of loading conditions (e.g., forces, pressures, thermal)

3. The potential hazards—ejection of parts and formation of shrapnel from failure tests, stored energy in pressure and hydraulic systems, and the handling, rigging, and assembly of test items and assemblies, possible temperature ranges and requirements for personnel protection. etc.

4. The types of equipment to be used, its general condition, state of repair, and calibration.

5. The general safety requirements to be met (usually part of company or organization safety program).

6. Ensuring that basic equipment operational training and safety practices are followed by test personnel (e.g., use of overhead cranes, forklifts and other motorized equipment, safety glasses and shoes, operation of hazardous equipment (grinders, drill presses, welding equipment, cutting torches, etc.), elimination of hazardous vapors and dust and/or ventilation requirements, etc.

7. Basic safety practices for conducting tests and test operations installation of safety equipment (e.g., safety shields which are portable clear screens made of impact resistant plastic, installation of pressure relief valves in pressure systems and vessels, etc.), access restriction by "extra" personnel to test area by closed doors, fences, etc., access restriction to test items while a test is in progress, handling of pressure and structural test items (pretest, during, and posttest), use of calibrated load cells, pressure transducers, and other diagnostic devices which indicate the loading conditions.

8. Basic restrictions on test personnel—tests that should be performed by pairs of qualified and properly trained test personnel, hazardous materials need to be clearly labeled, protected, and properly stored, appropriate warning signs need to be displayed.

9. Basic requirements for review and compliance by each member of a test laboratory, design or project group involved, or others who may participate, approval (by laboratory management, cognizant safety organizations, etc.) for responsibilities assigned, for resolution of safety concerns and differences of opinions. Compliance is typically indicated by a signed cover sheet.

10. Basic requirements for unusual situations—test items and assemblies, hazardous materials involved (ra-

dioactive, toxic, explosive, etc.), loading conditions, operating conditions (extremes of temperature, humidity, etc.).

13.6 SUMMARY OF SAFETY, HEALTH, AND ENVIRONMENTAL CONSIDERATIONS

A brief but accurate summary of the S/H/E considerations is given in this list of questions below:

1. Are the people qualified to perform structural tests?

2. What are the safety hazards involved? Have they been evaluated? Have the effects of these hazards been eliminated or sufficiently mitigated? Is adequate personnel training available?

3. What are the health hazards involved? Have they been evaluated? Have the effects of these hazards been eliminated or sufficiently mitigated? Is adequate personnel training available?

4. Will chemicals be used? Are the MSDS available for the chemicals? Have personnel read the MSDS's and understood the hazards associated with these chemicals? Is the proper training and equipment available for mitigating or eliminating the hazards associated with working with these chemicals?

5. Is help needed from other organizations to perform the test properly and safely? Is that help available?

6. Is the proper personal protective equipment available? Is this equipment worn properly?

7. Are environmental problems or hazards involved? How will the environment be protected and preserved?

8. Are the right tools and equipment available for the tests? Can this equipment be modified to be acceptable and usable?

9. Does any of the equipment and instruments used need to be calibrated? Are these calibrations current?

10. Are formal procedures required? Have they been read and understood? Are they practiced in the laboratory, field locations, and other sites? Do they apply to all locations where personnel will be located while testing?

It should be assumed that in responding to each question above that when deficiencies are found that they will be addressed and corrected. It is also assumed that test procedures and operations will be in compliance with federal, state, local, and industrial/university requirements and regulations.

13.7 REFERENCES

13-1. _____, *Code of Federal Regulations. Title 29. Parts 1900 to 1910*, U.S. Department of Labor, July 1990, (Two parts).

13-2. Commerce Clearing House Editorial Staff, *Occupational Safety and Health Standards for General Industry with Amendments as of September 5. 1989 Promulgated by the Occupational Safety and Health Administration. U.S. Department of Labor*. Commerce Clearing House, Inc., Chicago, 1989.

13-3. _____, *OSHA Basic Guide to Voluntary Compliance*, OSHA Training Institute, Des Plaines, IL, current edition.

13-4. _____, *Accident Prevention Manual for Industrial Operations*, National Safety Council, Chicago, IL, latest edition.

13-5. _____, *Fire Protection Handbook*, National Fire Protection Association, Quincy, MA, latest edition.

13-6. _____, *American National Standards Institute*, various standards listed below:
ANSI A10.8-1969, Safety Requirements for Scaffolding
ANSI A11.1-1965, Practice for Industrial Lighting
ANSI A12.1-1967, Safety Requirements for Floor and Wall Openings, Railings, and Toeboards
ANSI A13.1-1968, Safety Requirements for Portable Wood Ladders
ANSI A13.2-1956, Portable Metal Ladders
ANSI A13.3-1956, Safety Code for Fixed Ladders
ANSI A58.1-1958, Minimum Design Loads in Building and Other Structures
ANSI A64.1-1958, Requirements for Fixed Industrial Stairs
ANSI A90.1-1969, Safety Code for Manlifts
ANSI A92.1-1971, Standard for Manually Propelled Mobile Ladder Stands and Scaffolds
ANSI A92.2-1969, Standard for Vehicle-Mounted Elevating and Rotating Work Platforms
ANSI B19, 1938, Safety Code for Compressed Air Machinery
ANSI B30.1-1943, Safety Code for Jacks
ANSI B30.16-1973, Overhead Hoists
ANSI B30.2.0-1967, Safety Code for Overhead and Gantry Cranes
ANSI B30.5-1968, Safety Code for Crawler, Locomotive, and Truck Cranes
ANSI B30.6-1969, Safety Code for Derricks
ANSI B30.9-1990, Slings
ANSI B30.10a-1990, Hooks
ANSI B30.20d-1990, Below-the-Hook Lifting Devices
ANSI B56.1-1969, Safety Standard for Powered Industrial Trucks
ANSI C2-1981, National Electrical Safety Code
ANSI Z41.1-1967, Men's Safety-Toe Footwear
ANSI Z49.1-1967, Safety in Welding and Cutting
ANSI Z87.1-1968, Eye and Face Protection
ANSI Z88.2-1958, Standard Practice for Respiratory Protection
ANSI Z89.1-1969, Standard Requirement for Industrial Head Protection

13-7. Erwin, H., *OSHA 1910 General Industry Standards Made Easy*, 2nd Ed., Erwin Training Institute, 104 James St., Edmonton, KY 42129, 1989.

13-8. Sandia National Laboratories, *Environmental, Safety, and Health Manual*, SAND88-1161, September 1988, Albuquerque, NM, 87185

13-9. Sabnis, G.M., Harris, H.G., White, R.N., and Mirza, M.S., *Structural Modeling and Experimental Techniques*, Prentice-Hall, Englewood Cliffs, NJ, 1983.

13-10. E.W. Shepherd and R.T. Reese, "Perception of Risks in Transporting Radioactive Materials," 6th Packaging and Transportation of Radioactive Materials (PATRAM), New Orleans, May 1984.

13-11. U.S. Department of Energy, DOE Order 5481.1B, Facility Safety Analysis and Review System, Sept. 23, 1986

13-12. U.S. Department of Energy, DOE Order 5480.19, Conduct of Operations Requirements for DOE Facilities, July 1990.

13-13. _____, *Metals Handbook. 9th Ed., Mechanical Testing Vol. 8*, American Society for Metals, Metals Park, OH, 1985.

13-14. R.D. Archibald, Managing High-Technology Programs and Projects, Wiley-Interscience, New York, 1976.

13-15. M.D. Rosenau, Jr., Project Management for Engineers, Van Nostrand Reinhold, New York, 1984.

Standards Organizations:

13-1. American National Standards Institute, 1330 Broadway, New York, NY 10018

13-2. American Society of Civil Engineer, United Engineering Center, 345 East 47th St., New York, NY 10017

13-3. American Society of Mechanical Engineers, United Engineering Center, 345 East 47th St., New York, NY 10017

13-4. American Society for Testing and Materials, 1916 Race St., Philadelphia, PA, 19103

13-5. American Welding Society, 550 NW LeJeune Road, P.O. Box 351040, Miami, FL, 33135

13-6. Compressed Gas Association, 500 Fifth Ave., New York, NY, 10036

13-7. National Association of Plumbing and Mechanical Officials, 5032 Alhambra Ave., Los Angeles, CA, 90032

13-8. National Board of Boiler and Pressure Vessel Inspectors, 1155 North High Street, Columbus, OH, 43201

13-9. National Fire Protection Association, 1 Batterymarch Park, Quincy MA, 02269.

13-10. National Institute for Standards and Technology, Gaithersburg, MD 20899

13-11. Society of Automotive Engineers, 485 Lexington Ave., New York, NY 10017

CHAPTER 14

MEASUREMENT UNCERTAINTIES IN STRUCTURAL TESTING

Richard B. Pettit, Ph.D.,
Measurement Standards Laboratory,
Sandia National Laboratories, Albuquerque, NM 87185-5800

14.1 INTRODUCTION

This chapter discusses the basic aspects of measurement uncertainties associated with structural testing. The purpose of this chapter is to document the basic description of uncertainty analyses, to show the importance of these analyses, to give basic guidelines for their application, and to present specific examples. There are four important prerequisites that must be fulfilled before the detailed uncertainties associated with a test can be determined: (1) all instruments and transducers involved in structural tests must be calibrated; (2) the instrument calibrations must be valid during the period of time covering the tests; (3) the calibrations must have traceability to national standards; and (4) the details of the structural testing techniques must be well understood (e.g., temperature effects, modeling assumptions and limitations, material properties, etc.).

While measurement uncertainties have their basis in science and can be quantified, errors, on the other hand, are mistakes which cannot necessarily be quantified. Examples of errors in structural testing might be substituting a heat-treated tensile specimen when an annealed one was to be tested or applying a tension load when a compression load was needed. If the error can be defined mathematically, then either the results can be corrected or the error can be treated as an uncertainty. It is appropriate in structural testing to consider both the uncertainties and the possibility of errors. The quality of a test depends on the elimination of errors, using calibrated instruments and transducers, and performing an uncertainty analyses of the test equipment, transducers, instruments, and data acquisition system.

Whenever a test is performed, there is usually a question regarding the quality of the data obtained. This quality is directly related to the uncertainties involved in each part of a test. With an increasing emphasis on performing more detailed and exacting structural tests, greater requirements have been placed on the quality of the data. These requirements result from the use of more sophisticated and precise instruments and data systems, improvements in analytical models and the correlation efforts involved in understanding structural response, and improvements in the basic uncertainty of the data systems and instruments.

Uncertainties have their basis in measurement standards and how the standards are used to calibrate working instruments and transducers. One basic aspect of these standards is their traceability to accepted national measurement systems or standards (e.g., those maintained by the National Institute for Standards and Technology—NIST, formerly the National Bureau of Standards—NBS). That is, uncertainties and traceability are not independent or separate issues. Brian Belanger gives the following definition (Ref. 14-2):

"Measurements have traceability to designated standards if and only if scientifically rigorous evidence is produced on a continuing basis to show that the measurement process is producing measurement results (data) for which the total measurement uncertainty relative to national or other designated standards is quantified."

Measurements are traceable, not because of an unbroken chain of calibrations, but because the quality (uncertainty) of the measurements is scientifically defendable. The underlying purpose for traceability is to ensure that measurements are "true" or, at least that they can be reproduced at different times, conditions, and places. An unbroken chain will not provide this assurance unless each step in the chain is correct (e.g., the uncertainty is correct). If the uncertainty is not true, then the next calibration or measurement made down the chain may not be valid as well as all subsequent calibrations and test measurements.

One cannot describe a set of rules for uncertainty analyses because there are far too many conditions, instruments, and metrology disciplines involved. This chapter discusses the principles involved, suggests guidelines, and recommends practices applicable to structural testing. It is anticipated that the result of this chapter will be an improved understanding of the concepts involved and of the associated science and its application to often complex measurements.

14.2 UNCERTAINTIES—TOP LEVEL AND WORKING STANDARDS

In earlier calibration and measurement standards work, there was a fixed, large ratio between the uncertainty of the basic standard and the uncertainty of the instruments in common use. For example, it was typical to use a high quality, relatively expensive transducer or instrument for the standard and then to use lower quality, less expensive devices for actual testing. A ratio of 10:1 was accepted as a desirable ratio between the uncertainty of the standard and the uncertainty to which an item could be calibrated with reference to that standard. This 10:1 ratio helps ensure that the contribution of the uncertainty of the standard to the total uncertainty for the working instrument is less than 10 percent for straight summation of the uncertainties and less that one percent for root-sum-square (RSS) combination of uncertainties. Under this system, an instrument or transducer manufacturer could produce devices with uncertainties of ± 1.0 percent through the use of a ± 0.1 percent in-house standard that was referenced to an NIST standard with an uncertainty of $+0.01$ percent or 100 parts per million.

However, as the quality of commercially available instruments and transducers has improved and the number of calibration steps between the national standard and the working instrument increased, the required uncertainty of the primary standards faced a compound assault on capabilities. For example, adding a calibration station in a production line can require an increase in the uncertainty in the primary standard of a factor of 10 if a 10:1 ratio is maintained at each calibration step. Eventually, when performance specified by an instrument manufacturer reached or exceeded the uncertainty levels of the existing primary standards, then the fixed ratio of 10:1 could not be maintained Today, there are many instruments which have capabilities comparable to top level standards, such as dc digital voltmeters with submicrovolt resolution, timers with picosecond resolutions, etc. (It should be noted that the accuracy of such instruments should be verified and one should not assume these capabilities exist because they are described in product information.)

Therefore, calibration and test measurement systems often utilize reduced ratios of uncertainties such as 4:1 or 2.2:1 and accomplish reasonable results. This apparent relaxation is due to the improvements in the quality and stability of transducers and instruments. *To require the continued use of the 10:1 ratio can place unnecessary requirements on the primary standard.* In determining an instrument's total uncertainty, the first consideration is the contribution of the standard used in the calibration process. The uncertainty of an instrument should be increased by some value depending on the ratio R of the uncertainties of the calibrated instrument and the standard used. The following relationship expresses the probable percentage increase in the total uncertainty for the instrument as influenced by the standard as the root-sum-square of the uncertainty of the instrument with the uncertainty of the standard:

$$\% \text{ INC} = [(1 + 1/R^2)^{1/2} - 1] \times 100. \qquad \text{Eq. 14-1}$$

Figure 14-1 shows the percent increase for R = 2.2, 4.0, and 10.0. A frequently used value for the ratio of the uncertainty of a working instrument to a standard is 4:1, e.g., by the military services.

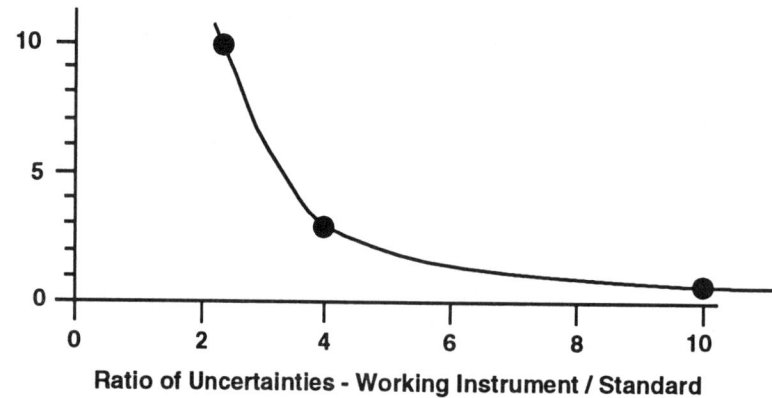

Figure 14-1. Percentage Increase in the Total Uncertainty of a Working Instrument as Affected by the Uncertainty Ratio, R.

14.3 UNCERTAINTY TYPES

Traditionally, uncertainties have been divided into two groups commonly called random and systematic. Random referred to data variability following some statistical distribution (generally Gaussian). Systematic referred to constant (though unknown) data offsets whose sign is not known and whose magnitude must be estimated. Specific rules were used to combine random uncertainties, to give an overall random uncertainty, and to combine systematic uncertainties, to give an overall systematic uncertainty (Refs. 14-1 and 14-8). The resulting overall random and overall systematic uncertainties could be further combined to provide a single overall uncertainty (Ref. 14-8).

It has been found that the random/systematic formulation of uncertainties leads to confusion concerning classifying uncertainties and to situations where an uncertainty changes from random to systematic depending upon the viewpoint of the experimenter (Refs. 14-4 and 14-11). For these and other reasons, recent deliberations have resulted in the recommendation to divide uncertainties into two new classifications called by the provisional names Type A and Type B (Ref. 14-9). Type A represents uncertainties that can be evaluated using the usual statistical techniques developed for random variables. Type B represents uncertainties that are evaluated using any other objective method. This recommendation was the result of a 1980 Bureau International des Poids et Mesures (BIPM or the International Bureau of Weights and Measures) working group on the statement of uncertainties. More details on the BIPM recommendation can be found in references 14-4, 14-6, and 14-9.

TYPE A UNCERTAINTIES

Type A uncertainties are those evaluated using objective statistical results from a measurement system in statistical control. A Type A uncertainty component should be reported as an estimated standard deviation (or variance) along with the number of degrees of freedom. For correlated uncertainties, an estimated covariance should also be given. Note that what were previously called random uncertainties will generally be Type A uncertainties under the new classification.

When gathering data for a statistical analysis, it is important to sample the short-term variability of the measurement process. Here, short-term variability is defined as the variations expected during a typical calibration as opposed to long-term variability, which is defined as the variations expected between separate calibrations (Refs. 14-6 and 14-7). For example, if the mounting of an instrument affects the calibration results (as often happens for strain gage transducers), then the instrument or transducers should be frequently remounted in order to sample this variability. As another example, 50 samples from a voltmeter obtained over a 2 second time interval in response to a single voltage input may well be a poor sample of the short-term signal variability if the real variability in the signal comes from environmental changes which have a time constant longer than 2 seconds. In any case, once the process is properly sampled, standard statistical procedures can then be applied to the data in order to calculate a mean and standard deviation (Ref. 14-3).

Type A standard deviations from each source (denoted by x, y, z, ..., which may represent temperature, voltage, time, etc.) should be determined. If the desired measured quantity is denoted by Z, then the effective standard deviations for Z will be represented by the variables $\Delta Z_A(x)$, $\Delta Z_A(y)$, ... All Type A uncertainty components are then combined into a single overall Type A uncertainty using the well-known root-sum-square (RSS) approach for combining standard deviations. Thus, if the overall Type A uncertainty is expressed as a standard deviation and denoted as ΔA_T, then it is calculated using the equation

$$\Delta A_T = \{[\Delta Z_A(x)]^2 + [\Delta Z_A(y)]^2 + ...\}^{1/2}. \qquad \text{Eq.14 -2}$$

TYPE B UNCERTAINTIES

Type B uncertainties are defined in the BIPM recommendation as those uncertainty components which are *not* evaluated by statistical methods. All components classified as Type B are characterized by *effective* variances or *effective* standard deviations, which may be considered as approximations to actual variances or standard deviations, the existence of which are assumed. These effective standard deviations are then treated like true statistical standard deviations. Where correlations appear, an effective covariance should also be estimated and treated in a way similar to statistically determined covariances.

The determination of Type B uncertainties involves mostly judgement by the experimenter. After identifying the specific factors (x, y, ...) that are important in affecting the uncertainty of the desired quantity, Z, the factor's range or some other measure of its dispersion must be estimated. This dispersion can be estimated from theoretical arguments, from model calculations, from published results or private communications of other experimenters, or from direct experimental measurements. For simplicity, we will restrict our discussion to estimated ranges ($\pm \Delta x, \pm \Delta y, ...$).

The next step is particularly difficult for Type B uncertainties since it involves converting information on the estimated range $\pm \Delta x$ into an estimation of the effective standard deviation of each x factor. While there are several approaches that can be taken, all approaches demand that the experimenter decide on a probability distribution for the factor x over the range in order to convert that range to an interval which contains the "true value," x_t, with a certain probability. In many cases, a normal distribution is an appropriate approximation since one generally expects the factors x, y, ... to have higher probabilities of being near their central or expected value than near the limits of the range (Ref. 14-7). In this case, the experimenter must decide if the estimated range $2\Delta x$ represents a 95% or 99% probability interval, for example. The effective standard deviation can then be estimated as $\Delta Z_B(x) = \Delta x/2$ for the 95% probability interval and $\Delta Z_B(x) \Delta x/3$ for the 99% probability interval. As another frequent choice for the probability distribution, the expected values may be uniformly distributed over the range, so that the effective standard deviation is given by

$$\Delta Z_B(x) = \Delta x/\vartheta^3 . \qquad \text{Eq. 14-3}$$

The estimate given by equation 14-3 is a common choice and is given special emphasis by the BIPM (Refs. 14-4, 14-5, and 14-9).

Since Type B uncertainties represent effective standard deviations, they, like Type A uncertainties, are combined using the RSS approach. If ΔB_T stands for the overall Type B uncertainty, then

$$\Delta B_T = ([\Delta Z_B(x)]^2 + [\Delta Z_B(y)]^2 + \ldots)^{1/2}. \qquad \text{Eq 14 4}$$

14.4 COMBINING UNCERTAINTIES

Since Type A and B uncertainties are expressed in terms of the variances ΔA_T^2 and ΔB_T^2, then adding the variances can be used to determine an overall variance as $[\Delta A_T^2 + \Delta B_T^2]$. The overall combined uncertainty, ΔZ_T, is given by

$$\vartheta Z_T = K^*[\Delta A_T^2 + \Delta B_T^2]^{\frac{1}{2}} \qquad \text{Eq. 14-5}$$

The range factor K converts the square root of the variance (or the standard deviation) to a probability interval which contains the true value with some defined probability. For instance, if the overall variance is well defined and the probability distribution is approximately Gaussian, then K = 2 gives a 95% probability that the interval will contain the true value, while if K = 2.6, the probability of containing the true value rises to 99%.

14.5 SOURCES OF UNCERTAINTY

This section outlines the sources of uncertainty and associated considerations that are important when assigning uncertainties. The following description is not all inclusive and the reader will need to interpret and apply the ideas to each specific situation.

1. Instruments: As previously described, the uncertainty associated with a measuring instrument is very important. An instrument's uncertainty should be taken from its certificate and should include corrections for particular environments and use conditions, any restrictions listed, time since the last calibration, and effects of the calibration processes.

2. Measurement Process: Among the factors to be considered are: (a) the analytical model used for data analysis (linear fit, least squares, etc.) and the limitations associated with the model; (b) evaluation of subsidiary experiments needed for the measurement; (c) use environments and conditions versus calibration conditions and their affect on uncertainties; (d) corrections and the uncertainty of the correction; (e) limitations of readout devices; and (f) appropriateness of interpolations and extrapolations of the calibration and measurement data.

3. Data Scatter and Statistical Techniques: Among the factors to be considered are: (a) valid statistical techniques can be applied only when all controllable sources of variation have been eliminated from the process; (b) use of the appropriate data analysis technique as dictated by the data and needs; (c) anomalies and outliers should be eliminated only with careful consideration; (d) resolution of instruments sets the lower bound on uncertainty; and (e) an objective evaluation of the process and understanding the variables involved.

4. Comparisons used to quantify shifts. drifts. and scatter: In order to ensure that a measurement process is in control, various comparison techniques are invaluable. These techniques identify shifts in calibration factors, point out long-term drift, and quantify long-term data scatter in a process or measurement. Among the techniques are: (a) comparisons with previous measurements; (b) direct comparisons of two similar (same) instruments or transducers; (c) comparisons with other measurement techniques; (d) comparisons with subsidiary measurements; (e) comparisons with the results obtained from other laboratories; and (f) before and after test comparisons which could include

the effects of transportation, handling, and exposure to potentially harmful environments.

5. Use Conditions: There are many aspects of use conditions which need to be known or estimated: (a) an adequate description or estimation of use conditions; (b) measure or establish the necessary correction coefficients if use and calibration conditions differ; (c) use the correction coefficient or appropriate estimates to calculate uncertainties; (d) impose appropriate restrictions on use in order to limit uncertainties; (e) identify other conditions which could affect the test results and determine what their effects could be on the uncertainties; (f) restrict use of instrument to calibration conditions if so needed; and (g) choose whether to make corrections, combine uncertainties, and other limits or analyses as needed.

6. Intervals: A time period should be assigned after which the measurements and the associated uncertainty are no longer valid. Estimates of possible drift should be made and factored into the uncertainty analyses.

7. Judgment: The actual process of test measurement is a science. However, there is considerable art and experience needed to perform many of the tests. Issues where judgments will be required are: (a) the requirements on the uncertainties (the cost effectiveness of improving the measurement capabilities); (b) the estimates of wear, abuse or rough handling, or other conditions which could affect the measurements; (c) changes due to environment such as temperature, humidity, vibration, corrosion, contamination, test environments such as radiation or magnetic fields; (d) relationships between the environments and the uncertainties; (e) prior experience with testing and calibration of similar structures and instruments; and (f) variability over the calibration time interval.

8. Anomalies: Sometimes measurements are made which contain anomalies or have data which lie outside expected uncertainty bounds. These measurements need to be evaluated carefully to determine whether the results reflect problems in the measurement system. For example, load cells calibrated in the vertical plane exhibited different results when used in a horizontal position.

9. Uncertainty Traps: Among the traps awaiting the unsuspecting tester are: (a) lack of understanding of the use and operation of an instrument or measurement system; (b) functioning of instruments in ways which were not accounted for in their original design (e.g., drift, affected by heat, power fluctuation, etc.); (c) test requirements which appear to be relevant to the function of a structure or instrument but which are not related; or when required tests are performed, they do not yield sufficiently accurate data. For example, the measurement of dynamic forces using a load cell could be performed with a strain gage based system or a crystal type cell. Calibration of these cells is usually performed statically using tension or compression loads. After considerable attempts to perform such a calibration (which was to simulate the expected use conditions), it was determined that the dynamic calibrations could not meet the accuracy limits imposed. That is, the standard used to measure the dynamic forces was not sufficiently accurate. A static calibration was performed and a confirmatory dynamic check calibration was performed instead of the more complicated dynamic force calibration.

14.6 UNCERTAINTY STATEMENTS

The main item of interest in a statement of the accuracy of test data is the numeric value of the quantity that has been measured. For example, in a test involving an accelerated component, the peak acceleration and velocity change delivered from a shock testing machine must be known. If the uncertainties of the shock testing equipment are not known, then it would be difficult to determine how to shape a shock pulse.

Some agencies of the federal government, as well as private laboratories, require that an uncertainty statement accompany test and calibration data. The criteria that an uncertainty statement must meet are difficult to quantify. Therefore, a simpler check-list of comments and questions was developed for things to consider; there is no significance in the order of the items in the list.

1. Are the units of the certified value, the uncertainty, or the test data clearly specified? Are they the same or is the uncertainty given in percent, parts per million, percent full scale, etc.? Is the uncertainty applicable over the full range or just over a portion of the range of data?

2. Will the user need the certified value in different units than the ones specified (torque in newton-meters rather than foot-pounds)? Chances for errors may be reduced if the conversions of both the value and the uncertainty are made to the same desired units.

3. Is the stated uncertainty equal to (or greater than) the appropriate combination of the uncertainties of the standards and measurement process used in the certification?

4. Does the present certified value for a standard fall outside the uncertainty bounds specified on the preceding certificate? That is, has the uncertainty increased from the previous calibrations or test data? Changes of this type can be important in some processes and test laboratories.

5. Often the tests are performed or instruments are used in situations which are different from the calibration conditions. If so, are these situations factored into the uncertainty statements? It is important to identify the information and factors which differentiate between calibration conditions and use situations.

6. Are the measurement and its uncertainty symmetrical or asymmetrical? For example, load cells may have different calibration factors in tension and in compression. Another example is torque transducers which have different sensitivities (millivolt/volt/unit torque) for clockwise and counterclockwise torques.

7. Does a single uncertainty value apply for all reported values of test data or do different uncertainties need to be given for different values, ranges, use conditions, environments, etc.?

8. What is the form of the uncertainty statement: worst case value, a 99.7 percent confidence value, root-sum-squares of several uncertainty components, etc.?

9. Are the uncertainties specified on a copy of the test data documented so that one can justify and defend them, if necessary, in the future?

10. In the use of a calibrated instrument or transducer:
    a. Does the uncertainty hold for a single reading, is it averaged over multiple readings, are there other factors to consider (e.g., power line cycles)?
    b. Does the same uncertainty apply to all ranges?
    c. Are there specific instrument settings that must be set for the uncertainty to be valid?
    d. Are there limitations on the characteristics of the quantity being measured that must be observed for the stated uncertainty to be valid (e.g., orientation, duty cycle, warmup period)?
    e. Are there preliminary operations (zeroing, nulling, battery checks, etc.) that must be completed before using a transducer or instrument so the uncertainty will be valid?
    f. Are there specific ancillary equipment that must be used for the uncertainty to hold for the ranges of the device?
    g. Does the uncertainty apply to the reading taken directly from the instrument or does it apply after corrections described elsewhere have been applied?
    h. Is the stated uncertainty valid only at a specific port, terminal, reference plane, etc. of the instrument?
    i. How should the uncertainty components be defined: as a fixed term, a value dependent term, a time dependent drift term, a random uncertainty term, a systematic uncertainty term, an interval uncertainty term, or an environmentally dependent term?

14.7 OUT OF RANGE CALIBRATIONS

Suppose a test is in progress and a calibrated load cell appears to be reading incorrectly. Upon checking the calibration, it is found that the load cell's sensitivity (millivolts/volt/unit force) is different from the sensitivity currently specified for the cell. This change affects the current performance, the uncertainty involved, and the previous measurements made with the load cell back to the point in time the last calibration was completed or when the cell was known to be operating with the original sensitivity. Various types of actions are needed, ranging from doing nothing if the difference is very small and has no appreciable effect on the test to sending notices to all organizations and users who have had work performed in which this load cell was a part of a test suggesting that the work performed may not be as accurate as originally stated.

There can be many reasons for the sensitivity of a load cell to change including: rough handling or dropping, excess power applied or voltage surges, moisture intrusion, effects of temperatures, etc. It may even not be known that the cell was exposed to conditions which changed the sensitivity. Nonetheless, changes in performance of instruments, transducers, and cells do occur; there should be procedures in place to account for such occurrences.

The alternatives to retroactive or repeat testing to account for changes in transducers or instruments include maintaining relatively short calibration intervals or performing periodic spot checks on the instrument's performance. It is always important to have the staff of the test lab protect the instruments as much as practical and to report problems when they occur. By performing operational or performance checks both prior to and after the tests, the proper operation of transducers or instruments can be verified.

14.8 VALUE OF UNCERTAINTY ANALYSES

An uncertainty analysis of a structural test set-up is essential for determining the validity of the measurements obtained. Often overlooked or postponed, the uncertainty analysis provides vital information on the operation, performance, and setting up of a test. The level of effort required for an uncertainty analysis may range from modest to very extensive. Experience shows that the investment of resources into a thorough uncertainty analysis will return significant dividends in guiding future operations, identifying equipment and computer system upgrades, and, above all, verifying the performance of a test or measurement system.

A situation may occur in which detailed structural analyses have been performed on a complicated structure or component before testing and the test results differ from the analytical models by some amount. When these differences become large (e.g., ten percent or factors of two), the analysts often question the test setup, the accuracy of the data systems, and other factors that may indicate discrepancies in the experimental analyses. In such situations, the detailed results of the uncertainty analysis are necessary in order to determine the cause of the large discrepancy. Once the analysts and requesters have confidence in the test results, test results which are markedly different from theory must be accepted.

As equipment and capabilities improve and better techniques become available, the uncertainty analyses become increasingly important. In part, this importance results from the increasing complexity of the instruments and methods involved. As equipment ages and becomes older, the uncertainty analysis may need to be updated. The same holds for situations where new or more accurate components are added to a measurement system. For example, a linear potentiometer displacement gage has been calibrated for many years using a precision micrometer accurate to 10^{-4} in. A new laser system is acquired which improves the accuracy to 10^{-6} in. If the calibrations using the micrometer take ten minutes per gage while an hour is required using the laser system, which system should be used? Suppose that fixtures and procedures are developed so that a calibration on the laser-based system takes five minutes. Which system should be used and how is the final measurement system uncertainty affected?

Another aspect of testing is that equipment and systems age and finally wear out. At some point in time, these systems will not perform as they once did and even careful maintenance and calibrations may not return them to their former capabilities. Because many structural tests are often destructive or performed at limits above use conditions (e.g., proof tests), transducers and load cells also tend to degrade more quickly than they would if they were not exposed to such rigorous test conditions. Because of these effects, all equipment must be periodically calibrated and its performance determined. Historical calibration data and the results of check standards are important for determining when to replace equipment or perform routine maintenance.

Experienced operators have an inherent feeling about how a testing machine or test equipment should be operating. This inherent feeling is very important because each structure or component, each type of test (Chapter 4), and each piece of equipment has unique features and characteristics which tend to develop an acceptable repeat history. However, it is only through a comprehensive uncertainty analysis that this inherent feeling is given substance and quantified. Without a documented uncertainty analysis, individuals performing tests and calibrations will develop their own feelings about the quality of the data obtained and accuracy estimates given. These feelings will likely contain some erroneous assumptions and rely on improper procedures.

The uncertainty analysis also points out the most critical parameters and instruments affecting the measurement process and thus aids in the optimization and upgrading of the measurement system. Improvements in the system aimed at providing a lower uncertainty in experimental test data can be directed toward the parameters and instruments which contribute most to the uncertainty. For example, if temperature measurement uncertainty contributes 0.01 percent and pressure uncertainty contributes 1.0 percent to an overall uncertainty of 2.0 percent, then system enhancements should be directed toward improving the pressure measurement and not the temperature measurement. A detailed uncertainty analysis readily shows the relationship of each uncertainty to the others. When a minor alteration is made in the system, an updated uncertainty can be easily determined by following the previous analysis.

In documenting an uncertainty analysis, the governing equations and analytic approach used for calculating the total uncertainty must be included. The derivation of each equations should include all assumptions leading to the final uncertainty equation. Each parameter included in the final uncertainty equation should have its uncertainty listed along with its working range. Factors considered insignificant should be mentioned, and it should be shown mathematically why these factors do not need to be considered. An appropriate statistical technique for combining uncertainties must be chosen, and the documentation should include why the particular combination method was used.

14.9 UNCERTAINTY ANALYSES OF STRUCTURAL TEST LABORATORY

Richard L. Crabb, Measurement Standards Laboratory, Sandia National Laboratories, Albuquerque, NM 87185-5800.

As technological improvements are made at ever increasing frequency, the need for uncertainty analyses becomes more and more critical. The proliferation of high-resolution, high-accuracy instrumentation available today can lead one to the euphoric, but inaccurate, belief that any measurement made with this instrumentation is unquestionably true. In fact, it is clear that sophistication and complexity merely contribute more sources of uncertainty, and make it even more essential that one quantifies the sources of uncertainties associated with the data that are collected. Familiarity with sources of uncertainty in a system, and some knowledge of how those uncertainties interact, is essential in all areas of measurement, and particularly in an area such as structural mechanics. An understanding of these factors enables one to quantify an overall uncertainty of measurements that is both practical and realistic. This section summarizes an investigation of the uncertainties associated with specific instrumentation used in the Structural Mechanics Laboratory in the Experimental Mechanics Department at Sandia National Laboratories in Albuquerque, New Mexico. Although this analysis is prepared for a specific laboratory, the general approach is applicable to any laboratory.

Traceability

It is important that the calibration of instruments used in structural testing is related to nationally accepted standards in order to assure measurement quality. This relationship is traceability. B. C. Belanger of NIST was quoted in section 14.1 with the emphasis that the standard used for reference must be traceable.

The next issue is one of practical traceability. The manufacturers of instrumentation and transducers provide statements that their devices are "Traceable to the National Institute of Standards and Technology," but more specific information is required to establish traceability. Traceability really requires a current or real-time unbroken chain of measurements with uncertainties that are repeatable and defendable (e.g., they are capable of being audited for quality assurance and control requirements). For example, sending a load cell to NIST once does not assure a 20-year chain of traceability. There must be a trail to a long-term series of NIST measurements which demonstrates a high repeatability (or precision) that consistently lies within a stated value of uncertainty. It follows, then that a statement of uncertainty which is realistic will result in a measure of confidence which is the goal of the study described in this section.

The Measurement System

The Structural Mechanics Laboratory described in this section uses several measurement systems. The laboratory's overall force measurement range is from 0.1 pound to 600,000 pounds. Computer-controlled systems by MTS Systems Corporation are used in the low and medium ranges of force and a Super L system from Tinius-Olsen is used in the high range. The study reported in this section is centered on an MTS 880 system with an upper range of 220,000 pounds. It consists of a Model 880 controller, and MTS LVDT built into the hydraulic actuator for displacement measurements, and an MTS Model 661-31A-02 load cell. Strain measurement data are taken with a Hewlett-Packard (HP) based Data Acquisition system which includes a HP 3456 Digital Voltmeter (DVM), HP 3495A scanners, and a bank of B&F signal conditioner/bridge completion networks for strain and transducer measurements. The signal conditioner/bridge completion networks are standardized with a Baldwin-Lima-Hamilton (BLH) Model 625 precision calibrator. The DVM and precision calibrator are calibrated by the Sandia Standards Laboratory to a combined uncertainty of $\pm 0.11\%$ of reading. The load cell and LVDT are calibrated by the manufacturer on an annual cycle with the load cell calibration traceable to NIST through a transfer standard and the LVDT calibrated using a precision micrometer calibrated by the Sandia Standards Laboratory. Additionally, the MTS load cell is calibrated in series with a load cell certified by the Sandia Primary Standards Laboratory. This cell has an uncertainty of $\pm 0.1\%$ of reading. The micrometer has an overall uncertainty of better than $\pm 0.1\%$ of reading. All calibrations are traceable to national standards.

Instrumentation Uncertainty

Instrumentation uncertainty is defined as the total uncertainty inherent in instruments and devices used in the measurement process. In structural testing, the factors determining the instrumentation uncertainty are: (1) randomness, (2) load cell calibration uncertainty, (3) LVDT calibration uncertainty, and (4) data acquisition system uncertainty. Instrumentation uncertainty is the limiting uncertainty in a measurement. While the precision of the measurement may be very good and the random uncertainty very low, the attainable measurement accuracy will never be better than the instrumentation uncertainty. In determining the instrumentation uncertainty, one approach is to assume the condition of worst-case uncertainty in all systems and that the worst-case uncertainties are all of the same sign or all in the same direction. In this case, the uncertainties become cumulative and may be simply added to determine overall instrumentation uncertainty. The addition of uncertainties represents a conservative approach to combining uncertainties. If effective standard deviations can be associated with each source of uncertainty following the BIPM recommendations discussed in Ref. 14-1, then the resulting standard deviations may be combined by the root-sum-square method.

The data involved in the analysis of uncertainties to randomness, the load cell, and the LVDT were gathered from several iterations of different measurements which included calibration of the MTS system and measurements made with bridge-connected strain gages on various test components. For the purpose of analysis, the strain gages were not considered in the determination of the instrumentation uncertainty; strain gage uncertainty will be discussed separately.

Table 14-1 lists the mean offsets derived from a number of measurements observed for the MTS system with the load cell and the LVDT. Mean offsets shown are percent of reading and represent the mean deviations from the standard. The offsets can be applied as corrections to the reading for the appropriate portion of the range. The tolerance listed with each offset in Table 14-1 represents three standard deviations of the mean offset and should be applied to the reading to which the offset was added. For example, if a tension force was measured with a load cell at 95% of range with a reading of 9489 pounds, the offset would be (9489 × 0.002) = 19 pounds. Thus the corrected reading would be (9489 + 19) = 9508 pounds with a three standard deviation value of ±14 pounds (±0.15% of reading).

These corrections just described apply to the cardinal points of the range. Interpolation would be required for intermediate portions of the range. While this will yield the best uncertainty of measurement, determining and applying these corrections can be a time-consuming effort. A more practical, but slightly less accurate, method is to use an assigned uncertainty which can be directly applied to all the readings. The assigned uncertainty is a conservative value of approximately three standard deviations plus the offset of the worst-case error encountered and applies to all values at or above 40% of range. Values for tension and compression are given in Table 14-1. Thus, for a load cell in tension the assigned uncertainty is ±0.35% and for compression is ±0.55%. In the example given above, the tension load would be reported as 9489 ±33 pounds (instead of 9508 ±14 pounds).

Please note that compression and tension both should be calibrated because of the reciprocity of those parameters is not a safe assumption, as illustrated by the values given. The assigned uncertainty values in parentheses in Table 14-1 indicate uncertainties to be applied if measurements are to be made at or near 20% of scale. Use of this region probably should be avoided due to its large uncertainties, particularly with the LVDT.

Now that the individual uncertainties have been

Table 14-1. Measurement System Uncertainties

Mode	Percent of Range	Load Cell		LVDT	
		Mean Offset	Assigned Uncertainty	Mean Offset	Assigned Uncertainty
Tension	95	0.20 ± 0.15%	0.35%	0.11 ± 0.08%	0.55%
	80	0.11 ± 0.13%		0.04 ± 0.09%	
	60	0.00 ± 0.01%		0.32 ± 0.23%	
	40	0.10 ± 0.12%		0.03 ± 0.10%	
	20	0.20 ± 0.15%		0.50 ± 0.15%	
Compression	−20	0.45 ± 0.10%	(0.55%)	0.75 ± 0.10%	(0.85%)
	−40	0.40 ± 0.10%		0.40 ± 0.15%	
	−60	0.33 ± 0.11%		0.13 ± 0.10%	
	−80	0.26 ± 0.11%		0.15 ± 0.12%	
	−100	0.17 ± 0.10%		0.22 ± 0.17%	

determined, the overall instrumentation uncertainty can be calculated. As an example, assume that displacement measurements were made with the MTS system with load cell and LVDT and all measurements were made above 40% of range. The tension uncertainties associated with the load cell and LVDT should be added:

0.35% of reading + 0.55% of reading = 0.90% of reading.

These combined or cumulative percentages refer only to the instrumentation uncertainty in the measurement system. Another uncertainty that should be considered is the uncertainty assigned to a device (e.g., load cell) that has been measured with the MTS system and which will be used outside the system under conditions beyond the control of the Structural Mechanics Laboratory. For this uncertainty additional factors should be included: for example, the uncertainty should be valid over a span of time and/or for specific measurement conditions. It has been the experience of metrology laboratories for some time that maintaining a given ratio of uncertainties between the standard and the working device can provide confidence in a measurement over a span of time and/or conditions. In the past, this ratio stood at 10:1 with the standard being 10 times better that the device under test. With the state-of-the-art in measurements and standards in use today, the 10:1 ratio produces a pronounced assault on capabilities. Experience has shown that ratios as low as 2:1 may be used if the uncertainties are broadened somewhat to accommodate the increase risk of error. For the purposes of this analysis, a ratio of 2.5:1 will be applied. As an example, if the MTS system is used to calibrate a device, then the instrumentation uncertainty determined above is used for the standard. The uncertainty given to the device, which will be used beyond the control of the control of the Structural Mechanics Laboratory, would be:

$0.9 \times 2.5 = \pm 2.25\%$ of reading

This uncertainty applies to displacement measurements made with MTS system on external devices.

The uncertainties in all measurement data examined fell well within the values calculated using the procedures and analyses summarized above. When measurements of strain are made with the system the uncertainty in strain data should also be included as discussed below.

Strain Measurements

Uncertainty statements are significantly more elusive when measurements are made using strain gages. Manufacturers usually provide an accuracy statement of ±1% of full scale as the uncertainty associated with strain gage calibration factors (microstrain/millivolt). Note that this equates to 2% at half scale, etc. However, this uncertainty figure does not and cannot account for variations caused in the installation of the gage. Uncertainties associated with the integrity of the adhesive used for mounting, the axis of mounting, and the location of mounting with relation to the strain within the item under test are situational and nearly impossible to determine with any reliability. In any case, strict quality measures should be taken with strain measurements. Measures that could, and perhaps should, be taken include multiple iterations of each measurement with the analysis of the resultant strain data for precision and accuracy; installation of strain gages with precisely the same technique for each measurement, perhaps always by the same technician; and careful maintenance of all strain data taken, with frequent analysis. A recorded history of repeatable data with consistent standard deviations will go far in establishing confidence in measurement of parameters with complex uncertainty components. Additional information on strain gage uncertainties is contained in Ref. 14-10. The recommendation, then, is to combine the data acquisition system uncertainty and the manufacturer's accuracy statement for the strain gage, together with any other uncertainty factors determined from the quality measured in each laboratory. The result should be an uncertainty value which is defendable with confidence.

Conclusions on Uncertainty Analysis of Structural Mechanics Laboratory

It is clear that measurement uncertainties within the Structural Mechanics Laboratory are very close to the 1-2% values that were expected from previous evaluations. Continued care in measurement will assure that these levels are maintained, and perhaps even improved upon, as technology improves. Strain measurements should be made with particular care, and the customer should be advised of the quantified uncertainties associated with the measurements. Careful measurement history should be maintained to provide evidence of justification for measurement confidence.

14.10 UNCERTAINTIES DUE TO STRAIN GAGE MISALIGNMENT
W.A. Kawahara, Sandia National Laboratories, Livermore, CA

If a strain gage rosette is bonded to a specimen but is misaligned from the desired orientation, the correct strain in the desired orientation can be calculated from the misoriented rosette strains through the strain tensor transformation equations.

In practice, there is a tendency to ignore this correction if the misalignment is only a few degrees, but uncertainties exceeding 20 percent can occur for misalignments under five degrees, depending on the relationship of the strain field to the rosette. The purpose of the discussion below is to impart a 'feel' for when such a correction may be significant.

Consider a tensile specimen along the "1" direction and let θ be the desired gage orientation along the 1' direction; then the strain components in the primed axes can be found from those in the unprimed axes from the transformation equations:

$$\varepsilon_{ij} = l_{ip}\, l_{jq}\, \varepsilon_{pq}$$

where $l_{ij} = e_i \cdot e_j$

From the present discussion, the 3- and 3'- axes are identical, and the 1'- and 2'- axes are rotated from the 1- and 2-axes by an angle θ. The matrix of direction cosines simplifies to:

$$l_{ij} = \begin{bmatrix} \cos\theta & \sin\theta & 0 \\ -\sin\theta & \cos\theta & 0 \\ 0 & 0 & 1 \end{bmatrix}$$

For example, then

$$\varepsilon_{11}' = l_{1p}\, l_{1p}\, \varepsilon_{pq}$$

$$= l_{11} l_{1q} \varepsilon_{1q} + l_{12} l_{1q} \varepsilon_{2q} + l_{13} l_{1q} \varepsilon_{3q}$$

$$= (l_{11} l_{11} \varepsilon_{11} + l_{11} l_{12} \varepsilon_{12} + l_{11} l_{13} \varepsilon_{13})$$

$$+ (l_{11} l_{11} \varepsilon_{11} + l_{11} l_{12} \varepsilon_{12} + l_{11} l_{13} \varepsilon_{13})$$

$$+ (l_{11} l_{11} \varepsilon_{11} + l_{11} l_{12} \varepsilon_{12} + l_{11} l_{13} \varepsilon_{13})$$

$$\varepsilon_{11}' = \cos^2\theta\ \varepsilon_{11} + \sin^2\theta\ \varepsilon_{22} + 2\cos\theta\ \sin\theta\ \varepsilon_{12}$$

Similarly,

$$\varepsilon_{22}' = \sin^2\theta\ \varepsilon_{11} + \cos^2\theta\ \varepsilon_{22} - 2\cos\theta\ \sin\theta\ \varepsilon_{12}$$

$$\varepsilon_{12}' = -\cos\theta\ \sin\theta\ (\varepsilon_{11} + \varepsilon_{22}) + (\cos^2\theta - \sin^2\theta)\, \varepsilon_{12}$$

For our tensile specimen along the 1-direction, we note

$$\varepsilon_{11} = \varepsilon_{11} \qquad \varepsilon_{22} = \upsilon \varepsilon_{11} \qquad \varepsilon_{12} =$$

and substituting into the equation for ε_{11}', we get:

$$\varepsilon_{11}' = \varepsilon_{11}\left(\cos^2\theta - \upsilon\ \sin^2\theta\right)$$

Figure 14-2 shows how ε_{11}' varies as a function of θ. For $\theta = 0$, $\varepsilon_{11}' = \varepsilon_{11}$, decreases to zero when $\tan\theta = \sqrt{1/\upsilon}$, and increases to $\upsilon\varepsilon_{11}$ at $\theta = \pi/2$ with the cloverleaf pattern shown.

Hence, for desired gage orientation angle θ, a small misalignment uncertainty $\pm\alpha$ will produce a percentage uncertainty in strain ε_{11}' that may be negligible near $\theta = 0$, $\pi/2$ or significant when θ is near 45°.

For a state of biaxial tension, the cloverleaf becomes more rounded in shape, the sharp re-entrant cusps disappear, and as a consequence one can see that the percentage uncertainty due to misalignment decreases.

Hence, we conclude that the percentage uncertainty due to misalignment depends on:

1. The ratio of maximum to minimum principal strains,

2. The angle θ between the desired gage orientation and the maximum principal stress direction, and

3. The misalignment angle α from the desired orientation.

Application: Unidirectional fiber composite specimens under uniaxial tension in the 1-direction. For $\theta = 0$, (i.e., desired gage orientation is along the tensile axis), the percentage uncertainty due to misalignment angle α can be significant depending on the angle between the fibers and the 1-direction. This is because, as the fiber angle deviates from the 1-direction, the cloverleaf pattern rotates and changes shape. For the ten-degree off-axis tensile specimen, the misalignment α of less than 4 degrees can produce an error in strain exceeding 20 percent if not corrected by the strain transformation equations.

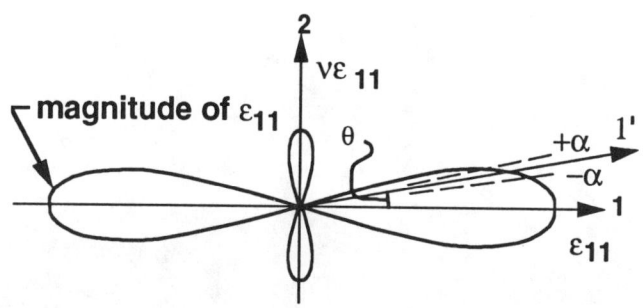

Figure 14-2. Variation of Strains Due to Strain Gage Misalignment

14.11. REFERENCES

14-1. R.B. Abernethy, et. al., *Measurement Uncertainty Handbook*, Rev. 1980, Instrument Society of America, Report AEDC-TR-73-5 (AD755356), Feb. 1973. See also J.W. Thompson, Jr., ASME Performance Test Codes Supplement of Instruments and Apparatus, Part 1, "Measurement Uncertainty," ANSI/ASME PTC 19.1-1985, ASME, New York, 1986.

14-2. B.C. Belanger, "Traceability: An Evolving Concept," ASTM Standardization News, Vol. 8, pp. 22-28, 1980.

14-3. A.H. Bowker and G.J. Lieberman, *Engineering Statistics*, Prentice-Hall, New York, 1972 and J. Mandel and L.F. Nanni, "Measurement Evaluation," NBS Special Publication 700-2, Industrial Management Series, March 1986.

14-4. E.R. Cohen, "The Unification of Random and Systematic Uncertainties," Measurement Science Conference Proceedings, pp. 1-10, Jan. 23-24, 1986.

14-5. A.R. Colcough, *"Two Theories of Experimental Error,"* Journal Research of the National Bureau of Standards, Vol. 92, pp. 167-185, 1987.

14-6. C. Croakin, "Measurement Assurance Programs Part II: Development and Implementation," NBS Special Publication 676-II, April 1984, see pp. 13-15. (also available from NTIS).

14-7. C. Eisenhart, "Realistic Evaluation of the Precision and Accuracy of Instrument Calibration Systems," Journal of the National Bureau of Standards, Vol. 67C, pp. 21-47, 1963.

14-8. R. Kaarls, "Report of the BIPM Working Group on the Statement of Uncertainties to the Comite International des Poids et Mesures," Proces-Verbeaux du Comite International des Poids et Mesures, Vol. 49, pp. 35-41, 1981 and P. Giacomo, "News from the BIPM," Metrologia, Vol. 17, pp. 69-74, 1981.

14-9. H.H. Ku, ed., *Precision Measurements and Calibrations*, NBS Special Publication No. 300, 1969. (See C. Eisenhart, "Expression of the Uncertainties of the Final Results," p. 69 and H.H. Ku, "Expression of Imprecision, Systematic Error, and Uncertainty Associated with a Reported Value," p. 73.) (This publication is also available from National Technical Information Service (NTIS), U.S. Department of Commerce, Springfield, VA, 22161.)

14-10. Pople, J., "Errors and Uncertainties in Strain Measurement," *Strain Gauge Technology*, Edited by A.L. Windows and G.S. Holister, Chapter 5, (pp 209-263), Applied Science Publishers, 1982.

14-11. R.B.F. Schumacher, "Systematic Measurement Errors," Journal of Quality Technology, Vol. 13, pp. 10-24, 1981.

CHAPTER 15
STRUCTURAL TESTING—EXAMPLES

15.1 INTRODUCTION

The purpose of this chapter is to provide examples of many different types of structural tests. The intent in providing these examples is to explain the test request or description, show how the tests were configured to obtain the responses needed, give ideas on how to interpret the results of structural tests, and to give basic insights into the science and art of testing. Included are some of the problems that can be encountered and how these problems were solved. Also presented are some examples of test data and suggestions on how to interpret different types of data. It is anticipated that by giving these examples it will help the reader to better visualize and understand the many possibilities and situations in which meaningful and valid structural tests are needed and can be performed. The tests range from relatively straightforward requests to those that can only be partially achieved.

When structural tests are performed, there are usually more than one way in which they can be accomplished. That is, there is usually more than one machine, one facility, or group of experienced people, or one set of instrumentation and data that can be taken to give the insights and understanding needed. The examples give one way that a test can be performed.

If a test laboratory or facility had unlimited resources and personnel, then virtually any test could be completed. However, the actual tests are often compromises between the data desired and the availability of test techniques and diagnostics. The examples are organized along the lines of the earlier discussions by types of tests with various types of loading conditions used.

The references indicated by a chapter number (15) and a following number are those found at the end of the chapter and indicate the source for many of the examples used. The references indicated by a number are those for a specific section and are found at the conclusions to that section.

15.2 LOADS

15.2.1 Thermal (Hydrocarbon Fire) (Ref. 15-4)

Dr. David W. Larson, Fluid and Thermal Sciences Department, Sandia National Labs, Albuquerque, NM 87185-5800.

Among the most difficult challenges for engineers is the design of structures, electrical components, and mechanical assemblies to survive and function after exposure to severe thermal environments. A severe fire is difficult to describe analytically. Combustion is a closely coupled, highly interactive, transient phenomena. The response of materials exposed to fire conditions indicates many different failure modes such as melting, burning, exploding, etc. The choices for engineering designs become much more limited at elevated temperatures.

Performing tests in severe thermal conditions is also a very difficult task. The temperatures within a fire are difficult to control and measurement of temperatures requires rugged and expensive transducers. There are few facilities available to perform fire tests. Information has been obtained from severe fires which have occurred.

Because designs in which thermal environments are the the major loads on a structure or component are usually accomplished on a case-by-case basis, the governing thermal conditions are based on the materials involved and how the heat is transferred. No specific loads can be defined. However, it is possible to define worst case conditions for thermal events which are typically associated with fires. An alternate approach to providing severe thermal environmental descriptions that approximate "worst case" conditions for design input is to carefully examine the results and effects of severe fires and develop information that accurately describes severe fires. That translation can occur through evaluation of the "experimental" evidence and associated corroborating analyses. A tunnel fire provides as close to an ideal set of conditions as can be expected in major fire accidents to yield data for examination and evaluation. The tunnel provides physical constraints and relatively well-defined boundary conditions not found in large open fires thus limiting the assumptions needed to analyze the fire environment. The thermal and chemical conditions are more severe in a tunnel fire than would generally be found in most other types of fires because the energy released is largely confined. In addition, the tunnel retains significant information on the behavior of the materials and structures involved in the fire.

One example of a severe fire is a fire which occurred in the Caldecott Tunnel in 1982. This tunnel is located between Oakland and Walnut Creek, California. The fire was investigated and analyzed to determine the following characteristics:

- Fire temperatures
- Fire duration
- Temperature distributions

The Caldecott Tunnel Fire

The Caldecott Tunnel complex consists of three nearly parallel tunnels located on California Highway 24 between Oakland and Walnut Creek. The east-west daily flow of traffic averages 110,000 vehicles, including 1220 tractor trailers with five axles or more. The north bore, in which the fire occurred, is dedicated to westbound traffic and was opened in 1964. The tunnel is constructed of a steel frame encased with reinforced concrete. The tunnel has a nearly constant downslope along its 1027 meter (3371 feet) length resulting in an elevation change of 49 meters (160 feet) between the east entrance and the west exit.

An isometric drawing of the cross-section of the north bore (an eastward view) is shown in Fig. 15-1. Features of the tunnel that are useful in determining the fire environment are the concrete ceiling, the tile lined concrete walls, the ventilation system, the emergency call boxes, the lighting system, and the tunnel dimensions, 5.5 meters (18 feet) high and 10.5 meters (34.5 feet) wide. Although the ventilation system was not in operation during the fire, it provided a source of air to support combustion and retained evidence (thermally deformed cover plates) of the environment after the fire was extinguished.

The fire was caused by nearly simultaneous collisions of a transit bus (no passengers) and a gasoline tank truck and trailer with a stalled automobile. The collisions caused the trailer of the tank truck to become unstable and overturn with the truck remaining upright. The over-

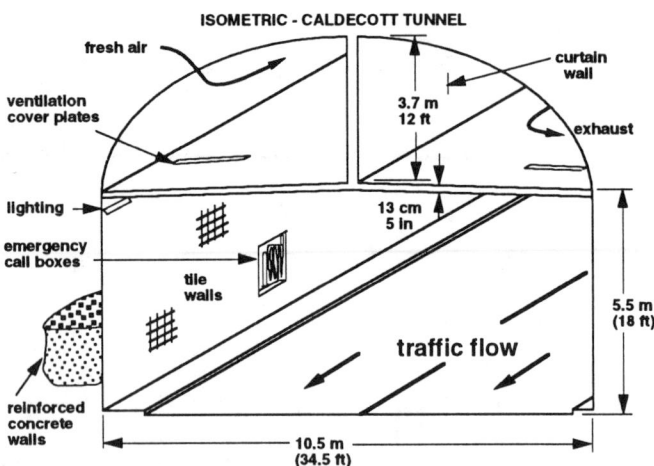

Figure 15-1. Cross-Section of North Bore of Caldecott Tunnel

turned trailer released gasoline at a rate of about 75 to 375 liters (20 to 100 gallons) per minute. The first people on the scene stated that within 2-4 minutes after the collisions burning gasoline flowed down the gutters of the roadway into the drop inlets for drainage. Approximately five minutes after the truck and trailer had stopped in the tunnel and the driver had made his way to safety by running downhill out of the tunnel, the fire characteristics changed dramatically, apparently as a result of the ignition of a vapor cloud from the spilled gasoline. The fire changed from a localized conflagration to one whose combustion products completely filled the tunnel from the tank truck to the east end of the tunnel 525 m (1721 feet) away. The air to support this combustion came primarily from the west (lower) end of the tunnel, creating a low-velocity blowtorch effect. The fire burned unabated for up to 40 minutes consuming the 18,120 liters (4800 gallons) of gasoline in the trailer and 16,140 liters (4000 gallons) of gasoline in the tank truck, except for approximately 750 to 1160 liters (200 to 300 gallons), which drained into the gutters in the first few minutes of the fire or remained in the damaged tanks after the fire was extinguished.

Figure 15-2 shows the locations of the vehicles involved and the extent of: spalled wall tiles, spalled concrete from the walls and ceiling, and thermally induced deformations of the ventilation cover plates. There were four main sources of information about the fire environment: (1) the damage to the tunnel, (2) the damage to the vehicles, (3) the gasoline fuel source, and (4) the observations of the first responders.

The thermally induced spall patterns of tile, grout, and concrete were uniform for about 230 meters (750 feet) east of the fire source. The concrete walls were spalled to

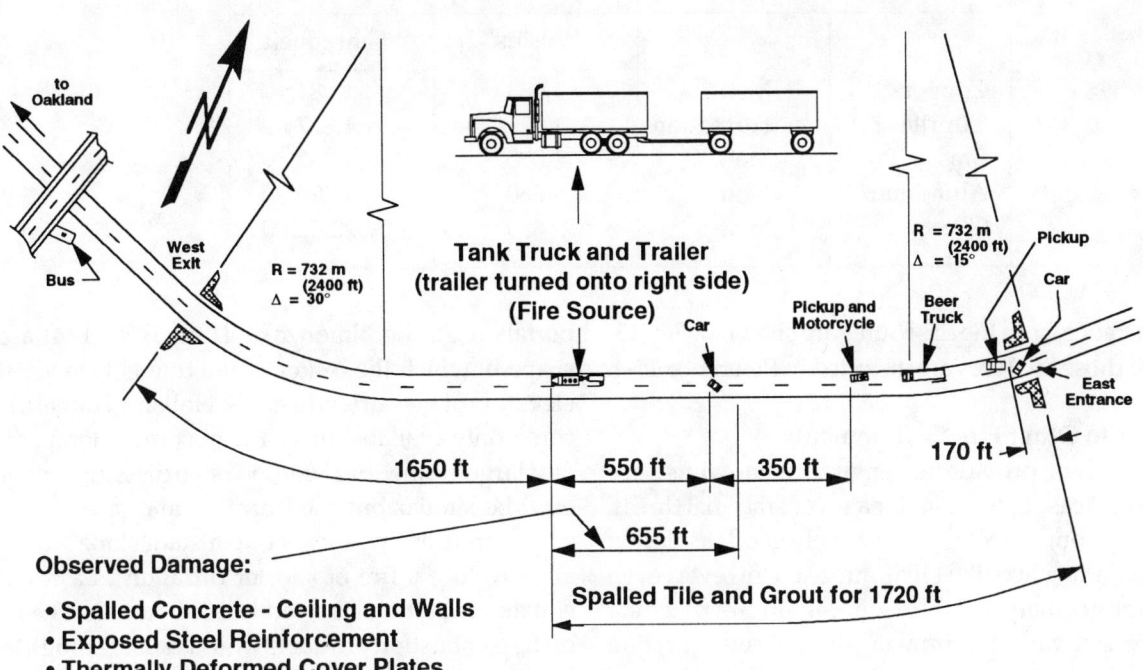

Figure 15-2. Plan View of North Bore of Caldecott Tunnel, Vehicle Locations, and Damage Description and Locations, April 7, 1982

the reinforcing steel, which normally is from 5 to 7.5 centimeters (2 to 3 inches) below the surface. The false ceiling concrete was spalled to a depth of 5 to 10 centimeters (2 to 4 inches) and exposed the double reinforcement steel pattern used in the ceiling. The steel ventilation cover plates (see Fig. 15-2) were buckled by the heat for a distance of 205 meters (675 feet) east of the tank truck and trailer.

The seven vehicles remaining in the tunnel were destroyed. The tank truck was constructed of aluminum except for the engine, parts of the frame, and the axles. The aluminum parts were melted or partially melted and included the front wheels, which were 2.5 centimeters (1 inch) thick, the rear dual wheels, the truck frame, the large gasoline tanks, the truck cab, fenders, and hood. Brass fittings on the fuel distribution system for the engine were melted. The copper core of the radiator was exposed but did not melt. The window glass had melted and pooled. All of the combustible materials in the other vehicles (tires, seat covers, and interior materials) were consumed.

The damage to the tunnel indicated that significant over-pressure had not built up at any time during the fire. The false ceiling, while extensively damaged, did not show any evidence of explosion.

Fire Temperatures and Durations

The extensive melting of aluminum, glass, and brass fittings indicate a severe thermal environment was produced by the burning gasoline. The copper wire in the lighting fixtures and vehicles showed the formation of cuprite, a red oxide of copper. Table 15-1 contains a summary of these melt and oxide formation temperatures.

The cuprite could have been formed directly in the fire or by the combustion products associated with the insulation surrounding the copper wire. In either event, a bound on the maximum temperatures in the tunnel was determined by the absence of melted copper (1,083°C) while evidence of melted brass indicated local temperature of at least 1,000°C.

Simplified thermal analyses using evidence from the accident resulted in estimates of fire durations that were consistent with reports from observers [2]. For example, the thermal environment required to heat and melt large structural members of the tank truck and energy requirements to vaporize the available fuel led to estimates of fire temperatures near 1000°C (1832°F) and fire durations of 28 to 40 minutes. The times and temperatures required to melt various thicknesses of aluminum are shown in Fig. 15-3. The profile of temperatures in the tunnel as functions of

Table 15-1. Melt or Formation Temperatures of Specific Materials

Material	Behavior	Temperatures	
		Celsius	Fahrenheit
Copper	Melt	1,083	1,981
Cuprite	Formation	1,025	1,877
Brass	Melt	1,000	1,832
Aluminum	Melt	660	1,220
Glass	Melt	550-650	1,022-1,200

time and distance from the fire source are shown in Fig. 15-4. Details of those analyses are included in Reference 15-4.

Comparison to Other Fire Environments

Tunnel fires provide an environment which approaches worst case conditions for a severe thermal threat. Generally, the confinement and relatively good insulating walls creates a furnace-like environment. However, even in tunnels the geometry can provide a significant potential for environment variation from one tunnel configuration to another. The length, cross-sectional area, and slope of the tunnel each play a role in determining whether a fire will burn fuel-rich, stoichiometric, or oxygen-rich, and thus affect temperatures, rates of energy release (combustion rate), and combustion products. Four basic tunnel geometries are common: (1) nearly level with little change in elevation between portals, (2) a relatively constant slope with the elevation of one portal well below the other and no pockets in between (e.g., the Caldecott Tunnel), (3) a convex shape with the center of the tunnel higher that the portals (e.g., the Nihonzaka Tunnel), and (4) a concave shape in which the center of the tunnel is lower than the elevation of the portals (e.g., the Holland Tunnel). For most commonly available fuels, the maximum temperatures in any large tunnel fire would (regardless of tunnel geometry) be similar but the duration and spatial extent will vary significantly. A level or constant slope would generally produce a fire of shorter duration as a result of adequate ventilation. The combustion rate and the velocities of the combustion products would cause the high temperature environment to extend over large dimensions as they did in the Caldecott. As a result of buoyancy forces, convex tunnels without forced ventilation would likely burn very fuel rich and very slowly leading to more localized effects and long duration fires as in the Nihonzaka Tunnel. It could also produce large quantities of unburned vapor as an additional hazard. A concave tunnel fire would generally respond between these two extremes.

There is substantial information in the literature reporting the thermal conditions produced by large hydrocarbon-fuel open pool fires (e.g., References 5 through 8). For these conditions, measured temperatures generally

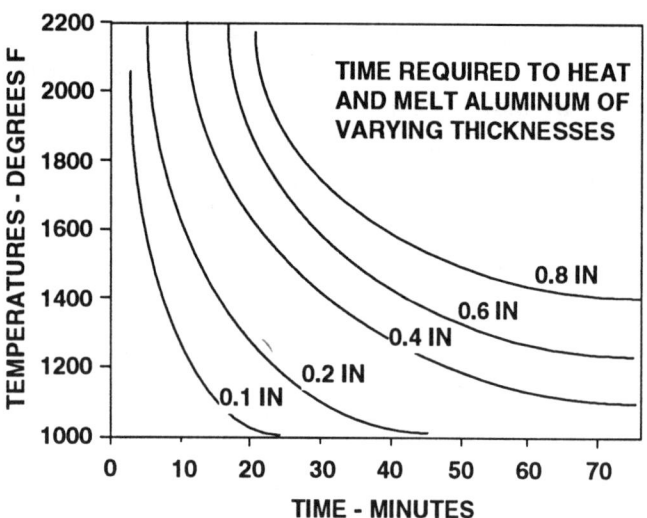

Figure 15-3. Time Required to Heat and Melt Aluminum of Varying Thicknesses (insulated back face in thermal radiation environment)

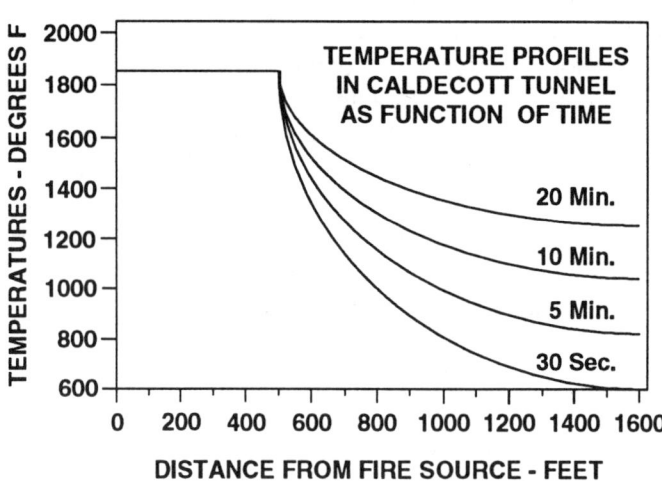

Figure 15-4. Approximate Temperature Profiles in Caldecott Tunnel

range from 800 to 1100°C (1472 to 2012°F), with the large majority of the flame and plume volume below 1000°C (1832°F). Therefore, systems designed to withstand the tunnel fire conditions should be safe in most other fire environments.

In the Caldecott fire, the damage to the tunnel was extensive. The damage to civil engineering structures (buildings, bridges, etc.) would be similar. Without insulation, the concrete cover protecting the reinforcing steel is expected to spall to a depth of 2.5 to 10. cm (1 to 4 inches) when temperatures approach 1000°C (1832°F) for durations of 30 minutes or more. Structures that need to maintain strength in a fire environment should, therefore, either be insulated, covered with an intumescent coating, or have concrete thickness exceeding this minimum depth to prevent exposure of the steel to these temperatures. Any bond lines will be opened rapidly in a fire including the bond lines in the tile and grout.

Components and systems required to survive and function in these environments will be severely threatened. Materials typically used for structural, mechanical, and electrical assemblies-steel, aluminum, wood, glass, copper, plastics-will either melt, decompose, or show significant reductions in strength and modulus of elasticity unless they are somehow shielded from these environments. Insulation, passive cooling systems, high temperature alloys and refractory materials are methods of coping with the environment with design specifies a function of the required survival time. Copper will be prone to oxide formation which will reduce its effectiveness for electrical circuitry. High temperature electrical insulation such as ceramic coatings are necessary to prevent electrical failures at these temperatures. Generally, most solid state electronic devices will not function much above 100°C (212°F) and they will either have to be cooled or extremely well insulated. Unfortunately, most electronic systems also produce thermal power and super insulation can be self-defeating.

Worst Case Thermal Design Conditions

In today's society there are a significant number of circumstances which require a designer to protect his system against extreme thermal environments. There is a wide base of hydrocarbon products such as paints, plastics, other petrochemicals, wood, gasoline, etc., than can lead to temperatures of approximately 1000°C (1832°F). The evidence associated with fires that have occurred in major highway tunnels indicates that these temperatures can readily endure for 30 minutes or more. While it is uncommon for any structure to have the concentration of fuel that was available in the Caldecott or other major tunnel fires, the damage and evidence from these fires indicate that severe and relatively long-term high temperature conditions must be given consideration in situations where structural and system failures could lead to a serious or catastrophic public hazard.

In application to structural tests involving thermal conditions, facilities required to support these tests will need to be capable of generating the 2000°F temperatures described and maintaining them for extended periods of time (e.g., at least 30 minutes). Personnel and projects requiring thermal tests at or near these conditions will need to review the capabilities of available facilities to ensure that the temperature magnitudes and durations can be attained. If the thermal environments are not known, then these worst case conditions could be used. If the prototype design structure or component passes tests involving this severe threat, then it would be able to withstand the very large majority (over 99 percent) of the fires that could occur.

References:
1. D.W. Larson, R.T. Reese, and F.L. Wilmot, "The Caldecott Tunnel Fire Environments, Regulatory Considerations, and Probabilities," 7th International Symposium on Packaging and Transportation of Radioactive Materials (PATRAM VII), New Orleans, LA, May 15-20, 1983.
2. Private communication with A.H. Lassiegne, National Transportation Safety Board, Washington, DC.
3. ——————, "The Holland Tunnel Chemical Fire," The National Board of Fire Underwriters, New York, May 13, 1949.
4. C.W. Richards, *Engineering Materials Science*, Wadsworth Publishing Co., San Francisco, 1961.
5. B.E. Bader, "Heat Transfer in Liquid Hydrocarbon Fires," Chemical Engineering Progress Symposium series, Vol. 61, No. 56, p. 7840, 1965.
6. L.H. Russell and J.A. Canfield, "Experimental Measurement of Heat Transfer to a Cylinder Immersed in a Large Aviation Fuel Fire," Journal of Heat Transfer, ASME, Series C, 95, pp. 397-407, August 1973.
7. R.K. Clarke, J.T. Foley, W.F. Hartman, and D.W. Larson, "Severities of Transportation Accidents," Sandia National Laboratories, SLA 74-0001, Albuquerque, NM, 1976.
8. J.J. Gregory, R. Mata, Jr., and N.R. Keltner, "Thermal Measurements in a Series of Large Pool Fires," Sandia National Laboratories Report, SAND85-0196, Albuquerque, NM.

15.2.2 Biomechanical Force-Platform Design Based on Strain Gages (Ref. 15-1)

Clarence A. Calder, Mechanical Engineering Dept., Oregon State University, Corvallis, OR and S. Smith, Freightliner Corp., Charlotte, NC

The activities of running, walking, and jumping consist of a series of force impacts between the body and the ground surface. These forces are transmitted into the body to be converted into mechanical energy in the form of muscle contractions and friction between the internal body parts. The magnitude of the transmitted forces has been shown to be affected by the compliance of the contact surface and the use of energy-absorbing materials in the soles of footwear (Refs. 1 and 2).

For years, efforts of biomechanical researchers have focused on determining the mechanisms involved in the human gait. Theoretical methods were developed along with experimental techniques for determining peak forces and the effects of various parameters on them. Studies involving running or walking have produced the largest wealth of information on the subject. In these instances, horizontal and vertical forces were usually obtained using commercially available force-platform systems costing upwards of $20,000. This example describes an inexpensive (less than $500) and easily constructed force platform designed to reliably measure vertical dynamic forces associated with jumping and running activities encountered in sporting events. The purpose of the force-measuring platform is to be able to compare shock absorption performance of various athletic shoe designs and styles.

Background

Biomechanical force platforms designed to measure the foot-ground forces during various activities are basically flat plates supported by a base frame. The support structures are instrumented to record the interaction forces. The plate may be supported by elements such as columns, or suspended by tension members.

Special attention to static and dynamic design considerations is necessary for force plates. In static design, if platforms of the support types are not rigid enough, possible nonlinearity between transducers and unwanted bending effects in the supports may occur. One solution would be to make the top plate sufficiently rigid resulting in an increased weight. Conversely, a light plate plate with supporting elements as stiff as possible is a dynamic requirement. This allows the natural frequencies of the plate to be sufficiently high so that measurements follow acceptably close to the actual dynamics of the applied load.

Force Plate Design

A force plate of the supported type was chosen for the design. The supported top plate consisted of a 50 cm square by 1.27 cm thick piece of aluminum. The size was determined to be convenient for accommodating the large feet often found in many athletes engaged in court sports. This plate was supported on each corner by four cantilever blocks machined from aluminum. The top plate rested on rounded aluminum supports, machined hemispherically at one end, and screwed to the cantilever blocks. The cantilever support beams (machined from 25 mm x 50 mm stock, 100 mm long) were mounted to an aluminum base plate which was 50 cm square and 1.6 cm thick. Strain gages were attached near the supported end of the cantilever on the top and bottom of each beam 38 mm from the rounded tip. The arrangement is shown in Fig. 15-5 for one of the four supports.

The top plate dimensions were chosen to provide a large enough surface to jump on, while keeping deflections sufficiently small to provide the rigidity required. The bottom plate was made more massive to provide a firm support. Vertical forces are transferred directly to the four instrumented-cantilever beams. The two plates were constructed from aluminum to minimize the total weight of the structure. At 220 N (50 lbs), the platform can be moved easily to any desired location for testing purposes.

From a static viewpoint, the cantilever support blocks required careful attention to stress, strain, deflection, and yield criteria. It has been well established that peak forces in running can be as high as two or three times body weight. Activities involving jumping could reach peak forces of five to six times body weight. Thus, design was based on a factor of safety of 2.0 assuming a peak force of 4.45 kN (1000 lbs) acting on a single support. At the same time, a sufficient magnitude of strain was required for sensitivity.

For a good dynamic response, the force plate should have a natural frequency well above any body damping and impact frequencies expected. Calculations showed a Mode I natural frequency of the plate on the order of 100 Hz while the fundamental frequency of the support beam was about 4000 Hz. Impact and body oscillations are known to be considerably below this level (Refs. 3 and 4). The design was believed to provide adequate dynamic response but, if the response is questioned, the dynamic response could easily be increased by using a thicker top plate.

Instrumentation

Two 120 ohm strain gages were bonded to each support with the four tensile (top) gages connected in series to form one leg of a half-bridge and the bottom gages likewise connected to for the adjacent leg. This arrangement provided increased sensitivity and cancelled any undesirable axial load which might be present. Accurate placement and alignment were verified using an optical

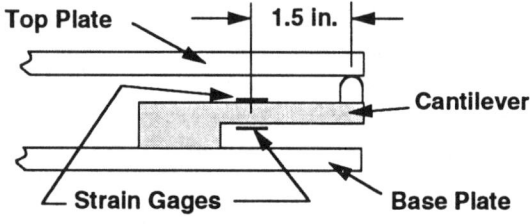

Figure 15-5. Platform Construction Showing Side View of Individual Support

comparator equipped with vernier table adjustments.

All experiments used a Vishay-Ellis VE-20A digital strain indicator connected to a Nicolet 204-A digital oscilloscope. Dynamic force histories were recorded on floppy diskettes and later hard copy was obtained by using the output from Nicolet to an attached x-y recorder. The instrumentation is shown in Fig. 15-6.

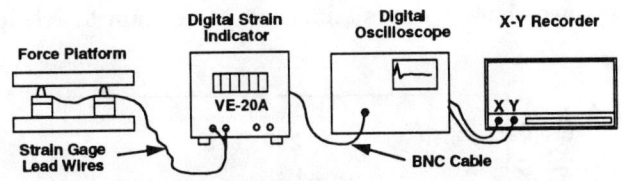

Figure 15-6. Schematic Showing Force Platform and Associated Instrumentation

Static Testing and Calibration

As an initial verification to the performance and alignment check of the strain gages, all eight gages were connected in series, in theory, when these are connected in a quarter-bridge, the tension and compression readings should cancel each other. Various weight combinations up to 900 N (200 lbs) were applied near the center of the force plate and the readings on the VE-20A were observed. In all cases, a reading of zero microstrain + 1 microstrain was indicated. Next, the four tension gages were wired in series as the leg of a quarter bridge. A reading of 85 microstrain was obtained for a centered load of about 900 N (200 lb). The same reading, within one microstrain, was obtained for a repeated test using the four compression gages.

To obtain reproducible loading situations, 22 interlocking weights and a weight-lifting bar were acquired. For each test, the weights were centered in the same order at each desired location and balanced to locate their center of gravity as close as possible over the intended location. Readings were recorded for each weight increment. First, each cantilever support was loaded individually while recording the corresponding strain readout of the tension and compression gage. Each gage was found to perform satisfactorily. Following this, similar testing was done on the assembled plate to verify a linear and equal response for loading any place on the plate surface. A satisfactory linear and equal response was found at all locations on the top plate.

An operational calibration of the force platform was accomplished with the use of approximately 3.3 kN (750 lbs.) of dead weights (lead in different sizes). The lead weights were measured and then stacked on the platform while recording the strain and voltage indicated by the strain indicator and oscilloscope. The calibration results are given in Fig. 15-7 for the strain indicator showing the highly linear nature of the force plate throughout the loading range. Similar results were obtained with the scope data. Using a linear-fitted regression routine, Eq. 15-1 was found for the strain indicator reading (static tests) and Eq. 15-2 for the scope reading (dynamic tests).

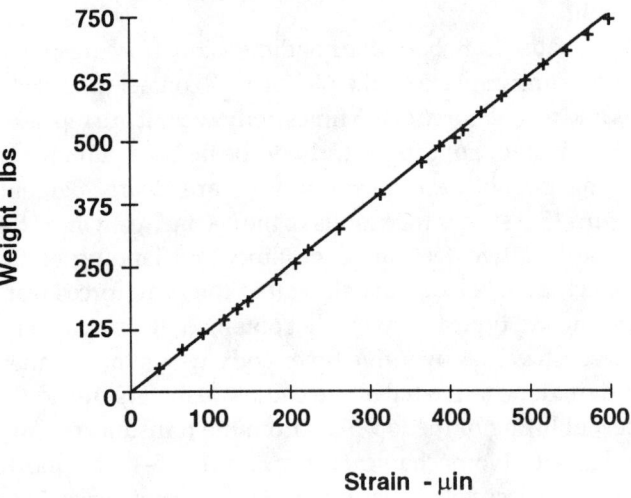

Figure 15-7. Weight Versus Strain Calibration for Force Platform

Force (N) = 5.6150 × microstrain − 9.8   Eq. 15-1

Force (N) = 41.281 × millivolts + 6.7   Eq. 15-2

Application

The force platform was used to study the dynamic forces transmitted into the leg structure due to common athletic activities such as running, jumping, and landing. Five subjects were selected to do three different activities each. The first was running in place. The second was a jump consisting of the test subject lowering his body, and with arm motion, pushing off with both feet straight up and landing back on the platform. This activity would represent what an individual might expect when performing a jump shot or blocking a volleyball. In the third activity, the subject would take several steps, place his left foot in the middle of the force platform, push up and land off the platform. This procedure would simulate a layup or blocked shot in basketball.

Considerable variability between subjects doing the same test was expected and did occur. Representative results from each activity are presented for the three tests. Running in place produced the most repeatable results with peak forces about 2.5 times the body weight. These results are consistent with earlier work on runners with commercial force plates (Ref. 2, 4, and 5). The second test is one that might simulate a slam dunk or rebound in basketball. The peak forces obtained in this activity ranged from 2.66 to 6.82 times the subject's body weight. In the third test, simulating a layup, the forces ranged from 2.29 to 3.49 times the body weight. For the fourth test, one representing a forward jump with two feet, the forces ranged from 1.67 to 3.22 times body weight. The last test, representing two foot landing, produced forces of 2.68 to 6.55 times body weight.

Figure 15-8 shows the nondimensional force response for running in place on the platform. A contact time of 250 ms and a peak force of 2.5 times body weight are repeated for each step. Foot impact, rise to peak force, fall to zero during pushoff, and airborne time are clearly evident. Figure 15-9 shows the results of the second test where the pushoff of two feet, airborne time, and landing of the subject are labeled. Data show that the peak forces from landing are typically twice that obtained by jumping and can reach values over five times body weight and higher. The small negative tail at the end of pushoff is due to the inertial lifting of the top plate from the transducer beams and is not a biomechanical force. Figures 15-10, 11, and 12 show, respectively, nondimensional force histories for a left foot jump, a two foot jump off, and a drop landing from two feet high. Each plot is only representative of each test performed and notable variability was found among the five subjects.

Figure 15-10 shows the nondimensional force history for a left foot jump. Here, one peak occurs at heel strike with a following higher peak during pushoff. This curve is very similar to running tests done on commercial force plates of the piezoelectric type (Ref. 2). A reduction in force is noted before pushoff in the standing jump from the platform (Fig. 15-11) due to arm motion and squatting before pushoff. A negative tail is again present in Figs. 15-10 and 11 for the same reason explained for Fig. 15-9. The two foot drop landing results indicate a double peak with maximum forces reaching over four times body weight.

Conclusions

A force platform of the supported type using strain gages attached to cantilever supports has been constructed and calibrated for use in dynamic testing of court athletic shoes. Calibration was accomplished using dead weight loading and showed a linear response of forces versus voltage or microstrain. A least square curve fitting technique was used to give the calibration constants. Representative testing by jumping onto, running in place on, and jumping from the back onto the platform showed good response characteristics and good correlation to results

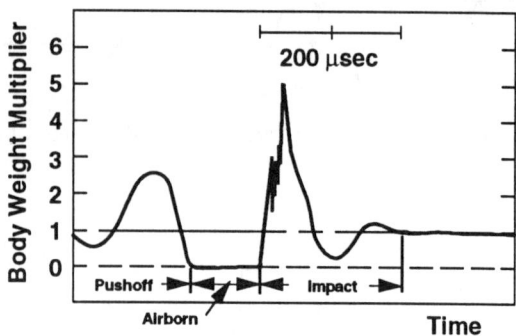

Figure 15-9. Nondimensional Force History for Two-Foot Jump and Land Test

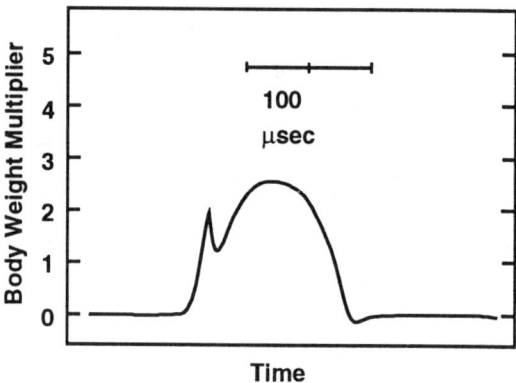

Figure 15-10. Nondimensional Force History for Left-Foot Jump Test

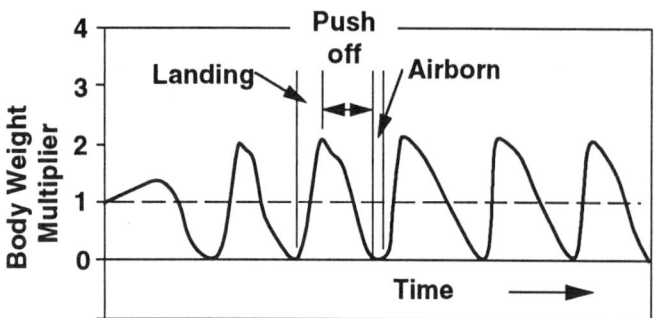

Figure 15-8. Nondimensional Force History for Running in Place with Jobbing Shoes

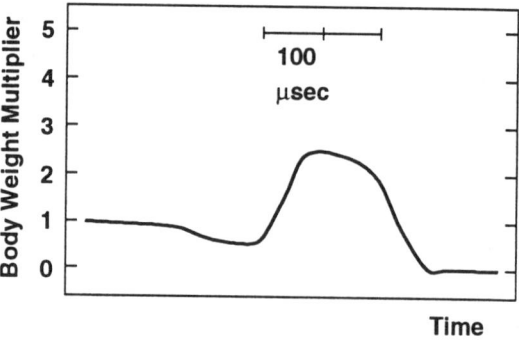

Figure 15-11. Nondimensional Force History for Two-Foot Jump from Platform Test

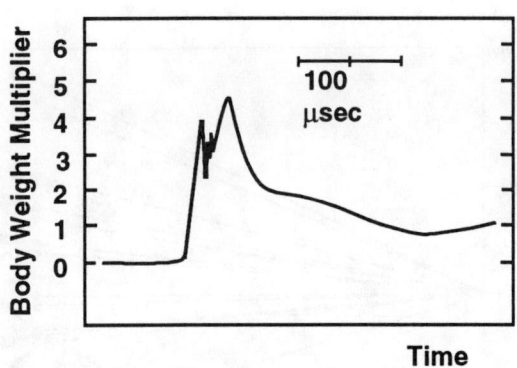

Figure 15-12. Nondimensional Force History for Two-Foot Drop from Elevation of Two Feet

obtained with commercial force plates. Peak dynamic forces transmitted into the leg structure in these activities reached values of over 1000 pounds. It was concluded that this inexpensive force platform performed similarly to commercial units and was quite satisfactory for testing dynamic loading in court athletic activities.

References
1. McMahon, T.A. and Greene, P.R., "The Influence of Track Compliance on Running," *Journal Biomechanics*, Vol. 12, pp. 893-904, 1979.
2. Cavanagh, P.R., "The Running Shoe Book," Arnold World Inc., Mountain View, CA, 1980.
3. Cavagna, G.A., "Elastic Bounce of the Body," *Journal Applied Physiology*, Vol. 29, pp. 279-281, 1970.
4. Alexander, R.McN. and Jayes, A.S., "Fourier Analysis of Forces Exerted in Walking and Running," *Journal Biomechanics*, Vol. 13, pp. 383-390, 1980.
5. Garhammer, J., "Forceplate Evaluation of Weightlifting and Vertical Jumping," *Medicine and Science in Sport and Exercise*, Vol. 11, No. 1, p. 106, 1979.

15.2.3 Field Measurement of Tension in a T-142 Tank Track

R.M. Trusty, M.D. Wilt, G.W. Carter, and D.R. Lesuer, Lawrence Livermore National Laboratory

Introduction

Computer models are being developed to simulate and evaluate the thermal and mechanical response of several tank track-pad designs on different tanks during field maneuvers. Hysteretic heating in the rubber has been addressed through laboratory experimentation and is fairly well understood. The mechanical behavior, however, is not easily simulated in the laboratory; thus, very little data is available. An important mechanical input to the rubber pad is the shear force in the pad caused by a differential tension in the steel links a across the pad (Fig. 15-13). The goal of this project was to determine this differential tension in the track while the tank traversed obstacles in a controlled field environment.

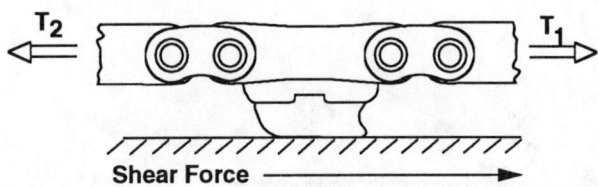

Figure 15-13. Differential Track Tension is Made up as a Shear Force in the Rubber Pad

Instrumentation

The measurement configuration selected consisted of a small track section instrumented with strain gages and calibrated as a tension load cell. High stresses occur in areas of small cross section or in areas of stress concentration such as corners and edges. The best choice for strain gage location was on the end connectors of the track. Several tests were performed to determine the location of the high stresses.

First, a two dimensional photoelastic model of an end connector was constructed and tested in tension. The stress patterns were visible when viewed through polarizing filters with each successive fringe representing one higher step in stress (Fig. 15-14). The highest tension levels were at the 4 o'clock position of the binocular tube hole, while zero stress levels were at the center area of the end link. Notice that there was one fringe of compressive stress at the 3 o'clock position.

In the second test, a brittle lacquer stress coat was painted on a T-142 end connector which was mounted on the section of the track and loaded in a tensile machine. The areas of highest stress were the same as indicated by the photoelastic model.

The final test was to place strain gages on an end connector at the high stress locations indicated by the previous tests (using photoelastic models and brittle lacquers) and to calibrate each location in tension. The results

Figure 15-14. Stress Patterns in a Photoelastic Model of an End Link While in Tension

---

Work performed under the auspices of the U.S. Department of Energy by the Lawrence Livermore National Laboratory under Contract No. W-7405-ENG-48 and supported by the U.S. Army Tank-Automotive Command, Warren, MI.

of the instrumentation efforts are summarized in the calibrations obtained and in the data measured during testing. An experiment was also conducted where the outside of a composite tube was gaged and then potted into the inside of the binocular tube to take advantage of the shear force in the pin at the pad end connector interface as shown in Fig. 15-15.

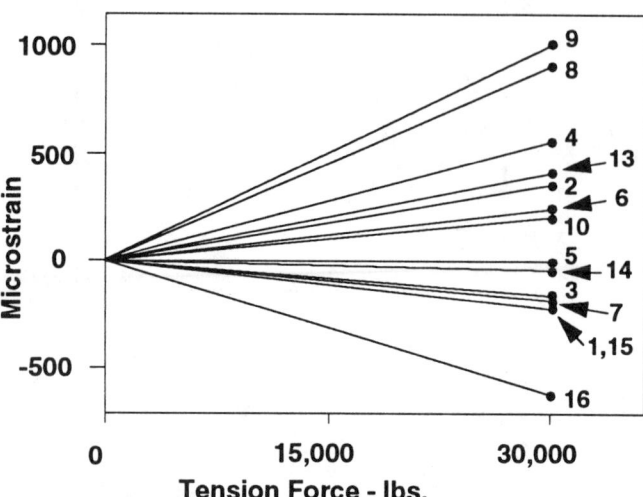

Figure 15-17. End-Point Approximation of Strain Gage Output as a Function of Applied Tension Force

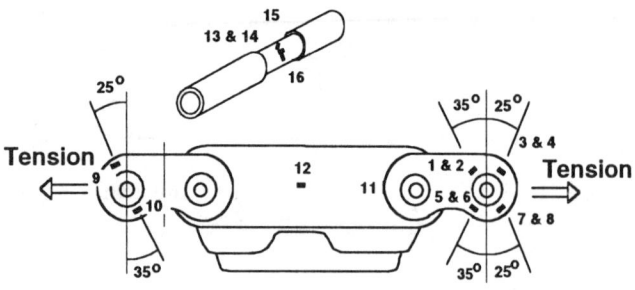

Figure 15-15. Strain Gage Locations on Test Section of T-142 Tank Track

Calibration

Two end connectors were instrumented and the entire section was mounted in a tensile machine and cycled slowly between zero and 40,000 lbs. (see Fig. 15-16). The output as a function of tension for several gage locations is shown in Fig. 15-17. Data indicates some of the highest repeatable stresses occurred at gage location 8.

Areas of high stress were natural choices for gage locations to optimize the signal output for the load cell design. Once these areas were found, two strain gages were installed at similar locations on six end-connectors.

A method to protect the strain gage wiring from rocks, dirt, and other elements was developed. The strain gages were located on the inside surface of the end connector and protected by the minimal size of the gap between the end connector and the track pad. Figure 15-18 shows the wire routing details. The junction box which routed all of the wiring from the inside of the track through the drive and idler sprocket teeth to the outside was designed to convert easily to telemetry in the future.

For the second time, the track section was mounted in the tensile calibrator. Each strain gage was connected to a Wheatstone bridge network integral to the calibration data acquisition system using three-wire circuitry to decrease

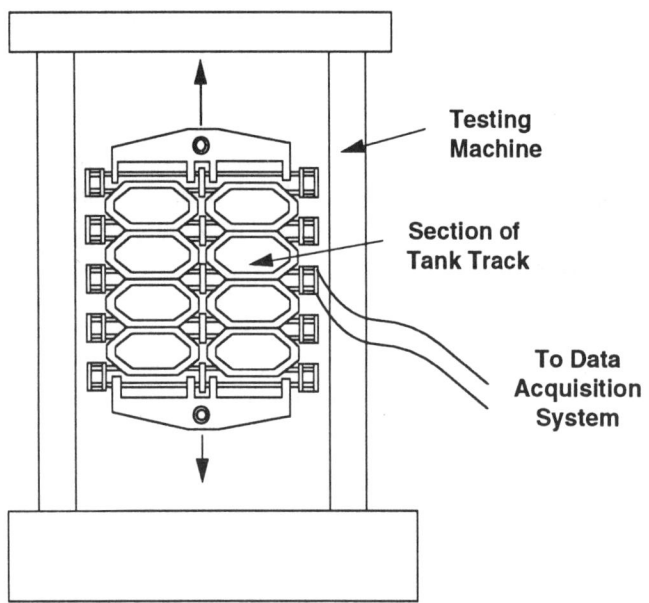

Figure 15-16. Test Section Mounted in Tensile Calibrator and Wired to Data Acquisition System

Figure 15-18. Mechanical Protection for Signal Cable Routing on Test Section

lead length effects. The 50 ft. long electrical tether was included in the calibration system. The track section was cycled slowly between zero and 40,000 lbs. tension. The output voltage of the 16 gages was plotted versus the tension force. The outputs were all reasonably linear and repeatable with force (Fig. 15-19). This design philosophy yielded a multi-output, high sensitivity load cell suitable for use in a tough field environment.

Data Acquisition System

The data acquisition system used in the field was a Hewlett Packard 3054 system and was located in an instrumentation van and connected to the tank via the 50 ft. long electrical tether (Figure 15-20). The maximum data rate achievable was approximately 0.1 second (5) per channel yielding a 5 Hz signal response for one channel and slower response as more channels were added. It was recognized that this system was not fast enough to keep up with the signal outputs when the tank was moving quickly; but under the controlled testing environment, the tank speeds were slow enough to be able to acquire meaningful data.

The data were stored on floppy disks and were later reduced to engineering units (lbs.) and plotted. In order to correlate the data plots with physical phenomena a portable video recording system was used with a clock superimposed on the video screen. The data acquisition clock was synchronized with the video clock.

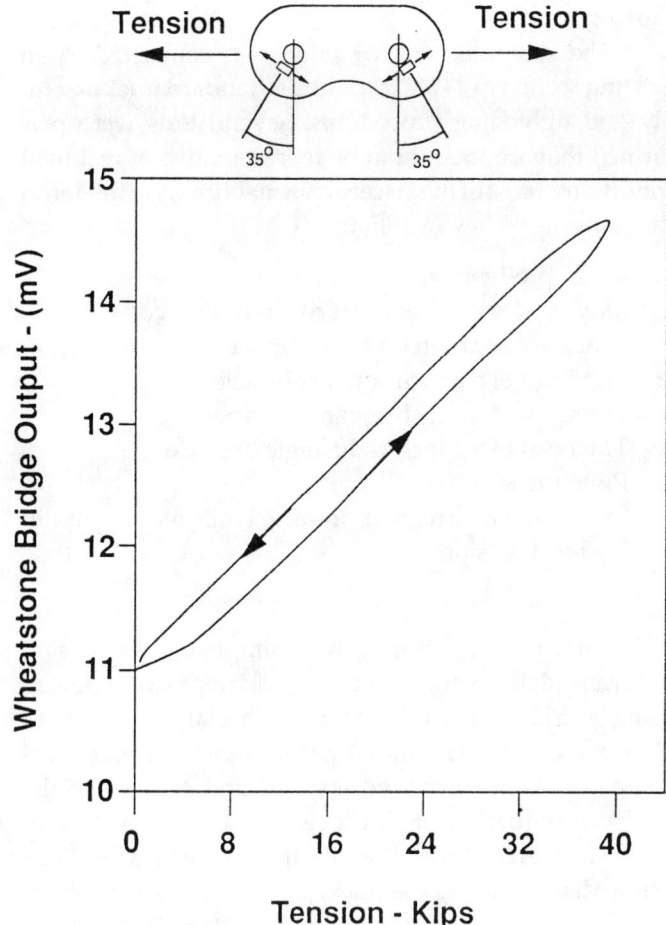

Figure 15-19. Calibration Curve of Typical Strain Gage

Figure 15-20. Electrical Tether Connecting Test Section to Instrumentation Van

Testing

The calibrated track section was connected to an existing section of T-142 track using standard track assembly and tightening procedures. Several tests were performed that are thought to be representative of real field conditions. Most of them were reasonably easy to model on the computer. They included:

1. Initial pre-tension
2. Moving slowly over a flat concrete surface
3. Quick acceleration on a flat surface
4. Traversal of a 2 inch square obstacle
5. Traversal of a 4 inch square obstacle
6. Traversal of a 4 inch right angle obstacle
7. Pivot turns
8. Traversing an irregular obstacle while moving uphill
9. Post-test tension

Data

After the field testing was completed, the tension data were plotted versus time. The video tapes were viewed using both slow and stop motion to correlate the data with physical stimuli. The initial pre-tension data was measured after the tank moved several hundred feet and the track was properly seated. The results range from 14.3 k-lb to 17.5 k-lb. After three days of running the tank, and just before disassembling the track, post test tension data was taken. Post-test tension averaged about 2000 lb. greater than the pre-test tension.

Data was acquired and reduced for several repetitions of each of the nine tests mentioned previously. An examination of typical data in Figs. 15-21 through 15-24 shows that large track tensions (greater than 40,000 lbs.) and track tension differences (greater than 10,000 lbs.) can be produced in simple, low velocity, obstacle contact situations. The accuracy is thought to be better than 10 percent of full scale due to the hysteresis loop in the calibration curves.

Conclusions

The field test at the Yuma Proving Grounds was a success as well as an invaluable learning experience. The gage installation, calibration, wire routing and protection worked well. Data was gathered successfully in a controlled manner. The use of video allowed the correlation of real time events to digital data. The results have been used in the latest version of the modeling codes.

The tank speeds were limited by the capability of the data acquisition system. Now that the logistics associated with this type of test are understood, the data system could be converted to a dynamic system capable of higher frequency response. The tank speed could then be increased to more realistically simulate field maneuvers.

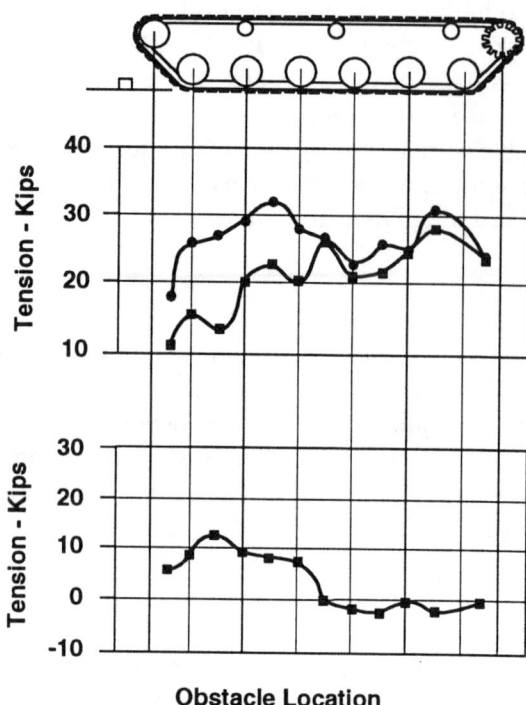

Figure 15-21. Traversal of a 4 in. by 4 in. Obstacle. The Circles and Squares are Output from Two Strain Gages Straddling Obstacle.

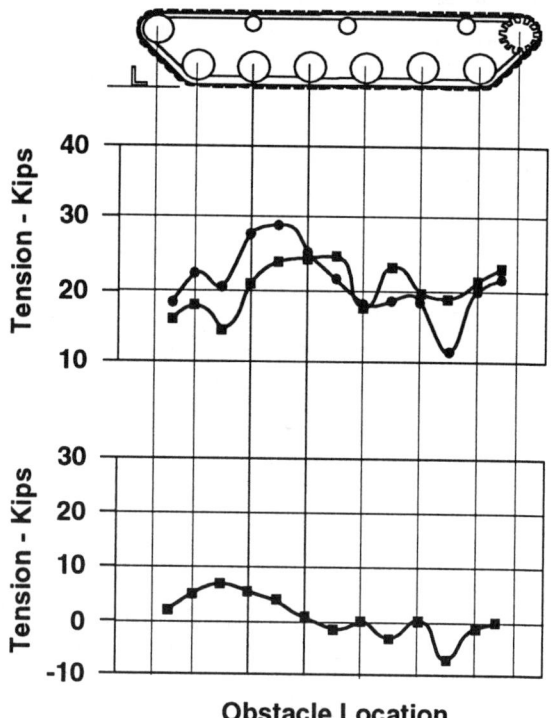

Figure 15-22. Traversal of a 4 Inch Right-Angle Bracket. The Circles and Squares are Output from Two Gages Straddling Obstacle

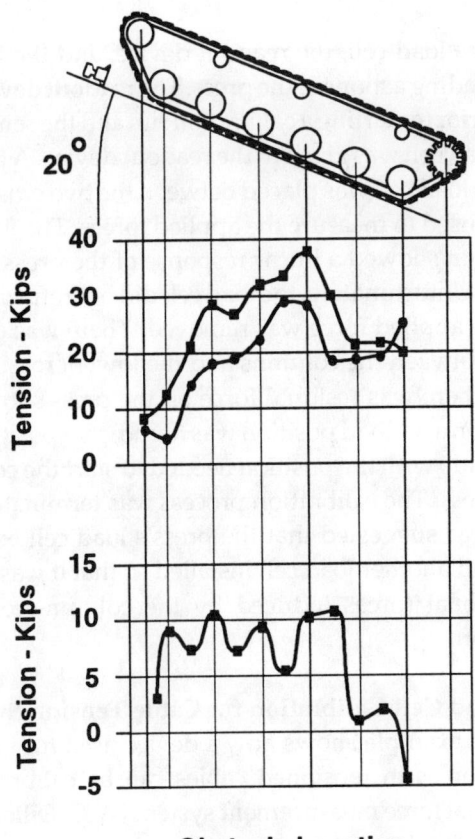

Figure 15-23. Uphill Traversal of a Railroad Rail. The Circles and Squares are Output from the Two Strain Gages Straddling the Obstacle

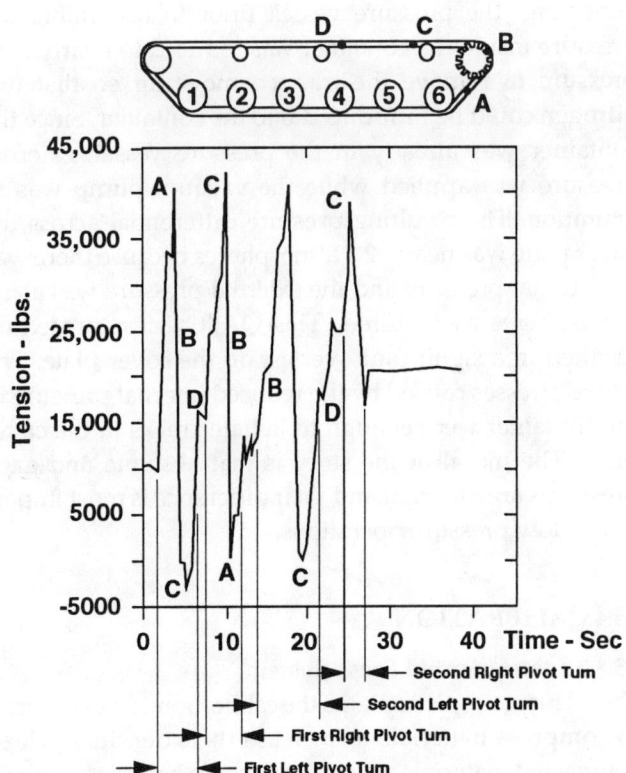

Figure 15-24. Consecutive Right/Left Pivot Turns. The Letters on the Plot and Tank Schematic Correlate Strain Gage Locations and Data.

While the limitations of the tethered method of acquiring data from moving vehicles are obvious, many advantages also exist. These include simplicity, cost effectiveness, and the ability to record more channels of data.

This test was divided into three parts: (1) sensor installation and calibration, (2) method of signal transmission, and (3) data acquisition system. We consider that the sensor installation and calibration are the most important and challenging aspect of this test. Our philosophy in the first field test was to gain experience in this area. This was effectively accomplished by keeping the signal transmission and data acquisition systems as simple as possible. Future field tests can now be performed with a high level of confidence in the sensor installation and calibration. The data acquisition system and data transmission method will be upgraded for future testing.

15.3 QUALITY ASSURANCE AND QUALITY CONTROL

Cover Plate Pressure and Vacuum Tests

This example describes the results of tests performed on a cover plate which was designed to resist one atmosphere of external pressure (14.7 psig) while the container was maintained at a vacuum. The plate was used to seal a container of electronic equipment used in flight testing. The plate had a series of holes in which connectors are installed to access the sensors, monitor components, and to connect to data systems contained within.

The development testing included the appropriate pressure proof testing to 1.5 times the maximum working differential pressure because the cover plate would be accessible to personnel while in operation. The cover passed development tests with no visible degradation.

A design decision was made to protect the electronic equipment by backfilling the volume with dry nitrogen at a low pressure (2 psig). During production, some decisions were made to reduce the number of operations needed. Since each cover plate and container had to be tested, it was decided to combine the pressure testing and backfilling operations. When cracks appeared in the 7075-T6 Aluminum cover plates after external pressure tests and backfilling operations had concluded, an investigation was initiated to determine the cause.

The container and cover plate were placed inside a pressure vessel. Plumbing connections were made for a vacuum pump and fill port through the wall of the container and the pressure vessel. Prior to backfilling the pressure inside the container was reduced to nearly zero pressure to remove the trapped moist air so that dry nitrogen could be introduced into the container. Since the container was already in the pressure vessel, external pressure was applied while the vacuum pump was in operation. The resulting pressure differential across the cover plate was nearly 2.5 atmospheres because there was no internal pressure and the external pressure was at 1.5 atmospheres as required. This QA/QC test procedure resulted in a significant overtest on the cover plate. The added stresses caused by the reduced internal pressure in the container were enough to initiate cracks in the cover plate. The moral of the story is that absolute and gage pressures are different and their difference is most important at low pressure operations.

15.4 CALIBRATION

15.4.1 Instrumented Press System

This example explains the calibration of a press used to compress materials to increase their densities. These compacted materials are made for use in batteries, ceramic insulators, and a variety of similar products. Calibration of the press used in making these specimens is needed for process control so that desirable characteristics of the compacted materials could be achieved.

The press consisted of beam type load cell attached to a crosshead as shown in Fig. 15-25. The press had a second crosshead used to react the forces imposed on the compacted specimens. The force was imposed by a mechanical crank which advanced the lower crosshead toward the upper crosshead with the applied force transmitted to the specimens. The lower crosshead was required to slide along the supporting columns.

The calibration procedure used focused on three things: the load cell, the readout device, and the loading and unloading actions of the press. The readout device was zeroed prior to starting a calibration run and the sensitivity for the load cell was given to the readout device. A transfer standard load cell was placed between the two crossheads and was used to measure the applied forces. The first calibration run showed a linear response of the press and its load cell. Unfortunately, the load cell did not return to zero when the applied force was removed. There was enough friction between the columns and the lower crosshead so that the there was residual force on the press's load cell. Even when a no-load position was found, that position did not coincide with the position needed to start the compaction process. The calibration process was terminated.

It was suggested that the press's load cell be abandoned and another load cell installed so that it was free of the frictional forces induced by the column-crosshead interaction.

15.4.2 Load Cell Calibration for Cable Tension System

This example shows how a device used for measuring the forces in tensioned cables can be calibrated. A commercial force measurement system (W.C. Dillon Running Line Tensionmeter) was calibrated on a structural test frame using a load cell in series with the cable. The calibration setup is shown in Fig. 15-26. The cable is positioned so that it contacts three pulleys in the force measurement system. The tension force in the cable is resisted by the load cell attached to the pivot arm in the tensionmeter. A hydraulic ram is used to apply the force. Table 15-2 gives the results of the calibration.

These data for the calibration of this tensionmeter were obtained using a digital voltmeter. A bridge excitation of 10 volts was applied, an increment of force was applied and indicated to the data acquisition system, the voltage change was measured by a digital voltmeter, and the measured sensitivity was calculated. The voltage change measurements were used to calculate the deviations from the applied force in terms of indicated pounds and per-

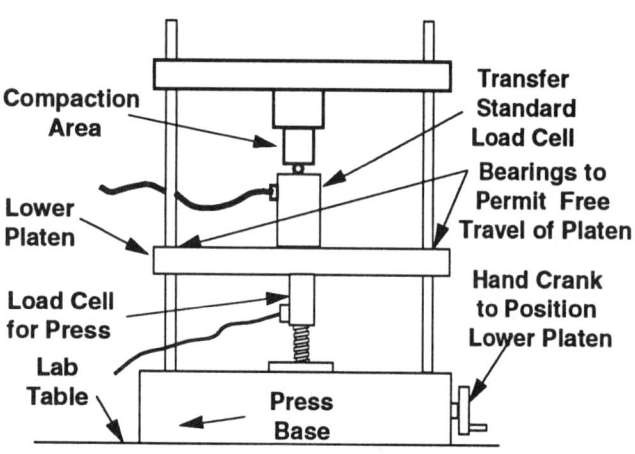

Figure 15-25. Compaction Press Calibration Setup

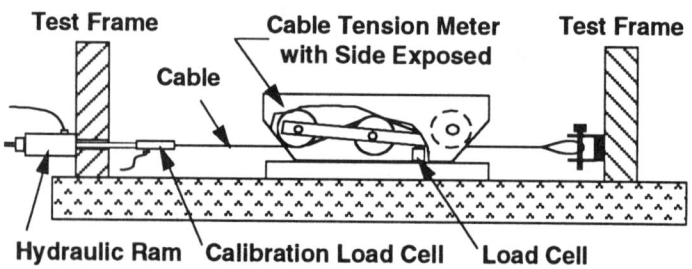

Figure 15-26. Calibration Setup for Cable Tension System

Structural Testing—Examples

cents. The data were analyzed using a binomial least squares fit.

Binomial Least Squares Fit for Force Calibration:

Force indicated by load cell = $-3.4 + 10 \times 103$ mV/V -140 (mV/V)^2

A gage factor (G.F.) was calculated by

G.F. = $4000 \times$ sensitivity = 4000×0.0945 µvolts/volt/lb = 0.3781 µin/lb.

The use of a gage factor for this load cell permits the use of a portable strain indicator to be used as a readout device. By use of interface connectors and cables, a load cell can be used in conjunction with a portable strain indicator.

15.5 CONNECTIONS AND JOINTS

Fixtures and Test Procedures for Miniature Threaded Fasteners (Ref. 15-6)

R.T. Reese and W.A. Kawahara, Sandia National Laboratories and L. J. Lazarus, Allied Signal Kansas City Division

Introduction

Generally accepted design guidelines are lacking for miniature threaded fasteners, with diameters of 6.35 mm (0.25 in.) or less, due in part to the difficulty in testing them in a precise, consistent, and repeatable manner. One consequence is that designers tend to call out increased thread depths with apparent excess threads in the hope that full strength and integrity of the threaded connection will be achieved. When threads are tapped into a part in excess of the length to diameter ratio (L/D) needed to achieve the full tensile strength of a screw, the incidence of rejected parts increases significantly. These added threads and rejected parts are due to broken taps or damaged and oversized threads associated with tapping such small holes. These parts are otherwise nearly completed and typically have many machining operations.

The basis for the study described in this article was to verify static design principles for miniature screws and to confirm or show where differences were found in applying these principles. This example includes: (1) a description of a portable tensile test fixture which provided precise, reproducible alignment, uniform loading, minimizes system compliance to permit more accurate displacement and energy-to-failure measurements, (2) outlines the experimental plan for testing small threaded fasteners, (3) a description of the testing procedure used, and (4) summaries of the results obtained to date. The test fixture presented allowed testing of screws shorter than the three

---

This work performed at Sandia National Laboratories supported by the U.S. Department of Energy under contract number DE-AC04-76DP00789.

Table 15-2. Force Calibration of Cable Tension Measurement System

Load Cell Capacity	= 20,000 lbs	Bridge Excitation	= 10 volts
Load Range Needed	= 10,000 lbs	Nominal Output	= 2 millivolts/volt
Load Cell Resistance	= 350Ω		
Nominal Sensitivity	= 0.1 microvolt/volt/lb		

Applied Force lbs	Bridge Output volts ×10⁻³	Output Change volts ×10⁻³	Measured Output millivolts per volt	Measured Sensitivity microvolts /volt/lbs	Indicated Force lbs	Deviation from Applied Force - Pounds	Deviation from Applied Force - Percent
0	0.791	0.0	0.0	0.0	0	0	0.0
1000	1.740	0.949	0.0948	0.0948	1002	2	0.21
2000	2.688	1.896	0.1896	0.0948	2003	3	0.16
3000	3.634	2.843	0.2842	0.0947	3003	3	0.11
4000	4.580	3.788	0.3787	0.0947	4002	2	0.06
5000	5.523	4.732	0.4731	0.0946	4999	-1	-0.02
6000	6.462	5.670	0.5669	0.0945	5991	-9	-0.16
7000	7.398	6.607	0.6605	0.0944	6980	-20	-0.29
8000	8.351	7.560	0.7559	0.0945	7987	-13	-0.16
9000	9.266	8.475	0.8473	0.0941	8954	-46	-0.52
10000	10.193	9.402	0.940	0.0940	9933	-67	-0.57
0	0.779	0.012	0.012	0.0	13	13	

length to diameter ratios (L/D) required by ASTM and permitted determination of screw response in static and dynamic conditions.

Test Fixtures

The test subassembly shown in Fig. 15-27 consisted of five parts: (1) the screw (fastener) tested, (2) the load washer made from heat treated tool steel (it can be made from the material being fastened if the whole joint is to be studied), (3) the engaged material which was threaded to accept the screw, (4) three hardened tool steel load pins transferred the applied force to the load washer and pushed the screw head away from the engaged material, and (5) the load collar which evenly distributed the applied force P among the load pins. The load washer was designed to apply the force uniformly to the head of the screw without bending it. The engaged material specimen may be made of high strength material (if the screw strengths are the desired test results) or of the material to be used in the actual threaded connection. Those reported in this article were made from round bar stock. The load pins were designed to be stronger than an individual screw. These pins were sized to slide through the holes in the engaged material and yet not wobble or shift the applied force while under load. The load collar was designed so that alignment could be ensured initially and preserved throughout the test.

The force applied to this five-part subassembly was resisted by a rigid cylindrical test housing mounted to the test system as shown in Figure 15-28. This housing contained a non-contact capacitance displacement gage (Capacitec Model 160-1260) aligned below the screw head.

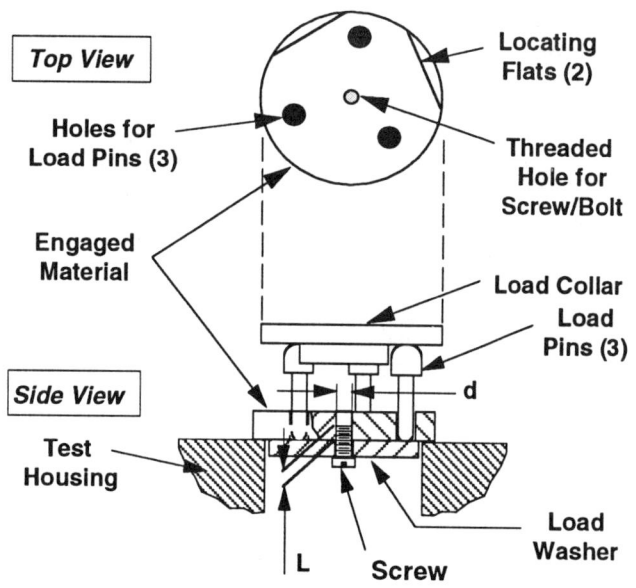

Figure 15-27. Test Subassembly

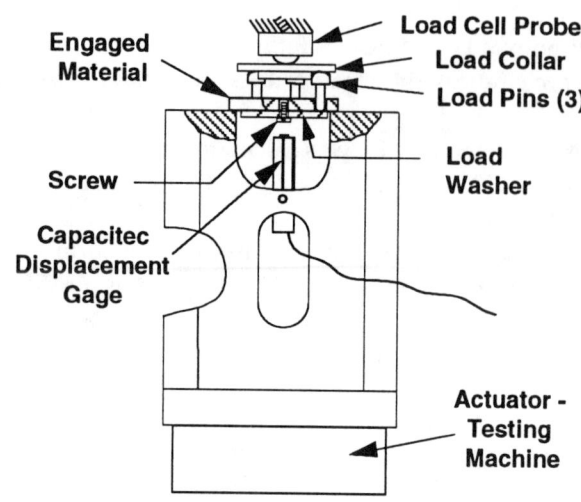

Figure 15-28. Test Housing

The position of this displacement gage minimized undesired displacement contributions from system compliance. Calibration of this displacement gage was performed with the screw, load washer, and gage coaxially mounted on a precision micrometer. The test housing could be modified to accommodate a precision micrometer providing truly in situ calibration. (Other types of displacement gages could be used; this particular gage offered the advantages of both small size and calibration. The fixture was designed for the capacitance gage and would require modification if other gages were used.)

The 1/4-20 screws were tested in a conventional fixture shown in Fig. 15-29. This fixture was modified with small circumferential grooves machined in the test fixture to prevent the extensometer from slipping.

Testing Procedure

The test housing was attached to the actuator of a servohydraulic testing machine. The actuator was moved toward the cross-head of the servohydraulic machine. The tests were performed using computer control although other control procedures (e.g., manual, function generator, or profiler) could have been used. The load cell and the displacement gage signals were fed into an x-y recorder so that plots of load vs. deflection were generated.

The testing sequence consisted of inserting an assembled screw/load washer/engaged specimen into the housing. Then the load pins were inserted and the load collar was positioned on top of the load pins. The housing was then positioned very close to the bearing at the probe tip; the probe was attached to the load cell. The load collar was aligned on the probe tip by rotating the collar on the heads of the load pins. (Experience gained in using this test

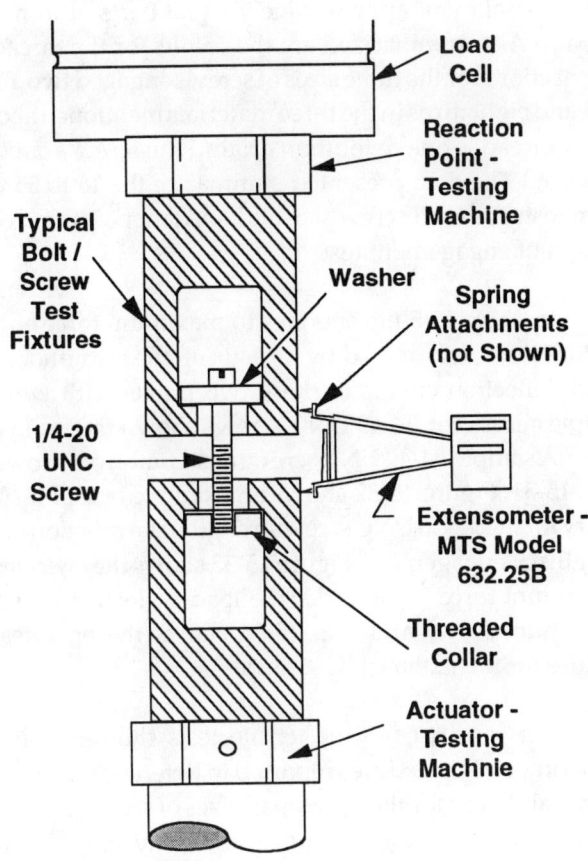

Figure 15-29. 1/4-20 Test Fixture

subassembly showed that the rotation of the load collar resulted in the applied force being uniformly distributed through the load pins to the load washer as required and eliminated bending in the screws.) The force was then applied until the threaded fastener failed in one of four modes: (1) failure of the screw, (2) stripping of the screw threads, (3) stripping of the threads in the engaged material, or (4) fracture of the engaged material sample which can occur in brittle materials. The load-deflection curve was obtained for each screw tested.

The test procedure for the 1/4-20 screws conformed to the ASTM requirements with the added use of the extensometer.

Experimental Plan

Three screw sizes (4-40 UNC, 2-56 UNC, and 1.0 UNM) were chosen for tensile tests using test subassembly and procedure described in this example. Data for 1/4-20 UNC screws were also obtained. Fifteen screws from each lot were tested to failure to form a reference data base for the other specimens. These screws were fully engaged (eight or more turns except for the shorter #4-40 screws which were engaged 5.5 turns or more) in stainless steel.

Because of the concerns of obtaining enough information for the technical basis for design guidance, individual screws were engaged in three different materials; 6061-T6 Aluminum, Hiperco 50, and 303 Stainless Steel (the 1/4-20 UNC screws were not tested in Hiperco 50 because no application could be found for this combination). These specimens were manufactured and assembled by Allied Signal Aerospace/Kansas City Division (KCD) and shipped to Sandia Laboratories in Albuquerque for testing. The screws selected for testing were from production lots at KCD and all were made from a 300 series stainless steel. All fasteners tested were obtained from production supplies and all met MS (Military Specification) quality. The internal threads on the engaged parts were made on KCD production machines (holes were reamed to a 0.0002 inch diameter tolerance and checked with pin gages before tapping).

The specimens consisted of four different thread engagements; 2 threads, 4 threads, 6 threads, and 8 threads

Table 15-3. Parameters Used in Small Screw Static Tests

Screw Size	Number of Tensile Tests	Engaged Materials-Number Tests			Number of Threads Engaged
		Aluminum 6061-T6	Hiperco 50	Stainless Steel 303	
1.0 UNM	15	24 †	24	24	2, 4, 6, 8
2-56 UNC	15	24	24	24	2, 4, 6, 8
4-40 UNC	35*	24	24	24	2, 4, 6, 8
1/4-20 UNC	15	24	—	24	2, 4, 6, 8

*Both pan and socket head screws were used for this test
†These totals come from the following specimen parameters: 3 specimens for each test step times 2 radial engagements (55% and 75%) times 4 different thread engagements (2, 4, 6, and 8 threads) = 24 tests.

and two values of percent thread (radial engagement) –55 and 75. This difference in radial engagement was accomplished by increasing the minor diameter of the engaged material. Three samples of each were tested to provide a sufficient data base for post-test evaluation on each size. Table 15-3 summarizes the different parameters used for the tests.

Sample Test Results

Tests results gave the tensile strengths of three sizes of stainless steel fasteners: #4-40 UNC, #2-56 UNC, and 1.0 UNM engaged in stainless steel. Two lengths of #4-40 screws were tested because the available socket head cap screws were not long enough to be engaged two diameters deep as was thought to be needed to generate the full strength of the screws. Table 15-4 summarizes the results for these tensile tests; note that shorter lengths of engagement for the #4-40 screws also achieved the full strength of the screws. (The coefficient of variation is 100 times the ratio of standard deviation to the mean with a low value indicating how repeatable the data are.)

Table 15-4 also lists the tensile strengths of the threaded connections for these four sizes of stainless steel screws engaged in three materials 6061-T6 Aluminum, Hiperco 50, and 303 Stainless Steel with different thread engagements. These materials are often used in connections involving small threaded fasteners. (The Hiperco 50 is a magnetic alloy of approximately equal parts of iron and cobalt.) Also summarized are the results for the threaded connections for the three sizes of screws engaged two, four, six, and eight turns in the three materials mentioned above. (One thread is one revolution around the screw's circumference.) The data presented summarize the 15 to 35 tension tests for each screw size and six data points for each length of engagement test.

In Table 15-5 the energies to maximum force and to failure were determined by measuring the area under the load-deflection curves to these two points with two example curves for #4-40 UNC screws shown in Fig. 15-30.

A sample #4-40 UNC screw tensile failure is shown in Fig. 15-31. Figure 15-32 shows an example of a plot summary for the #2-56 UNC screw strengths as a function of the lengths of engagement. Figure 15-33 shows the energies to maximum force (area under the load-deflection curve to maximum force) and Fig. 15-34 shows the energies to failure for the #2-56 UNC screws.

The results from the tensile tests showed that the experimental procedures reported in References 1 through 3 are valid even for the very small sizes of screws. The data are tightly grouped with more repeatability than expected. The coefficients of variation do not increase uniformly with decreasing size as was originally expected.

Table 15-4. Summary of Tensile Test Data and Results of Engagement Test Data

Screw Size		Measured Tensile Strength [Military Required Strength-lbs]	Standard Deviation-lbs [Coefficient of Variation]	Engaged Material	Mean Tensile Strength-lbs vs. Number Threads Engaged			
					2	4	6	8
1/4-20 UNC		4129 [2540]	57.4 [1.4]	Aluminum St. Steel	1876 3303	3699 4142	4099 4135	4114 4104
#4-40 UNC	Pan	715	9.5 [1.31]	Aluminum Hiperco 50 St. Steel	413 612 638	672 676 677	701 691 681	721 725 734
	Socket	743 [480]	14.7 [2.0]					
#2-56 UNC		462 [300]	2.7 [0.6]	Aluminum Hiperco 50 St. Steel	238 396 397	437 449 443	449 451 448	457 460 467
1.0 UNM		101 [none]	3.7 [3.7]	Aluminum Hiperco 50 St. Steel	42 88 102	75 100 108	83 99 99	98 100 100

Table 15-5. Summary of Results for Energies to Maximum Force and Failure

Screw Size	Engaged Material F.E.-Full Engagement	Average Energies to Max. Force by Number of Threads				Average Energies to Failure by Number of Threads			
		2	4	6	8	2	4	6	8
1/4-20 UNC	Aluminum	20.8	73.3	157.7	175.6	53.5	177.5	502.3	519.2
	St. Steel	66.0	181.8	163.5	132.1	173.1	539.6	497.2	460.4
	F. E.				141.8				459.9
#4-40 UNC	Aluminum	1.2	5.5	5.6	4.4	7.0	28.7	30.6	26.9
	Hiperco	3.0	5.1	4.8	2.9	17.0	28.9	28.7	24.1
	St. Steel	5.1	4.6	4.6	3.0	19.3	27.9	28.0	24.8
	F. E.		4.1		3.0			26.1	25.2
#2-56 UNC	Aluminum	0.5	2.3	2.6	2.8	2.7	10.4	13.8	12.2
	Hiperco	1.5	3.0	2.6	2.2	7.5	14.8	14.5	11.3
	St. Steel	1.6	3.0	2.8	2.2	6.0	15.2	13.7	12.0
	F. E.				2.3				11.7
1.0 UNM	Aluminum	0.07	0.27	0.46	0.71	0.3	0.9	1.1	1.5
	Hiperco	0.55	0.64	0.62	0.67	0.8	1.6	1.0	1.2
	St. Steel	0.72	0.89	0.59	0.60	1.2	1.6	1.2	1.4
	F. E.				0.57				1.2

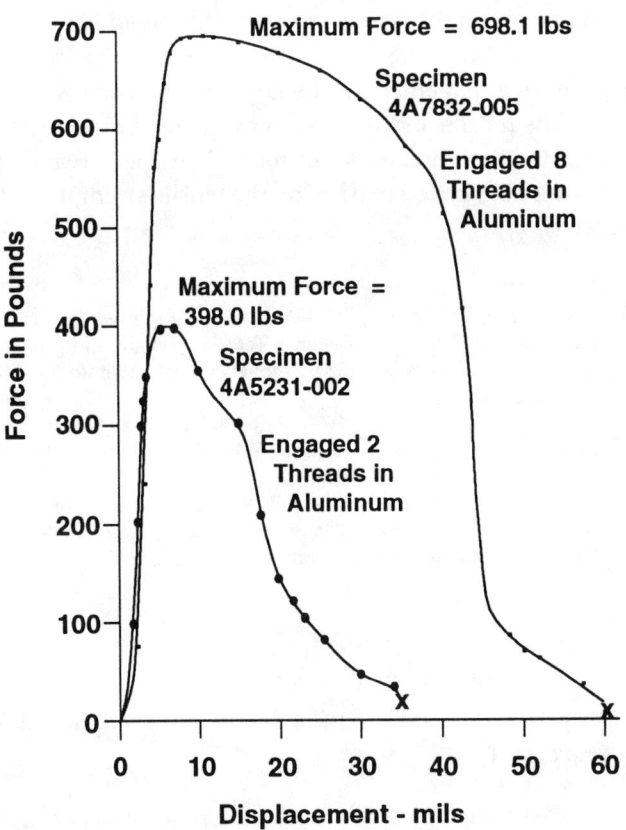

Figure 15-30. Two Sample Load-Deflection Curves

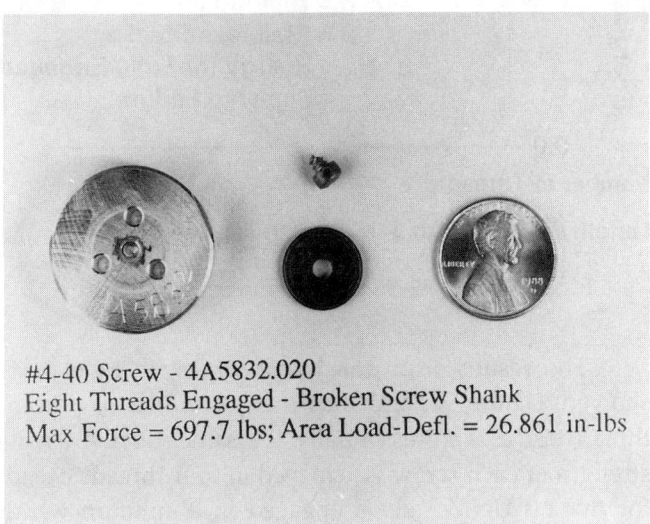

#4-40 Screw - 4A5832.020
Eight Threads Engaged - Broken Screw Shank
Max Force = 697.7 lbs; Area Load-Defl. = 26.861 in-lbs

Figure 15-31. #4-40 UNC Screw Failure

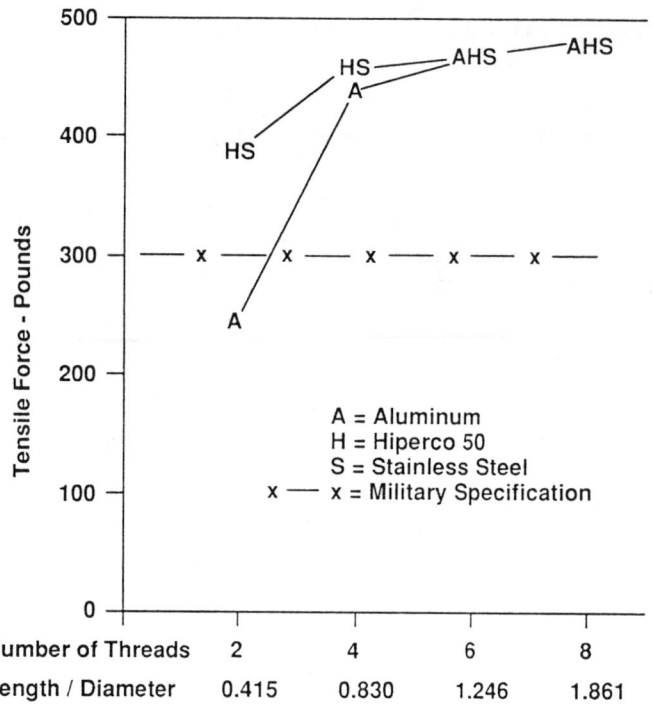

Figure 15-32. #2-56 UNC Screw Strengths

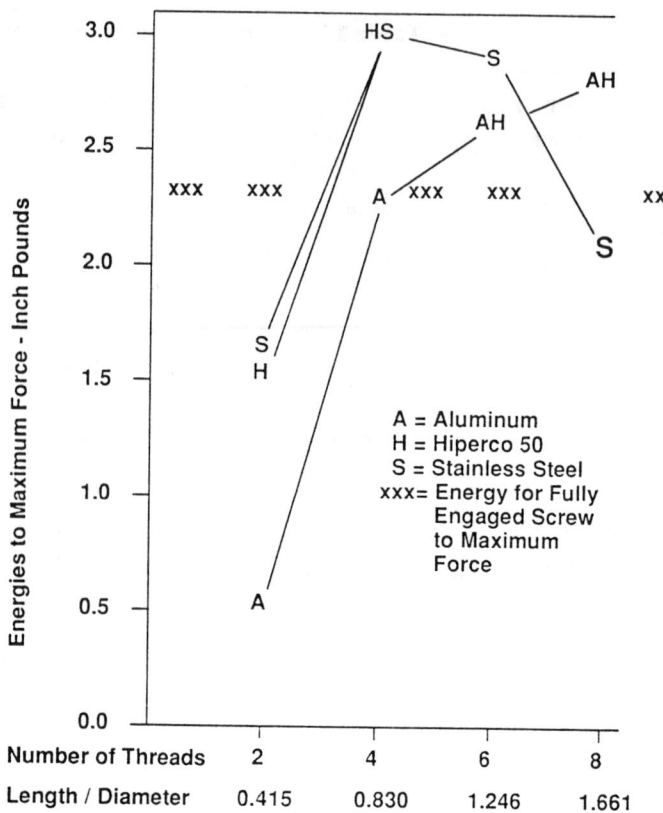

Figure 15-33. #2-56 Screws-Energies to Maximum Force

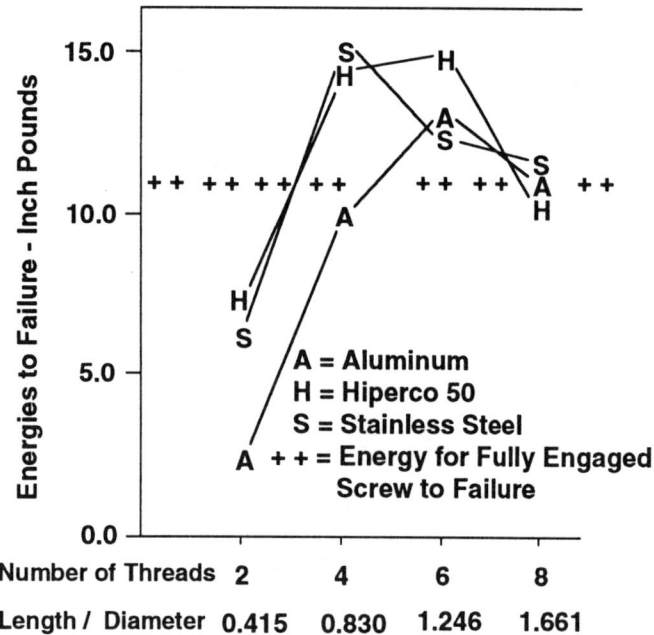

Figure 15-34. #2-56 Screws-Energies to Failure

The results from the length of engagement tests showed: (1) the strengths for the two values of percent thread tested were essentially the same; (2) the tensile strength of each screw is achieved in four threads except for the 1.0 UNM screws engaged in aluminum which requires eight threads; (3) the tensile strengths of the screws exceeded the military specifications by 49 percent or more; and most important (4) small threaded fasteners transfer forces in the same manner and require the same lengths of engagement as the larger sizes of screws.

The tensile and length of engagement tests strongly suggested that thread depths more than one to two diameters are not needed to develop the tensile strengths of the screws.

References:

1. K.V. Diegert, L.R. Dorrell, R.T. Reese, and L.J. Lazarus, "Small Screw Study-Interim Report on Fastener Tensile Strength and Optimum Thread Depth," SAND89-1320, Sandia National Laboratories, Albuquerque, NM 87185.
2. Dorrell, L.R. and Laing, J., "Test Methods for a Small Screw Study," Proc. SEM Spring Conference, June 5-7, 1990, Society for Experimental Mechanics, Bethel, CT., 06801.
3. J.L. Cawlfield, T.L. Ernest, and R.T. Reese, "Data and Conclusions from Tests on Small Screws," Proc. Spring Conference, June 4-6, 1990, Society for Experimental Mechanics, Bethel, CT., 06801.

15.6 ENERGY ABSORPTION— SUPPORT STRUCTURE

Robert T. Reese, Ph.D., Sandia National Laboratories, Albuquerque, NM

A structural test lab received a request to perform a load-deflection test to failure on a small support structure for a component. The support structure is intended to absorb energy upon severe impact and then limit the forces transmitted into the component. A preliminary impact test into a concrete wall was conducted and the results indicated that this support structure had a higher than desired failure load and there was too much area under the load deflection to failure curve-the structure absorbed too much energy when it failed. In this case the area under the load-deflection curve to failure was about two times too large. Unfortunately, about 90 of the support structures were already fabricated. The project engineer wanted to make some simple modifications to the support structure to reduce the area under the load-deflection curves to failure and he wanted the behavior of the redesign support structure confirmed by tests. The support structure is shown in Fig. 15-35 and the test set-up is shown in Fig. 15-36 with load P applied horizontally (in shear) at the top of the structure as this application of load most closely approximated the actual manner in which loads were applied. Although the support structure was filled with low-density foam, the major load carrying member was the aluminum shell structure and the foam did not make a significant contribution to the stiffness or strength of the support structure.

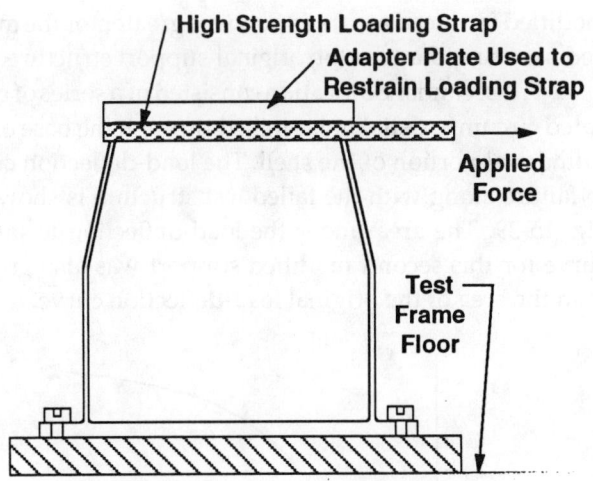

Figure 15-36. Loading System

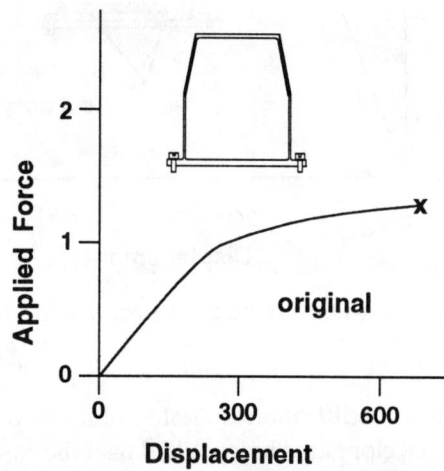

Figure 15-37. Load-Deflection Curve for Original Structure

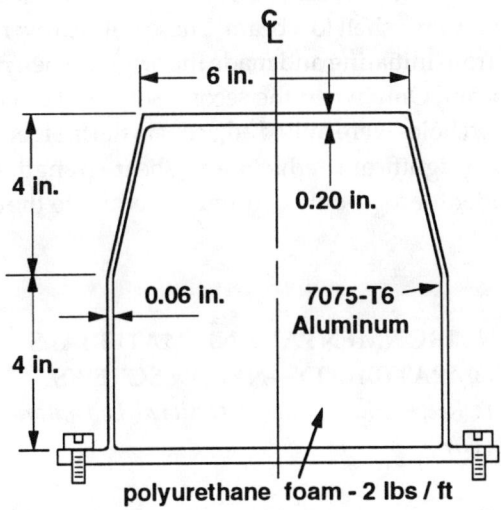

Figure 15-35. Support Structure

The load-deflection curve for the original (not modified) support structure along with the failed test unit is shown in Fig. 15-37. The first modification to the completed support structure consisted of a series of elongated holes milled through the aluminum shell structure at 30 degree increments parallel to its axis of revolution in both the cylindrical and conical sections as shown in Fig. 15-38 along with the load-deflection curve to failure. (This choice was partly decided by constraints for the simplest modifications in the machine shop.)

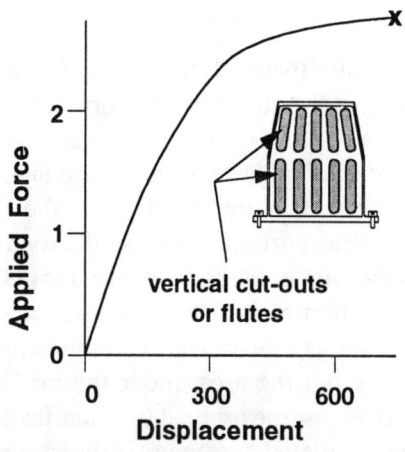

Figure 15-38. Load-Deflection Curve for First Modification

The area under the load-deflection curve for the modified (reduced area) support was greater for the modified structure than for the original support structure.

The second modification consisted of a series of elongated circumferential holes milled through the base of the cylindrical portion of the shell. The load-deflection curve to failure along with the failed test structure is shown in Fig. 15-39. The area under the load-deflection to failure curve for this second modified support was also greater than the area of the original load-deflection curve.

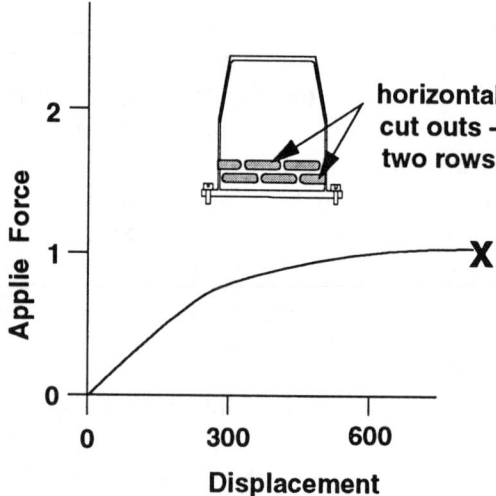

Figure 15-40. Load-Deflection Curve for Third Modification

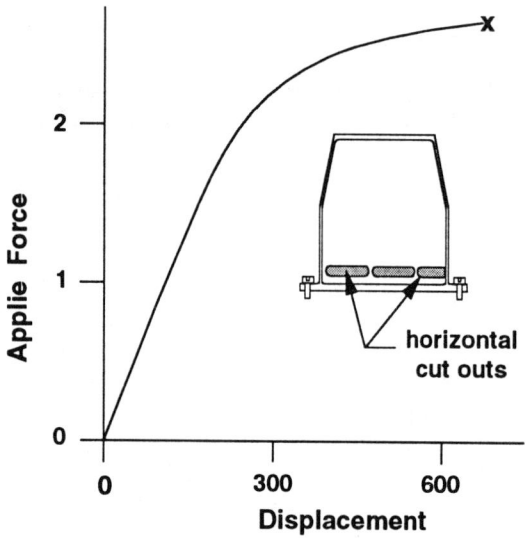

Figure 15-39. Load-Deflection Curve for Second Modification

The third modification consisted of a second set of circumferential elongated holes milled near the base of the structure. This modified structure along with its load-deflection to failure curve is shown in Fig. 15-40. This third modification was successful in reducing the area under the load-deflection curve.

Solution:

The removal of material from the walls of the support structure changed the structural response of the support structure from a shell-type structure to a beam-type structure. When the original support structure failed, a small buckle was initiated at the interface of the cylindrical section and conical portion of the shell. Attempts to increase the applied force resulted in continued plastic deformation of the aluminum shell structure without significant increase in the applied force. The maximum applied force did not increase but the area under the load-deflection curve did increase for the original (unmodified) structure until failure was initiated by fracture of the base of the shell. When the elongated milled holes were cut parallel to the axis of revolution of the shells and in the walls of the structure, its behavior was described by beam action and it did not exhibit a buckling at the same interface and the structure could accommodate greater load to failure until the same basic failure mode was initiated.

When the circumferential holes were milled at the base of the structure, the result was that the behavior also changed from a shell to a beam. These holes prevented the buckle from initiating and made the original shell respond like a beam. Only when the second set of circumferential elongated holes were milled adjacent to the first set of holes causing a significant reduction in shear strength did the area under the load-deflection curve reduce to the desired levels.

15.7 ENVIRONMENTAL AND MATERIALS COMPATIBILITY—NYLON SCREWS
Robert T. Reese, Ph.D., Sandia National Laboratories, Albuquerque, NM

This particular example was a test request to determine the shear strength of nylon screws. These screws were needed for an application in which lower strengths were needed than for even the lowest grade metal fasteners. The application required these screws to fail at a predictable applied force.

This test engineer had designed a special test fixture to test eight screw simultaneously. The test plan and analysis procedure was to test these screws and use average values for their strengths. The screws were radially engaged in fixture shown in Fig. 15-41. The upper portion of the fixture attached to a swivel attached to the fixed crosshead in a screw-driven testing machine. The lower portion

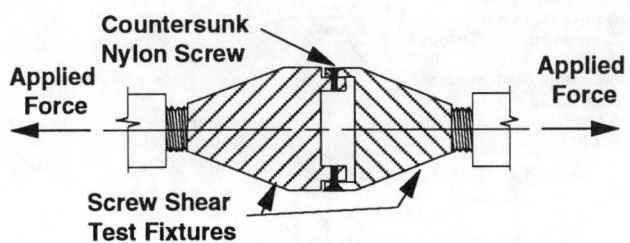

Figure 15-41. Test Setup for Nylon Screws in Single Shear

of the test fixture attached directly to the load cell in a testing machine.

One day (warm and dry) the maximum force determined in testing in shear was 1240 lbs. Testing was resumed the next morning after a summer thunderstorm had occurred in a desert climate with the test results giving a maximum force in shear of 1088 lbs. The next day (warm and dry) the maximum force was 1270 lbs. A fourth set of tests were performed during another summer thunderstorm with the maximum force of 1055 lbs.

The test results indicated that the strength of these nylon screws were very susceptible to the ambient conditions in the laboratory. Since the laboratory did not have a controlled atmosphere (e.g., temperature and humidity), the test results for shear tests on nylon screws could not be performed repeatably. The test fixture was retained for use in other tests and proved satisfactory for many types of shear tests.

It is known that the strength of nylon is a function of temperature and humidity. Other factors in the environment also affect nylon (e.g., presence of steam, acids, corrosives). Relationships are known for the effects some of the environmental conditions have on nylon. These environmental conditions must be defined prior to and during testing so their effect can be measured.

15.8 MATERIALS TESTING

Tensile and Compressive Materials Testing with Sub-Sized Specimens. *Wendell A. Kawahara, Sandia National Laboratories, Livermore, CA*

Other considerations being equal, there is little reason to prefer a miniature test specimen geometry over a larger, standard one. Smaller specimens have potential disadvantages including:

- Higher machining and inspection costs due to closer tolerances

- Susceptibility to unintentional damage in fabrication (for example, surface work hardening) or test setup (bending of small diameter tensile bars)

- In tension, the need for special extensometry

- Grain size effects on apparent material behavior for excessively small cross sectional areas, and unrepresentative failure levels in brittle materials having volumetric flaw distributions.

Materials testing to support constitutive modeling efforts, however, often necessitates the use of miniature tensile and compression specimens. This section describes some the these applications and test techniques at Sandia National Laboratories (SNL). Standard specimen geometries and testing procedures are prescribed elsewhere (Refs. 1 and 2).

We are most often driven to miniature specimens by lack of material. SNL applications have included:

- Investigating the degradation of parent metal mechanical properties in the heat-affected zones of welds

- Verifying the heat treatment strengthening or cold-work hardening on a forging after it has been machined to a thinner section

- Investigating the through-thickness behavior of a plate to assess material isotropy relative to in-plane properties.

- Investigating the residual effects of shock waves on plate impact targets

- Testing precious metals

Miniature specimens may also be chosen to accommodate test facilities. For example;

- Large strain-compression tests can require high forces as both flow stress and specimen cross sectional area increase during the test

- Defining F, σ, ε, and D_o as the maximum force, maximum true stress, maximum true strain, and initial specimen diameter (all quantities but the last are negative), one can verify that for incompressible materials:

$$F = (\pi/4)\, \sigma \cdot D_o^2 \cdot \exp(-\varepsilon) \qquad \text{Eq. 15-3}$$

or

$$D_o = [4 \cdot F / (\pi \cdot \sigma \cdot \exp(-\varepsilon))]^{\frac{1}{2}} \qquad \text{Eq. 15-4}$$

- For example, if we wish to compress a specimen with initial diameter 0.2 in. to a strain of -1.0 with an anticipated true stress of about 200,000 psi, Eq. 15-3 shows that will require 17,000 lbs. of force. Conversely, Eq. 15-4 can be used to size a specimen given a test system's force capacity, the anticipated true stress, and the desired true strain.

- High strain-rate test systems (e.g., Hopkinson bar apparatus) require short specimens to maximize the number of waves traversing the specimen length per microsecond so that a near-equilibrium stress state can be presumed during the test duration; this can require miniature specimens, especially in tension.

- Less frequent examples include a moderate increase of the maximum attainable strain rate for a given test systems' peak actuator/crosshead velocity, or reducing the time for specimen temperature conditioning prior to testing.

Figures 15-42 and 15-43 show a threaded-end microtensile specimen and axial extensometer with a clip-on feature to reduce the opportunity for unintentionally bending the specimen during installation. This system has been qualified by comparing stress-strain curves with that obtained from larger standard specimens extracted from the same material stock. The specimen ends thread into short rods which in turn thread into spherical seats; if the specimen threads are not chased to specification, some bending can enter to affect the apparent elastic modulus and near-yield behavior.

To support constitutive modeling, uniaxial compression testing has proven essential in addition to tensile testing for several reasons:

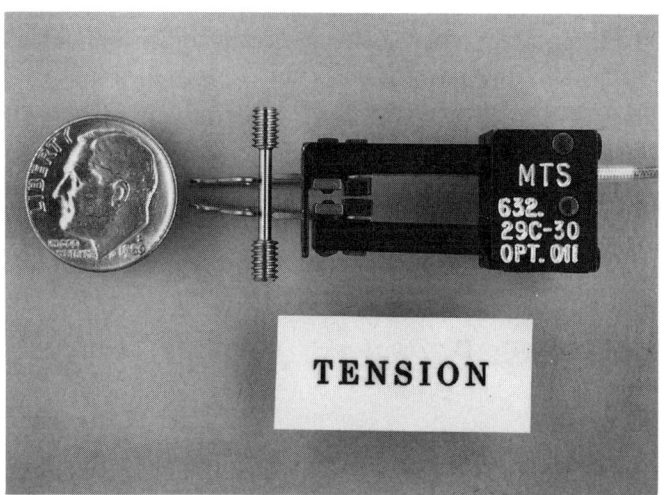

Figure 15-42. Microtensile Specimen and Extensometer with Clip-on Attaching Feature

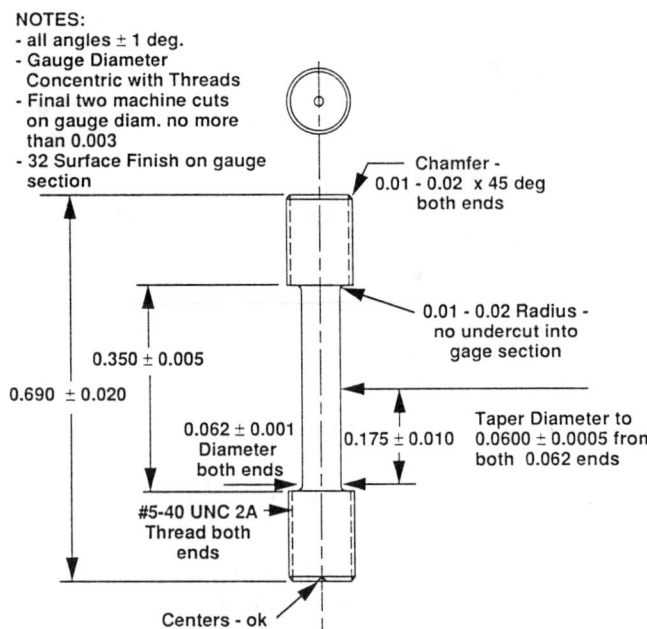

Figure 15-43. Dimensions of Microtensile Specimen

- The strain attainable in a tensile test is limited by necking (the Considere condition). Linear extrapolation of tensile data to larger strains can severely overestimate the stress-strain curve for many metals (e.g., 6061-T6 aluminum, 4340 steel).

- Materials may have significantly different tensile versus compression stress-strain curves (e.g., Ti-3A1-2V). This strength differential must be quantified for accurate modeling.

- Ductile rupture resulting from nucleation, growth, and coalescence of voids has been approximated as an ensemble of dispersed voids growing within a deforming undamaged matrix material. Compression testing, which prevents void growth due to lack of tensile triaxial stress, is useful in modeling this matrix material.

To promote uniform deformation, specimen ends are lightly grooved concentric pattern to entrap lubricant as the specimen is compressed between two hard, polished platens as shown in Fig. 15-44. The grooves also serve as fiduciary marks, testifying whether the specimen has expanded uniformly or has barrelled. Barreling results in erroneously high stress-strain curves calculated from load and overall specimen height reduction; a study of the effects or specimen aspect ratio, and lubrication is documented in Ref. 3. The specimens are compressed in a test system which permits monotonic or strain-rate jump test-

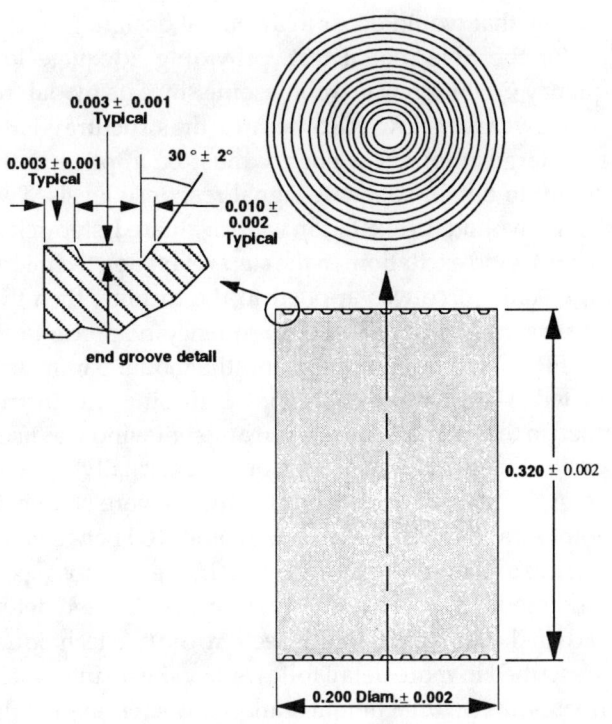

Figure 15-44. Dimensions of Uniaxial Compression Specimen (Reface after Grooving; Dimensions in inches)

ing at constant true strain rates; the control software also permits corrections for load frame compliance, and supports compressive creep and stress relaxation testing (Ref. 4).

Figure 15-45 two unacceptable extremes of lubrication in the top row. The left is an unlubricated specimen with obvious barreling; the right specimen has been excessively grooved and has anti-barrelled (note hourglass profile) as the lubricant extrudes excessive specimen end material outwards. The bottom row consists of three specimens-undeformed and at two states of uniform compression.

References:
1. ———, *Annual Book of ASTM Standards. 03.01*, American Society for Testing and Materials, Philadelphia, PA.
2. ———, *Metals Handbook Vol. 8. Mechanical Testing*, American Society for Metals, Metals Park, OH, 1985
3. W.A. Kawahara, "Effects of Specimen Design in Large-Strain Compression," Experimental Techniques, March/April 1990, pp. 58-60.
4. W.A. Kawahara, S.L. Brandon, and J.S. Korellis, "SNLL Materials Testing Compression Facility," SAND86-8219, Sandia National Laboratories, April 1986.

15.9 MODAL ANALYSIS

Modal Testing of a Very Flexible 110 m Wind Turbine Structure
T.G. Carne, J.P. Lauffer, and A.J. Gomez, Sandia National Laboratories, Albuquerque, NM and H. Benjannet, Shawinigan Consultants, Canada.

Introduction
The structural dynamics of a very large, flexible structures has recently received increased attention. This is due in part to an interest in large space structures. The modal testing of these structures presents some challenging problems. The subject of this example is the modal test of the 110 meter tall Eole wind turbine which, in spite of its very substantial structure, had a number of modes below one Hz.

The Eole wind turbine is located approximate 800 km northeast of Montreal on the St. Lawrence seaway, in an area of very high mean wind speed. The turbine is rated at four megawatts. It is a vertical axis turbine, as compared to the more traditional propeller machines. Vertical axis turbines can accept wind from any direction without yawing into the wind. They have all their power generation equipment at ground level rather than atop a tower, and they can use guy cables for support. Figure 15-46 is a photograph of the Eole turbine just after construction was finished in December, 1986. One can still see the truss structure used as the construction crane.

The objective of the Eole modal test was the verification of the analytical model. In the design of large flexible structures, which are subjected to dynamic loads, knowledge of the modal frequencies and mode shapes is essential in predicting structural response and fatigue life. For large rotating wind turbines, these modal parameters are particularly important since the applied loads acting on the turbine have large periodic components at integral mul-

Figure 15-45. Examples of Unacceptable (top row) and Acceptable (bottom row) Compressions

Figure 15-46. The Eole Vertical Axis Wind Turbine

tiples of the fixed rotation speed. During the design process, analytical or finite element models must be relied upon for estimates of the modal parameters. However, when the turbine hardware becomes available for testing, then the actual modal parameters can be measured and compared to the analytical predictions. The model is used for design, for redesign if required, for choosing the fixed operating speeds, for setting limits on the operational wind speeds, and for computation of predicted fatigue life. Consequently, it is of the utmost importance to have a test verified model.

Vertical axis wind turbines are excited by the applied aerodynamic loads which have primary spectral content at integral multiples of the rotation speed. If these discrete frequencies are close to any modal frequency, then a resonance can result in which very high strains reduce the fatigue life of the structure. Figure 15-47 shows typical mode shapes of a vertical axis turbine. Displayed in this figure are the top, side, and front views of the deformed shape. They are not necessarily in increasing order of frequency that would occur in an actual design.

For the Eole modal test, providing adequate low frequency excitation was troublesome since its modal frequencies were very low. Also, because the structure is large in size, large amounts of energy must be applied to the structure to obtain adequate signal response levels. Two different methods of excitation were evaluated, step-relaxation and wind excitation. In the step-relaxation technique a large static force was applied to the turbine, then this force was suddenly released. Frequency response functions (FRF's) can be computed and the modal parameters extracted. With wind excitation, the turbine was instrumented in the normal manner, but ambient wind was used to excite the structure instead of an externally applied force. A number of response transducers were chosen as references, and cross spectra were computed between the references and the other transducers. Using the cross spectra, the modal frequencies and mode shapes were determined. In the remainder of this section, the two techniques are described in more detail followed by the results which compares the two techniques and compares the results with analysis.

Excitation Using Step-Relaxation

The step-relaxation method of structural excitation involves applying a static force to the structure and then suddenly releasing this force. This process is analogous to plucking a violin string. Step-relaxation has been discussed in References 1 and 2. It is a technique that is seldom used because it can be mechanically difficult to implement; also there are problems involved in performing fast Fourier transforms (FFT) of the step function force signal. However, the low frequency weighting of the step function, which rolls off as 1/frequency, and the large amount of strain energy that can be input to the structure make it ideal for testing large structures with very low modal frequencies.

In using the step-relaxation on the Eole turbine, two force application points were used, one on the tower and one on a blade. This resulted in two sets of FRF's referenced to these two driving points. The forces were applied to the turbine rotor with a high strength steel cable and loaded with a diesel powered winch, which was located on the ground approximately 100 m from the base of the turbine. For the tower driving point, a force of 135 kN (30,000 lbs.) was applied at mid-span on the tower in a direction slightly out of the plane of the rotor. For the blade driving point, a 45 kN (10,000 lbs.) force was applied at a point between the lower horizontal strut and mid-span on the blade. (See Fig. 15-46). The required force magnitude, direction, and location of application were calculated as part of the pre-

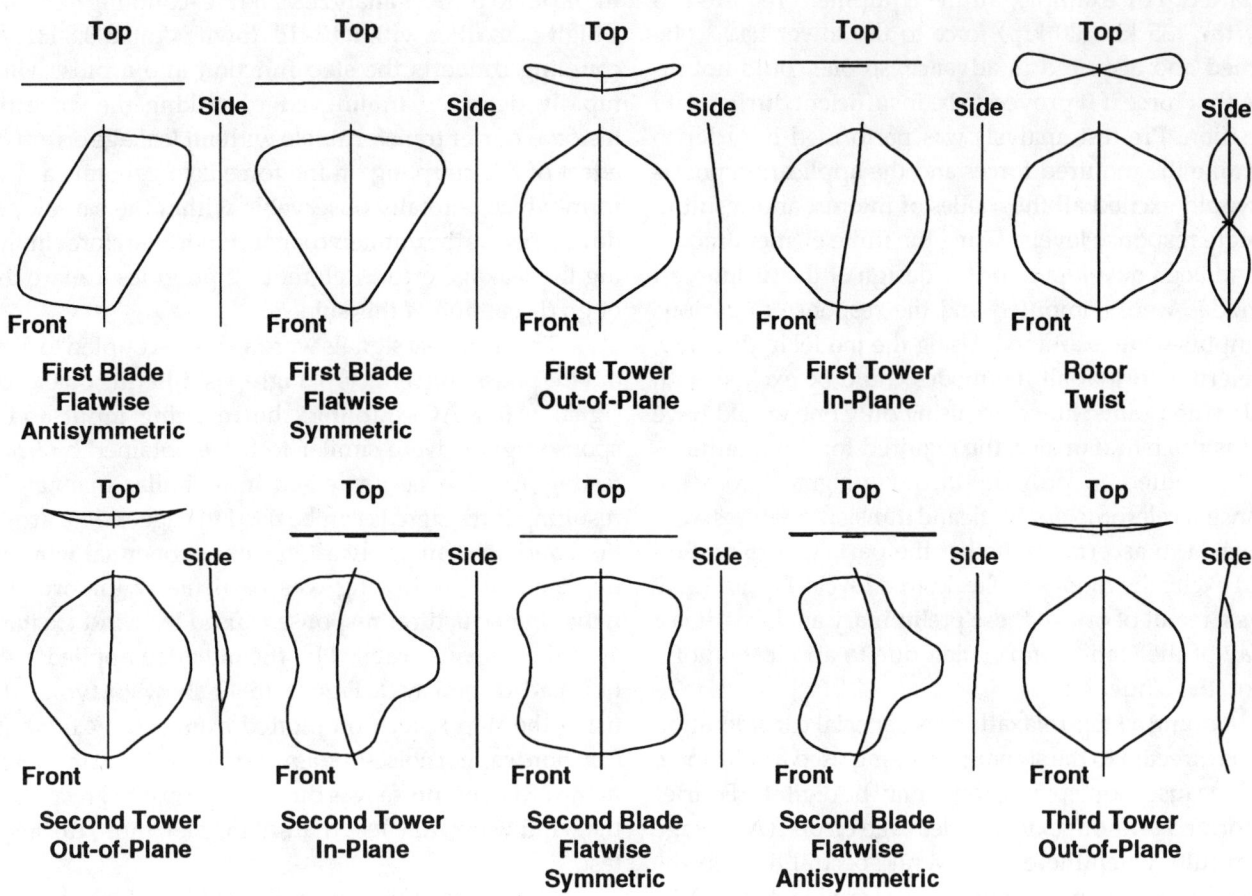

Figure 15-47. Typical Vertical Axis Wind Turbine Modes (Front, Top, Side Views)

test planning and will be discussed later in this example.

A quick-release device was used between the winch and the cable to allow an immediate relaxation of the force. For the forces required by this structure, the quick-release device utilized an explosively driven cable cutter which cut a small piece of replaceable steel cable. A load cell was placed in-line with the cable to measure the force signal. It is fairly important that the load cell be close to the structure so it senses the force actually applied to the structure. A total of 45 accelerometers measured the response on both the tower and two blades. The entire setup is pictorially displayed in Fig. 15-48. Not shown in Fig. 15-48, however, is a very important element in the design of step-relaxation hardware. Depending on the forces used and the length of the pull-down cable, a tremendous amount of strain energy can be stored in the cable and then suddenly converted into kinetic energy in the cable. To prevent the cable rebound from striking and damaging the turbine, a nylon restrain strap was designed which absorbs the strain energy stored in the steel pull-down cable. A maximum deformation, the nylon strap experience approximately ten percent strain.

Another of the difficulties in testing very large structures is that the test plan must be well established before

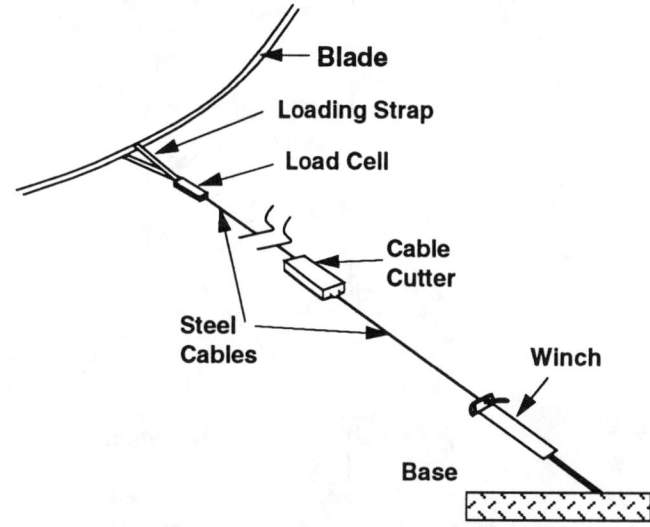

Figure 15-48. Step-Relaxation Hardware

the test since changes may be impossible once the testing has started. For example, all the equipment required to apply the 135 kN (30 kip) force to the tower had to be designed and acquired in advance; so one could not increase that force if it proved to be insufficient during data acquisition. Pre-test analysis was performed in order to determine the required forces and the application points that would excited all the modes of interest and result in adequate response levels. Using the finite element model that had been developed for the design of the turbine, all the modes were computed and the response to various load inputs were examined. Using the model in this way, we determined that all the modes could be excited with two driving points; however, using only one would have been insufficient. Further, the required force magnitudes were computed to provide adequate signals from the response accelerometers. Static and transient analyses were performed to ascertain whether the particular excitation would excite the modes to the desired levels. Figure 15-49 shows a result of one of these preliminary analyses; it is a display of the static deformation due to application of a load on the blade.

In doing a step-relaxation test, special consideration must be directed to the signal processing used for the force signal because the step function cannot be digitally Fourier transformed without extensive leakage errors. (A leakage error results when there are several spectral lines in the discrete Fourier transform over the sampling period.) To alleviate this difficulty, the force signal was AC-coupled at the input to the FFT analyzer. The AC-coupling network is a high-pass filter with its 3-dB down point at 0.8 Hz. AC-coupling converts the step function into a pulse with a rapidly decaying trailing edge, making the force time history Fourier transformable without leakage errors. The effect of AC-coupling on the force is to generate a waveform which is totally observable within the sample window, whereas the unfiltered signal is not, therefore eliminating the leakage error. Reference 2 provides a more thorough discussion of this subject.

The response signals were also AC-coupled to cancel out the phase shift effects of high-pass filtering on the force signal. After AC-coupling, the resulting input and response signals were similar to those obtained by impact testing and can be processed in a similar manner. The resulting force signals can be used to trigger data acquisition, and one can apply a force or exponential windows. Exponential windowing was particularly important for removing structural response caused by wind excitation after the response caused by the intended applied excitation had diminished. Figure 15-49 shows a typical FRF using the step-relaxation plotted from 0.4 to 4.0 Hz. The function was not noise-free as was manifested at the notches in the FRF. This noise was due to wind excited response as the wind was rarely less than 10 m/s (30 mph) during the test.

Using one of the sets of the 45 FRF's for a particular driving point, the mode shapes and frequencies were extracted from these data using standard techniques. Figures 15-50 and 15-51 show two of the mode shapes from this step-relaxation data. Referring to fig. 15-47 for reference, they are the first blade flatwise anti-symmetric (front view) and the second rotor out-of-plane (side view). More

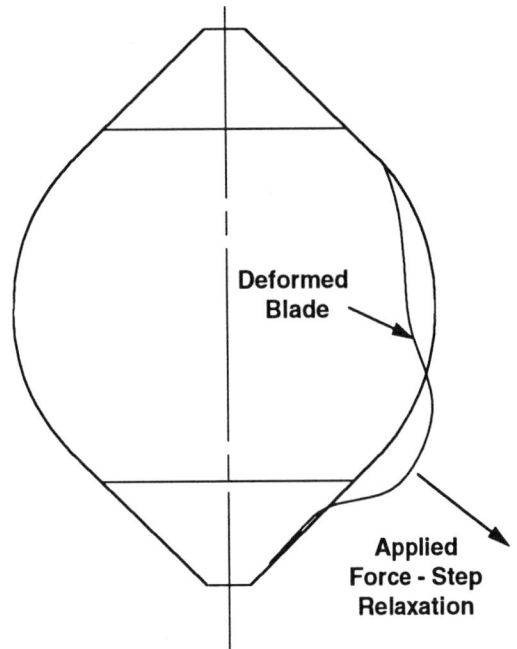

Figure 15-49. Deformation Due to Static Force Applied to Blade

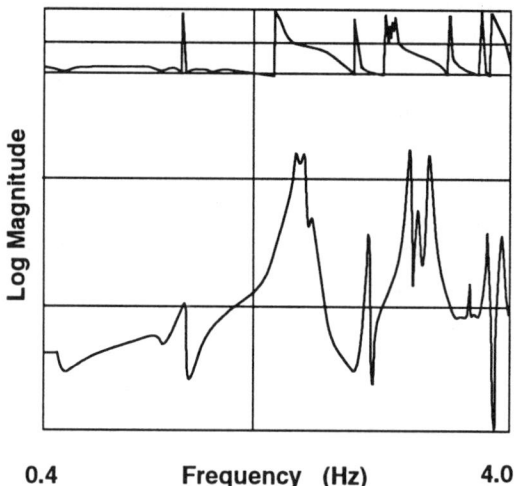

Figure 15-50. Frequency Response Function Acquired from Step-Relaxation

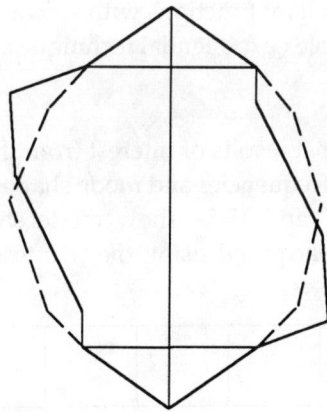

Figure 15-51. First Blade Anti-Symmetric Flatwise Mode Shape from Step-Relaxation

discussion of the test results along with a comparison with the analytical predictions is given in the results.

Wind Excitation Testing

During previous wind turbine tests, high winds have induced vibratory response levels larger than those resulting from the step-relaxation excitation, resulting in poor estimates of the FRF's. Waiting for the winds to cease, however, was not a reasonable alternative, since test scheduling on a prototype was extremely tight. Consequently, an alternative method of testing was devised to complement step-relaxation excitation testing. It was decided to measure the wind induced vibration of the Eole to determine its modal parameters.

References 3, 4, and 5 have indicated that for broadband excitation, response data alone could be used to determine modal parameters. No measurement of the force is required. Reference 3 indicates that it is possible to extract modal parameters from transmittance functions which are defined as the complex ratio between Fourier Transforms of response points. This was the approach followed in this test. The method used to calculate the transmittance functions was to take the ratio of the cross-spectrum to the auto-spectrum.

The procedure for performing wind excitation is similar to that used in performing artificial excitation testing. One significant difference is that the forces acting on the structure are not measured. Reference degrees-of-freedom (DOF's) are selected based upon their degree of participation in each of the mode shapes. The complete set of references should strongly participate in all of the modes of the structure within the frequency band of interest. For each set of response measurements, auto-spectra of the reference DOF's are evaluated, and cross-spectra are measured between the response DOF's and the reference DOF's.

The reference auto-spectra provided appropriate scaling of the mode shape to account for different levels of wind excitation for different sets of response measurements. Because 16 data acquisition channels were available for acquiring data and 42 response locations were selected, three separate measurement sets were required to acquire response at all of the locations of interest.

For this test, time histories of the vibrational response of the turbine were digitized and recorded on disk. For each measurement set, the vibrational response of the three reference DOF's were also recorded. The time histories were then processed to generate power spectra as described above. This process was performed for each of the three measurement sets. Shown in Figs. 15-52 and 15-53 are typical auto- and cross-spectra.

Figure 15-52. Second Rotor Out-of-Plane from Step-Relaxation

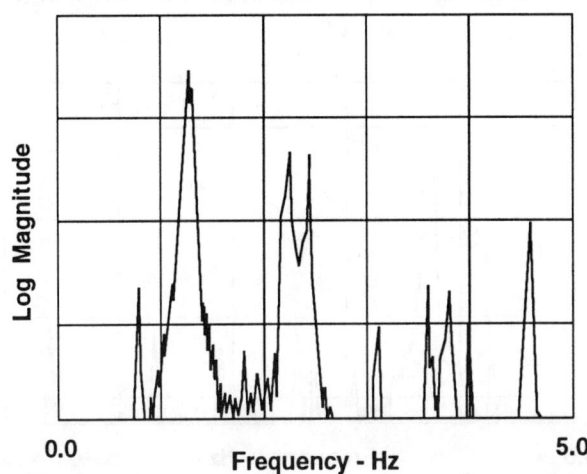

Figure 15-53. Acceleration Auto-Spectrum from Wind Excitation

Modal frequencies were determined from the peaks in the auto-spectra of the reference DOF's and the peaks of the indicator function. The indicator function was created by summing the magnitude squared of the power spectra. Particular modes, within the indicator function bandwidth, were enhanced by selecting response DOF's based upon a knowledge of the mode shape. Shown in Fig. 15-54 is a typical mode indicator function calculated to enhance the flatwise blade modes.

In the vicinity of a resonance, where the response is dominated by a single mode, the transmittance function is flat and its value can be taken as the mode vector entry for that mode at that DOF. If the transmittance function is directly calculated using block floating point arithmetic, dynamic range problems can exist because of zero and near zero values in the denominator. To avoid this problem, the spectra was zeroed at all frequencies except those corresponding to a narrow band about the resonances. The zero values in the reference spectra (the denominators) were replaced by a small number to prevent a division by zero. Transmittance functions were then calculated by taking the ratio of the cross-spectra to the reference auto-spectra for each data set. A typical transmittance is shown in Fig. 15-55.

Mode shapes can be calculated from the transmittance data base by moving the value of the transmittance, at resonance, into the corresponding location in the mode vector. Two typical mode shapes are shown in Figs. 15-56 and 15-57. These are the same two mode shapes shown in the step-relaxation part. Because damping was not a principal consideration in this test, no attempt was made to estimate it from the response data. However, damping values could have been extracted from the auto-and cross-spectra using the definition of sharpness of resonance. A second method of damping estimation would be to use the random decrement approach, outlined in Ref. 4, and to curve-fit the resultant functions with any one of a variety of damped complex exponential techniques.

Results

The principal results of interest from the modal test were the modal frequencies and mode shapes. Figures 15-51, 15-52, 15-57, and 15-58 show mode shapes of two different modes acquired using the two test techniques.

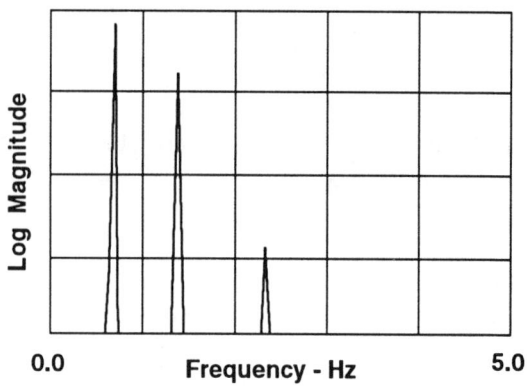

Figure 15-55. Summation Mode Indicator Function

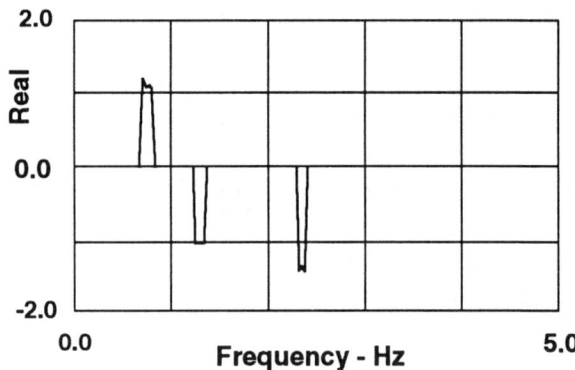

Figure 15-56. Processed Transmittance Function

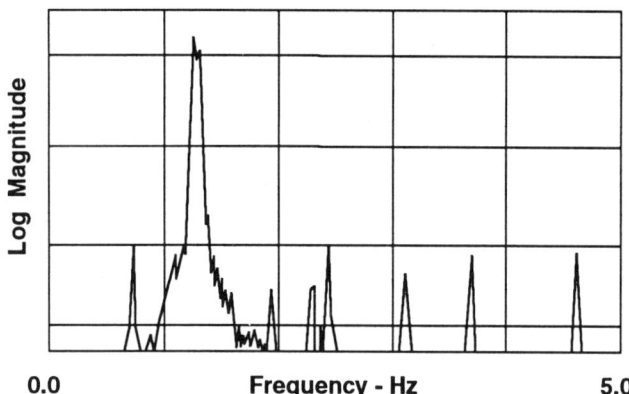

Figure 15-54. Acceleration Cross-Spectrum from Wind Excitation

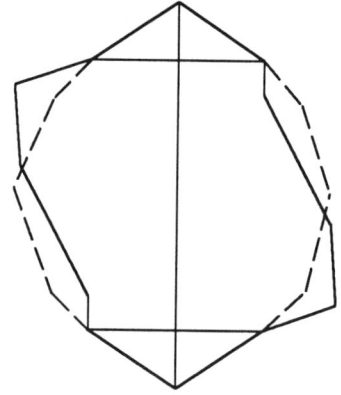

Figure 15-57. First Blade Anti-Symmetric Flatwise from Wind Excitation

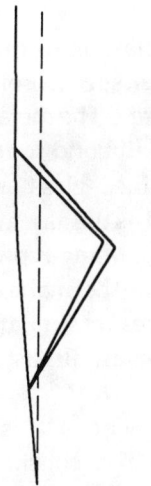

Figure 15-58. Second Rotor Out-of-Plane from Wind Excitation

The shapes are virtually the same. An examination of the other mode shapes show similar comparisons.

Table 15-6 compares the modal frequencies from the two test techniques for the first fourteen modes. Agreement between the two sets of frequencies is excellent, with all the differences less that the resolution of the transmittance function (0.016-Hz), except for mode 6 where the difference is only 1.5 percent. The two excitation techniques have produced virtually identical results.

As indicated earlier, the objective of this modal test was to verify the finite element model. Although a discussion of the accuracy of the model is not included in this example, a comparison between the predicted frequencies and those from the test shows very fine agreement and establishes the accuracy of the model. The average deviation between the test and analysis frequencies is less than two percent for modes 2 through 12. This is extremely close agreement. The first mode (propeller) was deleted from the comparison because the turbine brakes had to be engaged during testing, and they were not adequately represented in the model.

Conclusions

Both wind and step-relaxation testing methods worked extremely well and yielded virtually the same mode shapes and frequencies. Step-relaxation methods required the following: significant pre-test analysis for sizing of excitation hardware, unusual hardware such as explosive cable cutters and a winch capable of applying 135 kN, expensive fixturing, and significant ground support. Both methods required a crane and personnel to mount accelerometers. Step-relaxation testing required a higher dependence on the site workers. However, damping information is not as readily available from the power spectra obtained from wind excitation as it is from FRF's obtained using step-relaxation testing. Finally, and most importantly, we were able to extract modal data from FRF's measured, using step-relaxation, during high winds. However, it would not be possible to obtain meaningful wind response measurements during completely calm days.

References

1. T G. Carne, et. al., "Finite Element Analysis and Modal Testing of a Rotating Wind Turbine, AIAA paper 82-0697, 23rd Structures, Struc-

Table 15-6. Modal Frequencies from Step-Relaxation and Wind Excitation Tests

Mode Shape Description	Step-Relaxation Hz	Wind Excitation Hz
1. Propeller	0.421	0.420
2. 1st Rotor-Out-of-Plane	0.628	0.625
3. 1st Rotor-In-Plane	0.738	0.734
4. 2nd Rotor-Out-of-Plane	0.930	0.937
5. 1st Blade-Flatwise Anti-Symmetric	1.304	1.296
6. 1st Blade-Flatwise Symmetric	1.321	1.342
7. 2nd Rotor-In-Plane	1.383	1.391
8. Blade Edgewise-Anti-Symmetric	1.546	1.547
9. Rotor Twist (Dumbbell)	1.928	1.938
10. 2nd Blade Flatwise-Symmetric	2.241	2.250
11. Blade Edgewise Bending	2.329	2.328
12. 2nd Blade	2.396	2.391
13. 3rd Rotor-In-Plane	3.084	3.101
14. 3rd Blade Flatwise-Symmetric	3.564	3.563

tural Dynamics and Material Conference, May 1982, pp. 334-347.
2. J.P. Lauffer, T.G. Carne, and A.R. Nord, "Mini-Modal Testing of Wind Turbines Using Novel Excitation," Proc. 3rd International Modal Analysis Conf., Jan. 1985, pp. 451-458.
3. R.J. Allemang, "Investigation of Some Multiple Input/Output Frequency Response Function Experimental Modal Analysis Techniques," Ph.D. Dissertation, University of Cincinnati, Dept. of Mechanical Engineering, 1980.
4. S.R. Ibrahim and G.L. Goglia, "Modal Identification of Structures from Responses and Random Decrement Signatures," NASA-CR-155321, 1977.
5. J.S. Bendat and A.G. Piersol, *Engineering Applications of Correlation and Spectral Analysis*, John Wiley and Sons, New York, pp. 183-186.

15.10 NONDESTRUCTIVE TESTING

15.10.1 Nondestructive Testing of Concrete
Dr. Albert Lin, NIST, Gaithersburg, MD

The use of nondestructive test (NDT) methods, such as radiographic and ultrasonic, to assess the condition of metals is widely accepted throughout industry. These methods are used to inspect steel pipe sections, structural steel welds, and aircraft composite assemblies. The use of NDT in concrete, however, is not as widely accepted. Yet, the argument can be made that because most concrete is site-mixed and cast-in-place, the degree of consistency and control is far less than control processes used in the formulation of metals, and the need for adequate testing is much greater.

Reinforced concrete poses additional challenges. Reinforced concrete is a composite, made up of concrete (rock, sand, cement paste, and air), and reinforcing steel. NDT can be used to estimate the location and amount of steel in place, data that are necessary for repair or retrofit of damaged or inadequately constructed buildings. Other qualities that can be assessed include concrete strength, voids, and cracks.

The NDT of concrete has received considerable attention in recent years. The NDT of concrete can be used for two general purposes: (1) estimation of strength by surface hardness, penetration, pullout, breakoff, or maturity techniques, and (2) the estimation of in-place properties, such as moisture content, density, thickness, and location of cracks, voids and delamination.

The detailed discussion of each test method is beyond the scope of this handbook. Readers interested in a concise summary of concrete NDT methods are referred to Ref. 15-5.

15.10.2 Glass-to-Metal Seals
Alan G. Beattie and Robert T. Reese, Sandia National Laboratories, Albuquerque, NM

Glass-to-metal seals are widely used in various electrical devices—batteries, connectors, insulators, and lighting devices. The glass is formulated so that its thermal expansion characteristics closely approximate the expansion of the metal used. The metal must be prepared so that the molten glass will bond to it. The thermal contraction behavior of the joined glass and metal is critical to the integrity of the seal as the assembled battery header or pin connector is cooled. Many types of glasses used for these seals have a bilinear thermal expansion curve. While the slopes of these curves are similar, they are enough different that residual stresses are almost always present in glass-to-metal seals.

The development of glass-to-metal seals relies on materials sciences and considerable skill in producing strong bonds which have a minimum of residual stress. Testing of glass-to-metal seals is a particularly difficult challenge because many of the seals are small, they are at high temperatures when they are formed, they are significantly affected by impurities such as oxides which can form on the bonding surfaces, and they are susceptible to tolerances because glass is brittle and has a low strain to failure value.

One approach to evaluating glass-to-metal seals is to perform a lot sample tests on seals. Using failure tests on random samples from a lot of similarly produced glass-to-metal seals will give evidence to accept or reject the lot. In the development stages for glass-to-metal seals, various formulations of glass are tried as the test variable. Test results indicate the relative strengths of the different glass formulations.

The test approach described here used gas pressure to load one side of a seal as a means of applying a known condition to the glass. The pressure was increased until failure was initiated. The pressure at failure indicated the relative strengths of the glass. Acoustic emission was used as the principal diagnostic technique. A failure can be heard using acoustic emission sensors long before a crack could be detected by a decrease in pressure. Other types of point techniques or full field methods do not have the advantages of acoustic emission.

The test sample was the same size and design as the prototype seal for the actual battery header. A battery header seals the end of a battery by joining the cylindrical case to the battery terminal. The problem with testing battery headers was to determine when they fail. Glass exhibits behavior in which cracks can occur at low pressure but will not allow the gas to leak through until the pressure has been raised considerably. However, the electrical behavior and functionality is affected by these early cracks. Detecting these early cracks is important in the evaluation of glass-to-metal seals.

A test setup using acoustic emission is shown in Fig. 15-59. The test setup consisted of a pressure fixture which had an opening on the bottom so a test seal can be inserted. The seal was captured by two O-ring seals and a retainer ring. One side of the seal was pressurized at a rate of 15 psi/sec until failure occurred. Failure is indicated by a sudden increase in the acoustic emission counts sensed by the microphone. An acoustic record for one test sample is given in Fig. 15-60. The ordinate in Fig. 15-60 is the sum of the acoustic emission events (counts) to that point in time and pressure. The somewhat cyclic appearance of the record is actually scale changes to allow the data to fit on one page. The actual number of counts is much greater than what is shown in Fig. 15-60. Comparative data for development is easily obtained in this manner and quality control and assurance practices can also use this method. A laboratory setup as shown resulted in the evaluation of seals at a rate of about one seal every five minutes.

15.11 SERVICE LIFE EXTENSION

Load Testing of the Huey P. Long Bridge
Gilbert T. Blake, Wiss, Janney, Elstner Associates, Northbrook, IL

The Huey P. Long Bridge over the Mississippi River near New Orleans, shown in Fig. 15-61 is 22,996 ft. long structure designed to carry both railroad and vehicular traffic. Completed in 1936, the bridge was designed by Modjeski and Masters of Mechanicsburg, Pennsylvania to carry steam engines and the typical automobile and truck traffic common in that era. Trains run inside a 750 ft. main span cantilever truss, and two-lane roadways cantilever out from each of the truss, as shown in Fig. 15-62. The 18 ft. wide roadways are functionally obsolete and due for replacement.

During a routine inspection, cracks were discovered in the webs of the outside roadway stringers. In order to allow time for retrofit planning, the cracks were arrested by drilling out the ends of the cracks and load limits were imposed on bridge traffic.

Controlled load tests were conducted on the roadway spans to determine the effectiveness of the crack-arrest techniques. The tests consisted of installing strain instrumentation at the crack sites and measuring strain levels while imposing controlled traffic loads.

Fourteen biaxial strain gages (Measurements Group type EA-06-125TA-120) were bonded to the damaged and undamaged stringers. The gages were attached with M-Bond 200 adhesive. Each biaxial strain gage was nominally 0.34 in. long by 0.213 in. wide and contained one horizontal

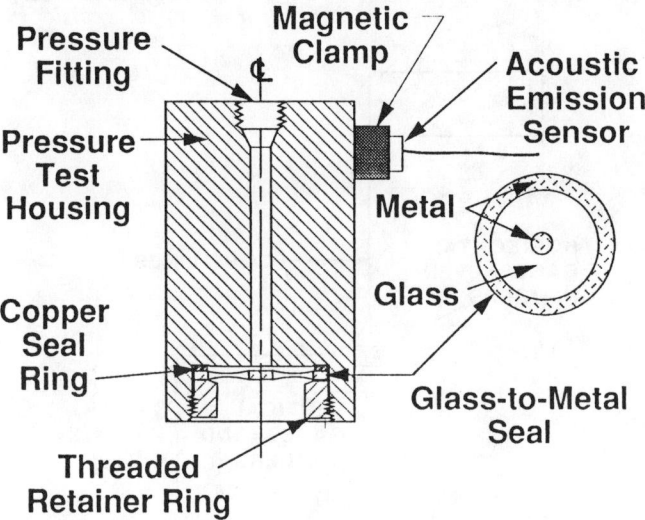

Figure 15-59. Glass-to-Metal Seal Evaluation Test Setup with Acoustic Emission

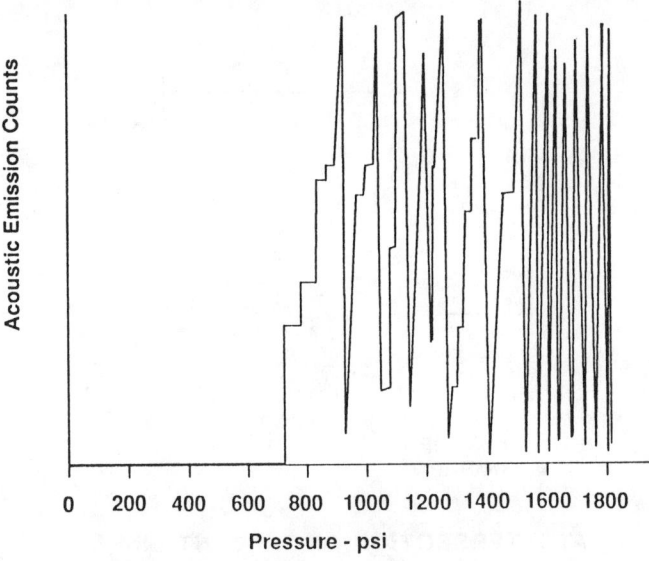

Figure 15-60. Test Record of Acoustic Emissions vs. Pressure for Glass-to-Metal Seal

Figure 15-61. Huey P. Long Bridge

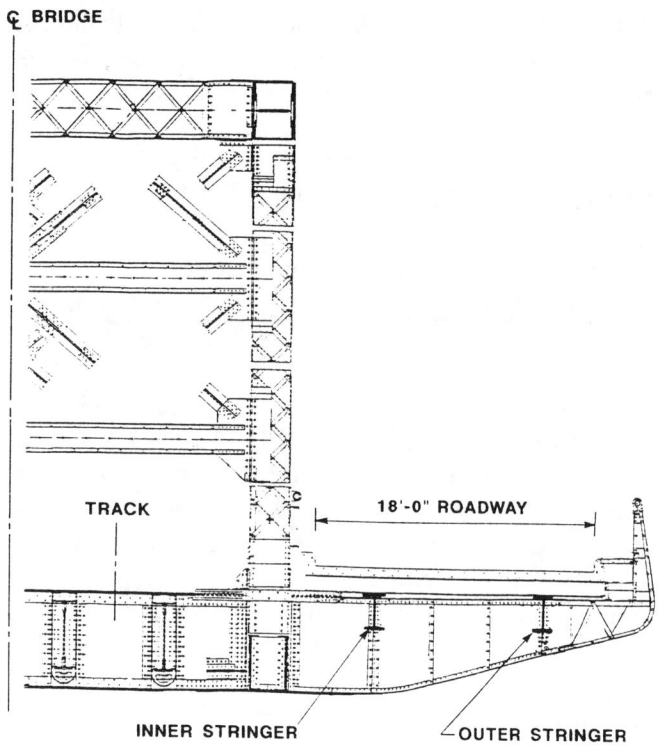

Figure 15-62. Typical Bridge Cross-Section

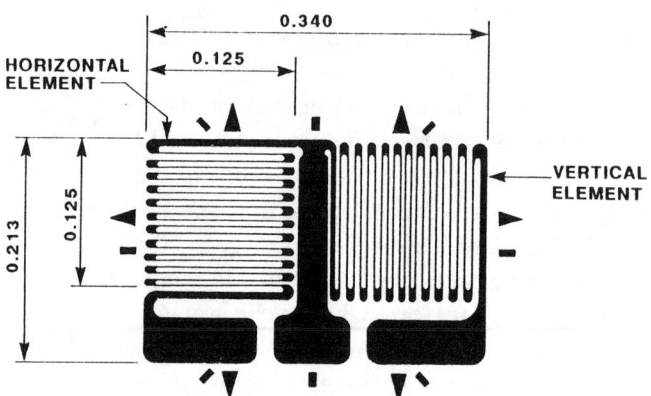

Fig. 15-63. Micro-Measurements Strain Gage Geometry

and one vertical strain gage, each with a gage length of 0.125 in. as shown in Fig. 15-63. The gages were mounted so that the center of the biaxial gage was centered at the intersection of either the projections of the horizontal and vertical cope lines or the horizontal and vertical tangents to the crack removal circle as shown in Fig. 15-64. In order to detect any bending strains, the gages were bonded on

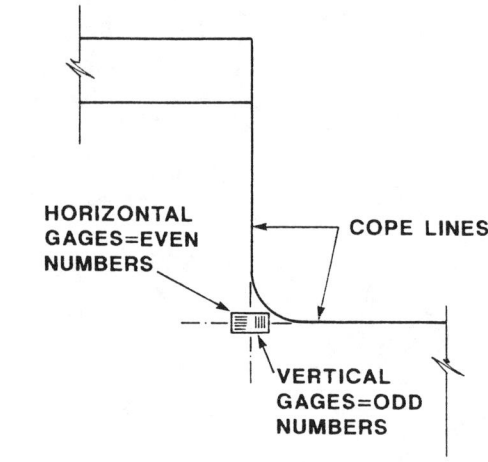

A.) AT INTERSECTION OF COPE LINES

Table 15-7. Strain Gage Locations on Huey P. Long Bridge

Panel Point	Location	Gage Numbers
1-9 Westbound East Side	Both Sides of Both Stringers	1, 2, 3, 4, 5, 6 7, 8
1-13 Westbound West Side	Both Sides of Both Stringers	9, 10, 11, 12, 13, 14, 15, 16
1-17 Westbound West Side	Both Sides of Inboard Stringer	17, 18, 19, 20
1-16 Westbound West Side	Both Sides of Outboard Stringer	25, 26, 27, 28
3-5 Westbound West Side	Both Sides of Inboard Stringer	21, 22, 23, 24

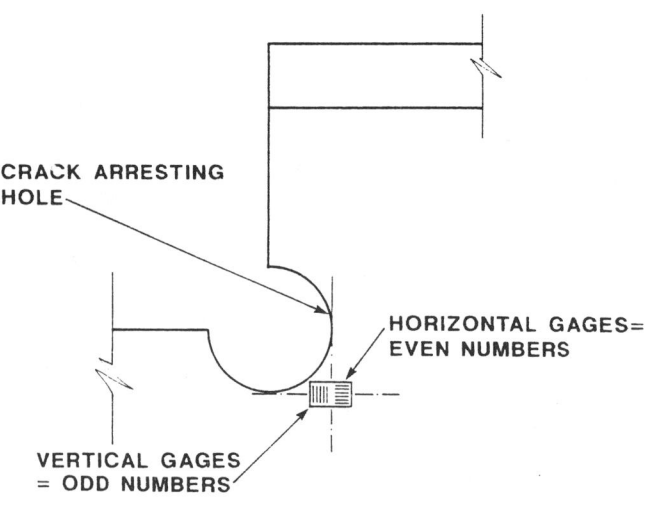

B.) AT INTERSECTION OF TANGENT LINES

Figure 15-64. Strain Gage Locations—(a) At Intersection of Cope Lines—(b) At Intersection of Tangent Lines

opposite sides of each stringer to form a mirror image arrangement. For instance, vertical gages were mounted back to back. The stringer locations and identification of all strain gages are shown in Table 15-7 and shown in Figures 15-65 through 68.

The data from the strain gages were obtained using dynamic recording instrumentation as shown in Fig. 15-69. An eight channel Hawkeye Model 1800 Signal Conditioner and Amplifier, a fourteen channel Kyowa Model RTP-600B Data Tape Recorder and eight-channel Western Graphtek Model WR3101 thermal strip chart recorder were used to record the data. All of the gages at each panel point were recorded simultaneously.

The strains due to the traffic loadings recorded at each panel point resulted from normal bridge traffic, a control vehicle, a variety of heavy vehicles, and train traffic. The control vehicle was a Louisiana Department of Transportation dump truck with front and rear axles loads of 4,950 and 17,950 lbs. respectively. The Huey P. Long bridge police provided traffic control which enabled the recording of single vehicles at different lane locations. Train traffic was monitored and measurements of train-induced strains were recorded as the movement of trains occurred.

All of the data were replayed in real time in the laboratory. The pertinent events and resulting strain traces were condensed into shorter records. Figures 15-65 through 68 show typical traces for a variety of vehicles recorded at each panel point. Each figure shows scale and time factors for every event.

Based on the analysis of these test results, interim repairs to the bridge were design and constructed pending the reinforcement and widening of the highway portions of the structure now scheduled for 1993.

15.12 SMART STRUCTURES
A Calibration Approach for Smart Structures Using Embedded Sensors (Ref. 15-7).
S.H. Smith, A.A. Boiarski, and D.G. Rider, Battelle Columbus Laboratory

Introduction

A "smart structure" is one which is sufficiently well instrumented so that the data can be synthesized to form an

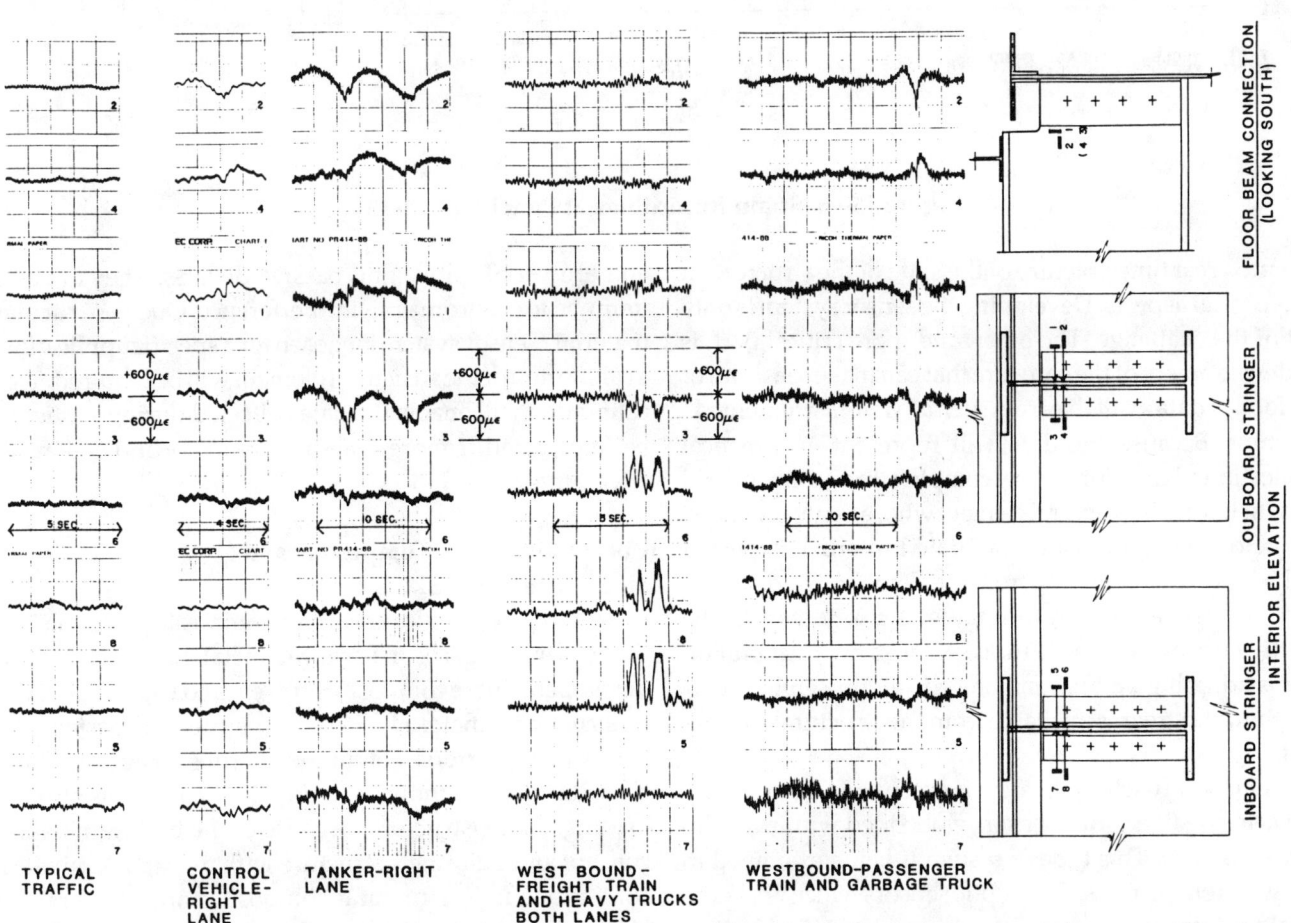

Figure 15-65. Strain Recordings at Panel Point 1-9

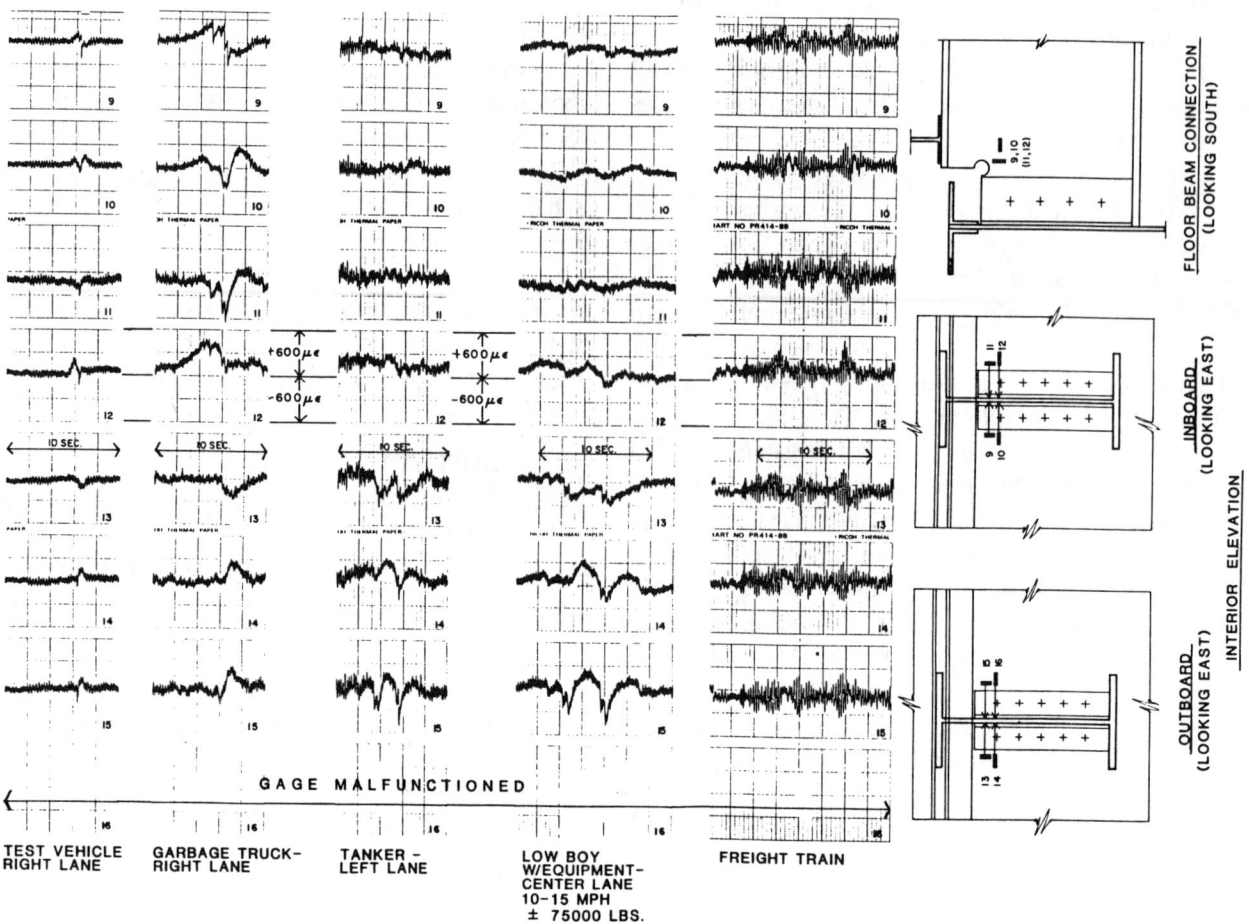

Figure 15-66. Strain Recordings at Panel Point 1-13

accurate "real time" picture of the state of the structure in all its critical aspects. Developing the sensor system is only part of the challenge. The other, and more critical part, is the development of the software that can make sense out of the flood of data available from such a "well-instrumented" structure. Because the data will represent, in practical applications, behavior in very complex three-dimensional stress fields, each structural element whose "state of health" will be of concern must be calibrated in order to permit analysis of the data coming from the element. The development of this calibration technique is a key step in the effective use of smart structures. As used here, calibration means adapting an analysis procedure to usefully interpret the data streams available from the structural component.

One approach to the development of a smart structure is to use fiber optic sensors embedded within a composite material. This type of system is being examined in many current projects. Fiber optic sensors which are being developed for many applications, can be embedded directly in the structure during the manufacturing process, and are very lightweight, passive devices. They are also immune to electromagnetic interference. Once a particular material/sensor system is chosen for a specific application, a calibration of the structural element will be required. This calibration will enable the data collected during service to be interpreted if more than just trends or indications of effects are desired.

Continuous Fiber Composites and Smart Structures/Skins

Recently, continuous fiber composite materials, either polymer matrix or advanced metal matrix, have been developed. These composites are referred to as engineered or structured materials. Based on structural design and basic loading criteria, continuous fibrous composite materials can be directionally tailored to carry the applied loadings in the structure. Because of the basic lamination concept and building a structure with multiple laminates, continuous fiber structural composites are a natural for using embedded sensors to monitor and measure the structural response. The combinations of these structural

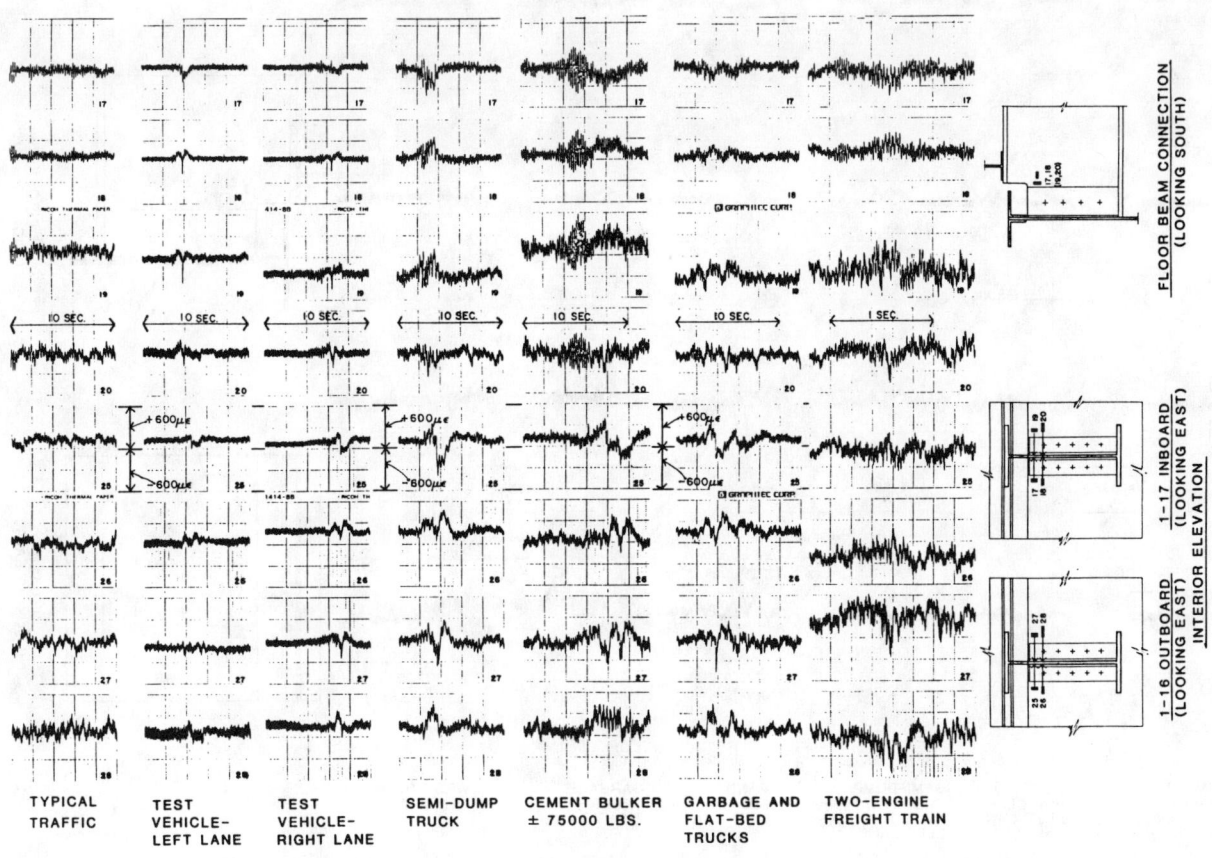

Figure 15-67. Strain Recordings at Panel Points 1-16 and 1-17

composites with embedded sensors is termed "smart structures," "smart skins," or "smart composites" (Refs. 1 through 7) and can be thought of as having an internal "nervous" system potentially capable of evaluating stress, strain, impact, fatigue, damping, and temperatures as well as other conditions such as fractures or delaminations. In addition, the purposes of smart structures and skins are to:

- Monitor the manufacturing process
- Measure service loadings and environmental parameters
- Assess the integrity of the structure by monitoring degradation with time

The Monitoring Hierarchy and Smart Structures

In order to determine the degree of sophistication of what is required in a smart structure in service, the monitoring hierarchy chart shown in Fig. 15-70 will be used. In the monitoring hierarchy, status and condition monitoring are the lowest level and would be functional testing of smart structures/skins with sensors. This is very similar to accept/reject criteria evaluation placed on mechanical and electrical equipment. Performance monitoring as applied to smart structures/skins can be thought of as loads, mechanical strain, temperature and humidity data obtained with sensors in the smart structures/skins. The highest levels of monitoring hierarchy are prognostic and diagnostic monitoring. For smart structures/skins, this would be in-service performance plus life and degradation evaluation. These high levels of monitoring and complex objectives with embedded sensors in smart structures/skins require accurate calibration of the sensors and efficient, rapid data analysis method.

Fiber Optic Sensors/Systems and Measurements

In the selection of an optical fiber and optical system for measuring the response of multilayered composites, it is import to be familiar with the three basic types of fibers and their light transmission characteristics. These basic fiber types are step-index single-mode, graded-index multi-mode, and step-index multi-mode. The light transmission and refractive index profile of each of these types of fibers is shown schematically in Fig. 15-71. The behavior of light

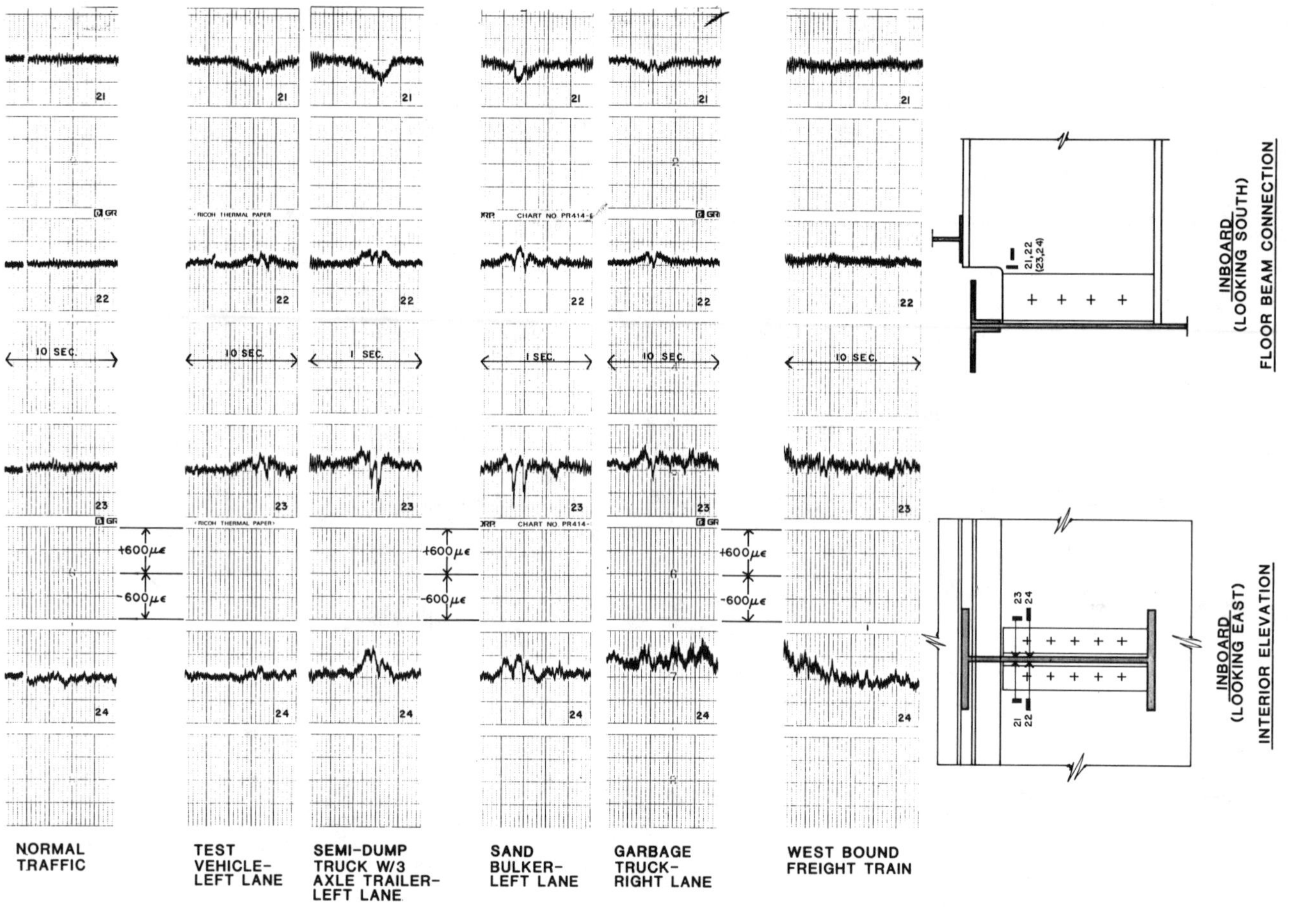

Figure 15-68. Strain Recordings at Panel Point 3-5

Figure 15-69. Dynamic Strain Recording Instrumentation

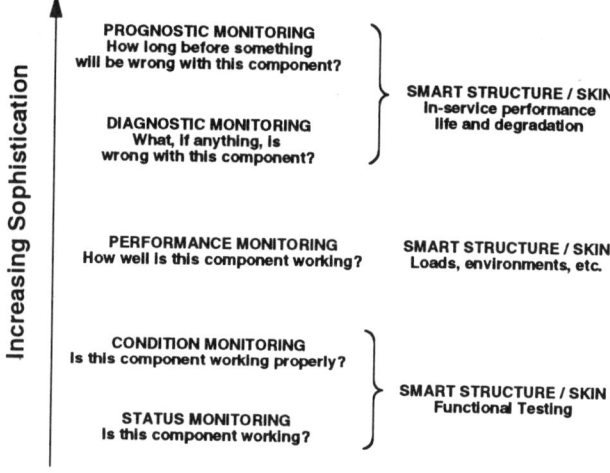

Figure 15-70. The Monitoring Hierarchy and Smart Structures/Skins Applications

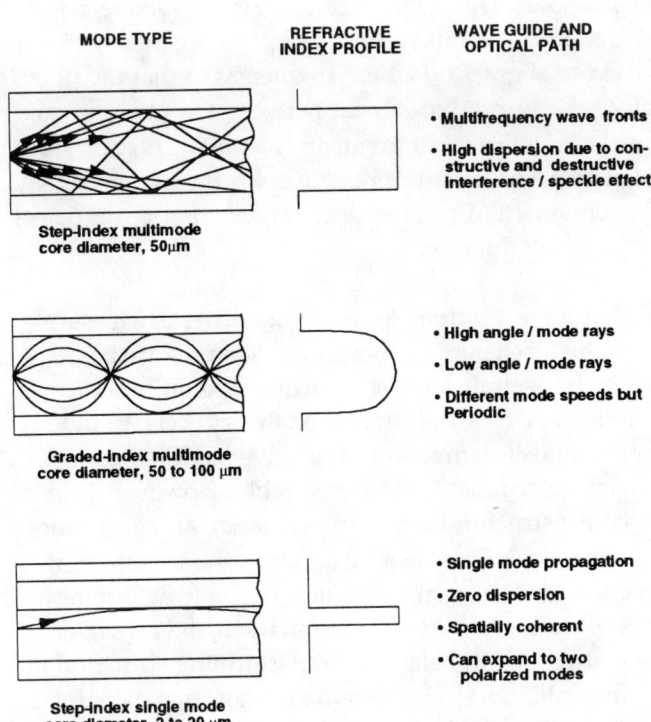

Figure 15-71. Three Basic Types of Optical Fibers and Resulting Waveguide

transmitting through each of these types of fibers is produced by the difference in the cladding and core glasses and their relative refractive indices. The cladding glass has a lower refractive index. Because of the differences between index of refraction of the core and the cladding glass, the light transmitted in the fiber refracts from the interface between the core and cladding producing different wave guide characteristics. The wave guide and optical path salient features are also listed in Fig. 15-71 and are important in selecting the type of fiber for experimental measurements of the mechanical response of the composites.

The next step is to select a fiber sensor system to act as a transducer for mechanical measurements. The three basic types of optical sensors which have potential for use as embedded sensors in composites are the intensity, polarization, and phase interferometric sensors. These are listed in Table 15-8 which also shows the properties measured based on industrial experience and the potential for use in composites as an embedded sensor.

Fiber optic sensing may be either intrinsic or extrinsic. The term "intrinsic" covers those types of fiber optic sensors that employ some change in the wave guide medium as a means of modulating the light in the wave guide. This is done by intensity or amplitude modulation or by frequency modulation. This behavior can be a bending loss, polarization, or phase modulation. The term "extrinsic" sensing applies to those cases where the fiber is used only to convey the light to and from a sensing tip or region and is not the sensor itself.

Three basic types of intrinsic-based fiber optic transducers are shown in Fig. 15-71. These transducers have been used in making mechanical measurements and the mechanisms measured along with the resolution and ranges of certain parameters are also shown in Fig. 15-72. The Y-guide or reflective bifurcated probe and the fiber-to-fiber variable coupling are extrinsic types of fiber sensors. The bending loss fiber optic transducer is an intrinsic type sensor. The use of the fiber optic transducers as embedded sensors in composites will be discussed later when the mechanics of stress, strain, and deformation of composites and the measurement techniques are presented.

Table 15-8. A Summary of Optical Sensors and Potential Use in Composites

Sensor Type	Properties Measured	Typical Uses (Industry)	Potential Use in Composites
Intensity	Mechanical Variables	Electro-optical Transducers	Very High
	Nuclear Radiation	Pressure Recorders	
	Temperature	Flow Meters	
	Chemical and Medical Variables	Spectrometers	
Polarization	Electrical Variables	Current and Voltage Transducers	Low
Phase (Interferometric)	Strain Pressure Temperature Changes	Hydrophones Magnetometers Gyroscopes Current Detectors	High

In order to determine the stress field within a structure, measurements must be made at multiple points. there are two methods of using fiber optic sensors to perform this multi-point measurement; distributed and quasi-distributed fiber optic sensing. With distributed sensors, the fiber is considered an intrinsic sensor and optical radar techniques (optical time domain reflectometry) are used to obtain data for arbitrary segments of the fiber along its entire length. In the quasi-distributed format, data from a set of extrinsic fiber optic sensors is multiplexed using one or more fiber cables.

In Fig. 15-72, the bending loss sensor can be made into a distributed sensor if the fiber distorting devices are installed along the entire fiber length. However, if short section of distorters are placed at intervals along the fiber, a quasi-distributed sensing format is produced.

Interferometric Fiber Optic Systems

One type of fiber optic sensor that can be embedded in the composite will require an interferometric system. There are four common systems used in fiber optics technology (Refs. 8 and 9). These systems are Mach Zehnder homo-dyne or heterodyne, Michelson, and the Sagnac. Figure 15-73 shows the Mach Zehnder fiber optic heterodyne receiver with frequency discriminator. The system consists of equal optical path lengths from beam splitter to beam splitter and the reference fiber (5) is used for calibration purposes. The signal fiber (s) are embedded sensors in the composite. In this system, an optical frequency shifter is used to slightly offset the frequency so that the photo detectors can produce a beat between the reference and signal beams. The beat frequency has all the phase modulation that has been impressed upon the optical signal. Conventional FM receiver techniques are used to detect the phase modulation.

Mechanics of Composites and Quantities to Measure

The mechanics of composite materials are based on elastic stress-strain analysis of orthotropic and anisotropic materials and classical lamination theory (Refs. 10 and 11). For a composite laminate, the effective mechanical properties are determinate and measurable. However, in performing a structural integrity evaluation and in monitoring service performance of a composite structure, it is desirable to measure the lamina behavior to determine the stresses and strain in each lamina and the interlaminar shear stress-strain behavior. In performing structural integrity evaluations, it is necessary to know the peak stresses and their distributions within each lamina of the laminate. These quantities are used in determining design margins and in predicting service life performance.

The fabrication of composite structures involves a temperature/time cure cycle for polymerization of multi-directional composites. This thermal processing will in-

Fiber Type / Sensor	Mechanism Measured	Resolution / Range
Y Guide or Reflective Bifurcated Probe 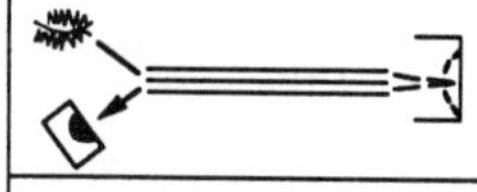	Vibration Surface Texture Pressure	Resolution: 0.1 μm (analog) Range: 50 μm - 10 mm Frequency Response > 1MHz Resolution : 0.1 Ångstrom (Interferometric)
Fiber-to-Fiber Variable Coupling 	Acceleration Pressure Transverse coupling Frustrated internal reflection	Pressure: 1000 μPa Resolution: 0.001 Ångstrom
Bending Loss 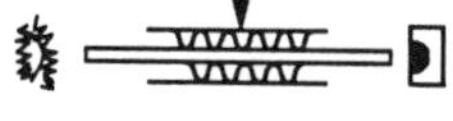	Pressure Strain Displacement	Resolution: 0.1 Ångstrom Pressure: 1 μPa Range: 0.1 μm

Figure 15-72. Fiber Optic Intensity Transducers for Mechanical Measurements

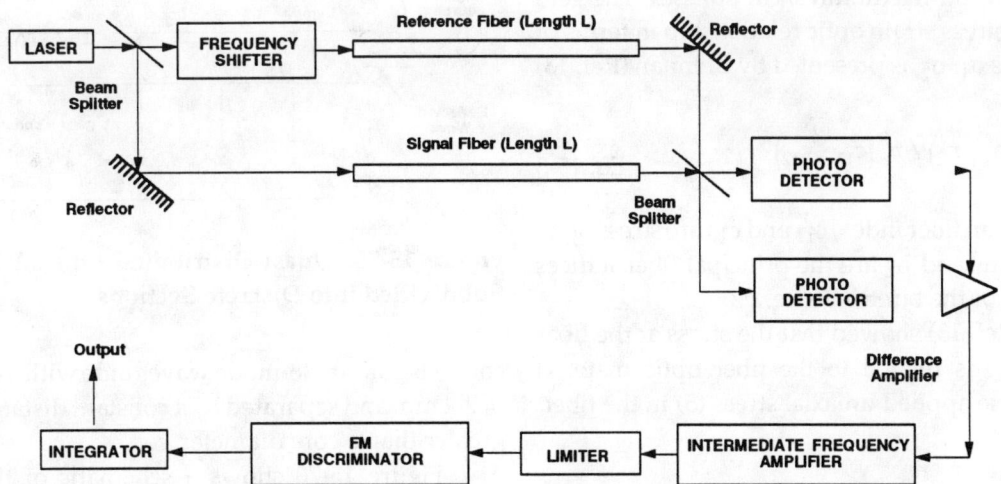

Figure 15-73. Mach-Zehnder Fiber Optic Heterodyne Receiver with Frequency Discriminator

duce thermal residual stresses in each of the lamina. The magnitude and direction of the residual thermal stresses (normal and shear) will effect the strength and service life performance of the composite structure.

Based on the objectives to measure lamina and interlaminate stresses and strains, a survey was conducted to determine the fiber optic sensors most suited for such an application. Figure 15-74 shows a bundle of fiber optic sensors using optical time domain reflectometry (OTDR). The flexible behavior of the multi-mode bundle allows the fibers to be placed in a composite for either sensing within a layer or from layer-to-layer. This quasi-distributed sensing system with each transmitting/receiving fiber has a different length resulting in a different round trip delay time. Therefore, a single input pulse from the OTDR laser will result in a train of return pulses whose amplitude is related to the strain at each of the sensing points.

Benchmark calibrations will be necessary to accurately calibrate the embedded fiber optic sensor (5) in the composite structure. The following is a suggested list of key elements in the calibration process:

- Reference fibers are needed for a baseline comparison.

- The measured strain optic behavior and relationship needs to be determined.

- Simple loading conditions of the composite that produce uniform remote stresses need to be used. Uniaxial tension and compression, four point bending, or constant stress beam are suggested.

- Residual stresses due to thermal cycle curing of the composite need to be accounted for in the calibration of the embedded fiber optic sensor.

Strain Optic Behavior of Embedded Fibers

The primary experimental objective of using embedded fiber optic sensors in composites is to measure the strain response of the fiber due to external application of loadings. Several studies (Refs. 12 through 16) have been conducted on correlating the optical transmission quantities of the fiber with the indicated deformation behavior of the sensor. For continuous layed up fiber optics, the response of the fiber will measure the average strain over the fiber length. Polarized light transmitted through the fiber

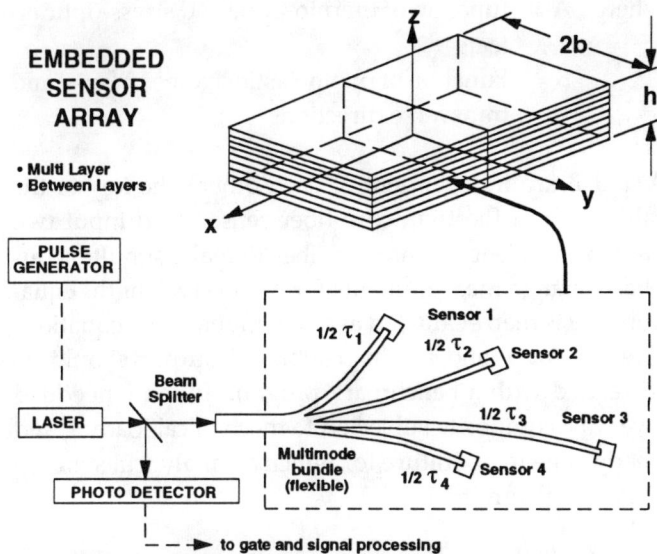

Figure 15-74. Quasi-Distributed Fiber Optic Sensor Using Multimode Fiber and OTDR Techniques for Laminate and Interlaminar Strains

optic will induce birefringence which is correlated with principal stresses and maximum shear stresses. The general form of the stress strain optic relationship in terms of principal stresses σ_1, σ_2 as presented by Brennan (Ref. 13) is:

$$\sigma_1 - \sigma_2 = 2(n_1 - n_2)/n^3 (c_{11} - c_{12}) \qquad \text{Eq. 15-5}$$

where n is the mean fiber index, c_{11} and c_{12} are stress optic coefficients and n_1 and n_2 are the principal fiber indices (e.g., magnitude of the birefringence).

Brennan (Ref. 13) showed that the stress at the fiber surface $(\sigma_1 - \sigma_2)$ is related to the fiber optic material properties and the applied uniaxial stress (5) in the fiber. That is,

$$(\sigma_1 - \sigma_2)/2 = cS \qquad \text{Eq. 15-6}$$

where c contains the geometrical and material constants. The main optical parameter measured over the fiber length, L, is the retardation $\Delta\phi$ which has been shown to be

$$\Delta\phi = 2\pi\, c\, \omega\, L\, n^3 (c_{11} - c_{12}) S/\lambda \qquad \text{Eq. 15-7}$$

where ω is light frequency and λ is the wavelength.

Experiments performed by Brennan (Ref. 13) showed that the experimental output is retardation ϕ versus applied stress and is linear in behavior.

Another approach for fiber embedding is to use subdivided fiber optics which are of equal length as shown in Fig. 15-75. Zimmerman, et.al. (Ref. 16) have shown that the strain, $\varepsilon_{f,\,i}$, at discrete sections is given as

$$\varepsilon_{f,i} = [(\Delta t_i - \Delta t_{i-1})/(t_i - t_{i-1})] \cdot [1/(1+a)] \qquad \text{Eq. 15-8}$$

where i is the segmented section number, t_{i-1} and t_i are near and far reflected pulses of the light reflected back to the source and a is the slope of the n versus ε_f curve. Here n is the average index of refraction of the fiber optic. Experiments performed by Zimmerman, et.al (Ref. 16) showed that the segmented optical fiber approach can measure strains at discrete locations.

Twin Core Fiber Sensor

The twin core fiber sensor has "built-in" optical characteristics for measuring microstrain and temperature with a single fiber-optic sensor. Reference 1 describes how twin core sensors are fabricated and explains experiments conducted using the sensor as an embedded sensor in composite materials. The basic concept of this sensor is that each

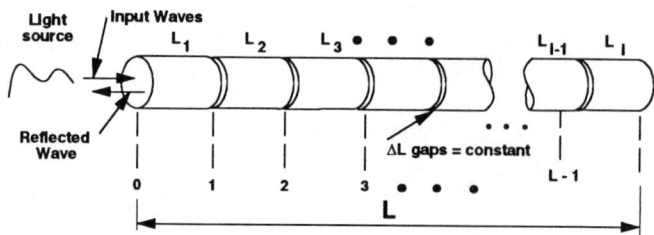

Figure 15-75. Quasi-Distributed Optical Fiber Sensor Subdivided into Discrete Sections

core acts as a single-mode waveguide with a core diameter of 2-4 mm and separated by a constant distance equal to or greater than a core diameter.

Figure 15-76 shows a schematic of the functional characteristics of the twin core fiber sensor. With laser output into one of the cores, two modes are excited and propagate with different phase velocities or different propagation constants. This interference sets up what is called "cross-talk" between the two fibers. The detection system determines the "beat-phase" by measuring the optical power coupled from each core. Core contrast is measured as the ratio $(R1 - P2)/(R1 + R2)$ and is a function of beat-phase b. The detected output can be correlated with microstrain and temperature. According to Ref. 1, the beat-phase shift, $\delta\beta$, where $b = p(L/\lambda)$, can be expressed as a function of wavelength (1), temperature (T), and strain (e) (Ref. 1):

$$\delta\beta = A(T, \lambda)\, T + B(T, \lambda)\, \varepsilon \qquad \text{Eq. 15-9}$$

where A = function of thermo-optic and stress-optic effects
      B = Function of photoelastic changes in axial and transverse directions

A and B are independent of each other. The key to the calibration of the twin core fiber sensor is to input two known wavelengths into the fiber to make simultaneous phase-change measurements for each wavelength. Equation 15-9 is then evaluated as two simultaneous equations to solve for T and ε. This calibrated output should be correlated with a benchmark program for an embedded layer in a composite subjected to mechanical loading and steady state temperature for which an analytical solution has been determined.

Experimental Setup and Data Processing

An experimental setup for calibration and testing of a particular embedded fiber optic sensor is shown in Fig. 15-77. Table 15-9 lists the basic components of the electrical

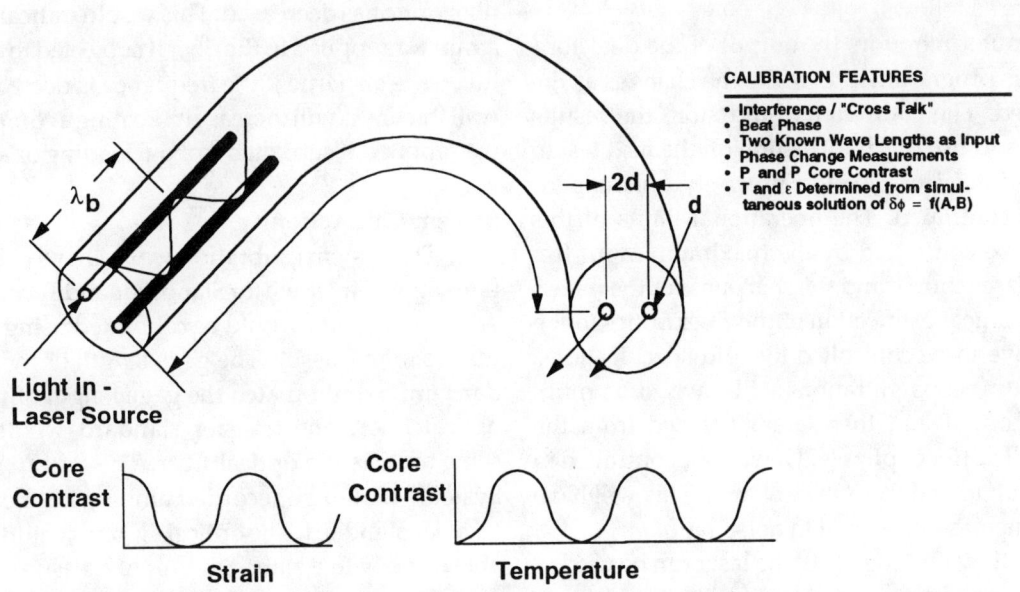

Figure 15-76. Twin Core Sensor for Measuring Mechanical Strain and Temperature

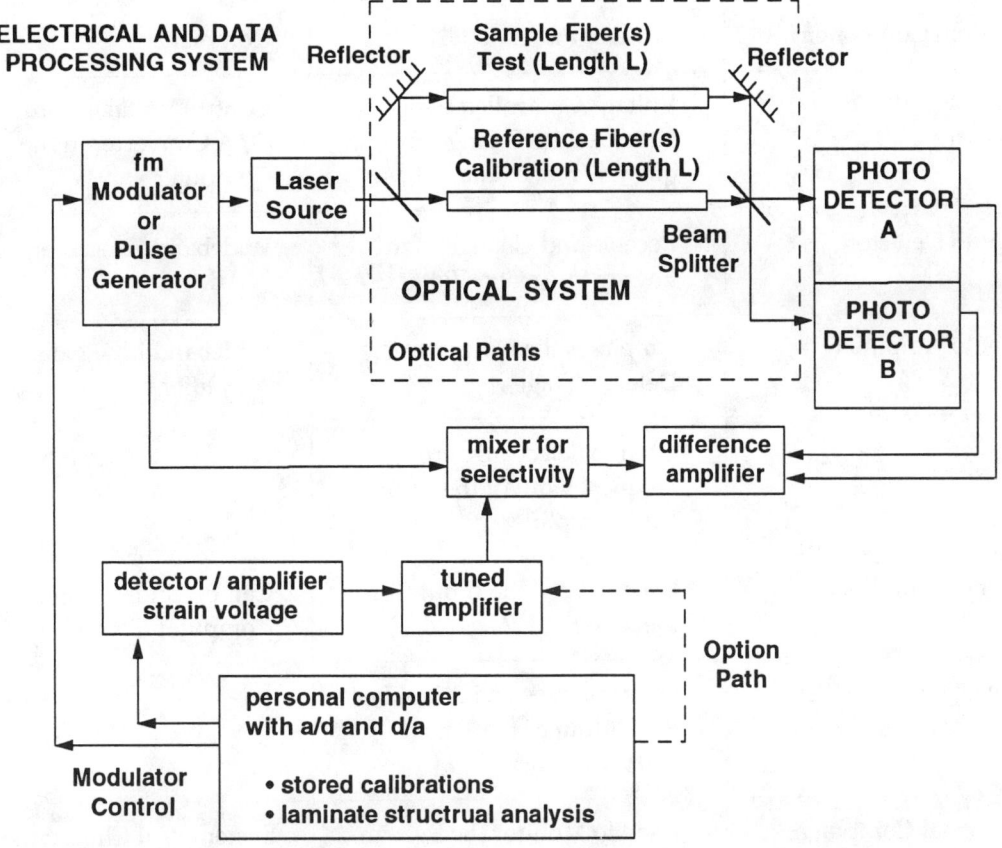

Figure 15-77. Smart-Structure Work Station—Experimental Setup and Data Processing

and data processing systems. This setup is called a "Smart Structure Workstation" and the salient features of the workstation are as follows.

The computer monitors the output of the detector/amplifier and is programmed to make the changes as the sample is tested. The computer would store the results from each test and use this information on the next test to collect information faster and more precisely as the experimental efforts continued. The operational limits of this system would be controlled by the maximum input frequency to the laser source and the sharpness of the mixer.

The mechanical connection of the fiber optics to the system will have to be controlled to rigid specifications. The outside extraneous vibrations will have to be minimized and recorded and then later analyzed from the recorded data. The fiber optics will have to be controlled in their optical path to and from the test area. This will have to be done to increase the signal to noise ratio.

The operational frequency of the laser can be used to determine the strain direction (e.g., tension or compression). This can be done with either a phase lock loop analog system or the computer. If the frequency has to be increased to meet certain criteria, the wall thickness of the fiber optic has decreased. This would indicate that a tensile strain was applied to the fiber if subjected to axial load. The inverse is also true if the frequency is decreased. The fiber wall thickness will increase indicating a compression strain was applied if subjected to axial loading or a bending loss.

System Calibration

The system calibration would have to be traceable to the National Institute for Standards and Technology (NIST). An extensometer would be calibrated using ASTM E83-67 no less than class A. The gage length of the transfer standard unit would match the gage length of the embedded optical fiber. The transfer standard would overlay the same area as the optical fiber. The output of the "Smart System" would be recorded along with the output of the transfer standard. The optical system would then be traceable directly to a standard unit of measure.

The areas to consider and some potential problems that could arise are:

Table 15-9. Basic Components of Electrical and Data Processing System

System Component	Primary Function	Comment
FM Modulator Pulsar	Voltage Controlled Oscillator Used to Change Frequency/Pulse Rate of Laser	Control Voltage from D/A Converter in the Computer
Photo Detector	Receive and Change Light Energy to Analog Voltages	Wideband Amplifier
Photo Amplifier	Amplifies the Output of the Detector Stage	Wideband Low Noise Amplifier
Mixer	Used to Select a Desired Frequency out of the Wide Band Amplifier	Helps to Reduce Noise
Tuned Amplifier	Used to Scan the Input from the Mixer	Can Be Under Manual or Computer Control
Detector/Amplifier	Discriminates and Amplifies the Output of the Tuned Amplifier	
Personal Computer	Used to Monitor the System and Record Data Taken During the Operation	System Will Choose the Best Frequency

- Inaccurate measurements of the gage lengths
- The coupling coefficient change of the laser to the fiber optics
- Readings taken of the output of the Smart System (less that 0.1%)
- The placement of the standard unit over the optical fiber
- The connection between the laser source and the fiber optics The interface between the fiber optics and the sample Preconditioning of the sample
- The cohesion of the fiber optic material to the test material is critical
- Light from outside sources
- Light coupling from one fiber to the other in the opposite plane causing unwanted modes.

Summary

This section has presented an overview of fiber optics technology, sensors, and interferometric sensing systems. The potential use of fiber optics with continuous (distributed) or segmented (quasi-distributed) have the potential for use as embedded sensors in "smart-structures/skins" made of composites. Mechanical stress and strain can be correlated to fiber optic indications. For continuous fibers, the indicated stress and strain is averaged over the embedded fiber length. Strains at discrete points can best be measured using embedded fibers which are segmented and contain initial gaps of equal length. A segmented fiber with a misalignment will show a change in (wavelength). If the fiber rotates and distance between the fiber remains constant there will apparently be no change in λ. The separation of the segment coupling coefficient will be the axial strain indicator.

References:

1. R. Mach, "Fiber Sensors Provide Key for Monitoring Stresses in Composites," *Laser Focus—The Magazine of Electro-Optics Technology*, May 1987.
2. A.K. Caglayan, S. M. Allen, and S.J. Edwards, "Hierarchial Damage Tolerant Controllers for Smart Structures," Charles River Analytics, Inc., AFWAL-TR-89-3009, March 1989.
3. J. Sirkis, "A Surface Application Technique for Optical-Fiber Strain Sensors," *Experimental Techniques*, SEM, Nov. 1988.
4. L. Leonard, "Smart Composites: Embedded Optical Fibers Monitor Structural Integrity," *Advanced Composites*, March/April 1989.
5. E. Udd, "Embedded Sensors Make Structures "Smart," *Laser Focus-The Magazine of Electro-Optics Technology*, May 1988.
6. ———, "Fiber Optic Smart Structures and Skins," *SPIE Proc.*, Vol. 986, Sept. 1988.
7. ———, "Materials that Think for Themselves," *Business Week—Science and Technology*, Dec. 5, 1988.
8. S.A. Kingsley and A H. Harmer, "Fiber Optic Sensors and Their Potential to Industry," Battelle Technical Inputs to Planning Report No. 48, Aug. 1986.
9. G.R. Tschulena and M. Selders, "Sensor Technology in the Microelectronics Age," Battelle Technical Inputs to Planning Report No. 40, June 1984.
10. J.M. Whitney, I. M. Daniel, and R.B. Pipes, "Experimental Mechanics of Fiber Reinforced Composite Materials," SEM Stress Analysis Monograph No. 4, 1982.
11. G.H. Staab, "Mechanics of Composite Materials," Short Course Notes, Ohio State University, 1984.
12. E. Udd, "Overview of Fiber-Optic Applications to Smart Structures," Review of Progress in Quantitative Nondestructive Evaluation, Vol. 7A, Plenum Press, New York and London, 1988.
13. B.W. Brennan, "Prototype Polarimetric-Based Fiber Optic Strain Gauge," Review of Progress in Quantitative Nondestructive Evaluation, Vol. 7A, Plenum Press, New York and London, 1988.
14. J.S. Shoenwald and P.M. Beckham, "Distributed Fiber-Optic Sensor for Passive and Active Stabilization of Large Structures," Vol. 7A, Review of Progress in Quantitative Nondestructive Evaluation, Plenum Press, New York and London, 1988.
15. R.O. Claus, K.D. Bennett, and R.G. May, "Optical Fiber Methods for the NDE of Smart Skins and Structures," Proc. of 1988 SEM Fall Conf., Indianapolis, IN, Nov. 7-8, 1988.
16. D.B. Zimmerman, K.A. Murphy, and R.O. Claus, "Embedded Optical Fiber Sensor for Internal Material Measurement," Proc. 1989 SEM Spring Conf. on Experimental Mechanics, p. 752, Cambridge, MA, May 29-June 1, 1989.

15.13 STIFFNESS TESTS

Stiffness of Pin Connectors
Joseph A. Kubas and Robert T. Reese, Sandia National Laboratories, Albuquerque, NM

This example describes load-deflection tests that were performed on electrical connectors. Each connector consisted of a pin (small diameter rod) positioned between two thin blades (bayonets). Electrical continuity is achieved when the pin maintains contact with either of the two bayonets. The concern in using this type of connector is the possible detrimental effects that high acceleration and vibration fields could have on the integrity of the connector. The bayonets, pins, and setup for static tests are shown in Fig. 15-78.

The connector is manufactured so that the bayonets have a preload on the pin when the pin is inserted. Integrity in the connector is ensured if the preload on the pin is maintained. Load-deflection tests were performed to evaluated the preloads. A typical connector had ten or more of combined individual pin-bayonet connectors. The second part of the load-deflection tests was to determine the force on the bayonet required to disrupt electrical continuity.

The tests were performed in two ways: statically and using a centrifuge. The static tests consisted of applying a force through a small diameter wire (foot) positioned at a right angle to a probe attached to a small (50 gm) load cell.

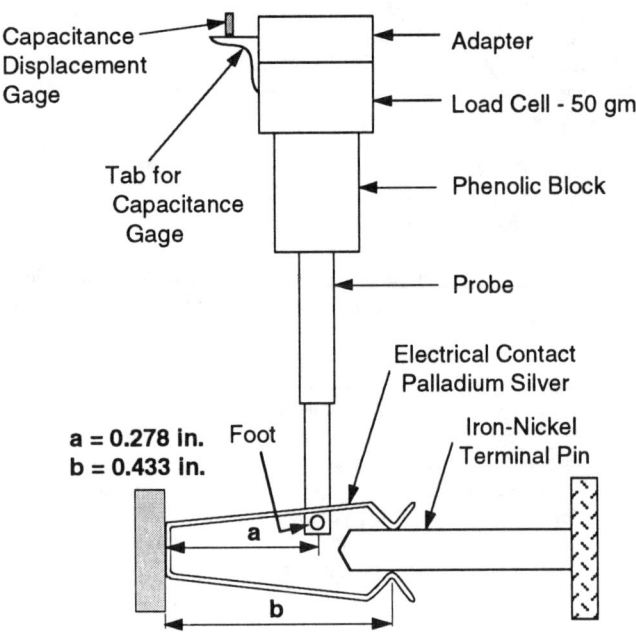

Figure 15-78. Pin/Bayonet Electrical Connector and Static Test Setup

The load cell was positioned in a small precision milling machine. The machining head was removed. The two-way movement of the machining table made precise alignment of the foot with the appropriate part of the bayonet possible. The deflection measurements were made with a non-contacting capacitive system. The setup for static tests is shown in Fig. 15-79. These static tests simulated a cantilever beam with a point load applied at a position on the bayonet away just outside the pin.

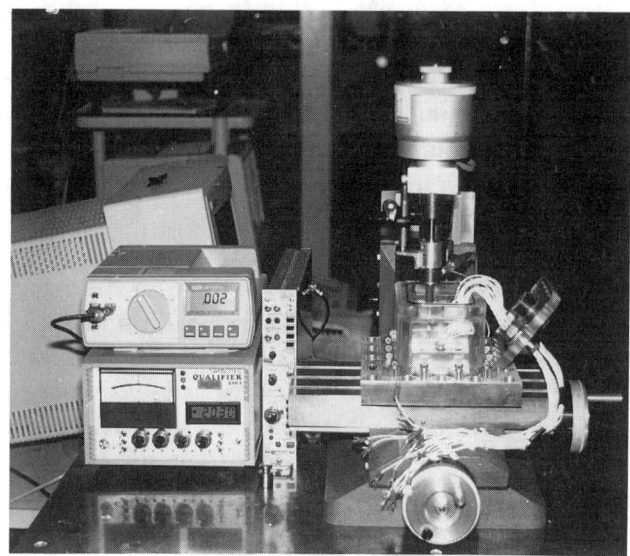

Figure 15-79. Setup for Static Tests on Pin/Bayonet Connector

The electrical continuity tests consisted of applying a current across the connector with one bayonet isolated electrically by means of a thin sheet of non-conductive material (e.g., mylar). A force was applied through the foot with the probe electrically isolated as well. The results showed that when the applied force was just greater than the preload, then the bayonet was separated from the pin and continuity was lost. The test plan was modified to determine the preloads from the bayonet to the pin for several connectors. The load deflection information indicated when separation of the bayonet and pin occurred with a resulting loss in electrical continuity.

The tests performed in the centrifuge simulated actual application of the connector because the real loading on the part consisted of a typical uniform acceleration over the length of the bayonet. The tests in the centrifuge also used a bayonet which was electrically isolated from the pin. Continuity was evaluated through application of a current through the centrifuge's slip rings into the connector. With a uniform acceleration applied to the connector as shown in Fig. 15-80, a measure of the preload on each individual bayonet/pin system in the connector was determined. A comparison of the governing equations and test results is given in Fig. 15-81 with the uniform inertial load on the left and the point load from the foot of the load cell on the right. From this correlation, it was determined that the two tests gave the same answers within the uncertainties associated with each test. Sample load-deflection data are given in Fig. 15-82. Electrical continuity was lost when the slope of the load-deflection curve changed.

It was determined that the preloads were not uniform. That is, the manufacturing process did not produce uniform preloads on each bayonet. The test procedures developed (both static and centrifuge) made evaluation of

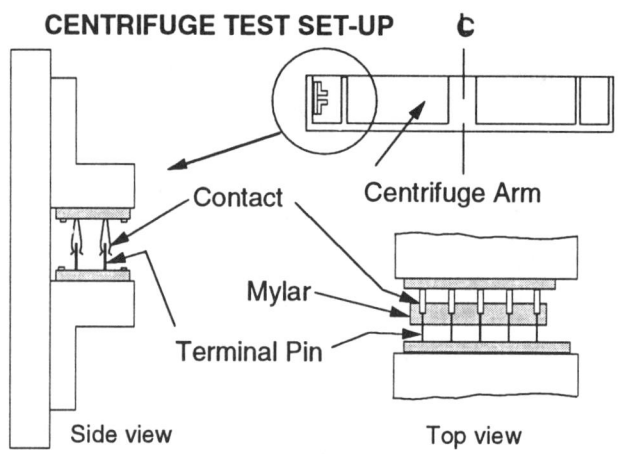

Figure 15-80. Pin/Bayonet Connector Subjected to Uniform Acceleration

COMPARISON OF LOAD-DEFLECTION MODEL AND UNIFORM LOAD MODEL

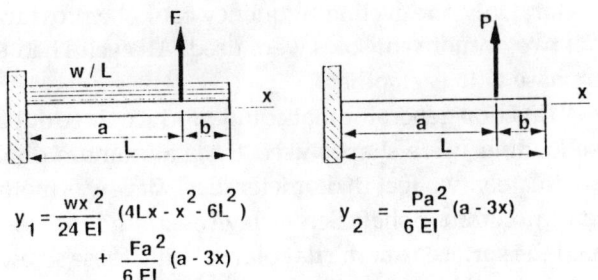

$$y_1 = \frac{wx^2}{24 EI}(4Lx - x^2 - 6L^2)$$
$$+ \frac{Fa^2}{6 EI}(a - 3x)$$

$$y_2 = \frac{Pa^2}{6 EI}(a - 3x)$$

For $y_1 = y_2$, and $x = a$ the above simplifies for the electrical contact to:
$$P = 23.5 \times 10^{-3} \text{ grams} / g \times \# g's$$

Figure 15-81. Comparison of Load-Deflection Model and Uniform Load Model for Pin/Bayonet Connector

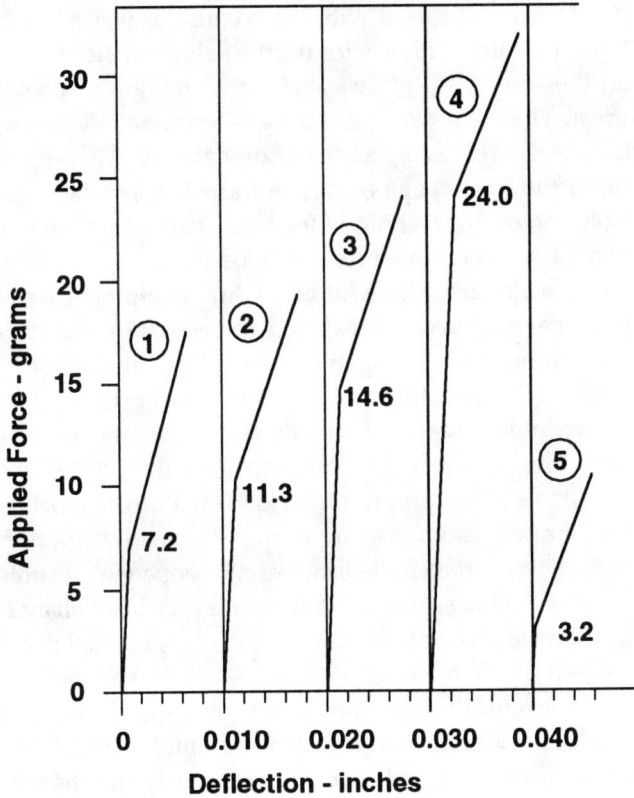

Figure 15-82. Load Deflection Test Results from Five Electrical Contacts

the required preloads possible. Further, acceptance tests based on the centrifuge test system were proposed for evaluating production lots of these connectors. The centrifuge test was selected for this evaluation because all the individual pin/bayonet connectors could be tested at one time which decreased test time.

15.14 TESTING OF AIRCRAFT STRUCTURES
Richard Grigat, Grumman Aerospace Corporation, Bethpage, NY

Since the late 1950's aircraft development and structural test validation has become increasingly complex. This has been brought about principally by the expansion in the knowledge of the true operational environment of the aircraft, improved knowledge of the effect of this environment on the structure, and the need for more reliable and efficient aircraft which will survive the operational environment for increasingly longer service lives. Paralleling the evolution of design capability has been the need for and the development of better fatigue testing methods, equipment, and facilities. This expanded test capability has made available to the aircraft manufacturer the ability to realistically evaluate the air frame fatigue life and to identify unanticipated deficiencies in the structure.

In the late 1950's the evolution of fatigue testing as a technique for ensuring a reliable service life was just beginning. So little was known of the significant factors affecting fatigue that arbitrary rules were established for fatigue test loads and their application to the structure. The tests performed were expected to yield conservative fatigue life results. Only simplistic test load systems were used which were within the then available capability of the testing equipment. Much of this test equipment was basically the same equipment used for static testing.

Early 1950 aircraft fatigue testing concentrated principally on the wing primary structure. Wing load applications were symmetrical and little consideration was given to other that flight loads although, for Navy carrier aircraft, catapulting and arresting fatigue testing was required. Pressure cabins were also extensively tested because of the untimely failures encountered in service use of the Comet aircraft. The wing test load distributions generally represented one design condition, usually the critical static condition, and variation in the loads for other than the maximum load were ratioed in proportion to load factor. This greatly simplified the test fixture requirements which was fortunate since the equipment capability to handle complex testing did not exist in that time frame. In recent years increasing demands for more sophistication in the type of testing considered the industry "state of the art" resulted in the acquisition and development of sophisticated test equipment, test controls, and test fixturing. The most significant changes in fatigue test requirements have been the introduction of flight by flight loading and the replication of time history loading conditions for both landing and flight.

Flight-by-flight testing applies loads to the aircraft structure in a realistic sequence of all loads (flight and non-flight) expected in service. To conduct a flight-by-flight test, the fixturing that formerly was used for each test condition, i.e., catapulting, flight, arresting, pressurization, and landing, must be placed on the vehicle. The number of load control channels needed to conduct this type of test is a large multiple of the number used on a simple condition test. Further, the loads for all these conditions must be sequenced properly and many strain readings were recorded. Because of the lack of access to and the difficulty in monitoring structure response, the test equipment has been made capable of monitoring anomalies measured under load since the troubled regions cannot be easily inspected. Automatic test shutdown beyond certain measured limits is also built into the equipment to prevent catastrophic failure. Clearly, the test fixtures, loading, and recording equipment needed for flight by flight testing is far more complex than that used in the early condition tests.

The introduction of load time history conditions into the test spectrum has added greatly to the number of test points to be covered in the test. Each landing condition for instance involves a number of critical load combinations which must be applied in sequence. The number of critical combinations of loads included the landing gear spin up, spring back, maximum vertical, and at rest conditions. Each combination must be applied in sequence without load relaxation between conditions. In order to accommodate both flight by flight testing plus time history conditions the number of individually controlled hydraulic loading cylinders and conditions to be used in the test has increased by an order of magnitude.

The development and sophistication of the testing equipment that has evolved since the late 1950's is described in this section.

Load Application Control Equipment

The first load control systems used for modern air frame testing were developed in the late 1950's to 1962 time frame. One example of an early system was a multi-channel servo controlled unit using closed loop servo amplifiers to control the pressure in the hydraulic system. A three channel disk function generator allowed three different hydraulic pressures to be used simultaneously by the synchronizing loading systems of each. The loading system was controlled by a tape reader which switched through the ten load channels in sequence and then repeated the sequence. The switch selected a set point voltage and a cycling voltage that were fed to the servo amplifier.

The tape reader was also used to count cycles. Each cycle advanced the tape one frame operating an electrical counter. At the required number of cycles a punched hole in the tape switched the console to the next load level.

The cycling signal was provided by a low frequency generator. Only one cycling frequency could be provided even if two or more consoles were used. All cycles had to be in phase with each other.

A function generator that could produce three out of phase loading wave shapes used three aluminum disks approximately two feet in diameter. Each disk was motor driven on a common shaft. Screw devices around each disk warped the surfaces out the flat plane. Using these screws each disk could be warped into a sine wave shape. When the disks were driven, followers connected to Linear Variable Differential Transformers (LVDT's) converted the warp of the disk into an electrical output. Each output was synchronized with the other but not necessarily in phase with them. Each disk provided an independent electrical output which could be scaled without affecting the other outputs.

The readout used with these consoles was a Visual Strain Indicator. This device used a galvanometer to convert the strain gage electrical signal to motion of a small mirror. The mirror reflected a light beam onto a ground glass screen displaying a bright dot that moved in proportion to the strain gage bridge output. The accuracy and resolution of this technique for loads near full scale was good but was very poor for small loads.

A major advancement in loading equipment was a multi-channel console developed and operational in 1963. This equipment was capable of handling random block loadings. The disk function generator used on the earlier equipment was replaced with an electronic function generator. The console had a load control accuracy of 4%, and 16 channels of pre-programmed loads with 16 channels of precalibrated readouts incorporated. The read-out system displays utilized a single beam oscilloscope which could be easily used to set the various load levels. Load channel selection was accomplished by using of a punched paper tape and a binary coded channel selector. The use of a binary system allowed random selection of channels so, if a load level was desired, the console could switch to the channel directly regardless of the previously run channel.

Error detection systems included protection against load variations both above and below the prescribed limits. Any error would stop the console and operate external safety relays.

In order to perform complicated tests individual consoles were connected to a master console to synchronize the cycling and switching of the consoles and their associated hydraulic systems. The master console included a static load feature that provided proportional static load outputs from each of the consoles so static strain surveys could be run without changing the fixture or console setup.

During the later part of the 1960's it became apparent that this type of fatigue test system would not be adequate to run future full scale fatigue tests.

An early generation digital control system was built by Fishbach and Moore of Dallas, TX. This system proved sixty servo amplifiers and sixty loading data channels and was controlled by a Varian G20/I computer.

This system was used extensively on the F-14 and B1 composite tail fatigue test programs. All load and spectrum data was on magnetic tape and the system could easily be reconfigured from test to test through operating system software.

During late 1978 and early 1979 Grumman developed the concept for the "Mark V" loading system. This system has 180 channels of load control. Each channel has its own servo amplifier, servo valve, load link (feedback), and function generator.

The software operating system permits four different tests to be run at the same time without interaction among tests. The 180 channels can be assigned to the four tests through the operating system. No hardware changes are ever required to reconfigure the system. It is all done in the software operating system.

To facilitate data input to the system, a test control language and compiler was developed. Using this language, a test is described in an "English Like" language that is then compiled into binary for the system. An operator need not know the inner workings of the system to use it.

The Mark V can run four tests simultaneously with any distribution of its 180 channels assigned to each test. The spectra for each test can have up to 2,000 end points and is controlled by a sequence table of up to 64,000 steps stored on a magnetic disk. Using this system of end points and sequence table any spectrum loading wave shape can be easily generated.

During 1984 an additional system, the Mark VB, was constructed for the E2C Fatigue Test Program. The Mark VB is identical to the Mark V except it has 120 channels whereas the the Mark V has 180 channels.

All fatigue tests at Grumman are now run using the Mark V control systems. They have proven to be accurate and reliable with load application accuracy of 3% and a MTBF of about four months.

Specimen Safety Provisions

Specimen safety and loading error detector systems are built into the Mark V. Sensors on the specimen can signal if the deflections become excessive. Any function that can be detected with a switch closure can be an input and tested by the system.

Loading is checked for accuracy by two analog and two digital error checks. The analog errors check for deviation from the required loading wave shape. A 5% error will slow the cycle rate while an error of 8% will unload the system and signal the operator.

Digital error checks for test deviation from the required wave shape are performed exactly as it is with analog error while the second digital error test is for an expedience of the maximum load that the channel should ever apply. This second digital error check, (called the limit error), will unload all channels and signal the operator.

Another system that checks the data from load links and strain gages on the test specimen runs whenever a fatigue test is running. This system acquired baseline data when a test is first started and then during the test compares newly acquired data to these baselines. If data deviates by more than preset limits allowed, the system will flag the out-of-limits data and notify the operator. The system can also be set up to stop the test automatically if data limits are exceeded.

These failure prevention provisions are able to prevent only a limited number of possible failures or anomalies which may occur in a fatigue test. The major reason for this is that failures in fatigue most often occur from small undetected fatigue initiated cracks which eventually propagate catastrophically through the part. A prerequisite for detection of the crack before rapid crack propagation occurs is to have strain gages very close to the crack initiation site. These initiations sites are, unfortunately, not always recognized prior to the test. In highly stressed structures, the critical crack size is small and the change in the over-all deflection of the parts prior to rupture is relatively insignificant.

Wing fatigue failures are generally more serious than fuselage failures since little redundancy is designed into the wings of military aircraft and large regions of the wing structure are under high tension stress when under load. The fuselage structure in contrast usually has a degree of redundancy so that, should one member crack or fail, others can carry the load and thus prevent catastrophic failure of the specimen.

Failures in the test equipment can also result in damage to the structure. The complex equipment used to test an airframe can be subject to breakdown from many sources.

Data Acquisition Systems

Early fatigue test instrumentation usually consisted of an oscillograph which was used to record the input and reaction loads and multi-channel strip chart point plotters to record the output of the installed strain gages. The strain gaged load cells, each with a minimum of two circuits with four strain gages per circuit and wired in a Wheatstone bridge configuration, were used to provide an electrical output which is directly proportional to applied load. These electrical signals drove a small galvanometer which was mounted in a magnetic block within the oscillograph. A tiny mirror which was attached to the galvanometer coil rotated in direct proportion to the energizing current. A light focused on the mirror was reflected onto moving photo-sensitive paper contained within a cassette. The trace deflection on the paper was directly proportional to the measured parameter and provided a hard copy record. This photo-sensitive paper was developed using a wet process similar to that which is used in developing photographs. This process consumed an enormous amount of elapsed time and man-hours for manual data reduction. Periodic static strain surveys were run to check for strain distribution changes within the test article. The electrical output of the strain gages was recorded on strip chart point plotters, in which the measured strain was plotted against a manual input in percent of applied load.

When Grumman Aerospace was awarded the F-14 contract, it was apparent that the available equipment was inadequate. To meet the F-14 requirements of schedule, accuracy, number of measurement channels, and real-time data processing the Myti-Acq, a 600 channel, multiple user data acquisition system, was purchased.

Myti-Acq, a computer-controlled data acquisition system, eliminated many of the manual operations required for setup, data acquisition, and processing. Measurement uncertainties caused by transducer non-linearity, balance network loading and lead resistance were corrected using computer techniques. Multiprocessing capability allowed acquisition and reduction of up to 600 data channels from two simultaneous fatigue tests. Separate control consoles gave each used independence and effective access to the computer facilities of the system. Elapsed time and manpower savings are achieved through automatic checkout and analysis of transducer malfunctions as well as computer controlled connection of strain gage bridge unbalances.

During the F-14 structural test program, the fatigue article, was instrumented with approximately 1700 strain gage circuits located at critical stress areas. These strain measurements were recorded and displayed during various phases of the test program. Since Myti-Acq could record only 500 measurements using a single console, only those measurements that were critical during a particular phase of the fatigue test program were recorded. Time savings were achieved by the almost real-time data processing capability of Myti-Acq. This processing provided for tabulated engineering units (EU) data for all data acquired for just minimum/maximum data presentations and also for load and/or stress time histories. These data presentations provided test analysis engineers with the necessary information to make go/no go decisions during the test.

The Myti-Acq acquisition system capability was expanded to support the F14 A/C 90 fatigue test program. This modification incorporated a Fatigue Monitoring System (FMS) to allow for continuous unattended data system operation by providing data comparisons to be made selectively during critical loading periods and a "rotating" temporary data storage buffer with a permanent magnetic tape storage of pertinent data only.

If comparisons automatically performed during selected critical loading periods exceed predefined limits, a shutdown command is transmitted to the load control system (LCS). The "rotating" temporary data storage buffer will always contain the last several minutes of data acquired by Myti-Acq assuring a record of applied loads and resultant stress, strain, deflection, etc., in the event of test article failure or shutdown. In the event of a test article failure or shutdown, the "rotating" temporary data buffer is automatically collected on magnetic tape. Provisions are included to write selective data to tape via LCS sequence table commands.

In 1982, a plan to replace Myti-Acq was initiated. An evaluation was undertaken to study new "State of the Art" techniques, future test programs, consideration of the safety of the test article and to reduce costs.

The result of this evaluation was to plan and implement a multi-year £our phase replacement program. The first phase, called the Structural Data Acquisition Systems (SDAS), is presently operational and is in use on the E-2C Fatigue Test Program. This system is dedicated, single user computer controlled 384 channel data acquisition system which is located in a ground floor room adjacent to the test article.

This system is capable of accepting various transducer inputs such as load, pressure, all types of strain gage circuits, potentiometric, temperature, and EMF. The transducers are connected via test cabling and aircraft type junction boxes to the input assembly. The input assembly contains a signal conditioner, amplifier, and A/D converter. The signal output of the input assembly is connected to the Data Control Unit (DCU). The DCU contains a Central Processing Unit (CPU), operator station, 2 line printers, magnetic tape recorder, mass storage (disk), and a high-speed disk emulator.

The SDAS includes the following capabilities:

- 384 channel of data acquisition
- automated system setup
- fatigue data acquisition
- real-time data processing

- min/max engineering units data
- automated decision making
- two-way communication with Mark VB LCS

The SDAS provides a two-way communication with the Load Control System via RS-232 aerial links and required handshake lines. The communication network passes sequence table information which defines up to 150 areas of interest (window) in the test profile requiring data comparisons with the baseline run. The SDAS compares min-max data to limits established from the baseline data. If the values are acceptable, i.e., within limits, acquisition of the data will continue with no other action being taken. If a value is beyond the first (inner) limit, the actual min-max for the offending channels will be printed out in engineering units (EU) and the acquisition of data will continue. The same action takes place for values beyond the second (or outer) limit along with the issuance of an "unload complete" signal to the LCS. Upon receipt of an unload complete signal from the LCS, or a timeout of thirty seconds, the SDAS will transfer the data in the temporary buffer to the magnetic tape. If this action is caused by the "unload complete" signal, testing will be halted, but the SDAS remains on line in a standby mode, waiting for further commands from the LCS. If data is written to tape due to a timeout, an exceptional status condition will have occurred. The appropriate error messages will be an output on the high speed line printer and both the SDAS and LCS consoles.

The rotating temporary data buffer will always contain the last two minutes of data based on a 984 channel test. This buffer will be configured as a first-in/first-out storage area. This will ensure in the event of test article failure or shutdown, that a permanent record is made of the applied loads, stresses, strains, deflections, etc. The rotating temporary data buffer may contain more than two minutes of data for tests smaller than 984 channels, i.e., 384 channels is approximately 5 minutes.

Phases II and III of SDAS are now operational. This improved SDAS capability provides for support of static tests with real-time displays in engineering units, including rosette solutions, graphics/hard copies, X-Y plotters, and on-line printers, as well as support of two concurrent tests with 256 data channels.

Fixturing Development
Wing Test Load Fixturing

The type of attachment of the load fixture to the wing is influenced by the wing substructure design (multibeam, rib stiffener or stiffened cover), the wing geometry (low or high aspect ratio), and the load intensity. Loading the wing through large numbers of tension pads bonded to the upper and lower surfaces at the beam rib intersections has generally been preferred. This fixturing has the advantage of distributing the net surface loads in a manner closely replicating the distributed load applied to the wing in flight. The rib load distributions can be reasonably matched with this type of pad system. In some wing designs such as Delta wings the matching of the surface load distribution can be very important in obtaining suitable internal structural member loads where there are no test alternatives. Since the lower cover is most important in fatigue tests, lower cover pads need to be kept to a minimum to provide adequate local lower cover visibility. Fortunately, a low number of lower surface pads is usually needed since down loads on the wing are generally not high.

Less costly fixturing which can be used in some cases employs what is called "clamshell" loading fixtures. This type of fixture is particularly advantageous if more than one wing is to be tested. A clamshell fixture is shown in Fig. 15-83.

This clamshell fixture loads the wing at a fewer number of discrete points, with loads generally applied at the front and rear beams. The correct net spanwise loads (net shear, bending moment, and torsion) for the wing are well represented with clamshell fixtures; however, the chordwise load distribution is not expect that the center of pressure (C.P.) of the wing loaded section by each fixture is correct. This leads to significant errors in the rib loads. However ribs are not generally too critical in the high aspect ration wings. Figure 15-84 shows a vertical fin being loaded by a "clamshell" fixture.

The massive weight of the clamshell fixtures is balanced to prevent unrealistic "at rest" loads when the load system is turned off. This is accomplished through two tension jacks per wing attached to fixtures over the wing and cabled to the wing clamshell fixtures. These jacks are hydraulically connected to an accumulator with a pressure

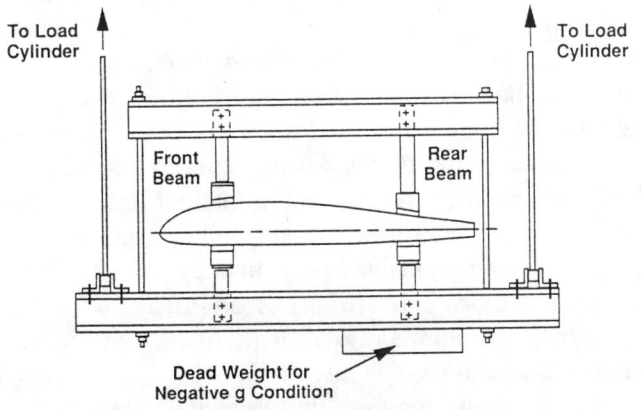

Figure 15-83. Clamshell Loading Fixture

Figure 15-84. Vertical Fin Being Loaded by Clamshell Fixture

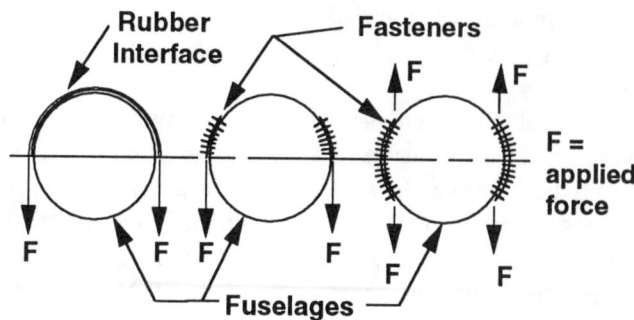

Figure 15-85. Examples of External Straps Used to Load Aircraft Fuselages

set to provide the correct lift load when the operating load system pressure is off. When the operating load system pressure is on the opposite side of the cylinder these tension jacks are pressurized which inactivates the jacks. This loading system is ideal in that it is easily reusable for a number of test wings without a problem of bonding tension pads to each tested wing.

Fuselage Test Load Fixturing

A fuselage does not usually carry high air loads as is the case for a wing. Therefore the running loads to be duplicated in test are those primarily due to inertia. Ideally many loads would be applied at or near the points of heavy load concentrations inside the body, e.g., the engine supports. If this were practical the bulkhead and frame members would be properly loaded to simulate the loading conditions. As this normally requires penetrations of the structural shell this approach is often not used except for the engine support loads. In addition, where fuel is carried in the fuselage the true load distributions are hard to duplicate and may in fact require internal fixturing for application of pressure loads to the bulkheads.

To obtain reasonable fidelity in body shear, bending and torsion, much of the fuselage load is usually applied through external straps attached to the fuselage at fames or bulkheads as shown in Figure 15-85.

Most internal forces on the fuselage are down so few if any upward acting forces are needed. The band system often used consists of a substrate of rubber bonded to the shell surface over a frame or bulkhead and the band loads the fuselage shell by normal pressure applied through the rubber as shown in Fig. 15-85 (a). The band system is easiest to apply on round or nearly round shells.

Although the load distribution on the internal structure is not correct using the band system it is usually less damaging to the fuselage substructure than the riveted-on strap systems shown in Figures 15-85 (b and c). Both types of strap systems have been used in the industry. The strap system shown in Fig. 15-85 (b) is the better of the two and can be attached either at a frame or on the shell surface between frames. The resulting internal frame structure, however, may be more severely loaded than by the band system. If a reasonable number of these straps are attached to the side of the body so as to minimize the load concentration and are whiffle-treed together, the resulting stresses in the internal frames can be reduced to an acceptable level. (Whiffle tree loading systems are shown in Chapter 6, Fig. 6-12). This type of strap system must be used if the type in Fig. 15-85 (a) is not suitable because of interferences with overhead structure, peculiar body shapes or interferences with other loading fixtures. The cost of the two systems, a and b of Fig. 15-85, is not too different. The third type (Fig. 15-85 (c)) has the advantage of application of both up and down loads through a single strap located at a frame or bulkhead. In order to prevent objectionable tension forces on the end attachment fastener (which can cause trouble in the substructure and/or the fastener itself), the straps must be installed tangent to the structure. Applications of fuselage loading through use of external strapping is exemplified in Fig. 15-86.

Pressurized Structural Loads

Early pressure testing of the structure for flight conditions were conducted in a large water tank. These tests followed the approach set by the British in pressure cycling of the Comet airframe. Early on testing of the cockpit

Figure 15-86. Aircraft Fuselage Loaded with Straps

section had cause considerable damage to the test facility and subsequent Comet (and other aircraft) tests at that time were conducted under water for safety reasons. The water tank test had two major advantages over pressure tests conducted using air:

1. The stored energy under pressure loads was minimal
2. The rate of load application was high.

The disadvantages of using water versus air are:

1. The large equipment investment in a water tank large enough for a fuselage.
2. The structure under test was hard to inspect.
3. Strain gage wiring often deteriorates in the water environment.
4. All flight loads on the fuselage shell had to be applied in self constrained fixtures and were, of necessity, quite rudimentary.
5. The extended time in the water tank could cause significant corrosion effects on the fuselage structure.

The overriding consideration which dictated the use of the water tank in earliest tests was safety.

Due to complexity of flight by flight testing coupled with landing time history load conditions it is impractical to test in water. The fuselage is filled with an inert substance, e.g., styrofoam blocks, to reduce the free volume. The remaining free volume was approximately 15 percent of the original volume. This reduces the stored energy under pressure by a large factor and also permits a reasonable cycle time between pressurizations. Periodic inspection is accomplished by removal of the styrofoam blocks as needed. This type of testing for pressure loaded structures is expected to be standard practice in the future.

In today's tight budget environment the approach to assessment of life through test has assumed a new level of importance. Fatigue testing is not used as an instrument with which the fatigue life of the airframe may be extended to the greatest degree practical. While this approach had been used in the past it had never been relied on as heavily as now. This has been brought about by the need for life extension of existing airframes due to tight budget constraints on government purchases of new design replacement aircraft.

While monitoring controls can and should be used in both the load application and the data acquisition it is by no means certain the onset of serious failure will be detected before complete part rupture occurs. The failure monitoring provisions in the equipment described above should be supplemented by careful fixture design.

Most aircraft under test are or will be subjected to loads well above the original design loads and to lives far beyond the original design or test lives. To ensure the test goals are met it is of the utmost importance that the fixture design minimize the effects of preventable structural failures on the rest of the test article so that the test can continue until a major failure develops. There should be a deliberate effort to anticipate likely failures and where practical, ensure that extensive secondary damage to the structure will not occur because of the fixture design.

Most of the loading methods previously described are depicted in the photograph of the complete test installation shown Fig. 15-87.

15.15 TEST CONTROLS

Direct Digital Control of Full-Scale Structural Tests (Ref. 15-9)

B.P. Perrine, B.A. Nakatani, and R.L. Hannah, Lockheed Aeronautical Systems Company, Kelly Johnson Research and Development Center, Valencia, CA.

Introduction

The development of materials and structures for increased performance, maintainability, and reliability of future advanced aircraft requires extensive testing of components and full-scale airframes. Present large multiloop

Figure 15-87. Complete Test Setup Showing Aircraft, Loading System, and Computer Control and Data Acquisition System

static and fatigue structural airframe loading test systems containing individual servo amplifiers are overly expensive and difficult to maintain due to their complexity and large parts count. Down time due to equipment failure can also contribute to unacceptable high costs during periods of testing. More stringent military requirements for aircraft along with test vehicle protection needs dictate new approaches to test system design in order to provide advance test techniques and equipment to simulate critical flight and ground loading environments for full-scale aircraft. A major system improvement and a cost savings can be achieved by merging the individual servo controllers into an all digital environment.

Specific Problem

Current systems employ individual analog servo controllers costing between $2,000 and $3,000 each and are controlled by a digital computer which provides the load spectrum generation and overall test control. This method has been used on large multiloop tests such as on wide-bodied transports, fighter aircraft, and bomber airframes, where as many as 256 channels of servo actuator control are required. These servo controllers are expensive and difficult to adjust for maximum stability in multiloop tests. An advanced 32 channel direct digital control (DDC) system, using the main computer to perform the functions for the analog servo controllers, was designed and built by Lockheed and is capable of small scaled structure tests. The same type system may be implemented on large scale structural tests, but the integration of DDC techniques into full-scale airframe fatigue testing is limited because of the timing constraints of the central processing unit (CPU). The present approach to DDC, the use of a single minicomputer, cannot be applied to systems requiring 256 channels or loops because of throughput limitations of the current processor.

Objectives

Lockheed's long-term objective for this continuing project is to develop a large multi-channel (200 plus) load control system using direct digital control techniques and capable of running both fatigue and static full-scale airframe structural tests.

The following overall objectives have been defined:
- Reduce problems associated with loop update rate, maximum cycle rate, number of steps per cycle, and number of loops.

- Investigate methods of reducing loop instability due to interchannel reactions.
- Improve test control and monitor capabilities.
- Develop a system for using DDC on medium size (128 channels) multiloop static structural tests.
- Develop a system for using DDC on large size (256 channels) multiloop structural fatigue tests.

Background

The first prototype DDC system we developed at Lockheed was based on a Hewlett Packard (HP) model 2100 and was written entirely in assembly language. Later systems were based on the HP-1000 series minicomputer. These systems made use of FORTRAN, assembly code, and micro-code. They were designed to run small (1-32 channels) multiloop tests as well as multiple tests (up to 32 individual tests per system). Larger systems were developed using the same computer front ends but with analog servo controllers for tests requiring from 128-256 channels.

In 1989, we took an existing 128 channel load control system that had utilized 128 individual analog servo controllers and converted it to a direct digital control system by eliminating the servo controllers and merging the servo control functions into the main computer through the use of micro-code. This gave us a system capable of handling 128 servo loops with a minimum loop update rate of 14 milliseconds. While this update rate is adequate for extremely slow moving static airframe tests of 128 channels or less, full-scale airframe fatigue and larger static tests require faster loop update rates. We then undertook the task of developing a multi-processor system using digital signal processors (DSP's) to overcome the throughput problem of a single CPU.

Overall System Design

The basic functional design of a DDC system remains the same regardless of whether it is implemented in microcode on a single CPU computer system or on a host computer with multiple DSP cards installed in it. This design consists of a user interface, a spectrum generator, a closed loop real time section and a test article protect system. The simplest form of a user interface is designed around an input command line system. This method has been implemented on the 128 channel static system described later in this example. A more user friendly method makes use of graphic pull-down menus. This method is being developed for the 256 channel fatigue test system but has not yet been implemented. All DDC systems require some form of spectrum generation. This simplest spectrum generator is a static generator. Fatigue testing however can use several different types of generation such as an externally generated disk based generator, a randomized spectrum generator running in real time on the host computer, or a block spectrum generator also running on the local computer (see Figs. 15-88 and 89).

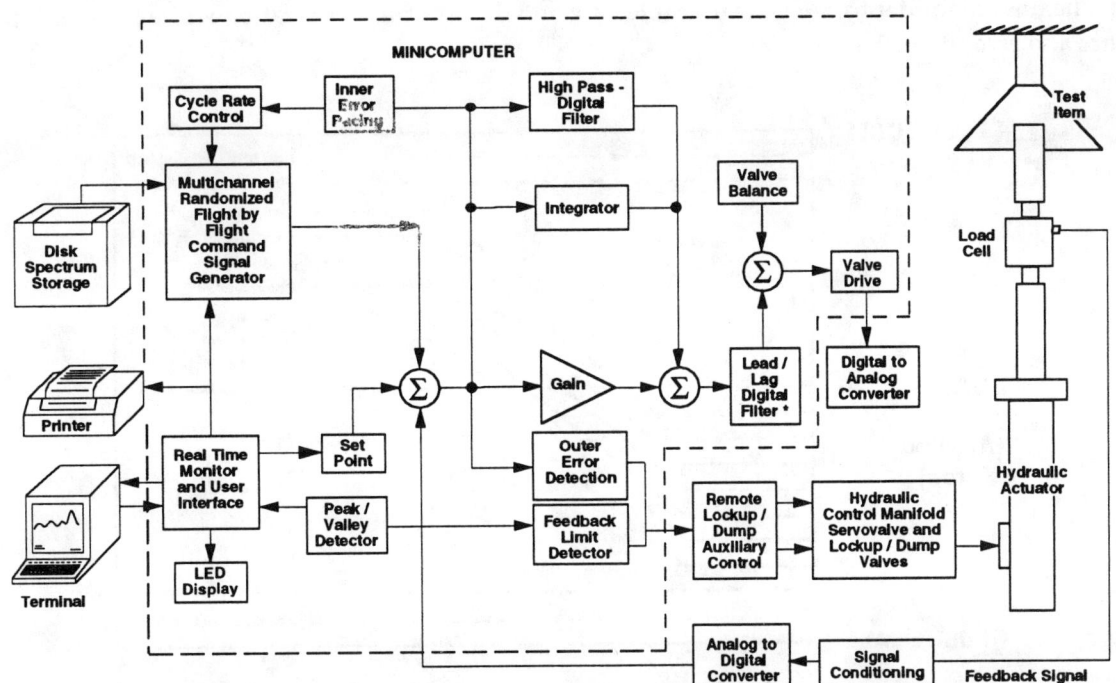

Figure 15-88. Functional Diagram of 128 Channel Test System. This diagram describes the overall functions required in a servo controlled loading system and which portions are performed digitally within the mini-computer.

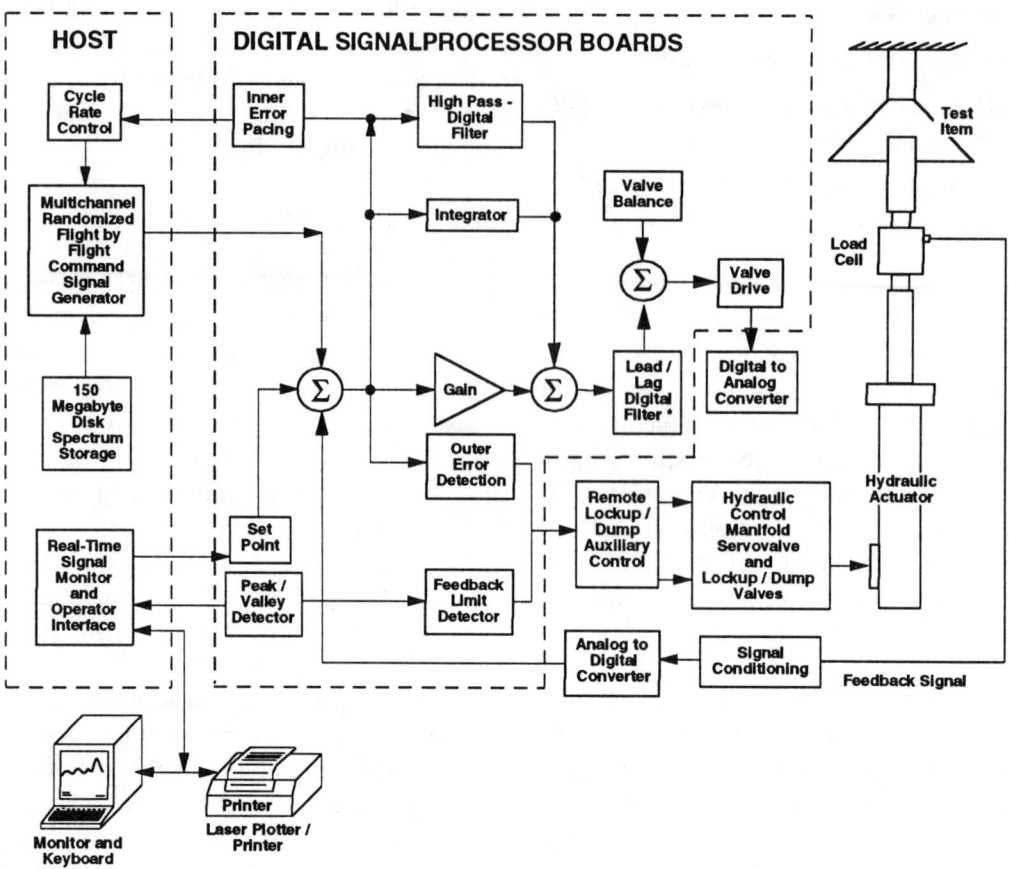

Figure 15-89. Functional Diagram of 256 Channel Test System. This diagram shows the overall division of functions among the host computer, the eight digital signal processor (DSP) boards, and the analog signal conditioning hardware. The host computer handles the user interface and the spectrum generation while the DSP's provide the loop closure and error detection functions.

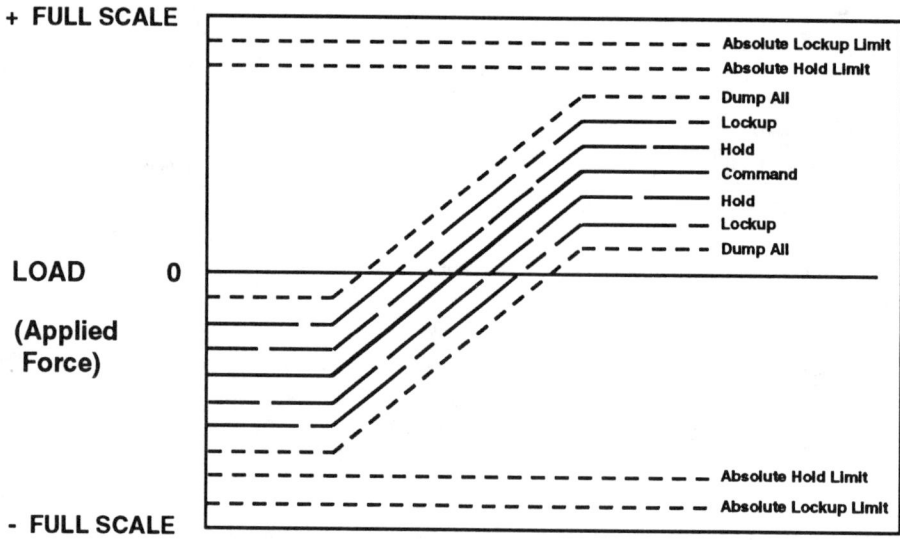

Figure 15-90. Load Protection System. This plot shows the relationship of the command signal to the various limits, both error and absolute. The feedback signal should track the command signal. If it strays through any of the indicated boundaries the designated protection level will be activated.

Test Article Protect System (Hold/Lockup/Dump)

There are three basic levels of load protection built into the computer software running the direct digital control loading system as shown in Fig. 15-90. These levels are explained as follows and use the system in Fig. 15-91:

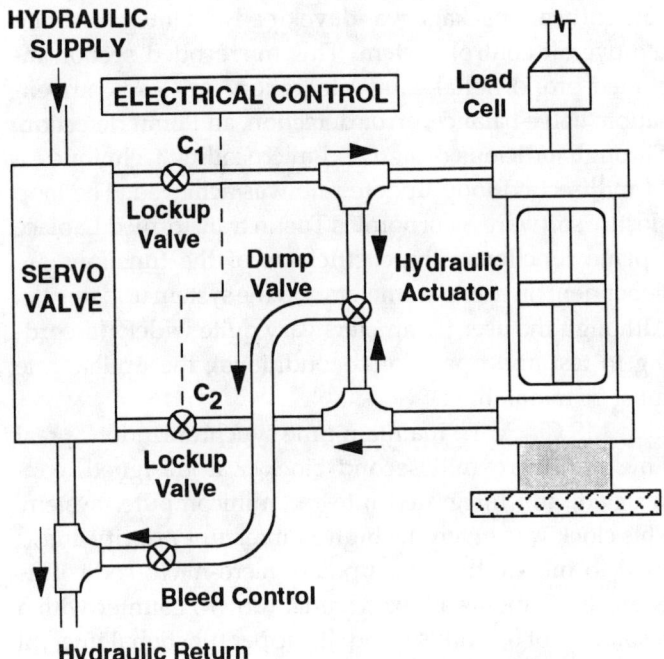

Figure 15-91. Hydraulic Lockup/Dump Diagram. The ideal hydraulic setup for Direct Digital Control loop consists of an equal area jack, two lockup valves, a dump valve, and a servovalve interface as shown.

Level 1. Command Hold of All Channels

"HOLD" is the process of freezing the command signal to all channels or loops in a given test. This will stop the ramp loading in a static test or the cycling of the loads in a fatigue test. A "HOLD" can be triggered by three different methods: (a) inner error pacing (IE), (b) absolute hold limit (AH), and (c) any one of several types of manual "HOLD" commands from the keyboard. Inner error pacing is activated whenever the delta between the command and feedback signal exceeds the IE percent value listed in the channel parameter (AM) table. Closed loop control is maintained while inner error pacing is in effect. The absolute hold limit activates a "HOLD" whenever the feedback signal crosses over the plus or minus AH limits listed in the channel parameter (AM) table. "Hold" is a non-latching mode and loading or cycling will automatically continue as soon as the inner error is reduced below the level of the IE pacing value, or the feedback signal drops below the AH limit. The "HOLD" commands from the keyboard will latch the "HOLD" until a "START" or "GOTO" command is issued.

Level 2. Lockup of Individual or Multiple Loading Jacks

"LOCKUP" is the process of locking the load applied by the hydraulic jack to the test specimen. To accomplish this, the system is put in "HOLD" and the pressure is trapped on either side of the jack piston by deactivating hydraulic valves leading directly into the jack ports. The supply valve is also de-activated, cutting off the supply of oil to the servovalve. "LOCKUP" can be triggered by either an outer error (OE) or an absolute limit (ABS) detection. Outer error detection is tripped whenever the delta between the command and the feedback signal exceeds the OE percent value listed in the channel parameter (AM) table. Absolute limit (ABS) detection is tripped whenever the feedback exceeds the ABS limit value listed in the channel parameter table. Closed loop control on the locked up channel is lost while "LOCKUP" is in effect. "LOCKUP" is a latching mode and the entire test is in a "HOLD" condition until the problem is resolved and an UNLOCK command is issued. A START command is also required to release the "HOLD" and continue the loading or cycling of the system.

Level 3. Dump of All Loading Jacks

"DUMP ALL" is the process of hydraulically dumping to zero the loads applied by the jacks to the test specimen. Individual jack dumps are not allowed in this system. A "DUMP ALL" is accomplished by putting the system in "HOLD" (if is not already), performing a "LOCKUP" on the jacks, and then performing a "DUMP" on those same jacks by de-activating the two dump valves that are plumbed across the piston of the jack. The command loads are ramped to zero.

There are six different events that can cause a "DUMP ALL" to occur:

1. The error signal exceeding the "DUMP ALL" outer limit.
2. A preset number of "LOCKUP's" tripping.
3. A manual "DUMP" from the keyboard.
4. A "dead-man" switch activated by a program failure of the computer.
5. An emergency manually activated external switch.
6. The loss of the 28 volt power supply.

Lockup/Dump Recovery

The lockup dump recovery system is designed to safely recover from either an individual lockup or a complete system dump. An "UNLOCK ch" command is issued from the keyboard to initiate the unlock of an individual jack. The command signal is set equal to the feedback signal in order to reduce the error signal to zero. The lockup and supply valve are activated to unlock the jack. The integrator is re-activated (if it has been assigned to this loop). The command signal is then ramped back to its original setting. An "UNLOCK ALL" command is issued from the keyboard to recover from a complete system dump. The dump valves and and the dump bleed valve for the first loop is activated to "undump" the jack. The previously described unlock sequence is then performed on that loop. The jacks of each loop are then "undumped" and "unlocked one by one. After a lockup recovery, a "START" command must be issued from the keyboard to continue the load transition or cycling.

Special Hydraulic System Safety Features.

Equal Area Jacks: The use of equal area jacks can greatly reduce the chance of damage to the test article during a dump situation. Unequal area jacks can sometimes apply a larger load after dumping than was applied prior to the dump. Loop tuning can also be more difficult with unequal area jacks.

Lockup Valves: Two lockup valves are located at the output ports (C1 and C2) of the servovalve. They lock off the flow of oil to the jacks during a lockup or dump condition. Normally closed valves are used in order to provide failsafe operation. (If power is lost to the coil of the valve, a lockup will occur.)

Dump Valves: The dump valve is plumbed across the ports on either side of the jack piston. It is used to equalize the pressure to each side of the equal area piston, thereby reducing the load to zero. Normally open valves are used in order to provide a failsafe dump of the load in case of power loss to the coil.

Pressure Relief Valves: Pressure relief valves are used to limit the maximum pressure that can be applied to either side of the jack piston. This provides the absolute load limit that can be applied to the test article.

Servovalve Size Selection: Special attention has been paid to the flow rate of the servovalves selected. Greater control of the servo loop can be maintained by selecting the minimum possible flow rate that will satisfy the maximum loading speed required for testing.

128 Channel Static Test System

Work was completed on the design, construction, and check out of a 128 channel static test system with a maximum loop update time of 14 milliseconds. An existing 128 channel load control system that had used 128 individual analog servo controllers was converted to a direct digital control system by eliminating the servo controllers and merging the servo control functions into the main computer through the use of micro-code in the operating diagram in Fig. 15-92. This required the addition of a writable-control store cored to the computer and the procurement of a micro-code software assembler.

Micro-Coded Loop Closure: A micro-coded loop closure software package was developed and integrated into the overall control system. This microcoded section included proportional gain, integration, lag/lead compensation, valve balance, error detection, and limit detection. Through judicious design and microcoding techniques, a 14 millisecond loop update rate was achieved. The loop closure software incorporates Tustin transformed Laplace s-plane functions. The coefficients of the functions are dependent on user parameters and the system update rate. Although the user parameters vary quite widely (according to test and operational conditions), the update rate must be maintained.

MS Clock: To maintain time synchronization, a real time hardware millesecond clock was designed, constructed, and integrated into the minicomputer system. This clock was given the highest interrupt priority and is used to initiate the loop update micro-micro-code. This board implements a hexadecimal (down) counter with a resolution of 0.1 millisecond. Its upper functional interval limit is bound at 65536 milliseconds. This board was installed and is in use in the 128 channel system. This gave a system capable of handling 128 servo loops with a minimum loop update rate of 14 milliseconds.

Power Failure Detection: If the power to the signal conditioning bays failed, the DDC computer could overload the test article. The solution was to design a unit that would monitor the signal conditioning power system and signal the failure to the host computer. A power system monitor using industry-standard opto-couplers was used to directly monitor each of the eight power supplies in the signal conditioning bays. The failure of one or more of the power units will trip the detection unit and signal the DDC computer.

Signal Conditioning: A 128 channel signal conditioning assembly was designed and interfaced with the overall control system. This two bay system includes excitation, amplification, bridge balance, and resistance calibration for all 128 channels as well as current conversion for the output signals to the servovalves.

Computer Input/Output (I/O) Interface System: The output of the signal conditioning system was connected to a 128 channel multiplexer and analog-to-digital converter (ADC) chassis which in turn is interfaced to the minicomputer through a 16 bit parallel I/O card. Two additional parallel I/O cards were required to interface a 128

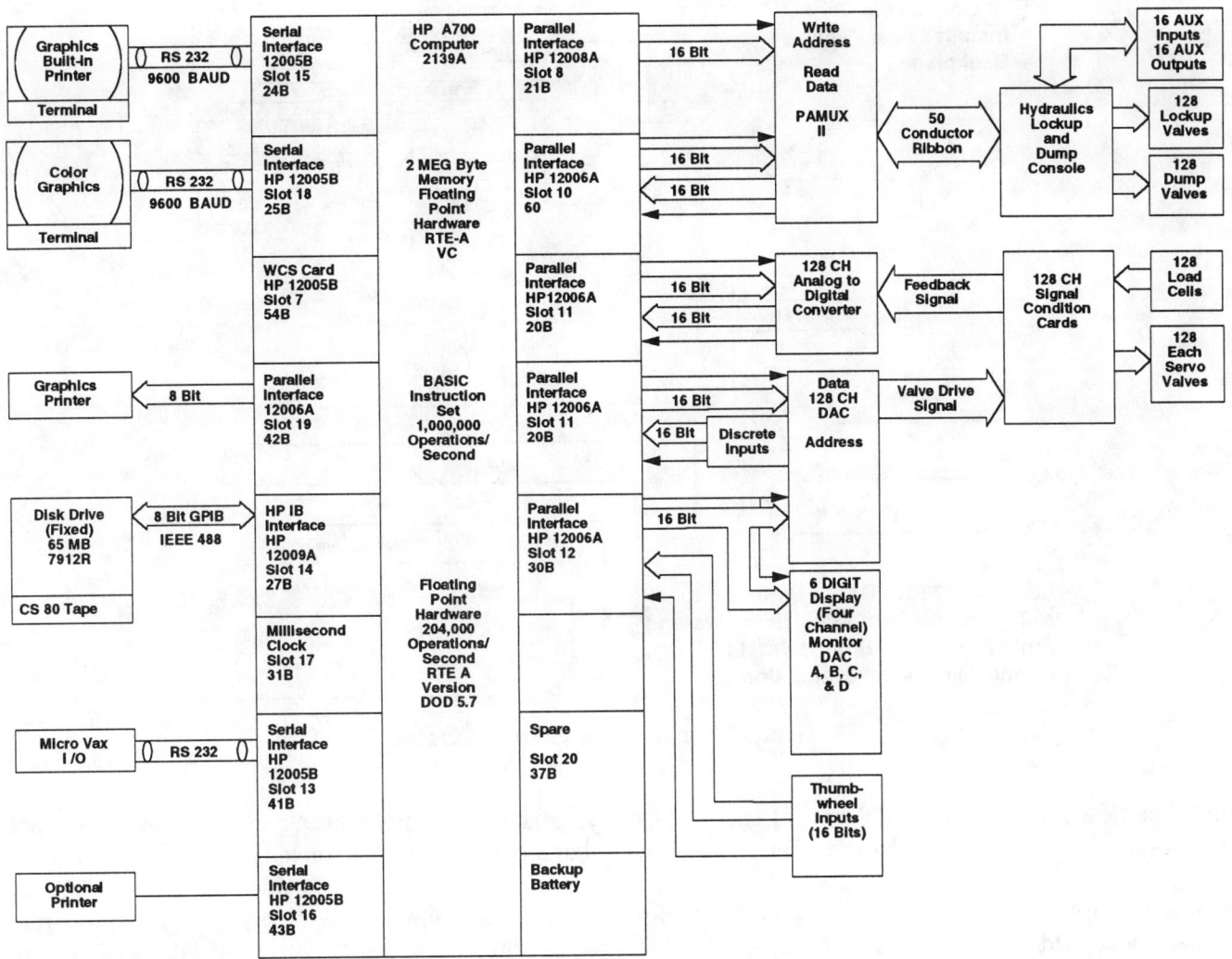

Figure 15-92. 128 Channel Static System Input/Output Interface

channel digital-to-analog converter (DAC) and four channel display chassis to the computer. The 256 channel Lockup/Dump chassis will also require two parallel I/O cards. Two graphic terminals and a remote computer link are interfaced to the computer through serial interface cards. The 65 M-byte fixed disk drive used an IEEE 488 parallel 8 bit I/O card. This system is shown in Fig. 15-93.

Circular Data Buffer: A multiple scan data buffer was developed that allows the operator to call up a plot of both the command signal and the feedback signal for evaluation after a lockup or system dump has occurred.

User to Host Computer Interface: The main users interface to the control system is through the host computer terminal. A large selection of command line instructions has been developed to input test and individual channel parameters, as well as spectrum generator data.

256 Channel Fatigue or Static Test System

While the 14 millisecond update rate is adequate for extremely slow moving static airframe tests of 128 channels or less, full scale airframe fatigue and larger static tests require faster loop update rates. Work was completed in 1991 on a more advanced DDC system capable of handling 256 or more channels with loop update rates of 2 milliseconds or less. This system is intended for larger fatigue and static airframe loading tests. A prototype and development system consisting of an 803865X host processor card installed in a special rack mounted "AT" type bus chassis capable of handling up to eight TMS320C25 digital signal processor cards was designed and constructed. A special I/O chassis was designed and constructed. This chassis can handle up to 128 channels of analog-to-digital output. A second chassis can be added for systems requiring up to 256 channels as shown in Fig. 15-94.

Digital Hardware System: A hardware and software prototype and checkout system was designed and constructed. This system consists of a 12 slot computer chassis with a passive backplane, an 80386SX based microcom-

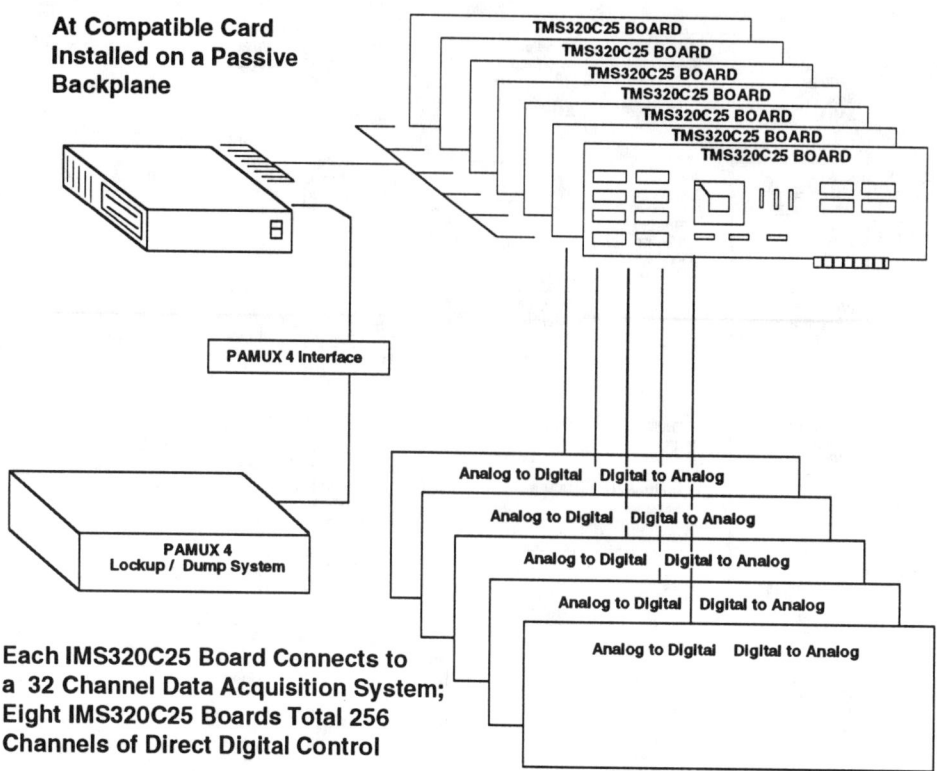

Figure 15-93. Channel Dynamic System Input/Output Interface

puter board, a 150 megabyte ESDI hard drive, a VGA graphics display, a keyboard, and room for up to eight TMS320C25 DSP boards. The TMS320C25 is a Harvard architecture (dual program memory structure), complementary metal-oxide semiconductor (CMOS), 16 bit, 40 MHz four phase clocked, 10 million instruction per second (MIP) DSP integrated circuit. The preliminary operating system is MS-DOS and the primary system development software is Microsoft C version 5.1 and Microsoft's Assembler version 5.1.

DSP to Analog Interface: A DSP to analog hardware interface chassis was designed and built as shown in Figs. 15-94 and 15-95. This system consists of a 32 channel analog-to-digital converter board and a 32 channel digital-to-analog converter board. Up to four pairs of boards can be added for a total of 128 channels per chassis. The new architecture required changes in the data acquisition system. Previously, the single computer approach required a single data acquisition system. In the new configuration, the eight DSP boards run snychronously and each handles 32 channels of loop closure. The eight boards could be interfaced to a single data acquisition system although a considerable amount of time is lost due to bus arbitration and scheduling. Because the data acquisition is extremely loosely coupled operation, each board could handle its own set of 32 channels. Therefore, we designed a two board system for each individual DSP that allow analog-to-digital and digital-to-analog conversion. The first board provides two parts: the DSP to analog converter interface and two analog-to-digital converters (ADC's). The interface section contains address registers and data bus drivers for the ADC's and DAC's. The ADC section is designed around two Burr Brown SDM-872 modules. The ADC module consists of a 12 bit, 50 kHz digitizer, a 16 channel multiplexer, a sample and hold amplifier, and an instrumentation amplifier. Each board has a maximum throughput of 100,000 samples per second, aggregate over two analog to digital converters. The DAC card design is based on the Analog Devices AD390 quad bipolar digital-to-analog converter chip. The DAC card contains eight of the DAC chips as well as a DAC address decoder section.

DSP Loop Closure: The loop closure software was developed and implemented on the DSP board. Proportional gain, integration, lag/lead compensation, valve balance, error detection, and load limit detection are included in this version of the software.

The task of loop closure was ultimately placed on the DSP's. Execution time is the most critical factor when dealing with these real-time routines. To maximize execution time, the entire set of code was written in TMS320C25 assembly language using COFF/Development Tools (version 5.10). Preliminary results have demonstrated that the

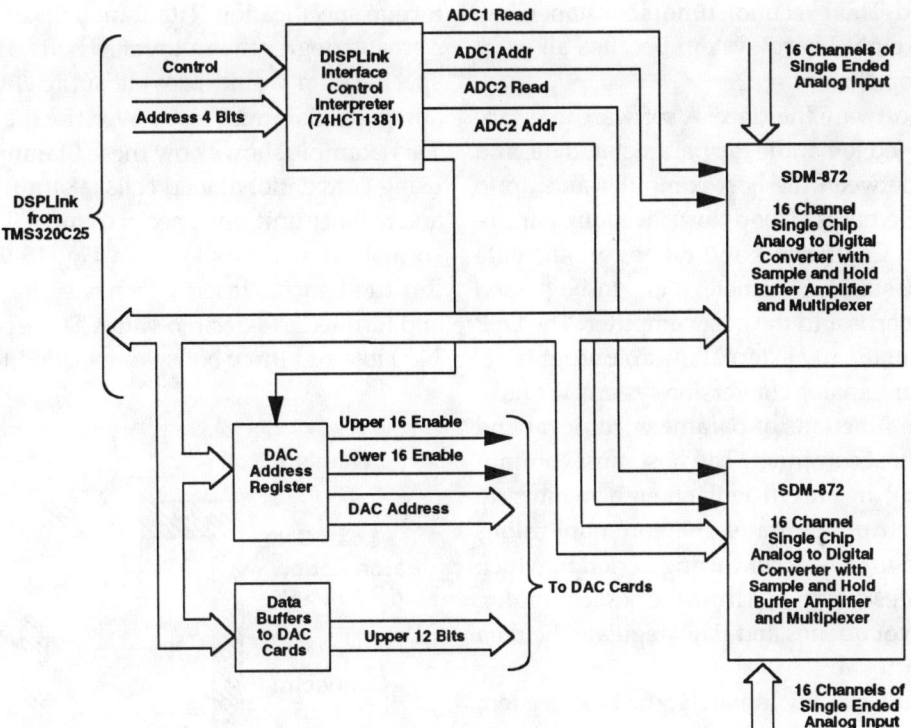

Figure 15-94. DSP Interface and Analog to Digital Conversion Card

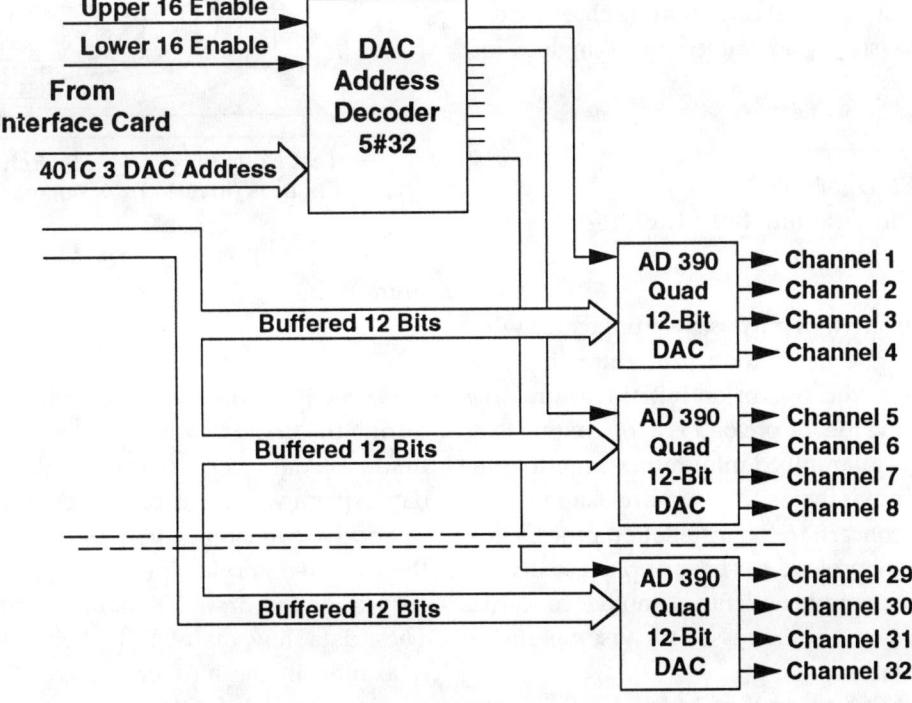

Figure 15-95. Digital to Analog Conversion

execution time for loop closure for 32 channels is approximately 1 millisecond. This execution time is constant when extrapolated to an eight board system because all eight boards calculate loop closure.

DSP to Host Software Interface: A software interface system was developed to handle the parameter, data, and error distribution between the host computer and up to eight DSP boards. Elements of loop closure require parameters such as Tustin Coefficients and other periodic data such as end points, status, parameters, etc. to be passed between the DSP boards and the host computer. The DSP system does not contain any external input/output function other that than the analog conversion system. We have task partitioned the functions of parameter retrieval and distribution to the host computer. The host must communicate with the DSP in an efficient enough manner to support the required update rate. The communications have been designed to only occur during period in which the DSP system requests servicing from the host computer. An intricate system of queues and flags regulate the data distribution and retrieval.

User to Host Interface: Ultimately, the DDC system must interface with the user. Since we were processed with the testing of real-time functions in this system, a preliminary command line user interface was developed and implemented for use in the checkout of the basic system. Basic functions to assign a test, assign spectrums, and initialize error limits were implemented.

The finalized prototype and development system consists of the host computer, an analog-to-digital/digital-to-analog I/O chassis, a signal conditioning chassis, and a power supply chassis all assembled in a single 6-foot console.

15.16 TIME DEPENDENT TESTS
Stress Relaxation in Titanium Bolts (Ref. 15-10)
Robert T. Reese

Bolts used for strength purposes are preloaded when they are installed. The installation procedures typically include a torque specification or an initial tension force. The recommended value for preload is a force equal to 80 to 90 percent of the guaranteed minimum strength of the fastener. After the nut is in position, some relaxation in the force occurs. One concern in the installation process is to ensure that sufficient preload has been applied so that the connection will not degrade with time. Another concern is that the bolt has not been overstressed to cause yielding or some other degradation.

In the case of this example of 1/4-28 UNF titanium bolts, tests were performed to evaluate their performance.

The purpose of the test was to determine the optimum torque specification. Titanium bolts are known to relax to a greater degree than alloy steel bolts. The optimum torque specification would provide sufficient preload to ensure adequate clamping force over the life of the connection. This example shows how these titanium bolts were tested using conventional load cells, a strain indicator, a switch and balance unit, and special fixtures. The test setup for the six bolts tested is shown in Fig. 15-96. A test bolt was threaded into an insert which was attached to the load cell and torqued to a desired value. Three bolts were loaded to 1450 lbs. and three bolts were loaded to 1350 lbs.

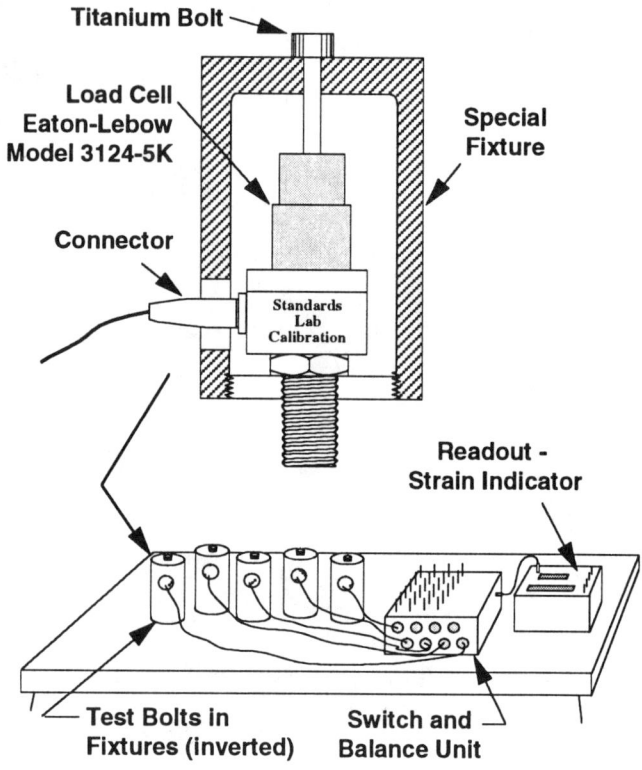

Figure 15-96. Test Setup for Measuring Forces in Titanium Bolts

The force data were recorded at frequent intervals during the first day following the installation. During the first week data were recorded at the start and end of each day. After a week, the forces were measured each day for a period of two months. The data were then extrapolated to the expected service life of 20 years. Two sample data curves for the first 1000 hours are shown in Fig. 15-97. These data show that at the higher preload there is a greater relaxation in the measured force. From these data and materials models for the behavior of titanium under load, estimates of the remaining preload at 20 years can be made.

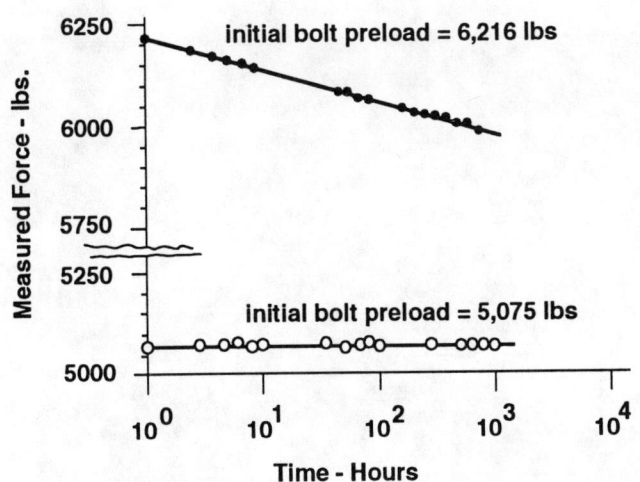

Figure 15-97. Two Sample Data Curves for Relaxation in Titanium Bolt Preload

15.17 REFERENCES

15-1. C.A. Calder, S.Smith, and P. Armony, "Biomechanical Force Platform Based on Strain Gages," Proc. SEM Spring Conference on Experimental Mechanics, June 1985, Las Vegas, NV, pp. 823-826.

15-2. T.G. Carne, J P. Lauffer, and A J. Gomez, "Modal Testing of a Very Flexible 110 M Wind Turbine," Proc. SEM Spring Conference on Experimental Mechanics, June 1987, Houston, TX, pp. 890-898.

15-3. W.A. Kawahara, "Tensile and Compressive Materials Testing with SubSized Specimens," Experimental Techniques, Nov./Dec. 1990, pp. 27-29.

15-4. D.W. Larson and R.T. Reese, "Design Considerations for Severe Thermal Environments Based on Tunnel Fires," Proc. SEM Fall Conference, Oct. 25-28, 1987 Savannah, GA., pp 138-143.

15-5. Malhotra, V.M. and Carino, N.J., Handbook on Nondestructive Testing of Concrete, CRC Press, Boca Raton, FL, 1991.

15-6. Reese, R.T., L J. Lazarus, and W.A. Kawahara, " Fixtures and Test Procedures for Miniature Threaded Fasteners," Experimental Techniques, Nov./Dec. 1990, pp. 30-34.

15-7. S.H. Smith, A.A. Boiarski, and D.G. Rider, "A Calibration Approach for Smart Structures Using Embedded Sensors," Proc. SEM Spring Conference on Experimental Mechanics, Albuquerque, NM, June 4-6, 1990, pp. 310-319.

15-8. R.M. Trusty, M.D. Wilt, G.W. Carter, and D.R. Lesuer, " Field Measurement of Tension in a T-142 Tank Track," Proc. SEM Spring Conference on Experimental Mechanics, June 1985, Las Vegas, NV, pp. 926-931.

15-9. B.P. Perrine, B.A. Nakatani, and R.L. Hannah, "Direct Digital Control of Full-Scale Structural Tests," Proc. Western Regional Strain Gage Committee, Summer Meeting, Overland Park, KS, Aug. 7-8, 1990, pp. 35-50.

15-10. J.J Stephens and J.W. Munford, "Stress Relaxation of a Titanium (Ti6A1-4V) Threaded Joint," Sandia National Laboratories, SAND87-1818UC25, May 1988.

INDEX

A

acceleration 14, 54, 56, 65, 66, 81, 87, 93, 103
accelerometers 15, 35, 36, 42, 48, 49, 154, 155
acoustic emission 135, 237
    acoustic waves 239, 242
    aliasing 257
    analysis methods 267
    applications 237, 362
    attenuation 244
    burst emissions 238
    characteristics of materials 240, 241
    comparison to ultrasonics 239, 272
    continuous emissions 238
    couplants 248
    data storage 259
    dispersion 243
    frequency spectra 252
    generating mechanisms 238
    instrumentation 254
        data storage 259
        preamplifier 254
        postamplifiers and signal processors 256
        recorder 259
        spatial discrimination 260
        spectrum analyzers 258
        transient recorders 256
        video systems 259
    interfaces 244
    principles 238
    sensors 245
        calibration 251
        sensitivity 250
    signals, signal characteristics 238, 251
        amplitude 264
        duration 265
        energy 253, 266
        event count 266
        mathematical models 253
        processing 261
            audio conversion 261
            count and count rate 262
            RMS voltage or signal level 262
            visual analysis 261
        rise time 265
    size effects 246
    spatial discrimination 260
    temperature effects 249
    wave motion 240
acoustic source 48
aerodynamic forces 58
aging (accelerated) 58
aircraft structures (testing) 1, 37, 52, 60, 72, 84, 89, 98, 99, 108, 153, 298, 377
Albree-Fraser airplane 18
amplifiers 36
Amsler 15
analysis and testing 1
approximations 88
ASME 50, 100
    pressure vessel safety code 51, 100, 138
    pressure vessels 51, 100, 141
    stored energy 54, 70
ASTM 41, 306
automobile testing 60, 72, 73, 76, 83, 104

B

Bailey bridges 42
Barlow, Peter 12
Bauschinger 16
beams/columns 24, 75, 88, 89, 94
Beggs 17
Belanger, Brian 319, 326
Beltrami 8
Bernoulli brothers 8
birefringent coatings 13
blast 92, 106
BLEVE 106
Boeing Aircraft Co. 143
Boltzman's constant 250
boundary conditions 110, 296, 304, 381
Bourdon tube 16, 17, 157
Bourns, Marlin 14
bridges 1, 82, 103, 108
Bridgman correction equations 47
brittle coatings 13, 14, 339
Brewster, David 13, 218
bronze 7
buckling 26, 27, 55
buildings/building codes 70, 83, 103

C

cables 83
Caldecott Tunnel 312
calibrations 31, 34, 35, 143, 167, 171, 212, 229, 231, 251, 337, 340, 344, 374
capacitance 14
Capp 12
centers of gravity 29, 45
centrifuges 44, 92, 102, 304
centers of gravity 29, 45
codes 10
code of federal regulations 308, 311
compliance 27, 110
compression specimens 47
compromises (in testing) 90
concrete 286, 290, 362
connections and joints 26, 345
copper 7
Coulomb 9
cracks 28, 138
cranes/hooks 83, 98
creep 27, 30, 57
Cross, Hardy 10

D

da Vinci 8, 11
data acquisition systems 88, 90, 287, 296, 304, 341, 372, 379
data systems 159
    A/D converters 178, 180
    aliasing 180, 257
    amplifiers 168, 170
    calibration 171
        shunt calibration 172
    CMOS switches 179
    common mode rejection 169
    computer based system 176
    converters 180
    D/A converters 178
    drift 169
    excitation 148, 167, 344
    filters 171
    ground loops 164
    multiplexing 176, 178
    power supplies 158, 166
        constant current 166
        constant voltage 166
    recorders, displays 173
        analog meters 173
        data loggers 176
        digitizing oscilloscope 175
        hybrid waveform recorders 175
        magnetic tape recorders 174
        oscillograph recorders 173
        oscilloscopes 174
        servo recorders 173
    sampling 179
    schematic 159
    signal conditioning 158, 166
        excitation 167
    transients 163
    voltmeters 172
data transmission 165
    pulse code modulation 165
    pulse duration 165
    slip rings 166
    telemetry 165
deconvolution 72
design margins 42
direct digital control 383
displacements 36, 41, 48, 54, 55, 56, 60, 66, 81, 87, 93, 138
    gages and transducers 36, 37, 135, 153, 287, 319, 346, 378
    variable resistance 153
Duleau, Alphonse 11
Dupin, Pierre 11
dynamic characteristics 27
dynamic response 138
dynamometers 17

E

El Centro, California earthquake 81
elastic behavior 27, 28
electrical assemblies and components 1, 24, 34
electrical components/connectors 24, 34, 298, 375
energy absorption 34, 39
energy deposition 92, 107
Eney 17
environments 1
environmental considerations 295, 304, 311
environmental incompatibilities 29, 40, 352
Environmental Protection Agency 310
Euler 8
experimental measurements, methods and techniques 2
    nondestructive evaluation 135, 136, 237, 362
    point 135
    whole field 135, 136
experimental uncertainties 93
explosives 54, 92, 101, 106
    line or strip charges 54, 83
explosive bolts 54
extensometers 150
    clip gages 150

F

fabrication 89
failure 26, 87
failure loads and modes 26, 41, 315
    non-structural 31
fatigue 21, 44, 54
federal regulations 307
feedback 123
fiber composites 367
fiber optics 367
    interferometric systems 370
finite element analysis 54
fire temperatures 73, 333
fixtures 381
force platform design 336
forces 35, 55, 66
fracture 27, 28
frequencies 66, 81, 102
fretting 30, 31
frequency response functions 49, 132, 356
full field techniques 138
full-scale tests 42

G

Gabor 14
gages 135, 141
Galileo 8, 11
gas guns 104
geometrical moiré 202
    in-plane strains 203
        application 209
        data processing 206
        data recording 203
        fringe multiplication 203
        specimen preparation 203
    out-of-plane strains 210
        application 216
        projection moiré 210
            test set-up and procedure 215
        sensitivity and calibration 212
        shadow moiré 211
            data reduction 215
            test set-up and procedure 215
Gerasmiov 183
glass-to-metal seals 362
Golden Gate Bridge 76
Grumman Aircraft Corp. 209, 210
Guild 183

H

Haigh 9

hazards evaluation 296, 302, 312, 314
health considerations 295, 306, 311
heat transfer 73
Hencky 9
highway bridges 70, 103, 108
history (structural testing) 2
Hoadley, G. B. 14
Hooke 11
Huggenberger 12
Huber 9
Huey P. Long Bridge 363
Hughes Aircraft Company 142
hydrogen embrittlement 28, 47

I

impact 48
impact-echo technique 136, 280
    applications 286
    data acquisition system 285
    instrumentation 283
    other concerns 284
    testing 282
    theory 281
impulse 68
inertial characteristics 27, 29, 376
Institute of Environmental Sciences 65
instrumentation system schematic 159
instrumented hammer 48, 102, 287
interlocks 300

J

joining 10
    soldering 10
    welding 10
    brazing 10
jumping forces 35

K

Kelvin, Lord 14
kips 66
Kirkaldy 16

L

laboratory courses 2
leaks, leakage 31, 100, 315
Liberty ships 22
light 184
lightning 65
limits, limit switches 90

linear variable differential transformers (LVDT) 91, 120, 153, 327, 378
load cells 116, 156, 157, 375
load test factors 30
loading conditions 38, 65, 93, 302
    rate/time dependency 40, 87, 159
loading systems 87
loads 65, 66, 311
    approximated live loads 68
    body forces 84
    bridge impact factor 70
    building code requirements 68
    dead loads 66, 67, 68
    densities of materials 69
    distributed 83
    dynamic 67, 68, 70, 75, 107
    earthquake (seismic) 69
    envelopes 74
    gravitational 66
    inertial 67, 84
    line 66, 83, 84
    live 67, 68
    jumping 335
    measured service 72
    moving 9
    point 66, 82, 87
    pressures 66
    probabilistic 76
    random static 76
    rates, time dependency 159
    relationships among 67
    running 335
    seismic 81
    spectra 74
    static 68
    stored energy 70, 71
    tank track 335
    temporary 68
    thermal, thermal gradients 66, 68, 73, 82, 311, 331
    variable 9
    walking 335
    wind 359
Lockheed Aeronautical Systems 384

M

Maney 10
Marten 13
mass/mass properties 29, 34, 45
material densities 69, 240
material safety data sheets (MSDS) 310
materials testing 47, 353
    misapplications 47

maximum allowable working pressure(MAWP) 42, 51
Maxwell, J C. 9, 218
Maxwell-Betti 17
measurements (experimental) 36, 50, 55, 160, 301, 319
    traceability 319, 326
measurands 36, 138, 172
mechanical triggers 54
metals 1, 7, 47, 346, 354
military standards 65
modal analysis 27, 48, 101, 355
    problems encountered 49
model tests 42, 43
Mohr 9
moiré 13
moiré interferometry 183
    basic interferometer 193
    diffraction gratings 188
        specimen grating preparation 197
    extraction of strains from displacement patterns 201
    fringes 183
    gratings 188, 197
    interference of light 186
    interferometers 193
        achromatic 196
        basic 193
        tuning procedure 196
    interpretations of fringes 200
        strains and displacements 201
    nature of light 184
    operation 195
    principles 191
    wave propagation 184
        diffraction 184
        interference 184
moments 111
moments of inertia 29, 45
Muller-Breslau 17
multistory structures 81

N

National Institute for Standards and Technology (NIST) 35
Navier 9
nylon screws 352

O

Occupational Safety and Health Administration (OSHA) 308, 309
operational tests 51
optical systems 365

P

parachute systems 107, 108
Parent 9
penetrants 137
Peters, O.S. 16
phases of (structural) engineering 1, 18, 19
photoelastic coatings 226
    bending correction 226
    calibration 229
    cementing coatings 232
    coating materials 230
    color sequence 228
    fundamental relationships 226
    measuring techniques 226, 227
    reflection polariscope 226, 227
    selection of coating thickness 231
    separation of principal stresses 232
photoelasticity 13, 218
    counting fringes 219
    fringes 220
        counting 219
        formation 222
        multiplication 221
        orders 220, 221
    history 218
    model 339
    patterns 14
    polariscopes 222
        alignment 223
        circular 223
        diffused light 224
        lens type 224
        lenses 225
        linear polarizers 222
        magnification 225
        plane 223
        wave plates 222
    polarizers (types) 223
    procedure 218
    Tardy method 221
piezoelectric 14
piezoresistive 14
pipeline 93
plastic deformation 27, 29
point techniques 141
Polytechnic Institute of Munich 16
power supplies 36
pressure relief devices 50, 100
pressure seals 83
pressure systems 70, 93, 98, 314, 382
pressure tests 42, 50, 66, 93, 304, 343
pressure transducers and gages 17, 35, 116, 138, 157
pressure vessels 42, 83, 84, 87, 98, 137, 304
proof test 51, 89
prototype hardware 33, 72, 90, 299
pseudodynamic testing 110
pseudovelocity 81

Q

quick release (step relaxation) testing 110, 357

R

radiography 32, 136, 137, 287
random processes 76
    narrowband 81
    probabilities 77
    probability density function 78
    random signal realization 78, 79, 80
    realization of random variables 77
    variables 77
    wideband 80
real-time radiography 287
    equipment 288, 291
    flaw detection 289
    image analysis 291
    image processing techniques 289
    internal strain measurements 290
    principles 287
    scattered radiation 288
    uncertainty analysis 291
residual stresses 27, 29
resistance temperature devices 36
response spectra 81
Ritter 9
rocket sleds 84
Roe 142
Romans 7
Ruge, Arthur 14
running forces 335

S

safe operating procedures 316
safety 33, 138, 295, 306, 308, 379
St. Venant 9
Sandia National Laboratories 45, 347
Saunders 142
Schaevitz, Herman 14
seismic 14, 42
sensors 135, 141
service life extension 52
servohydraulic test systems 115
    accumulators 124

    actuators 94
    bulk modulus 119
    characteristics 116
    command signal sources 127
        block programming 125
        constant amplitude 128
        variable amplitude 130
            artificial excitation 131
            double integrated acceleration 131
            real time direct excitation 130
            reconstructed peak valley 130
            shaped random noise 130
            time history reproduction 131
    components 118
    contamination control 125
    control electronics 126
    control response 120
    controls 115
    electric power loss 127
    electronics 126
    load control systems 121
    moving load cells 122
    oil column compliance 119
    operating pressures 118
    pressure control 117
    rotary 118
    safety and interlocks 125
    shut down 127
    supply and return 123
    valves 90
Shepherd 13
ships 1, 22
shock/shock tests 34, 53, 66, 67, 89, 93, 104
shock spectra 68
Simmons, Edward 14
smart structures 365
Society for Experimental Mechanics (SEM, formerly SESA) 18, 237
spall 27, 28, 332
spinners 102
static tests 90
steel 8
Stefan 15
Stevin 9
stiffness 27, 29, 33, 54, 110
storage-readiness-use-retirement cycle 72
stored energies 70
    gases 71
    liquids 71
strains/measurements 56, 87, 93, 137, 240, 328
strain gages 14, 15, 48, 88, 94, 116, 123, 135, 141, 167, 234, 363, 364
    adhesives 146, 152, 363
        curing times 151, 152
    alloys 145
    applications 144
    apparent strain 145
    applications 145
    backing materials 146, 152
    capacitance 142
    early developments 14, 141
    electrical noise 163
    equations 141
    factors to consider in gage installations 149
    free filament 143
    gage construction 145
    geometries 152
    hostile environments 148
    lead wires 152
    metal foil 142
    metal foil gage construction 145
    protective coatings 152
    semiconductor 143
    sensing elements 152
    shunt calibration 172
    strain ranges 147, 148, 149, 354
    strain sensitive alloys 145
    suppliers and manufacturers 149
    temperature ranges 147, 148
    three lead wires 163, 340
    transverse sensitivity 146
    weldable 142
    wire 142
strain rates/sensitivities 40
strength 27, 30
stress waves 67
structural analysis 88
structural design 72
structural engineering 1, 18
structural frames and test beds 97
structural responses 26, 81, 300
structural tests/testing 1, 21, 24, 34, 40, 84
    basic goals 2
    quality 2, 22
structural tests, types of 34
    acceptance/quality assurance 34, 343
    calibrations 35, 344
    connections and joints 36, 345
    instrumented test units 37
    dynamic tests 38
    energy absorption 39, 350
    environmental and materials compatibility 40, 352
    failure loads and modes 41
    inertial 43

INDEX

    centrifuges 102
    scaling relationships 44
    spinners 102
lightning 45
mass properties 46
materials testing 47, 353
    anisotropy 47
    compression specimens 47
    equation of state 48
    isotropy 48
    misapplication of test data 48
    necking 48
    tensile specimens 47
    tensile strength 47
    torsional testing 47
modal analysis 48, 355
    problems encountered 49
pressure tests 50, 141, 382
residual stresses 52
service life extension 52, 363
shock testing 53
    cable facility 105
    drop tower 105
    gas gun 105
    systems 103
stiffness testing 54, 375
smart structures 365
structure control interaction 56, 383
thermal testing 56, 331
    temperatures 56
    ranges 56
    cycling 56
    chambers 56
    burn facilities 56, 331
    test systems 101
time dependent tests 57, 392
    creep 57
    fatigue 58
    stress relaxation 392
vibrations 59
    systems 103
    tests 109, 110
wind tunnel 60
structures 1, 33, 298
structure-control interaction 94

T

Tacoma Narrows Bridge 22
tanks (battle) 72
Tatnall, Charles 14, 142
technology transfer 3
temperatures 55, 56, 161, 304, 333
    controls 93
    measurements 36, 55
    sensors 154, 155
tensile specimens 47
    necking 47
test beds 97
test conditions 33
test evaluation criteria 24
test facilities and laboratory evaluation 302, 304
test insights 33, 301
test perspectives 3
    conductors 3
    requesters 3, 4
    developers 3, 4
test planning 295, 296, 302, 347
test preparation 295, 302
test purposes 21, 24, 33, 295, 296, 305
test quality 2, 22
test request form 297
test setup 15, 305
test simulations 87, 88, 298, 302
testing concepts, controls, and procedures 56, 90, 115, 383
    closed loop 56, 89, 90, 120, 346, 378
    feedback 56, 90, 120
    open loop 56, 88, 90
testing systems and machines 15, 17, 92, 93
    dead weight 15, 95
    hydraulic 15, 96
    screw driven 15, 96
    servohydraulic 97
thermal systems and tests 56, 73, 101
    gradients 66
    radiation 82
thermography 137
threaded fasteners (bolts/screws) 39, 345, 352
    in single shear 42
    quick release 54
time dependent behavior 27
Timoshenko 9
Tollenaar 13
torque 38, 93
torrs 50
training 2
transducers and sensors 35, 135, 141, 160, 169
    offsets 168
truss 9

U

ultrasonic methods 136, 137, 272
    applications 272

couplants 279
　　equipment 276
　　　　couplant 279
　　　　instrumentation 276
　　　　reference standards 279
　　　　transducers 277
　　principles 272
　　standards 279
　　testing techniques 279
　　　　attenuation measurement 280
　　　　flaw detection 279
　　　　velocity measurement 279
uncertainty analysis 93, 291, 301, 319
　　accuracy 138
　　combining uncertainties 322
　　example 326
　　out of range calibrations 324
　　repeatability 138
　　sources of uncertainties 322
　　standards 370
　　statements 323
　　strain gage misalignment 328
　　top level 320
　　type A 321
　　type B 321
　　value 325
　　working standards 320
Uniform Building Code 70

V

vacuum tests 50
velocity 55, 66
velocity change 89
velocity gages 30
videos 137
vibration tests 59, 66, 92, 103, 298, 304
vibrations 38, 48, 68, 76
voltmeters 30

W

walking forces 335
wear 30
welds 36
Weller 13
Western Regional Strain Gage Committee 146, 165
Wheatstone Bridge 161, 167, 340
　　completions 161
　　precision resistors 172
　　wiring 163
　　　　standardized color coding 165
whiffle trees 16, 99
Whipple, Squire 9
wind tunnel balances 60
Winkler 9
Wright Brothers 17